Fishes of the Western North Atlantic

MEMOIR
SEARS FOUNDATION FOR MARINE RESEARCH
Number I

Fishes of the Western North Atlantic

PART ONE

LANCELETS

Henry B. Bigelow, *Museum of Comparative Zoology and*
Isabel Pérez Farfante, *Museo Poey,*
University of Havana

CYCLOSTOMES

Henry B. Bigelow *and* William C. Schroeder
Museum of Comparative Zoology

SHARKS

Henry B. Bigelow *and* William C. Schroeder

NEW HAVEN

SEARS FOUNDATION FOR MARINE RESEARCH, YALE UNIVERSITY

Publications of the Sears Foundation for Marine Research, Yale University

The Sears F●undati●n f●r Marine Research at Yale University was established in 1937 by Albert E. Parr, direct●r ●f Yale's Bingham Ocean●graphic Lab●rat●ry, thr●ugh a gift fr●m Henry Sears, t● pr●m●te research and publicati●n in marine sciences. The F●undati●n's Mem●irs, inaugurated in 1948, remain imp●rtant references. In 1959 the Bingham Ocean●graphic C●llecti●n was inc●rp●rated int● the Yale Peab●dy Museum ●f Natural Hist●ry.

Distributed by Yale University Press
www.yaleb●●ks.c●m | yaleb●●ks.c●.uk

MEMOIR I

Fishes of the Western North Atlantic

Part One
Lancelets, Cyclostomes, Sharks

Part Two
Sawfishes, Guitarfishes, Skates and Rays, Chimaeroids

Part Three
Soft-rayed Bony Fishes: Orders Acipenseroidei, Lepisostei, and Isospondyli
Sturgeons, Gars, Tarpon, Ladyfish, Bonefish, Salmon, Charrs,
Anchovies, Herring, Shads, Smelt, Capelin, et al.

Part Four
Soft-rayed Bony Fishes: Orders Isospondyli and Giganturoidei
Argentinoids, Stomiatoids, Pickerels, Bathylaconids, Giganturids

Part Five
Orders Iniomi and Lyomeri
Lizardfishes, Other Iniomi, Deepsea Gulpers

Part Six
Orders Heteromi (Notacanthiformes), Berycomorphi (Beryciformes),
Xenoberyces (Stephanoberyciformes), Anacanthini (Gadiformes)
Halosauriforms, Killifishes, Squirrelfishes and Other Beryciforms,
Stephanoberyciforms, Grenadiers

Part Seven
Order Iniomi (Myctophiformes)
Neoscopelids, Lanternfishes, and Atlantic Mesopelagic Zoogeography

Part Eight
Order Gasterosteiformes
Pipefishes and Seahorses

Part Nine, Volume One
Orders Anguilliformes and Saccopharyngiformes

Part Nine, Volume Two
Leptocephali

Part Ten
Order Beloniformes
Needlefishes, Sauries, Halfbeaks, and Flyingfishes

MEMOIR II

The Elementary Chemical Composition of Marine Organisms
by A. P. Vinogradov

Fishes of the Western North Atlantic

Editorial Board

Editor-in-Chief JOHN TEE-VAN
New York Zoological Society

CHARLES M. BREDER
American Museum of Natural History

SAMUEL F. HILDEBRAND
U. S. Fish and Wildlife Service

ALBERT E. PARR
American Museum of Natural History

WILLIAM C. SCHROEDER
Museum of Comparative Zoology

NEW HAVEN

SEARS FOUNDATION FOR MARINE RESEARCH, YALE UNIVERSITY

Yale

ISBN 978-1-933789-11-8 (pbk.)
ISBN 978-1-933789-24-8 (e-book)

Issued in paperback by the
Peabody Museum of Natural History, Yale University,
New Haven, Connecticut 06511 USA

Part One first published in hardcover

ISBN 978-0-912532-82-0 (cloth)
Library of Congress Control Number: 49000120

Distributed by Yale University Press
NEW HAVEN AND LONDON

Table of Contents

Preface

THE inhabitants of the waters of the earth have fascinated human beings ever since "God created great whales, and every living creature that moveth." Our interests have by no means been confined to the aesthetic or the gustatory; the reflections of Isaac Walton are an earnest of the composure and rapport with the universe that exists when fishes and their surroundings are contemplated; the mental relaxation of fly fisherman or surf caster needs no defense or explanation; the life of fishes, their migrations, their evolution, and the incredibly diverse facets of their activities, afford infinite opportunities for study by the scientist. In latter years man's curiosity about the inhabitants of "the water in the seas" has been increased and stimulated by his ever greater penetration into the deeps. Improved apparatus has enabled him to widen his sphere of effort and to capture fish for his markets farther from shore and deeper down than heretofore. With goggles and rubber fins he has pushed beneath the surface for momentary glimpses of those which live below; with diving helmet and diving suit he has gone deeper and investigated more closely; in the bathysphere he has dangled in the sea half a mile down and checked on the lives of the strange fishes which make their home in that dark and cold portion of the world.

Expeditions have gone forth with fishes as their prime consideration, and ichthyologists have studied what the expeditions brought back. Men and women in numerous laboratories have worked upon fisheries problems, while countless numbers of fishermen, professional and amateur, have added their bit to the knowledge of the whys and wherefores of our fishes. All this has produced an enormous quantity of information and lore which lies scattered in countless publications. The reason for the present series of volumes is to correlate the contents of the rich storehouse of knowledge relating to the fishes that live in the waters of the western North Atlantic.

This volume, the first of a series, describes the lancelets, the hagfishes and the lampreys, and those most interesting animals, the sharks. It has been written on the premise that it should be useful to those in many walks of life—to those casually or vitally interested in the general phenomena of life in our waters, to the sportsman whose interests are closely associated with pleasure and relaxation, to the fisherman whose livelihood depends upon knowledge of where fishes are gathered together, as well as to the amateur ichthyologist and the professional scientist. Special stress has been given to the relationship of the fishes to ourselves—in most cases this relationship is to man's advantage, but the present volume also carries this theme in reverse—some sharks will attack man!

Introduction

HALF a century ago Jordan and Evermann's *Fishes of North and Middle America* was published, and up to the present time these volumes have continued to be the only comprehensive descriptive account dealing with western Atlantic fishes. With the progression of years this work has become less available and more obsolete, which is understandable in view of the scientific advances made during the intervening decades.

Vast numbers of papers, both scientific and popular, have appeared since 1896–1900—the dates of issuance of Jordan and Evermann's work. Numerous new genera and species have been described; many groups of fishes have been subjected to detailed study and revision, especially within the last two decades; new viewpoints on classification and phylogeny have been presented; much additional information has been published on life histories and habits of many species, and some regional studies of the fish faunas have been made. However, this new information remains widely distributed in numerous books and periodicals.

Since our knowledge of the fishes on this side of the Atlantic has reached a point of relative stability, particularly with regard to purely descriptive accounts, the present time seems especially suitable for a publication which embraces all of our knowledge of the fish fauna of this region. To bring together and synthesize this scattered ichthyological information and to make it available to both the public and to marine biologists is the primary purpose of this work.

The first volume of FISHES OF THE WESTERN NORTH ATLANTIC brings to fruition, at least in part, a plan which was conceived at New Haven some years ago. With the establishment of the Sears Foundation for Marine Research at Yale University in 1937, funds became available for publication, and a group of interested ichthyologists met to discuss the preparation of a work such as is here presented. To lay a firm groundwork and to initiate production, the Editorial Board was formed, the members of which are Charles M. Breder, Jr., Samuel F. Hildebrand, Albert E. Parr, William C. Schroeder, John Tee-Van, and, until his death in 1944, the late J. R. Norman of the British Museum (Natural History). Assisting the Editorial Board is an Advisory Committee: William Beebe (New York Zoological Society), Rolf L. Bolin (Hopkins Marine Station), William K. Gregory (American Museum of Natural History), Carl L. Hubbs (Scripps Institution of Oceanography), Daniel Merriman (Bingham Oceanographic Laboratory), George S. Myers

(Stanford University), John T. Nichols (American Museum of Natural History), Luis Howell-Rivero (University of Havana) and Leonard P. Schultz (U.S. National Museum).

The articles in this and subsequent volumes, which will be co-operatively produced by many ichthyologists, are intended to be critical reviews or revisions of each group rather than perfunctory compilations or mere reprintings of previously published works. An outline of the general classification has been prepared, based on widely accepted schemes of classification (such as that used at the British Museum). Standards for both the text and the illustrations have been formulated so as to achieve a fairly uniform treatment for all volumes. Under each species will be found both the distinctive characters which set it apart from its nearest relatives, a detailed description, as well as discussions of its color, size, general habits, abundance, range, relation to man (that is, its economic importance, danger to man, sporting qualities, etc.), and its occurrence in the western Atlantic. Since the publication will be used by lay persons as well as by ichthyologists and marine biologists, the use of highly technical words and phrases has been avoided as far as possible. Because of the large number of references which are included in a study of this nature, particularly in the "Synonyms and References," abbreviations have been used throughout. References to periodicals are listed and abbreviated in accordance with the standards established in *A World List of Scientific Periodicals, Published in the Years 1900–1933* (Oxford University Press, Second Edition, 1934), and an approximate consistency has been developed for books and periodicals not listed in that publication. The final volume will contain a complete and extended bibliography. Common names which are most generally used have been included; for future volumes it is possible that the recommendations of the Committee on Common Names of the American Fisheries Society will be available.

The geographical range of FISHES OF THE WESTERN NORTH ATLANTIC embraces the western half of the North Atlantic, including the adjoining gulfs and seas, from Hudson Bay southward to the Amazon River. But this range is not strictly adhered to in all instances; a number of species living close to the outer borders of the region covered by this publication are included, particularly when their inclusion assists in a more adequate understanding of the group under consideration. Brackish water species are included, and naturally those which are cosmopolitan. As far as oceanic forms are concerned, pelagic species are treated in full, while the strictly deep-sea (bathypelagic) fishes are referred to only in keys and by references to the more recent reports describing these animals. Two factors dictate this decision: 1) The relative paucity and incompleteness of our knowledge of these animals, and 2) the fact that they rarely, if ever, come within the provenance of the nonspecialist in fishes, since special vessels and gear are required to effect their capture.

The map which accompanies this first volume is by no means complete. Since it was prepared before the manuscript was finished, all the localities given in the text could not be included, particularly in such heavily worked areas as New England. However, it will serve to give at least a general idea of locations; in future volumes there will be a closer relationship between the localities given in the text and those included on the map.

The expense incurred in the preparation of this volume has been extensive, and due appreciation and thanks are extended to the Sears Foundation for its share in making publication possible and to the institutions that supported the work of the authors and editors. Income derived from the sales of the volume will be used for the production of the remainder of the publication.

The Editorial Board would like to express its appreciation and gratitude to Yngve H. Olsen, Assistant Editor of the Sears Foundation for Marine Research, for his diligent and able editing of the manuscripts and for the guidance of the publication through the press.

To Henry Sears the members of the Editorial Board owe a personal and collective debt of gratitude for his understanding and for his unswerving continued support.

JOHN TEE-VAN
New York Zoological Society

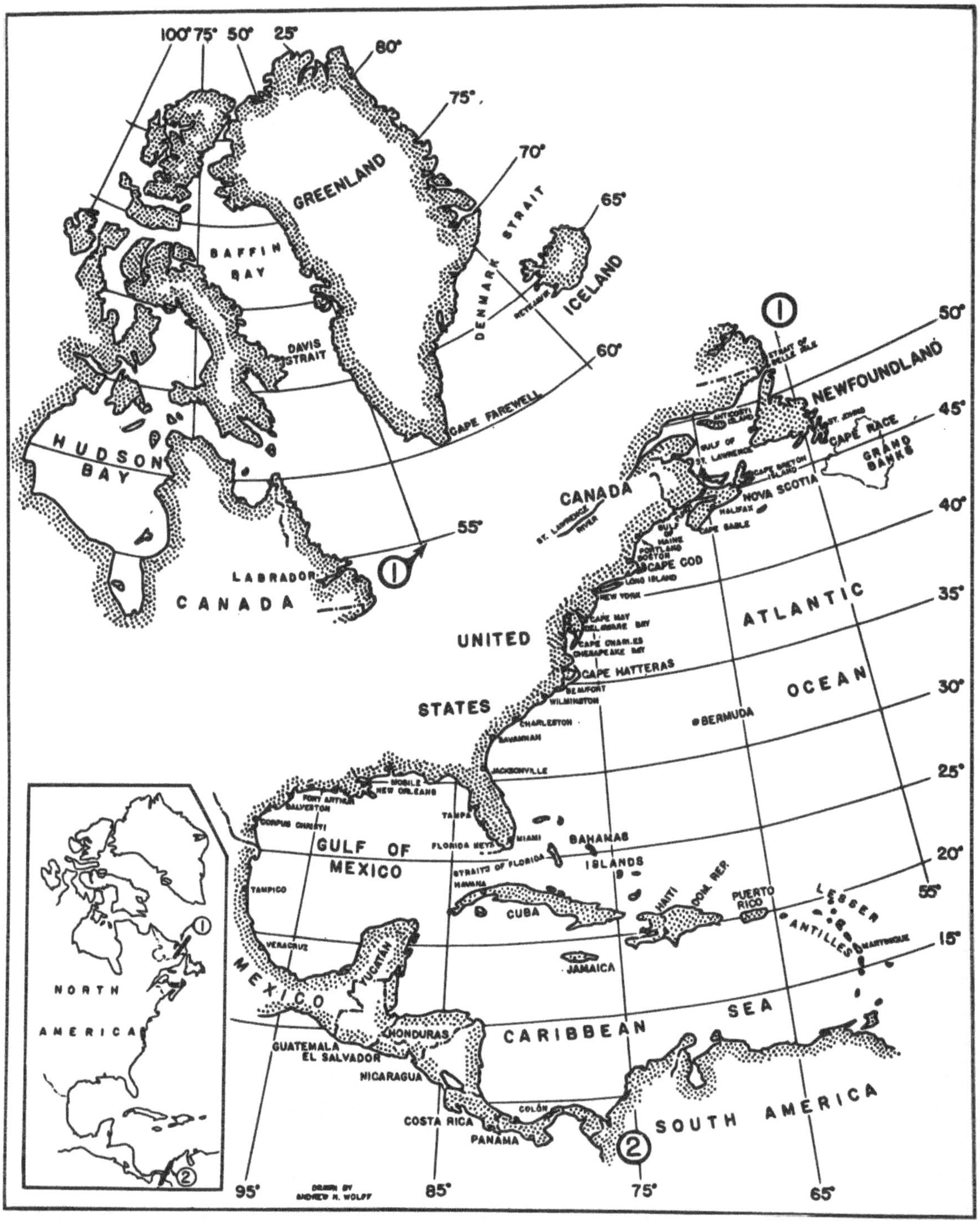

North America

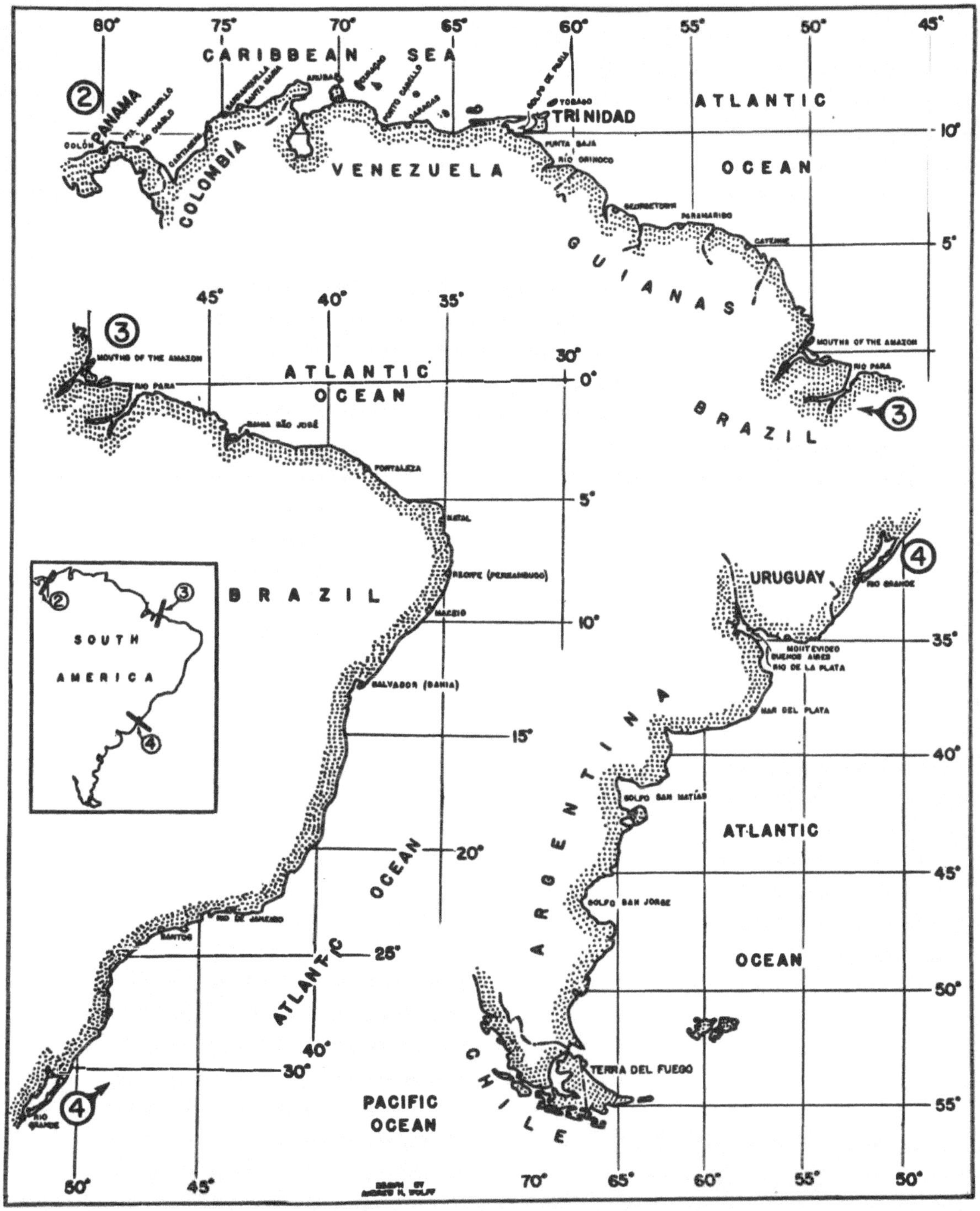

South America

CHAPTER ONE

Lancelets

BY

HENRY B. BIGELOW *and* ISABEL PÉREZ FARFANTE

ACKNOWLEDGMENTS

We are indebted to Thomas Barbour and Leonard P. Schultz for putting the Lancelet collections of the Museum of Comparative Zoology and of the United States National Museum at our disposal for study. Also, hearty thanks are due to Gerardo Canet for preparing all the original drawings included here.

GENERAL DISCUSSION

The Lancelets of the western Atlantic Ocean are included in the present volume for convenience, following the precedent established in existing manuals of the fishes of various parts of the world. Actually they are not fishes at all, although fish-like in appearance, but belong to a separate subphylum (Cephalochordata) of the Chordata, since they are much simpler in structure than are any of the true vertebrates of the subphylum Euchordata, or Vertebrata.

Class LEPTOCARDII

The notochord, extending the entire length of the body and persisting throughout life, is surrounded by a resistant sheath, this notochord and sheath forming a firm but flexible supporting structure. But there is neither protective skeleton nor cranium for the anterior part of the neural tube, no bony structures of any sort, and no jaws. The pharynx in the adult is surrounded by an atrial chamber, formed by the outgrowth and coalescence of two ridges (the metapleura) of the body wall; the pharynx opens into the atrium by a double series of gill slits, the number of which continues to increase throughout life; posteriorly, the atrial cavity opens to the exterior by a small aperture, the atriopore. The dorsal nerve tube terminates anteriorly some distance behind the anterior end of the notochord; it is much compressed laterally, and the only suggestion of a brain is that its axial canal widens anteriorly into a cerebral vesicle. The nerves given off by the neural tube (except for the first two) are dorsal and ventral in origin, but the dorsal and ventral roots do not join, and there are no ganglia on the dorsal roots. The muscular system is segmented, the successive muscle blocks, or myotomes, being separated one from the next by septa of connective tissue, or myocomma. The final number of myotomes is established early in life, but the number is somewhat variable in every species. The gonads are segmented. The circulatory system is very simple; there is no heart, but the larger blood vessels are peristaltically contractile. There is a well developed coelom, or body cavity. The outer surface of the body is clothed with an epidermis consisting of a single layer of columnar epithelial cells, without scales or other hard epidermal structures, and without cilia except in the mouth, pharynx, atrial cavity and intestine. There are no eyes and no limbs. The sexes are usually separate although similar in external appearance, but hermaphrodites have been reported on several occasions. Development is described below.

The Lancelets differ from all the higher groups of fish-like animals—cyclostomes, elasmobranchs, chimaeroids, and bony fishes—in the following important morphological features.

A. Their epidermis consists of a single layer of cells of ectodermal origin in contrast to several layers of cells in all higher groups.

B. They have no hard epidermal or tooth-like structures of any sort.

C. They have no eyes, no external nostrils and no true ears.

D. When adult, the pharyngeal region with the gill clefts is enclosed, on the ventral side, in a so-called atrial cavity.

E. The gill clefts increase in number throughout life whereas in all the higher groups their number is fixed.

F. They have no specialized internal respiratory structures, no true brain, no heart, no trace of a cranium and no hard vertebral structures, cartilaginous or bony.

G. The notochord extends forward beyond the anterior end of the dorsal nerve tube.

H. Their blood is colorless, without red corpuscles.

I. The neural canal, entirely closed dorsally in higher vertebrates, extends through the dorsal wall of the nerve tube as a longitudinal fissure, reminiscent of the ectodermal infolding by which the tube is formed.

J. The excretory organs are nephridia-like rather than kidney-like, consisting of numerous (up to 91) pairs of tubules in the pharyngeal region, each discharging independently into the atrial cavity.

K. The gonads are numerous, compared to only a single pair in higher groups, and segmentally arranged; each discharges its products directly into the atrial cavity, there being no permanent genital ducts.

L. The lining of the intestine bears cilia.

The relationship that the Lancelets bear to the Cyclostomes and to higher fishes has been actively discussed, one view being that they represent the specialization of some primitive prevertebrate stage in evolution, another that they are degenerate descendants of some early type of vertebrate comparable to the Cyclostomes that have developed peculiar adaptations for a very special mode of life. Perhaps the most that can be said at present is that possibly they may be "fairly close to the primitive types from which the vertebrates have arisen,"[1] although their atrial cavity has no parallel among the vertebrate series.[2]

Order AMPHIOXI

Description. This order includes all known representatives of the subphylum. They are slender, fish-like in external appearance, the body tapers at both ends and varies in length from one to eight cm. at maturity; they inhabit tropical and temperate seas. In the adult the buccal cavity, which leads into the mouth proper, opens on the ventral surface of the body a little behind the anterior end. It is bounded laterally by a pair of expanded muscular membranes, the so-called oral hood, the free edge of which bears 20 to 30 slender oral tentacles or cirri, each supported by a cartilaginous rod arising from a cartilaginous ring situated immediately behind the margin of the hood. Proximally, the inner surface of the oral hood bears a series of finger-like projections of ciliated epithelium, jointly forming the wheel organ, the ciliary action of which drives water inward through the buccal cavity to the mouth, and so to the pharynx. The mouth, at the bottom of the buccal cavity, is very small and surrounded by a vertical membrane, the so-called velum, from which several short velar tentacles project inward into the capacious pharynx. The linings of the pharynx, and of the vertical gill clefts that pierce its two sides, are clothed with cilia (those of the former having a complex pattern), the joint action of which is to drive the water from the mouth, along the pharynx, through the gill clefts and so out through the atrial cavity and atriopore. The pharynx serves chiefly as a feeding organ, as described below.

The integument is expanded as a single continuous finfold which extends along the ventral surface from close behind the atriopore, around the posterior end of the body, thence forward along the dorsal surface and around the anterior end of the latter, where it forms a snout or rostrum. The finfold thus surrounds the anterior end of the notochord and contains a lymph space; in the dorsal fin this is segmentally divided by vertical septa into a series of compartments known as fin-ray chambers and this is sometimes true of the ventral fin as well. These chambers are partially subdivided by so-called fin rays, the lateral and apical surfaces of which are free but the bases of which are connected with the continuous ridge of connective tissue that is derived from the roof of the neural sheath. The final number of rays and of ray chambers is established early in life, *i.e.*, at a small size, but is somewhat variable in all species. Anterior to the ventral fin the ventral surface of the body also bears a pair of prominent longitudinal ridges called the metapleura. As a result of their presence, the anterior part of the body is roughly triangular in cross section in adults, the dorsal fin forming the apex of the triangle, the two metapleura its other two

1. Romer, Man and Vert., 1941: 10.
2. The atrium of the Lancelets, while analogous to that of the tunicates, cannot be regarded as homologous with the latter, for the method of formation is very different.

angles, and the space between the latter forming its base, which is also the floor of the atrial chamber.

There is a rather conspicuous pigment spot at the anterior end of the nerve cord, which has been called an eye spot or median eye, but which appears not to be a light receptor. Also, an olfactory function has been ascribed to a small diverticulum from the cerebral vesicle, but it is doubtful whether this is correct.

Habits. Lancelets spend most of the time buried in the sand, in an oblique position, with the anterior end alone protruding.[3] If removed from the sand they swim actively, bending the body from side to side with a sinuous eel-like motion; it is with this same motion that they bore into the sand, which they do very rapidly. In most cases they burrow tail foremost, but they have been seen to do this with the anterior end foremost, in which case they then assume a U-contour to bring the anterior end out again from the sand. It seems that adults of the genus *Branchiostoma* seldom emerge spontaneously from the sand, or only for very brief periods, except at spawning time, for we find no record of their capture in tow nets.[4] But *Asymmetron* has been so taken (p. 21).

It has long been known that they feed on microscopic organisms which they strain out from the current of water that is drawn in through the mouth and driven by ciliary action through the gill apertures to the atrium, to be expelled through the atriopore. The buccal tentacles, folding over one another, prevent larger objects from entering. Particles small enough to pass through this screen are carried inward to the pharynx, where they become mixed with mucus and are driven against the gill bars. The cilia on the inner faces of the latter, beating in a ventro-dorsal direction, then drive the mingled food and mucus to the dorsal pharyngeal groove, along which it is swept to the oesophagus.[5] Feeding appears to be a continuous process. No doubt the diet includes whatever kinds of microscopic organisms may be available at any given time and place. The intestines of the European *Branchiostoma lanceolatum* have been found to contain diatoms chiefly, but also desmids, Foraminifera, Infusoria, Radiolaria, Cladocera and the eggs of various small invertebrates, as well as plant detritus.[6] Diatoms have also been reported from the intestines of Lancelets from Ceylon[7] and were again the most abundant item in the diet of young *Branchiostoma belcheri* at Amoy, China, although the adults also contained the larvae of tunicates, echinoderms and crustaceans.[8] At another time[9] this same species in the same general local-

3. For an excellent photograph of the European *Branchiostoma lanceolatum* in this situation, see Hagmeier and Hinrichs (Senckenbergiana, *13*, 1931: fig. 3b, 4b, facing p. 258).

4. Hensen (Ergebn. Plankton-Exped. Humboldt Stiftung, *1* A, 1892: 24–25) reported the capture of young Lancelets up to several centimeters long in plankton nets. But the fact that none so large were to be found subsequently in the collections (Goldschmidt, Dtsch. Sud-polar Exped., *11* Zool. 3, 1909: 235) suggests that the stated size was an error.

5. Condensed from a detailed account of the feeding mechanism in *Branchiostoma lanceolatum*, by Orton (J. Mar. biol. Ass. U.K., *10* [1], 1913: 19). For an account of the passage of food material through the gut, see Barrington (Philos. Trans., [B] *228*, 1937: 271).

6. For a list of the food of *B. lanceolatum* compiled from various sources, see Franz (in Grimpe and Wagler, Tierwelt N- u. Ostsee, Lief 7, 12b, 1927: 26).

7. Tattersall, in Herdman, Rep. Gov't. Ceylon Pearl Oyster Fish., Gulf of Manaar, pt. 1, suppl. 6, 1904: 221.

8. Chin, Philip. J. Sci., 75, 1941: 393.

9. Reeves, Ginling Coll. Mag. for Jan. 1931: 29.

ity was found feeding chiefly on bacteria, with a few protozoa also. The intestines of some of the specimens were filled with sand, showing that the oral tentacles do not always bar entrance to inedible particles.

Development. The larval development of the Lancelets has been the subject of several major investigations and has been much discussed in relation to the problem of the ancestry of the vertebrates. The process in the European *Branchiostoma lanceolatum*, which may serve as representative of the group, is briefly as follows.

Spawning takes place at sunset. The eggs are minute (0.1 mm. in diameter) and float freely in the water. Segmentation is not only complete but nearly equal and affords one of the classic examples of endoderm formation by invagination. About twelve hours after fertilization the embryo, now oval in shape and clothed externally with cilia, breaks out from the vitelline membrane and swims near the surface by ciliary action. By about the thirty-sixth hour the yolk is entirely absorbed; the mouth has appeared on the left-hand side; the first gill opening has been formed in the midline, soon to shift to the right side, however; and the anus has formed at the hinder end of the body a little to the left of the midline. During subsequent larval development, which may occupy as much as three months, the larvae live pelagically some distance below the surface of the sea, hanging for the most part in a vertical position which is maintained by the action of the long cilia, or flagellae, one of which is borne by each cell of the ectoderm. The larvae (Fig. 1), which

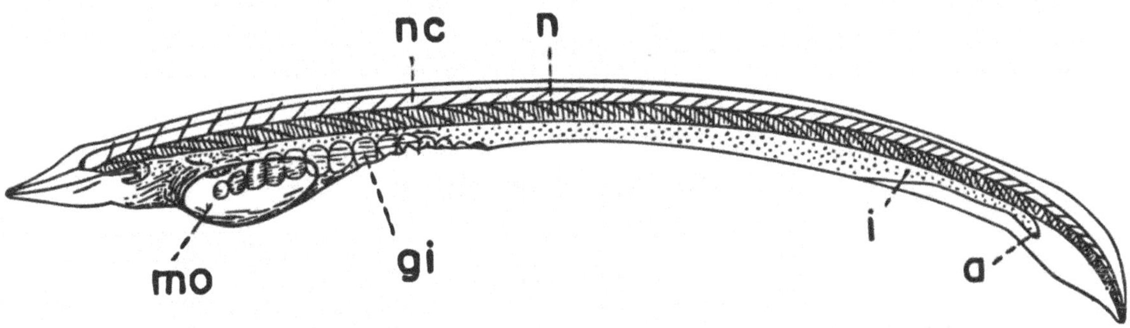

FIGURE 1. *Branchiostoma lanceolatum* Pallas; larva, with 61 myotomes, after Franz. *a* anus. *gi* gills. *i* intestine. *mo* mouth. *n* notochord. *nc* nerve cord.

have a very characteristic appearance because of the swollen gill region in an otherwise slender body, gradually assume the characters of the adult without any abrupt metamorphosis. The most striking of the external accompanying changes are in the numbers and locations of the gill openings, and the formation of the atrium, of the atriopore and of the adult mouth. The latter, at first on the left side and forming a most conspicuous feature of the larva because of its enormous size, shifts to the midline and decreases in relative size toward the end of larval life, while the preoral hood then develops above it. Additional gill openings, up to the number of 14 or more, are formed successively along the mid-ventral line, corresponding at first in number and location to the myotomes in that part of the body but later losing this relationship. Of these primary gill openings, only the second to ninth persist, however.

After the formation of the primary series of gill openings the number of segments increases at the posterior end of the body, the final number being attained early in larval life. In the meantime the embryonic tail fin, a simple ridge of columnar ectoderm cells, is replaced by the adult fin; this forms as an ectodermal fold, enclosing serial expansions of the body cavity, the ray chambers; the fin rays develop as columnar outgrowths of mesoderm upward into these chambers. A secondary series of gill openings, eight or nine in number, appear on the right side of the body, dorsal to the primary series; and each member of each set, except the first, becomes U-shaped and then entirely subdivided by a dorso-ventral bar. The primary series of gill openings then shift to the left side of the body, so that from then on the larva is bilaterally symmetrical so far as the location of its gills is concerned.

The metapleural ridges first appear in larvae with eight to ten gill openings of the second series. The atrial cavity results from the union of the median sides of these ridges, commencing posteriorly and progressing anteriorly. The canal so enclosed expands laterally in the pharyngeal region to the dimensions of the atrium of the adult, while it continues open posteriorly as the atriopore. During the formation of the metapleura the larva abandons its pelagic habit and comes to lie on one side or the other on the bottom. By the time the mouth has moved to the median position the oral hood has formed and the gills have assumed the final symmetrical arrangement. The little Lancelet, now resembling the adult in general appearance, buries itself in the sand; the only further change is the formation of pairs of tertiary gill openings, a process that continues throughout the life of the individual. The curious asymmetry of the larval Lancelet has been much discussed, but in our opinion none of the explanations which have been offered for it is adequate.

Gonads are formed in the second or third year, and the oldest noted among a large collection of *Branchiostoma belcheri* was four years old.[10]

Relation to Man. Lancelets are neither large enough or numerous enough to be of any commercial value anywhere in the western Atlantic, except as subjects for biological investigation; nor are they ever likely to be. However, near Amoy in southern China there has long been a fishery for Lancelets. Recently this employed about four hundred men in two hundred boats who fished with shovel- or scoop-shaped dredges from two to four hours each day on the ebb tide from August until April. This fishing ground is only about six miles long and less than one mile wide, but it has been estimated that the annual catch is in the neighborhood of 35 tons, or more than one billion Lancelets. Some of these are consumed in the near vicinity, while others are dried and shipped to Java and Singapore.[11] Lancelets are also used occasionally as food in Naples and Sicily.[12]

10. See Wells (Science, N.S. *64*, 1926: 188) for *Branchiostoma caribaeum* from Florida; Chin (Philip. J. Sci., *75*, 1941: 400) for *B. belcheri* from Amoy, China.

11. For more detailed accounts, from which the foregoing is condensed, see Light (Science, N.S. *58*, 1923: 57) and Chin (Philip. J. Sci., *75*, 1941: 369).

12. Franz, in Grimpe and Wagler, Tierwelt N- u. Ostsee, Lief 8, b, 1927: 44.

Families. The order includes two well defined families, Branchiostomidae and Epigonichthyidae, separated as indicated in the following key. A third assemblage of pelagic forms, usually grouped together as the genus *Amphioxides,* have sometimes been classed as a third family, Amphioxididae. But their chief distinguishing characters—mouth on the left side, atrial chamber unclosed and gill slits in an unpaired medio-ventral series—are those of larval Lancelets in general at an early stage of development (p. 23), and it now seems established in fact that they are larvae that have continued their pelagic existence for one reason or another until much larger and much further advanced in development than is usually the case, rather than taking to the bottom at a smaller size, as most of them do.[13] In fact, we think it is likely that these *Amphioxides* larvae never do descend to the bottom once they are carried out over deep water, but that they simply continue to exist for an indefinite period as they are swept along with the currents, finally perishing without producing offspring. On the other hand, it has been suggested that their existence may provide a means for the dissemination of the species. Up to the present time, none of them has been positively connected with any particular parent species.

Key to Families[14]

1a. Mouth nearly median, with oral cirri; closed atrial chamber and atriopore; a series of gill clefts on either side.

- 2a. Series of gonads developed on each side; both metapleura terminating close behind atriopore. Branchiostomidae, p. 7.
- 2b. Gonads developed on right side only; the right metapleuron continuous with ventral fin, the left-hand metapleuron terminating behind atriopore. Epigonichthyidae, p. 18.

1b. Mouth on left side without oral cirri; no closed atrial chamber; gill clefts in a single series along ventral side. Amphioxididae, p. 23.[15]

Family BRANCHIOSTOMIDAE

Description. Mouth nearly in midline, surrounded by oral cirri; tentacles with lateral sensory papillae, giving them a toothed appearance; closed atrial chamber; a series of gill slits on each side; gonad pouches developed on both left and right sides; both metapleura terminate close behind atriopore, including between their posterior ends the anterior end of the ventral fin; rostral fin continuous with right side of oral hood, but not with left side; posteriorly the median fin is expanded both dorsally and ventrally in lancet form as a distinct caudal fin, with its ventral lobe lying to the right of anus; ventral fin-ray chambers, except for the more anterior and more posterior, each contain a pair of fin rays in most species, although perhaps only a single fin ray in some;[16] but dorsal fin-ray chambers con-

13. For a recent discussion, see Goldschmidt (Biol. Bull. Wood's Hole, *64*, 1933: 321).

14. Amphioxididae included to facilitate identification. 15. See above discussion of these.

16. Willey (Quart. J. micr. Sci., *44*, 1891: 270) stated that in his *Dolichorhynchus indicus* the ventral fin rays are single; but they appear as paired in his illustration.

tain a single series of fin rays only; rostral fin, with anterior part of dorsal fin, lacks fin rays; dorsal fin-ray chambers much more numerous than myotomes, with four or five chambers to each myotome; the atrial chamber extends posterior to atriopore as a single blind sac as far as the anus; olfactory pit present.

Genera. Two genera, *Branchiostoma* and *Dolichorhynchus,* are commonly recognized in the family and are separated as indicated in the following key. In addition, a new subgenus of *Branchiostoma* has recently been proposed under the name *Amphipleurichthys*[17] for a species in which "the form is more elongated and less robust" than in *Branchiostoma,* "with the myotomes more acutely tapering at each end of the animal," and in which the "caudal fin is reduced to a low fold."[17] But the differences appear to us specific, rather than generic.

Key to Genera

1a. Rostral process, including anterior end of notochord, extends far beyond preoral hood. *Dolichorhynchus* Willey, 1901
Ceylon.

1b. Rostral process, including anterior end of notochord, extends only a short distance beyond preoral hood. *Branchiostoma* Costa, 1834, p. 8.

Genus *Branchiostoma* Costa, 1834

Branchiostoma Costa, Ceni. Zool., 1834: 49; type species, *B. lubricum* Costa. Naples.

Generic Synonyms:

Limax Pallas, Specil. Zool., Fasc. 10, 1774: 19, pl. 1, fig. 11; for *L. lanceolatus* Pallas, Cornwall; not *Limax* Linnaeus, 1758.

Gasterobranchus ? Rasch, Mag. Naturvid., Physiogr. Foren. Christiania, *12* (2) 2, 1836: 325, footnote; evidently *Branchiostoma*, from the excellent account, but only provisionally identified by that author; western Norway; not *Gasterobranchus* Bloch, 1795, which is a synonym for the cyclostome *Myxine* Linnaeus, 1758.

Amphioxus Yarrell, Brit. Fish., 2, 1836: 468; type, *Limax lanceolatus* Pallas, 1774.

Amphipleurichthys (subgenus) Whitley, Aust. Zool., 7 (3), 1932: 256; type, *A. minucauda* Whitley. Queensland.

Generic Characters. The rostral process, including the anterior end of the notochord, projects for only a short distance beyond the preoral hood; the characters are otherwise those of the family.

Range. European coasts from northern Norway to the Mediterranean, the Black Sea and tropical West Africa; western Atlantic from Chesapeake Bay[18] to the Rio de La Plata (including Bermuda); Pacific coast of the Americas from Middle California to Chile; Japan; China; East Indies; Philippines; Queensland; India; Ceylon; Madagascar; East and South Africa.

17. Whitley, Aust. Zool., 7 (3), 1932: 256.

18. Lancelets were said by Garman (in Kingsley, Stand. Nat. Hist., 3, 1885: 62) to range as far north as New York; but we find no positive record of any member of the group in the western Atlantic farther north than Chesapeake Bay.

Species. The characters that have been used chiefly in the classification of the species of the genus are: (1) number of ventral fin-ray chambers; (2) number of dorsal fin-ray chambers; (3) height of dorsal fin in relation to height from its crest to the margins of metapleura; (4) shape of caudal fin; (5) location of anus in lower lobe of caudal fin; (6) number of preatrial myotomes; (7) total number of myotomes. The five species that have been described from the western Atlantic (*B. caribaeum* Sundevall, 1853; *B. bermudae, B. floridae, B. platae* and *B. virginae* Hubbs, 1922) with the anus near, or posterior to, the midpoint of the ventral lobe of the caudal fin differ sharply from *B. lanceolatum* and *B. africae* of the eastern Atlantic, in which it is considerably farther anterior to it. Among this western Atlantic group, *B. bermudae* and *B. platae* are set apart by the fact that the lower lobe of the caudal fin originates considerably anterior to the origin of its upper lobe (Fig. 2 A, F), whereas in the others the two lobes originate opposite one another. *B. bermudae* is sharply separated from *B. platae* by a considerably smaller number of dorsal fin-ray chambers (200–242 vs. 278–330), and fewer myotomes (not more than 56 vs. at least 58). But *B. floridae* and *B. virginae* agree with *B. caribaeum* in the position of the anus, while counts of fin-ray chambers and myotomes in the specimens we have studied (Study Material, p. 13), together with those previously published, fail to show any clear distinctions among the populations of Virginia, North Carolina, Florida (including the Tortugas) or Porto Rico (representing the West Indian region). The most that can be said is that some Florida and West Indian specimens have fewer precaudal fin-ray chambers than have yet been recorded for more northerly localities. But this is not always true, since the maximum recorded counts are in fact for one specimen from Florida and for one from North Carolina. Therefore it cannot be invoked as a basis for specific separation.

Key to Species of *Branchiostoma*

1a. Caudal fin hardly higher than dorsal and ventral fins. *minucauda* Whitley, 1932.
Queensland.

1b. Dorsal or ventral lobe of caudal fin, or both, considerably higher than dorsal and ventral fins.

2a. Caudal fin not clearly marked off from ventral fin. *capense* Gilchrist, 1902.
South Africa.

2b. Caudal fin clearly marked off from ventral fin.

3a. Anus about at point of origin of caudal fin.

4a. Distance from anus to tip of caudal fin only ½ distance from anus to atriopore. *bazarutense* Gilchrist, 1923.
East Africa.

4b. Distance from anus to tip of caudal fin about as great as from anus to atriopore. *haeckeli* Franz, 1922.
Ceylon.

3b. Anus clearly posterior to origin of caudal fin.

5a. Anus far in advance of midpoint of lower lobe of caudal fin.

6a. 77 or more myotomes. *elongatum* Sundevall, 1852.
West coast of South America, Chile to Galapagos Islands.

6b. Not more than 73 myotomes.

7a. 42 to 44 myotomes anterior to atriopore.
africae Hubbs, 1927.
Tropical West Africa.

7b. Not more than 41 myotomes anterior to atriopore.

8a. 68 to 72 myotomes in all. *tattersalli* Hubbs, 1922.[19]
Ceylon.

8b. Not more than 66 myotomes.

9a. Not more than 62 myotomes; ventral lobe of caudal fin a little longer than distance from its origin to atriopore; anus clearly anterior to origin of dorsal lobe of caudal fin.
lanceolatum Pallas, 1778.
Northern Norway to Mediterranean and Black Sea.

9b. At least 63 myotomes; ventral lobe of caudal fin only as long as distance from its origin to the atriopore; anus below origin of dorsal lobe of caudal fin. *belcheri* Gray, 1847.
Japan, China, the East Indies, Philippines, India, and Ceylon to East Africa.

5b. Anus near midpoint of lower lobe of caudal fin or posterior to it.

10a. Origin of lower lobe of caudal fin considerably anterior to origin of its upper lobe.

11a. Not more than 242 dorsal fin-ray chambers or 56 myotomes. *bermudae* Hubbs, 1922, p. 11.

11b. At least 278 dorsal fin-ray chambers.

12a. Rostrum not marked off from dorsal fin by a notch; 65–74 myotomes. *californiense* Andrews, 1893.
Monterey, California to Gulf of California.

12b. Rostrum marked off from dorsal fin by a notch; 59–65 myotomes. *platae* Hubbs, 1922, p. 16.

10b. Origin of lower lobe of caudal fin about opposite origin of its upper lobe. *caribaeum* Sundevall, 1853, p. 13.[20]

19. Including *gravelyi* Prashad (Rec. Indian Mus., *36*, 1934: 333).
20. Including *floridae* Hubbs, 1922, and *virginiae* Hubbs, 1922.

Branchiostoma bermudae Hubbs, 1922

Figure 2 A–D

Study Material. Nineteen specimens, 29 to 49 mm. long, from Bermuda (U. S. Nat. Mus. and Harv. Mus. Comp. Zool.).

Distinctive Characters. Among Atlantic species, *B. bermudae* differs noticeably from *B. lanceolatum* and from *B. africae* in that its anus is about opposite the midpoint of the lower lobe of its caudal fin. In this respect it closely resembles *B. platae* and *B. caribaeum*, but it is separable from both of these by a smaller number of myotomes (56 at most) as well as by generally fewer precaudal fin-ray chambers (9 to 24, usually less than 16). The average number of dorsal fin-ray chambers also is smaller.

Additional Description. Anterior end of notochord extending forward in rostrum in a straight line; rostral fin marked off from dorsal fin by a subtriangular notch; origin of lower lobe of caudal fin anterior to origin of its upper lobe by a distance about ½ as great as length of lower lobe; dorsal fin ⅙ to ⅐ as high as distance from its base to margin of metapleura in the midregion of body; anus a little behind the midpoint of lower lobe of caudal fin; origin of lower caudal lobe about midway between its tip and atriopore; distance from tip of caudal to anus about 0.4 of distance from anus to atriopore; dorsal fin-ray chambers 204 to 242, the highest 3 to 4 times as high as long; precaudal fin-ray chambers 9 to 24; 35 or 36 myotomes anterior to atriopore; 12 to 14 between atriopore and anus, 5 to 7 posterior to anus, total number 54 to 56; gonads, 22 to 28 pairs.

Color. Living specimens are semitransparent and iridescent, but they become opaque after preservation.

Size. Maximum recorded length, 53.5 mm.[21]

Habits. The Bermuda Lancelets are usually found in one-half to six fathoms of water on coarse sandy bottom into which they burrow tail first and there remain most of the time with only the anterior part of the body exposed. If disturbed they swim vigorously for a short time but soon return to the sand. Observations in aquaria have shown that normally they are no more active by night than by day. Under experimental conditions they usually swim with the anterior end foremost. If a stimulus is applied to the anterior end, the Lancelet may dart backward for a short distance, or it may turn end for end. But this reversal in direction is of short duration, for it soon turns again and proceeds at only a slight angle from its original course. In taking to bottom after swimming, Lancelets usually sink quietly through the water to the sand; when in contact with the latter they may either lie there, passive for some time; or they burrow at once, usually tail first, or head first on rare occasions. When buried they usually are tortuous in outline, probably from being crowded among the grains of sand.[22]

Specimens adapted to the summer temperatures of Bermuda (about 31° C.) dart

21. Hubbs, Occ. Pap. Mus. Zool. Univ. Mich., 105, 1922: 10.

22. For detailed accounts of the experiments on which the above is based, see Arey (J. exp. Zool., 29 [1], 1915: 37) and Parker (Proc. Amer. Acad. Arts Sci., 43 [16], 1908: 413).

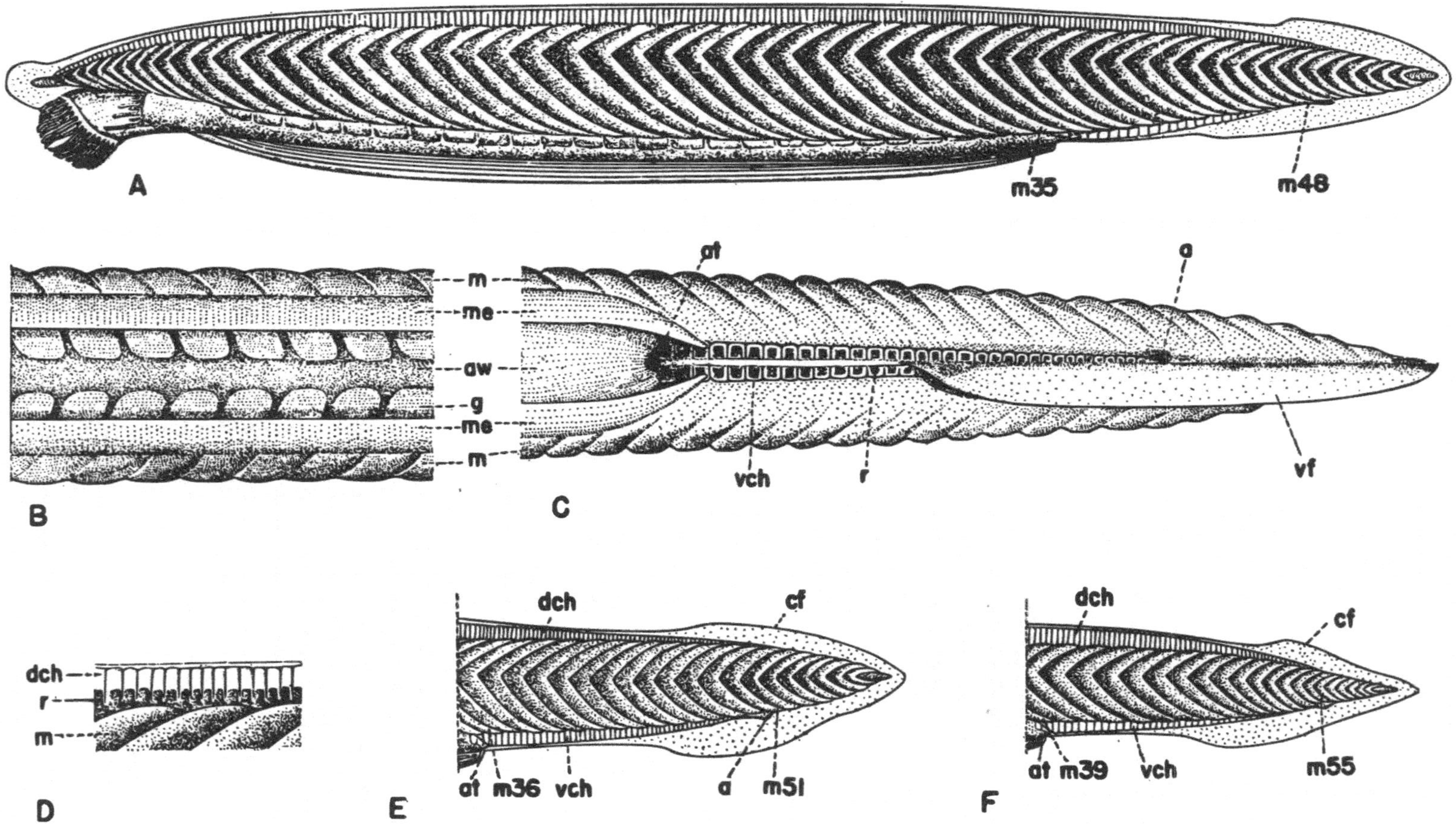

FIGURE 2. *A Branchiostoma bermudae* Hubbs, specimen 47 mm. long, from Bermuda. *B* Ventral view of same, 20th to 27th myotomes, to show gonads as seen through atrial wall. *C* Ventral view of posterior part of same. *D* Part of dorsal fin of same to show fin ray chambers and fin rays. *E Branchiostoma caribaeum* Sundevall, posterior part of specimen 55 mm. long, from Tampa Bay, Florida. *F Branchiostoma platae* Hubbs, posterior part of specimen 50 mm. long, from off mouth of Rio de la Plata ("Albatross" Station 2765). *a* anus. *at* atriopore. *aw* atrial wall. *cf* caudal fin. *dch* dorsal fin ray chambers. *g* gonads. *m* myotomes. *me* metapleuron. *r* fin rays. *vch* ventral fin ray chambers. *vf* ventral lobe of caudal fin. The fin rays have been omitted from *A*, *E* and *F* for the sake of clarity.

rapidly about for a short time if the temperature be either raised or lowered. If heated to 40° C. or higher they die; if chilled to 10° C. they become inactive and may die, as they invariably do if kept in a temperature of 4° C. for half an hour. But the thermal reactions are not known for specimens adapted to the winter temperatures that prevail at Bermuda.

It has been found that *B. bermudae* tends to swim away from a source of light; also it is stimulated to activity by the presence of light, *i.e.*, it is photokinetic,[23] and hence it may be expected to bore deeper into the sand if strongly illuminated, as by the sun. But it is more sensitive to mechanical than to photic stimulation, as is the European *B. lanceolatum.*[24] This is especially true of the preoral tentacles and of the outer fringes of the oral hood, which close and open with a sudden winking motion if touched. It is through this reaction that the Lancelet rids itself of the debris that may accumulate on its preoral tentacles, for when these become laden they contract sharply to loosen any waste particles, which are then swept away by water that is expelled simultaneously from the cavity of the oral hood.

Presumably it spawns chiefly in late spring, for the peak of the breeding season is passed before June–July.

Range. Bermuda.

Synonyms and References:

Branchiostoma lubricum Goode, Amer. J. Sci., *14*, 1877: 293 (Bermuda); not *B. lubricum* Costa, 1834.

Amphioxus (no specific name) Brooks, 3rd Annu. Rep. Johns Hopk. Univ., 1878: 54 (Bermuda).

Branchiostoma caribaeum Bristol and Carpenter, Science, N.S. *11*, 1900: 170 (Bermuda); Verrill, Trans. Conn. Acad. Arts Sci., *11*, 1901: 55 (Bermuda); Bean, Field Mus. Publ. Zool., 7 (2), 1906: 29 (Bermuda); Kutchin, Proc. Amer. Acad. Arts Sci., *49* (10), 1913: 571 (peripheral nervous system).

Amphioxus caribaeus Mark, Science, N.S. *20*, 1904: 179 (Bermuda).

Branchiostoma caribbaeum Barbour, Bull. Mus. comp. Zool. Harv., *46*, 1905: 110 in part (specimen from Bermuda); Parker, Proc. Amer. Acad. Arts Sci., *43* (16), 1908: 413 (sensory reactions, Bermuda); Arey, J. exp. Zool., *29* (1), 1915: 37 (swimming habits, Bermuda).

Branchiostoma carribaeum Mark and Crozier, Anat. Rec., *11* (6), 1917: 520 (photo receptors); Conklin, J. Morph., *54* (1), 1932: 70 (breeding season at Bermuda); not *B. caribaeum* Sundevall, 1853.

Branchiostoma bermudae Hubbs, Occ. Pap. Mus. Zool. Univ. Mich., 105, 1922: 9 (descr., discus., Bermuda); Jordan, Evermann and Clark, Rep. U. S. Comm. Fish. (1928), 2, 1930: 7 (Bermuda); Beebe and Tee-Van, Field Bk. Shore Fish. Bermuda, 1933: 2 (descr., ill., Bermuda); Goldschmidt, Biol. Bull. Wood's Hole, *64* (3), 1933: 321 (Bermuda); Pratt, Manual Common Invert. Anim., 1935: 757 (no. of myotomes and gonads, Bermuda).

Branchiostoma caribaeum Sundevall, 1853

Figure 2 E

Study Material. Numerous specimens, 12 to 66 mm. long, from Maryland, Chesapeake Bay, Virginia, North Carolina, eastern and western Florida, the Tortugas, Florida, and Vieques Island, Porto Rico.

23. Parker, Proc. Amer. Acad. Arts Sci., *43* (16), 1908: 426.
24. Franz, Wiss. Meeresuntersuch. Helgoland, *15* (14), 1924: 6.

Distinctive Characters. B. caribaeum differs from *B. lanceolatum* and from *B. africae* in that its anus is about in the middle of the lower lobe of the caudal fin. It is distinguished from *B. bermudae* by the shape of the caudal fin and by the origin of the ventral lobe below that of the dorsal lobe; by the position of the anus, in advance of the midpoint of the lower lobe of the caudal fin; and by the generally greater number of myotomes and dorsal fin-ray chambers (at least 230 of the latter). It is separated from *B. platae* by the shape of the caudal fin, as well as by the position of the anus and by its tendency to have fewer myotomes and dorsal fin-ray chambers.

Additional Description. Anterior end of notochord in rostrum extending forward in a straight line; rostrum marked off from dorsal fin by a subtriangular notch; caudal fin symmetrically lanceolate with narrowly rounded tip, its lower lobe considerably higher than ventral or dorsal fins, its origin opposite origin of its upper lobe and about midway between tip of caudal fin and atriopore; distance from tip of caudal to anus about ⅓ distance from anus to atriopore; dorsal fin ⅛ as high as distance from its crest to margins of metapleura in midregion of body; highest dorsal ray chambers 5 to 8 times as high as long; dorsal ray chambers 230 to 320; precaudal (ventral) fin-ray chambers 18 to 37; 35 to 38 myotomes anterior to atriopore, 13 to 17 between atriopore and anus, and 6 to 9 posterior to anus, recorded totals, 57 to 64; gonads 22 to 29.

Recorded counts for specimens from different localities.

Locality	*Dorsal fin-ray chambers*	*Precaudal fin-ray chambers*	*Myotomes*
Virginia and North Carolina	256–320	33–42	58–64[25]
Florida, including Tortugas	274–330	18–28	57–61
Porto Rico; Bahamas	227–300	15–37	58–61

Color. Live specimens are flesh-color or semitransparent, with a metallic iridescence; those kept in alcohol become opaque and whitish.

Size. The greatest length so far recorded is 66 mm. (see Study Material, p. 13).

Developmental Stages. In the Chesapeake region the pelagic larvae, of the sort usual for the group (p. 5), tend to settle to the bottom by the time they have reached a length of about 7.5 to 8 mm.[26]

Habits. The adults, like those of other species, live buried in coarse or fine sand. In Florida, and presumably elsewhere also, they are most numerous along the edges of sand bars just below the low tide mark where their presence is indicated by small holes in the sand. If the sand is laid bare by a low run of tides it appears that they simply burrow more

25. Andrews (Stud. Biol. Lab. Johns Hopk. Univ., *5*, 1893: 241) reports a total of only 48 myotomes for a specimen from Jamaica. But this is so much fewer than any other recorded count that we judge it to have been erroneous; see also Franz (Jena Z. Naturw., *58*, 1922: 399).

26. See Rice (Amer. Nat., *14*, 1880: 17, pl. 1, fig. 5) for a good illustration of the late larva.

deeply for the time being, instead of moving down the slope.[27] If driven out of their holes, as when a shovel is thrust into the sand close by, they shoot upward into the water and swim vigorously for a brief period with either the ventral or dorsal side uppermost, but always with the anterior end foremost. However, they soon sink to the bottom again. "Generally as soon as they touched the sand, they would half-arch their bodies and almost instantaneously disappear from sight . . . after their disappearance, they very rarely entirely emerged"[28] but continue buried in an oblique position, ventral side uppermost, either with the opening of the oral hood at the surface of the sand or with the anterior portion of the body protruding. Aquarium observations suggest that they protrude and feed chiefly at night. No specific information is available as to the diet of this species (p. 4). *B. caribaeum* has been recorded from the low tide zone down to a depth as great as 24 fathoms. In Florida, sexually mature males, and females "heavy with eggs,"[29] have been reported in March; they are to be expected perhaps two months or so later in the Chesapeake Bay region, where pelagic larvae are to be found in July and August. Sexual maturity is attained in the second or third year.

Range. Atlantic coast of America from Chesapeake Bay to the West Indies. Recorded localities are: several localities in Chesapeake Bay; North Carolina; many localities in Florida, both on the west coast north to Pensacola and on the east coast; the Tortugas; the Snapper Banks; Gulf of Mexico; Bahamas; Porto Rico; Jamaica. It is so common in Florida that one collector reports taking 5,000 of them.

Synonyms and References:

Branchiostoma caribaeum[30] Sundevall, Öfvers. Vet. Akad. Forh., Stockholm, *10*, 1853: 12 (in part, specimens from St. Thomas, West Indies); Gill, Rep. U.S. Comm. Fish. (1871–1872), 1873: 814 (listed, C. Hatteras to Fla.); Yarrow, Proc. Acad. nat. Sci. Philad., 1877: 218 (Bird Shoal, N. Carolina); Jordan and Gilbert, Bull. U.S. nat. Mus., *16*, 1883: 3 (Chesapeake Bay to West Indies); Günther, Rep. Zool. Coll. "Alert," Brit. Mus., 1884: 32 (in part, specimens from St. Thomas, West Indies, characterization, discus.); Garman, in Kingsley, Stand. Nat. Hist., *3*, 1885: 64 (Gulf of Mexico); Andrews, Stud. Biol. Johns Hopk. Univ., *5*, 1893: 240 (in part, specimens from Florida, Gulf of Mexico and West Indies, myotome formula); Jordan and Evermann, Bull. U.S. nat. Mus., 47 (1), 1896: 3 (in part, but *B. platae* also included);[31] Evermann and Kendall, Rep. U.S. Comm. Fish. (1899), 1900: 48 (Snapper Banks, Gulf of Mexico, and Tampa, Florida); Evermann and Marsh, Bull. U.S. Fish Comm., *20* (1), 1902: 59 (in part, Porto Rico, but not the ill., which probably is *B. platae* because of shape of caudal); Tattersall, Trans. Lpool. Biol. Soc., *17*, 1903: 271, 280 (comp. with *lanceolatum*); Lönnberg, Bronn's Klassen., *6*, Abt. 1, Buch 1, 1904: 339 (descr., distrib.); Barbour, Bull. Mus. comp. Zool. Harv., *46*, 1905: 110 (in part, but *bermudae* and *platae* also incl. because of distrib.); Fowler, Proc. Acad. nat. Sci. Philad., *59*, 1908: 461 (Vieques Island, Porto Rico); Smith, N. C. Geol. econ. Surv., Fishes, *2*, 1907: 27 (N. Carolina, but not the ill., which appears to be of *B. platae* because of shape of caudal); Ribeiro, Arch. Mus. nac. Rio de J., *14*, 1907: 146, 199 (in part, but *B. platae* probably included also); Fowler, Proc. biol. Soc. Wash., *33*, 1920: 143, footnote (mentioned); Hubbs, Occ. Pap. Mus. Zool.

27. For an interesting account of their occurrence in Florida, and of methods of collecting them, see Wells (Science, N.S. *64*, 1926: 187).
28. Rice, Amer. Nat., *14*, 1880: 8.
29. Wells, Science, N.S. *64*, 1926: 188.
30. Sometimes spelled *caribbaeum*.
31. The illustration of *caribaeum* by Jordan and Evermann (Bull. U.S. nat. Mus., 47 [4], 1900: pl. 1, fig. 1) appears from the shape of the caudal to have been based on a specimen of *B. platae*.

Univ. Mich., 105, 1922: 6 (descr.); Ribeiro, Fauna brasil., Peixes, *2* (1), Fasc. 1, 1923: 4 (in part, but *B. platae* included because of loc.); Meek and Hildebrand, Field Mus. Publ. Zool., *15* (1), 1923: 1 (descr., but *B. platae* included because of loc.); Nichols, Ann. N. Y. Acad. Sci., *10* (2), 1929: 180, fig. 1 (descr., distrib., Porto Rico); Jordan, Evermann and Clark, Rep. U.S. Comm. Fish. (1928), 2, 1930: 7 (West Indies); Pratt, Manual Common Invert. Anim., 1935: 757 (no. of myotomes, West Indies); Longley and Hildebrand, Pap. Tortugas Lab., No. 34, 1941: 1 (listed for Tortugas, Florida); Fowler, Monogr. Acad. nat. Sci. Philad., 7, 1945: 262 (Sanibel, Florida).

Branchiostoma lanceolatum Günther, Cat. Fish. Brit. Mus., *8*, 1870: 513 (in part, specimens from Caribbean Sea); Rep. Zool. Coll. "Alert," Brit. Mus., 1884: 32 (in part, specimens from N. America); Jordan and Gilbert, Bull. U.S. nat. Mus., *16*, 1882: 867 (in part, specimens from east coast of N. America); Adams and Kendall, Bull. U.S. Fish Comm., *9*, 1891: 292, 293, 298 (SW. Florida); Andrews, Stud. Biol. Johns Hopk. Univ., *5*, 1893: 239 (myotome formula of specimens from Chesapeake Bay); not *Limax lanceolatus* Pallas, 1774.

Amphioxus caribaeus Jordan and Gilbert, Proc. U.S. nat. Mus., *1*, 1879: 388; also, Smithson. misc. Coll., *19*, 1880: 388 (Bird Shoal, N. Carolina); Kirkaldy, Quart. J. micr. Sci., *37*, 1895: 313 (in part, descr., and distrib., but *B. platae* also included); Lönnberg, Bronn's Klassen: 6, Abt. 1, Buch 1, 1904: 239 (descr., distrib.).

Branchiostoma lubricum Goode and Bean, Proc. U.S. nat. Mus., 2, 1880: 121 (E. Florida, name only).

Amphioxus lanceolatus Rice, Amer. Nat., *14*, 1880: 1, 73, pl. 34, fig. 1, 2 (habits, struct., develop.); not *Limax lanceolatus* Pallas, 1774.

Branchiostoma lanceolata Gill, Proc. U.S. nat. Mus., *5*, 1883: 515 (Atlant. coast of U.S.); not *Limax lanceolatus* Pallas, 1774.

Amphioxus (no specific name) Wright, Amer. Nat., *24*, 1890: 1085 (Port Tampa, Florida); Andrews, Circ. Johns Hopk. Univ., *11*, 1892: 75 (young stages recorded from Jamaica); Wells, Science, N.S. *64*, 1926: 187 (ecology, habits, breeding season, age at sexual maturity and coll. methods, Florida).

Branchiostoma caribbaeum Tattersall, Trans. Lpool. Biol. Soc., *17*, 1903: 241, 280 (cf. with *B. lanceolatum* and *B. belcheri*).

Branchiostoma floridae Hubbs, Occ. Pap. Mus. Zool. Univ. Mich., 105, 1922: 7 (descr., Tampa and other Florida loc.); Breder, Field Bk. Mar. Fish. Atlant. Coast, 1929: 4 (Florida); Jordan, Evermann and Clark, Rep. U.S. Comm. Fish. (1928), 2, 1930: 7 (Florida); Pratt, Manual Common Invert. Anim., 1935: 757 (no. of myotomes, gonads, Florida).

Branchiostoma virginiae Hubbs, Occ. Pap. Mus. Zool. Univ. Mich., 105, 1922: 8 (descr., Chesapeake Bay); Hildebrand and Schroeder, Bull. U.S. Bur. Fish., *43*, 1928: 42 (descr., habits, Chesapeake Bay); Breder, Field Bk. Mar. Fish. Atlant. Coast, 1929: 4 (Chesapeake Bay); Jordan, Manual Vert. Anim. NE. U.S., 1929: 4 (descr., Chesapeake Bay to N. Carolina); Jordan, Evermann and Clark, Rep. U.S. Comm. Fish. (1928), 2, 1930: 7 (Chesapeake Bay to Florida); Cowles, Bull. U.S. Bur. Fish., *46*, 1931: 367 (Chesapeake Bay); Pratt, Manual Common Invert. Anim., 1935: 757 (no. of myotomes, Chesapeake Bay to Florida).

Branchiostoma platae Hubbs, 1922

Figure 2 F

Study Material. Thirty-six specimens, 31 to 51 mm. long, from the vicinity of Rio de Janeiro and San Sebastiao I., Brazil, and off the mouth of the Rio de La Plata, Argentina (Lat. 36° 43′ S.; Long. 56° 23′ W.), in the collection of the United States National Museum.

Distinctive Characters. B. platae differs from the two eastern Atlantic species of this genus (*lanceolatum, africae*), and from *caribaeum* as well, in having its anus considerably posterior to the midpoint of the lower lobe of its caudal fin; it differs further from *caribaeum* in that the lower lobe of its caudal fin originates considerably anterior to the origin of the upper lobe. The number of myotomes and dorsal fin-ray chambers is often larger also in *B. platae,* although there is no clear distinction between the two in these respects. *B. platae* differs from *B. bermudae* (which it resembles in the shape of the caudal fin) in its more numerous myotomes (at least 59) and dorsal fin-ray chambers (at least 275).

Additional Description. Anterior end of notochord in rostrum extends forward in a straight line; rostrum marked off from dorsal fin by a shallow notch; caudal fin lanceolate but asymmetrical, the origin of its lower lobe anterior to origin of its upper lobe by a distance ⅓ to ¾ as great as length of upper lobe, about midway between its tip and atriopore; anus considerably posterior to midpoint of lower lobe of caudal; distance from tip of caudal to anus ⅓ as long as from anus to atriopore; dorsal fin ⅕ to ⅐ as high as distance from its crest to the margins of the metapleura; highest dorsal fin-ray chambers 3 to 6 times as high as long; dorsal fin-ray chambers 280 to 330; precaudal fin-ray chambers 19 to 33; myotomes 37 to 40 anterior to atriopore, 13 to 17 between atriopore and anus, and 6 to 9 posterior to the anus, the recorded totals from 58 to 65; gonads 26 to 31.

Color. Presumably as in *B. caribaeum* (p. 14), but no specific information is available.

Size. Recorded specimens have ranged from 28 to 56 mm. in length.

Developmental Stages. Presumably as in other members of the genus.

Habits. Nothing is known of the habits of *B. platae* to differentiate it from its relatives.

Range. Specimens positively identified as *B. platae* are known up to the present time only from off the mouth of the Rio de La Plata and from southern Brazil (San Sebastiao I., the vicinity of Rio de Janeiro). But it seems probable that the Lancelets that have been recorded as *B. caribaeum* from Santos, from Santa Catharina at the mouth of the Amazon, from the Rio de La Plata and from Buenos Aires, were *B. platae.*

Synonyms and References:

Branchiostoma platae Hubbs, Occ. Pap. Mus. Zool. Univ. Mich., 102, 1922: 10 (descr., off Rio de La Plata); Jordan, Evermann and Clark, Rep. U.S. Comm. Fish. (1928), 2, 1930: 7, footnote.

Probable References:

Amphioxus mülleri Moreau, Bull. Acad. Roy. Belg., (2) *39*, 1875: 312, 1 pl., 12 figs. (micr. anat. of notochord, Rio de Janeiro, Brazil).[32]

Branchiostoma caribaeum[33] Günther, Rep. Zool. Coll. "Alert," Brit. Mus., 1884: 32 (in part, specimen from Botafogo, near Rio de Janeiro, Brazil); Jordan and Evermann, Bull. U.S. nat. Mus., 47 (1), 1896: 4 (in part, Brazil incl. in range); Bull. U.S. nat. Mus., 47 (4), 1900: pl. 1, fig. 1 (probably *B. platae* because of shape of caudal); Evermann and Marsh, Bull. U.S. Bur. Fish., *20* (1), 1902: 59 (ill., after Jordan and Evermann 1900, as above, but account is of *B. caribaeum,* N. Carolina); Barbour, Bull. Mus. comp. Zool. Harv., *46*, 1905: 110 (in part, because Rio de La Plata incl. in distrib.); Ribeiro, Arch. Mus. nac. Rio de J., *14*, 1907: 146, 199; Fauna brazil., Peixes, *2* (1), Fasc. 1, 1923: 4 (in part, because Brazil incl. in

32. The name *mülleri* would have priority over *platae* if the specimens in question actually were identical with the latter, as the locality suggests. But Moreau gave no account of their external characters, nor is it likely that the sections on which his studies of microscopic anatomy were based are still in existence.

33. Sometimes spelled "*carribaeum.*"

range); Schreiner and Ribeiro, Arch. Mus. nac. Rio de J., *12*, 1903: 77 (Santa Catharina, Brazil, probably *B. platae* because of loc.); Marelli, Elenc. Sist. Fauna B. Aires (Procord. Vert.); Mem. Ministr. Obras Publ. B. Aires (1922–1923), Fishes, 1924: 543 (no descr., Rio de Janeiro, Rio de La Plata, Buenos Aires, ident. probably because of loc.); Luderwaldt, Rev. Mus. paul., *16*, 1929: 40 (San Sebastiao I., Brazil); Sawaya and Carvalho, Bol. biol. Fac. Med. S. Paulo, N.S. *2*, 1938: 43 (no. of myotomes, behavior in aquarium, Santos, Brazil, at 25 meters).

Amphioxos (no specific name) Luderwaldt, Rev. Mus. paul., *16*, 1929: 11, 15 (in plankton, and from bottom in shallow water, San Sebastiao I., Brazil).

Family *EPIGONICHTHYIDAE*

Description. Gonads developed on right side only; right-hand metapleuron continuous with preanal fin. Characters otherwise those of the order.[34]

Genera. The family includes two well defined genera: in *Epigonichthys* the caudal fin does not extend as a long narrow process, and the oral tentacles are united, one to the next, by a uniformly low intertentacular membrane; in *Asymmetron* the caudal fin, as well as the notochord, is much prolonged posterior to the myotomes as a narrow process, and the intertentacular membrane is much higher ventrally than laterally.

Key to Genera

1a. Caudal fin prolonged as a narrow process; the intertentacular membrane much higher ventrally than laterally (Fig. 3). *Asymmetron* Andrews, 1893, p. 18.[35]

1b. Caudal fin not prolonged as a narrow process; the intertentacular membrane but little higher ventrally than laterally. *Epigonichthys* Peters, 1876.[36]
New Zealand; Australia; East Indies; Ceylon; Maldive and Laccadive Archipelagos, to East Africa.

Genus *Asymmetron* Andrews, 1893

Asymmetron Andrews, Stud. Biol. Lab. Johns Hopk. Univ., *5*, 1893: 237; type species, *A. lucayanum* Andrews. Bahamas.[37]

Generic synonyms:

Branchiostoma (in part) Willey, Amphioxus and Ancest. Vert., 1894: 41; including *A. lucayanum* Andrews, 1893; not *Branchiostoma* Costa, 1834.

Epigonichthys Fowler, Proc. Acad. nat. Sci. Philad., *59*, 1907: 461; including *A. lucayanum* Andrews, 1893; not *Epigonichthys* Peters, 1876.

34. Whitley (Aust. Zool., 7, 1932: 257, 260) divides this family into Epigonichthyidae and Asymmetrontidae.

35. Including *Notasymmetron* Whitley (Aust. Zool., 7, 1932: 260, pl. 13, fig. 6). Whitley mentions, as characters distinguishing this genus from *Asymmetron*, only that it is larger, with the origin and termination of the dorsal fin farther forward in relation to the myotomes.

36. Including *Bathyamphioxus* and *Merscalpellus* Whitley, 1932. The differences on which Whitley (Aust. Zool., 7, 1932: 257–259) has separated these two new genera from *Epigonichthys* are so small that we hesitate to judge their validity, not having seen specimens of them. *Paramphioxus* Haeckel, 1893, is clearly a synonym of *Epigonichthys* Peters, 1876, in our opinion.

37. The characters of this new genus were given also, but without a generic name, by Andrews (Johns Hopk. Univ. Circ., *12*, 1893: 104).

Generic Characters. Median finfold extending far beyond last myotome as a narrow urostyloid process, with notochord reaching nearly to its tip; intertentacular membrane much higher ventrally than laterally; ventral fin-ray chambers lacking in type species, but perhaps present in others;[38] caudal sector of median fin not demarked from more anterior portions dorsal or ventral; gonad pouches begin at myotomes 13 to 15; rostrum continuous ventrally with both right and left sides of oral hood, and these in turn with each metapleuron; atrial chamber extending behind atriopore as a pair of blind sacs; preoral tentacles lack sensory papillae; no olfactory pit.

Species. The type species of the genus is *A. lucayanum* Andrews, 1893, of the West Indian region and Bermuda with which *A. macricaudatum* Parker, 1904, of Florida is doubtless identical (pp. 19, 22); it is also reported from the Philippines. Our examination of its type specimens leads to this same conclusion for *A. orientale* Parker, 1904, of the Maldive Islands, Indian Ocean; nor does *A. caudatum* Willey, 1896, from the Louisiade Archipelago, southeast of New Guinea, appear to have any better claim to specific recognition.[39]

Asymmetron lucayanum Andrews, 1893

Figure 3 A–E

Study Material. Twelve specimens, from North Bimini I., Bahamas, and from Vieques I., Porto Rico (U. S. Nat. Mus.). Five specimens (all of them types of *A. macricaudatum* Parker, 1904) from Salt Key, Florida (Harv. Mus. Comp. Zool., No. 26282). Seven specimens (all of them types of *A. orientale* Parker, 1904) from the Maldive Is., Indian Ocean (Harv. Mus. Comp. Zool., No. 32816).

Distinctive Characters. The long, narrow caudal process marks this species off at a glance from all other Atlantic Lancelets, from which it differs further in the still more important morphological respects stated above (Key, p. 7).

38. Whitley's (Aust. Zool., *7*, 1932: pl. 13, fig. 6) illustration of a specimen identified by him as *caudatum* Willey, 1896, and on which he based the new genus *Notasymmetron*, shows ventral fin-ray chambers, although he made no mention of them in his description.

39. *A. caudatum* Willey (Quart. J. micr. Sci., *39*, 1896: 219, pl. 13, fig. 1–4) supposedly differs from *A. lucayanum* in that its rostrum is marked off by definite notches or constrictions both dorsally and ventrally. But Goldschmidt (Biol. Bull. Wood's Hole, *64*, 1933: 323, fig. 1a) has recently pictured the rostrum as of this same shape for a specimen of *A. lucayanum* from Bermuda, while we have seen one from the latter locality and another from Porto Rico with a notch on the dorsal side, although with none on the ventral side. *A. orientale* Parker (Bull. Mus. comp. Zool. Harv., *46*, 1904: pl. 1, fig. 4) was separated from *A. lucayanum* on the basis of a supposedly narrower caudal fin. But no sharp line can be drawn in this respect between its type specimens, which we have examined, and *A. lucayanum* of Florida and the West Indies (Fig. 3). We may also point out that the tail region of one specimen, a male, described by Willey (1896) as *A. caudatum* was what may be termed the "*lucayanum*" shape, that of the other, a female, of the "*orientale*" shape. It is possible, however, that the Australian form identified by Whitley (Aust. Zool., *7*, 1932: 260, pl. 13, fig. 6) as *caudatum*, and on which he founded the genus *Notasymmetron*, may represent a distinct species, in which case a new specific name would be needed for it; he has pictured it as having ventral fin-ray chambers, although these are not mentioned in his description of it. We may further note that ventral fin-ray chambers are also indicated in the illustration of *A. lucayanum* from the Maldives, by Forster-Cooper (in Gardiner, Fauna Geol., Maldive Laccadive Archip., *1*, 1903: pl. 18, fig. 1). But no trace of such is to be seen in the Maldive specimens that we have examined; nor are they indicated in Franz' (Jena Z. Naturw., *58*, 1922: 426, fig. 30) figure of a Philippine specimen.

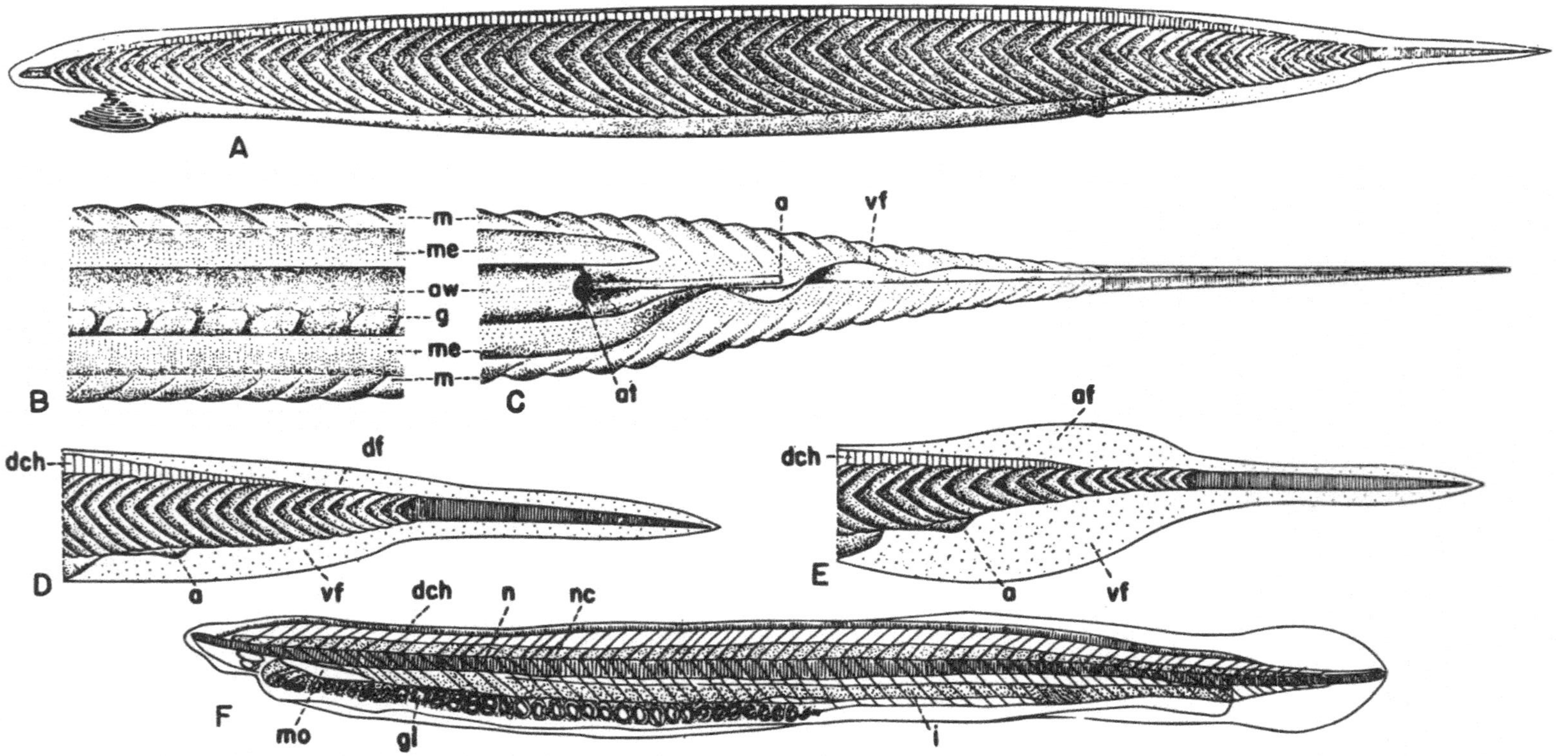

FIGURE 3. *A Asymmetron lucayanum* Andrews, 17 mm. long, from off Culebra, Porto Rico. *B* Ventral view of same from 27th to 33rd myotome showing gonads as seen through atrial wall. *C* Ventral view of posterior part of same to show continuity of right-hand metapleuron with ventral fin. *D* Lateral view of posterior region of same to show shape of caudal fin. *E* Posterior part of another specimen, 16 mm. long, from the same locality, showing the variation in the shape of the caudal fin. *F Amphioxides valdiviae* Goldschmidt, after Goldschmidt. *a* anus. *at* atriopore. *aw* atrial wall. *dch* dorsal fin ray chambers. *df* dorsal fin. *af* anal fin. *g* gonads. *gi* gills. *i* intestine. *m* myotomes. *me* metapleuron. *mo* mouth. *n* notochord. *nc* nerve cord. *vf* ventral fin.

Additional Description. Rostrum, continuous with dorsal fin, varies in shape from very narrow both above and below the notochord to more rounded in shape, and marked off by definite notches both dorsally and ventrally; dorsal fin-ray chambers from 170 to 180; preoral tentacles 21 to 29; intertentacular membrane much higher around the ventral side of oral hood than laterally, where the tentacles on either side are interconnected only near their bases.[40] Median fin (dorsal and ventral), posterior to atriopore, paddle-shaped in some specimens (wider ventrally than dorsally), narrowing rather abruptly between anus and last few myotomes; however, it is narrower in some, with a more gradual transition to the caudal process, there being a wide range of variation in this respect, even among specimens of a single lot, as illustrated in Fig. 3, D, E; the distance from the anus to tip of caudal process nearly twice as great as from last myotome to anus; myotomes 42 to 46 anterior to atriopore, 8 to 9 between atriopore and anus, 11 to 14 posterior to anus, total number 62 to 68;[41] gonads 26 to 29, in a single series on the right-hand side.

Color. This has not been described for living specimens.

Size. Nineteen mm. is the greatest length yet recorded for Atlantic specimens.[42] If, however, the Lancelets recorded as *A. lucayanum* from the Philippines are actually identical with the western Atlantic form, then the species grows larger in the Far East waters, for lengths up to 30 mm. have been reported there.

Developmental Stages. In larvae of 6 mm., with only 22 pairs of gill openings, the caudal extremity is expanded as a rounded fin; by the time the number of gill openings has increased to 27 pairs it has become pointed, after which it elongates to the adult form.[43]

Habits. This species, like other Lancelets, lives much of the time buried in the sand. But apparently it emerges more freely to swim about, for large numbers have been taken in tow nets at or near the surface in Bahaman waters; they are taken most abundantly during the early part of the ebb when the tide has been high about nine o'clock in the evening; rarely are they taken in the daytime, or late at night. In aquaria they seldom leave the sand in the daytime. Experiments have shown them to be negatively phototropic. The posterior part of the body has considerable power of regeneration if cut off just posterior to the

40. In the original account of *A. lucayanum*, Andrews (Stud. Biol. Lab., Johns Hopk. Univ., Zool., *5*, 1893: pl. 13, fig. 6) pictures the median ventral tentacle as considerably shorter than those next to it, with the membrane joining it to them lower than that which joins the next three or four tentacles; Kirkaldy (Quart. J. micr. Sci., *37*, 1895: 318, pl. 34, fig. 3), on the other hand, describes and pictures it as entirely free from the neighboring pair. Forster-Cooper (in Gardiner, Fauna Geogr., Maldive Laccadive Archip., *1*, 1903: 348, fig. 76) shows the membrane as notched where it connects with the ventro-median tentacle. But the membrane is higher there in a specimen from that same region that we have examined; it is so described and pictured also by Franz (Jena Z. Naturw., *58*, 1922: 429, 430, fig. 321) for one from the Philippines. Evidently, then, the difference in this respect is not geographic. Unfortunately, however, our West Indian series are not in good enough condition to clarify this point.

41. Parker (Bull. Mus. comp. Zool. Harv., *46*, 1904: 48) reported only four or five between atriopore and anus for the Florida specimens which he named *A. macricaudatum*. But re-examination of these same specimens yielded counts of eight to nine.

42. Kirkaldy, Quart. J. micr. Sci., *37*, 1895: 319.

43. Larval development is described by Andrews (Stud. Biol. Lab., Johns Hopk. Univ., *5*, 1893: 219, pl. 13, fig. 3, 5).

anus, although it is not known how far regeneration of the tail can proceed. The feeding habits are as described for the group in general (p. 4). The time occupied by the passage of food pellets through the digestive tract, as indicated by carmine particles, may be much less than an hour. In Bahaman waters sexually mature specimens have been taken in June and less often in July.[44]

Range. Circumtropical, with widely separated centers of distribution, and perhaps with local races; known in the western Atlantic from Bermuda, the Florida Keys, North and South Bimini in the Bahamas, Vieques I., Culebra I. and Humacao, Porto Rico, and off Pernambuco, Brazil. Known also from the Maldive Islands (Indian Ocean), the Philippines, the Louisiade Archipelago southeast of New Guinea, Zanzibar, and perhaps from North Australia. Evidently it is abundant locally in the tropical belt of the western Atlantic in suitable situations, for large numbers have been taken both in the Bahamas and at Castle Harbor, Bermuda.

Synonyms and References:

1. Atlantic:

Asymmetron lucayanum Andrews, Stud. Biol. Lab., Johns Hopk. Univ., *5*, 1893: 213, pl. 13, 14, fig. 1–25 (descr., anat., habits, sensory reactions, Bahamas); Kirkaldy, Quart. J. micr. Sci., *37*, 1895: 319, pl. 34, fig. 3 (descr., Bahamas); Jordan and Evermann, Bull. U.S. nat. Mus., 47 (1), 1896: 4 (descr., Bahamas); Evermann and Marsh, Bull. U.S. Fish Comm., *20* (1), 1902: 60 (descr., Porto Rico); Tattersall, Trans. Lpool. Biol. Soc., *17*, 1903: 291, 297, 302 (descr., discus., distrib.); Parker, Bull. Mus. comp. Zool. Harv., *46*, 1904: 49, pl. 1, fig. 2 (myotome formula, number of gonads); Mark, Science, N.S. *20*, 1904: 179 (Bermuda); Bean, in Shattuck, Bahama Islands, Fish., 1905: 296 (Bahamas); Barbour, Bull. Mus. comp. Zool. Harv., *46*, 1905: 110 (Bahamas and Bermuda); Lönnberg, Bronn's Klassen, *6*, Abt. 1, Buch 1, 1905: 244 (descr., Bahamas); Bean, Field Mus. Publ. Zool., 7, 1906: 29 (Bermuda); Gibson, Trans. Linn. Soc. Lond., Zool., (2) *13*, 1910: 241, 242 (number of myotomes; Bahamas, *Amphioxides pelagicus* perhaps the neotenic larva of this species); Hubbs, Occ. Pap. Mus. Zool. Univ. Mich., 105, 1922: 16 (off Brazil); Franz, Jena Z. Naturw., *58*, 1922: 377, 426, fig. 30–32 (descr., discus., distrib.); Breder, Field Bk. Mar. Fish. Atlant. Coast, 1929: 4 (Bahamas); Nichols, Mem. N. Y. Acad. Sci., *10*, 1929: 181, fig. 2 (descr., Bahamas, Porto Rico); Goldschmidt, Biol. Bull. Wood's Hole, *64*, 1933: 231, fig. 1 A, B (relation to *Amphioxides* larvae, Bermuda); Beebe and Tee-Van, Field Bk. Shore Fish. Bermuda, 1933: 21 (descr., distrib., Bermuda); Goodrich, Quart. J. micr. Sci., 75, 1933: 723 (nephridia; Bermuda); Pratt, Manual Common Invert. Anim., 1935: 757 (no. of myotomes and gonads); Andrews, Bigelow and Morgan, Sci. Mon., 61 (5), 1945: 341, 343 (habits, ill., Bimini, Bahamas).

Branchiostoma lucayanum Willey, Amphioxus and Ancestr. of Vert., 1894: 41 (Bahamas).

Asymmetron macricaudatum Parker, Bull. Mus. comp. Zool. Harv., *46*, 1904: 47, pl. 2, fig. 7 (descr., discus., Salt Key, Florida); Pratt, Manual Common Invert. Anim., 1935: 757 (no. of myotomes and gonads).

Epigonichthys leucayanum Fowler, Proc. Acad. nat. Sci. Philad., *59*, 1907: 461 (Bermuda).

2. Indo-Pacific:

Asymmetron caudatum Willey, Quart. J. micr. Sci., *39*, 1896: 219, pl. 13, fig. 1–4 (descr., ill. of head and tails of male and female; size; Louisiade Archipel.); Quart. J. micr. Sci., *44*, 1901: 271 (*caudatum* a subspecies of *lucayanum*); Willey's Zool. Res., *6*, 1902: 725 (ill., *caudatum* a subspecies of *lucayanum*); Hubbs, Occ. Pap. Mus. Zool. Univ. Mich., 105, 1922: 16 (refs.); Lönnberg, Bronn's Klassen, *6*, Abt. 1, Buch 1, 1905: 244 (doubts if distinct from *lucayanum*).

Asymmetron lucayanum Forster-Cooper, in Gardiner, Fauna Geogr., Maldive Laccadive Archip., *1*, 1903:

44. The foregoing account is based on observations by Andrews, 1893.

348, pl. 18, fig. 1 (descr., ill., size, Maldive Is., see footnote 39, p. 19); Punnett, in Gardiner, as above, 1903: 362 (number of myotomes, sizes); Tattersall, in Herdman, Rep. Govt. Ceylon Pearl Oyster Fish., Gulf of Manaar, suppl. 6, 1903: 222 (listed for Maldives and Zanzibar); Gibson, Trans. Linn. Soc. Lond., Zool., (2) 13, 1910: 241 (Maldives, by ref. to Forster-Cooper, 1903, *Amphioxides pelagicus* perhaps its neotenic larva); Raff, Zool. Res. "Endeavour," Austral. Dep. Trade. Customs, 1 (3), 1912: 305 (listed for Louisiade Archipel., Maldives, Zanzibar, Torres Strait); Franz, Jena Z. Naturw., *58*, 1922: 426, 427, fig. 30 (ill., myotome counts, size, Philippine specimens).

Asymmetron orientale Parker, Bull. Mus. comp. Zool. Harv., *46*, 1904: 46, pl. 1, fig. 4 (descr., ill., Maldive Is.); Hubbs, Occ. Pap. Mus. Zool. Univ. Mich., 105, 1922: 16 (ref.).

Epigonichthys caudatus Jordan and Evermann, Bull. U.S. Bur. Fish., *25*, 1906: 191 (name only, Louisiade Archipel.); Fowler, Mem. Bishop Mus., *10*, 1928: 17 (name only, Louisiade Archipel.).

Probable References:[44a]

Heteropleuron (*Asymmetron*) *lucayanum* Haswell, Rec. Aust. Mus., 7, 1908: 35 (Murray I., Torres Str., specimen subsequently named *Notasymmetron* by Whitley, 1932).

Notasymmetron caudatum Whitley, Aust. Zool., 7, 1932: 260, pl. 13, fig. 6 (descr., ill., Torres Str. spec.); Fish. Aust., *1*, 1940: 250, fig. 290 (N. Queensland, Murray I., Torres Str.).

Amphioxides Larvae

Synonyms:

Branchiostoma (in part) Günther, "Challenger" Rep., Zool., *31* (2), 1889: 43, for *B. pelagicum* Günther; not *Branchiostoma* Costa, 1834.

Amphioxides Gill, Amer. Nat., *29*, 1895: 458; type species, *Branchiostoma pelagicum* Günther, 1889.

"Pelagic larvae," Forster-Cooper, in Gardiner, Fauna Flora, Maldive Laccadive Archip., *1* (4), 1903: 354, pl. 6, fig. 3–6.

Asymmetron (in part) Pietschmann, in Kükenthal and Krumbach, Handb. Zool., *6* (1), Lief 1, 1929: 110, fig. 107, for *Branchiostoma pelagicum* Günther, 1889.

Group Characters. Small Lancelets, living pelagically, in which (as in larval Lancelets in general) the mouth is on the left side, without oral tentacles, the metapleural folds are separate, one from another, so that there is no closed atrial cavity, and in which the gill clefts are in a single row on the ventral side, but which grow to a greater length (up to 21 mm.)[45] and develop a greater number of gill clefts than is usual for Lancelet larvae before metamorphosis and which may show at least the rudiments of gonads.

As pointed out above (p. 7), these *Amphioxides* are juvenile specimens that retain their larval characteristics not only to a greater size than is characteristic of their parent species[46] but to a more advanced stage in their own development; they are not a primitive group as was originally supposed.[47] While they may develop gonads, as just stated, there is no evidence that Lancelets ever become mature sexually as *Amphioxides*.

44a. See footnote 35, p. 18.

45. Forster-Cooper (in Gardiner, Fauna Geogr., Maldive Laccadive Archip., *1*, 1903: 352, pl. 4) reports as *Branchiostoma pelagicum* a 21-mm. specimen from the central Indian Ocean that appears to be an *Amphioxides* because no trace of oral tentacles was to be seen.

46. Larvae of this sort are known, technically, as "neotenic."

47. Their larval nature, first suggested by Goldschmidt (Zool. Anz., *30*, 1906: 443) and accepted by Gibson (Trans. Linn. Soc. Lond., Zool., [2] *13*, 1910: 239), was substantiated by Goldschmidt (Dtsch. Sud-polar Exped., *11*, Zool. 3, 1909: 237), who discovered *Amphioxides* in which the secondary series of gill openings had begun to form, *i.e.*, which had commenced their metamorphosis.

Range. Amphioxides larvae have been reported from localities so generally distributed and so widely separated[48] that they are to be expected anywhere on the high seas, within the latitudinal belt where Lancelets of the family Epigonichthyidae occur in any abundance.

Species. One specimen of *Branchiostoma lanceolatum* has been reported in the *Amphioxides* stage, *i.e.*, it still retained its larval characters at a length of 5.5 mm., although this species usually undergoes its metamorphosis at about 4.5 mm.[49] The other *Amphioxides* larvae that have been described fall in two categories; the dorsal fin-ray chambers of the one extend forward well beyond the first myotome, those of the other terminate at the dorsal margin of the first myotome. Among the specimens of the second group, some agree in number of myotomes with the *Branchiostoma pelagicum* of Günther, 1889, and have been identified with the latter for this reason.[50] Other specimens of the group with a larger number of myotomes (70) for only two recorded specimens have been described as a distinct species, *Amphioxides stenurus* Goldschmidt, 1905. But it is doubtful whether the distinction between it and *pelagicus* is valid. The other category, with dorsal fin-ray chambers extending far forward, has been named *valdiviae* Goldschmidt, 1905.

None of these has been definitely connected with any particular parent species. In the few specimens in which the rudiments of gonads were to be seen, however, these were in a single series and on the left-hand side, suggesting an *Epigonichthys* or an *Asymmetron* parentage; *i.e.*, that they belong to the family Epigonichthyidae.

Key to Species of *Amphioxides*

1a. Dorsal fin originates opposite 21st to 25th myotome; ventral fin about opposite 40th myotome; dorsal fin-ray chambers do not extend forward beyond dorsal edge of 1st myotome.

- 2a. Not more than 68 myotomes. — *pelagicus* Günther, 1889, p. 25.
- 2b. 70 myotomes. — *stenurus* Goldschmidt, 1905. Indian Ocean.

1b. Dorsal fin originates opposite 32nd to 33rd myotome or even farther back; ventral fin about opposite 43rd myotome; dorsal fin-ray chambers extend forward considerably beyond dorsal edge of 1st myotome. — *valdiviae* Goldschmidt, 1905, p. 27.[51]

48. Reported from the English Channel, Bermuda, off the Amazon and at several other localities in the equatorial and south tropical Atlantic; mouth of the Red Sea; widespread in the tropical Indian Ocean; from the vicinity of the Hawaiian Islands. For a distribution chart, to which might be added a few more recent records, see Goldschmidt (Dtsch. Sud-polar Exped., *11*, Zool. 3, 1909: pl. 11).

49. Goldschmidt, Zool. Anz., *30*, 1909: 443.

50. So identified by Goldschmidt (Wiss. Ergebn. 'Valdivia,' *12*, 1905: 46). But this identification cannot be regarded as final until the type specimen of Günther's *pelagicum* is re-examined, because his illustration of it ("Challenger" Rep., Zool., *31* [2], 1889: pl. 6, fig. B) does not show the anterior termination of the fin-ray chambers clearly.

51. Previous accounts (Goldschmidt, Wiss. Ergebn. 'Valdivia,' *12*, 1905: 46, pl. 1, fig. 3, 4; Gibson, Trans. Linn. Soc. Lond., Zool., [2] *13*, 1910: 217, pl. 15, fig. 1) base the distinction between *valdiviae* and *pelagicus* chiefly on the shape of the tail fin, which is supposedly more sharply marked off and blunter at the tip, with the notochord ending more bluntly in the former than in the latter. But the two supposed species appear to intergrade in this respect.

Amphioxides pelagicus (Günther), 1889

Study Material. None.

Distinctive Characters. Amphioxides larvae differ from adult Lancelets in that they have neither atrial cavity nor oral cirri; their mouths are on the left-hand side, and their gill clefts are in a single series. *Pelagicus* is separable from *valdiviae* by the facts that its dorsal fin-ray chambers do not extend forward past the first myotome, and that the dorsal fin originates about opposite the 21st to 25th myotome, while in *valdiviae* it commences opposite the 32nd to 33rd myotome, or even more posteriorly; *pelagicus* is separated from *stenurus* by fewer myotomes (not more than 68 in *pelagicus*).

Additional Description.[52] Caudal fin usually not sharply marked off, although sometimes more definitely so, its tip lancet-shaped, usually pointed; notochord tapering to a narrowly pointed tip; dorsal finfold originates about opposite the 21st to 25th myotome; the ventral farther posterior in some (opposite the 40th myotome), but farther forward in others; the dorsal fin-ray chambers extend forward only to the posterior edge of the first myotome, anterior to which they are replaced by an undivided tapering canal; two chambers per myotome anteriorly, increasing to 3 or 4 per myotome posteriorly; gill clefts 16 to 18 in specimens of 5 to 6 mm., with 24–26 reported for Bermuda specimens of 8 to 10 mm. or longer, and up to 30 for the Indian Ocean form; myotomes usually 63 to 64 (50 or 51 preanal and 13 postanal) with totals of 67 also reported from Bermuda, and 62 to 68 from Indian Ocean.

Color. No information available.

Size. Pelagicus has been recorded up to 16 mm. in length from Bermuda; up to 10 mm. from the Indian Ocean.[53]

Parentage. It is probable that the *pelagicus* of the Atlantic is the neotenic larva of *Asymmetron lucayanum*, the *pelagicus* of the Indian Ocean that of the local representative of *lucayanum.*[54]

Habits. Nothing positive is known of the habits of this or of any other *Amphioxides*, except that it is planktonic. In the Indian Ocean *Amphioxides* of the *pelagicus* type have been taken in abundance at or near the surface and similarly at several localities in the tropical Atlantic. On the other hand, many of the records have been from nets fished at considerable depths.[55] In most instances, however, there is no certainty that the specimens were actually taken at the depth at which the major part of the haul was made, because the nets also fished while being lowered and hauled up again. Consequently, the depth of chief abundance is still to be learned. We think it probable that the odd speci-

52. Based on previous descriptions and illustrations.
53. A 21-mm. specimen from the Indian Ocean, reported and pictured by Forster-Cooper (in Gardiner, Fauna Geogr., Maldive Laccadive Archip., *1*, 1903: 352, pl. 4) was in such poor condition that its specific identity is doubtful.
54. This identity has been maintained by Gibson (Trans. Linn. Soc. Lond., Zool., [2] *13*, 1910: 241). Although Goldschmidt (Biol. Bull. Wood's Hole, *64*, 1933: 324) has questioned it, the number of myotomes that he records for the Bermudian *pelagicus* (50 to 51 preanal and 13 postanal, as well as a stated total of 67) falls within the limits reported for *lucayanum* from the Bahamas (62 to 68).
55. Ostensibly down even to 1,000 fathoms (1,829 meters).

mens that have been brought up from as deep as 250 to 500 fathoms in closing nets[56] were taken while in the process of sinking into the oceanic abyss, as may be the eventual fate of all the *Amphioxides* that drift out into deep water.

The frequency with which *pelagicus* has been reported from deep hauls makes it likely that it can exist for a time in considerably cooler water, although it is primarily tropical in its thermal relationships. But we have yet to learn how low a temperature may be fatal to it, and how rapidly.

Nothing is known of its feeding habits, nor of those of any *Amphioxides*.

Range. Specimens showing the characters of *pelagicus* have been reported from the vicinity of the Hawaiian Islands, the type locality; from numerous localities distributed across the tropical belt of the Indian Ocean between latitudes 10° 8′ S. and 9° 6′ N.; from five stations between the St. Helena and Ascension Islands and the African Coast (Lat. about 14° S. to about 4° N.); from one station off the mouth of the Amazon; and from the vicinity of Bermuda, whence 87 specimens were recorded from 27 townet hauls;[57] perhaps also from the Bahamas.[58]

Synonyms and References:

Branchiostoma pelagicum Günther, "Challenger" Rep., Zool., *3* (2), 1889: 43, pl. 6, fig. B (descr., ill., N. Pacific near Honolulu); Kirkaldy, Quart. J. micr. Sci., *37*, 1895: 320 (mention); Tattersall, Trans. Lpool. Biol. Soc., *17*, 1903: 296 (distrib.); in Herdman, Rep. Govt. Ceylon Pearl Oyster Fish., Gulf of Manaar, Snppl. 6, 1903: 214, plate not numbered, fig. 16 (descr., Indian Ocean); Lönnberg, in Bronn's Klassen, *6*, Abt. 1, Buch 1, 1904: 245 (ref. to type specimen); Franz, Jena Z. Naturw., *58*, 1922: 433 (refs., discus., incl. *valdiviae*); Pietschmann, in Kükenthal and Krumbach, Handb. Zool., *6* (1), Lief 1, 1929: 109 (discus.).

Amphioxides pelagicus Gill, Amer. Nat., *29*, 1895: 458 (name); Tattersall, Trans. Lpool. Biol. Soc., *17*, 1903: 275 (diagn.); Goldschmidt, Wiss. Ergebn. 'Valdivia,' *12*, 1905: 45, pl. 1, fig. 3, 4 (descr., ill., Indian Ocean and trop. Atlantic); Willey, Quart. J. micr. Sci., *50*, 1906: 581 (ref. to Goldschmidt, 1905); Goldschmidt, Dtsch. Sud-polar Exped. (1901–1903), *11*, Zool. 3, 1909: 234, pl. 27 (discus., trop. Atlant., chart of distrib.); Gibson, Trans. Linn. Soc. Lond., Zool., (2) 13, 1910: 217, pl. 15, fig. 1 (descr., discus., ill., Indian Ocean); Hubbs, Occ. Pap. Mus. Zool. Univ. Mich., 105, 1922: 4 (listed); Goldschmidt, Biol. Bull. Wood's Hole, *64*, 1933: 324 (meas., no. of myotomes, discus., Bermuda).

Asymmetron pelagicum Pietschmann, in Kükenthal and Krumbach, Handb. Zool., *6* (1), Lief 1, 1929: 110, fig. 107 (ill.).

Doubtful References:

Branchiostoma pelagicum Forster-Cooper, in Gardiner, Fauna Geogr., Maldive Laccadive Archip., *1* (4), 1903: 352 (21 mm., Indian Ocean; ident. doubtful because of poor condition).

Not *Branchiostoma pelagicum* Parker, 1904, Bull. Mus. comp. Zool. Harv., *46*, 1904: 40, pl. 1, fig. 1 (this was *valdiviae* in reality; see below, p. 28).

56. Gibson (Trans. Linn. Soc. Lond., Zool., [2] *13*, 1910: 214) lists two such instances from the Indian Ocean.

57. For a list of Bermuda records, see Goldschmidt (Biol. Bull. Wood's Hole, *64*, 1933: 322).

58. A six-mm. specimen from the Bahamas, pictured by Andrews (Stud. Biol. Lab., Johns Hopk. Univ., *5*, 1893: pl. 13, fig. 5) as *Asymmetron lucayanum*, is classed by Gibson (Trans. Linn. Soc. Lond., Zool., [2] *13*, 1910: 241) as *Amphioxides*. But Andrews' statement (p. 219) that it had "22 branchial clefts on a side" suggests that it was a specimen in the process of metamorphosis.

Amphioxides valdiviae Goldschmidt, 1905

Figure 3 F

Study Material. One specimen, 9 mm. long, with 33 gill clefts, from the Maldive Islands.[59]

Distinctive Characters. Amphioxides larvae of the *valdiviae* type are separated from those of the *pelagicus-stenurus* type by the following features: their dorsal fin-ray chambers extend forward well past the first myotome, and the dorsal fin originates about opposite the 32nd or 33rd myotome (in *pelagicus* about opposite the 21st to 25th myotome). Differences in the shape of the tail that have been given specific weight appear not to be constant.

Additional Description.[60] Caudal sector of fin paddle shaped with blunted tip and rather definitely marked off from more anterior portion (dorsal and ventral) by a constriction, about opposite anus; notochord blunt-tipped posteriorly; dorsal finfold originates opposite 32nd to 33rd myotome, the ventral finfold about opposite 43rd myotome; dorsal fin-ray chambers extend forward beyond first myotome; about 5 dorsal fin-ray chambers per myotome; gill clefts 25 to 35 in specimens of 5.7 to 8 mm., 33 to 35 in those of 8 to 9.25 mm.; myotomes 55 to 58 anterior to anus, 11 to 15 posterior to it, with recorded totals of 67 to 70.

Color. No information available.

Size. The maximum recorded length is 9.25 mm.

Parentage. If *Amphioxides* of this type are the neotenic larvae of species of *Epigonichthys,* as seems probable,[61] the parentage of *valdiviae* of the Atlantic presents an interesting question, because *Epigonichthys* is not yet known to occur there.

Habits. Nothing is known of the thermal or bathymetric occurrence of *valdiviae* to separate it from *pelagicus* (p. 25).

Range. Tropical Atlantic and Indian Oceans. While *valdiviae* has not yet been reported from the western Atlantic, it is to be expected in this section of the tropical belt, many specimens having been taken at the surface off tropical West Africa (Portuguese Senegal), some of them showing the beginnings of metamorphosis.[62] It has been reported also off the African Coast, south of Tenerife, and at a number of localities in the tropical Indian Ocean, including the vicinity of Sumatra, Bay of Bengal, Maldive Islands, near the Chagos Archipelago, southeast of the Seychelles, and in the vicinity of Farquhar Islands.

59. This is the specimen described and pictured by Parker (Bull. Mus. comp. Zool. Harv., *46*, 1904: 40, pl. 1, fig. 1, 2). The gonads credited to it in the original account prove actually to have been the gill bars.

60. Based on descriptions by Goldschmidt (Wiss. Ergebn. 'Valdivia,' *12*, 1905: 47, pl. 1, fig. 1), Gibson (Trans. Linn. Soc. Lond., Zool., [2] *13*, 1910: 217), and on the specimen listed above.

61. Gibson (Trans. Linn. Soc. Lond., Zool., [2] *13*, 1910: 241) suggests this parentage for *Amphioxides valdiviae* of the Indian Ocean.

62. Goldschmidt, Dtsch. Sud-polar Exped., *11*, Zool. 3, 1909: tab. p. 11.

Synonyms and References:

Branchiostoma pelagicum Parker, Bull. Mus. comp. Zool. Harv., *46*, 1904: 40, pl. 1, fig. 1 (ill. showing blunt notochord, rounded tail, and dorsal fin-ray chambers extending well beyond the first myotome; this is clearly *valdiviae*; see also Study Material, p. 27, and footnote 59, p. 27); not *B. pelagicum* Günther, 1889.

Branchiostoma pelagicum (in part) Franz, Jena Z. Naturw., *58*, 1922: 434 (*valdiviae* incl. in synonymy).

Amphioxides valdiviae Goldschmidt, Wiss. Ergebn. 'Valdivia,' *12*, 1905: 47, pl. 1, fig. 1 (descr., ill., trop. Atlant. and trop. Indian Oceans); Gibson, Trans. Linn. Soc. Lond., Zool., (2) *13*, 1910: 217 (descr., comp. with *pelagicus*); Goldschmidt, Dtsch. Sud-polar Exped., *11*, Zool. 3, 1909: 234, pl. 27 (specimens commencing metamorphosis, trop. Atlant., chart of distrib.); Hubbs, Occ. Pap. Mus. Zool. Univ. Mich., 105, 1922: 4 (listed); Goldschmidt, Biol. Bull. Wood's Hole, *64*, 1933: 321 (ref. to spec. showing beginning of metamorphosis).

CHAPTER TWO

Cyclostomes

BY

HENRY B. BIGELOW *and* WILLIAM C. SCHROEDER

ACKNOWLEDGMENTS

Many persons have contributed information used in the preparation of this paper; to all of them we are grateful. Our thanks are due in particular to A. A. Blair, Wilfrid Templeman, A. G. Huntsman, R. A. McKenzie, Raymond T. Kinney, and Ralph H. Osborn for assembling information as to the status of the Sea Lamprey in the rivers of Newfoundland, the Maritime Provinces of Canada and Massachusetts; to John Tee-Van for photostatic copies of bibliographic citations; to A. S. Romer, who assisted in the classification; to Ludlow Griscom and James A. Peters for assistance on scientific nomenclature; to H. W. Fowler and Leonard P. Schultz for assistance in numerous ways, and finally to the late Thomas Barbour, who placed the collections of the Museum of Comparative Zoology at our disposal.

GENERAL DISCUSSION

Scope of Study. The following pages give descriptions, life histories, geographic distribution so far as known, and lists of published citations for the genera and species of marine Cyclostomes that are known to occur on the western side of the North Atlantic. The characterizations of the orders, families and genera cover the Cyclostomes as a whole, as does the key to the species of the only genus in question that includes more than a single marine representative.

Descriptions. These are based on the Study Material listed under each species. The accounts of the habits and geographic distribution are taken partly from the published records, partly from data of our Study Material, and partly on information from numerous correspondents, supplemented by our own observations.

Keys. The keys, from the higher groups down to species, have been prepared solely for ease of identifying any Cyclostome that may come to hand; for that reason we have selected characters which are most easily used.

References. All citations listed among the references, with the few exceptions noted, were consulted in the original; for a list of co-operating libraries, see the general discussion for the section on Sharks.

Sources of Material. The collection of Cyclostomes in the Museum of Comparative Zoology has been the chief basis of our studies, but the collections at the United States National Museum and the Academy of Natural Sciences at Philadaelphia have also been drawn upon.

Class AGNATHA

Subclass CYCLOSTOMATA

Fish or fish-like vertebrates,[1] eel-like in form, the skeleton cartilaginous or fibrous, without bone; no definitely developed jaws or bony teeth; at least the rudiments of a cranium present in the form of a simple trough below the brain in some, but partially roofed in others; notochord not constricted at all segmentally; vertebral column represented by a simple notochordal sheath, without vertebral centra, but with rudimentary neural arches (not joined above) in some; no shoulder or pelvic girdles, no paired limbs and no true ribs; 6 to 14 pairs of gill pouches opening either directly into the pharynx internally or into a separate respiratory tube, which in turn opens into the mouth below the gullet, and opening to the exterior either separately or by a single aperture on each side; skin without scales; nostril single, either opening into the mouth or not; intestine with internal longitudinal ridges, or with a slight spiral fold; ear with either 1 or 2 semicircular canals only; no sympathetic nervous system, or spleen; heart without conus arteriosus; no cloaca, the genital apertures being separate from anus. Development oviparous, with or without a definite larval stage; the sexes separate or not.

The Cyclostomes are generally considered the most primitive of true vertebrates; structurally they are the simplest. They are easily distinguishable from all the higher fishes by their peculiar jawless mouths, by the fact that there is only one nostril, and by the very primitive cranium.[2]

Key to Orders

1a. Snout with prominent barbels; no separate dorsal fin; eyes not visible externally; nasal opening at tip of snout; mouth not funnel- or disc-like. Myxinoidea, p. 31.

1b. Snout without prominent barbels; one or more dorsal fins separate from caudal; eyes in adult well developed, and visible externally; nasal opening on upper side of head; mouth opens as a funnel or disc. Petromyzonida, p. 43.

1. Opinions differ as to whether the Cyclostomes are to be regarded as a class distinct from the true fishes, or as a subclass of the latter.

2. For detailed accounts of the anatomy of the Cyclostomes, see especially Lönnberg, Favaro, Mozejko and Rauther in Bronn's Klassen, *6*, Abt. 1, Buch 1, 1905–1924: 16–39, pl. 13–32; also Pietschmann, in Kükenthal, Handb. Zool., *6*, 1st half, 1929–1935: 2–5.

Order MYXINOIDEA

Description. Six to 15 pairs of gill pouches, opening internally into the pharynx, those on each side opening either separately to the exterior or by a single common aperture; 2 pairs of barbels on side of nostril and 1 or 2 pairs at side of mouth; single continuous fin running posteriorly around tail and anteriorly on lower surface; fin rays restricted to tail region; nostril at tip of snout opening into mouth and serving as the entrance for water in respiration; mouth not funnel-like; tongue evertible, with two rows of horny, rasp-like teeth; prominent row of segmentally arranged mucous pores along each side; anus near posterior end; eye, without lens or iris, not visible externally, and apparently degenerate; cranium a simple, unroofed trough below brain; barbels and tongue supported by cartilaginous bars; branchial basket reduced to a vestige; ear with one semicircular canal only; a pancreas-like gland well developed; notochordal sheath without rudimentary neural arches; intestine with internal longitudinal folds, but without spiral valve.

Development. According to recent studies (see discussion and footnote 14, p. 35) the myxinoids, although structurally hermaphroditic, are not functionally so. Development is direct, without a larval stage.[3]

Habitat. Exclusively marine.

Families. Only one, Myxinidae, is known.

Family MYXINIDAE

Hags

Characters. Those of the order.

Discussion of Genera. The members of the family fall in two sharply alternative groups, depending on whether the gill pouches of each side open to the exterior by a single common orifice, or separately. By common consent, members of the first group fall in one genus, *Myxine*. But the members of the second group have been divided, depending on the importance given by different students of classification to the number of gills and the grouping of their openings. Since none of the latter group occur in the western North Atlantic we need only point out that the use of the number of gills for generic separation does not seem permissible, for species occur with 5, 6 to 7, 8, 10, 11 to 12, and 14. But the difference between the close grouping of the gill openings in *Paramyxine*, and their wide spacing in all the others, does seem worth generic recognition, as indicated in the following key.[4]

3. Dean's detailed description of the early development of *Eptatretus stouti*, in "Festschrift von Kupfer's" (1899: 221–277, pls. 15–26) has formed the basis for subsequent accounts in many textbooks.

4. Holly (in Schultze, Kükenthal, *et al.*, Tierreich, Lief 59, 1933: 45) includes the shape of the gill openings as an additional generic character, but Matsubara (J. Imp. Fish. Inst. Tokyo, 32 [1], 1937: 13) has recently shown that this varies so widely in *Paramyxine* as not to be reliable.

Key to Genera

1a. Gill pouches on each side connect with exterior by single common aperture.
Myxine Linnaeus, 1758, p. 32.
Atlantic and Pacific Oceans.

1b. All gill pouches on each side open independently to exterior.
2a. Gill openings on each side (16 in number) are close together.
Paramyxine Dean, 1904.
Japan.

2b. Gill openings on each side (5–14 in number) separated by interspaces of considerable width. *Eptatretus* Cloquet, 1819.[5]
Pacific Ocean.

Genus *Myxine* Linnaeus, 1758

Hags

Myxine Linnaeus, Syst. Nat., 1758: 650; type species, *M. glutinosa* Linnaeus. Atlantic Ocean.

Generic Synonyms:

Petromyzon (in part) Walbaum, P. Artedi Genera Pisc. Emend. Ichthyol., Pt. 3, 1792: 500, for *M. glutinosa*; not *Petromyzon* Linnaeus, 1758.

Gastrobranchus Bloch, Naturg. Ausländ. Fische, Pt. 9, 1793: 66, pl. 413; type species, *G. coecus* Bloch. Denmark, Sweden, Norway and Iceland.

Muraenoblenna Lacépède, Hist. Nat. Poiss., *5*, 1803: 652; type species, *M. olivacea* Lacépède. Straits of Magellan.

Pholis Oken, Lehrb. Naturg., *3* (2), 1816: 122; alternative name for *Myxine*.

Gasteobranchus Buckland, Nat. Hist. Brit. Fish., 1881: 144; evident misspelling for *Gastrobranchus*.

Generic Characters. Five or 6 gill pouches on each side opening to exterior by a single aperture on ventral surface, close in front of origin of ventral finfold, the left-hand gill opening, which receives the oesophago-cutaneous duct, being much the larger; fleshy flap ("rostrum" or "labrum") overhanging nostril anteriorly; nostril close to tip of snout; snout with 6 barbels, flanking both nostril and mouth; each side with a series of large mucous pores, segmentally arranged, extending from a short distance behind the mouth rearward nearly to the caudal extremity. Characters otherwise those of the family and order.[6]

Range. Continental shelves and slopes of the North Atlantic in north temperate and

5. This genus has been called most commonly *Bdellostoma* Müller, 1835. It has been argued by Apstein (Sitzber. Gesellsch. Naturf. Berlin, 1915: 187) and Rauther (in Bronn's Klassen, *6*, Abt. 1, Buch 1, Lief 39, 1924: 685) that it would be well to accept this as a *nomen conservandum*. But, awaiting action by the International Commission on Zoological Nomenclature, it seems to us wiser to use the older name. As Rauther points out, the original description of *Eptatretus* was based by Cloquet (Dict. Sci. Nat., *15*, 1819: 135) on a combination of two species—the Chilean *dombei* and an unnamed species from the South Seas. But even if *Eptatretus* were to be abandoned as a generic name on that account, *Bdellostoma* is long antedated by *Homea* Fleming (Phil. Zool., *2*, 1822: 374) and by M'Murtrie's *Heptatremus* (Anim. Kingd. [after Cuvier], *2*, 1831: 298).

6. For an excellent account of the general morphology of *Myxine*, see Smitt (Hist. Scand. Fish., *2*, 1895: 1196).

subarctic latitudes, including the Mediterranean (Adriatic) in moderate depths; coasts of southern Argentina, Chile, Japan, and South Africa; Gulf of Panama in deep water (1,335 meters), the latter being the only locality where the genus is known to occur in tropical or subtropical latitudes.

Species. The representatives of the genus fall in two well defined groups, according to whether the first three lingual teeth of the anterior row are fused at the base, or only the first two, which is more usual. One member of the first of these groups, *M. circifrons*, is further set apart from all others in the genus by the fact that it has only five pairs of gill sacs. Unfortunately this feature is not apparent from the exterior, and other characters that have been used to separate supposed species, such as relative length of head and number of mucous pores, overlap to such an extent that it is doubtful how many of the named forms will finally stand. For further discussion, see p. 38.

Key to Species of *Myxine*

1a. First 3 lingual teeth in anterior series fused together at base.
- 2a. Lingual teeth $\frac{13}{11}$; head nearly or quite 33.3% of total length.
  circifrons Garman, 1899.[7]
  Gulf of Panama.
- 2b. Lingual teeth only $\frac{12}{11}$ or fewer; head not more than 29% of total length.
  - 3a. 26 or 27 mucous pores anterior to gill openings, and 12 or 13 posterior to anus. *garmani* Jordan and Snyder, 1901.
    Japan.
  - 3b. Only 22 mucous pores anterior to gill openings and 9 posterior to anus.
    tridentiger Garman, 1899.
    Straits of Magellan.

1b. Only 1st 2 lingual teeth in anterior row fused together at base.
- 4a. Lingual teeth only $\frac{6}{7}$. *paucidens* Regan, 1913.
  Japan.
- 4b. Lingual teeth $\frac{7}{7}$ or more.
  - 5a. 10–11 teeth in anterior series in adult. *affinis* Günther, 1870.
    Straits of Magellan.
  - 5b. Not more than 7–9 teeth in anterior series in adult.
    glutinosa Linnaeus, 1758, p. 34.[8]
    Both sides of North Atlantic.

7. This species is set apart from all others of the genus by the fact that it has only five pairs of gill sacs. This, however, is not apparent externally.

8. Including *capensis* Regan, 1913, South Africa, and *australis* Jenyns, 1842, Chile and southern Argentina; these species and *glutinosa* so overlap one another in the number of teeth and mucous pores and in the relative length of head that we have not been able to construct a key by which individual specimens could be identified with certainty. Neither can the presence of seven pairs of gill pouches in *capensis* be regarded as a unique specific character, since occasional specimens of *glutinosa* may have this same number (footnote 11, p. 35). Information on the number of teeth ($\frac{10}{10}$) and gill pouches of *capensis*, which was not included in the original description of the species (Regan, Ann. Mag. nat Hist., [8] *11*, 1913: 398), has been obtained subsequently (Barnard, Ann. S. Afr. Mus., *21* [1], 1925: 15).

Myxine glutinosa Linnaeus, 1758

Hagfish

Figure 4

Study Material. Forty-seven specimens of various sizes up to 610 mm. in length, from the Grand Banks and localities on both sides of the Gulf of Maine, north slope of Georges Bank, outer part of the continental shelf off Nantucket Island and off Cape Lookout. Also 13 specimens from the eastern Atlantic—Norway, Denmark, Kattegat, the Adriatic and Liverpool, England.

Distinctive Characters. The combination of jawless mouth, single nasal aperture, only a single pair of external gill openings, no operculum or covering fold of skin, wormlike form and lack of paired fins separate the Hag from all other fish-like vertebrates of the western North Atlantic.

Description. Trunk cylindrical throughout most of its length, its diameter about 1/24 to 1/25 of its total length, tapering rearward from dorsal origin of finfold to narrowly

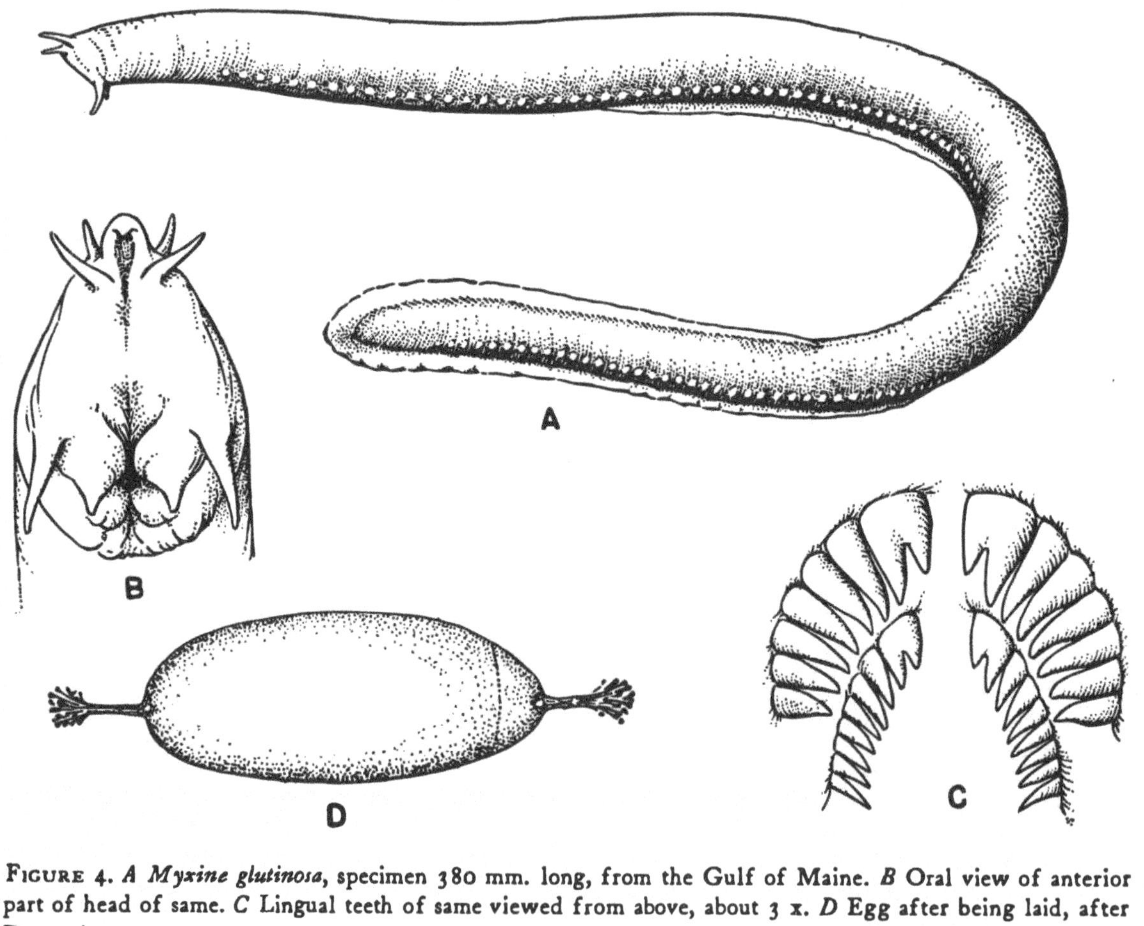

FIGURE 4. *A Myxine glutinosa*, specimen 380 mm. long, from the Gulf of Maine. *B* Oral view of anterior part of head of same. *C* Lingual teeth of same viewed from above, about 3 x. *D* Egg after being laid, after Dean, about 2 x.

rounded caudal extremity; a segmentally arranged row of mucous pores low down on each side, extending from about 1/18 the way back from snout to beyond anus; 26 to 33 pores in front of gill openings, 57 to 66 between gill openings and anus in those seen (53 to 70 recorded), and 11 to 13 posterior to anus in 9 specimens examined from Grand Manan Island, New Brunswick.[9]

Length of head to gill openings about 25 to 29% of total length (3.4 to 4 inches total length); snout obliquely truncate; fleshy rostrum a little higher than wide and broadly rounded in well preserved specimens, but sometimes more narrowly pointed, possibly due to contraction in the preservative; nostril an open pore on ventral surface near tip of snout; 2 pairs of slender, flexible barbels flanking either side of nostril, with a third pair, about twice as large, flanking the anterior part of mouth; mouth irregularly stellate when closed, without definite lip, but with a prominent, conical projection on either side of its margin;[10] gill openings close in front of origin of ventral finfold; usually 6 pairs of gill pouches, not visible externally, but sometimes 7 pairs.[11]

Lingual teeth comb-like, with swollen bases and sharp tips, moderately curved rearward, close together, decreasing in size from front to rear, of a strong orange color; those of anterior series about twice as large as those of posterior series, and partially overlapping the latter when tongue is retracted within mouth; 7 to 9 on either side in the anterior series and 8 to 10 in the posterior series; the first 2 in each series fused together at the base.

Ventral finfold originates about 1/3 of distance back from snout to caudal extremity, the dorsal fin about 2/3 the distance back and slightly anterior to anus,[12] both fins about 1/3 to 1/4 as high as the trunk is deep; ventral fold unsupported anterior to anus, but posterior to the latter it has a series of very slender, tapering cartilaginous rods, which extend around caudal extremity and forward along dorsal finfold (decreasing in length) nearly or quite to origin of latter.

Color. Grayish or reddish brown above, either plain, variously suffused, or mottled, with darker or paler gray, brown or bluish; whitish, or pale gray below. The variations in color may correspond more or less closely with the local color of the sea bottom.

Size. In American waters, on the coast of Maine, Hags are recorded up to 790 mm. in length, with one series of adults averaging 620 mm.[13] Apparently this is a greater size than they reach on the opposite side of the Atlantic, where the maximum recorded length is only 420 mm. (see discussion, p. 38).

Developmental Stages. The Hag was at first believed to be a functional protandrous hermaphrodite, its single unpaired sex organ first developing sperm in the posterior portion, then eggs later in the anterior portion.[14] However, recent detailed studies of the sex

9. 24–34, 54–64 and 10–14 respectively are recorded for European specimens.
10. These projections have sometimes been interpreted as a fourth pair of barbels.
11. Specimens with seven gill pouches on one side, or on both, are recorded by Cole (Anat. Anz., 27, 1905: 326).
12. Its origin is not clear-cut. The first indication of it is nearly as far anterior to the anus as the latter is distant from the tip of tail, in both American and Norwegian specimens.
13. Conel, J. Morph., 29, 1917: 77.
14. For a summary of earlier studies, see Smitt (Hist. Scand. Fish., 2, 1895: 1205) and Conel (Dean Memor. Vol., Amer. Mus. nat. Hist., Art. 3, 1931: 70).

organ[15] appear to show that this is not the case; either the male portion of the common sex organ matures in each individual, with the female portion remaining rudimentary, or vice versa. It has long been known that the eggs are few in number (only 19 to 30 having been counted in any one female) and large (up to 25 mm. in length), the horny shell with a cluster of anchor-tipped filaments at each end very characteristic in appearance.[16] But it was not until 1900 that any were found which had been laid naturally.[17] The eggs are deposited on the bottom, where they stick firmly in clusters to some fixed object[18] by means of their filaments and by threads of slime. The newly hatched Hag has not been seen as yet, but inasmuch as the smallest described, which is about two and one-half inches long and probably not long out of the egg, resembled the adult, there is no reason to suppose that the Hag passes through a larval stage.

Habits. The Hag is found chiefly, if not exclusively, where the bottom is soft mud or clay; its actions in aquaria[19] suggest that it spends most of its time imbedded in the clay or mud, with only the tip of its snout and the nasal barbels projecting, although it swims actively by an undulating motion in the horizontal plane when disturbed or when aroused by food in the vicinity; it is most active in the dark. Its depth range is considerable, extending commonly from 15 to 20 fathoms down to 250 fathoms or so, and it has been taken as deep as 524 fathoms.[20] The fact that it seldom, if ever, attacks hooked or netted fishes unless they are close to the bottom suggests that it never rises much above the latter.

In aquaria Hags die soon if the salinity is as low as 2.0 to 2.5 per cent;[21] survive for some weeks but do not feed if it is 2.9 to 3.1 per cent;[22] feed and thrive if it is as high as 3.2 to 3.4 per cent.[21] Also, it appears to be rather definitely limited in its dispersal toward the surface by high temperature, since it is rarely if ever found in water warmer than about 50 to 55 degrees, which in all but the most northerly part of its range would confine it to depths of 15 to 20 fathoms or more, except in the cold season. On the other hand, polar temperatures are probably a barrier to its northward dispersal (p. 40).

By its preference for soft bottom, comparatively high salinity (p. 37) and low temperature (see above), the Hag is confined within its area of regular occurrence to the deeper furrows and troughs on the Nova Scotian slope and in the Gulf of Maine, to the outer parts of the deeper bays, such as Fundy, Passamaquoddy, Massachusetts and prob-

15. Schreiner (Biol. Zbl., *24*, 1904: 91–104, 121–159, 162–173); Schreiner and Schreiner (Arch. Biol., *21*, 1905: 183–314, 8 pls., 315–355, 2 pls.; Arch. Zellforsch., *1*, 1908: 152); Conel (J. Morph., *29*, 1917: 75–163, 12 pls.; Dean Memor. Vol., Amer. Mus. nat. Hist., Art. 3, 1931: 70).
16. For reference to early accounts of eggs, see Smitt (Hist. Scand. Fish., *2*, 1895: 1206).
17. Dean, Mem. N. Y. Acad. Sci., *2* (2), 1900: 34, pl. 2.
18. To a Bryozoan in one case; see Jensen (Vidensk. Meddel. dansk. Naturhist. Foren., Copenhagen, 1900: 1).
19. For an interesting account of the habits of the Hag in aquaria, see Gustafson (Arkiv. f. Zoologi, Stockholm, *28A* [2], 1935).
20. Southeast slope of Georges Bank, Lat. 41° 32′ N., Long. 65° 55′ W. (Goode and Bean, Smithson. Contr. Knowl., *30*, 1895: 3; Spec. Bull. U.S. nat. Mus., 1895; Mem. Harv. Mus. comp. Zool., *22*, 1896).
21. Gustafson (Arkiv. f. Zoologi, Stockholm, *28A* [2], 1935), in western Sweden.
22. This is the usual summer range for surface water in Passamaquoddy Bay, where Hags were kept in captivity by Coonfield (Trans. Amer. micr. Soc., *59*, 1940: 398).

ably Penobscot, and offshore on the continental slope to the zone deeper than about 100 fathoms.

The Hag is not a parasite, as has sometimes been suggested, there being no reason to believe that it ever attacks living, uninjured fish. But it is a scavenger, feeding largely on dead or disabled fish of any sort, into which it bores by means of its rasp-like tongue. It is best known for its habit of penetrating the body cavities of hooked or gilled fishes, eating out first the intestines and then the meat, leaving nothing but a bag of skin and bones, inside of which the Hag itself is often hauled on board; or it may be captured clinging to the side of a fish it has just attacked. In Norwegian waters as many as six Hags have been reported in a single haddock.[23] It is also known to prey on marine polychaete worms, at least in Norwegian waters, and it has been suggested that these may be its normal diet.[24]

Being blind, the Hag evidently finds its food by scent, and so successfully that large numbers are sometimes taken in pots baited with dead fish or other offal; a local instance is mentioned below.

The fact that the eggs of the Hag have been found off southern Newfoundland at the mouth of the Bay of Fundy and on Georges Bank on one side of the Atlantic, and on the other side, near the Faroes, in Norwegian waters and off Morocco, shows that it spawns throughout its range; also, it spawns throughout the year, for females nearing ripeness, and others nearly spent, have been recorded for various months, winter and spring, as well as summer and autumn; in Norwegian waters eggs have been taken from November to May. The few eggs so far reported have been from depths of 50 to 150 fathoms, and most of them have been trawled on mud, clay, or sandy bottoms.[25]

Numerical Abundance. In American waters the Hag has usually been noted as being not very common. Actually it occurs in very considerable numbers on suitable mud bottoms at the appropriate depths, though rarely elsewhere, if at all. Thus, in the spring of 1913 the Hag was so plentiful on the Boon Island–Isles of Shoals fishing grounds that three to five per cent of all the haddock that we saw taken in gill nets had been attacked by them. Similarly, fishermen report that in certain areas of soft bottom in the northern part of the Gulf of Maine they damage a large proportion of the fish caught on long lines, unless the latter are tended frequently. The vicinity of Grand Manan Island at the mouth of the Bay of Fundy, and the trough with mud bottom between Jeffrey's Ledge and the coastline on the western side of the Gulf of Maine, are centers of abundance with which local fishermen have long been familiar. And evidently they are plentiful locally on the upper part of the continental slope off southern New England as well, for we took 11 large ones in an hour or less with one set of the Monaco trap off Nantucket at 260 fathoms on July 8, 1908. But we question whether they ever occur in American waters in such numbers as in the fjords of western Sweden and southern Norway, where catches of 100 are

23. S. Nilsson, Prod. Ichthyol. Skand., 1832: 124.
24. Gustafson, Arkiv. f. Zoologi, Stockholm, *28A* (2), 1935.
25. Hjort, Rep. Norweg. Fish. Invest., *1* (1), 1900: 75.

usual in eel pots set overnight on suitable bottom, with 1,400 recorded as captured in one set of 24 hours.[26]

Relation to Other Species. The American form has been considered specifically distinct from the European by some authors (*M. limosa* Girard, 1859), but not by others. However, the American form falls well within the limits of the European *M. glutinosa* in numbers of lingual teeth and slime pores. Its rostrum is also of the same obtuse shape in the better preserved specimens we have examined, although it has been pictured as more acutely pointed in some.[27] Nor has our own comparison of specimens from the two sides of the Atlantic revealed any significant differences in other respects. While the American form may grow larger than the European (p. 35), we hesitate to use size as a basis for specific separation unless accompanied by other differences of a sort that could allow any given individual to be referred to the one species rather than to the other. *M. atlantica* Regan, taken off Nova Scotia, seems also clearly referable to *glutinosa*.

The relationship of *glutinosa* of the northern hemsiphere to *australis*, *affinis* and *capensis* of the southern hemisphere is not so clear, but is a question of interest from the standpoint of geographical distribution. The only clear-cut difference between *capensis* on the one hand and the *australis-affinis* group on the other (the former overlaps the latter in number of teeth and slime pores) is that *capensis* is described as having seven gill pouches while there are only six in *australis* and *affinis*. However, we doubt whether or not this apparent difference is of specific importance, for while in *glutinosa* the usual number is six, seven also have been recorded (p. 35).

According to Norman's[28] recent comparison of *australis* with *affinis*, the number of teeth is less and the average number of abdominal slime pores is smaller in the former (8 teeth in first series, 8 or 9 in second; 56 to 64 abdominal pores) than in the latter [10 or 11 (9 in young) teeth in first series, 9–11 in second; 63 to 69 abdominal pores]; and its rostrum or labrum is longer and more acutely pointed. But this last character, being somewhat variable in *glutinosa*, may be equally so in the southern hemisphere forms. However, although the number of pores overlap in the two species, it appears that individual specimens can be referred to the one or the other, depending on the number of teeth. The large number of teeth in its anterior series also appears to mark *affinis* apart from *glutinosa* (7–9), although it overlaps the latter in the number of teeth in the posterior series, and falls within the range of variation recorded for *glutinosa* in the number of abdominal pores; however, *australis*, by Norman's definition, falls within the limits recorded for *glutinosa*, both in numbers of teeth and in numbers of pores. Neither have we been able to separate individual specimens of the one from those of the other by shape of rostrum. But since none of the considerable series of *australis* that we have examined are in good condition, we hesitate to unite the two species, in view of their widely separated areas of distribution.

26. Lyngnes, Z. Wiss. Biol., Abt. A, Z. Morph. Ökol., *19*, 1930: 591.
27. Garman, Mem. Harv. Mus. comp. Zool., *24*, 1899: pl. 68, fig. 7.
28. "Discovery" Rep., *16*, 1937: 4; see this publication also for the somewhat confused synonymy of the two.

Relation to Man. The Hag, being of no value itself, is only a nuisance to the fishermen because of its habit of damaging better fish, and a loathsome one, owing to its ability to discharge slime from its mucous sacs out of all proportion to its size. One Hag, it is said, can fill a two gallon bucket, and we think this no exaggeration.[29]

In American waters the commercial fishes most often damaged by it are the haddock and the hakes (*Urophycis*), these being the species most often fished for with long lines or with gill nets over the particular type of bottom that the Hag frequents. But it sometimes damages cod also, and European authors describe it as attacking ling (*Molva*) and other gadoids, herring, mackerel, sturgeon, and even mackerel sharks (*Isurus*) under similar circumstances.

Range. Both sides of the northern North Atlantic. In the eastern North Atlantic it occurs on the Murman coast and in northern Norway[30] southward in abundance to the northern part of the North Sea, the Kattegat (not known from the Baltic) and the Irish Sea; less commonly to the English Channel (Cornwall); occasionally to Portugal. There are two records of it off Morocco, one just outside the Straits of Gibralter,[31] the other just inside in the Mediterranean.[32] It has been credited to the Adriatic[33] also, no doubt on the strength of the fact, reported by Garman,[34] that there are three specimens labelled "Trieste" in the collection of the Museum of Comparative Zoology (see Study Material, p. 34). But so far as we can learn it is not included otherwise in any of the general surveys of Mediterranean fishes[35] that have appeared. This makes it much more probable that the specimens in question were mislabelled, and that *Myxine* is actually not a regular member of the fauna of the inner parts of the Mediterranean.

On the western side of the Atlantic it occurs at least occasionally as far north as the northern part of Davis Strait (see p. 40), and southward as far as the latitude of Cape Fear in North Carolina. It is represented in the corresponding thermal belt in the southern hemisphere (Chile, southern Argentina, Straits of Magellan, Tierra del Fuego, South Africa) by a form, or forms, so closely allied that it is doubtful whether any sharp line can be drawn between them (see discussion, p. 33).

Occurrence in the Western Atlantic. While not known for certain along the west coast of Greenland,[36] so far as we can learn, the Hag has been taken on one occasion in the

29. Linnaeus (Sys. Nat., 1758: 650), referring to this habit, wrote "aquam in glutem mutat."

30. Apparently it does not occur around Iceland, for it is not included by Saemundsson (Skr. Komm. Havunders. Kbh., No. 5, 1900) in his survey of Icelandic fishes.

31. Eggs; Koefoed (Rep. Sars N. Atlantic Deep Sea Exped., Zool., *4* [1], 1927: 18).

32. Roule, Result. Camp. sci. Monaco, *52*, 1919: 129.

33. Schnakenbeck in Grimpe and Wagler, Tierwelt N- u. Ostsee, Lief 7, Teil 12d, 1927: 3; Cons. explor. Mer., Faune Ichthyol. N. Atlant., 1931.

34. Mem. Harv. Mus. comp. Zool., *24*, 1899: 348.

35. Müller (Vergl. Anat. Myxinoiden, Pt. 1, 1835: 17, footnote) long ago rejected Bloch's (Schr. Ges. Naturf. Freunde Berlin, *10*, 1792: 251) suggestion that *Myxine* is in the Mediterranean, which was based on Aristotle's account of the slime-producing habit of his Pholis.

36. It has been credited repeatedly to Greenland on the strength of Fabricius' (Fauna Groenl., 1780: 344) characterization of it as "rari in mari Groenlandico." But we find no other record of it among the many subsequent lists of fishes of Greenland, east or west, except as noted above.

northern part of Davis Strait, just south of the Greenland–Baffin Land Ridge.[37] But there is no report of it either in the region of Hudson Bay, along the Atlantic coast of Labrador, or on the east coast of Newfoundland; nor did any of the many cod that we saw caught by hook and line or nets in the summer of 1900 along the outer Labrador coast show any evidence of attack by Hags. Apart from the Davis Strait record just mentioned, the most northerly known stations for it on the American coast are the Grand Banks and the south coast of Newfoundland, where its eggs have been trawled.[38] Type of bottom, temperature and salinity are such that it is also to be expected in the deep trough of the Gulf of St. Lawrence, though we found no definite record of it there.

To the southward it is generally distributed at appropriate depths wherever the bottom is suitable: over the continental shelf and down the continental slope along Nova Scotia, throughout the Gulf of Maine, along the seaward slope of Georges Bank, and off southern New England and New York, where specimens have been taken at many localities by trawl or otherwise, at depths of 100 to 250 fathoms and deeper. Apparently this marks the limit of its common occurrence in this direction, however, for the only records of its occurrence south of the latitude of New York are: one specimen taken off Delaware Bay in 126 fathoms, and one or more in 178 fathoms off Cape Fear, North Carolina, many years ago.[39]

Synonyms and References:[40]

Myxine glutinosa Linnaeus, Syst. Nat., 1758: 650 (Atlant. Oc., grouped among the worms); Müller, O. F., Prod. Fauna Danica, 1776: 227 (Denmark); Pennant, Brit. Zool., *4*, 1777: 39, pl. 20, fig. 5 (habits, ill.); Fabricius, Fauna Groenl., 1780: 344 ("rari in mari Groenlandico"); Retzius, Fauna Sueciae, 1, 1780: 302 (refs., habits, west. seas); Gmelin, Syst. Nat., *1* (6), 1790–1791: 3082 (descr., Atlant. Oc.); Retzius, Svenska. Vet. Akad. Handl., *11*, 1790: 110, pl. 4 (*Myxine* and *Petromyzon* considered more worm-like than fish-like; plate referred to is not in copy seen); Abildgaard, Schr. Ges. naturf. Freunde, Berlin, *10*, 1792: 193, 244, pl. 4 (descr., ill.; a fish, not a worm); Bloch, Schr. Ges. naturf. Freunde, Berlin, *10*, 1792: 244 (disc. of earlier accounts; believed same as Pholis of Aristotle, therefore in Mediterranean, Greece); Fleming, Hist. Brit. Anim., 1828: 164 (descr., habits, England); Cuvier, Règne Anim., 2nd Ed., 1829: 406 (North Sea); Nilsson, Prod. Ichthyol. Skand., 1832: 123 (habits, type of bottom, feeding, north. Norway); Johnston, London's Mag. Nat. Hist., *6*, 1833: 15 (Scotland); Jenyns, Manual Brit. Vert. Anim., *1*, 1835: 413 (Ireland); Müller, J., Vergl. Anat. Myxinoiden, 1, 1835: 3 (history), 15 (class., diag., North Sea, Norway, Sweden, Greenland), 17 (footnote considers Bloch's reference of it to Mediterranean on basis of Aristotle incorrect); Templeton, Charlesworth's Mag. Nat. Hist., *1*, 1837: 413 (Ireland); Cuvier, Règne Anim., Poiss., 1838–1843: 383, pl. 120, fig. 3 (ill.); Krøyer, Danmarks Fisk., *3*, 1838–1853: 1068 (descr., habits, Denmark); Swainson, Nat. Hist. Fish. Amphib. Rept., *2*, 1839: 338; Fries, Ekstrom and Sundevall, Skand. Fisk., 6, 1840: 121, pl. 28 (descr., ill., Scandinavia); Gray, List Fish. Brit. Mus., Chondropt., *1*, 1851: 147 (Norway, Gt. Brit.); White, List Spec. Brit. Mus., Fish., *8*, 1851: 145 (north. England, Scotland); Nilsson, Skad. Fauna Fisk., *4*, 1855: 750 (not seen); Thompson, M., Nat. Hist. Ireland, *4*, 1856: 267 (Ireland, Scotland); Thomson, A., Art. "Ovum," in Todd's Cyclop. Anat. Physiol., *5* (suppl. vol.), 1859: 50, fig. 33, c, d (earliest descr. of egg, ill.);

37. Lat. 66° 37′ N., 450 meters, temp. 3.12° C.; Jensen, Rapp. Cons. explor. Mer., *39*, 1926: 98.
38. Dean, Mem. N. Y. Acad. Sci., *2* (2), 1900: 34.
39. One specimen from this lot is in the Harv. Mus. Comp. Zool.
40. *Myxine*, as representative of its order (subclass in some schemes of classification), has been the subject of many anatomical accounts and discussions, in reports of original observations as well as in general textbooks, etc. This list is confined to such citations as bear directly on its classification, on its habits or on its distribution.

Gill, Proc. Acad. nat. Sci. Philad., Addend., 1861: 63 (Polar regions to Cape Cod); Steenstrup, Overs. danske Vidensk.-Selsk. Forh. (1863), 1864: 233 (eggs); Günther, Cat. Fish. Brit. Mus., *8*, 1870: 511 (descr., coasts of Europe and N. Amer.); Gill, Rep. U.S. Comm. Fish. (1871–72), 1873: 814 (listed Greenland, Polar regions to Cape Cod, Mass.); Putnam, Proc. Boston Soc. nat. Hist., 1873: 135 (dimens. and no. of teeth of Grand Manan specimens); Collett, Vidensk.-Selsk. Forh. Christiania, 1874; also Norges Fisk., 1875: 220 (habits, distrib., depth, Norway); Lütken, Cat. Fish. Greenl., in Manual Instr. for Arctic Exped. by T. R. Jones, Manual Nat. Hist. Geol. Greenl., 1775: 122 (Greenland, by ref. to Fabricius, Fauna Groenl., 1780); Gervais and Boulart, Poiss., *3*, 1877: 258, pl. 100 (descr., ill., England to Scandinavia); Malm, Göteborgs och Bohusläns Fauna, 1877: 637 (habits, food, west. Sweden); Goode and Bean, Bull. Essex Inst. Salem, *11*, 1879: 31 (occur., depth, off Massachusetts); Winther, Prod. Ichthyol. Dan. Mar. in Natur. Tidsskr. Copenhagen, (3) *12*, 1879: 62 (Skagerrak, north. Kattegat, not in Baltic); Day, Fish. Gt. Brit., *2*, 1880–1884: 364, pl. 179 (refs., descr., ill., habits, Gt. Brit.); Buckland, Nat. Hist. Brit. Fish., 1881: 145 (ill.); Mela, Vert. Fennica, 1882: 372, pl. 10 (not seen); Jordan and Gilbert, Bull. U.S. nat. Mus., *16*, 1883: 5 (descr., Europe and America); Storm, K. norske Vidensk.-Selsk. Skr., Trondh., 1883: 48 (Trondh. Fjord); Bean, Rep. U.S. Comm. Fish. (1882), 1884: 344 (off Woods Hole); Goode, Fish. Fish. Industr. U.S., Section 1, 1884: 681, pl. 252 (Atlant. coast, U.S.); Lilljeborg, Sverig. Norg. Fisk., *3*, 1884: 730 (descr., anat., habits, refs., distrib., Sweden, Norway); Kingsley, Stand. Nat. Hist., *3*, 1885: 67 (habits, Eastport, Maine and Grand Manan); McIntosh, 3rd Annu. Rep. Fish. Bd. Scotland, App. F, 1885: 66, 204 (Scotland, not seen); Honeyman, Proc. N.S. Inst. Sci., *6* (1), 1886: 230 (off Nova Scotia); Cunningham, Quart. J. micr. Sci., N.S. *27*, 1887: 49, pl. 6, 7 (habits in aquarium, breathing, reprod. organs, abund., Scotland); Jordan, Rep. U.S. Comm. Fish. (1885), 1887: 791 (in N. Amer. list); Nansen, Bergens Mus. Aarb. (1887), 7, 1888: 5–34, pl. 1, 2 (sex organs, abund. near Bergen, Norway); Vaillant, Exped. Sci. "Travailleur" et "Talisman," Poiss., 1888: 384 (off Portugal, 460 meters); Beard, Rep. Brit. Ass. Adv. Sci. (1892), Edinburgh, 1893: 789 (Scotland); Goode and Bean, Smithson. Contr. Knowl., *30*, 1895; Spec. Bull. U.S. nat. Mus., 1895; Mem. Harv. Mus. comp. Zool., *22*, 1896: 3, pl. 1, fig. 1 (descr., ill., loc., depth, Grand Banks, off south. New England and off N. Carolina); Smitt, Hist. Scand. Fish., 2nd Ed., 2, 1895: 1208, pl. 53, fig. 5 (refs., descr., habits, breeding, distr., Scandinavia); Jordan and Evermann, Rep. U.S. Comm. Fish. (1895), 1896: 211 (Amer. coast south to Delaware); Bull. U.S. nat. Mus., 47 (1), 1896: 7 (descr., Amer. coast south to Cape Cod); Smith and Kendall, Rep. U.S. Comm. Fish. (1896), 1898: 169 (off Delaware, Lat. 39° N., Long. 72° W., 75 fath.); de Braganza, Result. Invest. Sci. "Amelia," 1899: 41 (off Portugal, not seen); Garman, Mem. Harv. Mus. comp. Zool., *24*, 1899: 342, 348, pl. 68, fig. 5 (compar. with other species, rep. on specimens in Mus. Comp. Zool. Coll., labelled "Trieste"); Dean, Mem. N. Y. Acad. Sci., *2* (2), 1900: 34, pl. 2 (descr. and ill. of eggs, taken off south. Newfoundland and Georges Bank, 103 and 150 fath., discus.); Hjort, Rep. Norweg. Fish. Invest., *1* (1), 1900: 75 (eggs taken in shrimp trawl, Norway, depth and type of bottom; also young); Jensen, Vidensk. Medd. Dansk Naturhist. Foren., Copenhagen, 1900: 1, pl. 1 (descr., ill. of eggs, off the Faroes); Bridge, Camb. Nat. Hist., 7, 1904: 422 (descr., habits, abund. in North Sea, species of fish preyed on); Werner, Zool. Jb., Syst. Abt. 1, *21*, 1904: 266 (Norway); Jordan, Guide to Study Fish., *1*, 1905: 490 (north. Europe); Fowler, Proc. Acad. nat. Sci. Philad., *59*, 1908: 461 (old record off Delaware and Bar Harbor, Maine); Kendall, Occ. Pap. Boston Soc. nat. Hist., *8* (7), 1908: 1 (off Maine and N. Hampshire); Fowler, Proc. Acad. nat. Sci. Philad., *63*, 1911: 5 (old record off Delaware); Seabra, Bull. Soc. portug. Sci. nat., 5, 1911: 205 (old record off Portugal); Regan, Ann. Mag. nat. Hist., (8) *11*, 1913: 397 (class., descr., Gt. Brit., Norway); Sumner, Osburn and Cole, Bull. U.S. Bur. Fish., *31* (2), 1913: 734 (off Cape Cod); Conel, J. Morph., *29*, 1917: 78, 12 pls. (urogenital syst., size up to 790 mm., off coast of Maine); Roule, Res. Camp. sci. Monaco, *52*, 1919: 129 (off Morocco, just inside Strait of Gibraltar); Bigelow and Welsh, Bull. U.S. Bur. Fish., *40* (1), 1925: 16 (descr., ill., habits, occur. in Gulf of Maine); Jensen, Rapp. Cons. explor. Mer, *39*, 1926: 98; Koefoed, Rep. Sars N. Atlantic Deep Sea Exped., Zool., *4* (1), 1927: 18 (eggs trawled, off Morocco, outside Strait of Gibraltar, 535 meters); Nichols and Breder, Zoologica, *9*, 1927: 9 (habits, off Cape Cod); Palingren, Acta zool., *8*, 1927: 135 (exper. with pressure on *Myxine* in aquarium, Drobäk, Norway); Schnakenbeck, in Grimpe and Wagler, Tierwelt N- u. Ostsee, Lief 7, Pt. 12^d, 1927: 3

(general); Rey, Fauna Iberica, Peces, *1*, 1928: 256 (descr., ill., off Portugal); Jordan, Manual Vert. Anim. NE. U.S., 1929: 5 (in synopsis, Newfoundland to Cape Cod); Breder, Field Bk. Mar. Fish. Atlant. Coast, 1929: 5 (descr., habits, Arctic to N. Carolina, and east. Atlant.); Lyngnes, Z. Morph. Ökol. Tiere, *19*, 1930: 591 (descr., and ill. of eggs, abund., type of bottom, Norway); Conel, Dean Memor. Vol., Amer. Mus. Nat. Hist., *3*, 1931: 70 (believed not hermaphroditic, see p. 35); Schnakenbeck, Cons. explor. Mer. Ichthyol. N. Atlant., 1931: plate not numbered (descr., ill., Greenland and Murman coast to Portugal and Adriatic); Rep. Newfoundland Fisher. Res. Comm., *1* (4), 1932: 107 (Newfoundland, no definite loc.); Holly, in Schultze, Kükenthal, *et al.*, Tierreich, Lief 59, 1933: 47 (class., refs., descr.); Bigelow and Schroeder, Canad. Atlant. Fauna, 12[d], 1934: 2 (descr., ill., comp. with European forms, depth, type of bottom, Greenland and Grand Banks to N. Carolina); Nobre, Faun. Marinh. Portugal, *1*, 1935: 497 (loc. off Portugal, depth); Gustafson, Arkiv. f. Zoologi, Stockholm, *28A* (2), 1935: 1 (habits in aquarium, feeding and food, rel. to salinity and light, abund., west. Sweden); Vladykov and McKenzie, Proc. N. S. Inst. Sci., *19* (1), 1935: 44 (Nova Scotia Banks and Bay of Fundy, depth); Bigelow and Schroeder, Bull. U.S. Bur. Fish., *48*, 1936: 321 (size of eggs and number, Georges Bank); Lübbert and Ehrenbaum, Handb. Seefisch. Nordeurop., *2*, 1936: 323 (habits, eggs, distrib.); Norman, "Discovery" Rep., *16*, 1937: 5 (comp. with *M. australis* and *M. affinis*); Coonfield, Trans. Amer. micr. Soc., *59*, 1940: 398–403 (in aquar., St. Andrews, Passamaquoddy Bay; skin pigment); Jensen, Vidensk. Med., 105, 1942: 55 (Greenland).

Sleep marken, Gunnerus, Trondh. Gesellsch. Schrift., *2*, 1766: 230–236, pl. 3 (descr., ill., habits; considered a worm; ref. to name *Myxine glutinosa* Linnaeus. Title page of copy seen is dated 1765, but date of Gunnerus' paper is given as 1766 by Dean, 1913, Bibliogr. Fishes).

Petromyzon myxine Walbaum, P. Artedi Genera Pisc. Emend. Ichthyol., 3, 1792: 500 (diagn., refs.).

Gastrobranchus coecus[41] Bloch, Naturg. Ausländ Fische, *9*, 1793: 67, pl. 413 (descr., ill., habits, Denmark, Sweden, Norway, and Iceland); Bull. Sci. Soc. philom. Paris, *1* (4), 1797: 26 (equivalent to *Myxine glutinosa* Linnaeus, 1758); Lacépède, Hist. Nat. Poiss., *2*, 1798: 406 (descr., habits); Bloch and Schneider, Syst. Ichthyol., 1801: 534, pl. 104 (descr., ill.); Shaw, Gen. Zool., *5* (2), 1804: 264, pl. 134 (descr., ill., habits); Turton, Brit. Fauna, 1807: 110 (brief descr.); Cuvier, Règne Anim., *2*, 1817: 406; Strack, Naturg. in Bildern, Fische, Lief 4, 1819–1826: pl. 33 (descr., ill., Norway, America); Yarrell, Hist. Brit. Fish., *2*, 1836: 462 (habits, descr., Gt. Brit.); *2*, 1841: 612 (same as foregoing); Hamilton, Brit. Fish., *2*, 1843: 424 (brief descr.); Buckland, Nat. Hist. Brit. Fish., 1881: 145 (ill., the descr., p. 144, is as "Gasteobranchus").

Gastrobranche aveugle, Lacépède, Hist. Nat. Poiss., 4° ed., *1*, 1798: 525, in Buffon, Hist. Nat. (descr., habits); in Sonnini, Hist. Nat. Poiss., *5*, 1802–1803: 145, pl. 17 (descr., ill., refs.).

Glutinous gastrobranchus, Shaw and Nodder, Naturalist Misc., *10*, 1798: pl. 362 (descr., ill.).

Glutinous hag, Pennant, Brit. Zool., *3*, 1812: 109 (descr., habits).

Myxina coeca Oken, Lehrb. Naturg., *3* (2), 1816: 127 (descr.; but loc. "Guinea" is no doubt in error); Blainville, in Vieillot, Faune Franc., *2*, 1825: 2 (descr., meas., north. seas; pl. 1a not included in copy seen).

Myxine limosa Girard, Proc. Acad. nat. Sci. Philad., 1852: 224 (descr., off Grand Manan, abund., type of bottom, depth); Gill, Rep. U.S. Comm. Fish. (1871–72), 1873: 814 (listed Nova Scotia to Massachusetts); Garman, Mem. Harv. Mus. comp. Zool., *24*, 1899: 343, pl. 68, fig. 7 (considered distinct from *glutinosa*); Dean, Science, N.S. *17*, 1903: 433 (*limosa* considered distinct from *glutinosa* on basis of egg case); Jordan, Guide to Study Fish., *1*, 1905: 490 (retained as distinct from *glutinosa*); Halkett, Check List Fish. Canad., 1913: 38 (Newfoundland south to Cape Cod); Regan, Ann. Mag. nat. Hist., (8) *11*, 1913: 398 (class., descr., Bay of Fundy); Fowler, Proc. Boston Soc. nat. Hist., *35*, 1917: 110 (Bar Harbor, Maine); Huntsman, Contr. Canad. Biol. (1921), 1922: 55 (Bay of Fundy, Passamaquoddy Bay, 18–60 fath., eggs); Holly, in Schultze, Kükenthal, *et al.*, Tierreich, Lief 59, 1933: 47 (class., refs., descr.); Jordan, Evermann and Clark, Rep. U.S. Comm. Fish. (1928), *2*, 1930: 8 (considered distinct from *glutinosa*, Newfoundland to Cape Cod).

Borer, Couch, Fish. Brit. Isles, *4*, 1867: 408 (descr., habits, distrib. in Europe).

Gasteobranchus coecus Buckland, Hist. Brit. Fish., 1881: 144 (descr., habits).

41. Sometimes spelled *caecus*.

Myxine (no specific name) Cole, Anat. Anz., 27, 1905: 323 (anat., good descr. of teeth, specimens recorded with 7 gills on one side, or both).

Myxine atlantica Regan, Ann. Mag. nat. Hist., (8) *11*, 1913: 398 (class., descr., off Nova Scotia); Holly, in Schultze, Kükenthal, *et al.*, Tierreich, Lief 59, 1933: 49 (class., descr.).

Order PETROMYZONIDA

Description. In the adult, seven pairs of gill pouches open separately to the exterior, but open inwardly into a special respiratory tube which is separate from the pharynx and which ends blind, posteriorly; however, this respiratory tube connects with the mouth anteriorly.[42] At the time of metamorphosis this tube loses its connection with the intestine, while a new pharynx develops above it to form a forward extension of the intestine which connects with the mouth. Snout without barbels; dorsal and caudal fins separate, supported by rays; nostril a blind sac, on dorsal surface of head, not opening into mouth; mouth opens as a funnel or disc surrounded by a circular lip with numerous horny teeth; sides of trunk without prominent rows of mucous pores; ear with two semicircular canals; eye well developed, with lens and iris in adult, although rudimentary in larva; cranium partially roofed over; notochordal sheath with rudimentary neural arches; a complex cartilaginous basket around gill pouches; intestine with slight spiral fold, apparently homologous with the spiral valve of the Chondrichthyes; pancreas represented by scattered follicles.

Development. Sexes separate; eggs small, numerous; development, with larval (Ammocoete) stage, different in appearance structurally from adult. In some of the fresh water species the growth stage that normally occurs between the times of metamorphosis and sexual maturity is omitted.[43]

Habitat. Fresh water, or entering fresh water to breed if marine.

Families. The single family Petromyzonidae, in which the various Lampreys have been grouped, has been divided recently into two subfamilies, which, in our opinion, may well be raised to the rank of families as follows:[44]

1a. Upper margin of central mouth with only one dental plate; margin of oral funnel with a series of fringed, as well as smooth, papillae. Petromyzonidae, p. 43.

1b. Upper margin of central mouth with two separate dental plates; margin of oral disc with only smooth papillae or cirri. Mordaciidae. Australia, Tasmania, Chile.

Family PETROMYZONIDAE

Characters. Upper margin of the central mouth with only one dental plate, usually toothed; margin of oral funnel or disc with a series of fringed lappets, as well as a series of smooth marginal papillae. Characters otherwise those of the order.

42. This respiratory tube represents the pharynx of the larva, into which the gill sacs then open.

43. For discussion, see Hubbs, Pap. Mich. Acad. Sci., *4* 1st half, 1924: 587.

44. Based on the definitions by Holly (in Schultze, Kükenthal, *et al.*, Tierreich, Lief 59, 1933: 12) and by Pietschmann (in Kükenthal and Krumbach, Handb. Zool., *6* [1], Lief 5, 1935: 540).

Discussion of Genera. Generic characters among the Petromyzonidae, as here limited, are afforded by the dentition and by the number of dorsal fins, *i.e.*, whether one or two. Seven genera are recognized in the most recent general synopsis of the family as limited above.[45] Five of these are known in the northern hemisphere, but only two, namely *Petromyzon* and *Lampetra*, occur in the North Atlantic. *Petromyzon* inhabits only the western North Atlantic, while *Lampetra*, which also is marine and anadromous along the coasts of Europe and northern Asia, is confined to fresh water in North America.

Key to Genera of the Northern Hemisphere[46]

1a. Only 1 dorsal fin. *Ichthyomyzon* Girard, 1858.
Eastern North America.[47]

1b. More than 1 dorsal fin.

2a. Teeth on oral disc, sometimes called labial teeth, close together, arranged in curvilinear radiating rows (Fig. 2 D).

3a. Supraoral dental plate with 2 large teeth; margin of anterior lingual dental plate deeply indented in the midline.
Petromyzon Linnaeus, 1758, p. 45.

3b. Supraoral dental plate with only 1 tooth; margin of anterior lingual dental plate not deeply indented in the midline. *Caspiomyzon* Berg, 1906.
Caspian Sea.

2b. Teeth on oral disc loosely spaced, not in radiating rows.

4a. Supraoral dental plate with a strongly developed sharp median tooth, as well as 2 still larger lateral teeth on each side. *Entosphenus* Gill, 1862.
Pacific Coast of North America, from California to Alaska.[48, 49]

4b. Supraoral dental plate without strong, sharp, median tooth, at most with 1 or more low, blunt, median denticles. *Lampetra* Oken, 1816.[50]
Both coasts of North Atlantic and western Pacific.[51]

45. Holly, in Schultze, Kükenthal, *et al.*, Tierreich, Lief 59, 1933: 12.
46. Somewhat amended from the synopsis by Berg (Annu. Mus. zool. Acad. Leningrad, 32 [1], 1931: 87). For synopses of the family as a whole, see Regan (Ann. Mag. nat. Hist., [8] 7, 1911: 193) and Holly (in Schultze, Kükenthal, *et al.*, Tierreich, Lief 59, 1933: 13).
47. In fresh water exclusively.
48. Some of the species that fall in *Lampetra* by this key are placed in *Entosphenus* by Creaser and Hubbs (Occ. Pap. Mus. Zool. Univ. Mich., 120, 1922: 6); if accepted, this would expand the range of the genus to northeastern United States, Mexico and Alaska in fresh water; to Japan and the White Sea.
49. Marine, but entering fresh water to breed.
50. Including *Eudontomyzon* Regan, 1911, which was classed as a subgenus of *Petromyzon* by Creaser and Hubbs (Occ. Pap. Mus. Zool. Univ. Mich., 120, 1922: 2), as a subgenus of *Lampetra* by Berg (Annu. Mus. Zool. Acad. Leningrad, 32 [1], 1931: 92) and by Holly (in Schultze, Kükenthal, *et. al.*, Tierreich, Lief 59, 1933: 22), with whom we agree.
51. Europe, northern Asia, Japan, North America, Mexico; some species confined to fresh water; others marine, but entering fresh water to breed.

Genus *Petromyzon* Linnaeus, 1758

Lampreys

Petromyzon (in part) Linnaeus, Syst. Nat., 1758: 230; type species, *P. marinus* Linnaeus. European seas.

Generic Synonyms:

A. Adult.

Bathymyzon Gill, Proc. U.S. nat. Mus., *6*, 1883: 254; type species, *B. bairdii* Gill. Continental slope off Cape Cod, Lat. 40° 02′ N., Long. 68° 51′ W., 547 fathoms.[52]

Oceanomyzon Fowler, Proc. Acad. nat. Sci. Philad., *59*, 1908: 461; type species, *O. wilsoni* Fowler. Atlantic Ocean.

B. Larva.

Ammocoetus (in part) Duméril, Dissert. Poiss., 1812: 16; generic diagnosis, no species mentioned.

Ammocoetes (in part) Cuvier, Règne Anim., *2*, 1817: 119; emended spelling for *Ammocoetus* Blainville, 1812; type species, *Petromyzon branchialis* Linnaeus, 1758: 230 (larva of *Lampetra fluviatilis* Linnaeus, 1758; however, the larva of *Petromyzon marinus* is not distinguishable from it).[53]

Ammocaetes (in part) Beithold, in Latreille's Natur. Famil. Tierreich, 1827: 109; emended spelling for *Ammocoetes* Cuvier, 1817, and *Ammocoetus* Blainville, 1812.

Ammocites (in part) Beithold, in Latreille's Natur. Famil. Tierreich, 1827: 564 (index); evidently a misspelling.

Generic Characters. Two dorsal fins, the 1st separated from 2nd by a definite interspace, the 2nd demarked from caudal by a deep notch, but continuous with it basally; teeth renewed periodically by growth, combined with a periodic sloughing off of the outermost horny layer, those on oral disc, also called labial teeth, close together in regular arrangement, the inner series much the largest, the outer series radiating outward in curved rows: supraoral dental plate small, with 2 teeth; infraoral dental plate broad, with 7 to 9 conical teeth (see footnote 57, p. 47); tongue with 3 denticulated plates, the anterior deeply indented anteriorly in the midline, its toothed margin biconcave; about 70 myomeres between rearmost gill opening and anus.

Larva worm-like in appearance, toothless, the oral disc of adult represented in young by a broad hood-like upper lip and very short lower lip; complexly branched papillae surrounding mouth and present on midzone of upper lip; eyes rudimentary and not visible externally; fins without rays; dorsals not at all, or only faintly, demarked from each other or from the caudal; muscular segmentation evident externally; gill sacs opening directly into the pharynx internally; cartilaginous branchial basket rudimentary; pharynx with a ventral ciliated pocket, a peripharyngeal ciliated groove anteriorly and a dorsal ciliated tract;[54] gall bladder and bile duct present.

52. This specimen, 275 mm. long, was made the basis of the new genus *Bathymyzon* because its supraoral and infraoral dental plates lacked distinct tubercles. But our own examination of the type specimen (U.S. nat. Mus., No. 33311) has shown that it simply represents a case where the tubercles have been worn down prior to their renewal, for a fresh set of very sharp tubercles is exposed when the outer layer of the suboral plate is lifted free at one end. This appears to apply equally to the type specimen of *Oceanomyzon* Fowler, 1908.

53. The parentage of the Ammocoete larva of *Lampetra fluviatilis* was established by A. Müller (Arch. Anat. Physiol. wiss. Med., 1856: 323).

54. For description of these, see Dohrn (Mitt. zool. Sta. Neapel, *6*, 1886: 59) and Shipley (Quart. J. micr. Sci., *27*, 1887: 325).

During metamorphosis the eyes become functional, the external segmentation disappears, the dorsal fin becomes subdivided, the oral disc and teeth are formed, the branchial basket and skull complete their development, the ventral pharyngeal ciliated pit or groove becomes the thyroid gland, the pharynx loses its connection with the alimentary tract, the latter forming a new union with the mouth, while the gall bladder disappears and the bile duct is obliterated.[55]

Range. Atlantic coasts of Europe and eastern North America; marine but entering fresh water to breed; also landlocked in certain lakes in the northeastern United States.

Species. It is now generally agreed that the marine Lampreys of this genus represent only a single species (*P. marinus* Linnaeus, 1758), and the landlocked form of *P. marinus* (*dorsatus*, Wilder)[56] appears to be merely a dwarfed race, without any distinguishing features other than its smaller size.

Petromyzon marinus Linnaeus, 1758

Sea Lamprey, Lake Lamprey, Stone Sucker

Figure 5

Study Material. Fifty-one American specimens, up to 710 mm. in length, from Eastport, Maine; Exeter, New Hampshire; various localities in Massachusetts and Massachusetts Bay; Havre de Grace, Maryland (Chesapeake Bay); and the Potomac R., including the types of *Bathymyzon bairdii* Gill, 1884 (U.S. Nat. Mus., No. 33311) and *Oceanomyzon wilsoni* Fowler, 1907 (Acad. Nat. Sci. Philad., No. 375). Also five Mediterranean specimens from Nice, Messina and Trieste.

Distinctive Characters. The eel-like appearance of the Lamprey, combined with its circular oral disc surrounding the jawless mouth and the large number of external gill openings, places it at a glance among Atlantic fishes.

Description. General form eel-like, the trunk about as thick as high anteriorly, but somewhat flattened dorsally, hence ovoid in midsection and strongly compressed toward tail; immature males with a faintly indicated mid-dorsal ridge from about opposite 6th or 7th gill opening to 1st dorsal fin, this much more prominent in large maturing males, even while still in salt water; females, at maturity, developing a fin-like crest between anus and caudal fin.

Head, to last gill opening, a little more than 1/5 of total length; nostril prominent, surrounded by a circular rim, about opposite anterior margin of eye, its distance back from tip of snout about 3/8 to 2/5 of length of head to last gill opening. Eye approximately circular, its diameter about 1/16 as great as length of head, its anterior margin a little posterior to posterior edge of oral disc; gill openings round or somewhat oval, about 1/2 as long as

55. For a more extensive account, see Bridge (Camb. Nat. Hist., 7, 1904: 429). It has long been realized that the small Lamprey, repeatedly reported in American waters by early authors as *P. nigricans* Lesueur, 1818, is merely the young of *P. marinus*.

56. In Jordan and Gilbert, Bull. U.S. nat. Mus., *16*, 1883: 869.

horizontal diameter of eye, about equally spaced, the interspaces about as wide as diameter of eye, the 1st gill opening behind eye by a distance about equal to diameter of eye, each gill opening successively lower on side of head from front to rear. Oral disc circular in outline when attached to a fish or other object, but at other times contracted transversely, leaving only a longitudinal fissure open, its diameter when expanded a little greater than greatest thickness of trunk, or about 1/3 as long as head, its margin with 2 to 4 rows of close-set fleshy papillae, the inner rows variously fringed and the outermost row also fringed around posterior part of disc, but smooth around anterior margin.

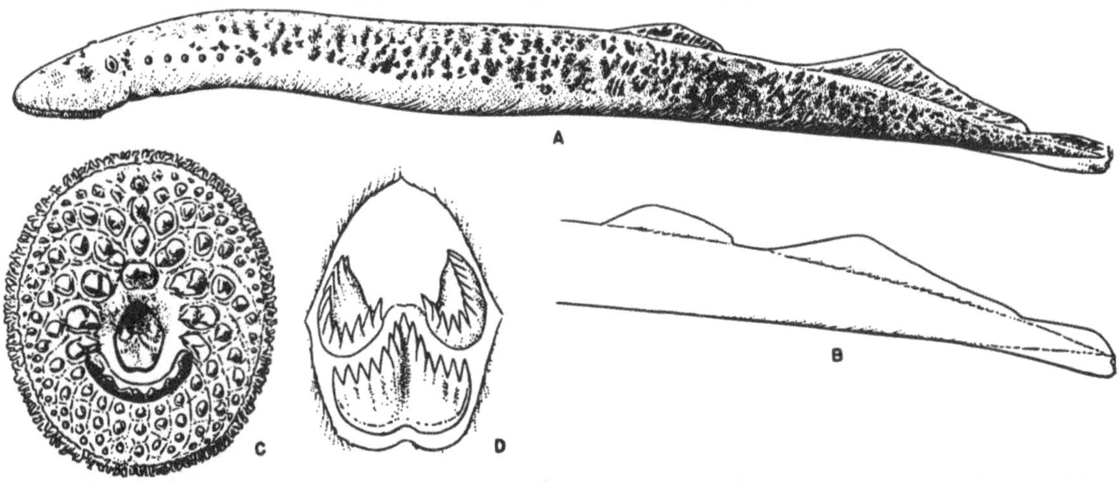

FIGURE 5. *A Petromyzon marinus*, adult about 450 mm. long, from Merrimack River, N. Hampshire (Harv. Mus. Comp. Zool., No. 35069). *B* Posterior portion of another specimen of about the same size to illustrate the variation in the length of the interspace between the dorsal fins. *C* Oral disc of another adult specimen from the Merrimack River (Harv. Mus. Comp. Zool., No. 24975), about natural size. *D* Central mouth of same with lingual teeth, about 4 x.

Teeth as described above for genus, those on disc about 112–125 in specimens counted, in curvilinear pattern as illustrated in Fig. 5 C, and marked off in a pavement-like arrangement by narrow furrows of the fleshy tissue, although actually their imbedded bases are separated, one from the next, by interspaces of considerable width; teeth varying in sharpness in different specimens according to the amount of wear, in extreme cases the supraoral and infraoral dental plates being nearly smooth.[57]

Origin of 1st dorsal fin a little posterior to midlength of trunk, its base about 1/2 as long as head, its height a little more than 1/5 as long as its base, with nearly straight but sloping margins and broadly rounded apex; interspace between 1st and 2nd dorsals varying from very short to about 1/3 as long as base of 1st dorsal; 2nd dorsal about twice as long as 1st dorsal basally, but similar in shape, its height a little more than 1/5 its base, and separated from caudal by a definite notch,[58] but continuous with latter at its base;

57. The genera *Bathymyzon* and *Oceanomyzon* were based on specimens in this condition. See footnote 52, p. 45.
58. Many of the earlier illustrations fail to show this notch, although others do show it.

caudal brush-shaped, with rounded corners, extending forward on ventral side of trunk for a distance about as long as base of 1st dorsal; no separate anal fin; anus anterior to ventral origin of caudal by a distance about ¾ as long as base of 1st dorsal.

Color. Small specimens, whether on their way downstream or in salt water, are white below and uniformly colored above, usually described as blackish-blue or lead-colored and as more or less silvery.[59] But large specimens, approaching maturity, are usually olive-brown above, or of varying shades of yellow-brown, green, red or blue, mottled with a darker shade of the ground color, although sometimes nearly black, the dark patches confluent; lower surface whitish, gray, or of a pale shade of the same hue as ground color of back. During the breeding season, at least in the landlocked form, the colors become still more brilliant, with the ground tint described as turning bright yellow.

Size. The length, at the time of transformation, ranges from about 100 to 200 mm. Sexually mature specimens, taken in American rivers, average about 2 to 2½ feet in length, the largest of a considerable series from the Navesink River being 33 inches long, weighing two pounds, four ounces; the maximum recorded length is about three feet.

Developmental Stages. The eggs are small, spherical. A female has been found to contain 236,000 ova. Segmentation is total, but slightly unequal. The larvae, which differ widely from the adult in external appearance and habits, as well as in internal morphology, are described above (p. 45).

Habits. Since Lampreys never take the hook and are seldom captured in nets, except close to the beach in pound nets or in estuarine situations with shad nets, they are not often seen in the open sea; consequently, little is known of their habits in the sea, except that they are rapid, vigorous swimmers, progressing by an undulating motion, as does an eel, and that they are exceedingly aggressive in their attacks on other fishes. Occasionally they are found attached firmly to driftwood and even to boats.

The fact that Lampreys, when encountered in salt water, are usually close to the land or even in estuarine situations, suggests that most of them remain in comparatively shallow water during their sojourn in the sea. But some stray far offshore and descend to considerable depths. Odd specimens have been caught on the Grand Banks at 86 fathoms north of Emerald Bank; on the seaward slopes of the Nova Scotian Banks off Nova Scotia, at 200 to 350 fathoms;[60] at 85 and at 100 fathoms on the western side of the Gulf of Maine; at 247 fathoms off Martha's Vineyard and at 547 fathoms off Nantucket, Massachusetts.

The geographic range of the species, combined with observations on the vertical distribution of temperature at different seasons, shows that it is tolerant of a wide range of temperature. It is equally tolerant of salinities ranging from fresh water to that of full oceanic saltness (3.5 per cent or even more).

The normal food of the Sea Lamprey is the blood of other fishes, which it attacks by sucking with the oral disc. Usually the Lamprey fastens to the side of its victim, where

59. Those that we have seen have lost all color in the preservative.
60. Specimens in the U.S. National Museum.

it rasps through skin and scales by means of its horny teeth and then sucks the blood. The secretion of its buccal glands has been found to have an anticoagulating action, thus helping the flow of blood.[61] Its prey sucked dry, it attacks another. After metamorphosis, young ones in aquaria attack any fish that may be available and doubtless older Lampreys do the same. In salt water they have been found preying in this way on mackerel, shad (*Alosa*), cod, haddock, American pollock (*Pollachius*), salmon, basking sharks, the various anadromous herrings, swordfish, hake (*Urophycis*), sturgeons and eels; as many as three or four sometimes have been found fast to a single shad. Near river mouths the shad and herring tribes suffer most from them. Judging from their landlocked relatives and from the frequency with which they have been found attached to marine fish, they must be extremely destructive to the latter when they are at all plentiful. So far as we are aware, nothing but fish blood has been found in the stomachs of Lampreys at sea, except fish eggs, of which they are said to be full occasionally.[62] But it is probable that they take in a certain amount of solid flesh also, for muscular tissue, as well as blood, has been found in the stomachs of fresh water Lampreys of another genus.[63]

Before its metamorphosis, the larval Lamprey in fresh water subsists entirely on such microscopic organisms as may be suspended in the constant stream of water that is drawn into the pharynx and discharged through the gill chambers, the oral papillae acting as a sieve to prevent the entrance of grains of sand, etc. When the sieve formed by these papillae becomes clogged, the gill openings are closed and the water is forced back through it.[64] How the food particles are separated from the water and carried into the oesophagus is not definitely known.[65]

It has been known from early times that the Sea Lamprey is anadromous.[65a] However, it does not enter all the streams within its range indiscriminately, but chooses certain ones and avoids others. As an illustration, we may cite outer Nova Scotia and the Bay of Fundy, where Lampreys run in the St. Marys, Sackville, Annapolis, Shubenacadie, Petitcodiac and St. John Rivers, but not in the Margaree, Moser or Apple Rivers, although these last are also "salmon" rivers.[66] For successful reproduction this selectivity is essential in order to obtain gravelly bottom in rapid water for spawning beds, as well as muddy or soft sandy bottom in quiet water for the larvae.

The mature Lampreys enter the rivers of the New England and middle Atlantic

61. Gage and Gage, Science, N.S. *66*, 1927: 282.
62. Goode, Fish. Fish. Industr. U.S., Sect. 1, 1884: 677.
63. Jordan, Guide to Study Fish., *1*, 1905: 491.
64. For a detailed account of the observations on the larva of the landlocked race, see Gage (Sci. Mon., N. Y., *28*, 1929: 401).
65. The food particles have been described as being entangled in strings of mucus and swept back with the latter to the oesophagus by the ciliated tracts on the pharyngeal walls (Bridge, Camb. Nat. Hist., 7, 1904: 429). But so far as we can learn this has not actually been observed.
65a. See Fontaine (Bull. Inst. Oceanogr. Monaco, No. 848, 1942: 2) for a recent study of the osmotic pressure of the body fluids of *Petromyzon marinus* in relation to sexual maturity and to its migrations from salt water into fresh.
66. The above statement is based on extensive observations made in connection with salmon investigations by the Biological Board of Canada, communicated to us by A. G. Huntsman.

states as early as the end of March or early April. In the rivers tributary to the Gulf of Maine the runs are at their maximum peak during May and early June. Few, if any, enter the rivers after that. In New Jersey and Pennsylvania the peak is from late April through May. Precise seasonal data are lacking for rivers farther south or farther north.[67] In many small streams, and in larger ones also, if their passage is blocked by dams or falls, they may spawn only a very short distance upstream, even within the influence of the tide, although invariably in fresh water. They are able to ascend falls, if not too high and steep, by clinging to the rocks with their oral discs and resting, but they do not leap as salmon do in similar circumstances. They may run up for long distances in large rivers. Such, for instance, was formerly their habit in the Merrimack and Hudson River drainage systems, while in the upper tributaries of the Delaware and Susquehanna systems they are still to be found 200 miles or more from the sea, and 150 miles upstream in the Savannah River system.

Since the breeding activities of the Sea Lamprey take place in fresh water, a brief account will suffice here. As the two sexes ripen they become dissimilar in appearance, the males developing a strong ridge along the back, the females a fin-like crest between the anus and the caudal fin (p. 46). Analogy with the landlocked form, and dates actually recorded, suggest that spawning is commenced when the temperature is about 10° C. and is completed by the time the water has warmed to about 20 to 21° C.

Spawning takes place in stretches of the stream where the bottom is stony or pebbly. Working in pairs, a male and a female, with a second female sometimes assisting, make depressions two to three feet in diameter and about six inches deep in the bed of the stream by dragging away the stones by means of their oral discs, leaving the stones in a pile downstream. They are able to move stones as large as one's fist. It is in these depressions that the eggs are deposited, not among the piles of discarded stones that have often been described as "nests."[68] To quote from Regan:[69]

> The female now secures herself by means of her sucker to some large stone near the upper end of the nest, and her mate attaches himself to her in the same way near her head, and winds himself partly round her; then the two together stir up the sand with vigorous movements whilst the eggs and milt are simultaneously deposited. The eggs are covered with an adhesive substance, and particles of sand stick to them, so that they sink to the bottom of the nest. The pair now separate and at once commence removing stones from above the nest and enlarging the pile at the lower end, the sand thus loosened being carried down and covering all the eggs. The process is repeated at short intervals until the spawning is completed. . . .

After spawning, it seems that the parents die, for not only have they been found dead

67. The landlocked form commences to "run" when the temperature has warmed to about 7 to 9° C. (Surface, 4th Rep. For. Comm. N. Y., 1899: 227).
68. Bigelow and Welsh (Bull. U.S. Bur. Fish., *40* [1], 1925: 20) fell into this same error. For an excellent account of the nesting and spawning of the Sea Lamprey, see Hussakoff (Amer. Nat., *46*, 1912: 729); for the landlocked form, see Surface (4th Rep. For. Comm. N. Y., 1899: 191), Coventry (Publ. Ont. Fish. Res. Lab., Biol. Ser. No. 20, 1922) and especially Gage (Sci. Mon., N. Y., *28*, 1929: 401).
69. Regan (Fresh Water Fish. Brit. Isles, 1911: 6), based on accounts of the American landlocked form.

repeatedly along the streams,[70] but their intestines atrophy, they are attacked by fungus, and they become so debilitated that recovery seems unlikely. The larval stage is believed to last from three to five years,[71] during which time the larvae live in burrows or under stones in the mud of the parent stream. Having reached a length of from four to six inches they undergo transformation to the adult form, an event occupying about two months (August to September in New England). They then descend the stream to the sea and are described as reaching salt water in late autumn or early winter in America. The length of life in the sea is not known, but large ones, not yet mature, are to be found there the year round.

Numerical Abundance. It is certain that along the American coast as a whole the Sea Lamprey is now far less numerous than it was, a decrease probably resulting from the construction of dams that it cannot pass in many of the streams that it enters to spawn. This decrease has been most severe in the larger rivers of New England. In the Merrimack River, for example, several cartloads were caught daily for a considerable period in 1847 after the dam was completed there. But so few, if any, now succeed in passing the dams at Lawrence and Lowell, Massachusetts, notwithstanding the fact that fishways are now maintained, that a recent survey yielded no evidence that any now breed in the upper stretches of the river.[72] Similarly, there is a recorded catch of 3,800 in one night at Hadley Falls in the Connecticut River in 1840, but by 1866 Lampreys had become nearly extinct in the Connecticut's upper reaches, although still plentiful in its lower part. However, Lampreys still continue numerous where suitable spawning areas are accessible to them. For example, we may quote catches of 18, 15, and 119 specimens at three localities on the Petitcodiac River system, Nova Scotia, during salmon investigations in May and June, 1942 and 1943;[73] of over 100 on several occasions recently in the lower Exeter River, New Hampshire;[74] and of 98 specimens collected in Swimming River, tributary to Sandy Hook Bay, New York.[75] While Lampreys, like other anadromous fishes, may seem plentiful when condensed within the narrow bounds of a river's banks, their numbers as a whole are in no wise comparable with those of the more common salt water fishes.

Relation to Man. In Europe, during the Middle Ages, Sea Lampreys were considered a great delicacy, and formerly, when they were more plentiful, large numbers were taken in the rivers of New England for human food, particularly in the Merrimack and Connecticut Rivers.[76] Many were also sold in fish markets in New Jersey as late as the

70. For example, Perley (Rep. Fish. Bay of Fundy, 1851: 156) saw dead Lampreys for miles along the Nerepis River, New Brunswick, in August, 1840; and he reports a similar situation in the Miramichi (Cat. Fish. N. Brunsw. and Nova Scotia, 1852).
71. We have no first-hand information to contribute on this point.
72. Bailey (Biol. Surv. Merrimack Watershed, New Hampshire Fish Game Dep., 1938: 155). For an account of early attempts to restore the Lamprey and other anadromous fish in the Merrimack, see Marston (Biol. Surv. Merrimack Watershed, New Hampshire Fish Game Dep., 1938: 193).
73. Personal communication from A. G. Huntsman.
74. Collected by R. Witter for the use of the Biological Lab., Harvard College.
75. Nichols and Breder, Zoologica, 9, 1927: 10.
76. For an account of the Lamprey fishery in New England during the first half of the 19th century, see Goode (Fish. Fish. Industr. U.S., Sect. 1, 1884: 680).

middle of the 19th century.[77] But so far as we can learn they were never valued in the more southern part of their American range. For the past half century the Lamprey fishery has been hardly more than a memory, even in New England, except in a small way for local home consumption or to supply the needs of biological laboratories. In salt water they have never been of any commercial importance; the average fisherman might not see one in a lifetime, nor is there any sale for the few picked up by chance. The larvae are taken in considerable numbers for bait, however, in the Susquehanna River, and perhaps in other streams.

Range. Both sides of North Atlantic; northern Norway; only occasional individuals from Iceland;[78] the Faroes in the east, and southward to Portugal along the coast of Europe, including the North Sea and the Baltic inward to the Finnish Gulf, the western Mediterranean (including Algeria),[79] and the Adriatic; also reported for West Africa;[80] southern Greenland, Gulf of St. Lawrence and Newfoundland in the west, south to Florida; breeding exclusively in fresh water, and landlocked in certain American lakes (p. 54).

Occurrence in the Western Atlantic. The Sea Lamprey has been listed recently for Greenland,[81] where it seems to have been unknown previously. However, apart from this the estuary and southern side of the Gulf of St. Lawrence (reported from Trois Pistoles,[82] Gaspé Basin, Bay of Chaleur and Prince Edward Island) are its northernmost outposts along the American coast,[83] the local stock evidently maintained by reproduction in the tributary streams, for Lampreys run up the St. Lawrence for at least 40 to 50 miles above Quebec City.[84] Adults are taken in large numbers also in the Restigouche[85] and the Miramichi, both in the salt estuary and upstream in fresh water during May and June.[86]

Lampreys have never been reported in the rivers of Newfoundland, although these are fairly well frequented by anglers and wardens.[87] But one specimen was taken 1½ miles off the Newfoundland coast near St. John (found attached to the bottom of a fishing boat) in November 1946;[87] one in the U.S. National Museum is recorded for the Grand Banks south of Newfoundland; also a swordfish, scarred by a Lamprey, was taken off Cape Breton. Earlier characterization of their presence in numbers along outer Nova Scotia is in line with their presence in the Sackville and St. Marys Rivers, Musquedoboit, Mersey

77. Abbott, in Cook, Geol. N. J., 1868: 830.
78. For list of Icelandic records up to 1909, see Saemundsson (Skr. Komm. Havunders. Kbh., No. 5, 1909: 127).
79. Seurat and Dieuzeide, Bull. Sta. Aquic. Pêche Castiglione, 2, 1931: 83; Algerian record; not seen.
80. Günther, Cat. Fish. Brit. Mus., *8*, 1870: 502.
81. Nordgaard, K. norske Vidensk.-Selsk. Aarsber., 1924: 65; Jensen, Rapp. Cons. explor. Mer, *39*, 1926: 101.
82. Personal communication from J. L. Tremblay, conveyed to us by A. G. Huntsman.
83. Bigelow and Schroeder (Canad. Atlant. Fauna, biol. Bd. Canad., 12[d], 1934: 4) based its presence on the north shore of the Gulf on the capture of a small Lamprey taken in the upper Bersimis River, a northern tributary to the St. Lawrence estuary (Low, Labrador Peninsula, 1896: 329). However, it was not specifically identified and may have been an *Ichthyomyzon.*
84. Personal communication from J. L. Tremblay, communicated to us by A. G. Huntsman.
85. Specimen in U.S. National Museum.
86. Personal communication from R. A. McKenzie, of the Biological Board of Canada.
87. Information contributed by A. A. Blair, of the Newfoundland Fishery Research Laboratory.

and Medway Rivers, and at the mouths of streams flowing into St. Margaret and Mahone Bays.[88] They have also been taken repeatedly as far offshore as the vicinity of Emerald Bank, the seaward slope of Banquereau Bank and Sable Island Bank, Lahave Bank, Browns Bank, in the deep gully between the latter and Georges Bank,[89] and on the continental slope off Nantucket and Martha's Vineyard. Lampreys are to be expected anywhere around the shores of the Bay of Fundy, they being recorded from salt water in the St. Andrew's region; adults were plentiful in the St. John River and its tributaries, formerly, and no doubt still are, for small ones were found in the stomach of a *Lota maculosa* in Grand Lake, St. John River system, in the winter of 1926–27. They spawn in the Annapolis and Petitcodiac River systems, as well as in the Shubenacadie River, where larvae have recently been reported as abundant.[90]

They have been reported as being present at many localities along the northwestern and western shores of the Gulf of Maine and as breeding not only in the Penobscot, Saco and Merrimack River systems, but in various smaller streams, including the Exeter River, where they still occur in large numbers, the Lamprey River, a tributary of Great Bay, New Hampshire, and the Parker River in northern Massachusetts;[91] no doubt they occur in other rivers for which there is no published record. In southern Massachusetts they still run in some numbers in several of the small streams tributary to Buzzards Bay,[92] and in the Taunton River system.[93] There is one record for Nantucket.

They are taken occasionally in pound nets in the Woods Hole region, in Narragansett Bay where a few breed in the Taunton River, and in Long Island Sound; they spawn in at least one of the small Long Island tributary rivers which empty into Long Island Sound.[94] The Connecticut and Housatonic Rivers were famous in past years for their runs of Lampreys, although their passage today is barred by dams. Some still enter the Hudson, and there are records of their presence in the Raritan drainage system. They are common in the Navesink and Swimming Rivers tributary to Sandy Hook Bay;[95] and within the Bay itself large and small specimens are taken from time to time in pound nets, or found there attached to fish; they are also taken in Gravesend Bay at the mouth of New York Harbor.

There are numerous recent records for Lampreys, large and small, all along the coast of New Jersey, north to south; also up the Delaware River system to the northern part of Pennsylvania in the Erie River. Although we find no published record of them for the coastal sector between the mouths of Delaware and Chesapeake Bays, Lampreys no doubt occur in this area, for the Bay is a center of abundance for them, with Lampreys recorded

88. Information gathered for us by A. G. Huntsman and R. A. McKenzie of the Biological Board of Canada.

89. Specimens in the U.S. National Museum.

90. Information gathered for us by A. G. Huntsman.

91. Personal communication from Q. A. Arlin, Coastal Warden.

92. Wareham River, Agawam River, Red Brook; also reported in Cape Cod Canal; personal communication from H. G. Smith, Coastal Warden.

93. Palmer River, personal communication from E. H. Trask, Coastal Warden.

94. The Nissiquague, Hussakoff (J. Amer. Mus. nat. Hist., *13*, 1913:323).

95. See especially the account by Hussakoff (Amer. Nat., *46*, 1912: 72) of the nest-building of the Sea Lamprey in the Navesink.

at many localities down to its mouth at Virginia Beach; they run up the Patuxent, Potomac and Susquehanna Rivers, the latter a productive spawning region with larvae reported in abundance in the flats near its mouth.

The next suitable spawning grounds, southward, are the streams discharging via Pamlico Sound. Correspondingly, Lampreys, both young and adult, are recorded as taken in shad nets in Albemarle Sound, while they did run up the Neuse River at least as far as Raleigh, North Carolina, and probably still do. They have been taken in Winyah Bay, South Carolina,[96] and are reported from the Pee Dee and Savannah River systems. Although unreported from Georgia, an early characterization of Lampreys as not uncommon in the St. Johns River system of northern Florida[97] is supported by specimens in the United States National Museum.[98] But it is not known from the Gulf of Mexico,[98a] nor from the drainage area of the latter.[99]

It has long been known that a dwarf, landlocked race of the Sea Lamprey occurs abundantly in Lake Ontario and the lakes tributary to it in northern New York State, where it is very destructive to other fishes. Formerly it was barred from the upper Great Lakes by the falls at Niagara. However, with the construction of the Welland Canal, a passage was opened for it and by 1921 it had reached Lake Erie, where it was unknown previously; by 1936 it was in Lake Michigan; and its spread to Lake Huron and Lake Superior is to be expected, if it has not already taken place.[100]

Synonyms and American References:[101]

Petromyzon marinus Linnaeus, Syst. Nat., *1*, 1758: 230 (descr., refs., European seas); Mitchill, Trans. Lit. phil. Soc. N. Y., *1*, 1815: 461 (descr., N. York); Williams, Hist. Maine, Fish., *1*, No. 13, 1832 (not seen); Holmes, 2nd Annu. Rep. Nat. Hist. Geol. Maine, 1862: 33, 63 (listed for Maine); Goode and Bean, Bull. Essex Inst. Salem, *11*, 1879: 31 (Salem, Massachusetts, specimen attached to American Pollock; also in Massachusetts rivers); Jones, List Fish. Nova Scotia, 1879: 11 (not uncommon, Nova Scotia); Goode, Proc. U.S. nat. Mus., *2*, 1880: 121 (listed for east. Florida); Proc. N. S. Inst. Sci., *5*, 1882: 97 (same as Jones, 1879); Bull. U.S. Fish Comm., *2*, 1883: 349 (American form ident. with European; Nova Scotia to Cape Hatteras; species of fish attacked); Jordan and Gilbert, Bull. U.S. nat. Mus., *16*, 1883: 11 (descr., Atlant. coast, Europe, N. Amer.); Bean, Rep. U.S. Comm. Fish. (1882), 1884: 344 (off Woods Hole); Proc. U.S. nat. Mus., *6*, 1884: 637 (mouth of Susquehanna R., spring); Goode, Fish. Fish. Industr. U.S., 1, 1884: 677, pl. 25, upper fig. (descr., habits, former abund., commercial utilization); Holder, Marvels of Anim. Life, 1885: 5 (abund. in lower Saco R., Maine); Jordan and Fordice, Ann. N. Y. Acad. Sci., *3*, 1885: 283 (class., synonymy, descr., concludes landlocked form not separable from *marinus*); Lee, Portland (Maine) Adviser (Mar. 3), and Brunswick (Maine)

96. Specimen in the U.S. National Museum.
97. Evermann and Kendall, Rep. U.S. Comm. Fish. (1899), 1900: 48.
98. Identification of Lake George, Florida specimens verified by Leonard P. Schultz.
98a. A specimen is listed as *Petromyzon castaneous* (Girard) by Goode and Bean (Proc. U.S. nat. Mus., *5*, 1883: 240), but this is an *Ichthyomyzon* and probably was taken in fresh water.
99. According to Creaser and Hubbs (Occ. Pap. Mus. Zool. Univ. Mich., 120, 1922) and Gudger (Copeia, No. 4, 1930: 146), a specimen earlier reported as from Muscatine, Iowa (Mississippi drainage system), probably was in reality from Lake Cayuga, New York, where *P. marinus* is landlocked.
100. For the history of this expansion of its range, see Hubbs and Pope (Trans. Amer. Fish. Soc. [1936], *66*, 1937: 172).
101. The Sea Lamprey is also mentioned in most of the larger works on European and American fishes as well as in a great number of anatomical and embryological papers, zoological textbooks and natural histories.

Telegraph (Mar. 13), 1885: (listed for Maine, not seen); Gage and Meek, Proc. Amer. Ass. Adv. Sci., *35*, 1886: 269 (nesting of landlocked form in Lake Cayuga, New York); Honeyman, Proc. N. S. Inst. Sci., *6*, 1886: 230 (Nova Scotia, no specific loc.); Jordan, Rep. U.S. Comm. Fish. (1885), 1887: 792; Meek, Ann. N. Y. Acad. Sci., *4*, 1889: 299 (Lake Cayuga, New York, size when adult); Cox, Bull. nat. Hist. Soc. New Brunsw., 1893: 42 (south. New Brunswick, Kennebecasis Bay; species of fish attacked); Gill, Proc. U.S. nat. Mus., *17*, 1894: 107 (larva); Cox, Bull. nat. Hist. Soc. New Brunsw., 13, 1895: 63 (listed for New Brunswick); Jordan and Evermann, Bull. U.S. nat. Mus., 47 (1), 1896: 10 (descr., synonyms, Atlant. coast, south to Chesapeake Bay); Rep. U.S. Comm. Fish. (1895), 1896: 212 (south to Chesapeake Bay); Evermann and Kendall, Rep. U.S. Comm. Fish. (1894), 1896: 584 (small specimens, Vermont); Bean, Bull. Amer. Mus. nat. Hist., *9*, 1897: 329 (Gravesend Bay, N. York); Mearns, Bull. Amer. Mus. nat. Hist., *10*, 1898: 311 (lower Hudson R., juveniles and adults); Smith, Bull. U.S. Comm. Fish., *17*, 1898: 88 (Buzzards Bay); Smith, E., Proc. Linn. Soc. N. Y., No. 9, 1898: 11 (fresh and brackish water near N. York); Smith and Bean, Bull. U.S. Comm. Fish., *18*, 1899: 180 (District of Columbia); Surface, 4th Ann. Rep. For. Comm. N. Y., 1899: 193 (Hudson and Susquehanna Rivers); Bean, Bull. N. Y. St. Mus., 60, Zool., *9*, 1903: 11 (descr., habits, Massachusetts, Connecticut, N. York, Delaware, and Susquehanna River systems); Fowler, Rep. N. J. Mus. (1905), 1906: 48 (descr., N. Jersey loc.); Roy, Nat. Canad., *33*, 1906: 33 (lower St. Lawrence R.); Tracy, 36th Annu. Rep. R. I. Comm. inl. Fish., 1906: 44 (Narragansett Bay, breeds in Taunton R.); Fowler, Amer. Nat., *41*, 1907: 5 (Penn. loc., Delaware R. system); Smith, Geol. Econ. Surv. N. C., *2*, 1907: 28 (Albemarle Sound and Neuse R. system, N. Carolina); Fowler, Rep. N. J. Mus. (1907), 1908: 48 (Delaware R. system, attacking shad, larva); Kendall, Occ. Pap. Boston Soc. nat. Hist., 7 (8), 1908: 1 (New England loc.); Fowler, Proc. Acad. nat. Sci. Philad., *61*, 1909: 407 (small specimens attached to anadromous herring, N. Jersey); Rep. N. J. Mus. (1908), 1909: 351 (descr., size, Delaware River system, species of fish attacked); Tracy, 40th Rep. R. I. Comm. inl. Fish., 1910: 58 (same as Tracy, 1906); Regan, Ann. Mag. nat. Hist., (8) 7, 1911: 198 (class., descr., ***Bathymyzon*** Gill, also the landlocked form, classed as synonyms); Fresh Water Fish. Brit. Isles, 1911: 4 (nesting habits, based on published descr. of landlocked forms); Cornish, Contr. Canad. Biol. (1906–1910), 1912: 79 (Prince Edward I., on mackerel); Fowler, Proc. Acad. nat. Sci. Philad., *64*, 1912: 42, 51, 57 (loc., Maryland, Delaware R. and Virginia); Hussakoff, Amer. Nat., *46*, 1912: 729 (abund., nesting habits, size at maturity, in Navesink River, N. York); Murray and Hjort, Depths of Ocean, 1912: 644 (Newfoundland Bank, at surface); Stafford, Contr. Canad. Biol. (1906–1910), 1912: 54 (Gaspé Bay, Gulf of St. Lawrence); Halkett, Check List Fish. Canad., 1913: 38 (Maritime Provinces, Prince Edward I. and Gaspé Bay); Hussakoff, J. Amer. Mus. nat. Hist., *13*, 1913: 323 (nest building, Nissiquague R., Long Island, N. York); Nichols, Abstr. Proc. Linn. Soc. N. Y., No. 20–23, 1913: 90 (off New York); Fowler, Proc. Acad. nat. Sci. Philad., *66*, 1914: 347 (Delaware R.); Kendall, Proc. Portland Soc. nat. Hist., *3* (1), 1914: 9 (loc., Maine); McAtee and Weed, Proc. biol. Soc. Wash., *28*, 1915: 9 (Potomac River, attached to shad); Fowler, Copeia, No. 31, 1916: 41 (Sandy Hook Bay, N. York season); Proc. Acad. nat. Sci. Philad., *69*, 1917: 122 (loc. for Chesapeake Bay); Latham, Copeia, 57, 1918: 56 (Long Island, N. York, season); Fowler, Proc. Acad. nat. Sci. Philad., *71*, 1920: 292 (N. Jersey); Proc. biol. Soc. Wash., *32*, 1919: 52 (Delaware and Susquehanna R. systems); Latham, Copeia, 71, 1919: 53 (Long Island Sd., color of young, season); Fowler, Proc. biol. Soc. Wash., *33*, 1920: 143 (N. Jersey loc.); Breder and Crawford, Copeia, 103, 1922: 11–15 (Potomac R.); Creaser and Hubbs, Occ. Pap. Mus. Zool. Univ. Mich., 120, 1922: 9 (class., descr., synonyms); Huntsman, Contr. Canad. Biol. (1921), 3, 1922: (7) 55 (Eastport, Maine, region of Passamaquoddy Bay, Kennebecasis Bay, St. John R., larvae in Shubenacadie R.); Breder, Copeia, 114, 1923: 2 (color of young, Sandy Hook Bay; adults in tributaries; nest building in Navesink R.); Hubbs, Pap. Mich. Acad. Sci., *4*, 1924: 590 (size of larvae and after transformation, Virginia Beach); Rauther, in Bronn's Klassen, *6*, Abt. 1, Buch 1, 1924: 677 (class., size, distrib.); Bigelow and Welsh, Bull. U.S. Bur. Fish., *40* (1), 1925: 18 (descr., habits, Gulf of Maine); Breder, Copeia, 138, 1925: 4 (Sandy Hook Bay, New York); Jensen, Rapp. Cons. explor. Mer, *39*, 1926: 101 (listed for Greenland, name only); Nichols and Breder, Zoologica, *9*, 1927: 10 (descr., habits, season, size at maturity, Sandy Hook Bay; breeding in tributaries); Fowler, Fish Culturist, 7 (10), 1928 (descr., habits, N. Jersey, Pennsylvania, Maryland, Delaware); Hildebrand and Schroeder, Bull. U.S. Bur. Fish., *43*, 1928: 43 (loc. for Chesapeake

Bay); Breder, Field Bk. Mar. Fish. Atlant. Coast, 1929: 6 (general account); Dymond, Hart and Pritchard, Publ. Ont. Fish. Res. Lab., 27, 1929: 37, in Univ. Toronto Stud. Biol. (abund., size, Lake Ontario; also in Lake Erie); Fowler, Proc. Acad. nat. Sci. Philad., *80*, 1929: 607 (N. Jersey); Hubbs and Brown, Trans. Roy. Canad. Inst., *17*, 1929: 17 (recent spread to Lake Erie); Jordan, Manual Vert. Anim. NE. U.S., 1929: 6 (general account); Truitt, Bean and Fowler, Md. Conserv. Dep., Conserv. Bull., *3*, 1929: 27 (Maryland loc.); Gudger, Copeia, 4, 1930: 146 (loc. for N. and S. Carolina); Jordan, Evermann and Clark, Rep. U.S. Comm. Fish. (1928), 2, 1930: 8 (in check list); Osburn, Wickliff and Trautman, Ohio J. Sci., *30*, 1930: 120 (listed for Ohio); Berg, Annu. Mus. zool. Acad. St. Petersb., *32*, 1931: 89 (class., descr., refs., distr., alterations at sexual maturity); Breder, Copeia, 2, 1931: 39 (Sandy Hook Bay); Creaser, Copeia, 3, 1932: 157 (Lake Ontario, Lake Erie); Holly, in Schultze, Kükenthal, *et al.*, Tierreich, Lief 59, 1933: 13 (class., refs., descr., distrib.); Bigelow and Schroeder, Canad. Atlant. Fauna, 12d, 1934: 3 (descr., distrib.); Vladykov and McKenzie, Proc. N. S. Inst. Sci., *19* (1), 1935: 45 (distrib. Nova Scotia and offshore); Bigelow and Schroeder, Bull. U.S. Bur. Fish., *68*, 1936: 321 (ref. to Greenland record); Fowler, Bull. Amer. Mus. nat. Hist., *70* (1), 1936: 21 (descr. Amer. and Italian specimens); Vladykov, Copeia, 3, 1936: 168 (N. of Emerald Bank, off Nova Scotia); Hubbs and Pope, Trans. Amer. Fish. Soc. (1936), *66*, 1937: 172 (history of its recent appearance in Lakes Erie and Michigan); Bailey, Biol. Surv. Merrimack Watershed, New Hampshire Fish Game Dept., 1938: 151, 155 (former abund. Merrimack R., recent record for Exeter R.); Breder, Bull. N. Y. zool. Soc., *41*, 1938: 28 (New York Harbor and lower Hudson R.).

Sea Lamprey, Pennant, Brit. Zool., *3*, 1776: 76, pl. 8, no. 27.

Ammocoetes branchialis (in part) Cuvier, Règne Anim., *2*, 1817: 119 (descr.); Gill, Proc. U.S. nat. Mus., *17*, 1894: 108. (This is as applicable to the larvae of *P. marinus* as it is to *Lampetra fluviatilis* Linnaeus, 1758, the two being indistinguishable at this stage in development; note, however, that *Petromyzon branchialis* Linnaeus, 1758, is a synonym of *L. fluviatilis* Linnaeus, 1758, after transformation.)

Petromyzon americanus Lesueur, Trans. Amer. phil. Soc., N.S. *1*, 1818: 382 (descr., Amer. rivers); Storer, Rep. Fish. Rept. Birds Mass., 1839: 195 (descr., nesting, Merrimack R., Boston Harbor); DeKay, Zool. N. Y., 4, 1842: 379, pl. 66, fig. 216 (descr., N. York, Hudson R.); Linsley, Amer. J. Sci., *47*, 1844: 79 (Housatonic R., Connecticut); Storer, Mem. Amer. Acad. Arts Sci., (2) *2*, 1846: 517 (descr., Massachusetts, Connecticut, N. York); Thoreau, Concord and Merrimack Rivers, 1849: 35 (former occur. in Concord R., abund. in Merrimack R.; also in subsequent edits.); Gray, List Fish. Brit. Mus., Chondropt., *1*, 1851: 139 (N. Amer. spec.); Perley, Cat. Fish. N. Brunsw. and Nova Scotia, 1852: 226 (New Brunswick, St. John, Miramichi R. systems); Gill, Proc. Acad. nat. Sci. Philad., Addend., 1861: 63 (listed for N. Amer.); Storer, Mem. Amer. Acad. Arts Sci., N.S. *9*, 1867: 251, pl. 38, fig. 4; also Fishes Mass.: 275, pl. 38, fig. 4 (descr., habits, Massachusetts Bay, Merrimack R.); Abbott, in Cook, Geol. N. J., 1868: 830 (N. Jersey, value); Woods, Amer. Nat., *3*, 1869: 20 (lower Connecticut R.); Baird, Rep. U.S. Comm. Fish. (1871–1872), 1873: 827 (Woods Hole); Gill, Rep. U.S. Comm. Fish. (1871–72), 1873: 814 (listed Cape Cod to Cape Hatteras); Abbott, Rep. U.S. Comm. Fish. (1875–1876), 1878: 827 (abund. in Delaware R. system); Goode, Bull. U.S. Fish Comm., *2*, 1883: 349 (considered ident. with *marinus*); Fish. Fish. Industr. U.S., Sect. 1, 1884: 677 (considered probably ident. with *marinus*).

Petromyzon nigricans Lesueur, Trans. Amer. phil. Soc., *1*, 1818: 385 (descr., color, near Philad.; this is young *marinus*); Storer, Rep. Fish. Rept. Birds Mass., 1839: 197 (size, species of fish attacked, off Massachusetts); DeKay, Zool. N. Y., 4, 1842: 381, pl. 79, fig. 247 (account from Storer, 1839); Linsley, Amer. J. Sci., *47*, 1844: 79 (listed for Connecticut); Storer, Mem. Amer. Acad. Arts Sci., N.S. *2*, 1846: 577 (descr., Massachusetts, Connecticut); Gray, List Fish. Brit. Mus., Chondropt., *1*, 1851: 139 (refs.); Thompson, Hist. Vermont, 2, 1857: 150 (descr., habits, small specimens, Vermont); Gill, Proc. Acad. nat. Sci. Philad., Addend., 1861: 63 (N. Amer.); Storer, Mem. Amer. Acad. Arts Sci., N.S. *9*, 1867: 253, pl. 39, fig. 6; also Fishes Mass., 1867: 277, pl. 39, fig. 6 (descr., size, Massachusetts and Connecticut, fishes attacked); Abbott, in Cook, Geol. N. J., 1868: 830 (N. Jersey, with *americanus*); Amer. Nat., *4*, 1870: 719 (N. Jersey, attached to an eel); Provancher, Nat. canad., *8*, 1876: 262 (descr., common in Gulf of St. Lawrence, species of fish attacked); Jordan and Gilbert, Bull. U.S. nat. Mus., *16*, 1883: 11 (descr., Atlant. coast U.S. and Cayuga Lake, N. York).

Petromyzon (no specific name) Perley, Rep. Fish. Bay of Fundy, 1851: 156 (abund., season, St. John R. system; death after spawning).

Ammocoetes unicolor DeKay, Zool. N. Y., 4, 1842: 383, pl. 79, fig. 250 (good descr. and ill. of larva, Connecticut R. at Northampton, Massachusetts).

Petromyzon maculosus Gray, Cat. Fish. Coll. Descr. by L. T. Gronow, in Brit. Mus., 1854: 2 (equivalent to *P. marinus*, England).

Lampetra marinus Malm, Vetensk. Handl. Göteborg, *8*, 1863: 87; Göteborgs och Bohusläns Fauna, 1877: 630 (occur. west. Sweden).

Ammocoetes fluviatilis Jordan, Ann. N. Y. Acad. Sci., *1*, 1879: 120 (Cayuga Lake, N. York); not *Petromyzon fluviatilis* Linnaeus, 1758.

Petromyzon marinus dorsatus Jordan and Gilbert, Bull. U.S. nat. Mus., *16*, 1883: 869 (compared with *marinus*, Cayuga Lake, Gage Mss.); Jordan, Rep. U.S. Comm. Fish. (1885), 1887: 792 (fresh water, east. U.S.); Rauther, Bronn's Klassen, *6*, Abt. 1, Buch 1, 1924: 678 (landlocked, L. Ontario, and lakes of northern N. York); Berg, Annu. Mus. zool. Acad. St. Petersb., *32* (1), 1931: 89 (ref. to landlocked form); Vladykov, Trans. Canad. Inst., *20*, 1935: 122 (Gt. Lakes).

Petromyzon (Bathymyzon) bairdii Gill, Proc. U.S. nat. Mus., *6*, 1884: 254 (continental slope off Nantucket, 547 fath.); Goode and Bean, Smithson. Contr. Knowl., *30*, 1895: 4 (by ref. to Gill, 1884).

Bathymyzon bairdii Jordan, Rep. U.S. Comm. Fish. (1885), 1887: 792 (by ref. to Gill, 1884); Jordan and Evermann, Bull. U.S. nat. Mus., 47 (1), 1896: 9 (descr., Gulf Stream).

Petromyzon concolor Wright, Prelim. Rep. Fish Fish. Rep. Ont. Game Fish Comm. (1892), 1892: 439 (Landlocked race, Gt. Lakes); not *Ammocoetes concolor* Kirtland, 1840, which is an *Ichthyomyzon.*

Petromyzon unicolor DeKay, or *P. dorsatus* Wilder; Gage, Wilder Quar. Century Bk., 1893: 430 (cf. landlocked form with *marinus*; descr. and photo of larvae).

Petromyzon marinus (*americanus*) McClure, Zool. Anz., *16*, 1893: 360 (segmentation of egg, N. Jersey, specimens, season, temp.).

Petromyzon marinus unicolor Jordan and Evermann, Bull. U.S. nat. Mus., 47 (1), 1896: 10 (descr., the landlocked race; lakes of north. N. York); Rep. U.S. Comm. Fish. (1895), 1896: 212 (lakes of north. and cent. N. York); Surface, Bull. U.S. Comm. Fish., *17*, 1898: 209, pl. 10, upper fig. (life hist., damage to fishes, Lake Cayuga, N. York); 4th Annu. Rep. For. Comm. N. Y., 1899: 194, 200 (extended account of life history of landlocked form, damage to fishes, lakes of N. York); Halkett, Check List Fish. Canad., 1913: 38 (Lake Cayuga, reported Lake Champlain); Bensley, Contr. Canad. Biol. (1911–1914), *2*, 1915: 10 (abund., Lake Ontario, species of fish attacked); Dymond, Publ. Ont. Fish. Res. Lab. 4, in Univ. Toronto Stud. Biol., Ser. *20*, 1922: 60 (first Lake Erie record); Evermann, Bull. U.S. Bur. Fish., *21*, 1922: 95 (listed for Gt. Lakes); Gage and Gage, Science, N.S. *66*, 1927: 282 (anticoagulation action of secretion of buccal glands); Gage, Biol. Surv. Oswego R. System, 1928: 163 (not seen). Probably not *Ammocoetes unicolor* DeKay, 1842, which appears to be an *Ichthyomyzon*; see Creaser and Hubbs, Occ. Pap. Mus. Zool. Univ. Mich., 120, 1922: 9.

Lamprey (no scientific name), Bumpus, Science, N.S. *8*, 1898: 850 (date of breeding in Taunton R.); Nichols and La Monte, Amer. Mus. Novit., 901, 1937: 1 (a swordfish marked by a Lamprey, Cape Breton, Nova Scotia); Stroud, Bull. Bowdoin Coll. (April 1), *6*, 1939: 22 (Kent Island, Bay of Fundy, in stomach of cod).

Oceanomyzon wilsoni Fowler, Proc. Acad. nat. Sci. Philad., *59*, 1908: 461 (descr., ill., Atlant. Oc.).

Petromyzon marinus var. *dorsatus* Huntsman, Ottawa Nat., *31*, 1917: 25 (Lake Ontario); Coventry, Pub. Ont. Fish. Res. Lab., *9*, in Univ. Toronto, Stud. biol., ser. 20, 1922: 133 (spawning, nests, Humber R., tributary to Lake Ontario).

Lake Lamprey, Gage, Sci. Mon., N.Y., *28*, 1929: 401 (habits, larvel devel., photos).

Lamper-eel, Marston and Gordon, Biol. Serv. Merrimack Watershed, New Hampshire Fish Game Dep., 1938: 193, 197 (hist. of attempts to restore Lampreys in upper Merrimack R.).

Probable Synonyms:

Ammocoetes bicolor Lesueur, Trans. Amer. phil. Soc., *1*, 1818: 386 (young, after transformation, Connecticut R.); DeKay, Zool. N. Y., 4, 1842: 383, pl. 79, fig. 248 (descr., ill. after Lesueur, 1818); Linsley, Amer.

J. Sci., *47*, 1844: 80 (listed for Connecticut); Storer, Mem. Amer. Acad. Arts Sci., N.S. *2*, 1846: 519 (brief descr., after Lesueur, 1818); Gray, List Fish. Brit. Mus., Chondropt., *1*, 1851: 146 (by ref. to Lesueur, 1818); Gill, Rep. U.S. Comm. Fish. (1871–72), 1873: 814.

Not ***Ammocoetes appendix*** DeKay, Zool. N.Y., *4*, 1842: 381, pl. 64, fig. 211 (included by Jordan and Evermann, 1896, as a synonym of ***Petromyzon marinus***, but probably a ***Lampetra*** because two dorsal fins are pictured as continuous at their bases).

Not ***Petromyzon lamotteni*** DeKay, Zool. N. Y., *4*, 1842: 382, pl. 79, fig. 249 (included by Jordan and Evermann, 1896, as a doubtful synonym of ***P. marinus***, but probably a ***Lampetra*** because two dorsal fins are pictured as continuous, at their bases).

Not ***Petromyzon lampetra*** Pallas, Zoogr. Rosso Asiat., *3*, 1814: 66 (name and loc. only; included in the synonymy of ***P. marinus*** by Holly, 1933, but probably a combination of a ***Lampetra*** with ***Caspiomyzon***, because of localities White Sea and Caspian).

CHAPTER THREE

Sharks

BY

HENRY B. BIGELOW *and* WILLIAM C. SCHROEDER

ACKNOWLEDGMENTS

In preparing the present paper we have received invaluable assistance and co-operation from many people, both here and abroad. Numerous correspondents have contributed information of various sorts, including photographs of freshly caught specimens, and these are noted under the accounts of the respective species. We are particularly grateful to Luis Howell-Rivero and Stewart Springer for contributing much-needed specimens, together with extensive notes on the occurrence of many species from Cuba and Florida. We wish also to express our gratitude to the following persons: J. L. Baughman for an extensive series of specimens from Texas; Maj. C. M. Duke, U.S. Army, for a specimen of the fresh-water Shark from Lake Nicaragua, and F. B. Richardson for arranging its transportation; Capt. James Whaley for sending us a "Mako" taken off Ocean City, Maryland; Richard Foster and John Huntington for a "Mako" from Cat Cay; Dr. Heloisa Alberto Torres for entrusting to us the type specimen of Scyliorhinus haeckelii *(Ribeiro) for study; Lieut.-Commander J. W. Lowes, U.S.N.R., for records of his own captures of* Carcharodon, *together with color notes, measurements and photographs; President Don Anastasio Somozo of Nicaragua, Capt. W. B. Brinker and Frank Fisher of the National Geographic Society for photographs of newly caught specimens of the fresh-water Shark of Lake Nicaragua; Carlos de la Torre for permitting us to have photographs taken of the late Filipe Poey's unpublished drawings of Cuban sharks, with copies of Poey's unpublished notes; A. Fraser Brunner and Lieut. Colonel W. P. C. Tenison for drawings of* Pseudotriakis *and* Echinorhinus *from specimens in the British Museum, and Miss Ethelwynn Trewavas who enlisted their kind assistance; W. H. Rich for records of recent captures of the Greenland Shark by Gulf of Maine fisher-*

men; John Tee-Van, who furnished bibliographic citations; A. S. Romer, who assisted us in determining classifications of major groups and who contributed summaries of the various genera which occurred in earlier geologic periods; Ludlow Griscom and J. A. Peters, who assisted us in the solution of puzzling questions regarding scientific nomenclature; H. W. Fowler, J. T. Nichols and L. P. Schultz, for their assistance; and to the late Thomas Barbour, not only for putting the collections of the Museum of Comparative Zoology so freely at our disposal, but for constant encouragement and personal help in many ways.

GENERAL DISCUSSION

Scope of Study. The following pages give descriptions, illustrations, life histories and geographic distribution, as well as lists of published citations, for all species of sharks so far known on the western side of the North Atlantic. In some genera represented within these geographic limits, a few additional species occurring in the western South Atlantic but not reported as yet north of the equator are included as addenda. The pertinent characterizations of the suborders and families, as well as the keys to major groups and genera, cover the sharks as a whole and in some cases this applies to the species keys within genera. However, it seems wiser in other cases to limit them geographically until the relationships of species from the western North Atlantic to those of adjacent parts of the ocean are clarified.

Descriptions. The descriptions are based on the Study Material listed under each species, except for *Pseudotriakis microdon* and *Echinorhinus brucus*, no specimens of which were available. The discussions of habits and geographic distribution are based on data of our Study Material, on information submitted to us through the co-operation of numerous correspondents, checked in many cases by our own observations, and on previously published accounts.

Keys. The keys, whether to higher groups or to genera and species, have been arranged solely for the purpose of facilitating the identification of any shark. Therefore, we have selected as alternative characters those that are most easily visible and measurable. Our personal views on phylogeny are not discussed. Species within a genus are presented in alphabetical sequence.

References. All citations listed in the references, with the few exceptions noted, were consulted in the original through the kind co-operation of the following libraries and institutions: Museum of Comparative Zoology and other departments of Harvard University; American Academy of Arts and Sciences; Yale University; United States Fish and Wildlife Service; Library of Congress; American Museum of Natural History; American Philosophical Society; and Massachusetts Institute of Technology.[1]

Sources of Material. The well-rounded collection of sharks from many parts of the world, preserved in the Museum of Comparative Zoology, has been our chief source of

1. For pre-Linnaean names and references, see Garman (Mem. Harv. Mus. comp. Zool., 36, 1913).

reference. The extensive collections of the United States National Museum have also been made available to us, as well as specimens from the Academy of Natural Sciences at Philadelphia, the American Museum of Natural History, the Bingham Oceanographic Collection at Yale University, the Carnegie Museum at Pittsburgh, the Chicago Natural History Museum, the California Academy of Sciences, the Museu Nacional in Rio de Janeiro and the Woods Hole Oceanographic Institution. Other specimens and data are acknowledged on page 59. We regret that war conditions have prevented us from examining the types of many species of sharks that are in the British and European museums.[1a]

Proportional Dimensions and Illustrations. The actual measurements from which the proportional dimensions of the several species have been calculated were taken on a horizontal line between perpendiculars at given points; for example, the distance from tip of snout to origin of first dorsal fin is the line BC in the accompanying illustration (Fig. 6), *not* AC; the length of snout in front of nostril is line ED, *not* DF. The illustrations have been drawn on this basis so that the proportions can be scaled from them directly, if

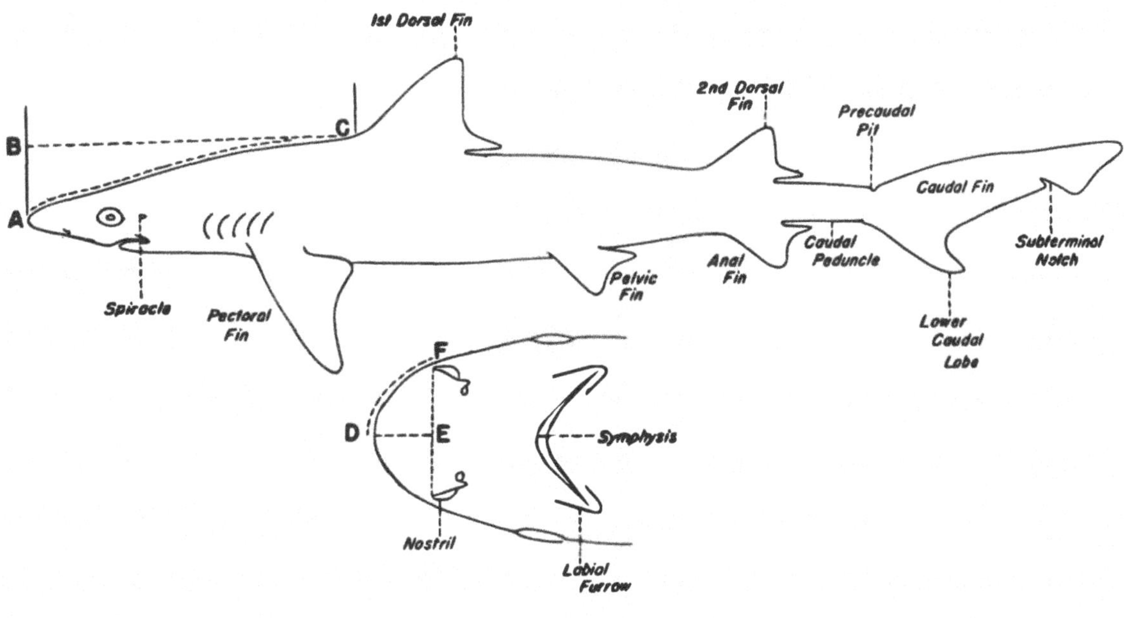

FIGURE 6. Outlines of a typical shark to illustrate terminology and methods of measurement.

desired. In the shark illustrations, the dermal denticles pictured are from high on the sides of the trunk, below the first dorsal fin, unless otherwise noted. All the illustrations are original, except as indicated; the great majority were prepared by the well known zoological artist E. N. Fischer. *Rhincodon* was drawn by Janet Roemhild, *Pseudotriakis* by A. Fraser-Brunner and *Echinorhinus* by Lieut. Colonel W. P. C. Tenison.

1a. For a list of type specimens of cyclostomes, elasmobranchs and chimaeroids in the Paris Museum, see Bertin (Bull. Mus. Hist. nat., Paris, [2] *11*, 1939: 65–93).

Class CHONDRICHTHYES

Elasmobranchs and Chimaeroids

Characters. Fish-like vertebrates with well developed lower jaws and bony teeth; 2 pairs of appendages supported by pectoral and pelvic girdles; a cartilaginous skeleton which, while more or less calcified, lacks any true bone; scales essentially tooth-like in structure, the ectoderm as well as the mesoderm sharing in their formation (placoid scales); two nostrils, each single, partially subdivided; olfactory sacs blind, not opening into mouth; posterior end of vertebral column either straight or heterocercal; sympathetic nervous system, pancreas, spleen and contractile arterial cone present; two, three or more series of heart valves; swim bladder absent.

Relation to Other Classes. Chondrichthyes are most obviously separated from the Cyclostomes by their well developed lower jaws and bony teeth, by their much more highly developed cranium and visceral skeleton, as well as by the presence of pectoral and pelvic girdles, paired limbs, spleen and a contractile conus arteriosus with two, three or more series of heart valves. The lack of true bone in the skeleton, which is most apparent in the skull and pectoral girdle, separates them from all so-called higher fishes, including the Lung Fishes (Dipnoi). Other features marking them apart from bony fishes are: (a) cranium, without sutures consequent on its lack of bone; (b) outer margins of fins supported by horny rays or filaments as contrasted with bony rays or spines among bony fishes; (c) first gill pouch with a row of gill filaments, which are lacking among bony fishes, and gill filaments attached to the interbranchial septa except at the tips (free for a greater or lesser part of their length in bony fishes); (d) no true operculum, but at most a fold of skin serving the same purpose (in Chimaeroids); (e) nostrils single; (f) teeth simply imbedded in the gums, not firmly attached to jaws or imbedded in the latter; (g) scales (placoid or dermal denticles) tooth-like in structure, consisting of a hollow cone of dentine of dermal origin surrounding a pulp cavity; externally this cone is covered with a layer of an enamel-like substance (vitro-dentine) or possibly true enamel (among Rays) formed at least partially by the epidermis; *i.e.*, it is of ectodermal origin,[1b] whereas in bony fishes the scales are formed by the dermis alone, *i.e.*, they are bone-like in origin; (h) fertilization is internal, and in all modern representatives is effected by cartilaginous appendages, commonly called claspers, which are developed from the inner margins of the pelvic fins of the males (among such of the bony fishes as have internal fertilization, the intromittent organs are developed either from the genital papilla or urogenital orifice, or in connection with the anal fin, or as a special structure situated on the chest, but never from the pelvic fins). Furthermore, the invariable presence among the Chondrichthyes of the so-called spiral valve in the posterior portion of the intestine separates them from most bony fishes,[2] as does the presence of a pair of spiracles in many of them (representing vestigial gill clefts) which open on the dorsal or dorso-lateral side of the head, frequently with a

1b. See Tomes (Philos. Trans., *190*, 1898: 460) for further discussion of this question.

2. In a vestigial form in sturgeons, *Amia*, lung fishes and some others.

number of branchial lamellae that probably aid in the oxygenation of blood to the eyes and brain.[3]

Key to Subclasses

1a. 5 to 7 pairs of gills and 5 to 7 pairs of gill clefts, each of the latter opening separately to exterior; dorsal fin or fins, and spines if present, rigid, not erectile; skin with or without dermal denticles; teeth numerous; upper jaw or palatoquadrate cartilage not fused to cranium, although it may be locally attached to it; rostral cartilage fused to cranium; vertebral centra more or less clearly differentiated, and the notochord more or less constricted segmentally; at least some of vertebrae of trunk region with articulated transverse ribs; the 2 halves of pelvic girdle fused into a single bar; anus and urogenital canals discharge into a common cloaca; males without prepelvic or frontal tenacula. Elasmobranchii; Sharks, Skates, Rays, p. 63.

1b. Only 4 pairs of gills and 4 pairs of gill clefts, with only 1 opening to the exterior on each side of head;[4] dorsal fin and spine erectile; skin in adult naked, without dermal denticles; teeth represented by 6 pairs of grinding plates; upper jaw or palatoquadrate cartilage fused with cranium; rostral cartilages articulated to cranium, not fused; no vertebral centra, and the notochord not constricted segmentally; ribs lacking; the 2 halves of pelvic girdle separate; no cloaca, the urogenital aperture being distinct from anus and posterior to it; males with an erectile prepelvic tenaculum, and usually with a frontal tenaculum on the head also.[5] Holocephali; Chimaeroids.

Subclass ELASMOBRANCHII

Sharks, Skates, Rays

Characters. Five to 7 pairs of gill clefts, all opening separately to exterior and not covered by an opercular fold of skin; dorsal fin or fins, and fin spines if present, rigid, not erectile; spiracle present or absent; skin armed with numerous placoid scales or "dermal denticles"; teeth numerous and in several series; no frontal or prepelvic tenacula in males; notochord more or less constricted segmentally, persisting only between the vertebrae in many cases; vertebral centra more or less well developed; at least some of vertebrae of trunk region with short articulated ribs; upper jaw or palatoquadrate cartilage not fused to cranium, although it may be firmly attached to the latter by 1 or 2 articular surfaces of limited extent; lower jaw, or Meckel's cartilage, articulated to upper jaw, and as a rule attached also to hyomandibular arch, which thus takes part in the suspension of jaws;[1] rostral cartilages (1–3) fused to cranium; the 2 halves of pelvic girdle fused into a single bar; inner margin of each pelvic fin in males modified to form a copulating organ or "clasper" grooved for the passage of the sperm and supported by an axial cartilage, the

3. Among the skates, the water that reaches the gills is inhaled through the spiracles, at least for the most part.
4. The true gill clefts open into a common branchial chamber (covered by an opercular fold of skin supported by cartilaginous rays) which opens to the exterior by a single secondary branchial aperture on each side.
5. The frontal tenaculum is lacking in the genus *Harriotta.*

1. Among notidanoids this arch is much reduced and has no attachment to the lower jaw.

latter a rearward extension of the basal cartilaginous element of the fin. Development either oviparous,[2] ovoviviparous,[3] or viviparous;[4] embryos with transitory external gills.[5]

For convenience, the modern representatives of this subclass may be grouped in two orders; the one to include all living sharks as well as the fossil group (hybodonts) from which they appear to have descended, the other to include the skates and rays, which have probably descended from the hybodonts also.

Key to Orders of Modern Elasmobranchs

1a. Gill openings at least partly lateral; edges of pectoral fins not attached to sides of head anterior to gill openings; upper margin of orbit free from eyeball (eyelid free).
Selachii; Sharks, p. 64.

1b. Gill openings confined to ventral surface; edges of pectoral fins attached to sides of head anterior to gill openings; upper margin of orbit not free from eyeball (no free eyelid). Batoidei; Skates, Rays.[6]

Order SELACHII

Modern Sharks

Characters. Gill openings at least partly lateral; edges of pectoral fins not attached to sides of head anterior to gill openings; upper edges of orbits free from eyeballs, as free eyelids. Other than as indicated above, no sharp lines can be drawn between the sharks on the one hand and the skates and rays on the other, so far as external characters are concerned; the gap between the prevalent cylindrical body shape of the former and the much flattened form of the latter is bridged by one group of true sharks (Squatinoidea, p. 533).

Skeletal differences between the two groups are considerable, however, corresponding chiefly to the highly specialized external features of the rays. Thus, to mention only the most obvious, among sharks the propterygial cartilage of the pectoral bears many fewer radials than the metapterygial and is smaller than the latter,[1] while the reverse is the case among rays. Among sharks the shoulder girdle is neither directly nor firmly attached to the vertebral column, nor are its elements united above, while in rays it is attached above by a separate scapular element, or elements. The suspension of the jaws differs also between the two groups; in sharks the ceratohyal cartilage is attached to the lower end of the hyomandibular as well as more or less intimately to the posterior end of the lower jaw (Meckel's cartilage), thus assisting to support the latter; in rays (typically,

2. Eggs laid before hatching.
3. Eggs hatching and embryos developing within the mother, but without placental attachment.
4. Embryos attached to the uterine wall of the mother by a yolk-sac placenta.
5. For an excellent and comprehensive account of the morphology of the elasmobranchs, see Daniel (Elasmobranch Fishes, Univ. Calif. Press, 1934).
6. The skates and rays are classed as a suborder only by some authors. But the skeletal differences between them and the typical sharks discussed above (p. 64) seem to us sufficient to set them apart as a separate order.

1. This applies even to the Squatinoidea (pp. 77, 533).

at least) it is connected by a ligament with the hyomandibular only, at the upper end of the latter, and is entirely separate from the lower jaw, hence it does not take a direct part in the suspension of the lower jaw. The cranium is also much less intimately connected to the vertebral column in sharks than in rays. It has been stated that while the upper jaw (palatoquadrate cartilage) bears a transverse process by which it is attached to the cranium by a ligament in sharks, this is not true in rays. Actually, however, the two groups intergrade in this respect, for the ligamentary attachment (but not the transverse process) is present in some skates at least,[2] while the transverse process may be represented by an articular area only, in some sharks (Heterodontidae), or altered to a rounded knob in others (Isuridae).

Replacement of Teeth. The number of series of teeth that are in actual use at any given time varies from one to four or even five in different sharks, and in different parts of the jaw of a given shark. There are also one to several additional reserve series lying in a reversed position (points up in the upper jaw, points down in the lower) against the inner surfaces of the gums, new series being developed in a deep dental groove along the inner margin of the jaw and covered over by a fold of the mucous membrane. As functional teeth are lost, whether by accident or by orderly migration to the outer anterior edge of the jaw, those of the next younger series move forward to replace them. This process of replacing older and smaller teeth by younger and larger ones continues throughout life, there being as many reserve rows in adult specimens of a given species as in the young. It is this process that provides for the increase in the size of the teeth, which accompanies the growth of the shark.

Among the majority of galeoid sharks the loss of older teeth is irregular, in part accidental, the older teeth being lost and replaced by younger ones individually. Thus, in *Carcharias*[3] two days to one week are required for a directly observable tooth to become detached; sometimes one may be seen dangling from one of the outer corners of its base; as a rule, too, the teeth are lost singly and not by entire series simultaneously. Evidently this last feature applies equally to the various carcharhinids, to the White Shark (*Carcharodon*), and to the Hammerheads, for their jaws commonly show various stages in the progression of teeth. There is no reason to doubt that the replacement is correspondingly irregular in those squaloids in which the teeth are slender, raptorial and spaced along the jaw. But in others of that group, in which they form a continuous cutting edge (Figs. 88 A, 97), the process of replacement involves a revolution of the younger series as a unit from the reversed to the erect position, otherwise gaps would occur in the series in use, which is seldom, if ever, the case.[4] The teeth of the older series, which are being replaced, do not loosen and fall out forthwith, but continue for some time attached to the outer side of the gum to which they have moved, although standing meantime at a somewhat lower

2. This is recorded and well illustrated by Parker (Trans. zool. Soc. Lond., *10*, 1879: 223, pl. 41, fig. 4) for *Raja clavata.*

3. Breder, Copeia, 1924: 42.

4. We have never detected such a gap in numerous specimens of *Squalus acanthias* except as a result of mutilation by the hook at time of capture; nor among specimens of *Centroscymnus, Dalatias, Isistius* or *Somniosus* that we have examined.

level than the series that has replaced them. Consequently, the jaw, as viewed from the outside, may show two series (perhaps even three or only one), while the series next younger than those in actual use (one, or possibly two) may be either oblique or may still lie in the reversed position (Fig. 7), depending on the momentary stage of replacement. It is probable, also, that replacement of teeth is similar (*i.e.*, by series rather than singly) in the Smooth Dogfishes (*Mustelus*), in which they are arranged in mosaic.[5]

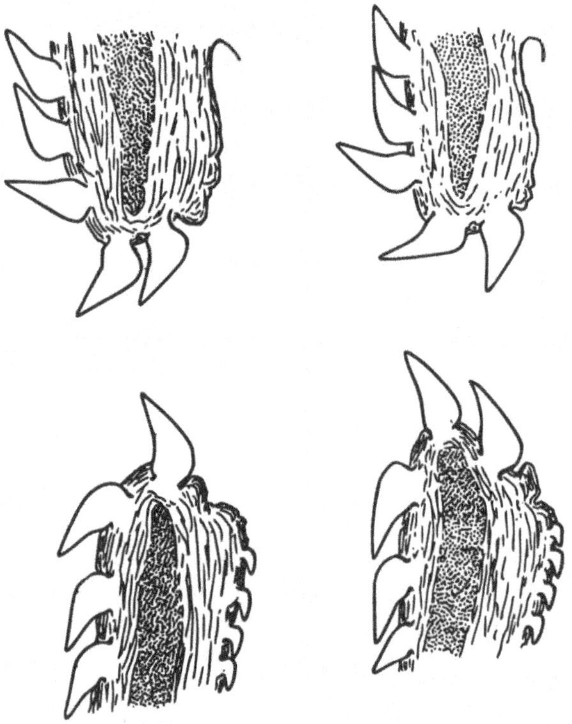

FIGURE 7. Semidiagrammatic cross-sections of upper jaws (left) and lower jaws (right) of two adult specimens of *Squalus acanthias* illustrating different stages in the replacement of teeth.

Form, Activity, Size. Most sharks are subcylindrical in form; some are as beautifully streamlined as the larger members of the mackerel tribe, giving rise to the vernacular name "Mackerel Sharks." On the other hand, a few are very much flattened dorso-ventrally and expanded laterally, so that they resemble skates or rays in general appearance. There is a wide variation in their swimming also. The Mackerel Sharks (Isuridae) are exceedingly active, swift and powerful, whereas others, such as the Greenland and Portuguese Sharks (*Somniosus*, *Centroscymnus*) are so sluggish and inert that it is a question of some interest how they succeed in capturing their prey. Sharks also vary widely in size.

5. Cawston, in a series of recent papers, has maintained that the reserve teeth in sharks come into use only when an individual tooth, lying in front, happens to be lost through injury (Brit. dent. J., *35*, 1938: 321; S. Afr. J. Sci., *35*, 1938: 321; Dent. Rec., *59* [10], 1939: 1; Dent. Rec., *60* [11], 1940: 435; S. Afr. dent. J., *14* [12], 1940: 312; Tidskr. Wetensch. en Kuns, 2, 1941; S. Afr. dent. J., *17*, 1943: 117; S. Afr. dent. J., *17*, 1943: 295; Copeia, 1944: 184). However, direct observations on the shedding of teeth in *Carcharias* (Breder, Copeia, 1942: 42) have proved that the shedding of the older teeth, and their replacement by younger teeth, is a normal process. Successive stages in the process, such as those illustrated in Fig. 7 for the Spiny Dogfish (*Squalus acanthias*), are

On the one hand the Whale Shark (*Rhincodon*) reaches a length of at least 45 and probably 50 to 60 feet, making it by far the largest of fishes, while certain scyliorhinids (p. 213) and triakids (p. 239) mature at lengths of only 300 to 400 mm. (less than 1½ feet).

Breeding and Development. Fertilization is internal (p. 62). The males have a pair of copulatory organs (claspers or myxopterygia) that are developed as appendages from the inner edges of the pelvic fins, supported by cartilages derived from the basipterygial cartilage of the latter, with a groove along the inner side for the guidance of the sperm. In copulation they are inserted through the cloaca of the female into her two sexual orifices.[6] In some species at least, as in the European *Scyliorhinus caniculus*, only one clasper is inserted at a time and coitus lasts about twenty minutes. As a rule the eggs are enclosed in horny cases, at least for a time, but the Greenland Shark (*Somniosus*) may be an exception (p. 520).

Development is oviparous in some sharks, ovoviviparous in the majority and truly viviparous in still others. In the first type the horny egg capsules usually (but not always) bear long tendrils at the corners at one or both ends, by which they are attached to algae, etc.; in one group (Heterodontidae) there is also a very prominent spiral flange, giving the egg a very distinctive appearance. In one species representing this category (*Scyliorhinus caniculus*) the period of incubation is 157–178 days.[7] Among the ovoviviparous species the embryos, early liberated from the capsule, develop in the oviduct of the mother; they are nourished from the original yolk alone which is chiefly in the yolk sac, or from yolk, as well as from nutritive fluids secreted by filaments which are developed from the walls of the maternal oviducts; these nutrients are absorbed both by the yolk sac of the embryo and in many cases by appendages borne by its stalk, the so-called umbilical or placental cord. The young are not born until fully formed and after the yolk sac has been absorbed. In the viviparous species the young lie in special uterine dilations of the oviducts during development; the yolk sac develops folds and processes that interdigitate with corresponding folds of the uterine wall, thus forming the so-called yolk-sac placenta. The number of young is small, as compared with many bony fishes; the maximum number so far reported in a gravid female of any ovoviviparous shark of which we have found record is 82.

Intelligence and Senses. It is recognized by common observation that the intelligence of sharks is of a very low order, although we cannot find that any significant tests have been made of their capacity for learning. Their indifference to injury of any kind is proverbial. In numerous recorded instances a shark, severely mutilated or even disembow-

also instructive, since in this case entire series are involved and not merely individual teeth alone; Spiny Dogfish are so commonly stocked by biological supply houses that large numbers are easily to be had.

6. The method of copulation with which Aristotle was acquainted and which is now a matter of common knowledge, was rediscovered by Louis Agassiz (Proc. Boston Soc. nat. Hist., *14*, 1871: 340); see also Garman and Putnam (Proc. Amer. Ass. Adv. Sci., *23* [2], 1874: 14) and Garman (Proc. Boston Soc. nat. Hist., *17*, 1875: 171, 172). For an eyewitness account of the act of coition in *Scyliorhinus caniculus*, see Bolaw (Z. Morph. Ökol. Tiere, *35*, 1888: 321); and Lübbert and Ehrenbaum (Handb. Seefisch. Nordeurop., *2*, 1936: pl. 20, fig. 249) show an excellent photograph of a pair so engaged.

7. Observations in the Hamburg Aquarium; see Bolaw (Z. Morph. Ökol. Tiere, *35*, 1888: 324).

elled, has returned to continue feeding on the carcass of a whale, or on offal thrown overboard, or even to take the hook a second time. However, some of their senses are of a much higher order than the foregoing might suggest, particularly their sense of smell. It has been shown by experiment that the Smooth Dogfish (*Mustelus*) seeks its prey chiefly by smell (p. 248) and it can be only because of their keen scenting ability that sharks gather so quickly around a whale that is being cut up, or around a dead horse or other carcass in some tropical harbor. As evidence of the ability of a large shark to scent a comparatively small object from a considerable distance, we might mention an occasion in the Gulf Stream, off Key West, Florida, when we saw a large carcharhinid tracking our bait (a Spanish mackerel) up-current, its dorsal fin cutting the surface as it tacked back and forth across the trail, and finally dashing forward on a direct line.

Experiments on the Smooth Dogfish (*Mustelus canis*)[8] have shown that it has at least fair vision for objects that are close at hand, and this no doubt applies to sharks generally. In experiments, however, they seldom responded to any object until the latter was within one foot of them,[9] thus bearing out the general concept that sight is of very little importance in the lives of sharks.

No evidence of any response by sharks to vibrations of high frequency (sound) has been reported, although it seems well established that their auditory (8th) nerves, as well as the nerves of the lateral-line system, are sensitive to water vibrations of low frequency.[10]

Luminescence. A few genera are luminescent, as noted below (p. 509), but the great majority are not.

Food. Sharks are carnivorous without exception. Seaweeds have often been found in the stomachs of one or another species, but no doubt these were taken with the animals on which they were preying, and the more voracious kinds are so indiscriminate in their feeding that they often swallow any kind of inedible rubbish.[11] A few that have crushing teeth (e.g., *Mustelus* and the heterodonts) feed largely on hard-shelled crustaceans (crabs, lobsters) or on mollusks; but the majority prey chiefly on fishes smaller than themselves, on squid and to some extent on pelagic crustacea. In general the size of the prey is relative to the size of the shark. However, some of the more fiercely predaceous species regularly attack other fish, including other sharks nearly as large as themselves, if they are in a position to do so; sea turtles and seals are a regular item in the diet of some sharks. On the other hand, the two largest species (Whale Shark and Basking Shark) subsist wholly on minute planktonic forms, chiefly crustacea, and on small schooling fishes.

Number of Species. In spite of the antiquity of the group, and in spite of the fact that they appear to be as numerous and as varied now as at any time in the past, there are many less species of sharks than of bony fishes; not more than 225 to 250 are now known.

Danger to Man. Dependable information on the danger of sharks to man is fragmentary; nevertheless, we think it necessary to discuss the subject briefly, since it is of

8. Parker, Bull. U.S. Bur. Fish., *29*, 1911: 46.
9. Parker, Bull. U.S. Bur. Fish., *33*, 1914: 64.
10. Parker, Bull. U.S. Bur. Fish., *24*, 1905: 201.
11. For instances, see p. 69.

interest to seamen, to fishermen and to seaside visitors who frequent shark-infested regions.

Most species of sharks are either too small, too sluggish, too weakly armed or normally live at too great a depth to be of any potential danger. This applies also to some of the larger and better-armed species which feed on small rather than on large prey. On the other hand, there are unquestionably a considerable number of species, proverbially voracious, which are large, active and armed with very effective teeth, and which habitually feed on large prey such as other sharks, large fishes and sea turtles; it is equally true that many persons in various parts of the world have been attacked by sharks. Notable among dangerous species are the White Shark (*Carcharodon*), the Tiger Shark (*Galeocerdo*), certain members of the genus *Carcharhinus*, the Lemon Shark (*Negaprion brevirostris*) and the larger Hammerheads. All these bear evil reputations as potential man-eaters and the charge seems to be sufficiently proved against them in one part of the world or another (see discussions below under the respective species). Perhaps the Makos (*Isurus oxyrinchus* and *I. glaucus*), which feed chiefly on small fish, may deserve a similar reputation, but we do not believe that the Blue Shark (*Prionace glauca*) does, unless attracted by blood to a wounded man in the water; under these conditions any shark more than five or six feet long would be a menace. Among the foregoing list the White Shark (*Carcharodon carcharias*) is beyond question the most dangerous. Fortunately, however, even the smaller sizes of this species appear not to be common anywhere, while large adults are very seldom seen, especially close inshore.

In estimating the risk, even from the more dangerous species, we should keep in mind that man is not the habitual prey of any shark; hence the scent of man in the water is not likely to prove especially attractive, since it is presumably by scent chiefly that sharks discover and track down their food. On the other hand, sharks soon learn to gather where dead animals or garbage are to be expected, as where refuse from a slaughter house drifts out to sea. When in a feeding mood, some of the more voracious kinds, especially the "Tiger," will gulp down wholly indigestible objects, such as boots, old clothes, a sack of coal, tin cans, etc., as readily as a chunk of salt pork or a dead dog. Nor is there any reason to suppose that the scent of man is repulsive to any shark.

In view of the foregoing it is not astonishing that many shark fatalities are on record, well attested by hospital reports or otherwise.[12] Shark attacks are much more frequent in warm waters than in cold, as might be expected. For example, from 1919 to 1933, 37 cases were reported for various parts of Australia on seemingly conclusive evidence, with many more for earlier years.[13] In fact, the shark menace is so real in New South Wales

12. The many reports of fatalities by sharks which are not so attested may be left out of the account; some have been based on rumor alone (even the individuals concerned may have been imaginary); others, involving the overturning of small boats, etc., or the disappearance of swimmers without trace, may have resulted from quite other causes; and in still others, an observed attack may have been by a Barracuda (*Sphyraena*) and not by a shark.

13. See Coppleson (Med. J. Aust., April 15, 1933: 449) for a list of these and other such happenings for Australia, with references for shark attacks in other parts of the world; see also, Whitley (Fish. Aust., *1*, 1940: 13, 259) for further details, discussion and list of attacks in Australian waters, with photographs of wounds suffered by victims.

that patrols are maintained on the more popular bathing beaches, some of which are further protected by wire netting; in some parts of Cuba bathing areas are similarly protected with closely spaced palmetto logs. Attacks have been reliably reported from South Africa, the Red Sea, India,[14] Ceylon, the East Indies, the Philippines, the Pacific coasts of Mexico and Panama, the coast of Ecuador, the Gulf of Mexico, the West Indies, the Guianas, the eastern coast of the United States (see below), tropical West Africa, the eastern Mediterranean, Port Said, and no doubt from other regions as well. However, the incidence of attacks is very irregular. Sharks, for example, although plentiful enough along the beaches of Florida, are so slight a menace that we have positive word of only one or two attacks in recent years (pp. 368, 408), despite the fact that many thousands of persons bathe there constantly throughout the year. The most recent instance was of a girl severely bitten while bathing in the surf, only waist deep, at Mayport, Florida, in late May or early June 1944. The size of the shark's jaws, as outlined by the wounds, showed that it was only 5½–6½ feet long, and other circumstantial evidence pointed to a *Carcharhinus maculipinnis* as responsible.[15] Shark attacks appear to be similarly unusual throughout the West Indian region in general; although local inhabitants in Porto Rico and among the Antilles have informed us that, while they would not hesitate to swim by day even if sharks were about, it would be hazardous in the extreme to do so at night.

Attacks occur from time to time, however, even to the northward along the Atlantic coast, although sharks of the dangerous sorts are progressively less numerous in that direction. Near Charleston, South Carolina, for example, several well-attested cases have been reported recently.[16] More widely heralded was a series of attacks on six bathers on the New Jersey coast in July 1916, probably by a small *Carcharodon* (see p. 139) that was caught nearby a few days later. More recent still was an attack in Buzzards Bay, Massachusetts, July 26, 1936, on a bather who was so badly injured that he died shortly afterward in the New Bedford Hospital. The shark, about six feet long but not identified as to species, was driven away by the victim's companions who came to his rescue in a boat. However, these last two instances are the only ones along our northeastern coast that have come to our attention in a lifetime experience. Such events are certainly no more common along the bathing beaches of the northern Mediterranean or of northwestern Europe, for we have not found a single definite case of recent date recorded in the literature of sharks, in natural history journals or in the press. It also happens that the few large sharks which are at all common close along the shore north of Cape Hatteras on the one side of the Atlantic, or of Portugal on the other,[17] are either wholly innocuous, as is the Basking Shark, or at least have never been proved guilty of attacks on bathers, whatever may be

14. The Augustine Friar, Sebastiao Manrique, in 1643, was an eyewitness to attacks by sharks on pilgrims wading out into the sea at Hugli, in Bengal; see translation by Collis, The Land of the Great Image, 1943: 76.

15. This case was reported to us by Stewart Springer and was mentioned in the local press. The victim was treated at the dispensary of a Naval Base near by.

16. Burton, Sci. Mon., N. Y. *40*, 1935: 279.

17. Sand Shark (*Carcharias taurus*); Basking Shark (*Cetorhinus maximus*); Tope (*Galeorhinus galeus*); Common Mackerel Shark or Porbeagle (*Lamna nasus*); Blue Shark (*Prionace glauca*); Brown Shark (*Carcharhinus milberti*); Dusky Shark (*Carcharhinus obscurus*).

true of them in warm seas. This applies also to the Greenland Shark (*Somniosus*) of Arctic seas, for while it preys habitually on living seals it is so sluggish that both Eskimos and whale fishermen look upon it with contempt (p. 522).

The general conclusion from the foregoing is that in continental waters in temperate and boreal latitudes on either side of the North Atlantic the danger to a swimmer of attack by a shark, although existent, is so exceedingly remote as to be wholly negligible, unless it be known that a shark of some dangerous kind has been seen in the vicinity recently. We believe this to be equally true of the coastwise waters of the North Pacific south to southern California on the one side and to northern Japan and northern China on the other, although our personal information is less extensive there than for the North Atlantic.

Categorical statement is not so safe for warmer seas, because reported attacks have been much more frequent there, because large sharks of the potentially dangerous kinds are far more numerous, and because local conditions differ widely between different regions. For coral-reef areas all our sources of information, including personal experience, agree that while dangerous sharks may be numerous offshore and along the seaward slopes of the reefs, they seldom enter the lagoons and are much less likely to enter any smaller pools among banks and coral heads. Large sharks do not often come into wading depth along open beaches, especially if the swell is breaking heavily a short distance out, as is so often the case, unless attracted by slaughterhouse wastes, etc., or by corpses, as in India. Shoal-water bathing is therefore reasonably safe in such situations, at least in the daytime, unless as just qualified or unless the local inhabitants advise against it. In deeper harbors, more open to the sea, it is wiser to err on the side of caution, unless the locality is declared safe by local report, which is usually reliable.

Under normal circumstances the danger of attack to a bather offshore, even in tropical seas, also appears very slight, for the chances are much against any dangerous sharks being close at hand or of their being in a feeding mood if present. But if persons in the water are bleeding from injuries the danger from shark attack may be imminent and the results may prove fatal. The more voracious of the larger sharks are excited by blood in the water to such a degree that they will make ferocious attacks, whether the object be fish, whale or man, dead or alive. Attempts to drive the attacker away by blows or splashing are likely to be futile, although success might be achieved if the swimmer were uninjured. Instances are on record, apparently on good evidence, of crews from capsized boats being attacked and pulled down in tropical seas; the southwestern Pacific is reported to have been the site of such events during the recent war.

So extensive is the resulting laceration likely to be that bites from any large sharks are extremely dangerous for they are followed by very rapid bleeding and severe shock, even if the wounds are not still more directly destructive. Thus, "so far as known, about one-half of Australia's shark attacks have ended fatally."[18]

18. Whitley, Fish. Aust., *1*, 1940: 16.

To class sharks "harmless" as a group, as some authors have done, is contrary to all the weight of evidence. On the other hand, the danger of attack to the ordinary bather is very small indeed, except in such special localities and under such circumstances as those mentioned.

Commercial Importance. Sharks are the objects of minor fisheries in the warmer parts of the world, largely for their liver oil and for their fins (considered a great delicacy by certain oriental races), and to a lesser extent for their hides and flesh.

Shark liver oil was formerly valued highly in combination with other fish oils for tanning, the yield from local fisheries being considerable, notably from the Greenland Shark. Recently a new demand for the liver oil of some species has developed because of the high vitamin content. This is notably the case in the northeastern Pacific, the California catch having risen from about 555,000 pounds in 1936 to about 7,800,000 pounds in 1940, although it dropped to 2,613,431 pounds in 1944. This increase has resulted from the oil of one vitamin-rich species, the Soupfin or Oil Shark (*Galeorhinus galeus*). Interest in the commercial possibilities of shark oil has given impetus to shark fisheries along the eastern coast of the United States also, but to date no western Atlantic shark that occurs in large numbers has been found to equal the California *Galeorhinus* in showing a consistently high Vitamin A content (nor do representatives of that same species in the eastern Atlantic), although individual specimens, such as the larger Hammerheads, may give a high yield. The following table, condensed from a more detailed one,[19] gives the maximum and minimum potencies in Vitamin A (stated in U. S. P. Units) for the liver oil of several Florida sharks.

Species	*Number of Specimens*	*Potency Maximum*	*Potency Minimum*
Carcharodon carcharias	6	7,350	750
Ginglymostoma cirratum	many	6,720	641
Galeocerdo cuvier	many	4,625	1,375
Carcharhinus leucas	many	20,875	1,812
Carcharhinus milberti	many	15,500	283
Carcharhinus obscurus	6	58,500	6,500
Carcharhinus limbatus	many	22,250	4,250
Negaprion brevirostris	many	11,425	3,000
Sphyrna diplana	34	137,000	5,400
Sphyrna tudes	many	340,000	8,250

19. Springer and French (Industr. Engng. Chem., *36* [19], 1944: 190). See Walford (U.S. Fish Wild Life Serv., Fish. Mkt. News, *6* [6], 1944: 4) for a detailed table giving the amounts of Vitamin A, both per gram of oil and per pound of liver, for several species taken in the Gulf of California. For methods of calculating amounts of Vitamin A in livers, see Sanford (U.S. Fish Wild Life Serv., Fish. Mkt. News, 7 [1], 1945: 6) and Bolomey and Tompkins (Fish. Bull., Sacramento, 64, 1946: 73); for relationship between liver yield of Vitamin A and the biology of the Soupfin Shark (*Galeorhinus galeus*) in California waters, see Ripley and Bolomey (Fish. Bull., Sacramento, 64, 1946: 39).

From earliest times the fins of certain sharks have been highly prized as food in China and Japan because of their gelatin content, and often the demand has exceeded the supply. We regret that statistics are lacking for the total amounts marketed. However, as long ago as 1850 not less than 40,000 sharks were caught yearly in the Arabian Sea, chiefly for the export of fins to China.[20] Until the recent war, supplies were regularly drawn from as far afield as California. In fact, the species from which the fins are taken there (*Galeorhinus galeus*) has been known locally as Soupfin Shark, although at present the name Oil Shark is more commonly applied to it.

The kinds of sharks which have firm meat are better food fish than is generally appreciated, and various species are regularly placed for sale in the fish markets of the temperate parts of the world. In Chile, for example, 2.7 million pounds of sharks (about 10 per cent of the total catch of fish in all categories) were landed in 1940 to be consumed locally.[21] Local consumption may be considerable in northern Europe also where the Spiny Dogfish (*Squalus acanthias*, p. 462) is in demand. Along the coasts of the United States the larger sharks have been increasingly marketable of late years. Efforts have also been made by the United States Bureau of Fisheries to promote the sale of canned meat from the Spiny Dogfish (p. 462) as "gray fish," but the project failed when discoloration and spoilage resulted from the generation of ammonia in the cans due to the high content of urea in shark flesh.

It has long been known that the hides of many of the larger sharks yield leather comparing favorably with cowhide, and minor fisheries for this purpose have been carried on in various parts of the world. In the western North Atlantic these fisheries have been located off southern Massachusetts, North Carolina, eastern Florida, Key West, Florida, the Bahamas, and among the Virgin Islands. Up to the present time, however, the amount of shark leather marketed has been very small, as compared with leather from domestic animals. In some cases the fisheries have been short-lived, because of depletion of the local stock of sharks which are large enough to be serviceable (for local instance, see p. 104). But in regions where a fishery may be expected to draw its supply of sharks from a wide area, as on the east coast of Florida with the Gulf Stream near at hand, the prospects of commercial success appear to depend chiefly on an expansion of the demand for shark leather.

The dermal denticles of many sharks are so sharp and so close set that the skins make an effective abrasive, and shark skin, often known as "shagreen," was formerly in wide use by cabinet makers the world over for polishing wood, but it has been almost entirely supplanted of late by other recently developed abrasives, except perhaps in remote parts of the world.

At the present time shark scrap, like other fish scrap, is in demand for feeding poultry and other livestock, and in sum total considerable amounts are marketed. However, we find

20. See Buist (Proc. zool. Soc. Lond., *18*, 1850: 100) for an account of the Karachi Fishery in India.
21. Fiedler, Geog. Rev., *34*, 1944: 104.

no statistics as to actual amounts of shark scrap for comparison with scrap from other fishes. Small amounts of shark refuse also find their way into commercial fertilizers, but here again definite statistics are lacking. Efforts have even been made in the Maritime Provinces of Canada (p. 462), as well as in the United States and possibly elsewhere, to develop this industry. But so far as we know all such attempts have been short-lived, because of irregularity in the supply of sharks.

Sharks are not as highly esteemed for food as are various bony fishes that support the great fisheries, partly because the available supply is only a fraction as great; hence, the landings of sharks are correspondingly smaller, especially in northern seas, and they are correspondingly less in value. Thus, the reported catch of sharks (4,417,700 pounds) was less than one-half of one per cent of the total catch of all kinds of fish (1,458,687,600 pounds) along the Atlantic and Gulf coasts of the United States in 1942, and about one per cent (10,171,900 pounds out of a total of 1,346,559,600 pounds of fish of all kinds) on the Pacific Coast of the United States. In warmer regions the shark catch may rank relatively higher, the catch of bony fishes being much smaller than it is in the northern seas, *e.g.*, the Chilean catch mentioned above (p. 73). But previous experience suggests that fisheries for large sharks, if intensive and on a large scale, are likely to be short-lived, seemingly through exhaustion of the local supply.

Recently commercial shark fishing in the western Atlantic has been carried on most actively off the southern part of the North Carolina coast (Morehead City), along eastern and southeastern Florida (Mayport, Salerno, Cortez and Key West) and off the Bahamas. The yield consists chiefly of Tiger Sharks (*Galeocerdo*), Sand Sharks (*Carcharias*), various species of *Carcharhinus*, Nurse Sharks (*Ginglymostoma*), Hammerheads, and Lemon Sharks (*Negaprion*); on the whole, the first two rank foremost in commercial importance, both in quantity and in value. Anchored gill nets with a stretched mesh of about 20 inches, and anchored set lines (best of chain) with snoods of wire rope every six to eight feet, both set at depths of 3 to 20 fathoms, are the types of gear chiefly used. The catches of Greenland Sharks that are made in the waters off Iceland and Greenland are mostly by long lines, or by hand lines. Basking Sharks have usually been harpooned because of their large size, and this applies equally to the Whale Sharks that have been fished from time to time in the Bay of Bengal and in the waters around India.

Habitat and Range. Sharks are marine for the most part, but a few members of the genus *Carcharhinus* run far upstream into brackish or even into fresh water in large rivers such as the Ganges, the Tigris and the Zambezi. We have received two specimens of *Carcharhinus leucas*, a well known west tropical Atlantic species, that had been taken in Lake Yzabal, Guatemala (p. 341), and one landlocked species is known in Lake Nicaragua (p. 381). Many are oceanic and roam the high seas,[21a] while others dwell on the ocean bottom or close to it. In warm latitudes they are often seen following ships for days

21a. A shark tagged off Ventura, southern California, was recaptured on the west coast of Vancouver Island, it having migrated about a thousand miles; see Ripley (Calif. Fish Game, [2] 32, 1946: 101).

at a time, feeding on garbage thrown overboard. However, the great majority are confined to comparatively shallow water. While a few, which are mentioned below in the appropriate connections, find their homes on the continental slopes at depths of some hundreds of fathoms, the greatest depth for which there is definite record of the capture of a shark of any species is about 1,500 fathoms. Nor is it likely that any shark is a regular inhabitant of the floor of the oceanic abyss. The group is cosmopolitan, but the great majority inhabit the tropical-subtropical belt. Characteristically temperate species are much fewer in numbers, and only one genus (*Somniosus*) is a regular inhabitant of truly polar waters.

Classification. The question of how to subdivide the modern sharks so as to illustrate the supposed phylogenetic relationship of different groups, which has been argued since the days of the early comparative anatomists, is one that we pass over briefly.

The paleontologic history of the groups of sharks that still exist throws little light on the matter. Groups as diverse as the heterodonts, orectolobids, galeoids and squatinoids were all in existence as far back as the Upper Jurassic, and the hexanchids were present in the middle Jurassic and the squaloids in the Cretaceous; while "by the beginning of the Tertiary period all of the living families of Elasmobranchs appear to have come into existence."[22]

Students of living sharks have agreed generally that the most primitive are those (Hexanchidae and *Chlamydoselachus*) in which the vertebrae are calcified but weakly, if at all, and in which the notochord is but little constricted segmentally. The hexanchids likewise appear to agree with the Mesozoic genus *Hybodus* both in these features, and further, in the mode of suspension of the upper jaws (p. 78). However, if these supposedly primitive groups were actually derived from the hybodoids, as has been suggested, they have diverged widely from the ancestral stem by a multiplication of gill arches (*Hybodus* had five only), by the loss of the second dorsal fin and of fin spines, and by modification in their dentition. On the other hand the heterodonts, which resemble the ancient hybodoids so closely in dentition, in number of gills and in the presence of two dorsal fins and fin spines that they have often been united with them in a single suborder, differ from the hybodoids in having the vertebrae well calcified, the notochord strongly constricted segmentally and the upper jaw (palatoquadrate cartilage) attached to the cranium in one region only, without the postorbital connection which has often been regarded as primitive.[23]

Among the remaining, and far more numerous, living members of the order, much weight in classification has been given to the degree and arrangement of the internal calcifications of the vertebral centra. These centra may consist of only a primary ring surrounding the notochord ("cyclospondylic"), or of a primary ring with secondary calcifications as well, either in concentric rings around the primary one or in bars (simple or branched)

22. Romer, Vert. Paleont., 1933: 54.
23. See De Beer (Devel. Vert. Skull, 1937: 421–425) for definitions of the rather complex terminology that has been employed to define the different methods of attachment of upper jaw to skull.

radiating out from the primary ring and which may or may not invade the four primary uncalcified areas that radiate out to the bases of the neural and haemal spines.[24]

Jordan and Evermann,[25] for example, followed in 1930 by Jordan, Evermann and Clark,[26] classed all sharks, other than the notidanoids, in two orders, Asterospondyli (corresponding to our Galeoidea and Heterodontoidea) and Cyclospondyli (including the squaloids, pristiophoroids and squatinoids).[27] The sharks have also been subdivided according to the external or the skeletal structure of the male copulatory organs. However, this results in grouping the notidanoids with the squaloids, and the squatinoids with the Batoidei in one case,[28] or *Chlamydoselachus* with the Holocephali in another.[29]

The majority of modern authors[30] have given primary consideration to characters that are visible externally in both sexes, such as the number of gill openings, the presence or absence of the anal fin, number of dorsal fins and the dentition.

White[31] classed the Selachii ("Antacea") as a superorder with four orders—Hexanchea, Heterodontea, Squalida and Galea, dividing the Squalida into the suborders Squalida and Rhinida, the Galea into the suborders Isurida and Carcharinida. Still more recently, Bertin[32] classed the skates and rays with the sharks as four suborders under the order Euselachii, and distributed among three suborders (Scylliformes, Musteliformes and Lamniformes) the assemblage of families that are united here as the suborder Galeoidea (White's order Galea).

In our opinion, however, the characters on which these subdivisions of the galeoid sharks are based—the presence or absence of a nictitating fold or membrane, the position of first dorsal relative to pelvics, the details of vertebral calcification and the morphology of the spiral valve—are of a lower taxonomic grade than are those by which the notidanoids, heterodontoids, squaloids, pristiophoroids and squatinoids can be defined.[33]

Apart from the names employed, the subordinal classification used in the present paper follows that of Rey,[34] which in turn is based in its essentials on Garman's[35] system,

24. Hasse (Nat. Syst. Elasm. Algem. Theil, 1879) proposed the names "cyclospondylic" for the vertebral type with primary annular calcification only, "tectospondylic" for that with secondary concentric rings of secondary calcification, and "asterospondylic" for that with radiating bars in addition to the primary ring. Regan (Proc. zool. Soc. Lond., 1906: 737), however, has more recently limited "asterospondylic" to the type with four radial bars only, which do not invade the four primary uncalcified areas, and has expanded "tectospondylic" to include all types that are not either "asterospondylic" as so limited, or "cyclospondylic," an emendation that has caused some confusion in nomenclature.

25. Bull. U.S. nat. Mus., 47 (1), 1896.

26. Rep. U.S. Comm. Fish. (1928), 2, 1930.

27. In the interim, Jordan (Class. Fish., Stanford Univ. Publ. Biol., 3 [2], 1923) had recognized five orders of sharks and employed the term "Tectospondyli" in place of "Cyclospondyli" for the order including squaloids, plus squatinoids.

28. Huber, Z. Wiss. Zool., *70*, 1901: 671.

29. Leigh-Sharpe, J. Morph., *42*, 1926: 336.

30. Notably Regan (Proc. zool. Soc. Lond., 1906: 722), Garman (Mem. Harv. Mus. comp. Zool., *36*, 1913), Rey (Fauna Iberica Peces, *1*, 1928), White (Bull. Amer. Mus. nat. Hist., 74 [2], 1937) and Bertin (Bull. Inst. oceanogr. Monaco, 775, 1939).

31. Bull. Amer. Mus. nat. Hist., *74*, 1937: 100, 101.

32. Bull. Inst. oceanogr. Monaco, 775, 1939.

33. For a recent discussion of inter-relationships of modern sharks, with resultant proposals as to classification, see Bertin (Bull. Inst. oceanogr. Monaco, 775, 1939).

34. Fauna Iberica Peces, *1*, 1928: 280.

35. Mem. Harv. Mus. comp. Zool., *36*, 1913.

except that *Chlamydoselachus* and the Pristiophoridae are each made a distinct suborder for the reasons stated below (pp. 94, 532).

Key to the Suborders of Modern Sharks

1a. Anal fin present.
  2a. 6 or 7 gill openings.
    3a. Margins of 1st gill openings not continuous across throat; upper and lower teeth notably unlike toward center of mouth. Notidanoidea, p. 77.
    3b. Margins of 1st gill openings continuous across throat; upper and lower teeth similar in center of mouth as well as along its sides. Chlamydoselachoidea, p. 92.
  2b. Only 5 gill openings.
    4a. Dorsal fins preceded by stout spines; teeth toward center of mouth in each jaw markedly different from those toward its corners. Heterodontoidea,[35a] p. 94.
    4b. Dorsal fins not preceded by spines; teeth toward center of mouth of same basic type as those toward its corners. Galeoidea, p. 95.
1b. No anal fin.
  5a. Snout of only moderate length, without lateral teeth or barbels.
    6a. Trunk subcylindrical; eyes lateral; anterior margins of pectorals not overlapping gill openings. Squaloidea, p. 449.
    6b. Trunk much flattened dorsoventrally; eyes dorsal; anterior margins of pectorals far overlapping gill openings. Squatinoidea, p. 533.
  5b. Snout greatly elongate, as a narrow beak, armed on either side with sharp teeth, and with a long fleshy barbel. Pristiophoroidea,[36] p. 532.

Suborder NOTIDANOIDEA

Characters. Anal fin present; only one dorsal fin, without spine; either 6 or 7 gill openings, all anterior to pectorals; margins of 1st gill openings not continuous across throat; snout not beak-like, without lateral teeth or barbels; upper and lower teeth toward center of mouth widely dissimilar, but essentially similar to those toward corners. Trunk subcylindrical (shark-like); eyes lateral; anterior margins of pectorals not expanded forward beyond 1st gill opening; nostrils separate from mouth, anterior margins without barbels; eye without nictitating fold or membrane; spiracles present; segmentation of vertebral column incomplete, but centra more or less differentiated, with axial canal somewhat contracted in its passage through them; notochord partially constricted segmentally in

35a. For a recent account of this group, see Smith, B. G. (Dean Memor. Vol., Amer. Mus. nat. Hist., *8*, 1942).

36. These "Saw Sharks" superficially resemble the true "Sawfishes" (Pristidae) which, however, fall among skates and rays (Batoidei), they having ventral gill openings as well as the edges of the pectorals attached to the sides of the head anterior to the gill openings.

correspondence, more strongly so posteriorly than anteriorly; vertebral centra either without calcification, or those in the tail region with calcareous lamellae radiating from a central ring in some forms; neural spines not attached to dorsals; cranium on each side with a well developed antorbital process, continuous, however, with the auditory capsule; rostral cartilage single; upper jaw (palatoquadrate cartilage) attached to cranium at two points (*i.e.*, to the suborbital region and to a postorbital process), but not to the hyomandibular arch which is much reduced; propterygial cartilage of pectoral without radials; heart valves in 4 or 5 rows. Development ovoviviparous.

Families. One modern family known, Hexanchidae.

Family HEXANCHIDAE

Characters. Either 6 or 7 gill openings; margins of all gill openings widely interrupted at throat; eyes without nictitating folds or membranes; spiracles present; upper teeth sharp, with slender, curved, primary cusps; lower teeth blade-like, quadrate or triangular, their margins with several small cusps; caudal fin with well marked subterminal notch, its axis raised but little; inner margins of pelvics either separated or briefly united posterior to cloaca; no precaudal pits; dorsal fin posterior to pelvics; lower jaw with either large or rudimentary labial furrows; no upper labial furrows; clasper of male largely enclosed by a leaf-like expansion of the pelvic fin, its axial cartilage small, simple distally, and attached to the basipterygial cartilage of fin by 2 small connecting segments.[1] Development ovoviviparous.

Key to Genera

1a. 6 gill openings. *Hexanchus* Rafinesque, 1810, p. 78.

1b. 7 gill openings.

  2a. Head narrow; snout tapering; horizontal diameter of eye considerably greater than distance between nostrils. *Heptranchias* Rafinesque, 1810, p. 87.

  2b. Head broad; snout broadly rounded; horizontal diameter of eye considerably smaller than distance between nostrils. *Notorynchus* Ayres, 1885.
  Mediterranean, South Africa, Argentina, California to Oregon, Japan, China, Australia–New Zealand, Indian Ocean.

Genus *Hexanchus* Rafinesque, 1810

Six-gilled Sharks, Cow Sharks, Mud Sharks

Hexanchus Rafinesque, Caratt. Gen. Spec. Sicil., 1810: 14; type species, *H. griseus* Rafinesque, Sicily, equivalent to *Squalus griseus* Bonnaterre, 1788.

Generic Synonyms:

Monopterhinus (in part) Blainville, Bull. Soc. philom. Paris, 1816: 121.

1. For illustrations of the cartilages of the clasper, see Huber (Z. Wiss. Zool., 70, 1901: pl. 27, fig. 1) and Daniel (Elasmobranch Fishes, 1934: 51).

Notidanus Cuvier, Règne Anim., 2, 1817: 128; type species, *Squalus griseus* Bonnaterre, 1788, designated by Jordan, Genera Fish., 1, 1917: 97.
Notidamus Münster, Beitr. Petrefak., 5, 1842: 66; evident misspelling for *Notidanus* Cuvier, 1817.
Hexancus L. Agassiz, Nomencl. Zool. Index, 1846: 181; equivalent to *Hexanchus* Rafinesque, 1810.

Generic Characters. Six gill openings, decreasing in length from front to rear; snout short, rounded; mouth very large, mostly lateral; lower labial furrow well developed; upper lip widely expanded posteriorly; spiracles small, situated far behind eye; fins of moderate size; anterior upper teeth slender, pointed; anterior lower teeth broad, quadrate; no median upper tooth; lower median tooth present or absent. Characters otherwise those of the family and suborder.

Range. Both sides of North Atlantic, Mediterranean, Argentina, southern Indian Ocean, Island of Reunion, Natal and Agulhas Bank, Japan, west coasts of North and South America. All known representatives of the genus appear to belong to a single wide-ranging species.

Fossil Teeth. Middle Jurassic to Pliocene, Europe; Upper Cretaceous, western Asia, New Zealand, Madagascar; Upper Cretaceous to Oligocene, South America; Eocene, Africa; Miocene, North America.

Species. The representatives of this genus from different seas resemble one another closely. But opinions have differed as to whether *H. corinus* Jordan and Gilbert[2] of the Pacific coast of the United States is separable from the well known Six-gilled Shark (*griseus*) of Europe. Supposedly, *corinus* is set apart from *griseus* by the fact that its lower teeth other than the median are finely serrate along their inner edges. Actually, however, no difference exists in this respect between the populations of the two geographic regions in question, for the lower teeth of the European *griseus* were long ago excellently pictured[3] and described[4] as having finely serrate inner edges, although this fact seems to have been overlooked in some of the more recent accounts.[5] On the other hand, it is expressly stated in the original account of *corinus*[6] that in a small specimen from Puget Sound the lower teeth were smooth-edged, as they are also in a 2½-foot *griseus* from the Mediterranean that we have examined;[7] and our comparison of the latter with a four-foot specimen from Puget Sound reveals no significant difference in any other respect. We therefore follow Regan[8] and Fowler[9] in referring *corinus*, as well as the Japanese *Hexanchus*, to *griseus*. There is nothing in the descriptions or obviously generalized illustration to suggest otherwise for the Chilean *edulis*.[10]

2. Proc. U.S. nat. Mus., *3*, 1880: 352.
3. L. Agassiz, Poiss. Foss., *3*, 1835: pl. e, fig. 2–4.
4. Müller and Henle, Plagiost., 1841: 81.
5. Garman (Mem. Harv. Mus. comp. Zool., 36, 1913: 17), for example, describes them without qualifications as smooth-edged, and Rey (Fauna Iberica Peces, *1*, 1928: fig. 83) so pictures them.
6. Jordan and Gilbert, Proc. U.S. nat. Mus., *3*, 1880: 352.
7. Similarly, the lower teeth are smooth-edged in a 429-mm. specimen from Cuba; the first large lower tooth is serrate but the others smooth in one of about five feet; all the large lateral lower teeth are more or less serrate along their inner margins in one of 11 feet (see Study Material, p. 80).
8. Ann. Mag. nat. Hist., (7) *16*, 1905: 57.
9. Bull. U.S. nat. Mus., 100 (*13*), 1941: 57.
10. Perez-Canto, Estud. Escual. Chile, 1886: 8; Philippi, An. Univ. Chile, 71, 1887: 554, pl. 6, fig. 1.

Hexanchus griseus (Bonnaterre), 1780

Six-gilled Shark, Cow Shark, Grey Shark, Mud Shark

Figures 8, 9

Study Material. Female, 830 mm., from Nice, France (Harv. Mus. Comp. Zool., No. 946); male, 429 mm., from off Havana, Cuba, apparently newborn, without trace of umbilical scar (Harv. Mus. Comp. Zool., No. 35630); male, 4 feet, Puget Sound (U.S. Nat. Mus., No. 104474); male, 1,167 mm., from off Pacific Beach, California (Harv. Mus. Comp. Zool., No. 36474); also jaws of an 11-foot and of a 5-foot specimen (Harv. Mus. Comp. Zool., No. 36217, 36216) and of one of 10 feet 2 inches from N. Carolina (U.S. Nat. Mus., No. 37790).

Distinctive Characters. The presence of six gill openings, combined with the facts that the lower ends of the members of each pair are widely separated one from the other in the region of the throat, and that the upper teeth are strikingly unlike the lowers, separates this species from all other sharks of the North Atlantic.

Description. Proportional dimensions in per cent of total length. Male, 429 mm., from Cuba (Harv. Mus. Comp. Zool., No. 35630). Female, 830 mm., from Nice, France (Harv. Mus. Comp. Zool., No. 946).

Trunk at origin of pectoral: breadth 10.7, 9.8; height 8.2, 7.8.

Snout length in front of: outer nostrils 1.6, 1.7; mouth 7.0, 5.0.

Eye: horizontal diameter: 4.7, 3.0.

Mouth: breadth 9.8, 10.0; height 6.3, 6.1.

Nostrils: distance between inner ends 4.4, 4.6.

Gill opening lengths: 1st 6.5, 7.5; 2nd 5.4, 6.5; 3rd 5.1, 5.8; 4th 4.7, 5.2; 5th 3.4, 4.3; 6th 3.5, 4.1.

First dorsal fin: vertical height 3.7, 4.8; length of base 5.8, 6.6.

Anal fin: vertical height 1.9, 3.5; length of base 4.4, 6.4.

Caudal fin: upper margin 35.2, 31.3; lower anterior margin 10.2, 8.3.

Pectoral fin: outer margin 13.0, 13.2; inner margin 5.2, 6.0; distal margin 10.0, 10.7.

Distance from snout to: 1st dorsal 46.7, 54.1; upper caudal 64.8, 68.7; pectorals 20.5, 21.3; pelvics 38.5, 44.6; anal 50.7, 57.2.

Interspace between: 1st dorsal and caudal 10.9, 9.0; anal and caudal 6.5, 5.2.

Distance from origin to origin of: pectorals and pelvics 20.0, 25.4; pelvics and anal 12.3, 14.4.

Trunk moderately stout anteriorly, its depth opposite origin of pectorals 1/7 to 1/8 as great as its length to origin of caudal fin, tapering rearward and strongly compressed laterally posterior to pelvics. Caudal peduncle without precaudal pits. Lateral line clearly visible as a pale streak from opposite last gill opening rearward out onto caudal and dipping downward abruptly on anterior sector of caudal. Denticles on sides of trunk (Fig.

8 E) usually tridentate (occasionally with one of the lateral points lacking), with a prominent axial crest and two (occasionally only one) lower lateral crests, loosely to moderately-closely spaced and overlapping but little; those along rear half of upper margin of caudal fin much larger than on trunk, smooth, ovoid in outline, forming a visible ridge.

Head flattened above. Snout broadly rounded and short, its length in front of mouth

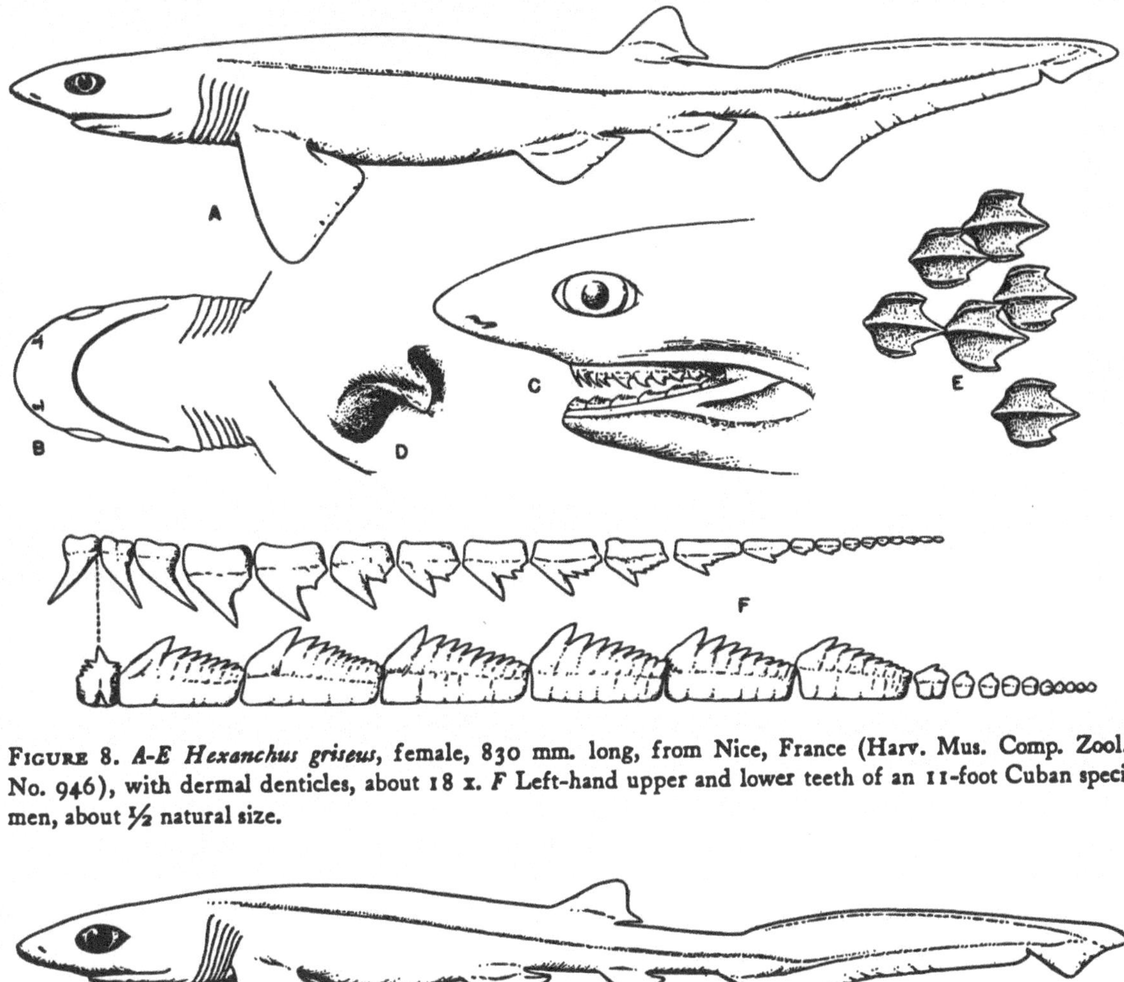

FIGURE 8. *A-E Hexanchus griseus*, female, 830 mm. long, from Nice, France (Harv. Mus. Comp. Zool., No. 946), with dermal denticles, about 18 x. *F* Left-hand upper and lower teeth of an 11-foot Cuban specimen, about ½ natural size.

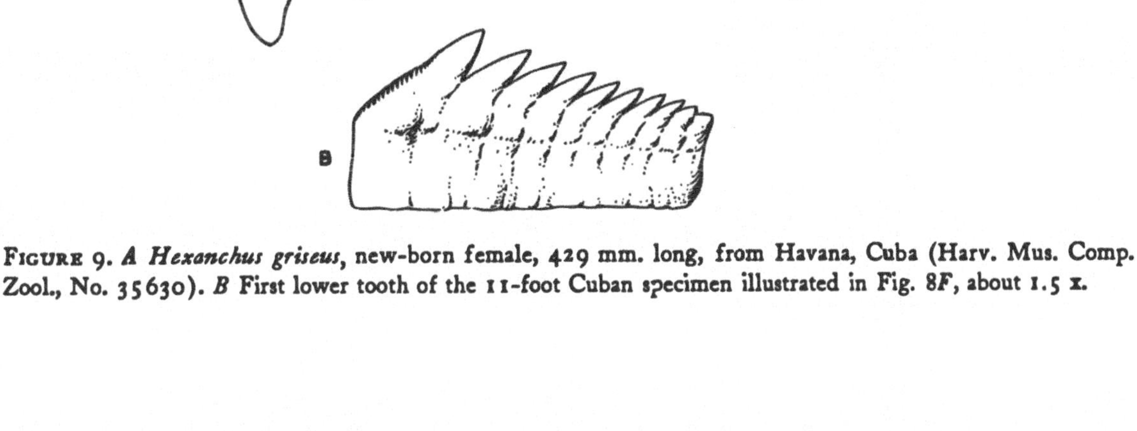

FIGURE 9. *A Hexanchus griseus*, new-born female, 429 mm. long, from Havana, Cuba (Harv. Mus. Comp. Zool., No. 35630). *B* First lower tooth of the 11-foot Cuban specimen illustrated in Fig. 8*F*, about 1.5 x.

only about ½ as great as breadth of mouth. Eye oval and noticeably large, its horizontal diameter about ⅐ as great as length of head in a 2½-foot specimen, but perhaps relatively smaller in adults.[11] Spiracle very small, at level of upper margin of eye, about opposite corner of mouth. Gill openings notably long, extending from high on sides far onto ventral surface of throat, and noticeably oblique, the 1st (longest) about 2½ times as long as horizontal diameter of eye, successively shorter, rearward, the 6th only a little more than ½ as long as the 1st; the inner margins of 2nd and 3rd gill arches with 2, and 4th to 6th with 3 to 5, fleshy tubercles, suggesting rudimentary rakers. Nostril much nearer to tip of snout than to mouth, small, strongly oblique, its anterior margin expanded as a subtriangular lobe with blunt tip. Mouth notably large, about ⅔ as high as broad, crescentic and inferior anteriorly, but extending along sides of head for most of its length, the gape reaching rearward about ⅔ of distance to origin of pectoral. Well developed labial furrow at corner of mouth on lower jaw, visible only when mouth is partly open; none on upper jaw. Upper lip enclosing posterior part of lower jaw as a free fold extending rearward past corner of mouth for a distance about equal to horizontal diameter of eye.

Teeth $\frac{16 \text{ to } 20 - 16 \text{ to } 20}{12 \text{ to } 16 - 1 \text{ (or 0)} - 12 \text{ to } 16}$; noticeably different in the 2 jaws; 1st 2 to 4 uppers simple with slender median cusp curved outward, the 1st noticeably smaller, the 2nd slightly smaller and with narrower bases than 3rd, the subsequent teeth to the 10th or 11th with 1, 2 or 3 short basal cusps on outer side (number increasing toward corner of mouth and with growth), the outermost 7 or 8 teeth rounded, with only very small cusp or none, and much lower than the others. Lower jaw usually with 1 symmetrical median tooth, having 1 median cusp and 1, 2 or 3 lower cusps on each side,[12] the next 6 (occasionally 5) teeth trapezoid, about twice as broad as high, with 7 to 8 pointed cusps in small specimens, increasing in number to 8 to 10 in large, the innermost cusp the longest, the others progressively shorter, the inner margins smooth in newborn specimens, but finely serrate in large, with intermediate sizes showing intermediate states,[13] the 7th lower tooth (6th in specimens which have only 5 large laterals) much smaller, with only 1 definite cusp, the outermost 4 to 6 teeth very low, rounded, without cusp; 2 or 3 series functional in center of upper jaw and 1 along its sides; 1 series functional in lower jaw.

Vertical fins small. Dorsal with rounded apex and weakly convex rear margin, free rear tip broadly triangular, about half as long as base, its origin slightly behind cloaca, the midpoint of its base about over origin of anal. Caudal about ⅓ of total length, with well marked subterminal notch, its lower anterior corner expanded as a low rounded lobe in newborn but not appreciably so in larger specimens (cf. Fig. 8 A and 9 A), its maximum breadth a little more than ⅓ its length. Anal about as long at base as 1st dorsal, rear margin nearly straight, free rear tip short. Interspace between rear end of base of anal and origin of caudal about ½ as long as between dorsal and caudal. Pelvics with nearly straight mar-

11. The eye, as in many sharks, is relatively larger in newborn than in older specimens.
12. This median tooth is lacking in the small Mediterranean specimen listed under Study Material, p. 80.
13. For further details, see discussion, p. 79.

gins and rounded apices, their bases a little less than 1½ times as long as base of dorsal, their inner margins united for a very short distance posterior to cloaca in female, but entirely separate in male, and partially enclosing the claspers, with their tips somewhat elongate. Pectoral about 1⅓ times as long as broad, with broadly rounded apex, nearly straight distal margin (moderately concave in smaller specimen) and very broad base, the inner margin nearly or quite ½ as long as anterior margin.

Color. Fresh specimens are variously described as coffee-colored (darkest along the midline of back) or as very dark gray above; at least some specimens with a pale streak along the side; lower surface of a paler shade of the same hue, or whitish. Of the preserved specimens we have seen, a 2½-foot Mediterranean specimen is dark chocolate brown above and paler below; one newborn from Cuba is mouse gray above and paler below; a 4-foot Puget Sound fish is very dark gray, hardly paler below than above.

Size. The fact that embryos of 650 mm. have been reported, with free-swimming young as small as 429 to 700 mm., points to a length of 16 to 26 inches at birth. Maturity is to be expected at a length of perhaps 6 to 6½ feet, large numbers of eggs having been found in females of about 7 feet in Cuban waters. The recorded lengths of the larger adults have ranged up to about 15½ feet (4.82 m.). One of 26 feet 5 inches was reported from Cornwall many years ago, a giant of its kind if its size was stated correctly.[14] Reported weights are about 220 lb. (100 kilo.) at 7 ft. 4 in. (2.25 m.); about 300–400 lb. at about 9 ft.; about 528 lb. (240 kilo.) at about 13 ft. (4 m.); 1,085 lb. at 14 ft.; and 1,300 lb. at 15 ft.[14a]

Developmental Stages. The litters are certainly large, for 47 embryos were counted in a female of 4.8 meters,[15] while fishermen have reported as many as 108 in a specimen slightly smaller (4.5 meters long).[16]

Habits. This shark is described as sluggish, hardly resisting when caught on hook and line, but we have had no experience with it alive. Characteristically, it is a "ground" species, usually living in at least moderately deep water. It is recorded from depths as great as 800 to 1,875 meters off Portugal, at a little deeper than 100 fathoms off the Irish slope, at about 90 to 560 fathoms in Scottish waters, including the Shetland-Faroe Channel, and usually from 75 fathoms down to 300 fathoms off Cuba, where, in fact, few are caught shoaler than 100 fathoms. And we have received photographs from Ollyandro del Valle of three large ones (922, 1,400 and 1,682 pounds) taken in the deep water shark fishery off the north coast of Cuba, said to have been hooked at 700 fathoms. On the other hand, one of the earliest recognizable reports of it[17] was of a specimen from the coast of Holland and therefore certainly from shoal water. Scattering specimens are caught in the North Sea, in depths certainly no greater than 15 to 20 fathoms, and they have even been seen swimming at the surface off Ireland. It has been suggested that this shark lies quiescent on

14. Day, Fish. Gt. Brit., *2*, 1880–1884: 308; this size has been frequently quoted since then for this species.
14a. Cuban specimens, reported by Luis Howell-Rivero. 15. Bolivar, Bol. Soc. esp. Hist. nat., *7*, 1907: 207.
16. Vaillant, Bull. Mus. Hist. nat. Paris, 1901: 202.
17. Spallanzani, Viag. Sicil., *4*, 1793: Chap. 31, pl. 2.

bottom by day, visiting the upper waters at night in search of food.[18] High temperatures probably act as a barrier to it toward the surface and inshore in the warmer part of its range, as in the Mediterranean and around Cuba.

The Six-gilled Shark was long ago reported as mating in spring and autumn and producing young at various seasons, but on how good evidence we cannot say.

Their food consists of fish and various crustaceans. In Spanish waters it feeds largely on hake (*Merluccius*); an entire torpedo has also been found in one. Off Cuba, dolphins (*Coryphaena*), small marlins (*Makaira*) and small swordfish (*Xiphias*) are reported from stomachs, as well as crabs, shrimps and parts bitten from other sharks that had been hooked.[19] They are described as coming to the surface on occasion to pick up fish thrown overboard.

Relation to Man. This species is not sufficiently abundant in American waters to be of any special importance, although such as are taken off Cuba are utilized for their oil. In the North Sea, any that are caught are marketable in Germany, even though the flesh has been credited with a purgative action. However, along the Iberian Peninsula, and in the Mediterranean, where it is much more plentiful, it is of no commercial importance, except as a nuisance to fishermen, since it drives away merchantable fishes.

Range. Continental waters on both sides of the Atlantic, including the Mediterranean; also Pacific coast of North America from southern California to British Columbia; Chile; Japan; Australia; southern Indian Ocean and South Africa.

Occurrence in the North Atlantic. On the eastern side of the Atlantic, although nowhere abundant, the center of population for this Shark appears to be in the Mediterranean, where it is widespread, and thence northward along the Atlantic coasts of the Iberian peninsula and France. It also enters the North Sea in numbers sufficient for fishermen to be familiar with it; it is taken from time to time on the south coast of England, along the Irish Atlantic slope, off western Scotland to the Faroe-Shetland Channel, and even as a stray off Iceland. To the southward it has been reported from Morocco to Mauritania.

Occurrence in the Western North Atlantic. It has long been known that the Six-gilled Shark occurs off the northern coast of Cuba, specimens being caught from time to time near Matanzas and Havana, and since the recent development of a hook and line fishery at 100 to 400 fathoms or deeper it has proved to be more plentiful there in deep water than was formerly supposed, large specimens being taken daily.[20] However, for it to stray northward must be a very rare event, the only record of its occurrence on the east coast of continental North America being a ten-foot two-inch specimen taken in March 1886 on the coast of North Carolina near Currituck Lighthouse. Neither is there any evidence of its presence anywhere in the Gulf of Mexico and Caribbean region, other than for Cuba. But it is to be expected there, at appropriate depths, and along the coast of South America generally, if a report of it from northern Argentina be well founded.[21]

18. Fraser-Brunner, Proc. R. Irish Acad., *42*, B-9, 1935: 519.

19. Communication from Luis Howell-Rivero.

20. Communication from Luis Howell-Rivero.

21. Lahille, An. Mus. nac. B. Aires, *24*, 1913: 26, 32 (identified by the teeth).

Synonyms and References:

1. Atlantic Ocean and South Africa:

Le Griset, Broussonet, Mem. math. phys. Acad. Sci. Paris, 1780: 663 (descr., Medit.).

Squalus griseus Bonnaterre, Tabl. Encyc. Meth. Ichthyol., 1788: 9 (descr., type loc. Medit.); Gmelin, in Linnaeus, Syst. Nat., 1789: 1495 (descr.); Bloch and Schneider, Syst. Ichthyol., 1801: 129 (Medit.); Latreille, Nouv. Dict. Hist. Nat., *24*, 1804: 72 (in table of contents); Risso, Ichthyol. Nice., 1810: 37 (descr., Medit.); Nacarri, Ittiol. Adriat., 1822: 24 (Adriatic, not seen); Nardo, Oss. Agg. Adr. Itiol., Giorn. Fis. Nat. Pavia, 7, 1824: 261 (not seen); Martens, Reise Vened., *2*, 1824: 408 (Medit., not seen); Nardo, Prod. Ichthyol. Adriat., 1827: 9 (Adriatic); Trapani, Cat. Fish Malta, 1838: 16 (Malta, not seen).

Squalo (not named), Spallanzani, Viag. Sicil., *4* (31), 1793: pl. 2 (jaws, Holland).

Le squale griset, Lacépède, Hist. Nat. Poiss., 4° ed., *1*, 1798: 167, 269, in Buffon, Hist. Nat.; in Sonnini, Hist. Nat. Poiss., *4*, 1801–1802: 96 (descr.).

Squalus vacca Bloch and Schneider, Syst. Ichthyol., 1801: 138.

Hexanchus griseus Rafinesque, Carratt. Gen. Nuov. Sicil., 1810: 14 (descr., Sicily); Indice Ittiol. Sicil., 1810: 47 (Sicily); Swainson, Nat. Hist. Fish. Amphib. Rept., *2*, 1839: 316 (descr.); Müller and Henle, Plagiost., 1841: 80 (descr., Medit., Atlant.); Heckel, Pesci Dalmat., in Carrara Opera, 1846: 91 (Medit., not seen); Busch, Selach. Ganoid. Encephalo., 1848: pl. 3, fig. 8 (brain); White, List Spec. Brit. Mus., Fish., 1851: 130 (Isle of Wight); Costa, Fauna Napol. Pesci Cat. Sist., *2–3*, 1854–1857: 15, 16 (39, 40) (Medit.); Canestrini, Cat. Pesci Genov., Arch. Zool. Anat. Fisiol., *1*, 1861: 267 (Medit.); Guichenot, Notes Isle Reunion, *2*, 1841: 30 (I. Reunion, not seen); Duméril, Hist. Nat. Poiss., *1*, 1865: 431, pl. 4, fig. 9–12 (teeth); Poey, Repert. Fisico-Nat. Cuba, *2*, 1868: 454 (Cuba); Gray, Ann. Mag. nat. Hist., (4) *1*, 1868: 76 (Cornwall); Brito Capello, J. Sci. math. phys. nat. Lisboa, *2*, 1870: 140 (Portugal); Miklucho-Maclay, Beitr. Vergl. Neurol. Wirbelt., *1*, 1870: 13, pl. 2, fig. 8–14 (brain); Ninni, An. Soc. Nat. Modena, *5*, 1870: 66 (Venice); Costa, La Pesca., Napoli, 1871: 87 (Medit., not seen); Gegenbaur, Unters. Vergl. Anat. Wirbelt., *3*, 1872: pl. 3, fig. 6, 7 (skelet.); Poey, An. Soc. esp. Hist. Nat., *5*, 1876: 398; also sep., Enumerat. Pisc. Cubens., 1875: 202 (Cuba); Brito Capello, Cat. Peix. Portugal, 1880: 45 (Portugal); Doderlein, Man. Ittiol. Medit., *2*, 1881: 76[22] (Medit., distrib. in Atlant.); Moreau, Hist. Nat. Poiss., France, *1*, 1881: 336 (descr., distrib., Medit. and Atlant.); Carus, Prod. Fauna Medit., *2*, 1889–1893: 499 (Medit.); Almeida and Roquete, Mammif. Peix. Costa e rios do Algarve, Inquerito Industr. Lisboa, 1892: 372 (Portugal, not seen); Jordan and Evermann, Rep. U.S. Comm. Fish. (1895), 1896: 213 (distrib.); Bull. U.S. nat. Mus., 47 (1), 1896: 19 (descr., Cuba); Carruccio, Boll. Soc. Roma Zool., *5*, 1896: 165, 1 pl. (anat., Medit.); Vieira, Ann. Sci. nat. Porto, *4* (2), 1897: 67 (Portugal, size); Sicher, Atti Accad. gioenia., (4) *11* (5), 1898: 16 (Medit.); Jordan and Evermann, Bull. U.S. nat. Mus., 47 (4), 1900: pl. 2, fig. 8: Huber, Z. wiss. Zool., 70, 1901: 600, pl. 27, fig. 1, 2 (claspers); Vaillant, Bull. Mus. Hist. nat. Paris, 7, 1901: 202 (Gulf of Gascony, embryos); de Braganza, Result. Invest. Sci. "Amelia," Ichthyol., *2*, 1904: 28 (Portugal); Regan, Ann. Mag. nat. Hist., (7) *16*, 1905: 571 (concludes *corinus* equals *griseus*); Smith, Bull. N.C. geol. econ. Surv., *2*, 1907: 30 (N. Carolina specimen); Fowler, Proc. Acad. nat. Sci. Philad., *60*, 1908: 52 (spec., no data); Diaz y Martinez, An. Acad. Habana, *46*, 1910: 82 (Cuba); Laboissière, Bull. Ass. Levall.-Perret, *5*, 1910: 6 (Gulf of Gascony); Seabra, Poiss. Port., 1911: 195 (Portugal, not seen); Roule, Bull. Inst. océanogr. Monaco, 243, 1912: 3 (Medit.); Garman, Mem. Harv. Mus. comp. Zool., *36*, 1913: 16 (descr.); Lahille, An. Mus. nac. B. Aires, *24*, 1913: 26, 32 (Argentina); Scharff, Irish Nat., *24*, 1915: 99 (Ireland); Smith, J. Amer. Mus. nat. Hist., *16*, 1916: 349, 351 (N. Carolina); Lahille, Physis B. Aires, *5*, 1921: 63 (spec. recorded, Argentina); Norman, Ann. Mag. nat. Hist., (9) *9*, 1922: 319 (Natal, S. Africa); Marelli, Elenc. Sist. Fauna, Mem. Ministr. Obras Publ. B. Aires (1922–1923), 1924: 544 (Argentina); Barnard, Ann. S. Afr. Mus., *21* (1), 1925: 22 (S. Africa); Rey, Fauna Iberica Peces, *1*, 1928: 292 (descr., off Spain); Hickling, Ann. Mag. nat. Hist., (10) *2*, 1928: 199 (NE. Atlant.); Jordan, Evermann and Clark, Rep. U.S. Comm. Fish. (1928), *2*, 1930: 11 (name, distrib.); Fowler, Proc. Pan-Pacif. sci. Congr. (1929), *3*, 1930: 484 (name); Sanchez-Roig, Revist. Agric. Peces Cubana, Commerc. Trabaj., *3* (9), 1931: 18 (Cuba); Baylis, Ann. Mag. nat. Hist., (10) *12*, 1933: 322 (parasites); Belloc, Rev. des. trav. Pêches Marit., 7,

22. See Doderlein, 1881, for extensive list of Mediterranean records, some in publications not accessible to us.

Fasc. 2, 1934: 151 (ill., Morocco, Rio de Oro, Mauritania); Borri, Atti Soc. Tosc. Sci. nat., *44*, 1934: 94 (Medit.); Fraser-Brunner, Proc. R. Irish Acad., *42* (B), 1935: 319 (Irish Atlant. slope); Pozzi and Bordale, An. Soc. cient. argent., 120, 1935: 149 (listed for Argentina); Nobre, Fauna Marinha Port. Vert., *1*, 1935: 413, pl. 61, fig. 189 (Portugal); Lübbert and Ehrenbaum, Handb. Seefisch. Nordeurop., *2*, 1936: 272, pl. 19, fig. 239 (general; E. Atlant.); Howell-Rivero, Proc. Boston Soc. nat. Hist., *41*, 1936: 42 (Cuba); Fowler, Bull. Amer. Mus. nat. Hist., *70* (1), 1936: 26 (descr., distrib.); Norman and Fraser, Giant Fishes, 1937: 5 (general); Desbrosses, Rev. des. Trav. Pêches Marit., *11*, Fasc. 1, 1938: 53 (growth, migrations); Tortonese, Atti Soc. ital. Sci. nat., 77, 1938: 286 (Medit.); Howell-Rivero, Torreya, *9*, 1941: 4, pl. 1, 2 (descr., Cuba); Fowler, Bull. U.S. nat. Mus., 100 (*13*), 1941: 11(part); Holmgren, Acta Zool. Arg., *22*, 1941: 9 (skelet.); Bigelow and Schroeder, Guide Comm. Shark Fish., Anglo Amer. Carib. Comm., Wash., 1945: 94, fig. 31 (descr., ill., habits, range).

Monopterhinus griseus Blainville, Bull. Soc. philom. Sci. Paris, 1816: 121.

Notidanus griseus Cuvier, Règne Anim., *2*, 1817: 128 (general); 2nd Ed., 1829: 390; Bory de St. Vincent, Dict. Class. Hist. nat., *15*, 1829: 597 (general), Voigt, in Cuvier, Tierreich, 2, 1832: 509 (descr.); Bonaparte, Fauna Ital. Pesc., *3* (2), 1835: pl. [55], fig. 1 (descr., Medit.); Agassiz, L., Poiss. Foss., *3*, 1835–1838: 92, pl. E, fig. 2–4 (teeth); Lowe, Trans. zool. Soc. Lond., *2* (3), 1837: 194 (Madeira); Bonaparte, Mem. Soc. neuchâtel. Sci. nat., *2* (8), 1839: 9 (in synopsis); Cuvier, Règne Anim., Poiss., 1843: 362, pl. 115, fig. 1 (teeth); Couch, Zoologist, *4*, 1846: 1337 (Cornwall); Bonaparte, Cat. Pesc. Europ., 1846: 17 (name); Sassi, Cat. Pesci Legur., 1846: 131 (Medit., not seen); Nardo, Sinon. Modern. Spec., descr. Pesci St. Chiereghin, 1847: 111 (name); Owen, Cat. Osteol. Roy. Coll. Surg., *1*, 1853: 91 (teeth); Machado, Peces Cadiz, 1857: 8 (near Cadiz); Nardo, Atti Ist. veneto., (3) *5*, 1859–1860: 787 (Medit.); Gemellaro, Atti Accad. gioenia, (*2*) *19*, 1864: 122 (Medit., not seen); Bocage and Brito Capello, Poiss. Plagiost., 1866: 15 (Portugal); de la Blanchère, Dict. Pêches, 1868: 371 (descr., Medit.); Günther, Cat. Fish. Brit. Mus., *8*, 1870: 397 (descr., distrib.); Canestrini, in Cornalia, *et al.*, Fauna d'Ital., 1872: 43 (Medit.); Lawley, Monog. Gen. Notidanus foss., 1875: 60, 65, pl. 4, fig. 1 (jaws); Gervais and Boulart, Poiss., *3*, 1877: 194, pl. 74, fig. 18 (descr.); Delfortrie, Act. Soc. linn. Bordeaux, (4) *2*, 1878: 253 (teeth); Doderlein, Prosp. Metod. Pesci Sicil., 1878: 30 (Sicily); Day, Fish. Gt. Brit., *2*, 1880–1884: 308, pl. 158, fig. 2 (descr., habits, Gt. Brit.); Stossich, Boll. Soc. adriat. Sci. nat., *5*, 1880: 69 (Adriatic); Perugia, Elenc. Pesc. Adriat., 1881: 55 (Adriatic); Graeffe, Arb. zool. Inst. Univ. Wien, 7, 1886: 447 (Medit.); Günther, Proc. roy. Soc. Edinb., *15*, 1888: 207 (Orkneys and Shetlands); Anderson, Ann. Scot. nat. Hist., 9, 1894: 182 (W. Scot.); Traquair, Ann. Scot. nat. Hist., *5*, 1896: 159 (W. Scot.); Philippi, Ann. Univ. Chile, 109, 1902: 304; Bolivar, Bol. Soc. esp. Hist. nat., 7, 1907 (Spain); Günther, Encycl. Brit., *24*, 1911: 809 (general); Ehrenbaum, in Grimpe and Wagler, Tierwelt N.- u. Ostsee, Lief 7 (*12*[e]), 1927: 7 (general, North Sea).

Notidanus vacca Cuvier, Règne Anim., *2*, 1817: 28.

Squalus (*Monopterhinus*) *griseus* Blainville, in Vieillot, Faune Franc., 1825: 77 (descr., Medit.).

Notidanus monge Risso, Hist. nat. Europ. merid., *3*, 1826: 129 (general, Medit.); Bory de St. Vincent, Dict. Class. Hist. nat., *15*, 1829: 597.

Grey Notidanus, Yarrell, Brit. Fish., 3rd Ed., *2*, 1859: 515 (not seen).

Notidanus sp. dubia Poey, Memorias, *2*, 1861: 359 (Cuba).

Six-gilled shark, Couch, Fish. Brit. Isles, *1*, 1867: 21, pl. 4 (England).

Notidanus (*Hexanchus*) *griseus* Werner, Zool. Jb., Syst. Abt., *21*, 1904: 286; Ehrenbaum, Faune Ichthyol., Cons. explor. Mer, 1930: plate not numbered (general).

2. Pacific:

Hexanchus corinus Jordan and Gilbert, Proc. U.S. nat. Mus., *3*, 1880: 352 (descr., Monterey Bay and Puget Sound); Bean, Proc. U.S. nat. Mus., *4*, 1882: 267 (listed for Puget Sound); Jordan and Gilbert, Proc. U.S. nat. Mus., *4*, 1882: 30 (listed for Soquel, Calif., and Puget Sound); Jordan and Gilbert, Bull. U.S. nat. Mus., *16*, 1883: 62 (descr.); Jordan, Rep. U.S. Comm. Fish. (1885), 1887: 792 (listed for Pacific coast, U.S.); Jordan and Starks, Proc. Calif. Acad. Sci., (2) *5*, 1895: 788 (listed for Puget Sound); Jordan and Evermann, Bull. U.S. nat. Mus., 47 (1), 1896: 19 (descr., Monterey, California, to Puget Sound); Rep. U.S. Comm. Fish. (1895), 1896: 213 (listed, Monterey Bay to Puget Sound); Starks,

Ann. Carneg. Mus., 7, 1911: 163 (cf. with *griseus*; color; San Francisco, Puget Sound); Garman, Mem. Harv. Mus. comp. Zool., *36*, 1913: 17 (descr.); Halkett, Check List Fish. Canad., 1913: 39 (listed Vancouver I. and Puget Sound to Monterey); O'Donoghue, J. Anat., Lond., *61*, 1926: 41 (descr., meas., ill., teeth, denticles; anat.; photo of Nanaimo, British Columbia spec.); Jordan, Evermann and Clarke, Rep. U.S. Comm. Fish. (1928), 2, 1930: 11 (listed Monterey Bay to Puget Sound); Townsend, Bull. N.Y. zool. Soc., *34* (6), 1931: 169 (photograph, Washington); Walford, Fish Bull. Sacramento, 45, 1936: 24 (diagn., photo., San Diego to Puget Sound); Barnhart, Mar. Fish south. Calif., 1936: 7 (off San Diego).

Notidanus vulgaris Perez Canto, Estud. Escual. Chile, 1886: 8 (descr., Chile); Philippi, An. Univ. Chile, *71*, 1887: 554, pl. 6, fig. 1 (descr., ill., Chile); Quijada, Bol. Mus. nac. Chile, *5*, 1913: 112 (listed for San Antonio, Chile).

Hexanchus griseus Regan, Ann. Mag. nat. Hist., (7) *16*, 1905: 571 (Japan, Pacific U.S. cf. with Atlantic); Starks, Calif. Fish Game, *3*, 1917: 146 (listed California to Puget Sound); Hubbs, J. Pan-Pacif. Res. Instn., *3*, 1928: 11 (listed for Oregon and Washington); Desbrosses, Rev. des Trav. Pêches Marit., *11*, Fasc. 1, 1938: 53 (migrations, rate of growth, breeding season); Fowler, Bull. U.S. nat. Mus., 100 (*13*), 1941: 11 (descr., Pacific refs.); Phillipps, Dom. Mus. rec. Zool., *1* (2), 1946: 5 (descr., photo, New Zealand).

Genus *Heptranchias* Rafinesque, 1810

Heptranchias Rafinesque, Carratt. Gen. Spec. Sicil., 1810: 13; type species, *H. cinereus* Rafinesque, Sicily, equals *Squalus perlo* Bonnaterre, 1788.

Generic Synonyms:

Monopterhinus (in part) Blainville, Bull. Soc. philom. Paris, 1816: 121.

Carcharias (in part) Cloquet, Dict. Sci. Nat., 7, 1817: 69.

Notidanus (in part) Cuvier, Règne Anim., *2*, 1829: 390.

Heptanchus (in part) Müller and Henle, Plagiost., 1841: 81.

Heptranchus Gray, List Fish. Brit. Mus., Chondropt., 1851: 68, equivalent to *Heptranchias* Rafinesque, 1810.

Heptrancus Costa, Fauna Napoli Pesci, 3, Chondropt., 1854–57: 5 (29), equivalent to *Heptranchias* Rafinesque, 1810.

Hexanchus (in part) Seabra, Bull. Soc. Portug. Sci. nat., *5*, 1911: 195.

Generic Characters. Seven gill openings, decreasing in length from front to rear; snout narrow, tapering; horizontal diameter of eye considerably greater than distance between nostrils. Characters otherwise those of the suborder and family.

Range. Eastern and western North Atlantic, South Africa, Australia, Japan.

Species. Our own comparison of medium-aged and small specimens from Japan, with others from Cuba and from the Mediterranean (see Study Material, p. 88) corroborates Garman's[1] conclusion that the North Pacific representative of the genus is identical specifically with the Atlantic form; such differences in proportionate dimensions as exist between them are no greater than might have resulted from the fact that the larger of the Japanese examples had been dried. However, the Australian form may be distinct, as indicated in the following key.[2]

1. Mem. Harv. Mus. comp. Zool., *36*, 1913: 23.
2. *Heptranchias haswelli* Ogilby (Proc. Linn. Soc. N. S. W., *22*, 1897: 62), *Notidanus medinae* and *N. wolniczkyi* Philippi (An. Univ. Chile, 109, 1901: 305, 307, Chile), and *N. ferox* Perez Canto (Estud. Escual. Chile, 1886: 7), which are included by Fowler (Bull. U.S. nat. Mus., 100 [*13*], 1941: 9) in the synonymy of *Heptranchias perlo*, all appear, from the original accounts, to fall in the genus *Notorynchus*.

Key to Species

1a. Origin of anal opposite rear base of dorsal. *perlo* Bonnaterre, p. 88.
1b. Origin of anal under middle of base of dorsal. *dakini* Whitley, 1931. Australia.[3]

Heptranchias perlo (Bonnaterre), 1788

Seven-gilled Shark

Figures 10, 11

Study Material. Female, 932 mm. long, containing 9 embryos, and also an adult male of 698 mm., both from Havana, Cuba (Harv. Mus. Comp. Zool., No. 36897); specimen of 732 mm., from Nice, France (Harv. Mus. Comp. Zool., No. 945); also specimens of 957, 980 and 255 mm. from Japan, the latter newborn with umbilical scar still faintly visible (Harv. Mus. Comp. Zool., No. 35070, 1040, 1299).

Distinctive Characters. The presence of 7 gill slits combined with narrow head and pointed snout separates *perlo* from all other Atlantic sharks.

FIGURE 10. *Heptranchias perlo*, male, about 689 mm. long, from the north coast of Cuba (Harv. Mus. Comp. Zool., No. 35897); *A* Anterior part of head, about ½ natural size. *B* Right-hand nostril, about 1.3 x. *C* Dermal denticles, about 37 x. *D* Lateral view of dermal denticle, about 64 x. *E* Apical view of dermal denticle, about 64 x.

3. An additional diagnostic character, according to Whitley (Aust. Zool., *6*, 1931: 310), is anal base as long as dorsal base in *perlo*, but shorter than the latter in *dakini*. Actually, however, no distinction can be drawn in this respect, the anal being appreciably shorter than the dorsal in two of the three Atlantic specimens of *perlo* that we have seen.

Description. Proportional dimensions in per cent of total length. Male, 698 mm., from Cuba (Harv. Mus. Comp. Zool., No. 35897). Female, 932 mm., from Cuba (Harv. Mus. Comp. Zool., No. 35897).

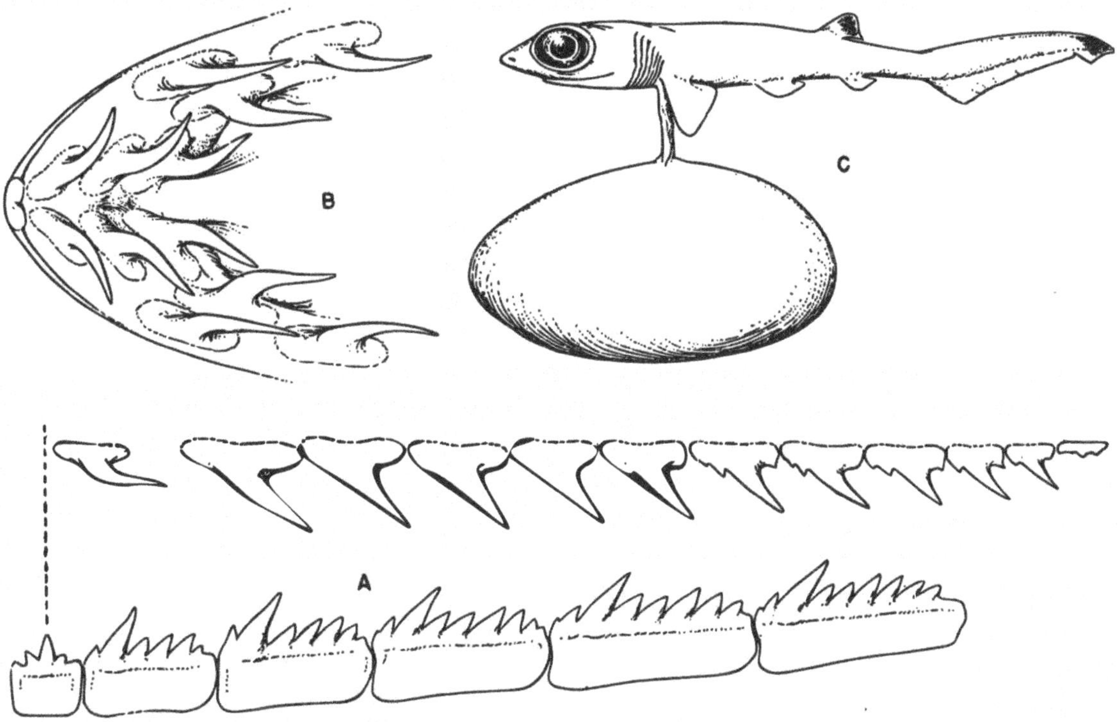

FIGURE 11. *Heptranchias perlo, A* upper and lower teeth of specimen pictured in Fig. 10, about 2.4 x. *B* Anterior part of upper jaw to show arrangement of teeth as viewed from below, about 2.4 x. *C* Embryo from Cuba with yolk sac attached (Harv. Mus. Comp. Zool., No. 35581), about 0.4 natural size.

Trunk at origin of pectoral: breadth 8.4, 9.1; height 9.6, 10.7.
Snout length in front of: outer nostrils 2.0, 2.0; mouth 5.0, 4.8.
Eye: horizontal diameter 4.0, 3.6.
Mouth: breadth 8.0, 7.0; height 6.4, 7.1.
Nostrils: distance between inner ends 2.6, 2.1.
Gill opening lengths: 1st 5.7, 7.2; 2nd 5.6, 6.5; 3rd 4.7, 5.9; 4th 4.2, 5.3; 5th 3.6, 4.6; 6th 3.1, 4.0; 7th 2.6, 3.1.
First dorsal fin: vertical height 4.3, 4.5; length of base 6.2, 6.6.
Anal fin: vertical height 2.2, 2.7; length of base 5.9, 5.9.
Caudal fin: upper margin 30.6, 30.4; lower anterior margin 9.0, 8.6.
Pectoral fin: outer margin 11.4, 11.4; inner margin 5.0, 5.5; distal margin, 9.6, 8.5.
Distance from snout to: 1st dorsal 49.0, 48.3; upper caudal 69.4, 69.6; pectoral 20.9, 19.1; pelvics 40.0, 38.4; anal 54.8, 52.2.

Interspace between: 1st dorsal and caudal 14.2, 14.8; anal and caudal 9.0, 9.7.

Distance from origin to origin of: pectorals and pelvics 18.9, 22.0; pelvics and anal 15.0, 15.0.

Trunk rather slender, compressed, its height at about midsection of body, where highest, 11.6 to 11.8% of total length, the body sector shorter than tail sector by a distance about equal to length of pectoral. Caudal peduncle about 75 to 80% as wide as deep; no precaudal pits. Dermal denticles on sides of trunk closely overlapping, a little longer than broad, each denticle with prominent median tooth, flanked by a pair of much smaller laterals, a strong median ridge and upturned lateral edges, the blades so thin and transparent that the pigment dots on the skin are visible through them; those along upper margin of caudal ovoid, without lateral marginal teeth, but with 3 longitudinal ridges, the median subdivided posteriorly, forming an ill-defined crest, much as in *Hexanchus griseus* (p. 81).

Head with dorsal profile slightly convex. Snout tapering, narrow, its tip slightly rounded. Eye notably large (as in *Hexanchus*), oval, its anterior edge about opposite front of mouth. Spiracle minute, about on level with upper edge of eye, its distance behind eye about equal to horizontal diameter of latter. Gill openings extending down onto throat, the 1st about 1½ times as long as horizontal diameter of eye, the 2nd to 7th decreasing successively in length, the 7th only about ½ as long as 1st. Nostril about equidistant between mouth and tip of snout, its anterior margin expanded as a broadly triangular, corrugated lobe (Fig. 10 B). Mouth narrowly rounded in front, notably long, the length about equal to breadth, with very extensive gape, lateral in position for most of its length, the margin of upper lip extending rearward past corner of mouth for a distance equal to ½ to ⅔ horizontal diameter of eye. An oblique labial furrow at angle of mouth, originating on upper jaw and extending downward and forward for a short distance onto lower jaw.

Teeth $\frac{12-12}{5-1-5}$ in grown specimens, $\frac{10-10}{5-1-5}$ in young of 257 mm., unlike in the 2 jaws; upper teeth fang-like, strongly oblique, the first 3 or 4 more or less sinuous in outline with base as well as cusp smooth-edged, but subsequent upper teeth with 1 or 2 small subsidiary cusps at base on inner side and 1 on the outer, the outermost upper tooth low, without definite cusp; lower jaw with 1 symmetrical tooth at symphysis with large median cusp, and 1 or 2 smaller on either side, the lateral lower teeth very broad and low, each with a series of 6 to 8 somewhat oblique triangular cusps in male, and 7 to 10 in female, the 2nd or 3rd of which is much the largest, their edges perfectly smooth; 2 to 3 series of teeth functional in front of upper jaw and 1 along sides; 1 series functional in lower jaw.

Origin of dorsal a little posterior to cloaca, its anterior margin straight or slightly convex, its apex broadly rounded, its rear margin concave, its free rear corner prolonged a distance equal to about ½ the horizontal diameter of eye, its vertical height about ⅓ as great as length of pectoral. Interspace between dorsal and caudal about as long as between axil of pectoral and origin of pelvics. Axis of caudal hardly raised, its upper margin moderately convex, lower margin with well marked subterminal notch, rather strongly concave anteriorly, the lower anterior lobe about 30% as long as upper. Anal with nearly straight margins and subacute corners, about as long as dorsal at base but only about ⅔ as high, its origin about under rear end of base of dorsal. Pelvics a little higher than anal and almost 1½ times as long at base, prolonged rearward in male, and partially enclosing the claspers, the inner margins entirely separate posterior to the cloaca in both sexes. Pectoral relatively small, ⅔ to ¾ as broad as long, with very broad base, the outer margin weakly convex, distal margin moderately concave, apex narrowly rounded and inner corner more broadly so.

Color. Fresh specimens from Cuba are described[4] as uniformly gray, sometimes shaded with brownish, somewhat paler below than above; pectorals bordered with white; pelvics and anal pale; dorsal black at apex, with two white spots, one midway of its anterior margin, the other near its rear base; caudal edged below with white, its apex with a black spot, edged with white. After preservation, these same specimens (see Study Material, p. 88) are dark mouse-gray above, grayish white below, with apex of dorsal and tip of caudal dusky, the latter pale-edged.

Developmental Stages. Gravid females have been taken off Cuba with as many as 18 embryos, ranging in size up to 150 mm.[5] Nine embryos, about 100 mm. long, taken from the female listed above (p. 88), are of approximately adult form, the chief differences being their much larger eyes, which is a common embryonic feature, relatively longer caudals, less deeply emarginate dorsal and pectoral fins, and relatively shorter body cavities. The large oval yolk sac shows no signs of any attachment to the wall of the oviduct of the mother.[6] Up to 20 embryos have been found in a female, in Cuban waters.[7]

Size. This Shark may be born at a length no greater than about 10 inches; males may mature at 2 to 2½ feet, and females at about 3 feet, or perhaps while even smaller. The few specimens for which sizes have previously been recorded in scientific literature have ranged from about one foot, two inches (350 mm.)[8] to a maximum of seven feet (about 2.14 m.).[9] Although the species has been credited repeatedly with reaching more than three meters, or 10 feet, we find no definite proof of so large a size for it.

Habits. Very little is known of its mode of life. It seems to be a bottom dweller chiefly, of coastal waters. Its depth range is wide, however, for on the one hand it is recorded from 380 to 460 meters depth off Portugal and from deep water off Cuba, while on the other hand it has been reported as common in the very shallow water of roadsteads and lagoon-like situations along tropical West Africa.[10] In Spanish waters it is classed as very voracious, destroying great numbers of food fish, especially hake (*Merluccius*). No precise information is available as to its stomach contents. Nothing is known of its breeding habits, other than as indicated above.

4. Howell-Rivero (Torreya, *9*, 1941: 8, and personal communication).
5. Personal communication from Luis Howell-Rivero.
6. See Lo Bianco (Mitt. zool. Sta. Neapel., *19*, 1909: 667) for an account of the egg capsules; Ranzi (Pubbl. Sta. zool. Napoli, *13*, 1934: 378, 417) for the structure of the uterine wall in the gravid female.
7. Personal communication from Luis Howell-Rivero.
8. Tortonese, Atti Soc. ital. Sci. nat., 77, 1938: 286.
9. Günther, Cat. Fish. Brit. Mus., *8*, 1870: 398.
10. Rochebrune, Act. Soc. linn. Bordeaux, (4) *6*, 1882: 46.

Relation to Man. It is not sufficiently plentiful anywhere to be of commercial importance.

Range. Atlantic, west and east, including Mediterranean; Cape of Good Hope; Japan in the North Pacific; it is represented in Australian waters by a relative (*dakini*) so close to *perlo* that it may finally prove identical (see discussion, p. 87). In the eastern Atlantic its chief center of population is apparently the Mediterranean, where it is widespread, although nowhere numerous. Its range extends thence southward to Senegambia, where it is reported from many localities. It is also caught occasionally and in small numbers to the northward along the Atlantic coasts of the Iberian Peninsula. The most northerly records are on the Portuguese coast and in the Gulf of Gascony (off Bayonne). It is also known from Madeira.

Occurrence in the Western Atlantic. The only published record of its presence in the western Atlantic is of the two specimens from Matanzas, Cuba, described above;[11] but Howell-Rivero writes us that specimens are now being taken occasionally in Matanzas Bay, including gravid females with embryos in all stages of development, suggesting that it is now experiencing an upswing in abundance in Cuban waters.

Synonyms and References:

Le Perlon, Broussonet, Mem. Math. Phys. Acad. Sci. Paris, 1780: 668 (descr., Medit.).

Squalus perlo Bonnaterre, Tabl. Encyc. Meth. Ichthyol., 1788: 10 (descr., type loc., Medit.).

Squalus cinereus Gmelin, in Linnaeus, Syst. Nat., 1789: 1497 (descr., Medit.); Bloch and Schneider, Syst. Ichthyol., 1801: 133 (descr., Medit.); Bosc, Nouv. Dict. Hist. Nat., *21*, 1803: 185 (diagn.); Latreille, Nouv. Dict. Hist. Nat., *24*, 1804: 72 (in table of contents); Risso, Ichthyol. Nice, 1810: 24 (near Nice, size); Desvaux, Essai Ichthyol. France, 1851: 24 (France, not seen).

Le squale perlon, Lacépède, Hist. Nat. Poiss., 4° ed., *1*, 1798: 220 (general); Sonnini, Hist. Nat. Poiss., *4*, 1801–1802: 14 (descr.).

Heptranchias cinereus Rafinesque, Carratt. Gen. Spec. Sicil., 1810: 13; Indice Ittiol. Sicil., 1810: 45 (Medit.); Swainson, Nat. Hist. Fish. Amphib. Rept., 1839: 314 (general); Bonaparte, Mem. Soc. neuchâtel. Sci. Nat., *2* (8), 1839: 9 (in synopsis); Icon. Faun. Ital., *3*, 1841: 1, index, No. 137 (name); Cat. Pesc. Europ., 1846: 17 (Medit.); Costa, Fauna Napol. Pesci, *3*, 1854–1857: No. 4, 11 (Medit.); Gill, Ann. N. Y. Lyc., 7, 1861: 404 (class.); Costa, La Pesca, Napoli, 1871: 87 (Medit., not seen); Rey, Fauna Iberica, Peces, *1*, 1928: 289 (descr., meas., Spain); de Buen, Faune Medit., Cons. Internat. Explor. Medit., 1930: plate not numbered (Medit., Atlant.); Ranzi, Pubbl. Sta. zool. Napoli, *13*, 1934: 378, 417 (struct. of uterine wall).

Monopterhinus cinereus Blainville, Bull. Sci. Soc. philom. Paris, 1816: 121 (name).

Carcharias cinereus Cloquet, Dict. Sci. Nat., 7, 1817: 69 (general).

Squalus (*Monopterhinus*) *cinereus* Blainville, in Vieillot, Faune Franc., Poiss., 1825: 80 (descr.).

Notidanus cinereus Cuvier, Règne Anim., 2nd ed., *2*, 1829: 390 (descr.); Lowe, Trans. zool. Soc. Lond., *2* (3), 1837: 194 (Madeira); Bonaparte, Icon. Faun. Ital., *3*, 1835: plate not numbered, Index, No. 127 (descr., Medit.); Cuvier, Règne Anim., Ill. Poiss., 1843: 364 (descr.); Van der Hoeven, Handb. Dierkunde, 2nd ed., *2*, 1855: 261 (general); Nardo, Atti Ist. veneto, (3) *5*, 1859–1860: 787 (Medit.); Fitzinger, Bild. Atlas Naturg. Fische, 1864: fig. 176 (good ill.); Günther, Cat. Fish. Brit. Mus., *8*, 1870: 398 (Medit., Atlant., Madeira); Gervais and Boulart, Poiss., *3*, 1877: 195 (Medit. and Atlant. coasts of France); Reguis, Ess. Hist. Nat. Provence, *1*, Fasc. 1, 1877: 53 (Medit.); Heldreich, Faune de Grèce, 1879: 91 (not seen); Giglioli, Elenc. Pesc. Ital., 1880: 52 (not seen); Perugia, Elenc. Pesc. Adriat., 1881: 55 (Adriatic); Rochebrune, Act. Soc. linn. Bordeaux, (4) *6*, 1882: 46; Faune Senegambie, *1*, Poiss., 1883–1885: 24 (Senegambia).

11. Howell-Rivero, Torreya, 9, 1941: 7.

Squalus (Notidanus) cinereus Voigt, in Cuvier, Tierreich, *2*, 1832: 509 (descr.).
Heptanchus cinereus Müller and Henle, Plagiost., 1841: 81, pl. 35, fig. 3 (teeth, descr., Medit.); Duméril, Hist. Nat. Poiss., *1*, 1865: 437, pl. 4, fig. 1–4 (teeth, descr., Medit., C. of Good Hope [?]); Bocage and Brito Capello, Poiss. Plagiost., 1866: 15 (no Portuguese record); Miklucho-Maclay, Beitr. Vergl. Neurol. Wirbelt., *1*, 1870: 12, pl. 2, fig. 1–6 (brain); Ninni, Ann. Soc. Nat. Modena, *5*, 1870: 66 (Medit.); Canestrini, in Cornalia, *et al.*, Fauna d'Ital., Pesc., *3*, 1871–1872: 43 (Medit.); Gegenbaur, Unters. Vergl. Anat. Wirbelt., *3*, 1872: pl. 21, fig. 5, 6 (skelet., Medit.); Hertwig, Jena Z. Naturw., *8*, 1874: 349, pl. 12, fig. 7, 10, 13 (develop. of denticles, teeth); Doderlein, Prosp. Metod. Pesci Sicil., 1878–1879: 30 (Medit.); Stossich, Boll. Soc. adriat. Sci. nat., *5*, 1880: 69 (Adriat.); Moreau, Hist. Nat. Poiss. France, *1*, 1881: 339 (descr., Medit. and Atlant. coasts, France); Doderlein, Man. Ittiol. Medit., *2*, 1881: 78 (Medit.);[12] Graeffe, Arb. zool. Inst. Univ. Wien, 7, 1886: 447 (Medit.); Carus, Prod. Fauna Medit., *2*, 1889–1893: 500 (Medit.); Vieira, Ann. Sci. nat. Porto, *4*, 1897: 67 (Portugal); Sicher, Atti Accad. gioenia, (4) *11* (5), 1898: 16 (Medit.); Steinhard, Arch. Naturgesch., *1* (69), 1903: 6, pl. 1, fig. 1–3 (oral and pharyngeal denticles); de Braganza, Result. Invest. Sci. "Amelia," Ichthyol., *2*, 1904: 30, 102 (Portugal); Borri, Mem. Soc. tosc. Sci. nat., *44*, 1934: 94 (Medit.); Belloc, Rev. des. Trav. Pêches Marit., 7, Fasc. 2, 1934: 152 (ill., Morocco to Senegal); Daniel, Elasmobranch Fishes, Univ. Calif. Press, 1934: 4, 24, 25, 51, 91 (scales, fin cartilages, musculature, thymus gland); Noronha and Sarmento, Peixes Madeira, 1934: 100 (Madeira, not seen); Nobre, Fauna Marinha Port., Vert., *1*, 1935: 414 (Portugal, size); Holmgren, Acta Zool. Intern. Tidskr. Zool., *22*, 1941: 3 (skull).
Heptranchus cinereus Gray, List Fish. Brit. Mus., Chondropt., 1851: 68 (Medit., ocean).
Heptrancus angio Costa, Fauna Napoli Pesci, *3*, 1854–1857: No. 4, 5, pl. 13, 14, fig. 3 (descr., Medit.).
Heptanchus (Heptranchias) cinereus Brito Capello, J. Sci. math. phys. nat. Lisboa, *2*, 1870: 141 (Portugal).
Notidanus (Heptanchus) cinereus, var. *pristiurus* (var. *aetatis*) Bellotti, Atti Soc. ital. Sci. nat., *20*, 1877: 60 (Medit.).
Notidanus (Heptanchus) cinereus Lo Bianco, Mitt. zool. Sta. Neapel, *13*, 1899: 542 (embryos); Imms, Proc. zool. Soc. Lond., *1*, 1905: 44 (denticles).
Heptranchias deani Jordan and Starks, Proc. Calif. Acad. Sci., (3) *2*, 1901: 384 (type descr., Japan); Jordan and Snyder, Annot. zool. jap., *3*, 1901: 128 (Japan); Jordan and Fowler, Proc. U.S. nat. Mus., *26*, 1903: 595 (descr., Japan); Pietschmann, S. B. Akad. Wiss. Wien, Math.-Naturn. Kl., *117* (1), 1908: 708 (descr., Japan).
Hexanchus cinereus Seabra, Bull. Soc. portug. Sci. nat., *5*, 1911: 195 (Portugal).
Heptranchias perlo Garman, Mem. Harv. Mus. comp. Zool., *36*, 1913: 21 (descr.); Hussakoff, Copeia, 67, 1919: 9 (descr., Japan); Fowler, Bull. Amer. Mus. nat. Hist., *70* (1), 1936: 27 (descr., W. Africa); Norman and Fraser, Giant Fishes, 1937: 5, fig. 8 (general); White, Bull. Amer. Mus. nat. Hist., *74* (2), 1937: 40, pl. 1e, 4m, 17c, 23b, 29e, f (anat.); Tortonese, Atti Soc. ital. Sci. nat., 77, 1938: 286 (Medit.); Fowler, Bull. U.S. nat. Mus., 100 (*13*), 1941: 9 (Medit. and Pacif. refs.); Norris, Plagiost. Hypophysis, 1941: 23, 37, pl. 1, fig. 4; pl. 20, fig. 79 (brain); Howell-Rivero, Torreya, 9, 1941: 7, pl. 3, 4 (descr., Cuba); Fowler, Revist. Chilen. Hist. Nat., anõs 46–47, 1944: 113, fig. 5 (Chile); Bigelow and Schroeder, Comm. Shark Fish., Anglo Amer. Carib. Comm., Wash., 1945: 92, fig. 30 (descr., ill., range).

Doubtful reference:
Monopterhinus ciliaris Blainville, Bull. Sci. Soc. philom. Paris, 1816: 121 (name only).

Suborder CHLAMYDOSELACHOIDEA

Characters. Anal fin present; only 1 dorsal fin, without spine; 6 gill openings, all in front of origins of pectorals; margins of 1st gill openings continuous across throat; snout

12. See Doderlein (1881) and Carus (1889–1893) for additional locality records for the Mediterranean in publications not accessible to us.

not beak-like, without lateral teeth or cirri; teeth alike in both jaws, those in front of mouth essentially similar to those toward corners; trunk subcylindrical (shark-like); eyes lateral; anterior margins of pectorals not expanded forward beyond 1st gill opening; mouth terminal, without distinct snout; nostril entirely separate from mouth, its anterior margin without barbel; eye without nictitating membrane; spiracles present; vertebral column only incompletely segmented, the notochord being somewhat constricted segmentally for a short distance back from head, but of uniform diameter thence rearward; a few of anterior vertebrae with primary calcifications (cyclospondylic), but the others not calcified; upper jaw (palatoquadrate cartilage) with transverse process attached to orbital region of cranium by a ligament (only attachment to cranium); also with ligamentary attachment to the hyomandibular arch, which is well developed and provides the chief suspension for both jaws; propterygial cartilage of pectoral fin bears no radial elements; heart valves in 6 or 7 rows; clasper of male not enclosed by margin of pelvic fin, its axial cartilage attached to basipterygial cartilage of fin by 1 small element only, its tip with 3 movable accessory cartilages. Development ovoviviparous.[1]

Remarks. The majority of recent authors have placed *Chlamydoselachus* (sole known representative of the group) among the notidanoids because of its large number of gill openings and the incomplete segmentation of its vertebral column. We believe a separate suborder is demanded for it,[2] because it differs so widely from *Hexanchus* and *Heptranchias* (representing the notidanoids) in the much less intimate attachment of its upper jaw to the cranium, as well as in the facts that its much larger hyomandibular arch affords the chief suspension for the jaws and that its notochord is of nearly uniform diameter throughout most of its length.

Families, Genera, Species. Only one modern species is known, *Chlamydoselachus anguineus* Garman; but teeth, apparently of this genus, have been described from the Pliocene of Tuscany.

Range. Japan, also eastern Atlantic off southern France, off the Iberian Peninsula, and near Madeira in moderately deep water; reported from New South Wales, but on doubtful evidence.[3]

Fossil Teeth. From Miocene, West Indies; Pliocene, Europe.

1. See Garman (Bull. Mus. comp. Zool. Harv., *12* [1], 1885), Goodey (Proc. zool. Soc. Lond., 1910: 540), Allis (Acta zool., *4*, 1923: 122) and Smith (Dean Memor. Vol., Amer. Mus. nat. Hist., Art. 6, 1937) for detailed accounts and illustrations of the skeleton and other anatomical features; Garman (Mem. Harv. Mus. comp. Zool., *36*, 1913: pl. 59, fig. 4, 5, pl. 61, fig. 7, 8) and especially Gudger (Dean Memor. Vol., Am. Mus. nat. Hist., Art. 7, 1940) for excellent illustrations of the egg capsule and of embryos in different stages of development.
2. Garman (Science, *3*, 1884: 117) proposed for *Chlamydoselachus* a new order, Selachophichthyoidi, a name based on the supposition that it "stands nearer the true fishes than do the sharks proper." Shortly afterward, however, Gill (Science, *3*, 1884: 346) united it with the fossil genus *Didymodus* (a pleuracanth) as the suborder Pternodonta, while Garman (Bull. Mus. comp. Zool. Harv., *12*, 1885: 30) united it, as Cladodonti, with the fossil *Cladodus* and its allies, of which he, by then, had come to consider it "the living representative." More recent studies of the fossil genera in question, however, make it so unlikely that *Chlamydoselachus* can be properly grouped with any pleuracanth or cladodont that we prefer to use for the suborder a name based on that of the modern genus.
3. See Whitley (Fish. Aust., *1*, 1941: 70).

Suborder HETERODONTOIDEA

Characters. Anal fin present; 2 dorsal fins with well developed spines; only 5 gill openings, the last 3 or 4 over base of pectoral; snout not beak-like, without lateral teeth or cirri; teeth similar in both jaws, those toward center of mouth smaller, with 3 to 5 cusps,[4] but those along outer parts of jaws much larger, rounded (molar), without cusps, several rows functional; midtrunk subcylindrical (shark-like), but head with snout strongly flattened both above and below; tail sector flattened below; anterior margins of pectorals not expanded forward beyond 1st gill opening; nostrils connected with mouth by a deep groove; eye without nictitating fold or membrane; spiracles present; inner margins of pelvics entirely separate, posterior to cloaca; vertebral column completely segmented throughout its length, its axial canal much contracted in the region of the centra, which are fully differentiated, and notochord greatly constricted segmentally in centra; vertebral centra with internal calcareous lamellae radiating from a central ring; skull without antorbital processes or separate antorbital bars; upper jaw (palatoquadrate cartilage) attached by a short ligament to hyomandibular arch as well as closely and much more extensively to sides of preorbital region of cranium;[5] rostral cartilage lacking; neural spines not attached to dorsals; propterygial cartilage of pectoral bears 1 radial element; heart valves in only 2 rows; claspers of males projecting freely from pelvics, their axial cartilages with 3 movable accessory cartilages at tip and attached to basipterygium of the fin by 2 small connecting elements. Development oviparous; egg cases horny with spiral flanges, but without tendrils.

Families and Genera. Only one modern family (Heterodontidae) and genus (*Heterodontus* Blainville, 1816) with the characters of the suborder.[6]

Range. East Africa, East Indies, New Zealand, Australia, China, Japan and eastern Pacific north as well as south; not known in Atlantic or Mediterranean.

Fossil Teeth. Upper Jurassic to Pliocene in Europe; Upper Cretaceous to Eocene in Africa; Miocene in South America, New Zealand, Australia.

Suborder GALEOIDEA

Characters. Anal fin present; 2 (rarely 1) dorsal fins, without spines; only 5 gill openings with rudimentary 6th arch; snout not beak-like, without lateral teeth or cirri; teeth of essentially the same type in front of mouth as near corners; trunk subcylindrical, not strongly depressed; eyes lateral; anterior edges of pectorals not extending forward past 1st gill openings; nostril either connected with mouth or separate from it; nictitating membrane and spiracles present or absent; inner margins of pelvics either separate posterior to cloaca, or more or less united; vertebral column completely segmented throughout its length, its axial canal much contracted or obsolete in regions of centra, the latter being fully differentiated; notochord greatly constricted segmentally in centra, or even completely obliterated there, but dilated in spaces between concave faces of adjoining vertebral centra;

4. Cusps may be entirely worn off in large specimens. 5. Firmly articulated there in fossil forms.
6. See Fowler (Bull. U.S. nat. Mus., 100 [*13*], 1941: 15) for list of generic synonyms.

vertebral centra with calcareous lamellae radiating from a central ring, or with latter alone calcified (genera *Galeus, Pseudotriakis*); neural spines not attached to dorsals; skull with antorbital processes more or less developed, but no separate antorbital bar; rostral cartilages 3 (united or separate at tip), 1 or none; upper jaw (palatoquadrate cartilage) not articulated with cranium, but connected with ethmoid region by a longer or shorter ligament;[7] its connection with hyomandibular arch also ligamentary only, at least in most cases.[8] Propterygial cartilage of pectoral much smaller than mesopterygium, with 1 to several radial elements; heart valves in 2 or 3 rows; claspers of male projecting freely from pelvics; axial cartilages either single or double, usually with a group of movable accessory cartilages at the tip when adult, and attached to basipterygium of fin by 1 small connecting element only.[9] Development oviparous, ovoviviparous, or viviparous.

Key to Families

1a. Only 1 dorsal fin. Scyliorhinidae (part), p. 195.

1b. 2 dorsal fins.

- 2a. At least ½ of base of 1st dorsal posterior to origin of pelvics.
  - 3a. Caudal lunate, large; gill arches connected one with the next by masses of spongy tissue, forming sieve-like structures. Rhincodontidae, p. 187.
  - 3b. Caudal not lunate, not very large; gill arches not connected one with the next by masses of spongy tissue.
    - 4a. Nostril connected with mouth by a deep groove, its anterior margin with a well developed barbel. Orectolobidae, p. 178.
    - 4b. Nostril not connected with mouth by a deep groove, or, if so connected, its anterior margin without a well developed barbel. Scyliorhinidae, p. 195.
- 2b. Base of 1st dorsal terminates over, or (usually) well anterior to, origin of pelvics.
  - 5a. Head greatly expanded laterally. Sphyrnidae, p. 407.
  - 5b. Head of normal shape, not expanded laterally.
    - 6a. Caudal fin lunate, its axis steeply raised.
      - 7a. Teeth large, few in number; gill arches without gill rakers. Isuridae, p. 109.
      - 7b. Teeth minute, very numerous; gill arches with well developed rakers. Cetorhinidae, p. 146.
    - 6b. Caudal fin not lunate, its axis raised only moderately at most.
      - 8a. 1st dorsal fin longer at base than caudal. Pseudotriakidae, p. 228.
      - 8b. 1st dorsal fin much shorter at base than caudal.
        - 9a. Caudal fin occupies nearly ½ total length, or even more. Alopiidae, p. 160.

7. This allows the jaws to be more or less protrusible in many cases.

8. Parker's (Trans. zool. Soc. Lond., *10*, 1879: pl. 38, fig. 2) illustration of the skull of *Scyllium canicula*, equals *Scyliorhinus caniculus* (Linnaeus), 1758, which shows these ligamentary connections well, has been copied in many subsequent textbooks of zoology.

9. For illustrations of the cartilages of the clasper in various galeoids, see especially Huber (Z. Wiss. Zool., 70, 1901: pl. 27); White (Bull. Amer. Mus. nat. Hist., *74*, 1937: pl. 46–50).

9b. Caudal fin occupies considerably less than ½ total length.

10a. 5th gill opening well in front of origin of pectoral; eye without nictitating fold or membrane.

11a. Jaws widely protrusible forward; snout greatly elongate. Scapanorhynchidae, p. 109.

11b. Jaws not widely protrusible; snout not greatly elongate. Carchariidae, p. 98.

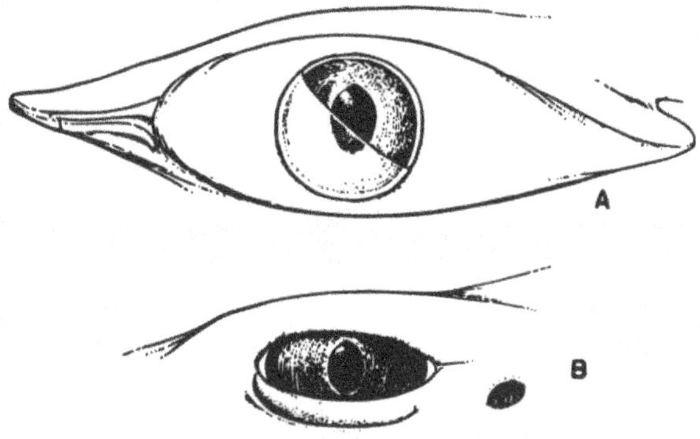

FIGURE 12. *A* Eye of *Sphyrna diplana*, about 1375 mm. long, to show nictitating membrane (U. S. Nat. Mus., No. 108452), about 2 x natural size. *B* Eye of *Mustelus canis*, about three feet long, to show subocular fold (Harv. Mus. Comp. Zool., No. 35245).

10b. 5th gill opening over or behind origin of pectoral; eye with a more or less strongly developed nictitating fold or membrane.

12a. Upper edge of nictitating fold continuous with edge of eyelid, or even arising outside latter posteriorly, although enclosing it anteriorly; teeth low, rounded or with 3 or more cusps, usually in mosaic arrangement, several series functional simultaneously in sides of jaws as well as in front. Triakidae, p. 233.[10]

12b. Upper edge of nictitating membrane arises far within edge of eyelid posteriorly, as well as anteriorly; teeth blade-like with 1 cusp only, not in mosaic arrangement, usually not more than 1 or 2 series functional in sides of jaws simultaneously. Carcharhinidae, p. 262.[10]

10. It may not be possible to draw a sharp line between Triakidae and Carcharhinidae with respect to the nictitating membrane or the teeth. However, the definition given above will serve to place any genus yet known from the Atlantic in the one family or in the other.

Family CARCHARIIDAE

Sand Sharks

Characters. Two dorsal fins, the 1st much shorter than caudal, the rear end of its base over or anterior to origin of pelvics; caudal not more than ⅓ of total length, not lunate, its lower anterior corner expanded as a distinct lobe, its axis raised but little; caudal peduncle not greatly depressed or expanded laterally; a precaudal pit above but none below; sides of trunk, anterior to anal, without longitudinal dermal ridges; inner margins of pelvics more or less united posterior to cloaca in male, less so in female; jaws not greatly protrusible; snout not greatly elongate; 5th gill opening anterior to origin of pectoral; gill arches without rakers, not interconnected by a sieve of modified denticles; nostrils entirely separate from mouth, their anterior margins without barbels; spiracles present; lower eyelid without nictitating fold or membrane; teeth large, awl-shaped, with or without lateral denticles and not very numerous (see counts, p. 102); skull of normal form (*i.e.*, not greatly expanded laterally); rostral cartilages united in one; mesopterygium of pectoral with about ½ as many radials as metapterygium, and nearly as large; meso- and metapterygia not separated by foramen. Development ovoviviparous.

Genera. Only one modern genus, *Carcharias*, so far known.

Genus *Carcharias* Rafinesque, 1810

Sand Sharks

Carcharias Rafinesque, Caratt. Gen. Nuov. Sicil., 1810: 10; type species, *C. taurus* Rafinesque. Sicily.[11]

Generic Synonyms:

Squalus (in part) Risso, Ichthyol. Nice, 1810: 38; for *S. ferox;* also subs. authors; not *Squalus* Linnaeus, 1758.

Galeorhinus (in part) Blainville, Bull. Soc. philom. Paris, 1816: 121; for *G. ferox*.

Odontaspis Agassiz, Poiss. Foss., 1838: *3*, 87; type species, *Carcharias ferox* Risso, 1826, equals *Squalus ferox* Risso, 1810.

Triglochis Müller and Henle, Arch. Naturg., 1837: 396; type species, *Carcharias ferox* Risso, 1826, equals *Squalus ferox* Risso, 1810.

Eugomphodus Gill, Proc. Acad. nat. Sci. Philad., Addend., 1861: 60; type species, *Carcharias griseus* Storer, 1846, equals *Carcharias taurus* Rafinesque, 1810; monotypic.

Synodontaspis White, Vert. Faun. Engl. Eocene, 1931: 51; type species, *Carcharias taurus* Rafinesque, 1810.

Paradontaspis White, Vert. Faun. Engl. Eocene, 1931: 63; type species, *Odontaspis platensis* Lahille, 1928.[12]

Generic Characters. Caudal peduncle with a well marked pit above (none below) and without lateral keels; dermal denticles with 3 broad longitudinal ridges; snout conical; jaws with or without labial furrows; anterior teeth in both jaws two-rooted, the posterior teeth less obviously so; spiracle small; 2nd dorsal about as large as 1st; caudal with small but definitely outlined lower anterior lobe and well marked subterminal notch. Characters otherwise those of the family.

11. Opinion 47 of the International Commission on Zoological Nomenclature (Smithson. Publ., 2060, 1912: 108; Copeia, 29, 1916: 28) confirms *Carcharias taurus* Rafinesque, 1810, as the type of *Carcharias* Rafinesque, 1810; it thus replaces *Odontaspis* Agassiz.

12. The fossil genus *Oxytes* Giebel, Fauna Vorwelt, Fische, 1847: 364, type species, *O. obliqua* Giebel (monotypic), is referred to the synonymy of *Carcharias* by Fowler (Bull. U.S. nat. Mus., 100 [*13*], 1941: 119).

Range. Both sides of warm temperate and tropical North Atlantic, including the Mediterranean; eastern South America south to northern Argentina; South Africa; India; Australia; China; Japan.

Fossil Teeth. Lower Cretaceous to Pliocene, Europe; Upper Cretaceous to Miocene, South America; Upper Cretaceous to Pliocene, North America, New Zealand; Upper Cretaceous, Asia; Paleocene to Pliocene, Africa; Miocene, Australia, West Indies.

Species. The members of this genus fall into two easily separable divisions, the one represented by a single well defined species (*ferox* Risso), the other by a group of named forms, so clearly allied one to another that it is still an open question how many of them deserve separate specific names. While awaiting comparison of specimens from different ocean areas, the accompanying key recognizes differences which may later prove merely varietal.

Provisional Key to Species

1a. 1st upper tooth notably smaller than 2nd, each tooth usually with 2 denticles on each side; 3rd upper tooth followed by 4 very much smaller teeth. *ferox* Risso, 1810.
Eastern Atlantic, Mediterranean.

1b. 1st upper tooth only slightly smaller than 2nd, if so at all; each tooth usually with 1 denticle only (rarely 2) on each side, or with none; 3rd upper tooth followed by 2 or 3 much smaller teeth at most.

2a. 3rd upper tooth followed by 2 or 3 much smaller teeth, no wide gap between these and the next large (5th or 6th) tooth. *platensis* Lahille, 1928.
Argentina.

arenarius Ogilby, 1911.
Australia.[13]

2b. 3rd upper tooth followed by 1 much smaller tooth only, the latter separated from the succeeding large tooth by a broad gap.

3a. Snout broadly rounded; inner margin of pectoral only ⅕ as long as outer; no labial furrow at angle of mouth. *tricuspidatus* Day, 1888.
India, China.[14]

3b. Snout pointed; inner margin of pectoral more than ⅓ as long as outer; well marked labial furrows at corners of mouth.

4a. Lateral denticles lacking on most teeth, minute on others; length of longest tooth less than ½ diameter of eye. *owstoni* Garman, 1913.
Japan.

4b. Most or all of teeth with a well developed lateral denticle on each side; length of longest tooth at least ⅔ diameter of eye.
taurus Rafinesque, 1810, p. 100.[15]

13. Published descriptions are not sufficiently detailed for critical comparison of *platensis* with *arenarius*.

14. Fang and Wang, Contr. Biol. Lab. Sci. Soc. China, *8*, 1932: 241.

15. It has been suggested recently that the American form (*littoralis*) differs from the European *taurus* in having no denticles on its first and fourth upper teeth (Giltay, Mém. Mus. Hist. nat. Belg., Hors Série, 5, Fasc. 3, 1933: 7). Our own examination of specimens of various sizes from southern New England and the vicinity of New York shows that while the teeth in question are smooth in some small specimens (about three feet long), they have

Carcharias taurus Rafinesque, 1810

Sand Shark, Sand Tiger

Figures 13, 14

Study Material. Five Massachusetts specimens, male and female, 943 to 1,081 mm. long (Harv. Mus. Comp. Zool.); jaws of a large specimen from New Jersey, and also of a female, 2,800 mm., from Englewood, Florida; also many medium-sized specimens, fresh caught at Woods Hole, Massachusetts.

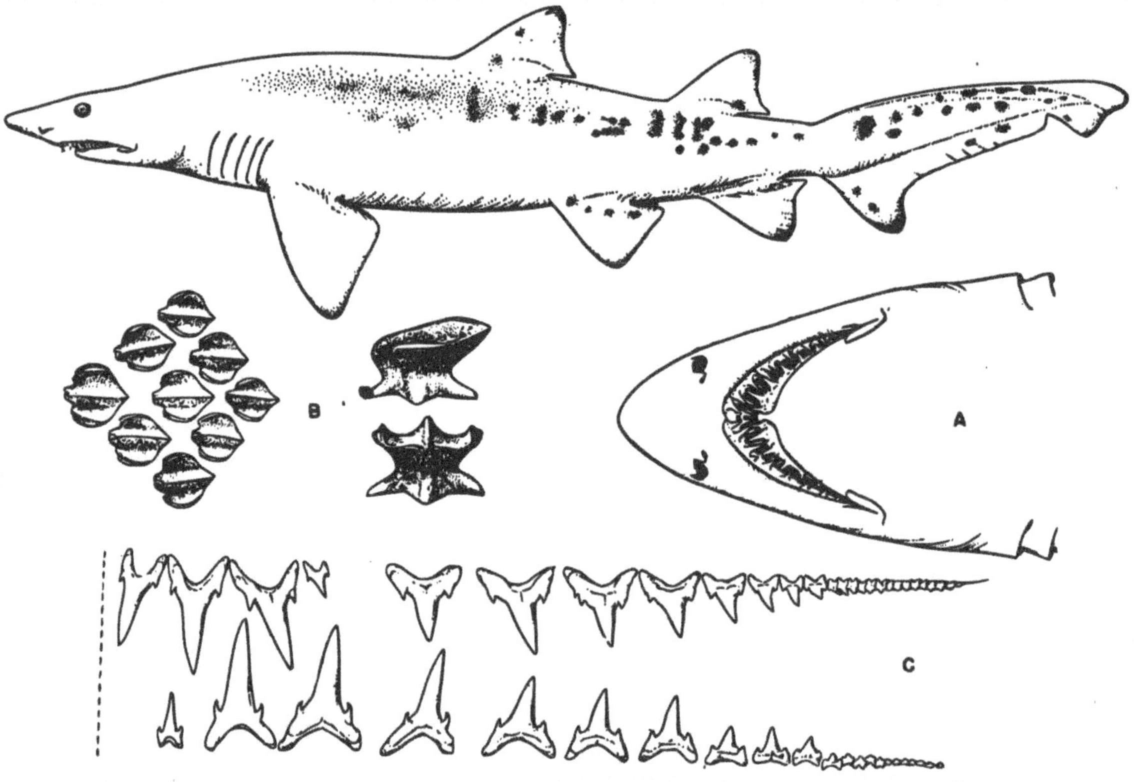

FIGURE 13. *Carcharias taurus*, young male, 1010 mm. long, from Cape Cod, Massachusetts (Harv. Mus. Comp. Zool., No. 351). *A* Anterior part of head of same from below. *B* Dermal denticles, general view, about 25 x; lateral and apical views, about 50 x. *C* Upper and lower teeth of larger specimen from New Jersey (Harv. Mus. Comp. Zool., No. 351), about 1 x.

Distinctive Characters. The five gill openings in front of the pectorals, the second dorsal about as large as the first, the position of the first dorsal entirely in front of the pelvics, the entire separation of the nostril from the mouth, and the highly characteristic teeth (Fig. 13 C) are diagnostic among sharks of our province.

denticles on one or on both sides in others and denticles on both sides in the two large specimens that we have examined. Neither do the supposed differences in the relative position of the rear end of the base of the first dorsal fin, or in the origin of the pelvics, invoked by earlier authors as a specific character, prove any more significant, for these vary considerably among American specimens (see p. 102).

Description. Proportional dimensions in per cent of total length. Female, 982 mm., from Mass. (Harv. Mus. Comp. Zool., No. 436). Male, 1,081 mm., from Mass. (Harv. Mus. Comp. Zool., No. 402).

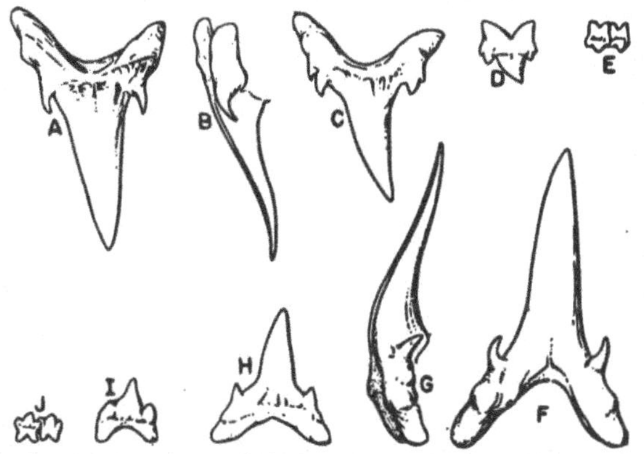

FIGURE 14. *Carcharias taurus*, showing teeth from jaws pictured in Fig. 13, about 2 x. *A*, *B* Third upper tooth. *C* Seventh upper tooth. *D* Twelfth upper tooth. *E* Seventeenth and eighteenth upper teeth. *F*, *G* Third lower tooth. *H* Seventh lower tooth. *I* Tenth lower tooth. *J* Seventeenth and eighteenth lower teeth.

Trunk at origin of pectoral: breadth 10.7, 10.2; height 12.2, 10.7.
Snout length in front of: outer nostrils 3.4, 3.3; mouth 3.7, 3.9.
Eye: horizontal diameter 1.5, 1.2.
Mouth: breadth 8.1, 8.0; height 4.7, 5.2.
Labial furrow length: upper (visible part) 0.9, 0.9; lower 2.4, 2.3.
Nostrils: distance between inner ends 3.1, 3.2.
Gill opening lengths: 1st 5.5, 5.1; 2nd 5.0, 5.1; 3rd 4.9, 4.6; 4th 4.9, 4.4; 5th 3.8, 3.4.
First dorsal fin: vertical height 7.2, 6.6; length of base 7.5, 8.4.
Second dorsal fin: vertical height 6.3, 6.1; length of base 7.0, 7.2.
Anal fin: vertical height 5.8, 5.9; length of base 7.1, 7.7.
Caudal fin: upper margin 30.1, 29.3; lower anterior margin 9.8, 10.0.
Pectoral fin: outer margin 13.5, 14.2; inner margin 5.8, 6.0; distal margin 9.4, 9.5.
Distance from snout to: 1st dorsal 40.8, 39.3; 2nd dorsal 56.8, 57.2; upper caudal 70.0, 70.5; pectoral 22.8, 22.8; pelvics 48.0, 49.5; anal 60.2, 61.7.
Interspace between: 1st and 2nd dorsals 11.2, 10.8; 2nd dorsal and caudal 7.1, 6.2; anal and caudal 3.8, 2.8.
Distance from origin to origin of: pectoral and pelvics 29.9, 27.5; pelvics and anal 14.4, 14.1.

Trunk moderately stout, its height abreast pectoral origin about 1/8, opposite origin of dorsal about 1/7, of the total length. Caudal peduncle relatively high and laterally compressed. Dermal denticles about 0.4 mm. broad, by 0.45 mm. long in a specimen of about 100 cm. length, loosely spaced, their blades ovoid lanceolate, their anterior margins entire or slightly indented between tips of the 3 ridges; axial ridge very prominent and sharp-edged anteriorly but usually flat-topped and subdivided posteriorly.

Head moderately flattened above. Snout short, its length in front of mouth about 1/4 to 1/5 the length to 1st gill slit, narrow ovoid, with rounded tip. Eye round and small, its diameter only about 2/5 as long as distance between nostrils. Spiracle minute, about on a level with upper margin of eye and behind latter by a distance about equal to length of snout in front of mouth. Gill openings relatively large, 4th about as long as snout in front of mouth, others slightly shorter, the 5th shortest. Nostril nearly transverse, its anterior margin with a small rounded flap near inner end. Distance from inner angle of nostril to mouth about equal to width of nostril. Mouth crescentic in front, about 3/4 to 4/5 as long as broad; angle of mouth with well marked labial furrow on lower jaw and a less prominent one on upper; upper furrow partially hidden when mouth is closed.

Teeth $\frac{44 \text{ to } 48}{41 \text{ to } 46}$ in specimens examined, 1st to 6th or 7th teeth in each jaw either with or without 1 or 2 small basal denticles on either side;[16] 1st upper tooth usually a little smaller, but sometimes of the same size as 2nd or 3rd, the 4th much smaller than 3rd or 5th, with a broad interspace between 4th and 5th; 1st lower tooth much smaller than 2nd to 6th, the teeth posterior to 6th or 7th successively smaller in each jaw and broader relative to length, with denticles successively larger relative to median cusp, the outermost 12 or 13 minute, close set, tricuspidate, about as broad as high; 3 or 4 series functional toward corners of mouth, but only 1 or 2 series toward center.[17]

Origin of 1st dorsal about midway between axil of pectoral and origin of pelvics, its base terminating a little anterior to latter, its apex subacute, its rear margin slightly concave, the free corner about 1/3 as long as its base, its vertical height about 1/2 as great as length of pectoral or about 1/4 as great as length of head. Second dorsal similar to 1st and only slightly smaller, its origin about midway between cloaca and origin of anal, 1/5 to 1/2 of its base overlapping base of latter. Caudal about 30% of total length, its axis only slightly raised, the subterminal notch well marked, the posterior outline of terminal sector concave, its lower anterior corner expanded as a definite lobe with rounded apex, its anterior margin about 1/3 as long as upper caudal margin; re-entrant corner, included between the 2 lobes, broadly rounded. Anal a little larger than 2nd dorsal in area and a little longer basally, its rear margin less deeply concave, its free rear tip about 1/3 as long as its base, the interspace between anal and caudal only about 1/2 as long as base of anal. Pelvics originating a little posterior to rear end of base of 1st dorsal, and about as large as latter, the inner margins entirely separate posterior to cloaca in female, but connected for a short

16. On a large New Jersey specimen every tooth, from the first to the seventh, is flanked by one or two denticles on each side; in another, from southern Massachusetts, the fourth upper tooth lacks a denticle, while on still other specimens from the same general locality some of the teeth have a denticle on each side, some have a denticle on one side only, and still others have no denticle on either side.

17. An account of the shedding of the teeth of specimens in an aquarium is given by Breder (Copeia, 1942: 42); see also p. 65.

distance in male. Pectoral a little more than ½ as broad as long, with nearly straight distal and outer margins, rounded corners, and wide base.

Color. Light gray-brown above, darkest along back, snout, and on upper sides of pectorals, paling on the sides to grayish white on belly and on lower sides of fins; sides of trunk rearward from pectorals as well as caudal and dorsals variously marked with roundish to oval spots, varying in color from yellowish brown to ochre yellow. In a specimen 100 cm. in total length these spots vary from less than ½ cm. to more than 1½ cm. in diameter, numbering upwards of 100. Posterior margins of fins edged with black on some specimens but perhaps not on all.

Size. In the northern sector of their American range, from Delaware Bay to Cape Cod, Sand Sharks are recorded from 3 feet to about 9 feet, the great majority of those caught being immature, of perhaps 4 to 6 feet. Large adults (7 to 8 feet or more) are also reported, not rarely, from widely scattered localities along the New Jersey coast, from the vicinity of New York, from Clinton, Connecticut (8 feet 10 inches), and especially from the vicinity of Nantucket, where commercial operations in the early nineteen-twenties are said to have yielded "a wealth of eight and nine foot Sand sharks."[18] From North Carolina southward, however, large ones alone have been reported, the recorded lengths ranging from about 8 to 9 feet in the Beaufort–Cape Lookout region; 6½ to 9½ feet for Charleston, South Carolina; 9 feet 2 inches to 10 feet 5 inches for southwestern Florida at Englewood, the last named being the greatest length yet positively recorded for *Carcharias taurus.* The recorded weight of about 250 pounds for an 8-foot 10-inch specimen from Clinton, Connecticut, shows how much lighter a fish this is, length for length, than the Mackerel Shark, Mako or White Shark. We have no firsthand information to contribute.

It appears, from the state of sexual development of the specimens we have seen, and from the sizes of the few females so far reported as containing eggs or embryos, that this Shark does not mature until it attains a length of perhaps seven feet or upward.

Developmental Stages. Females have been reported containing many eggs as well as embryos.

Habits. In spite of its trim appearance and voracious appetite (see below) this is a comparatively sluggish shark, living mostly on or close to bottom, being more active and biting the hook more freely by night than by day. It is a coastwise species, as contrasted with pelagic, most of those caught being taken in depths of not more than two to five fathoms; and it is often encountered close in to the tide line in only two to six feet of water, hence its frequent capture in pound nets. It has not been reported from the fishing banks off Nantucket or at the mouth of the Gulf of Maine. To the southward, however, it may not be so strictly confined, witness its presence on the North Carolina Banks.

Knowledge of its breeding habits is confined to the facts that a large female, taken at Beaufort, North Carolina, in April contained many large eggs; also that specimens a little

18. Young and Mazet, Shark, Shark, 1933: 132.

more than eight feet long, at Cape Lookout, North Carolina, contained many eggs and embryos more than nine inches long in July; and that females with unripe eggs have been reported at Woods Hole in the same month in different years. Since no embryos have been found in large females in Florida, and since immatures three to five feet have been reported so far only from the section north from Delaware Bay (where these constitute the majority of the local stock, p. 103) this is probably the chief center for the production of young, but information is still lacking as to the seasonal occurrence of gravid females there, or of newborn young.

Proverbially voracious, the Sand Shark feeds chiefly on smaller fishes, for the capture of which its slender raptorial teeth are admirably adapted. Large specimens have been taken with as much as 100 pounds of fish in their stomachs, and by eyewitness accounts, schools of them may surround other fish or even those imprisoned in fishermen's nets. On the east coast of North America the recorded diet, depending on the geographical locality, includes alewives (*Pomolobus*), black drum (*Pogonias*), bluefish (*Pomatomus*), bonito (*Sarda*), butterfish (*Poronotus*), cunner (*Tautogolabrus*), eels (*Anguilla*), flatfishes, menhaden (*Brevoortia*), mullet (*Mugil*), scup (*Stenotomus*), sea bass (*Centropristis*), sea robin (*Prionotus*), small sharks (species?), shark sucker (*Echeneis*), silver hake (*Merluccius*), spadefish (*Chaetodipterus*), spot (*Leiostomus*), tautog (*Tautoga*) and the weakfishes, spotted (*Cynoscion nebulosus*) and gray (*C. regalis*). No doubt a complete list for any given locality would include practically all the local species that were not too large. Squid have been found in their stomachs at Woods Hole, likewise crabs and lobsters, although the latter are perhaps only exceptionally eaten, for they were not found among the stomach contents of many more which were recently examined at Woods Hole on different occasions. There is no reason to suppose that this species ever attacks large prey.

Relation to Man. Although plentiful, the Sand Shark is of little commercial importance at present. A few are included in the catch of the Florida shark-fishery; occasional specimens are sold at a low price in fish markets. There were local fisheries for it for leather in Nantucket Sound, in the first quarter of the present century, but these were short-lived, reportedly because of exhaustion of the stock. However, it is of some interest to sport anglers, considerable numbers being caught by them yearly, both as objects of special pursuit or incidentally while surf-casting for other fish. But its resistance when hooked is so much less vigorous for its size than that of the more active pelagic sharks, such as the Mako or White Sharks (pp. 128, 139), that few would rate it as in the game class.

There is no record of attack by a Sand Shark on human beings in North American waters, although bathers often come close to them, our own experience bearing this out. Its relative (or relatives) in East Indian waters bears a sinister reputation, however.

Range. Mediterranean, tropical West Africa, Canaries and the Cape Verdes in the eastern Atlantic; South Africa; western Atlantic from the Gulf of Maine to Florida and southern Brazil; represented in Argentine waters and in the Indo-Pacific by close allies (see Species, p. 99).

Occurrence in the Western Atlantic. Next to the Smooth and Spiny Dogfishes (p. 466), the Sand Shark is probably the most abundant shark in season from Delaware Bay northward to Cape Cod; in this region it is far more plentiful than it is anywhere in the eastern Atlantic. Considerable numbers are caught all along the coast of New Jersey both in the bays and outside; it is a common visitor yearly to the vicinity of New York, along Long Island and presumably within Long Island Sound.[19] It is common in summer in Rhode Island waters, and it is fairly so around Block Island. So general is its occurrence along the southern shores of Massachusetts, including Martha's Vineyard and Nantucket, that every local fisherman knows it well. As an example of its local numbers we may cite the fact that a catch of about 1,900 sharks, made by three boats on Horseshoe Shoal in Nantucket Sound from June to September, 1918, consisted chiefly of this species.[20] Similarly, a catch of 350 sharks, made near Nantucket in the early 1920's, consisted of this species with few exceptions.[21] It is also taken in some numbers yearly along the outer shores of Cape Cod. But this marks the eastern boundary of its center of abundance, for while it is recorded at various localities around Massachusetts Bay, these are occasional specimens only. Only as a stray does it wander north of Cape Ann; it was reported once from Casco Bay and once from St. Andrews, New Brunswick, at the mouth of the Bay of Fundy.

Our data are not adequate to describe its status from Delaware Bay southward. It is reported from the Bay itself, both near the mouth at Bowers Beach and even from the vicinity of Philadelphia at its head; likewise from the coast of Maryland, and from Chincoteague and Smith Island in Virginia. However, these reports do not suggest any great numbers. The survey of the fishes of Chesapeake Bay by the United States Bureau of Fisheries[22] did not yield even a single record, although it has been reported there more recently. Nor does it appear with any regularity along North Carolina, although large schools appear at times off Cape Lookout, and it rarely enters the local sounds.[23] On the other hand, it is described as one of the commonest summer sharks on the South Carolina coast, near Charleston, with as many as six large specimens recorded from a single net haul. It is taken on the east coast of Florida at all seasons, as at Salerno, near Jupiter Inlet, where it appears irregularly in considerable numbers. However, it apparently reaches the west coast of Florida as a stray only, but two specimens being known from Englewood, where the shark stock has been the subject of special investigation.[24] It has been taken off the northern Bahamas.[25] We find no published report of it anywhere else for the Gulf of Mexico, the Bahamas, Cuba, the Antilles, or for the Caribbean region, although it is so easily recognizable and usually comes so close inshore that it could hardly have been overlooked if it occurred with any regularity within these general areas. However, its

19. The only published record of it in the Sound is for Clinton, Connecticut.
20. Identity established by excellent photographs by R. H. Bodman, who reported this catch to us.
21. Young and Mazet, Shark, Shark, 1933: 132.
22. Hildebrand and Schroeder, Bull. U.S. Bur. Fish., *43*, 1928.
23. Only two specimens, both large, reported from Beaufort, N.C.
24. Springer, Proc. Fla. Acad. Sci., *3*, 1939: 34.
25. Wise, Nat. Hist. N.Y., *38*, 1936: 322 (photo).

presence, at least as a stray, has been proved recently at Bermuda by the capture of typical specimens, in 1927 and 1942.[26] It apparently has a second center of occurrence on the coast of Brazil, since Sand Sharks, seemingly identical with the northern *taurus*, are plentiful in October and November near Rio de Janeiro, where many are placed for sale in the market; they are recorded as far south as the Rio Grande do Sul. But it is not yet possible to define the boundaries of this southern population, owing to the uncertainty of identity (whether *taurus* or *platensis*) of the nominal records from Brazil, Uruguay and Argentina. Nor does any explanation suggest itself for the apparent discontinuity between the areas of distribution of the North Atlantic and South Atlantic populations.

On the east coast of Florida *C. taurus* is taken irregularly at all seasons. From South Carolina northward, however, it has been reported only during the warm half of the year. Thus, at Charleston, South Carolina, it is reported for summer only; off North Carolina it may appear from late April on through the spring and equally early in the season at the mouth of Delaware Bay, as in 1921, when eight were taken at Cape May on April 21. However, May 27 appears to be the earliest recorded date for it on Long Island at Orient, with its season of maximum abundance extending from June into early October all along the coast from New York to Cape Cod. It withdraws from the neighborhood of New York in autumn, when the temperature of the water falls below about 67° F. (19–20° C.), and departs from the coasts of southern New England and New Jersey by November at the latest.

The winter home of the Sand Sharks that summer along the northeastern United States is not known. No increase in their numbers in autumn or early winter has been noted along North Carolina or Florida, coincident with their disappearance from the North. Like various bony fishes, it is possible that they move offshore, and possibly southward, to escape winter chilling.

Synonyms and References:

Carcharias taurus Rafinesque, Caratt. Gen. Nuov. Sicil., 1810: 10, pl. 14, fig. 1; Indice Ittiol. Sicil., 1810: 45 (type loc., Sicily); Garman, Mem. Harv. Mus. comp. Zool., *36*, 1913: 25, pl. 6, fig. 1–3, pl. 41, pl. 51, fig. 7 (descr.); Radcliffe, Trans. Amer. Fish. Soc., 1914: 35; Bull. U.S. Bur. Fish., *34*, 1916: 244, pl. 38, fig. 1, 2 (N. Carolina, teeth); Nichols and Murphy, Brooklyn Mus. Sci. Bull., *3* (1), 1916: 21, pl. 2 (occur.); Fowler, Copeia, 30, 1916: 36; Copeia, 31, 1916: 41 (N. Jersey); Copeia, 35, 1916: 69 (Long Island, N. York); Jordan, Copeia, 29, 1916: 281 (name); Smith, Amer. Mus. J., *16*, 1916: 347 (general); Thorne, Copeia, 35, 1916: 69 (N. York); Fowler, Proc. Boston Soc. nat. Hist., *35*, 1917: 110 (Nantucket, Woods Hole, Massachusetts); Proc. Acad. nat. Sci. Philad., *69*, 1917: 108 (N. Jersey, size); Occ. Pap. Mus. Zool. Univ. Mich., *56*, 1918: 15 (Virginia); Copeia, 68, 1919: 13 (N. Jersey); Proc. biol. Soc. Wash., *32*, 1919: 72 (Delaware R.); Proc. Acad. nat. Sci. Philad., *71*, 1919: 292 (N. Jersey); Roule, Result. Camp. sci. Monaco, *52*, 1919: 116 (St. Lucia I., C. Verde); Fowler, Proc. biol. Soc. Wash., *33*, 1920: 143 (N. Jersey); Sherwood, Copeia, 100, 1921: 77 (Connecticut, large size); Fowler, Proc. Acad. nat. Sci. Philad., *72*, 1921: 385 (N. Jersey, small size); Proc. Acad. nat. Sci. Philad., *74*, 1922: 2, 5, 7 (N. Jersey, Delaware); Breder, Bull. N. Y. zool. Soc., *25*, 1922: 137 (in aquarium); Huntsman, Contr. Canad. Biol. (1921), *3*, 1922: 8 (St. Andrews, New Brunswick); Linton, Proc. U.S. nat. Mus. *64*, 1924: 13, 14 (parasites); Breder, Copeia, 127, 1924: 27 (Sandy Hook Bay, New York, and New Jersey); Barnard, Ann. S. Afr. Mus., *21* (1), 1925: 36 (Medit. and Cape Seas); Bigelow and Welsh, Bull. U.S. Bur. Fish., *40* (1), 1925: 34 (descr., food, Gulf of Maine); Fowler, Copeia, 143, 1925: 41,

26. Personal communication from Louis Mowbray, director of the Bermuda Aquarium.

42 (N. Jersey, Delaware Bay, size); Breder, Copeia, 153, 1926: 122 (near N. York); Fowler, Copeia, 156, 1926: 146 (N. Jersey); Chabanaud and Monod, Bull. Etud. Hist. Sci. Afr. Occid. Franc. (1926), 1927: 228 (tropical West Africa); Barnard, Ann. S. Afr. Mus., *21* (2), 1927: pl. 2, fig. 2 (S. Africa); Fowler, Copeia, 165, 1927: 89 (Virginia); Rey, Fauna Iberica Peces, *1*, 1928: 394 (descr., Spain); Fowler, Proc. Acad. nat. Sci. Philad., *80*, 1929: 607 (N. Jersey, size); Truitt, Bean and Fowler, Bull. Md. Conserv. Dept., 3, 1929: 30 (Chesapeake Bay); Uriarte and Mateu, Notas Inst. esp. Oceanogr., 53, 1931: 10 (Canary Is., size); Wilson, Bull. U.S. nat. Mus., 158, 1932: 464, 469 (parasites, Woods Hole); Giltay, Mem. Mus. Roy., Hors Ser., *5*, Fasc. 3, 1933: 8 (class.); Young and Mazet, Shark, Shark, 1933: 132, 271 (commercial catch, near Nantucket); Bigelow and Schroeder, Canad. Atlant. Fauna, biol. Bd. Canad., 12e, 1934: 12 (descr., distrib.); Budker, Bull. Mus. Hist. nat. Paris, (2) 7, 1935: 184 (Dakar, W. Africa); Fowler, Bull. Am. Mus. nat. Hist., *70* (1), 1936: 29 (W. Africa); Wise, Nat. Hist. N. Y., *28*, 1936: 322 (photo, Bahamas); Fowler, Proc. Acad. nat. Sci. Philad., *89*, 1937: 303 (N. Jersey); Wise, Tigers of the Sea, 1937: 178 (general); Tortonese, Atti Soc. ital. Sci. nat., 77, 1938: 288 (Medit.); Fowler, Arch. Zool. Estado São Paulo, *4*, 1942: 127 (Brazil); Monogr. Acad. nat. Sci. Philad., 7, 1945: 262 (Florida); Bigelow and Schroeder, Guide Comm. Shark Fish., Anglo Amer. Carib. Comm., Wash., 1945: 100, fig. 34 (descr., ill., habits, range).

Squalus americanus Mitchill, Trans. Lit. Phil. Soc. N. Y., 1815: 483 (descr., N. York), not *S. americanus* Gmelin, 1789; DeKay, Zool. N. Y., 4, 1842: 366 (descr., N. York); Gill, Proc. Acad. nat. Sci. Philad., Addend., 1861: 60 (named).

Squalus littoralis Lesueur, J. Acad. nat. Sci. Philad., *1*, 1817: 224 (descr., N. York); Mitchill, Amer. Month. Mag. Crit. Rev., *2*, 1817: 328 (N. York, large size).

Squalus macrodous Mitchill, Amer. Month. Mag. Crit. Rev., *2*, 1817: 328 (substitution for his earlier *S. americanus*).

Carcharias littoralis Bory de St. Vincent, Dict. Class Hist. Nat., *15*, 1829: 597 (ref.); DeKay, Zool. N. Y., 4, 1842: 351 (descr., N. York); Storer, Mem. Amer. Acad. Arts Sci., *2*, 1846: 503; DeKay, Rep. State Cab. Nat. Hist. N. Y. (1855), 8, 1858: 64 (in N. York list); Gill, Proc. Acad. nat. Sci. Philad., Addend., 1861: 60 (name) and also subsequent eds.; Jordan, Rep. U.S. Comm. Fish. (1885), 1887: 797 (name, distrib.); Moore, Bull. U.S. Fish Comm. (1892), *12*, 1894: 358 (N. Jersey); Linton, Bull. U.S. Fish Comm. (1893), *13*, 1894: 103 (parasites); Jordan and Evermann, Bull. U.S. nat. Mus., 47 (1), 1896: 46 (descr., distrib.); Linton, Proc. U.S. nat. Mus., *20*, 1897: 445, 450 (Woods Hole, parasites); Bean, T. H., Bull. Amer. Mus. nat. Hist., *9*, 1897: 329, 371 (near N. York); Smith, Bull. U.S. Fish Comm., *17*, 1898: 89 (Woods Hole region); Jordan and Evermann, Bull. U.S. nat. Mus., 47 (3), 1898: 2747 (N. Carolina, large size); Parker and Davis, Proc. Boston Soc. nat. Hist., *29*, 1899: 165, pl. 1, fig. 1; pl. 2, fig. 4; pl. 3, fig. 7, 10 (heart veins); Linton, Bull. U.S. Bur. Fish., *19*, 1901: 273, 428 (Woods Hole, food, parasites); Bean, Rep. For. Comm. N. Y., 1901: 379 (abundance, season, near N. York); Gregg, Where to Catch Fish, 1902: 17 (named); Bean, T. H., Bull. N. Y. St. Mus., *60*, Zool. 9, 1903: 34 (general); Daniel, Elasmobranch Fishes, Univ. Calif. Press, 1934: 178 (veins); Sharp and Fowler, Proc. Acad. nat. Sci. Philad., *56*, 1904: 505, 506 (Nantucket); Wilson, Proc. U.S. nat. Mus., *28*, 1905: 573 (parasites); Fowler, Rep. N. J. Mus. (1905), 1906: 52 (N. Jersey); Tracy, Rep. R. I. Comm. inl. Fish., 1906: 46 (Rhode Island); Fowler, Rep. N. J. Mus. (1906), 1907: 254 (New Jersey); Wilson, Proc. U.S. nat. Mus., *33*, 1907: 414, 423, 430, 431 (Woods Hole, parasites); Smith, Bull. N. C. geol. econ. Surv., *2*, 1907: 37 (C. Lookout); Field, Rep. U.S. Comm. Fish. (1906), 1907: 16 (food, Woods Hole); Sullivan, Bull. U.S. Bur. Fish., *27*, 1907: 10 (digest. syst.); Fowler, Proc. Acad. nat. Sci. Philad., *60*, 1908: 54 (Nantucket and N. Jersey); Kendall, Occ. Pap. Boston Soc. nat. Hist., 7 (8), 1908: 6 (New England records); Fowler, Proc. Acad. nat. Sci. Philad., *61*, 1909: 407 (C. May); Fowler, Rep. N. J. Mus. (1908), 1909: 352 (N. Jersey, food); Tracy, Rep. R. I. Comm. inl. Fish., 1910: 60 (Rhode Island); Linton, Bull. U.S. Bur. Fish., *28* (2), 1910: 1200 (parasites); Sumner, Osburn and Cole, Bull. U.S. Bur. Fish., *31* (2), 1913: 736 (Woods Hole, food); Gudger, Proc. biol. Soc. Wash., *26*, 1913: 98 (N. Carolina, sizes); Nichols, Abstr. Linn. Soc. N. Y., 20–23, 1913: 91 (season near N. York); Fowler, Copeia, 13, 1914: not numbered (N. Jersey); Coles, Proc. biol. Soc. Wash., *28*, 1915: 91 (C. Lookout, N. Carolina; eggs, embryos, food); De Nyse, Bull. N. Y. zool. Soc., *19*, 1916: 1425 (in aquarium); Latham, Copeia, 41, 1917: 17 (off N. York); Wilson, Proc. U.S. nat. Mus., *53*, 1917: 69 (N. Jersey, parasites); Latham, Copeia, 99, 1921: 72 (near N. York); Huntsman, Contr. Canad. Biol. (1921), 3, 1922: 56 (Passamaquoddy Bay); Latham, Copeia, 118, 1923: 61 (Long Island, N. York); Nichols and Breder, Zoologica, N. Y., *9*, 1927:

18 (N. York region); Breder, Field Bk. Mar. Fish. Atlant. Coast, 1929: 20 (general); Wilson, Proc. U.S. nat. Mus., *60* (5), 1922: 23, 24, 67 (Woods Hole, parasites); Breder, Copeia, 1932: 31 (Block Island); Giltay, Mém. Mus. Hist. nat. Belg., Hors Ser., *5* (3), 1933: 8 (class.); Burton, Sci. Mon. N. Y., *40*, 1935: 283 (in abundance near Charleston, S. Carolina); Smith, H. W., Biol. Rev., *11*, 1936: 52 (blood serum); Breder, Copeia, 1942: 42 (shedding of teeth).

Triglochis taurus Müller and Henle, Arch. Naturg., (3) *1*, 1837: 396 (name); S. B. Akad. Wiss. Berlin, 1837: 114 (not seen); Gill, Ann. N. Y. Lyc., 7, 1862: 398 (name); Canestrini, in Cornelia, *et al.*, Fauna d'Ital., Pesci, 3, 1872: 44 (Medit.).

Odontaspis taurus Müller and Henle, Plagiost., 1841: 73 (Medit., Atlant., C. Good Hope); Bonaparte, Cat. Pesc. Europ., 1846: 17 (Medit.); Gray, List Fish. Brit. Mus., 1851: 63 (Cape seas); Duméril, Hist. Nat. Poiss., *1*, 1865: 417, pl. 7, fig. 3 (Medit., descr., teeth); Gervais and Boulart, Poiss., *3*, 1877: 187 (Medit., Atlant.); Doderlein, Prosp. Metod. Pesci Sicil., 1878–1879: 30 (Sicily); Giglioli, Elenc. Pesc. Ital., 1880: 52 (Medit., not seen); Doderlein, Man. Ittiol. Medit., *2*, 1881: 58 (Medit.); Moreau, Hist. Nat. Poiss. France, *1*, 1881: 291 (Algiers, descr.); Rochebrune, Act. Soc. linn. Bordeaux, (4) *6*, 1882: 46; Faune Senegamb., Poiss., *1*, 1883–1885: 23 (Senegambia); Carus, Prod. Fauna Medit., *2*, 1889–1893: 506 (Medit.); Carrucio, Boll. Soc. Zool. Ital., *71*, 1910: 267–272 (Medit., not seen); Roule, Bull. Inst. océanogr. Monaco, 243, 1912: 9 (St. Lucia I., C. Verde); Metzelaar, Trop. Atlant. Visschen, 1919: 189 (trop. W. Africa); Jordan, Copeia, 140, 1925: 20 (name); Copeia, 166, 1928: 4 (name); Belloc, Rev. des Trav. Pêches Marit., 7 (2), 1934: 135 (Morocco, Senegal, Canaries); Norman and Fraser, Giant Fishes, 1937: 8 (general); Fowler, Bull. U.S. nat. Mus., 100 (*13*), 1941: 120 (descr., synon. in part).

Carcharias griseus Ayers, Boston J. Nat. Hist., *4*, 1843: 288, pl. 12, fig. 4 (Long Island, N. York, descr.); Linsley, Amer. J. Sci., *47*, 1844: 76 (same specimen as Ayers, 1843); Storer, Mem. Amer. Acad. Arts Sci., *2*, 1846: 504; Proc. Boston Soc. nat. Hist., *2*, 1848: 256 (teeth); Mem. Amer. Acad. Arts Sci., *9*, 1867: 219, pl. 36, fig. 1; also Fishes Mass., 1867: 241, pl. 36, fig. 1 (descr., Massachusetts).

Odontaspis griseus Desor, Proc. Boston Soc. nat. Hist., *2*, 1848: 264 (teeth).

Carcharias ferox Guichenot, Explor. Alger. Poiss., 1850: 124; not *Squalus ferox* Risso, 1810 (Algiers).[27]

Odontaspis americanus Abbott, Proc. Acad. nat. Sci. Philad., 1861: 400 (N. Jersey, size); Duméril, Hist. Nat. Poiss., *1*, 1865: 419 (descr.); Abbott, in Cook, Geol. N. J., 1868: 828 (N. Jersey); Günther, Cat. Fish. Brit. Mus., *8*, 1870: 392 (descr., synonyms in part); Ribero, Arch. Mus. nac. Rio de J., *14*, 1907: 159, 202 (south. Brazil); Ribeiro, Fauna brasil. Peixes, *2* (1), Fasc. 1, 1923: 15 (south. Brazil).

Eugomphodus griseus Gill, Proc. Acad. nat. Sci. Philad., Addend., 1861: 60 (name).

Eugomphodus littoralis Gill, Proc. Acad. nat. Sci. Philad., 1863: 333 (name); Proc. Acad. nat. Sci. Philad., 1864: 260 (name, syn.); Verrill and Smith, Rep. U.S. Comm. Fish. (1871–1872), 1873: 521 (Woods Hole, food); Gill, Rep. U.S. Comm. Fish. (1871–1872), 1873: 813; Baird, Rep. U.S. Comm. Fish. (1871–1872), 1873: 827 (Woods Hole); Bean, Proc. U.S. nat. Mus., *3*, 1881: 115 (east coast, U.S.); Wilson, Proc. U.S. nat. Mus., *31*, 1907: 711 (Woods Hole, parasites); Proc. U.S. nat. Mus., *60* (5), 1932: 67 (Woods Hole, parasites).

Carcharias americanus Jordan and Gilbert, Bull. U.S. Nat. Mus., 16, 1883: 27 (east coast, U.S.).

Odontaspis littoralis Jordan and Gilbert, Bull. U.S. nat. Mus., 16, 1883: 874 (name); Proc. U.S. nat. Mus., *5*, 1883: 581 (S. Carolina); Stevenson, Trans. Vassar Bros. Inst., *2*, 1884: 92, 1 pl. (Nantucket); Rathbun, Proc. U.S. nat. Mus., 7, 1885: 490 (Vineyard Sound, parasites); Nelson, Rep. N. J. geol. Surv. (1888), 1890: 662 (N. Jersey record); Jordan, Guide to Study Fish., *1*, 1905: 534 (general); Man. Vert. Anim. NE. U.S., 1929: 12 (general); Jordan, Evermann and Clark, Rep. U.S. Comm. Fish. (1928), 2, 1930: 19 (distrib.); Norris, Plagiost. Hypophysis, 1941: pl. 18, fig. 69 (brain); Lunz, Bull. S. C. St. Planning Bd., 14, 1944: 27 (S. Carolina, Florida).

Doubtful References:

Odontaspis taurus Rochebrune, Act. Soc. linn. Bordeaux, (4) *6*, 1882: 46 (Senegambia; identity doubtful because color, as described, does not agree with that of *taurus*).

27. Not seen; according to Duméril (Hist. Nat. Poiss., *1*, 1865: 418) this specimen, preserved in Paris, was *taurus*, not *ferox*.

Odontaspis americanus Ihering, Rev. Mus. paul., *2*, 1897: 35 (listed, Rio Grande do Sul, Brazil); Thompson, Mar. biol. Rep. Cape Town, *2*, 1914: 146 (S. Africa, not seen); Lahille, Physis. B. Aires, *5*, 1921: 63; also Enum. Peces Cartilag. Argent., 1921: 15 (Argentina); Devincenzi, An. Mus. Montevideo, (2) *2*, 1926: 202 (size, weight, no descr., Uruguay).

Family SCAPANORHYNCHIDAE

Goblin Sharks

Characters. Two dorsal fins, the 1st very much shorter than caudal, the rear end of its base anterior to origin of pelvics; caudal slightly more or less than ⅓ of total length, its axis raised only very slightly, its lower anterior corner not expanded as a distinct lobe; caudal peduncle not greatly depressed or expanded laterally; sides of trunk without longitudinal dermal ridges; jaws greatly protrusible (much more so than in any other sharks) and widely expansible; snout greatly elongate; 5th gill opening over or anterior to origin of pectoral; gill arches without rakers and not interconnected by a sieve of modified denticles; nostrils entirely separate from mouth, their anterior margins without barbels; spiracles present; lower eyelid without nictitating membrane or subocular fold; teeth similar in the two jaws, with thorn-like cusps, smooth-edged, with or without lateral denticles, on broad bases; skull of normal shape (*i.e.*, not widely expanded laterally); rostral cartilages 3, united anteriorly as a long rod; radials of pectoral mostly borne on mesopterygium and on metapterygium; meso- and metapterygia not separated by a foramen; heart valves in 3 rows. Development not known, but probably ovoviviparous.

Genera. Only one genus, *Scapanorhynchus* Woodward, 1889, is known.

Range. Modern representatives of *Scapanorhynchus*[1] are known from Japan, the coast of Portugal, and perhaps from Australia.[2]

Fossil remains of the genus, mostly from the Cretaceous, have been found at many localities in Europe, North and South America, Asia, Africa and New Zealand.

Family ISURIDAE

Mackerel Sharks, Man-eater Sharks

Characters. Two dorsal fins, the 1st much shorter at base than length of caudal, the rear end of its base far in advance of origin of pelvics; 2nd dorsal and anal much smaller than 1st dorsal; caudal less than ⅓ of total length, lunate in form, its axis steeply raised; caudal peduncle strongly depressed dorso-ventrally and widely expanded laterally, forming a prominent keel on each side, extending well out on the caudal, with a less definite longitudinal keel close below it on the anterior part of caudal in some species; sides of trunk, anterior to anal, without longitudinal dermal ridges; upper and lower precaudal pits well developed; inner margins of pelvics entirely separate, posterior to cloaca; snout not very elongate and jaws not greatly protrusible; 5th gill opening in front of origin of

1. How many species these represent still remains unsettled.
2. According to Whitley (Fish. Aust., *1*, 1940: 136), the report of this Shark from Murray River (Zietz, Trans. roy. Soc. S. Aust., *32*, 1908: 291) is open to doubt.

pectoral; gill arches without rakers and not interconnected by a sieve of modified denticles; nostrils entirely separate from mouth, without barbels; spiracles present or absent; lower eyelid without nictitating fold or membrane; both jaws with labial furrows at corners; teeth large, few in number, awl- or blade-like, with 1 cusp; head of normal shape (not widely expanded); rostral cartilages 3, united at tip; metapterygium of pectoral with about 3 times as many radials as mesopterygium, but the latter nearly as large as former; meta- and mesopterygia not separated by a foramen; heart valves in 3 rows. Development ovoviviparous.

Genera. One of the members of this family, set apart from all the others by its triangular, serrate teeth, has long been considered as representing a well marked genus, *Carcharodon.* The remaining isurids fall in two groups: (A) Very stout-bodied; first dorsal originating over or anterior to inner corner of pectoral when latter is laid back; first two teeth in each jaw similar in shape to subsequent teeth; caudal fin (so far as known) with a secondary longitudinal keel on either side below the primary keel formed by the lateral expansion of the caudal peduncle. (B) More slender-bodied; first dorsal originating definitely posterior to inner corner of pectoral; first two teeth in each jaw noticeably more slender and more flexuous than the others; without secondary keels. It seems reasonable to accept the difference between the two groups, and especially the presence or absence of the secondary caudal keels, as sufficiently important for generic separation. This course is followed here. Fortunately there has been no need to coin a new generic name in either case.

Key to Genera

1a. Upper teeth broadly triangular with serrate edges.
Carcharodon Agassiz, L., 1838, p. 133.

1b. Upper teeth slender, with smooth-edged cusps.

- 2a. First 2 teeth in each jaw similar in shape to the succeeding teeth; most or all of teeth with lateral denticles in most species, and perhaps in all (lateral denticles in young specimens may be so small as to be difficult to recognize; they may even be lacking on some of the teeth); origin of 1st dorsal about over or anterior to inner corner of pectoral when latter is laid back; anterior part of caudal fin with a secondary caudal keel on either side below the primary keel formed by the lateral expansion of the caudal peduncle. *Lamna* Cuvier, 1817, p. 111.
- 2b. First 2 teeth in each jaw noticeably more slender and more flexuous than the others; no lateral denticles on any of the teeth; origin of 1st dorsal definitely posterior to inner corner of pectoral when latter is laid back; caudal fin without secondary keels, with only the primary keels formed by the lateral expansion of the caudal peduncle. *Isurus* Rafinesque, 1810, p. 123.

Genus *Lamna* Cuvier, 1817

Lamna Cuvier, Règne Anim., *2*, 1817; 126, 127; type species, *Squalus cornubicus* Gmelin, 1789, equivalent to *Squalus nasus* Bonnaterre, 1788.

Generic Synonyms:

Lamia Risso, Hist. Nat. Europe merid., *3*, 1826: 123; type species, *Squalus cornubicus* Gmelin, 1789; not *Lamia* Fabricius, 1775, for Coleoptera.

Selanonius Fleming, Hist. Brit. Anim., 1828: 169; type species, *Selanonius walkeri* Fleming, same as *Squalus nasus* Bonnaterre, 1788.

Exoles Gistel, Naturg. Tier., 1848: 9; to replace *Lamia* Risso, 1826; preoccupied.

Generic Characters. Teeth slender, awl-shaped, smooth-edged, with lateral basal denticles in most cases and perhaps in all, the first 2 teeth in each jaw similar in shape to those succeeding, the anterior ones with two widely divergent roots, the third upper tooth much smaller than second or fourth, but third lower tooth about same size as fourth; origin of 1st dorsal over or anterior to inner corner of pectoral when latter is laid back; trunk robust (Fig. 15); snout conical, pointed; caudal pits in the form of transverse furrows; a less distinct secondary longitudinal keel, broadly triangular in cross section, on anterior part of caudal on each side, close below the primary keel formed by the expanded caudal peduncle, in all species so far known; upper jaw very slightly protrusible. Characters otherwise those of the family.

Range. Widespread in boreal to warm temperate belts of the oceans in both hemispheres; not known from tropical seas.

Species. The genus *Lamna* is represented in the North Atlantic by the well known Porbeagle (*L. nasus*, p. 112); in the North Pacific by a form that has usually been considered identical with *nasus*, but which has recently been found to be a distinct and well marked species (*L. ditropis* Hubbs and Follett, 1947);[4] in Australian–New Zealand waters[5] and off Argentina[6] by close relatives whose precise relationships to *nasus* remain to be determined; and in the eastern side of the South Pacific by a form (*L. philippii* Perez Canto, 1886)[7] resembling *nasus* in general appearance and in the position of the first dorsal fin, but described and pictured as lacking lateral denticles on the teeth. Until it is known whether this is actually the case, and whether or not *philippii* has the secondary caudal keels (none are shown on the only published illustration of it), its status must remain problematical.

4. Hubbs and Follett, Copeia, 1947: 194.
5. L. *whitleyi* Phillipps, N. Z. J. Sci. Tech., *16*, 1935: 239, fig. 3; secondary caudal keels clearly shown in the photograph.
6. Reported as *nasus* by Lahille (An. Mus. nac. B. Aires, *34*, 1928: 310, pl. 4); teeth described as with denticles; secondary caudal keels clearly shown on the illustration.
7. Philippi, Anal. Univ. Chile, *71*, 1887: 549, pl. 3, fig. 2.

Key to Species of the Northern Hemisphere[8]

1a. Distance from tip of snout to anterior edge of eye at least ½ as great as from posterior edge of eye to 1st gill opening, each measurement taken between perpendiculars; lower surface plain-colored, without dark blotches.
nasus (Bonnaterre), 1788, p. 112.

1b. Distance from tip of snout to anterior edge of eye less than ⅓ as great as from posterior edge of eye to 1st gill opening; lower surface conspicuously marked with dark blotches, except perhaps in very young specimens. *ditropis* Hubbs and Follett, 1947.
Warm Temperate to Subboreal Belt, Both Sides of North Pacific.

Lamna nasus (Bonnaterre), 1788

Mackerel Shark, Porbeagle; Blue Shark (in Gulf of Maine)

Figures 15, 16, 17

Study Material. Male, 935 mm., from Nahant, Mass. (Harv. Mus. Comp. Zool., No. 209); specimens of 660, 963 and 966 mm., from New England, North Atlantic and Continental Slope off Sable I., Lat. 42° 37′ N., Long. 60° 55′ W. (U.S. Nat. Mus., No. 47528, 24288, 44057); head, from 70 miles SE. of Cape May (U.S. Nat. Mus., No. 125884); two embryos, no doubt of this species, about 180 mm. long, from Barnstable, Mass., taken in October 1942 (Harv. Mus. Comp. Zool., No. 35901); also eight specimens fresh caught in Gulf of Maine (not preserved); jaws from several of same; photograph of 180-mm. embryo from female caught off Portland, Maine, at "Mistaken Ground" in January 1927 by Capt. D. C. Train.

Distinctive Characters. Easily separable from the Sharp-nosed Mackerel Shark by its teeth (cf. Fig. 16 D with 19); from *Carcharodon* by the teeth (cf. Fig. 16 D with 20 B) and by the relative positions of the second dorsal and anal fins.

Description. Proportional dimensions in per cent of total lengths. Male, 935 mm., from Nahant, Mass. (Harv. Mus. Comp. Zool., No. 209).

Trunk at origin of pectoral: breadth 15.0; height 14.5.
Snout length in front of: outer nostrils 6.0; mouth 7.1.
Eye: horizontal diameter 2.4.
Mouth: breadth 8.1; height 5.5.
Nostrils: distance between inner ends 3.6.
Labial furrow length: upper 2.2; lower 1.3.
Gill opening lengths: 1st 7.5; 2nd 7.1; 3rd 6.5; 4th 6.5; 5th 6.5.
First dorsal fin: vertical height 11.0; length of base 9.1.
Second dorsal fin: vertical height 2.7; length of base 1.7.
Anal fin: vertical height 2.7; length of base 1.6.

8. It would be premature to expand the Key to the southern hemisphere for the reasons stated above.

Caudal fin: upper margin 24.2; lower anterior margin 15.4.
Pectoral fin: outer margin 17.7; inner margin 5.6; distal margin 14.3.
Distance from snout to: 1st dorsal 33.4; 2nd dorsal 66.3; upper caudal 76.0; pectoral 27.8; pelvics 50.7; anal 67.0.
Interspace between: 1st and 2nd dorsals 24.3; 2nd dorsal and caudal 9.2; anal and caudal 9.2.
Distance from origin to origin of: pectoral and pelvics 26.3; pelvics and anal 16.4.

Trunk fusiform, much stouter than in *Isurus oxyrinchus*, its height opposite origin of 1st dorsal (where highest) about equal to distance from eye to 4th gill slit, or about 18% of total length, tapering to a very slender caudal peduncle. Sides smooth, lateral line not apparent. Caudal peduncle very strongly flattened dorsoventrally, widely expanded laterally and sharp-edged, with a less distinct longitudinal keel, broadly triangular in cross section, on anterior part of caudal close below the rearward extension of the expanded peduncle; this is more obvious in large specimens than in small; upper and lower precaudal

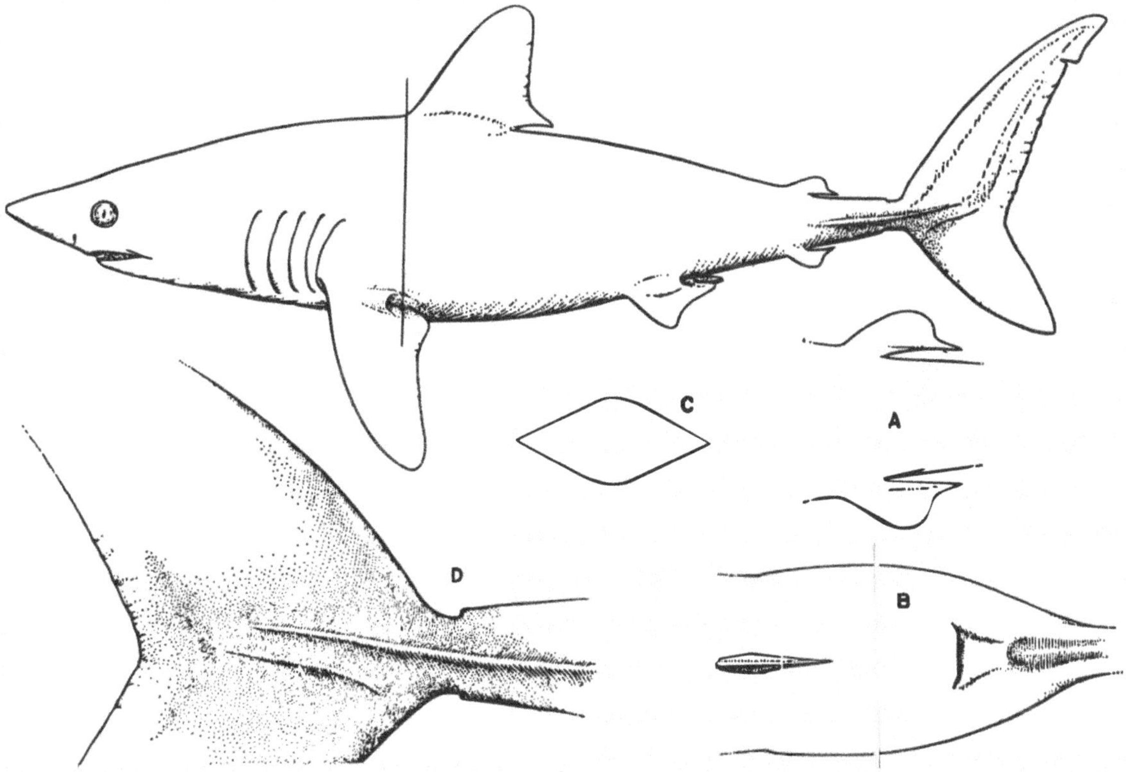

FIGURE 15. *Lamna nasus*, young male, 935 mm. long, from Nahant, Massachusetts (Harv. Mus. Comp. Zool., No. 209). *A* Second dorsal and anal fins, about 0.4 x. *B* Caudal peduncle viewed from above, about 0.4 x. *C* Cross section of caudal peduncle at region indicated by transverse line in *B*. *D* Caudal peduncle and base of tail, from the photograph of a fresh Gulf of Maine specimen about four feet long, to show secondary caudal keel.

pits both strongly developed as transverse furrows, straight or only slightly curved, with convexities directed rearward, the lower pit a little in advance of the upper. Dermal denticles so small and flat that the skin is velvety to the touch, each a little broader than long, with 3 teeth, the median a little longer than others, and a corresponding number of low, sharp-topped ridges, separated by broad valleys; pedicels moderately long, on broad bases.

Head and snout conical, the length of head to pectoral a little less than ⅕ of total length. Snout pointed, its length in front of mouth about ¼ of head to origin of pectorals. Eye circular, its diameter about 30% as long as snout in front of mouth. Spiracles lacking in specimens studied but described as sometimes present as minute pores behind eyes. First gill opening slightly longest, about as long as snout in front of eye, the lower end of 5th curving rearward and ventrad for a short distance around origin of pectoral, the space between 4th and 5th only about ½ that between 1st and 2nd, the 5th more oblique than others. Nostrils approximately transverse, hardly ⅓ as long as distance between them, their inner corners about ¼ as far from mouth as from tip of snout, the anterior margin with a low rounded lobe (Fig. 16 B). Mouth broadly rounded, about 1½ times as broad as high (thus somewhat shorter, relatively, than in *oxyrinchus*). Upper labial furrow about 0.3 as long, the lower 0.2 as long, as distance to symphysis of the respective jaws, the upper partly, and the lower almost entirely, concealed when mouth is closed.

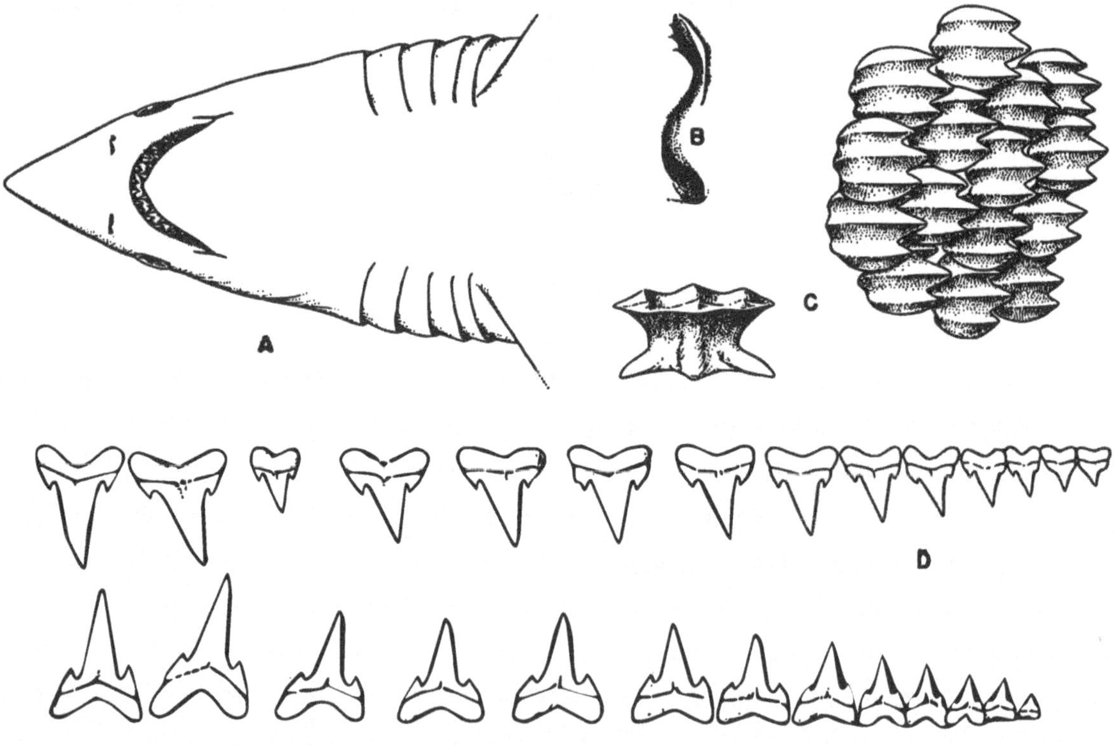

FIGURE 16. *Lamna nasus. A* Lower view of head of young male from specimen shown in Fig. 15. *B* Right-hand nostril, about 2.5 x. *C* Dermal denticles—general view, about 40 x; apical view, about 80 x. *D* Upper and lower teeth of a larger specimen from Platts Bank, Gulf of Maine, about 1.3 x.

Teeth alike in the two jaws, $\frac{24\text{ to }32}{14\text{ to }20}$ in specimens counted; no median tooth in either jaw; all teeth except those next to corners of jaw slender, narrow, straight, sharp-pointed, with a small sharp basal denticle on either side in adults, which, however, may be lacking on some or all of teeth in young specimens, or at least so small that their detection is difficult; 1st and 2nd teeth in each jaw largest, 3rd upper tooth much smaller than 2nd or 4th and sometimes lacking, but 3rd lower tooth about equal to 4th; 4th to 8th or 9th about equal in size in each jaw, but 10th and subsequent teeth progressively smaller; lateral teeth in both jaws, as well as anterior teeth in lower jaw, erect, but 1st, 2nd and sometimes 3rd upper teeth directed sharply inward; 1, or rarely 2, rows functional along sides of mouth, but 2, or rarely 3, rows near the center.

Origin of 1st dorsal over or very slightly posterior to axil of pectoral (thus relatively much farther forward than in *oxyrinchus*), its anterior margin slightly convex, its apex broadly rounded, its rear margin straight toward apex but moderately concave toward base, its free rear corner about ⅓ as long as its base, its vertical height nearly equal to distance from eye to 1st gill or about 60% as great as length of pectoral. Second dorsal about ¼ as high as 1st, its origin over origin of anal, its apex broadly rounded, rear margin deeply concave, its free rear corner about 1½ times as long as base, but only moderately slender. Lower lobe of caudal about 64 to 75% as long as upper, relatively shorter in young than in older specimens (about 68% in Fig. 15), each measured from the respective precaudal pit (thus somewhat shorter, relatively, than in *oxyrinchus*), the subterminal notch strongly marked, the posterior outline subangular, with rounded corner. Anal slightly larger than 2nd dorsal, similar in shape. Pelvics with rounded corners and moderately concave outer margins, their origins posterior to rear tip of 1st dorsal by a distance about ⅔ to ¾ as long as from tip of snout to mouth. Pectoral nearly or quite as long as from posterior margin of eye to 5th gill opening, about ½ as broad as long, the anterior margin moderately convex, the tip and inner corner rounded, the distal margin only moderately concave (less so than in *oxyrinchus*).

Color. Dark bluish-gray above, changing abruptly on the lower sides to the white of the lower surface; pectorals dusky on outer half or third, the anal white or slightly dusky.

Size. While *nasus* has repeatedly been reported to reach a length of 12 feet, a 10-foot female from Monhegan, Maine,[9] is the largest of which we find a definite record. However, at least one other of eight feet has been positively reported from the Gulf of Maine, and a number from seven to nine feet (up to 2,800 mm.) at different times from northern European waters. Very few, however, of those caught in the western side of the Atlantic are more than six feet long, with four to five feet perhaps the commonest size. For example, none of those that we have hooked has been longer than five feet, apart from one of perhaps eight feet hooked and lost over Cashe's Ledge on September 30, 1927. At the other extreme the smallest on record is 29 inches. Information as to the relationship be-

9. Hubbs, Copeia, 123, 1923: 101.

tween length and weight in *nasus* is scant. Reported weights of about 400 pounds at nine to ten feet, and 305 pounds at eight feet three inches would suggest that this is a much lighter fish than *oxyrinchus*. But since the stoutness of its trunk suggests rather the reverse, it seems more probable that the few reported weighings have been of fish that had been gutted, which, in the case of a shark with so large a liver, means the loss of a large part of the total weight. Females may contain embryos at a length of five feet.

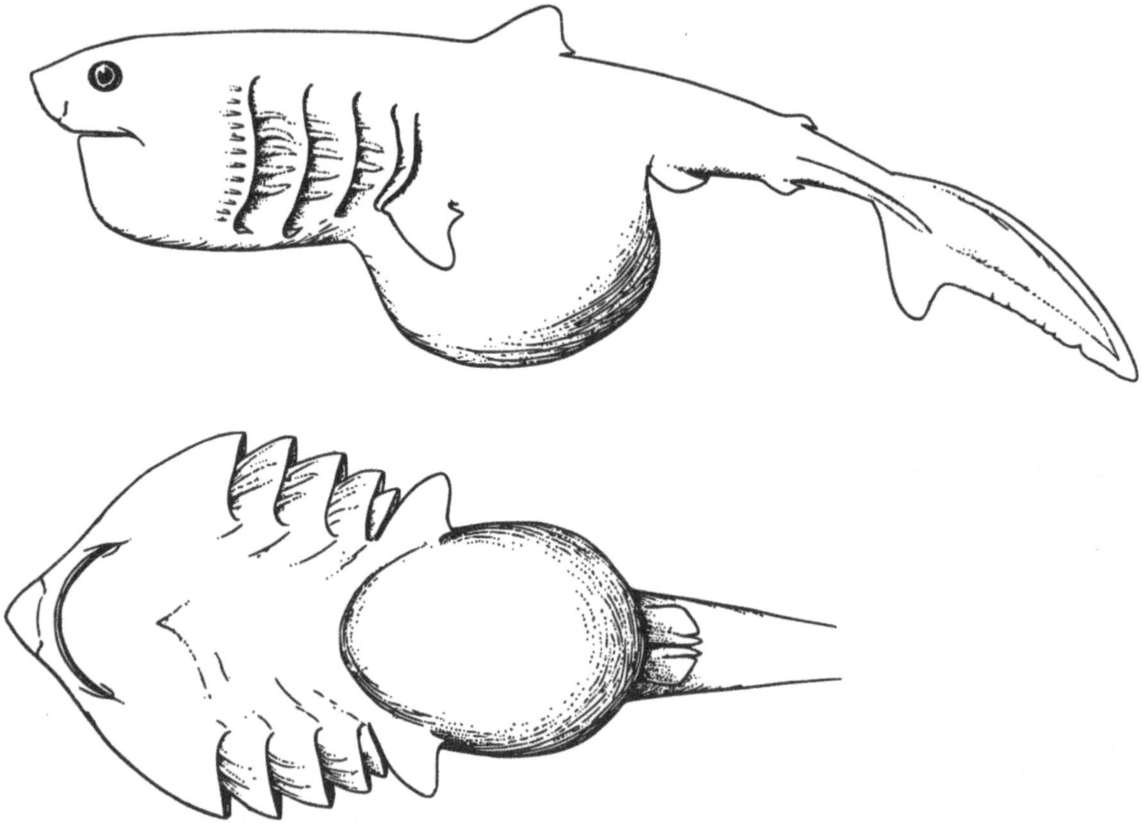

FIGURE 17. *Lamna nasus*, embryo, about 180 mm. long, from Barnstable, Massachusetts (Harv. Mus. Comp. Zool., No. 35901), about 0.8 natural size.

Developmental Stages. It has long been known that this is an ovoviviparous species, the young lying free in the uterus without connection with the mother. It also seems established that in *nasus*, contrary to the rule among most other ovoviviparous sharks, the yolk sac is absorbed and the umbilical cord entirely obliterated while the embryo is still very small (55 to 60 mm.) and still with well developed external gills. Thereafter, the embryo nourishes itself by swallowing the unfertilized eggs which lie close to it in the uterus, the result being that its stomach becomes enormously swollen by the masses of yolk so swallowed, forming a so-called "yolk stomach";[10] as the embryo grows this increases in relative

10. For more detailed accounts see Swenander (Zool. Stud. Tullberg, Uppsala, 1907: 283, pl. 1); also, on the North Pacific form, see Lohberger (Abh. bayer. Akad. Wiss., Suppl., 4, Abt. 2, 1910).

size from about 45 mm. in length (in one of 180 mm., Fig. 17) to about 235 mm. (in one of 400 mm.[11]) or to more than half the total length. The throat region of the embryo too is enormously expanded, giving it a most grotesque appearance. Also, the caudal fin is at first much more asymmetrical than in the adult, assuming the lunate form with growth of the embryo, and the young are very large at birth, witness embryos of 19, 24 and 18 inches in a five-foot female.[12] A Gulf of Maine female of 10 feet contained a 20-pound embryo.[13] Corresponding to the large size of the embryo, gravid females normally contain only one to four young (0–2 per oviduct), although five have been reported.[14]

Habits. This has been described repeatedly as an active, strong-swimming species when in pursuit of its prey. When hooked, however, it puts up only a very feeble resistance, as we have experienced. We have never seen or heard of one jumping in its attempt to escape, as the Mako does (p. 128). Nor is there any difficulty in landing specimens of four to five feet on an ordinary hand-line; in fact, it is as proverbial for its sluggishness under such circumstances as is the Mako for its activity.

Mackerel Sharks are often seen finning at the surface on calm days; on the other hand, many have been caught on bottom with cod and halibut lines as well as at mid-depths now and then in drift nets in northern European waters, while one is occasionally entangled in a mackerel net. Evidently, then, their depth range is from the surface down to bottom; on the cod fishing grounds that would be to some 70 to 80 fathoms at least; it is not known how much deeper they descend.

In the waters of northern Europe gravid females have been taken from localities so widely scattered as to show that the species produces young throughout its East Atlantic range. Presumably this is true in the western Atlantic also, although embryos have actually been recorded only from the vicinity of Monhegan Island, Maine, in August, from off Portland, Maine, in November and in January (see p. 119) and from Barnstable, Massachusetts, in October (see Study Material, p. 112). In Europe, females with embryos have been reported for the winter months as well as for summer. But the fact that the largest embryos have been found in summer indicates the latter as the chief season of production.

Lamna nasus preys largely on schools of mackerel, herring and (in the eastern Atlantic) pilchards; also on such ground fish as cod, hake, cusk, and other gadoids, flounders, or any other fish that may be available, and on squid. In the eastern Atlantic its diet also includes whiting (*Gadus merlangus*), spiny dogfish (*Squalus acanthias*) and john dory (*Zeus faber*). It also has the troublesome custom of foraging on cod, etc., that have been hooked on long lines, biting off the snoods in the process.

Relation to Man. During the first quarter of the last century the liver oil of this species, mixed with other fish oils, was in considerable demand (chiefly for tanning pur-

11. Nordgård, K. norske Vidensk-Selsk. Skr. Trondh. (1923–1924), 1925: 38, fig. 22.
12. Shann, Rep. Fish. Bd. Scot., 1911: 73, pl. 9.
13. Hubbs, Copeia, 123, 1923: 101.
14. Lübbert and Ehrenbaum, Handb. Seefisch. Nordeurop., *2*, 1936: 278. A shark with 10 embryos reported long ago as this species (Wilder, Science, *1*, 1880: 236) probably was some other.

poses). Provincetown was the center of activity. However, this local demand for shark-liver oil had almost entirely died before 1850. Of late years the only commercial importance of *nasus* in the western Atlantic (except as a nuisance to fishermen) has been its salability in the larger fish markets, for its range does not extend southward far enough to bring it within the scope of the shark-leather industry that is now operating there. In northern Europe, on the west coast of France, as well as in the Mediterranean, the flesh of *nasus* is in much greater demand; this is especially so in Germany, where the local supply was regularly augmented by considerable imports from Norway[15] before the war. The Norwegian catch has been made chiefly on long lines, the German catches chiefly in herring trawls. The few that are caught in American waters are taken incidentally either on hand-lines when fishing for cod, etc., or in mackerel nets. It is not game enough to be of interest to sport-anglers.

Range. Continental waters of the northern North Atlantic, from the Mediterranean and northwestern Africa to the North Sea, Scotland, Orkneys, and southern Scandinavia, on the eastern side; less common north to Iceland, northern Norway and the Murman Coast; from the Newfoundland Banks and Gulf of St. Lawrence in the west south to New Jersey, and perhaps to South Carolina.[16] It is represented in the North Pacific from northern California to southern Alaska, Kamchatka and Japan, as well as in the Australian–New Zealand region,[17] by forms very closely allied, but not identical.

Occurrence in the Western Atlantic. The area of regular occurrence for *nasus* is confined to a much narrower latitudinal belt in the West than in the East, *i.e.*, from southern New England to the outer coast of Nova Scotia and the Gulf of St. Lawrence, with the chief center of population lying in the western side of the Gulf of Maine. Thus, while there are but two records of it from the Newfoundland Banks,[18] and one, except for vague reports, from the Gulf of St. Lawrence, fishermen report it as the commonest large shark in summer along the Atlantic coast of Nova Scotia, including Cape Breton. Apparently it tends to shun the cold waters of the Bay of Fundy region, there being but one positive record for it in Passamaquoddy Bay. Farther west, however, in the Gulf of Maine, it is so numerous on occasion that there is record of incidental catches of 19 in one night by six men working hand-lines, with about 150 taken on cod-lines by a crew of fishermen on a three weeks' trip near Monhegan Island, Maine. During the cruises of the United States Bureau of Fisheries vessels we have seen and caught them most often in the immediate vicinity of Platts Bank off Cape Elizabeth. It is certainly the most often seen of the larger sharks around the Isles of Shoals and near Cape Anne, while in Massachusetts Bay it has

15. Exact amounts are not available, because the landings of all species of sharks are combined in the published statistics.

16. A shark reported under the name *Lamia nasus* from Argentina by Lahille (An. Mus. nac. B. Aires, *34*, 1928: 310) appears actually to have been a Mako (see p. 130).

17. See p. 111.

18. This includes one of 7 feet 10½ inches, taken June 28, 1946, in Lat. 44° 27′ N., Long. 50° 00′ W., reported by Dr. A. M. Ramalho of Lisbon, who sent us one tooth.

been described repeatedly as "common."[19] We have hooked or sighted on an average about one per three or four days on the cod fishing grounds, generally in the western side of the Gulf of Maine and on Nantucket Shoals during the summers of 1924 to 1930.[20] However, the fact that such large numbers have been caught in the past within brief periods (see above) is sufficient evidence that their numbers vary widely from year to year, or over a period of years, at least locally.

To the westward *nasus* is described (we have no first-hand information) as comparatively common in the vicinity of Woods Hole (more so in autumn) and it has been reported on several occasions from Rhode Island coastal waters. However, it appears only as a stray along Long Island, New York (one record), or along the New Jersey coast; the only evidence of its presence farther south is one somewhat doubtful report of it off Charleston, South Carolina. From this it appears that the isothermal belt of about 65° F. limits its normal range to the southward.

It seems equally certain that its on-and-off-shore range is similarly narrow, for we find no record of it (nor report of it by fishermen), from the offshore fishing banks abreast of the Gulf of Maine (Georges and Browns Banks); only one is reported from the Nova Scotian slope (see Study Material, p. 112), and two from the Grand Banks. On the other hand, few venture close enough to land to be picked up in the pound nets. There is, however, record of a Mackerel Shark, probably this species, which was entangled in the eel grass (*Zostera*) in Barnstable Harbor, Massachusetts, many years ago.[21] In the western Atlantic all published records of it, and those that we have observed, have been for the warm half of the year, but its presence in the Gulf of Maine in winter is proved by our receipt of a photograph of an embryo, certainly of this species, from a female caught off Portland, Maine, in January of 1927. Similarly, it is taken in winter as well as in summer off northern Europe, but less commonly. This, together with the absence of any evidence of migration southward along the middle Atlantic coast of the United States, suggests that in winter they simply descend into deeper water to avoid low surface temperatures, apparently feeding little then, otherwise more of them would be picked up by the winter fishery for hake.

Synonyms and References:

Porbeagle, Borlase, Nat. Hist. Cornwall, 1758: 265, pl. 26, fig. 4 (Cornwall); Pennant, Brit. Zool., *3*, 1769: 92 (descr., Cornwall); also later eds.

Squalus glaucus Gunnerus, K. norske Vidensk.-Selsk. Skr. Trondh., 1768: 1, pl. 1 (descr., embryo, Nor-

19. Actually, no sharks, other than the Spiny Dogfish, are ever common off the northeastern coast of the United States or Canada, in the sense in which that term can be applied to such fish as the cod, mackerel, etc., but only as relative to other sharks of corresponding sizes.

20. Cod-tagging cruises of the U.S. Bureau of Fisheries.

21. Goode, Fish. Fish. Industr. U.S., 1884: 670, footnote.

way, not seen); Skr. Norg. Gesellsch. Wiss. Kbh. and Leipzig, 1770: *1*, pl. 1, fig. 1–5 (translation of foregoing); Ascanius, Icon. Rerum Natur. Kbh., 1777: pl. 31; Strøm, K. norske Vidensk.-Selsk. Skr., N.S. *2*, 1788: 340, pl. 2, figs. (Norway, not *Squalus glaucus* Linnaeus, 1758).

Beaumaris Shark, Pennant, Brit. Zool., *3*, 1776: 104, pl. 17 (descr., Wales).

Le Nez, Broussonet, Mem. Math. Phys. Acad. Sci. Paris, 1780: 667; Sonnini, Hist. Nat. Poiss., *4*, 1801–1802: 5 (descr., Cornwall).

Touille-boeuf ou Loutre de Mer, Duhamel, Traité Gén. Pêches, *4* (2), sect. 9, cap. 5, 1782: 298, pl. 20, fig. 4 (general).

Squalus nasus Bonnaterre, Tabl. Encyc. Meth. Ichthyol., 1788: 10, pl. 85, fig. 350 (descr., type local., Wales, by ref. to Pennant, 1776).

Squalus cornubicus Gmelin, in Linnaeus, Syst. Nat., *3*, 1789: 1497 (descr., Cornwall); Goodenough, Trans. Linn. Soc. Lond., Zool., *3*, 1797: 80, pl. 15 (English Channel); Bloch and Schneider, Syst. Ichthyol., 1801: 132; Sonnini, Hist. Nat. Poiss., *4*, 1801–1802: 5 (descr.); Bosc, Nouv. Dict. Hist. Nat., *21*, 1803: 185 (diagn.); Latreille, Nouv. Dict. Hist. Nat., *24*, 1804: 72 (in table of contents); Shaw, Gen. Zool., *5* (2), 1804: 350 (general); Donovan, Nat. Hist. Brit. Fish., *5*, 1807: pl. 108 (ill.); Turton, Brit. Fauna, 1807: 113 (color); Risso, Ichthyol. Nice, 1810: 29 (Medit.); Neill, Mem. Werner. Soc. Edinb., *1*, 1811: 549 (Scotland); Couch, Trans. Linn. Soc. Lond., Zool., *14*, 1825: 91 (Cornwall); Nilsson, Prod. Ichthyol. Scand., 1832: 116 (mention); Jenyns, Manual Brit. Vert. Anim., 1835: 500 (Gt. Brit., size); L. Agassiz, Poiss. Foss., *3*, Atlas, 1835: pl. G, fig. 3 (teeth); Wright, Fries and Ekstrom, Skand. Fisk., *5*, *6*, 1838–1840: 135, pl. 30 (descr., Scandinavia); de la Blanchère, Dict. Pêches, 1868: 747 (descr.); Gatcombe, Zoologist, (3) *5*, 1881: 425 (English Channel).

Squalus pennanti Walbaum, P. Artedi Genera Pisc. Emend. Ichthyol., *3*, 1792: 517 (descr., Atlant.).

Squale Long-nez, Lacépède, Hist. Nat. Poiss., 4° ed., *1*, 1798: 216, pl. 2, fig. 3 (descr.).

Beaumaris, Sonnini, Hist. Nat. Poiss., *4*, 1801–1802: 8 (descr., Gt. Brit.).

Squalus monensis Shaw, Gen. Zool., *5* (2), 1804: 350 (general); Pennant, Brit. Zool., *3*, 1812: 154, pl. 20 (descr., Wales); Cuvier, Règne Anim., *2*, 1817: 127 (footnote).

Squalus cornubiensis Pennant, Brit. Zool., *3*, 1812: 152 (descr., Cornwall).

Squalus selanonus Leach, Mem. Werner. Soc. Edinb., *2*, 1814: 64, pl. 2, fig. 2 (Scotland).

Carcharhinus cornubicus Blainville, Bull. Soc. philom. Paris, 1816: 121 (name).

Lamie nez., Cloquet, Dict. Sci. Nat., Atlas Poiss., 1816–1830: pl. 28.

Lamna cornubica Cuvier, Règne Anim., *2*, 1817: 127; Cloquet, Dict. Sci. Nat., *25*, 1822: 183 (general); Fleming, Hist. Brit. Anim., 1828: 168; Ekstrom and Sundevall, Skand. Fauna, 1834: 67, pl. 30 (descr., Scandinavia, not seen); Bonaparte, Icon. Faun. Ital., *3* (2), 1835: plate not numbered (descr., Medit.); Yarrell, Brit. Fish., *2*, 1836: 384 (descr., Gt. Brit.); Krøyer, Danmarks Fisk., *3*, 1st ed., 1838: 852 (around Denmark); Parnell, Mem. Werner. Soc. Edinb., *7*, 1838: 413 (Gt. Brit.); L. Agassiz, Poiss. Foss., *3*, 1838: 86 (ref. to *3*, pl. A, should read pl. G, fig. 3, "Lamna"); Bonaparte, Mém. Soc. neuchâtel Sci. Nat., *2* (8), 1839: 9 (in synopsis); Swainson, Nat. Hist. Fish. Amphib. Rept., *2*, 1839: 314 (general); Owen, Odontogr., 1840–1845: 28 (teeth); Müller and Henle, Plagiost., 1841: 67 (descr.); Yarrell, Brit. Fish., 2nd ed., 1841: 515 (descr., Gt. Brit.); Mayer, Neue Notiz. Natur. u. Heilk. (Froriep) Weimar, *21* (446), 1842: 81; Agassiz, L., Poiss. Foss., *3*, 1835–1843: 287, pl. G, fig. 3 (teeth); Hamilton, Brit. Fish., *2*, 1843: 309 (general); also subsequent eds.; Bonaparte, Cat. Pesc. Europ., 1846: 18 (Medit., N. Europ.); Thompson, Zoologist, *8*, 1850: 2970 (English Channel); Baikie, Zoologist, *11*, 1853: 3846 (Orkneys); Krøyer, Danmarks Fisk., 2nd ed., *3*, 1852–1853: 852 (Denmark); Owen, Cat. Osteol. Roy. Coll. Surg., 1853: 95 (cranium, vertebrae); Nilsson, Skand. Fauna Fisk., *4*, 1855: 718 (Scandinavia); Thompson, Nat. Hist. Ireland, *4*, 1856: 251 (Ireland); Machado, Peces Cadiz, 1857: 9 (Spain); Jouan, Mem. Soc. Imp. Sci. Nat. Cherbourg, 7, 1860: 139 (English Channel); Nardo, Atti Ist. veneto, (3) *5*, 1859–1860: 787 (Medit.); Fitzinger, Bild. Atlas Naturg. Fische, 1864: fig. 172 (ill.); Perugia, Cat. Pesc. Adriat., 1866: 7 (Adriatic, not seen); Duméril, Hist. Nat. Poiss., *1*, 1865: 405 (descr.); Bocage and Brito Capello, Poiss. Plagiost., 1866: 12 (Portugal); Couch, Fish. Brit. Isles, *1*, 1867: 41, pl. 8 (descr., Gt. Brit.); Brito Capello, J. Sci. math. phys. nat. Lisboa, *2* (7), 1869: 139 (Portugal); Günther, Cat. Fish. Brit. Mus., *8*, 1870: 389 (descr., synon., distrib.); Van Beneden, Poiss. Côtes Belg. Parasit., 1870: 8 (parasites, Belgium); Ninni, An. Soc. Nat. Modena, *5*, 1870: 66 (Medit.); Canestrini, in Cornalia, *et al.*,

Fauna d'Ital., 1872: 45 (Medit.); Gulliver, Quart. J. Micr. Sci., N. S. *12*, 1872: 40 (size); Collett, Norges Fisk., 1875: 208 (Norway); Turner, J. Anat., Lond., *9*, 1875: 301 (spiracle); Gervais and Boulart, Poiss., *3*, 1876: 180, pl. 68 (descr.); Clement, Bull. Soc. Sci. nat. Nîmes, *4*, 1876: 109 (not seen); Reguis, Ess. Hist. Nat. Provence, (1) *1*, Fasc. 1, 1877:48 (Medit.); Doderlein, Prosp. Metod. Pesci Sicil., 1878: 30 (Medit.); Goode and Bean, Bull. Essex Inst. Salem, *11*, 1879: 29 (Gulf of Maine); Campbell, Sci. Gossip, *16*, 1880: 277 (Scotland, not seen); Brito Capello, Cat. Peixes Port., 1880: 44 (Portugal); Stossich, Boll. Soc. adriat. Sci. nat. Trieste, *5*, 1880: 67 (Adriatic); Day, Fish. Gt. Brit., *2*, 1880–1884: 297, pl. 156 (descr., Gt. Brit.); Buckland, Hist. Brit. Fish., 1881: 217 (descr.); Moreau, Hist. Nat. Poiss. France, *1*, 1881: 296 (descr., distrib.); Doderlein, Man. Ittiol. Medit., *2*, 1881: 60 (Medit.);[23] Hasse, Naturl. Syst. Elasm. besond. Theil., 1882: 215, pl. 28, fig. 1–6 (vertebrae); Jordan and Gilbert, Bull. U.S. nat. Mus., *16*, 1883: 30 (descr., N. Atlant., Woods Hole); Goode, Fish. Fish. Industr. U.S., 1, 1884: 670, pl. 249 (ill., habits, fishery); Storm, K. norske Vidensk-Selsk. Skr. Trondh. (1883), 1884: 45 (Trondhjem, Norway); Pilliet, Bull. Soc. zool. Fr., *10*, 1885: 283 (anat.); Rathbun, Proc. U.S. nat. Mus., 7, 1885: 488 (Vineyard Sound); Ehlers, Nachr. Ges. Wiss. Göttingen, *18*, 1886: 547 (Friesland, size); Whiteaves, Cat. Canad. Pinnep. Cetacea Fish., 1886: 5 (Gulf of St. Lawrence); Jordan, Rep. U.S. Comm. Fish. (1885), 1887: 797 (distrib.); Calderwood, 6th Ann. Rep. Fish. Bd. Scot., (3), 1888: 263 (embryo, Scotland); Bean, T. H., Bull. U.S. Fish Comm., 7, 1889: 256 (off N. Jersey); Carus, Prod. Faun. Medit., 1889–1893: 505 (Medit.); Ewart, J. Anat., Lond., *24*, 1890: 227 (spiracle); Bellotti, Atti Soc. ital. Sci. nat., *33*, 1891: 110 (Atlant. coast, France); Weber, Tijdschr. ned. dierk. Ver., (2) *4*, 1893: IV (Holland); Duerden, Irish Nat., *3*, 1894: 264 (Ireland); Dean, Fishes, living and extinct, 1895: 90 (general); Jordan and Evermann, Rep. U.S. Comm. Fish. (1895), 1896: 218 (distrib.); Bull. U.S. nat. Mus., 47 (1), 1896: 49 (descr., distrib.); Kermode, Zoologist, (4) *1*, 1897: 597 (Isle of Man); Knipovich, Annu. Mus. zool. Acad. St. Pétersb., *2*, 1897: 154 (Murman coast); Jungersen, Danish "Ingolf" Exped., *2* (1), 1899: 37, pl. 2, fig. 22, 23 (claspers); Yung, Arch. Zool. exp. gén., (3) 7, 1899: 140 (physiol.); Bassett-Smith, Proc. zool. Soc. Lond., 1899: 464, 469, 479 (parasites); Jordan and Evermann, Bull. U.S. nat. Mus., 47 (4), 1900: pl. 6, fig. 22 (ill.); Warren, Irish Nat., *9*, 1900: 293 (Ireland); Bean, T. H., Rep. For. Comm. N. Y., 1901: 380 (Long Island, N. York); Huber, Z. wiss. Zool., *70*, 1901: 617, pl. 27, fig. 6 (claspers); Prince and Knight, Contr. Canad. Biol., 1901: 55 (fins, New Brunswick); de Braganza, Result. Invest. Sci. "Amelia," *2*, 1904: 54 (Portugal); Bean, Bull. N. Y. St. Mus., 60, Zool., 9, 1903: 39 (New York); Drzewina, Arch. Zool. exper. gén., (4) *3*, 1905: 226, 243 (anat.); Evans, Ann. Scot. nat. Hist., 1905: 56 (Scotland); Borca, Arch. Zool. exp. gén., (4) *4*, 1905: 204, 207 (France); Collett, Medd. Norges Fisk., Suppl., *3*, 1905: 77 (size, food, embryos, season, Norway); Fowler, Rep. N. J. Mus. (1905), 1906: 56 (New Jersey record); Swenander, K. norsk. Vidensk.-Selsk. Skr. Trondh. (1905), 9, 1906: 73; Fowler, Science, N. S. *24*, 1906: 596 (N. Jersey); Rep. N. J. Mus. (1906), 1907: 254 (N. Jersey); Sullivan, Bull. U.S. Bur. Fish., *27*, 1907: 10, 15, 23 (anat.); Swenander, Zool. Stud. Tullberg, Uppsala, 1907: 283, pl. 1 (embryo); Wilson, Proc. U.S. nat. Mus., *33*, 1907: 386, 410, 423 (parasites, Woods Hole region); Winther, Hansen and Jensen, in Schiødte and Hansen, Zool. Danica, *2*, 1907: 296, pl. 27, fig. 2 (descr., distrib.); Kendall, Occ. Pap. Boston Soc. nat. Hist., 7 (8), 1908: 7 (off New England); Fowler, Proc. Acad. nat. Sci. Philad., *60*, 1908: 55 (Italy); Tracy, Rep. R. I. Comm. inl. Fish., 1910: 60 (Rhode Island); Günther, Encycl. Brit., 11th ed., *24*, 1911: 807 (general); Shann, Rep. Fish. Bd. Scot., *3*, 1911: 73, pl. 9 (embryos, Scotland); Ariola, Atti Soc. ligust. Sci. nat. geogr., 1913: 3 (Medit., not seen); Sumner, Osburn and Cole, Bull. U.S. Bur. Fish. (1911), *31* (2), 1913: 736 (Woods Hole region); Halkett, Check List Fish. Canad., 1913: 40 (Canada); Le Danois, Ann. Inst. océanogr. Monaco, *5* (5), 1913: 18 (English Channel); Leigh-Sharpe, J. Morph., 35, 1921: 367 (claspers); Shann, Proc. zool. Soc. Lond., 1923: 161 (embryos); Burne, Philos. Trans., (B) *212*, 1923: 209 (blood system); Jenkins, Fish. Brit. Isles, 1925: 311 (Gt. Brit.); Elmhirst, Scot. Nat., *161*, 1926: 151 (Scotland); Ehrenbaum, in Grimpe and Wagler, Tierwelt N.- u. Ostsee, Lief 7 (12e), 1927: 14 (general); Nordgaard, K. norske Vidensk.-Selsk. Skr. Trondh. (1925–1926), 1, 1927: 38 (embryo); Ehrenbaum, Rapp. Cons. Explor. Mer., Faune Ichthyol. N. Atlant., 1929: plate not numbered (descr., ill., distrib.); Nordgaard, K. norske

23. See Doderlein, 1881, for additional references for the Mediterranean in publications not accessible to us.

Vidensk.-Selsk. Skr. Forh., Trondh., *1*, 1929: 206 (embryos); Wilson, Proc. U.S. nat. Mus., 60 (5), 1932: 23, 59 (parasites); Belloc, Rev. des Trav. Pêches Marit., 7, Fasc. 2, 1934: (Morocco to Senegal); Borri, Mem. Soc. tosc. Sci. nat., *44*, 1934: 92 (Medit.); Daniel, Elasmobranch Fishes, Univ. of Calif. Press, 1934: 9 (teeth, body form); Nobre, Fauna Marinha Port., Vert., *1*, 1935: 430, pl. 58, fig. 182 (descr., Portugal); Norman and Fraser, Giant Fishes, 1937: 12 (general).

Lamia cornubica Risso, Hist. Nat. Europ. Merid., *3*, 1826: 124 (Medit.); Bory de St. Vincent, Dict. Class. Hist. nat., *15*, 1829: 596.

Selanonius walkeri Fleming, Hist. Brit. Anim., 1828: 169 (Gt. Brit.).

Lamia monensis Bory de St. Vincent, Dict. Class. Hist. nat., *15*, 1829: 596.

Squalus (*Carcharias*) *cornubicus* Blainville, in Vieillot, Faune Franç., 1825: 96, pl. 14,[24] fig. 2 (descr.).

Lamna monensis Yarrell, Brit. Fish., *2*, 1836: 387 (Wales); Bowerbank, Zoologist, (2) *8*, 1873: 3617 (English Channel, size); Zoologist, (2) *9*, 1874: 4202 (English Channel).

Lamna punctata Storer, Mass. Zool. Bot. Surv., Rep. on Fish., 1839: 185, pl. 3, fig. 2 (ill., Mass. Bay); Boston J. Nat. Hist., *2*, 1839: 534, pl. 8, fig. 2 (descr., size, habits, utilization, Massachusetts); Linsley, Amer. J. Sci., *47*, 1844: 76 (Conn.); Storer, Mem. Amer. Acad. Arts Sci., N. S. *2*, 1846: 504 (part, *L. nasus* and *I. oxyrinchus* combined; refs. also incl. *Squalus punctatus* Mitchill, 1815, which was perhaps *Scoliodon terrae-novae*); Mem. Amer. Acad. Arts Sci., N. S. *9* (1), 1867: 225, pl. 37, fig. 1; also in Fishes Mass., 1867: 249, pl. 37, fig. 1 (descr., ill., Massachusetts); Smith, S. I., Rep. U.S. Comm. Fish. (1871–1872), 1873: 576 (parasites, Woods Hole region); Norman and Fraser, Giant Fishes, 1937: 12 (east coast, U.S.).

Squalus (*Lamna*) *cornubicus* Cuvier, Règne Anim., Poiss., 1849: 362, pl. 114, fig. 3 (general).

Lamna pennanti Desvaux, Essai Ichthyol. France, 1851: 23 (not seen).

Isurus cornubicus Gray, List. Fish. Brit. Mus., Chondropt., *1*, 1851: 58; Smitt, Hist. Scand. Fish., 2nd ed., *2*, 1895: 1138, pl. 51, fig. 1 (descr., habits, Scandinavia); McIntosh, Ann. Mag. nat. Hist., (7) *10*, 1902: 254 (food, Scotland); Rey, Fauna Iberica Peces, *1*, 1928: 383 (descr.); Lübbert and Ehrenbaum, Handb. Seefisch. Nordeurop., *2*, 1936: 278 (general, North Sea).

Oxyrhina daekayi Gill, Proc. Acad. nat. Sci. Philad., Addend., 1861: 60.

Isuropsis dekayi Gill, Rep. U.S. Comm. Fish. (1871–72), 1873: 813 (part); Baird, Rep. U.S. Comm. Fish. (1871–1872), 1873: 827 (Woods Hole).

Oxyrhina punctata Duméril, Hist. Nat. Poiss., *1*, 1865: 409 (descr., N. Amer.).

Isurus punctatus Garman, Bull. Mus. comp. Zool. Harv., *17*, 1888: 79, pl. 1 (lateral line syst.); Mem. Harv. Mus. comp. Zool., *36*, 1913: 36, pl. 6, fig. 4–6, pl. 56, fig. 5, pl. 58, fig. 3, pl. 62, 63 (descr., anat., east coast U.S.); Nichols and Murphy, Brooklyn Mus. Sci. Bull., *3* (1), 1916: 23 (near N. York); Fowler, Copeia, 30, 1916: 36 (mid-Atlant., U.S.); Smith, J. Amer. Mus. nat. Hist., *16*, 1916: 342 (Atlant. coast U.S.); Bigelow and Welsh, Bull. U.S. Bur. Fish., *40* (1), 1925: 36, 551 (habits, distrib., Gulf of Maine); Jordan, Evermann and Clark, Rep. U.S. Comm. Fish. (1928), 2, 1930: 19 (distrib.); Burton, Sci. Mon. N. Y., *40*, 1935: 283 (off Charlestown, S. Carolina, but ident. not certain); Anonymous, Newfoundland Fish. Res. Lab. (1934), *2* (3), 1935: 79 (Grand Banks); Springer, Proc. Fla. Acad. Sci., *3*, 1939: 36 (no Florida record given).

Isurus dekayi Smith, Bull. U.S. Fish Comm., *27*, 1898: 89 (Woods Hole region, ident. probable because of local.); Tracy, Rep. R. I. Comm. inl. Fish., 1906: 46 (Narragansett Bay, ident. probable because of local.); Kendall, Occ. Pap. Boston Soc. nat. Hist., 7 (8), 1908: 7 (New England, ident. probable because of loc.); Tracy, Rep. R. I. Comm. inl. Fish., 1910: 60 (Rhode Island, ident. probable because of loc.); Sumner, Osburn and Cole, Bull. U.S. Bur. Fish., *31* (2), 1913: 736 (Woods Hole region).

Lamna (*Oxyrhina*) *cornubica* Imms, Proc. zool. Soc. Lond., (1), 1905: 44 (pharyngeal denticles).

Isurus nasus Garman, Mem. Harv. Mus. comp. Zool., *36*, 1913: 34 (descr., Gt. Brit., Medit., W. Atlant., Japan); Fowler, Copeia, 30, 1916: 36 (mid-Atlant. U.S.); Smith, J. Amer. Mus. nat. Hist., *16*, 1916: 342 (England); Fowler, Proc. biol. Soc. Wash., *33*, 1920: 143 (off N. Jersey); Huntsman, Contr. Canad. Biol. (1921), *3*, 1922: 56 (Passamaquoddy Bay); Bigelow and Welsh, Bull. U.S. Bur. Fish., *40* (1), 1925: 37 (European), 551 (*punctatus*, identical with *nasus*); Bigelow and Schroeder, Bull. Mus. comp.

24. The reference on p. 96 is to pl. "24"—actually the plate is No. 14.

Zool. Harv., *68*, 1927: 239 (discus., meas., Gulf of Maine); Breder, Field Bk. Mar. Fish. Atlant. Coast, 1929: 21 (general); Schroeder, Bull. Boston Soc. nat. Hist., *58*, 1931: 3 (Gulf of Maine); Bigelow and Schroeder, Canad. Atlant. Fauna, biol. Bd. Canad., *12e*, 1934: 13 (descr., distrib.); Vladykov and McKenzie, Proc. N. S. Inst. Sci., *19*, 1935: 47 (Nova Scotia); Bigelow and Schroeder, Bull. U.S. Bur. Fish., *48*, 1936: 322 (Grand Banks of Newfoundland); Fowler, Bull. Amer. Mus. nat. Hist., *70* (1), 1936: 32 (Italy).

Lamna nasus Jordan, Copeia, 49, 1917: 87 (name); Hubbs, Copeia, 123, 1923: 101 (embryo, Gulf of Maine); Jordan, Manual Vert. Anim. NE. U.S., 1929: 12 (general); Jordan, Evermann and Clark, Rep. U.S. Comm. Fish. (1928), 2, 1930: 19 (distrib.); Piers, Proc. N. S. Inst. Sci., *18* (3), 1934: 202 (Nova Scotia).

Lamna sp.(?) Gudger, Sci. Mon., N. Y., *34*, 1932: fig. 412, 413 (teeth).

Isurus (*Lamna*) *nasus* Tortonese, Atti Soc. ital. Sci. nat., 77, 1938: 289 (small size, Medit.).

Doubtful Reference:

Lamna cornubica Lowe, Trans. zool. Soc. Lond., *2* (3), 1837: 194 (Madeira).[25]

Not *Lamna punctata* DeKay, Zool. N. Y., 4, 1842: pl. 63, fig. 206, 207 (these ill. were evidently for *I. oxyrinchus*, see p. 131).

Not *Lamna punctata* Wilder, Science, *1*, 1880: 236 (Wilder's report of 10 embryos, in this case, makes it likely that the reference was actually to some genus other than *Isurus*).

Not *Lamia nasus* Lahille, An. Mus. nac. B. Aires, *34*, 1928: 310 (the specimen in question seems in all probability to have been *Isurus oxyrinchus*, see p. 130).

Not *Lamia nasus* Pozzi and Bordale, An. Soc. cient. argent., *120*, 1935: 150 (this reference is evidently to the same specimen reported by Lahille, see above; no doubt *I. oxyrinchus*).

Genus *Isurus* Rafinesque, 1810

Isurus Rafinesque, Carratt. Gen. Nuov. Sicil., 1810: 11; type species, *I. oxyrinchus* Rafinesque, Sicily.

Generic Synonyms:

Oxyrhina L. Agassiz, Poiss. Foss.,[26] Feuill. Addit., 1835: 71, 86; type species, "*Lamna oxyrhina* Cuvier and Valenciennes Mss.," equals *Isurus oxyrinchus* Rafinesque, 1810.

Oxyrrhina Bonaparte, Cat. Pesc. Europ., 1846: 17; type species, *Oxyrhina gomphodon* Müller and Henle, 1841, equals *Isurus oxyrinchus* Rafinesque, 1810.

Plectrosoma Gistel, Naturg. Tier., 1848: 10; to replace *Oxyrhina* L. Agassiz, 1835.

Isuropsis Gill, Ann. Lyc. Nat. Hist. N. Y., 1, 1862: 397; type species, *Oxyrhina glaucus* Müller and Henle, 1841.

Generic Characters. Teeth without lateral denticles; the first 2 in each jaw noticeably more slender and more flexuous than the others; origin of 1st dorsal definitely posterior to inner corner of pectoral when latter is laid back; trunk slender (Fig. 18); caudal fin without secondary caudal keels. Characters otherwise as in *Lamna* (p. 111).

Species. This genus includes: the "Sharp-nosed Mackerel Shark" or "Mako" of the Atlantic (*I. oxyrinchus*, p. 124); the closely allied Pacific Mako (*I. glaucus*), with which *bideni*[27] from South Africa and *mako*[28] from New Zealand and Australia appear to be

25. Being nominal only, and from a region where *oxyrinchus* is to be expected, there is no way of knowing to which species this record actually referred.
26. For dates of publication of the separate parts of the "Poissons Fossiles," see Jeannet (Bull. Soc. neuchâtel. Sci. nat., *52*, 1928: 102; *53*, 1929: 197).
27. Phillipps, N. Z. J. Sci. Tech., *13*, 1932: 227.
28. Whitley, Rec. Aust. Mus., *17*, 1929: 101.

identical; also, if the original account be correct, a third Indian Ocean species (*guntheri* Murray, 1884) that differs from *glaucus* in having about twice as many teeth and a much more prominent lateral line.[29]

Key to Species

1a. Teeth about 12–14 on each side on each jaw; lateral line not forming a prominent ridge along the side.

- 2a. Height of 1st dorsal about one-half as great as distance from eye to 4th gill opening and a little greater than length of its base; length of head to origin of pectoral about as great as from axil of pectoral to rear ends of bases of pelvics (or distance from axil of pectoral to origin of pelvics only about as long as from tip of snout to 2nd gill opening). *oxyrinchus* Rafinesque, 1810, p. 124.
- 2b. Height of 1st dorsal only about one-half as great as distance from eye to 2nd gill opening and a little less than length of its base; length of head to origin of pectoral only about as great as distance from axil of pectoral to origin of pelvics (or distance from axil of pectoral to origin of pelvics about as long as from tip of snout to origin of pectorals). *glaucus* Müller and Henle, 1841.[30] Indo-Pacific.

1b. Teeth 22–28 on a side in each jaw; lateral line forming a prominent ridge along the side, rearward to caudal peduncle. *guntheri* Murray, 1884. India.[31]

Isurus oxyrinchus Rafinesque, 1810

Sharp-nosed Mackerel Shark, Mako

Figures 18, 19

Study Material. Skin, with head attached, of adult male, 7 feet 6 inches long, from Bimini, Bahamas (Harv. Mus. Comp. Zool., No. 35367); young male (entire), 1,640 mm., from Ocean City, Maryland (Harv. Mus. Comp. Zool., No. 35899); skin, about 6 feet long, from Miami, Florida (Harv. Mus. Comp. Zool., No. 35366); also female (mounted) from Miami, Florida, about 6 feet 8 inches long; jaws from medium-sized and large specimens taken off New Jersey, New York and Cape Cod (U.S. Nat. Mus., No. 110881, Amer. Mus. Nat. Hist., No. 567 and 9220, Harv. Mus. Comp. Zool., No. 816).

Distinctive Characters. The Mako is separable from the common Mackerel Shark by its teeth and more slender form; from *Carcharodon* by its teeth, its slender form and by the relative position of the second dorsal and anal fins.

29. Only one specimen of this sort has ever been reported (Murray, Ann. Mag. nat. Hist., [5] *13*, 1884: 349).
30. Including *bideni* Phillipps (N. Z. J. Sci. Tech., *13*, 1932: 227, fig. 2, S. Africa); *mako* Whitley (Rec. Aust. Mus., *17*, 1929: 101).
31. Only one specimen of this sort has ever been reported (Murray, Ann. Mag. nat. Hist., [5] *13*, 1884: 349); the validity of this species is very doubtful.

Description. Proportional dimensions in per cent of total length. Male, 1,598 mm., from Ocean City, Maryland (Harv. Mus. Comp. Zool., No. 35899). Male, 2,337 mm., from Bahamas (Harv. Mus. Comp. Zool., No. 35367).

Trunk at origin of pectoral: breadth 10.3, 12.4; height 11.3, 11.9.
Snout length in front of: outer nostrils 4.2, ——; mouth 6.1, 6.5.
Eye: horizontal diameter 1.8, 1.8.
Mouth: breadth 6.2, 6.1; height 5.0, 5.6.
Nostrils: distance between inner ends 3.4, 3.8.
Labial furrow length: upper 1.1, ——; lower 0.6, ——.
Gill opening lengths: 1st 6.9, 6.9; 2nd 6.6, ——; 3rd 6.4, ——; 4th 6.3, ——; 5th 6.8, 7.2.
First dorsal fin: vertical height 9.2, 10.0; length of base 8.8, 9.1.
Second dorsal fin: vertical height 1.6, 2.1; length of base 1.6, 1.1.
Anal fin: vertical height 1.9, 2.1; length of base 1.5, 1.4.
Caudal fin: upper margin 20.6, 21.8; lower anterior margin 15.5, 16.7.
Pectoral fin: outer margin 17.1, ——; inner margin 4.8, ——; distal margin 14.0, ——.
Distance from snout to: 1st dorsal 36.6, 34.8; 2nd dorsal 69.0, 68.5; upper caudal 79.4, 78.2; pectoral 26.1, 25.0; pelvics 53.2, 50.0; anal 70.5, 68.5.

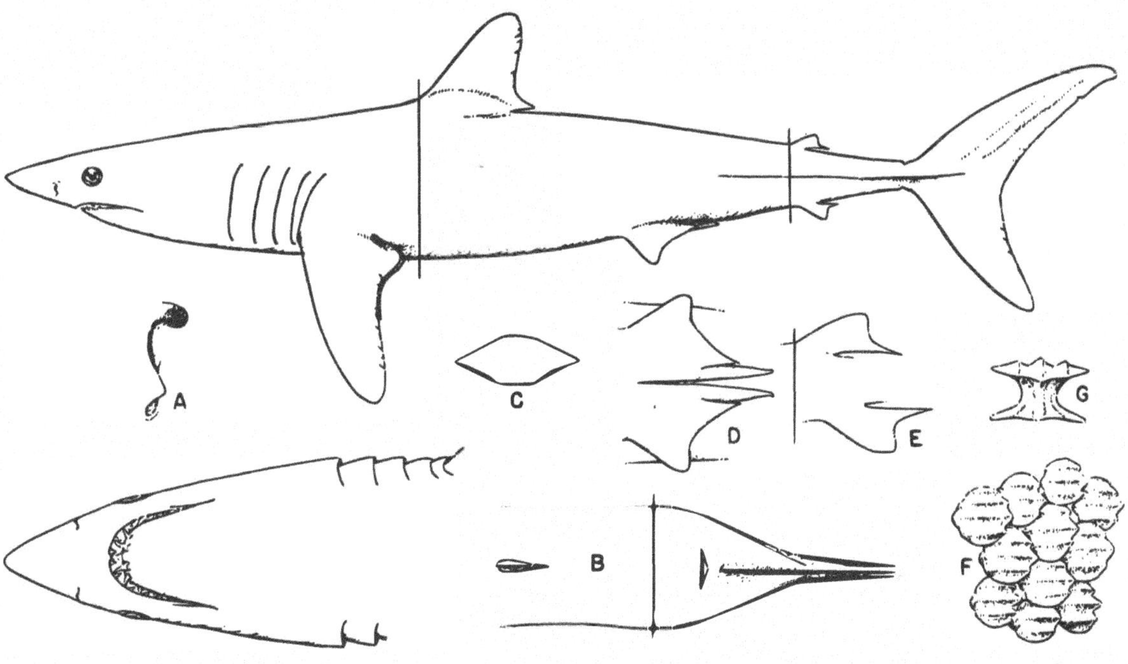

FIGURE 18. *Isurus oxyrinchus*, young male, 1,640 mm. long, taken off Ocean City, Maryland (Harv. Mus. Comp. Zool., No. 35899). *A* Left nostril, about 1.1 x. *B* Caudal peduncle viewed from above, about 1/7 natural size. *C* Cross section of caudal peduncle at point indicated by the transverse line on *B*. *D* Pelvic fins and claspers. *E* Second dorsal and anal. *F* Dermal denticles, about 32 x. *G* Apical view of dermal denticle, about 65 x.

Interspace between: 1st and 2nd dorsals 24.8, 25.5; 2nd dorsal and caudal 8.7, 8.7; anal and caudal 8.0, 7.2.

Distance from origin to origin of: pectoral and pelvics 27.9, 28.3; pelvics and anal 16.9, 17.2.

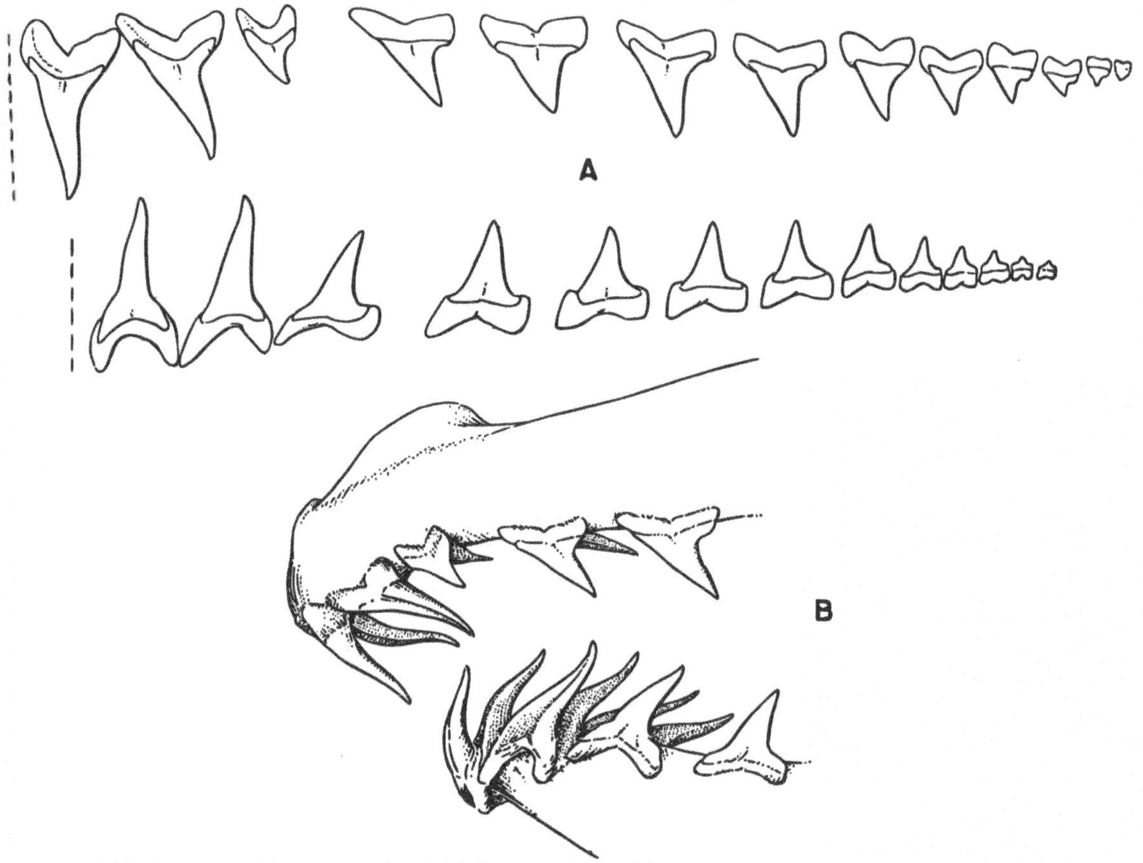

FIGURE 19. *Isurus oxyrinchus. A* Upper and lower teeth of specimen pictured in Fig. 18, about natural size. *B* Side view of anterior part of jaws of a large Cape Cod, Massachusetts, specimen (Harv. Mus. Comp. Zool., No. 816), about ½ x.

Trunk fusiform, considerably more slender than in *Lamna nasus*, its height at origin of 1st dorsal (where highest) about equal to distance from eye to 2nd gill opening, or about 15% of total length, tapering both rearward and forward. Caudal peduncle very much flattened dorso-ventrally, but broadly expanded laterally and sharp-edged as in other Isuridae (Fig. 18 B), but without the secondary keel below it that is characteristic of *nasus*. Sides smooth. Lateral line not prominent. Upper and lower caudal pits strongly developed as deep furrows, nearly transverse to peduncle, or perhaps slightly arcuate (convexity rearward) in some specimens. Denticles small, closely imbricate, with 3 to 5 ridges, and 3 marginal teeth, the median the longest but often worn down.

Head conical but somewhat flattened dorsally. Snout pointed. Eyes round, their diameters about $\frac{1}{3}$ as long as snout in front of mouth. Nostrils nearly transverse, about $\frac{1}{5}$ as long as distance between them, their inner margins without definite lobe, the distance from inner corner of nostril to mouth between 33 and 50% as great as to tip of snout. Spiracle a minute pore or slit, about at same level as upper margin of eye, and situated behind the latter by a distance equal to about 3 times the eye's diameter. Gill openings noticeably large, the 1st to 4th about as long as snout in front of mouth, the 5th slightly longest, the 1st nearly straight, but lower outlines of others increasingly flexuous, that of the 5th most strongly so, the 5th close in front of origin of pectoral and extending ventrally around the latter for a distance about $\frac{1}{2}$ as long as diameter of eye, the distance between 4th and 5th only about $\frac{1}{2}$ as great as between 1st and 2nd. Mouth very broadly rounded in front and notably long, about 1.15 times as broad as long. Upper labial furrow about 25% as long as distance (around the curve) from corner of mouth to symphysis of upper jaw, ending about opposite 7th tooth, the lower furrow slightly more than $\frac{1}{2}$ as long as upper and entirely concealed when mouth is closed, ending opposite 6th tooth.

Teeth $\frac{12 \text{ or } 13 - 12 \text{ or } 13}{12 \text{ or } 13 - 12 \text{ or } 13}$; alike in the 2 jaws; slender, somewhat flexuous in outline, smooth-edged, without lateral denticles; the 1st 2 in each jaw much the largest, the 1st 2 in each jaw recurved at base, but with curve reversed at tips, their outer faces flat but inner faces rounded; subsequent teeth relatively broader and increasingly blade-like, their outer margins varying from very strongly convex to very slightly concave, their inner margins slightly more concave; 3rd upper tooth much smaller than 2nd or 4th to 7th, but 3rd lower tooth about as large as 4th to 6th; 9th to 13th teeth in each jaw successively smaller and with cusps shorter relative to breadth of base, the 2 or 3 outermost minute; 1 or 2 rows functional along sides, but 2 or 3 in front of jaws.

Origin of 1st dorsal about over inner corner of pectoral when latter is laid back, or perhaps slightly behind it in some specimens (thus relatively farther back than in *nasus*), the midpoint of its base slightly nearer to anterior margin of eye than to origin of caudal, its length at base about equal to $\frac{1}{2}$ the distance from posterior margin of eye to 5th gill opening or to about $\frac{1}{2}$ the length of pectoral, its anterior margin slightly convex, its apex moderately rounded, its rear margin rather strongly concave, its free rear corner broadly triangular and about $\frac{1}{4}$ as long as base. Origin of 2nd dorsal slightly but unmistakably anterior to origin of anal, the rear end of its base about over midpoint of base of latter, its length at base only about $\frac{1}{7}$ to $\frac{1}{8}$ as great as that of 1st dorsal, its apex rounded, its posterior margin deeply concave, its free rear corner about $1\frac{1}{2}$ times as long as its base. Lower lobe of caudal slightly more than 75% as long as upper (77 to 80% in 2 specimens studied), each measured from its respective precaudal pit, the upper lobe about as long as from front of mouth to origin of pectoral, or slightly longer than latter, the upper anterior and lower anterior outlines of caudal only slightly convex, the tips subacute, posterior contour deeply and nearly evenly concave, with well marked subterminal notch. Anal similar to 2nd dorsal, but with relatively longer free rear corner (about twice as long as base), and of about the same size. Pelvics originate posterior to rear tip of 1st dorsal by a distance about equal to distance from tip of snout to mouth, or slightly greater, their corners rounded, their outer margins concave. Claspers of male long and slender, reaching about 3/4 of the distance to origin of anal. Pectoral about as long as distance from posterior margin of eye to 5th gill opening, or about twice as long as vertical height of 1st dorsal, a little less than 1/2, or about 45%, as broad as long, the outer margin slightly convex, the tip and inner corner rounded, the rear margin moderately concave.

Color. Described as deep blue-gray above when fresh caught,[32] but appearing cobalt or ultramarine blue in the water; snow-white below; dark slate gray above after preservation, and bluish white to pale dirty gray below, on head and body, and on lower surface of pectoral, with gradual transition from one shade to the other along the middle of the trunk.

Size. While the Mako is said to reach a length of 13 feet (4 m.), the maximum length reported for an actual specimen of this species is only about 12 feet.[33]

The largest West Atlantic specimen of which we find definite record, taken off St. Petersburg, Florida, was about 10 feet 6 inches long; one nearly as large (10 feet, 2 inches) was taken off New York Harbor many years ago. Males are sexually mature, as indicated by the claspers, at perhaps six feet, females perhaps not until somewhat larger. Recorded weights at different lengths are about 135 pounds at 6 feet; 230 pounds at 7 feet 8 inches; about 300 pounds at 8 feet; 1,009 pounds at 10 feet 6 inches. A weight of 700–800 pounds may be expected at about 9 feet, depending on condition. The largest specimen so far caught on rod and reel was one of 786 pounds taken off Bimini, Bahamas, by Ernest Hemingway in 1936. The largest Pacific Mako (*glaucus*) yet taken on rod and reel, by E. White-Wickham off New Zealand, weighed 798 pounds.

Developmental Stages. Embryos, like those of other members of the family (p. 116), are provided with a voluminous yolk stomach, and before birth they reach a very large size relative to that of the mother. Presumably the number of young in a brood is correspondingly small, but no definite information is at hand.[34]

Habits. This is one of the most active and strongest swimming of sharks, famous for its habit of leaping clear of the water under natural conditions and when hooked. It appears to be typically a near-surface fish, often seen swimming on sunny days with the tips of first dorsal and caudal fins above the water. Around the Canary Islands it is often hooked at depths of from five to eight meters, but we have no definite information as to how deep it may descend. Nothing is definitely known of its breeding habits, but presumably it is similar in these to its more familiar relative, *nasus* (p. 117).

Very little is known of its diet other than that it is a fish-eater, preying upon the

32. Shown as dark slaty blue above and grayish white below in colored sketch of a fresh 8-foot 4-inch specimen, by J. Henry Blake, Provincetown, Mass., October 1868.

33. 3.7 mm. calculated from the size of the jaws (Uriarte and Mateu, Notas Inst. esp. Oceanogr., *53*, 1931: 12); specimen from the Canaries.

34. Vaillant (Bull. Soc. philom. Paris, [8] *1*, 1889: 38) reports an embryo of this species from the Mediterranean, 50 cm. long, including caudal, with yolk stomach 23 to 24 cm. long; size of the mother is not known.

schools of scombroids, clupeids, or other small fishes, of which it destroys great quantities. Around Bermuda, for example, it is seen most often when in pursuit of scombroids; off the coast of the United States it has at least the reputation of following schools of mackerel. It also feeds on much larger fish. A 120-pound swordfish (*Xiphias gladius*) nearly intact with sword still attached was found in the stomach of a 730-pound specimen taken near Bimini. Another Mako of about 800 pounds, harpooned off Montauk, Long Island, had been seen attacking a swordfish and was found, when landed, to contain about 150 pounds of its flesh. These instances illustrate its capabilities, and one well known angler described it as the only marine enemy of the broadbill swordfish.[35] But there is no reason to suppose from the nature of its teeth, or from repute, that it attacks sea turtles. Probably, like most other pelagic sharks, it feeds also on squids when opportunity offers itself.

Relation to Man. The flesh is sold in limited quantities, but the chief importance of the species is as a game fish because of its famous habit of leaping when hooked, as mentioned above. In this respect, as well as in the fierceness of its resistance to capture, it falls little or not at all behind its better known relative, *glaucus*, of New Zealand waters.

Range. An oceanic species of the tropical and warm-temperate Atlantic, north and south; it is replaced in the Pacific (including New Zealand and Australian waters) by the closely allied but easily distinguishable *I. glaucus* (p. 124). The fact that the ranges of *glaucus* and *oxyrinchus* appear to be continuous around the Cape of Good Hope, although widely separated off the southern part of South America, lends special interest to the identity of any specimens that may be caught off the Cape.

Occurrence in the Eastern Atlantic. In the eastern side of the Atlantic *oxyrinchus* is known as far south as St. Helena and Ascension and northward to northern France; to northern Scotland and southwestern Norway as a stray. Coastwise, however, its zone of reasonably frequent occurrence appears to extend only from tropical West Africa to the Iberian Peninsula, including the Mediterranean, whence it has been recorded repeatedly as "common" or "abundant" from many localities. It is also known from the Azores, Madeira and the Canaries, where it is said to be one of the commoner sharks, occasionally numerous enough to be a great annoyance to net fishermen. This, together with its long known presence in at least small numbers around Bermuda, shows that it is to be expected anywhere in the middle Atlantic.

Occurrence in the Western Atlantic. For the western Atlantic only a very fragmentary picture of the occurrence of this offshore shark could be derived from the captures reliably reported in scientific literature, since these[36] total not over 20 to 25 distributed as follows: off the tip of Cape Cod; 10 miles N.E. of Nantucket Lightship, Mass.; Long Island, New York; vicinity of New York Harbor; coast of New Jersey; off Cape Hatteras, North Carolina; western and northwestern Florida; east coast of Florida; Santa Rosa Island near

35. See K. Farrington (Field and Stream, 47, February 1943) for the instances mentioned above and for other interesting notes on the Mako.

36. It seems certain that at least most of the nominal records for this species in the Gulf of Maine and for the vicinity of Woods Hole actually referred to *L. nasus* (p. 130).

Pensacola, Florida; Cuba; Gulf of Mexico; Havana, Cuba; Rio de Janeiro, Brazil; Bermuda. A shark taken in 1927 off Mar del Plata, in northern Argentine waters,[37] probably belonged to this species also.

Fortunately, however, there is now available a much more extensive source of information in the published and verbal reports of anglers, since *oxyrinchus* is a favorite game fish. From these it is well known to be tolerably plentiful in the winter on the Bahaman side of the Straits of Florida, where many are caught off Bimini, Cat Key and Nassau, but it is less frequent on the Florida side, although it is a familiar fish there. Also, in the summer considerable numbers journey northward along the continental shelf as far as the offings of Maryland, New Jersey, New York and southern New England, although they rarely, if ever, come close enough inshore to be picked up in the pound nets. Perhaps they never penetrate far into inlets. However, to keep offshore is not an invariable part of its behavior pattern, for on the tropical coast of West Africa it has been reported from sundry estuarine situations. During the past few summers we have heard repeatedly of "Makos" seen jumping, or occasionally hooked, near the tip of Cape Cod. Recently a large one (about 9 feet long) was caught on the southern side of Massachusetts Bay a few miles off Plymouth.[38] It is thus evident that at least scattered individuals enter the southwestern part of the Gulf of Maine, probably in pursuit of the schools of mackerel, but it appears that this is its extreme northerly outpost in inshore waters on this side of the Atlantic. The sundry early reports that ostensibly referred to it farther north in the Gulf of Maine all appear to have been based on its close relative, *nasus* (p. 118). Apparently it rarely if ever occurs in water colder than about 60°. But it would not be astonishing if it were encountered farther north, offshore, in the sweep of the Gulf Stream, although there is as yet no positive record of it either from the Nova Scotian Banks or from the Banks of Newfoundland.

Except for its presence in Bahaman waters, knowledge of it in the southern part of its western Atlantic range is very scant, but the records for western and northwestern Florida and Cuba,[39] together with evidence from recently-received photographs of one from southern Texas (Cameron County), is evidence that it ranges over the Gulf of Mexico generally, and in all probability over the entire Caribbean region. But information as to its occurrence off the South American seaboard is limited to the one positive record for Rio de Janeiro, and one probable record for northern Argentina.[37]

Synonyms and References:

Isurus oxyrinchus Rafinesque, Caratt. Gen. Nuov. Sicil., 1810: 12, pl. 13, fig. 1 (type loc., Sicily); Fowler, Bull. Amer. Mus. nat. Hist., *70* (1), 1936: 33 (West Africa, descr.); Tortonese, Atti Soc. ital. Sci. nat.,

37. Pictured and described by Lahille (An. Mus. nac. B. Aires, *34*, 1929: 310) as "*Lamia nasus.*" But his illustration (p. 311, fig. 10) shows the origin of the first dorsal as being over the inner corner of the pectoral, the second dorsal as slightly in advance of the anal, and the lower caudal lobe as only slightly shorter than the upper, as in *oxyrinchus*, while his statement that the teeth have lateral denticles appears to refer to the species *nasus* as a whole, rather than to the particular specimen.
38. Personal communication from W. J. Mixter in the late summer of 1941.
39. Luis Howell-Rivero writes us that it is always taken offshore there.

77, 1938: 290 (Medit.); Springer, Proc. Fla. Acad. Sci., *3*, 1939: 35 (Florida); Bigelow and Schroeder, Guide Comm. Shark Fish., Anglo Amer. Carib. Comm., Wash., 1945: 102, fig. 35 (descr., ill., habits, range).

Isurus spallanzanii Rafinesque, Indice Ittiol. Sicil., 1810: 45, 60 (Sicily); Jordan and Gilbert, Bull. U.S. nat. Mus., *16*, 1883: 874 (name, descr., C. Cod, West Indies); Nobre, Fauna Marinha Port. Vert., *1*, 1935: 431, pl. 58, fig. 184 (descr., Portugal).

Lamia oxyrhincus Bory de St. Vincent, Dict. Class. Hist. Nat., *15*, 1829: 596 (ref.).

Squalus cepedii Lesson, Voy. "Coquille," Zool., *2*, 1830: 93 (equatorial Atlantic).

Lamna oxyrhina L. Agassiz, Poiss. Foss., *3*, 1838: 86, ref. to pl. A, Lamna (teeth); Owen, Odontogr., 1840–1845: 28, pl. 3, fig. 1 (teeth).

Isurus oxyrynchus Swainson, Nat. Hist. Fish. Amphib. Rept., *2*, 1839: 313 (general).

Oxyrhina spallanzanii Bonaparte, Icon. faun. Ital., *3* (2), 1839: pl. [136], fig. 1; Agassiz, L., Poiss. Foss., *3*, 1839: 276, ref. to pl. G, fig. 2, Lamna; Nardo, Atti Ist. veneto, (3) *5*, 1859–1860: 787 (Medit.); Duméril, Hist. Nat. Poiss., *1*, 1865: 408, pl. 7, fig. 4 (descr., teeth); Miklucho-Maclay, Beitr. Vergl. Neurol. Wirbelt., *1*, 1870: 26, pl. 5, fig. 3A, B (brain); Ninni, Ann. Soc. nat. Modena, *5*, 1870: 66 (Medit.); Canestrini, in Cornalia, *et al.*, Fauna d'Ital., 1870–1872: 45 (Medit.); Poey, An. Soc. esp. Hist. nat., *5*, 1876: 391, pl. 14, fig. 1; Enumerat. Pisc. Cubens., 1876: 185, pl. 9, fig. 1 (spec. 2,585 mm., ill., tooth, descr., discus., Cuba); Gervais and Boulart, Poiss., *3*, 1877: 182, pl. 69 (descr.); Stossich, Boll. Soc. adriat. Sci. Nat., *5*, 1880: 68 (Adriatic); Moreau, Hist. Nat. Poiss. France, *1*, 1881: 298 (descr., France); Doderlein, Man. Ittiol. Medit., *2*, 1881: 62 (Medit.); Vaillant, Bull. Soc. philom. Paris, (8) *1*, 1889: 38 (embryo); Huber, Z. Wiss. Zool., *70*, 1901: 619 (claspers); Sicher, Atti Accad. gioenia, (4) *11* (5), 1898: 16 (Medit.); Belloc, Rev. des. Trav. Pêches Marit., 7 Fasc. 2, 1934: 137 (ill. after Bonaparte; Morocco, Senegal).

Oxyrrhina glauca Bonaparte, Mém. Soc. neuchâtel. Sci. nat., *2* (8), 1839: 9 (in synopsis); Heckel, Peix. Dalmaz. in Carrara, 1864: 91 (Medit., not seen); not *Oxyrhina glauca* Müller and Henle, 1841.

Oxyrhina gomphodon Müller and Henle, Plagiost., 1841: 68, 191, pl. 28 (descr.); Gray, List Fish. Brit. Mus., *1*, 1851: 60; Bocage and Brito Capello, Poiss. Plagiost., 1866: 13, pl. 3, fig. 3 (Portugal); Brito Capello, J. Sci. math. phys. nat. Lisboa, *2*, 1869: 139; Hasse, Naturl. Syst. Elasm. besond. Theil, 1882: 230, pl. 31, fig. 36–38, 41 (skelet.); Rochebrune, Act. Soc. linn. Bordeaux, (4) *6*, 1882: 8; Faune Senegamb. Poiss., *1*, 1883–1885: 22 (Senegambia).

Lamna punctata DeKay, Zool. N. Y., 4, 1842: 352, pl. 63, fig. 206 (descr., good ill., teeth, size, N. York); not *Lamna punctata* Storer, 1839.

Lamna cornubicus Cuvier, Règne Anim. Poiss., ill. ed., 1843: pl. 114, fig. 3 (teeth).

Lamna punctata (?) Goode, Bull. U.S. nat. Mus., *5*, 1876: 73; not *Lamna punctata* Storer, 1839 (Bermuda).

Oxyrrhina spalanzanii Bonaparte, Cat. Pesc. Europ., 1846: 17 (Medit.).

Isuropsis dekayi Gill, Ann. N. York Lyc., 7, 1862: 409; Poey, Repert. Fisico-Nat. Cuba, *2*, 1868: 446 (Cuba); Yarrow, Proc. Acad. nat. Sci. Philad., *29*, 1877: 217 (Ft. Macon, North Carolina). Not *Oxyrhina daekayi* Gill, Proc. Acad. nat. Sci. Philad., Addend., 1861: 60.

Lamna latro Owen, Cat. Osteol. Roy. Coll. Surg., *1*, 1853: 96 (teeth, ident. by ref. to Owen, Odontogr., 1840–1845: pl. 5, fig. 1).

Carcharias tigris Atwood, Proc. Boston Soc. nat. Hist., *10*, 1865: 81; *12*, 1869: 268 (descr., G. Mexico, C. Cod).

Lamna spallanzanii Günther, Cat. Fish. Brit. Mus., *8*, 1870: 380 (descr., distrib., synon.); Reguis, Ess. Hist. Nat. Provence, *1* (1), 1877: 49 (Medit.); Perugia, Elenc. Pesc. Adriat., 1881: 53 (Adriatic); Graeffe, Arb. zool. Inst. Univ. Wien, 1886: 446 (Medit.); Helgendorf, Arch. Naturgesh., *54* (1), 1888: 213 (Azores); Carus, Prod. Fauna Medit., *2*, 1889–1893: 505 (Medit.); Steindachner, S. B. Akad. Wiss. Wien, *100* (1), 1891: 363 (Canaries); Taylor, Ann. Scot. nat. Hist., 1910: 250 (off Scot.); Günther, Encycl. Brit., 11th ed., *24*, 1911: 807 (general); Metzelaar, Trop. Atlant. Visschen, 1919: 189.

Lamna glauca Günther, Cat. Fish. Brit. Mus., *8*, 1870: 391; not *Oxyrhina glauca* Müller and Henle, 1841.

Isuropsis sp. Goode, Amer. J. Sci., (3) *14*, 1877: 293 (Bermuda).

Isurus glaucus Jordan and Gilbert, Bull. U.S. nat. Mus., 16, 1883: 28 (Atlant. Oc., Cuba); Barnard, Ann. S. Afr. Mus., *21* (1), 1925: 33 (part, S. Afr.).

Isurus dekayi Jordan and Gilbert, Proc. U.S. nat. Mus., *5*, 1882: 241 (descr., Santa Rosa I., near Pensacola, Florida, color); Goode and Bean, Proc. U.S. nat. Mus., *5*, 1882: 240 (Gulf of Mexico); Jordan and Gilbert, Bull. U.S. nat. Mus., 16, 1883: 874 (descr., C. Cod to W. Indies); Jordan, Rep. U.S. Comm. Fish. (1885), 1887: 797 (distrib.); Nelson, Rep. N. J. Geol. Surv., *2* (2), 1890: 663 (N. Jersey record, feeding habits); Bean, Rep. For. Comm. N. Y., 1901: 379 (near N. York, but perhaps confused with *L. nasus*); Kendall, Occ. Pap. Boston Soc. nat. Hist., 7 (8), 1908: 7 (part); Jordan and Evermann, Bull. U.S. nat. Mus., 47 (4), 1900: pl. 6, fig. 21; Evermann and Kendall, Rep. U.S. Comm. Fish. (1899), 1900: 49 (Florida); Gregg, Where to Catch Fish, 1902: 17 (Florida); Bean, Bull. N. Y. St. Mus., 60, Zool. *9*, 1905: 38 (off N. York); Fowler, Rep. N. J. Mus. (1905), 1906: 54, pl. 1 (N. York); Tracy, Rep. R. I. Comm. inl. Fish., 1906: 46; Rep. R. I. Comm. inl. Fish., 1910: 60 (Rhode Island, same as foregoing); Nichols, Abstr. Linn. Soc. N. Y., 20–23, 1913: 91 (off N. York); Nichols and Murphy, Brooklyn Mus. Sci. Bull., *3* (1), 1916: 22 (descr., W. Indies-C. Cod).

Isurus (*Isurus*) *oxyrhynchus* Jordan and Evermann, Rep. U.S. Comm. Fish. (1895), 1896: 218 (distrib.); Bull. U.S. nat. Mus., 47 (1), 1896: 48 (descr., distrib.).

Isurus (*Isuropsis*) *dekayi* Jordan and Evermann, Rep. U.S. Comm. Fish. (1895), 1896: 218 (distrib.); Bull. U.S. nat. Mus., 47 (1), 1896: 48 (descr., C. Cod, W. Indies).

Isurus oxyrhynchus Schreiner and Ribeiro, Arch. Mus. nac. Rio de J., *12*, 1903: 79 (Brazil); de Braganza, Result. Invest. Sci. "Amelia," 2, 1904: 52 (Portugal); Collett, Norges Fisk., Suppl., *3*, 1905: 76 (Bergen, Norway); Fowler, Proc. Acad. nat. Sci. Philad., *60*, 1908: 55 (jaws); Garman, Mem. Harv. Mus. comp. Zool., *36*, 1913: 37 (descr.); Jordan, Copeia, 49, 1917: 87 (name); Ribeiro, Arch. Mus. nac. Rio de J., Fauna Brasil., Peixes, *2* (1), Fasc. 1, 1923: 18, pl. 6 (Rio de Janeiro); Rey, Fauna Iberica Peces, *1*, 1928: 376 (descr., habits, Spain); Jordan, Evermann and Clark, Rep. U.S. Comm. Fish. (1928), 2, 1930: 19 (distrib.); Uriarte and Mateu, Notas Inst. esp. Océanogr., (2) *53*, 1931: 12 (Canaries); Bigelow and Schroeder, Canad. Atlant. Fauna, biol. Bd. Canada, *12e*, 1934: 15 (general); Lübbert and Ehrenbaum, Handb. Seefisch. Nordeurope, *2*, 1936: 279 (Bergen, Norway; Medit.).

Isurus (probably *dekayi*) Bean, Field Mus. Publ. Zool., 7 (2), 1906: 30 (Bermuda).

Isurus tigris Garman, Mem. Harv. Mus. comp. Zool., *36*, 1913: 36 (descr.); Fowler, Copeia, 30, 1916: 36 (off U.S.); Radcliffe, Bull. U.S. Bur. Fish., *34*, 1916: 247 (teeth); Fowler, Proc. Acad. nat. Sci. Philad., *69*, 1917: 109 (N. Jersey); Jordan, Copeia, 49, 1917: 87 (name); Murphy, Copeia, 69, 1919: 32 (descr., photo., N. York); Bigelow and Welsh, Bull. U.S. Bur. Fish., *40* (1), 1925: 38 (descr., Gulf of Maine);[40] Breder, Copeia, 153, 1926: 121 (Sandy Hook Bay, New York); Nichols and Breder, Zoologica, N. Y., *9*, 1927: 19 (Long Island, N. York); Bigelow and Schroeder, Bull. Mus. comp. Zool. Harv., 68, 1927: 240 (discuss.); Jordan, Manual Vert. Anim. NE. U.S., 1929: 12 (general); Breder, Field Bk. Mar. Fish. Atlant. Coast, 1929: 22 (general);[41] Fowler, Proc. Acad. nat. Sci. Philad., *80*, 1929: 607, pl. 31; (N. Jersey); Jordan, Evermann and Clark, Rep. U.S. Comm. Fish. (1928), 2, 1930: 19 (distrib.); Firth, Bull. Boston Soc. nat. Hist., *61*, 1931: 8 (off Nantucket); Pearson, Invest. Rep., U.S. Bur. Fish., (10) *1*, 1932: 18 (North Carolina); Beebe and Tee-Van, Field Bk. Shore Fish. Bermuda, 1933: 23 (Bermuda);[42] Bigelow and Schroeder, Canad. Atlant. Fauna, biol. Bd. Canad., *12e*, 1934: 14 (general); Bull. U.S. Bur. Fish., *48*, 1936: 322 (near Nantucket); Fowler, Proc. Acad. nat. Sci. Philad., *89*, 1937: 303 (N. Jersey); Breder, Bull. N. Y. zool. Soc., *41*, 1938: 28 (near N. York).

Isurus cepedii (in part) Fowler, Bull. Amer. Mus. nat. Hist., 70 (2), 1936: 34 (St. Helena, Ascension, but *glaucus* also included).

Lamna oxyrhynchus Borri, Mem. Soc. tosc. Sci. nat., *44*, 1934: 92 (Medit.); Norman and Fraser, Giant Fishes, 1937: 12 (general).

Mako Shark (*Lamna*), Kaplan, Big Game Fisherman's Paradise, Dep. Agric. Fla., 1936: 104 (Florida).

Lamna tigris Norman and Fraser, Giant Fishes, 1937: 12 (general).

Isurus oxyrinchus Fowler, Arqu. Zool. Estado São Paulo, *3*, 1942: 127 (Brazil).

Isurus cepedii Fowler, Monogr. Acad. nat. Sci. Philad., 7, 1945: 43, fig. 3, 4 (Bimini, Bahamas, and Worcester Co., Maryland).

40. At least some of the Gulf of Maine records listed there probably referred to *nasus*, p. 118.

41. The illustrations of *oxyrinchus* ("*tigris*") and *nasus* are transposed.

42. Their illustration (p. 24) actually represents *nasus*.

Doubtful References:

Squalus cepedii Lesson, Voy. "Coquille," Zool., *2*, 1830: 93.[43]

Isurus dekayi Linton, Bull. U.S. Bur. Fish., *19*, 1901: 429 (Woods Hole); Sharp and Fowler, Proc. Acad. nat. Sci. Philad., *56*, 1904: 506 (Nantucket).

Lamia nasus Lahille, An. Mus. nac. B. Aires, *34*, 1928: 310 (Mar de la Plata, Argentina).

Isurus oxyrhynchus Wilson, Proc. U.S. nat. Mus., *60* (5), 1932: 24 (parasites).

Isurus tigris Wilson, Bull. U.S. nat. Mus., 158, 1932: 427, 446 (parasites).

Not *Isurus dekayi* Smith, Bull. U.S. Bur. Fish., *17*, 1898: 89;[44] Kendall, Occ. Pap. Boston Soc. nat. Hist., 7 (8), 1908: 7;[45] Sumner, Osburn and Cole, Bull. U.S. Bur. Fish., *31* (2), 1913: 736.[45]

Genus *Carcharodon* Agassiz, 1838

Carcharodon Agassiz, L., Poiss. Foss., *3*, 1838: 91; type species, *verus* Agassiz, L., equals *Carcharias verus* Cloquet (Dict. Sci. Nat., 7, 1817: 69) and *Squalus carcharias* Linnaeus, 1758.[1]

Generic Synonyms:

Squalus (in part) Linnaeus, Syst. Nat., 1758: 233 and subsequent authors.[2]

Carcharias (in part) Cloquet, Dict. Sci. Nat., 7, 1817: 69; Cuvier, Règne Anim., *2*, 1817: 125, and subsequent authors; not *Carcharias* Rafinesque, 1810 (see p. 98).

Carcharocles Jordan, Stanford Univ. Publ. Biol., *3*, 1923: 99; type species, *Carcharias auriculatus* Blainville, fossil.

Carcharhinus Whitley, Mem. Qd. Mus., *10*, 1934: 199; not *Carcharhinus* Blainville, 1816 (see p. 320).

Generic Characters. Teeth triangular, with slightly concave margins and coarsely serrate edges, but without lateral denticles; lower teeth smaller and more slender than uppers; 3rd upper tooth nearly as large as 2nd and 4th; snout conical, flattened above, only moderately acute; anterior part of caudal without secondary longitudinal keel below rearward extension of expanded caudal peduncle. Characters otherwise those of the family.

Range. Pelagic; cosmopolitan in tropical, subtropical and warm temperate seas, including the Mediterranean.

Fossil Teeth. From Upper Cretaceous to Pleistocene, Europe; Eocene to Pliocene,

43. Tortonese (Atti Soc. ital. Sci. nat., 77, 1938: 291) revives this name to replace *glaucus* Müller and Henle for the Indo-Pacific form. But Lesson (1830: 93) expressly states that his specimen was harpooned "dans la ocean Atlantique," in Lat. 6° S., though the longitude as given, 27° E., is patently in error, which accords with the general location of the ship on the stated date of capture, Sept. 28, 1822.

44. The account of occurrence near Woods Hole makes it highly probable that these citations actually referred to *nasus*.

45. From the authorities cited, from the widespread distribution and from the abundance credited to it in the Gulf of Maine, this evidently referred to *nasus*.

1. The early history of the generic name *Carcharodon* is confused. Proposed in 1838 by Müller and Henle (Charlesworth Mag. nat. Hist., [2] 2: 37) with diagnosis but without mention of any particular species, its type species was designated in the same year by L. Agassiz (Poiss. Foss., *3*: 91) as "*C. smithii* Müller and Henle." However, since this is a *nomen nudem*, not used by Müller and Henle, the genus must be credited to L. Agassiz, its type being *Carcharodon verus* Agassiz (Poiss. Foss., *3*, 1838: 91), the account of which, added to his illustration of its teeth printed three years earlier simply as "*Carcharias*" (Poiss. Foss., *3*, 1835: pl. F, fig. 3), leaves no doubt as to its identity. We may point out that the specific name *verus* (equivalent to *Squalus carcharias* Linnaeus, 1758) actually dates from Cloquet, 1817, for as used earlier by Blainville (Bull. Soc. philom. Paris, 1816: 121) it was a *nomen nudem* also, since it lacked any indication as to identity.

2. See under References, *Carcharodon carcharias*, p. 142.

Africa; Eocene to Pleistocene, North America; Miocene, South America, West Indies; Miocene to Pliocene, West Indies, New Zealand; Pliocene, eastern Asia.

Species. It is probable that all published accounts of this genus, whether Pacific or Atlantic, belong to a single species.[3] Since final conclusion must await critical comparison of specimens from the two oceans, or at least comparable measurements, the Pacific references are segregated below (p. 144) from those for the Atlantic and Mediterranean.

Carcharodon carcharias (Linnaeus), 1758

White Shark, Man-eater

Figures 20, 21, 22

Study Material. Jaws from specimens of about 6 feet from Long Island, New York (Amer. Mus. Nat. Hist., No. 14773), 8½ feet from Woods Hole, Mass. (U.S. Nat. Mus., No. 10899) and of about 12 feet from an unknown locality (Harv. Mus. Comp. Zool.); a mounted specimen about 6 feet long from Woods Hole (in New England Mus. Nat. Hist.); two fresh caught specimens, about 9 and 10 feet long, from Massachusetts Bay, but not preserved; good photographs of several fresh specimens, of about 5 to 10 feet long, taken off the tip of Cape Cod, off Rhode Island and off Sarasota, Florida.[4]

Distinctive Characters. The combination of strongly lunate caudal with very large triangular and coarsely serrate teeth is diagnostic. The more rearward position of the anal relative to the second dorsal and the blunter snout further separate it from its relatives of the genera *Isurus* and *Lamna.*

Description. Proportional dimensions in per cent of total length. Female, immature, 4,700 mm. total length,[5] from Florida.

Snout length in front of: mouth 6.3.
Mouth: height 1.1.
Nostrils: distance between inner ends 3.8.
Gill opening lengths: 1st 9.0; 5th 9.7.
First dorsal fin: anterior margin 12.8; length of base 9.7.
Second dorsal fin: anterior margin 2.8; length of base 1.4.
Anal fin: anterior margin 2.6; length of base 1.4.
Caudal fin: upper margin 20.0; lower margin 13.5.
Pectoral fin: outer margin 18.9; inner margin 4.6; distal margin 16.6.
Distance from snout to: 1st dorsal 37.5; upper caudal 80.0; pectoral 27.7.
Interspace between: 1st and 2nd dorsals 21.6; 2nd dorsal and caudal 10.1.

3. Whitley (Aust. Zool., *9* [3], 1939: 240) proposes the new name *albimors* for the Australian *Carcharodon* earlier described by McCoy (Prod. Zool. Victoria, Decade *8*, 1885: pl. 74) as *Carcharodon rondelletii.* But there is nothing apparent, either in McCoy's account or in his measurements, or in Whitley's subsequent illustrations (Fish. Aust., *1*, 1940: 126) to set it apart from the *Carcharodon* of the Atlantic.

4. Contributed by J. W. Lowes, James Miller and Stewart Springer.

5. From measurements by Springer (Copeia, 2, 1939: 115).

Trunk fusiform, moderately stout, broadest and highest opposite 1st dorsal fin and tapering to caudal peduncle, the girth in specimens 6 to 7 feet long about 58 to 62% of total length. Caudal peduncle strongly depressed dorso-ventrally and widely expanded

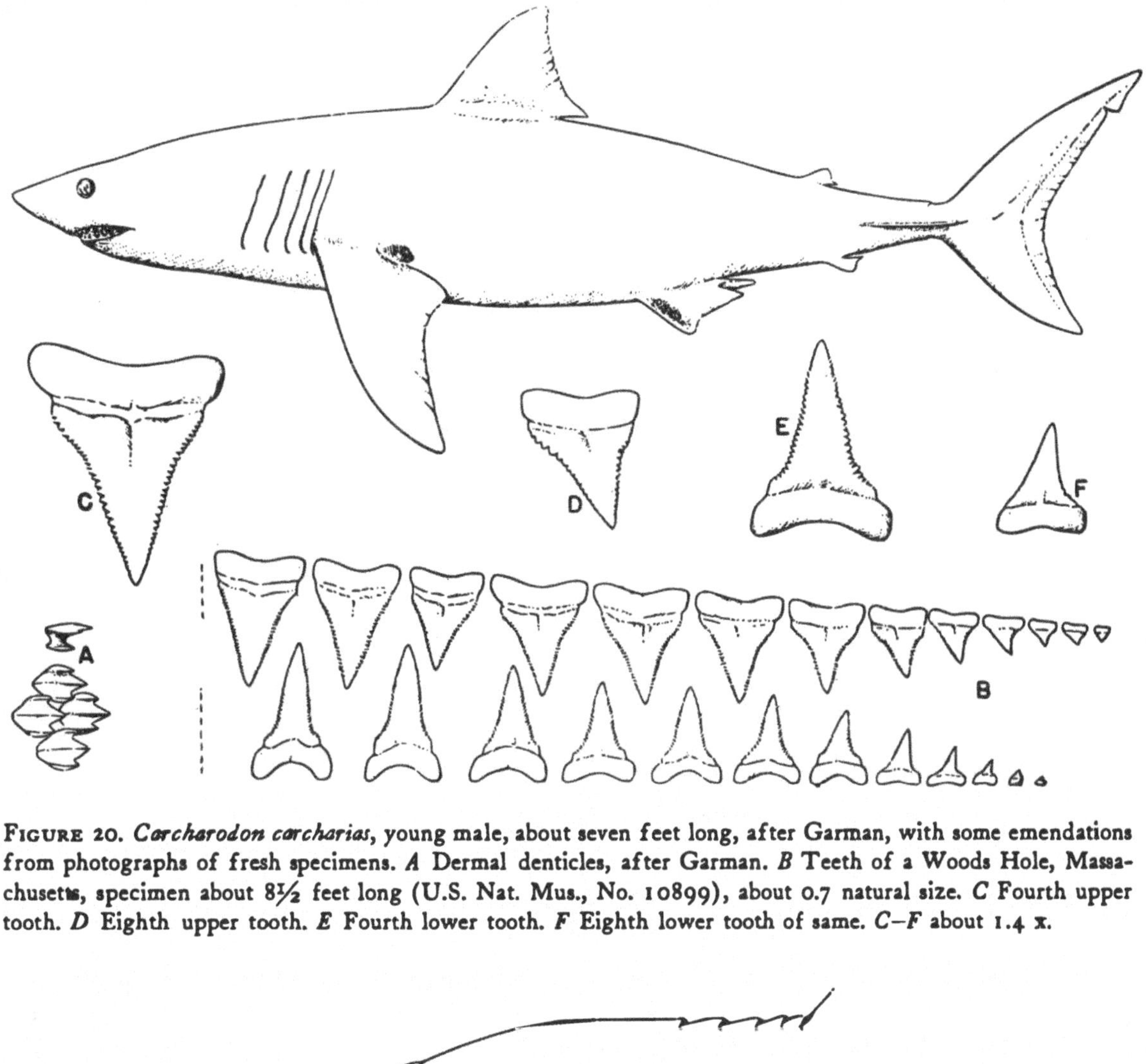

FIGURE 20. *Carcharodon carcharias*, young male, about seven feet long, after Garman, with some emendations from photographs of fresh specimens. *A* Dermal denticles, after Garman. *B* Teeth of a Woods Hole, Massachusetts, specimen about 8½ feet long (U.S. Nat. Mus., No. 10899), about 0.7 natural size. *C* Fourth upper tooth. *D* Eighth upper tooth. *E* Fourth lower tooth. *F* Eighth lower tooth of same. *C–F* about 1.4 x.

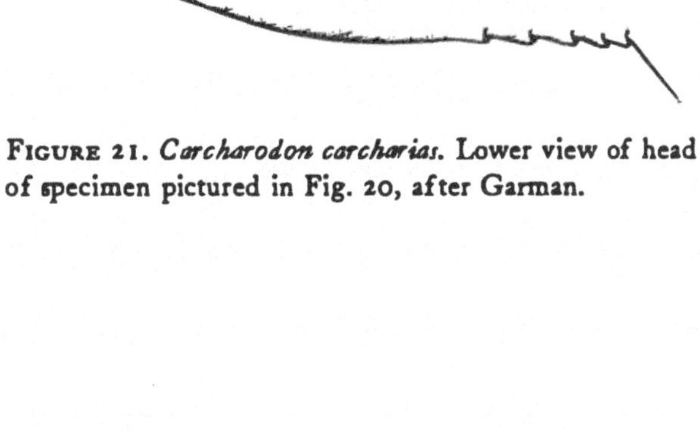

FIGURE 21. *Carcharodon carcharias*. Lower view of head of specimen pictured in Fig. 20, after Garman.

laterally, as in *Isurus,* its breadth, including its lateral keel-like extensions, about 3 times its depth, with a prominent transverse furrow above and below just in front of origin of caudal. Dermal denticles minute, as in *Isurus,* 3-ridged, their free margins correspondingly indented, the blades so nearly flat that the skin is hardly rough to the touch.

Head 25 to 30% of total length. Snout shorter than in *Isurus,* its length in front of mouth a little less than 1/4 of length of head, obtusely conical, but somewhat flattened dorsally, with blunt tip; but in large, heavy specimens, suspended or dragged up on the beach by the front of the mouth, the head is often greatly distorted in appearance as seen in photographs, since the upper jaw is slightly protrusible. Eye small, circular, its anterior margin a little posterior to front of mouth. Spiracle lacking in fresh specimens seen by us, pore-like if present, behind eye by a distance about equal to length of snout in front of mouth. Gill openings as in *Isurus,* noticeably long, the 5th a little the longest, between 1.5 to 2 times as long as snout in front of mouth, the 1st shortest, the spaces between them successively narrower from front to rear, that between 4th and 5th being only about 1/2 as great as that between 1st and 2nd, the 5th close in front of origin of pectoral and curving posteriorly around latter. Nostril narrow, transverse, near side of head, nearer to mouth than to tip of snout, its anterior margin with very low subtriangular lobe. Mouth broadly rounded, a little more than twice as broad as high. Labial furrows very short, the lower concealed except when mouth is open.

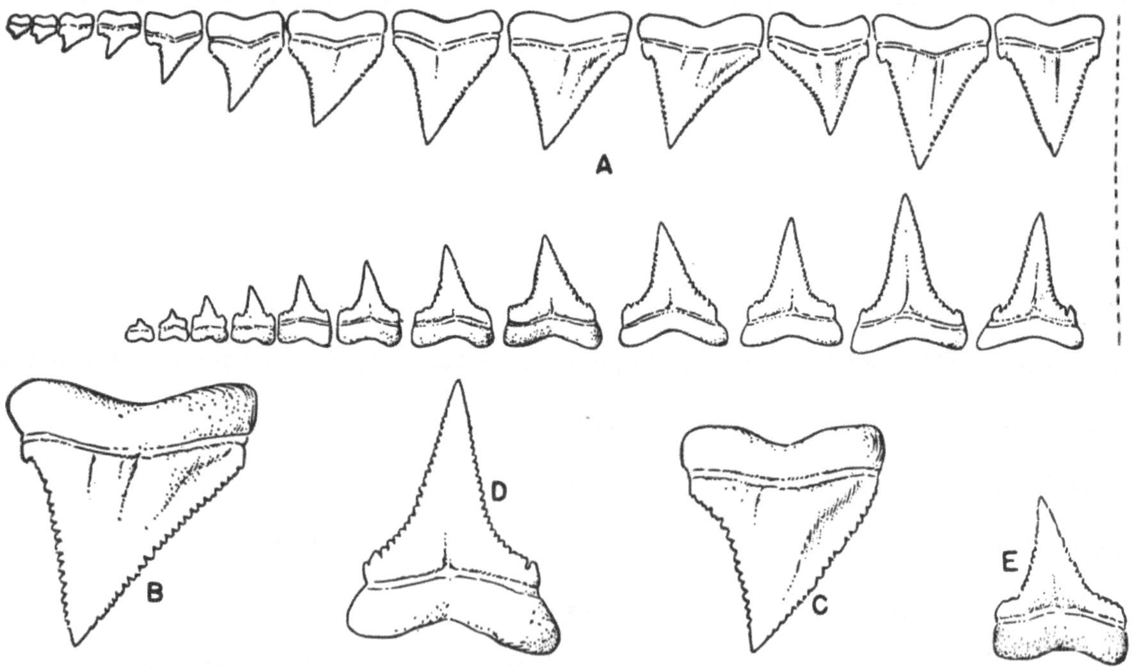

FIGURE 22. *Carcharodon carcharias,* about six feet long, from Long Island, New York (Amer. Mus. Nat. Hist., No. 14773). *A* Upper and lower teeth, right-hand side, about natural size. *B* Fourth upper tooth. *C* Seventh upper tooth. *D* Third lower tooth. *E* Seventh lower tooth. *B-D,* about 2 x.

Teeth $\frac{13}{11 \text{ or } 12}$ in each side of mouth, large,[6] subtriangular, erect or very slightly oblique, their edges coarsely and regularly serrate; uppers about as high as broad, 1st with inner margin nearly straight, but others with both margins usually slightly concave, the outer edge the more so;[7] lowers narrower than uppers, their margins more concave; 1st and 2nd teeth the largest in each jaw, those toward corners of mouth successively smaller, the outermost 2 or 3 minute; 1st and 2nd lowers in small specimens (Fig. 22 A, D) with basal serrations considerably the largest; 1, or at most 2, series functional in each jaw. Gap at symphysis wider in lower jaw than in upper.

First dorsal nearly an equilateral triangle, its apex rounded, its rear margin only slightly concave, its free rear tip only about 1/4 as long as its base, its origin opposite or a little anterior to inner corner of pectoral.[8] Second dorsal only 1/5 to 1/6 as large in linear dimensions as 1st, its apex rounded, its margins nearly straight, the rear end of its base over, or a little anterior to, origin of anal. Upper anterior and lower anterior outlines of caudal moderately convex, posterior outline lunate, with strongly marked subterminal notch, the tips subacute, the lower anterior margin about 4/5 (76 to 92%) as long as upper anterior margin, each measured from precaudal furrow. Anal similar in size and shape to 2nd dorsal, and wholly behind latter. Pelvics much larger than 2nd dorsal or anal, their anterior margins about 1/2 as long as anterior margin of 1st dorsal, their distal margins concave, their corners rounded. Pectoral noticeably larger in area than in *Isurus*, a little less than 1/4 as long as from tip of snout to origin of caudal, and considerably less than 1/2 as broad as long, with convex anterior and concave posterior margins, subacute tip and rounded inner corner.

Color. Specimens up to 12 to 15 feet long, including those seen by us, are slaty-brown, dull slate-blue, leaden gray, or even almost black above, shading more or less abruptly to dirty white on the lower surface with a black spot in the axil of the pectoral; the tips of the pectorals also black, usually with some adjacent black spots; the dorsals and caudal dark along rear edges, but the pelvics darkest (olive) along anterior edges, fading rearward to white. Large specimens (perhaps some smaller ones also) are described as dun-colored above, or even leaden-white. They may also lack the black axillar spot.[9]

Size. This shark has been credited repeatedly with reaching a length of 40 feet. Actually, however, the stated length of the Australian specimen on which the foregoing has been based, the jaws of which are now in the British Museum, was 36½ feet.[10] The next largest, the actual capture of which is authentically recorded, was reported as of about 30 feet, seemingly not measured.[11] However, these appear to have been giants of their kind, for while 20 to 25-footers have been reported as seen on several occasions, the three next

6. The largest teeth of a specimen 36½ feet long were about two inches long.
7. Individual teeth vary in this regard, irrespective of their positions along the jaws.
8. Sometimes shown as a little behind inner corner of pectoral in photographs of specimens suspended by mouth, and hence more or less distorted.
9. Personal communication from Stewart Springer, from his observations on about a dozen large Florida specimens.
10. Günther, Cat. Fish. Brit. Mus., *8*, 1870: 392; Guide to Study Fish., 1880: 321.
11. Jordan and Evermann, Bull. U.S. nat. Mus., 47 (1), 1896: 50.

largest actually measured have been 21 feet[12] and 17 to 19 feet in length. We should perhaps caution the reader that estimates of the size of the larger sharks are frequently much too high; *e.g.*, an Australian specimen, reported in the local newspapers as 16 feet long, actually measured only eight feet six inches.[13] On the other hand, the smallest free-living specimen of which we find record was about 5 feet long.[14] Among 44 other specimens from various localities, of which measurements are available, 15 were between 6 and 8 feet; 7 between 8 and 10 feet; 9 between 10 and 12 feet; 7 between 12 and 14 feet; 4 between 14 and 16 feet; 2 between 16 and 18 feet. The two gravid females on record were 14 feet 9 inches and 18 feet 4 inches (5.7 m.) in length; similarly the gonads of a male of 14 feet 6 inches, taken off Salerno, Florida, were much enlarged, but other males of 12 to 12½ feet showed no signs of approaching maturity.[15] The fact that females of 8 feet 6 inches and 15 feet 6 inches have been reported as containing neither embryos nor even enlarged ova suggests that sexual maturity is not usually reached at a length less than perhaps 13 to 14 feet. That so few adults are captured anywhere is no doubt due to their large size, great strength and formidable nature.

Recorded weights of Atlantic specimens in relation to length are: 600 pounds at 8 feet 3 inches; 960 pounds at 9 feet 8 inches; 998 pounds at 12 feet; 940 pounds at 12 feet 2 inches; about 1,300 pounds at about 13 feet; and 7,100 pounds, with a liver of 1,005 pounds,[16] at 21 feet (Cuban specimen mentioned above, see footnote 12, page 138); also an estimated weight of 1,200 pounds for a specimen 12 feet 8 inches long. Weights of Pacific specimens taken on the coast of the State of Washington are: 342 pounds at 8 feet 2 inches; between 800 and 1,000 pounds at about 12 feet; up to 2,000 to 2,400 pounds at 13 feet.[17] A 5 foot 4 inch specimen from Catalina Island weighed 87 pounds.[18] Australian data[19] show: 928 pounds at 11 feet 3 inches; 910 pounds at 12 feet 6 inches; 1,291 pounds at 13 feet 6 inches; 1,334 pounds at 13 feet 5 inches; and 1,720 pounds at 15 feet 2 inches; a South African specimen of only 13 feet 3 inches weighed 2,176 pounds.[20] The variation in weight at given lengths with differences in the condition of the individual specimens is thus very wide, and increasingly so with growth.

Developmental Stages. No account of the developmental stages has yet appeared. The few embryos so far reported have ranged in length from about 20 to 61.6 cm. A Mediterranean specimen, probably of this species, contained nine young, each about two feet long.[21]

12. Taken recently off Havana, Cuba, and reported to us by Luis Howell-Rivero.
13. Macleay, Proc. Linn. Soc. N. S. W., *4*, 1880: 459.
14. Doderlein (Man. Ittiol. Medit., *2*, 1881: 66) reports a specimen of .63 m., or about 2 feet, but this may have been an embryo.
15. Personal communication from Stewart Springer.
16. We have received a good photograph, apparently of this specimen, with weight stated at 7,302 pounds, from Ollyandro del Valle.
17. Bonham, Copeia, 1942: 264.
18. Personal communication from W. I. Follett.
19. Whitley, Fish. Aust., *1*, 1940: 127.
20. London Illustr. News, July 14, 1928: 53.
21. Norman and Fraser, Giant Fishes, 1937: 18; but the stated weights of these embryos (about 100 pounds at a length of two feet) were evidently in error.

Habits. This is an active, strong-swimming species, putting up a dogged and savage resistance to capture. The reports of it attacking boats, when harpooned or hooked, are too numerous and too circumstantial to be dismissed. However, it does not have the leaping habit of the Mako. So few are seen that nothing is known of its life apart from the foregoing and the fact that it is voracious. The great majority of records have been of specimens taken at the surface or close to it. But it appears that they may descend to considerable depths, for a large one caught off the north coast of Cuba, of which we have a photograph, was said to have been hooked at a depth of 700 fathoms. Nothing is known of its breeding habits.

Characterization of this Shark by an earlier student as "the most voracious of fish-like vertebrates,"[22] is no doubt well deserved. The frequency with which it captures large prey, which it devours practically intact, is illustrated by the presence of other sharks from four to seven feet long, as well as a young sea lion of 100 pounds, in the stomachs of White Sharks; also seals, sturgeons and tuna have been found in specimens no larger than eight to nine feet. Sea turtles are also described as a regular item in its diet in southern waters. On the other hand, it also preys on a wide variety of smaller fishes and marine animals, including chimaeroids and squids. The mouth of a Massachusetts Bay specimen recently examined by us was festooned with hooks and snoods from a long line, while its stomach contained a spiny dogfish evidently torn off a hook. This, together with similar reports by others, including the report of a large Florida specimen containing two *Carcharhinus milberti* six to seven feet long which were evidently torn from hooks on the setline on which the *Carcharodon* itself was taken,[23] shows that when White Sharks stray in on the fishing grounds they find a convenient source of food.

It has been described also as a scavenger when occasion offers; for example, the stomach of a shark said to be this species, caught in Sydney Harbor, New South Wales, contained a variety of garbage, including horse meat, legs of mutton, parts of a pig, a dog, etc.

Relation to Man. This is perhaps the only shark against which the charge of unprovoked attack on small boats is proved through identification of the teeth left imbedded in the sides of the boat. It has borne an unsavory reputation from the earliest times as a man-eater. It is so classed, for example, in Australia, where attacks by sharks on bathers, especially near Sydney, are of such common occurrence that most of the bathing beaches are protected by wire-netting enclosures.[24] It is not possible to tell whether men, reported by earlier authors to have been found in the stomachs of White Sharks, were alive or dead when eaten; but it is probable that a seven-foot specimen, taken a few days later in Sandy Hook Bay at the mouth of New York Harbor, was responsible for four shark fatalities

22. Jordan, Guide to Study Fish., 1905: 538.
23. Springer, Copeia, 2, 1939: 114.
24. See Coppleson (Med. J. Aust., April 15, 1933: 449) and Whitley (Fish. Aust., *1*, 1940: 259) for circumstantial accounts (many of them recent) of shark fatalities in Australia. In most cases the species of shark responsible was not determined.

that occurred on the bathing beaches of New Jersey from July 6 to 12, 1916.[25] A *Carcharodon* also may have been responsible for the fatal attack on a swimmer at Mattapoisett on Buzzards Bay, Massachusetts, on July 25, 1936;[26] in this case the shark was driven away and not identified. However, these are the only recently recorded instances anywhere on the eastern seaboard of the United States in which *Carcharodon* is under suspicion. Hence, while the possibility of attack by it on bathers is always present, since White Sharks do occasionally come close inshore near populous sectors of the coast line, it is exceedingly remote. The most recent report of an attack by this species (fatal in this instance) was of a 6- to 7-foot specimen on a swimmer in Panama Bay, the species being identified by a well known ichthyologist on the basis of fragments of its teeth taken from wounds by the surgeon attending the victim.[27]

In spite of its ferocity and its muscular power, the White Shark does not put up as spectacular a resistance as the Mako when hooked (p. 129), not having the habit of jumping. Nor does it seem to make as strong a fight, pound for pound, as the tuna or the swordfish. For example, it is recorded that a 1,329-pound specimen was landed on rod and reel by an angler after fifty-three minutes in Australia;[28] another of 2,176 pounds was landed in South Africa from the shore in five hours,[29] the latter one of the largest, if not the largest, fish ever landed on rod and reel.[30]

Range. Oceanic; widespread in tropical, subtropical and warm temperate belts of all the oceans, including the Mediterranean; exceedingly irregular in its occurrence; apparently most numerous in Australian waters, but nowhere abundant.

Occurrence in the Atlantic. Although this shark has been so long known and so much discussed because of its ill repute, very little detailed information is available as to its geographic distribution anywhere. While repeatedly reported from the Mediterranean and from many other localities, it certainly is not common there. It appears to be decidedly scarce on the eastern side of the open Atlantic, it being positively recorded, so far as we can learn, only from the Cape of Good Hope region, from Morocco, Rio de Oro, Mauritania, Senegal, the Canaries, and from the coast of the Iberian Peninsula, with nominal records from the vicinity of Teneriffe and Madeira.

The list of positively identified captures for the tropical-subtropical belt in the west is limited to one record for Brazil (several times repeated by subsequent authors); one from St. Lucia in the West Indies; one from the vicinity of Nassau in the Bahamas; four from the west coast of Florida; and one or two from the east coast. Reputedly, however,

25. Nichols and Murphy, Brooklyn Mus. Quart., 4, 1916: 157.
26. The victim was taken to the hospital in New Bedford, where he died.
27. Reported in J. Amer. med. Ass., July 22, 1944, and in Science News Letter, July 29, 1944: 73. Identification by J. T. Nichols.
28. Whitley, Fish. Aust., *1*, 1940: 126.
29. London Illustr. News, July 14, 1928: 53, photograph; recorded as a Mako, but identifiable by the teeth as a *Carcharodon*.
30. For a graphic account of the capture of one 9 feet 2 inches long off Virginia by an angler, see Wise (Tigers of the Sea, 1937: 61).

it is considerably more plentiful among the West Indies than the paucity of the published records would suggest; this is certainly true along the east coast of Florida, where one correspondent (a well known student of sharks) reports the recent capture in the shark fishery of about a dozen fair-sized ones.[31] To the northward it is either more plentiful or at least more often caught or reported. Thus, four were taken near Cape Lookout, North Carolina, during the summer of 1918, with others reported as seen in that and previous summers; one is recorded off Smith Island, Virginia; three or four from the coast of New Jersey, with others reported by sport fishermen.[32] Occasional specimens are encountered off New York; a small one of about five feet was taken in a pound net at Sakonnet, Rhode Island, May 30, 1939.[33] Nine or ten are definitely listed and several additional ones are reported from the Woods Hole region and Nantucket, with two at the most, however, in any one year. While it is generally considered a warm water species, reliable reports of its presence have been received more often from the southwestern part of the Gulf of Maine than from any other coastal sector of comparable length on the American seaboard. In Massachusetts Bay alone at least nine were either actually captured or harpooned and lost during the period from 1935 to 1940, with stray specimens taken for earlier years back to 1848, most of them in the vicinity of Cape Cod. Still farther north there are scattered records for the vicinity of Portland, Maine (2),[34] the most recent a 13-foot specimen, taken in a gill net off Casco Bay in November 1931; from Eastport at the mouth of the Bay of Fundy (1), and from Digby, Nova Scotia, within the Bay (1). It may visit the outer coast of that Province more often than formerly supposed, there being several reliable records for St. Margaret Bay, and perhaps for Halifax also. The most northerly record for American waters is St. Pierre Bank, south of Newfoundland, where one attacked a fisherman many years ago in a dory, leaving in the sides of the boat fragments of its teeth, by means of which Dr. Garman was able to identify it.[35]

The fact that all records of its presence off the northeast coast of the United States and Canada are for the warm half of the year suggests that it is an oceanic visitor, but nothing whatever is known of its status offshore in the western Atlantic, there being no record of its presence around Bermuda.

Although typically an inhabitant of the high seas, it frequently comes inshore and even into very shallow water, as in the following cases: one taken inside Sandy Hook Bay, New York, in 1916; a considerable number that have been picked up at different times in the fish traps within a few yards of the beach in the vicinity of Woods Hole and on Cape Cod; one harpooned in 10 feet of water in Provincetown Harbor many years ago; two specimens caught close to Boston Harbor in 1839; one harpooned about two miles off one of the most popular bathing beaches at the mouth of Boston Harbor in 1937; another simi-

31. Personal communication from Stewart Springer.

32. A recently received photograph, supposedly of a Mako taken off New Jersey in October 1935, unmistakably represents a *Carcharodon* of 11 to 12 feet.

33. Photograph received from James Miller.

34. Received from Walter H. Rich.

35. Putnam, Bull. Essex Inst., *6*, 1874: 72.

larly harpooned within half a mile of the land off Cohasset, on the southern side of Massachusetts Bay in August 1940.[36]

Synonyms and References:

1. Atlantic:[37]

Squalus carcharias Linnaeus, Syst. Nat., *1*, 1758: 235 (type loc., Europe); Brunnich, Ichthyol. Massil., 1768: 5 (Adriatic, food); Bloch, Naturg. Ausländ. Fische, *4*, Atlas, 1758: pl. 119; Gmelin, in Linnaeus, Syst. Nat., 13th ed., *3*, 1789: 1498 (part); Walbaum, P. Artedi Genera Pisc. Emend. Ichthyol., *3*, 1792: 514 (descr., Medit., Atlant.); Bloch and Schneider, Syst. Ichthyol., 1801: 132 (descr., ref.); Blumenbach, Handb. Naturg., 7, 1803: 263 (ref.); Bosc, Nouv. Dict. Hist. Nat., *21*, 1803: 185 (general); Latreille, Nouv. Dict. Hist. Nat., *24*, 1804: 72 (in table of contents); Nardo, Prod. Itiol. Adriat., 1827: 9 (Adriatic); Voigt, in Cuvier, Tierreich, *2*, 1832: 505 (descr.); Griffith, in Cuvier, Anim. Kingd. with Adds., *10*, 1834: 599 (general); Bonaparte, Mém. Soc. neuchâtel Sci. nat., *2* (8), 1839: 10 (in synopsis); Nardo, Atti Ist. veneto, (3) *5*, 1859–1860: 787 (Medit.); Gemellaro, Sagg. Itiol. Catania, 1864: 120 (Medit., not seen); Vieria and Clavijo, Dicc. Hist. Nat. Isl. Canaries, Real Soc. Econ. Las Palmas, 1866–1869, Mss. of 1799 (Canaries, not seen); Buckland, Hist. Brit. Fish., 1881: 211 (general).

White Shark, Brookes, Nat. Hist., *3*, 1763: 28 (general); Pennant, Brit. Zool., *3*, 1769: 82 (general); also subsequent ed.; Pennant, Arctic Zool., Suppl., 1787: 105 (America).

Cane carcharia Cetti, Amfib. Pesci Sardegna, 1777: 69 (Sardinia).

Squale requin, Lacépède, Hist. Nat. Poiss., *1*, 4° ed., 1798: 165, 169, in Buffon, Hist. Nat. (part, combined with *Carcharhinus*), not pl. 8, fig. 1, which is *Carcharhinus*; in Sonnini, Hist. Nat. Poiss., *3*, 1802: 332 (part, combined with *Carcharhinus*, see p. 320), pl. 4, fig. 2, 3 (teeth), not pl. 4, fig. 1, which is *Carcharhinus* (see p. 320).

Carcharias lamia Rafinesque, Indice Ittiol. Sicil., 1810: 44 (substituted for *Squalus carcharias* Linnaeus, 1758, Sicily); Fitzinger, Bild. Atlas Naturg. Fische, 1864: fig. 169 (ill.).

Squalus (*Carcharias*) *carcharias* Cuvier, Règne Anim., *2*, 1817: 126 (general, size).

Carcharias verus Cloquet, Dict. Sci. Nat., 7, 1822: 69 (general); Reguis, Ess. Hist. Nat. Provence, *1* (1), 1877: 46 (Medit.).

Squalus (*Carcharhinus*) *carcharias* Blainville, in Vieillot, Faune Franc., 1825: 89 (descr.).

Carcharias carcharias Bory de St. Vincent, Dict. Class. Hist. nat., *15*, 1829: 596; Cuvier, Règne Anim., Ill. ed., Poiss., 1842–1843: 360, pl. 114, fig. 2a (tooth); Jordan, Copeia, 140, 1925: 20 (name).

Carcharias rondeletti Bory de St. Vincent, Dict. Class. Hist. nat., *15*, 1829: 596 (general); Reguis, Ess. Hist. Nat. Provence, *1* (1), 1877: 47 (Medit.).

Squalus (*Carcharias*) *vulgaris* Richardson, Fauna Boreal. Amer., *3*, 1836: 288 (ref.).

Carcharias (no specific name), Agassiz, L., Poiss. Foss., *3*, 1835: pl. F, fig. 3 (teeth, see footnote 1, p. 133).

Carcharodon smithii Agassiz, L., Poiss. Foss., *3*, 1838: 91 (name only);[38] Bonaparte, Mém. Soc. neuchâtel. Sci. nat., *2* (8), 1839: 9 (in synopsis).

Carcharodon verus Agassiz, L., Poiss. Foss., *3*, 1838: 91 (teeth).

Carcharias vulgaris Hamilton, Brit. Fish., *2*, 1843: 304, and subsequent eds. (Gt. Brit., general).

Carcharodon lamia Bonaparte, Icon. Faun. Ital., *3* (2), 1839: pl. [52] (descr., colored ill., general); Cat. Pesc. Europ., 1846: 17 (Medit.); Sassi, Cat. Pesci Liguri., 1846: 123 (Medit., not seen); Nardo, Atti Ist. veneto, (3) *5*, 1859–1860: 787 (Medit.); Ninni, An. Soc. Nat. Modena, *5*, 1870: 66 (Medit.); Moreau, Hist. Nat. Poiss. France, *1*, 1881: 302 (descr., France); Bellotti, Atti Soc. ital. Sci. nat., *33*, 1891: 111 (name); Belloc, Rev. des Trav. Pêches Marit., 7, Fasc. 2, 1934: 138 (ill. after Bonaparte; Morocco, Rio de Oro, Mauritania, Senegal).

Carcharodon rondeletti Müller and Henle, Plagiost., 1841: 70 (Brazil, Australia, descr., meas.); Gray, List Fish. Brit. Mus., 1851: 61; Gill, Proc. Acad. nat. Sci. Philad., Addend., 1861: 60 (name); Duméril, Hist. Nat. Poiss., *1*, 1865: 411 (descr., size, Medit., Algeria, Atl., C. Good Hope), pl. 7, fig. 7 (tooth);

36. We had the opportunity of examining two of these Massachusetts Bay specimens soon after they were landed, and have received photographs of others.

37. For Indo-Pacific references, see p. 144.

38. *Nomen nudem.*

Bocage and Brito Capello, J. Sci. math. phys. nat. Lisboa (1868–1869), *2*, 1870: 140 (Portugal); Günther, Cat. Fish. Brit. Mus., *8*, 1870: 392, 518 (descr., Cape Seas, jaws of 36½-ft. spec., Australia); Canestrini, in Cornelia, *et al.*, Fauna d'Ital., 1872: 45 (Medit.); Doderlein, Prosp. Metod. Pesci Sicil., 1878–1879: 30 (Medit.); Stossich, Boll. Soc. adriat. Sci. nat., *5*, 1880: 68 (Adriatic); Doderlein, Man. Ittiol. Medit., *2*, 1881: 66 (Medit.);[39] Perugia, Elenc. Pesc. Adriat., 1881: 53 (sizes, Adriatic spec.); Hasse, Naturl. Syst. Elasm. besond. Theil, 1882: 224, pl. 30, fig. 24–33 (vertebrae, dermal dent.); Rochebrune, Act. Soc. linn. Bordeaux, (4) *6*, 1882: 46; Faune Senegambie, Poiss., *1*, 1883–1885: 23 (Senegambia); Jordan and Gilbert, Bull. U.S. nat. Mus., *16*, 1883: 30 (American coast); Graeffe, Arb. zool. Inst. Univ. Wien, 7, 1886: 446 (Medit.); Carus, Prod. Fauna Medit., *2*, 1889–1893: 506 (Medit.); Sicher, Atti Accad. gioenia, (4) *11*, 1898: 16 (Medit.); Bridge, Camb. nat. Hist., 7, 1904: 451 (general); Werner, Zool. Jb., Syst. Abt., *21*, 1904: 285 (embryo); Günther, Encyc. Brit., 11th ed., *24*, 1911: 807 (general); Metzelaar, Trop. Atlant. Vischen, 1919: 190 (Senegambia); Norman and Fraser, Giant Fishes, 1937: 14 (general, embryos).

Carcharias atwoodi Storer, Proc. Boston Soc. nat. Hist., *3*, 1848: 72 (Provincetown, Massachusetts); Gill, Proc. Acad. nat. Sci. Philad., Addend., 1861: 59 (listed); Storer, Mem. Amer. Acad. Arts Sci., *9*, 1867: 222, pl. 36, fig. 4; also as Fishes of Mass., 1867: 246, pl. 36, fig. 4 (descr., ill., Massachusetts); Gill, Rep. U.S. Comm. Fish. (1871–1872), 1873: 813 (Newfoundland to Florida); Smith, in Verrill, Rep. U.S. Comm. Fish. (1871–1872), 1873: 576 (parasites, Vineyard Sound).

Carcharodon atwoodi Lyman, Rep. Mass. Comm. inl. Fish., 1872: 53 (Massachusetts); Smith, S. J., Rep. U.S. Comm. Fish. (1871–1872), 1873: 576 (Vineyard Sound); Baird, Rep. U.S. Comm. Fish. (1871–1872), 1873: 827 (Woods Hole); Goode, in Verrill, Rep. U.S. Comm. Fish. (1871–1872), 1873: 812 (listed C. Cod to Florida); Goode and Bean, Bull. Essex Inst. Salem, *11*, 1879: 38 (Massachusetts); Rathbun, Proc. U.S. nat. Mus., 7, 1885: 488, 489 (parasites).

Carcharodon (without specific name) Owen, Odontogr., 1840–1845: 30 (size of teeth in 37-ft. spec.).

Carcharias (*Prionodon*) *lamia* Putnam, Bull. Essex Inst. Salem, *6*, 1874: 72 (St. Pierre Bank, S. of Newfoundland).

Carcharodon carcharias Perez-Arcas, An. Soc. esp. Hist. nat., 7 (2), 1878: 15 (Valencia, Spain; history); Jordan and Gilbert, Bull. U.S. nat. Mus., 16, 1883: 875 (name); Stephenson, Trans. Vassar Bros. Inst., *2*, 1884: 83, pl. 1, 2 (meas., Nantucket); Amer. Nat., *18*, 1884: 940, pl. 31 (same specimen as preceding); Goode, Fish. Fish. Industr. U.S., 1, 1884: 670 (habits, ferocity, distrib.); Jordan, Rep. U.S. Comm. Fish. (1885), 1887: 797 (distrib.); Jordan and Evermann, Rep. U.S. Comm. Fish. (1895), 1896: 218 (distrib.); Bull. U.S. nat. Mus., 47 (1), 1896: 50 (descr., distrib., 30-ft. Calif. spec.); Smith, H. M., Bull. U.S. Bur. Fish., *17*, 1898: 89 (Woods Hole); Bean, Rep. For. Comm. N. Y., 1901: 380 (old C. Cod records); Bull. N. Y. St. Mus., 60, Zool. *9*, 1903: 40; also in Rep. N. Y. St. Mus. (1902), *3* (5), 1904: 40 (general); Jordan, Guide to Study Fish., *1*, 1905: 538 (general); Ribeiro, Arch. Mus. nac. Rio de J., *14*, 1907: 160, 203 (Brazil, synonyms); Wilson, Proc. U.S. nat. Mus., *33*, 1907: 371, 414, 423 (Woods Hole, parasites); Kendall, Occ. Pap. Boston Soc. nat. Hist., 7 (8), 1908: 7 (New England); Berg, Faune Russie Poiss., *1*, 1911: 55 (not seen); Sumner, Osburn and Cole, Bull. U.S. Bur. Fish., *31* (2), 1913: 737 (Woods Hole region); Halkett, Check List Fish. Canad., 1913: 40; Garman, Mem. Harv. Mus. comp. Zool., *36*, 1913: 32, pl. 5, fig. 5–9 (descr., Massachusetts Bay spec.); Radcliffe, Trans. Amer. Fish. Soc., 1914: 37 (teeth); Coles, Proc. biol. Soc. Wash., *28*, 1915: 91 (N. Carolina); Smith, H. M., J. Amer. Mus. nat. Hist., *16*, 1916: 341, 343 (Martha's Vineyard, Mass.; Florida; danger to man); Fowler, Copeia, 30, 1916: 36 (Middle Atlant., U.S.); Hussakoff, Copeia, *37*, 1916: 86 (N. Jersey); Nichols, Copeia, 37, 1916: 87 (N. Jersey); Nichols and Murphy, Brooklyn Mus. Quart., 3 (4), 1916: 145–157 (N. Jersey, shark fatalities of 1916); Nichols and Murphy, Brooklyn Mus. Sci. Bull., *3* (1), 1916: 24 (Woods Hole); Radcliffe, Bull. U.S. Bur. Fish., *34*, 1916: 247 (N. Carolina, teeth, Woods Hole); Coles, Copeia, 69, 1919: 34 (descr., habits, ill., N. Carolina records); Fowler, Proc. biol. Soc. Wash., *33*, 1920: 144 (N. Jersey, fatalities); Proc. Acad. nat. Sci. Philad., *72*, 1921: 385 (N. Jersey); Huntsman, Contr. Canad. Biol. (1921), *3*, 1922: 56 (Bay of Fundy); Meek and Hildebrand, Field Mus. Publ. Zool., *15*

39. For locality references from the Mediterranean in publications not accessible to us, see Doderlein.

(1), 1923: 63 (general); Ribeiro, Fauna Brasil., Peixes, Mus. nac. Rio de J., *2* (1) Fasc. 1, 1923: 18 (Brazil); Wilson, Proc. U.S. nat. Mus., *64* (17), 1925: 12 (Woods Hole); Bigelow and Welsh, Bull. U.S. Bur. Fish., *40* (1), 1925: 39 (Gulf of Maine); Nichols and Breder, Zoologica, *9*, 1927: 19 (general); Hildebrand and Schroeder, Bull. U.S. Bur. Fish., *43*, 1928: 46 (general); Rey, Fauna Iberica Peces, *1*, 1928: 788 (Spain); Breder, Field Bk. Mar. Fish. Atlant. Coast, 1929: 22 (general); Truitt, Bean and Fowler, Bull. Md. Conserv. Dept., *3*, 1929: 29 (off Maryland); Jordan, Manual Vert. Anim. NE. U.S., 1929: 12 (general); Jordan, Evermann and Clark, Rep. U.S. Comm. Fish. (1928), 2, 1930: 20 (distrib.); Townsend, Bull. N.Y. zool. Soc., *34* (6), 1931: 169 (St. Lucia, W. I., Australia); Bellon and Mateu, Notas Inst. esp. Océanogr., *2* (53), 1931: 11 (Canaries); Wilson, Bull. U.S. nat. Mus., *20*, 1932: 428, 432, 461, 464 (Woods Hole, parasites); Young and Mazet, Shark, Shark, 1933: 267 (general); Piers, Proc. N.S. Inst. Sci., *18* (3), 1934: 192 (Nova Scotia, Bay of Fundy); Bigelow and Schroeder, Canad. Atlant. Fauna, biol. Bd. Canad., *12e*, 1934: 15 (descr., distrib.); Vladykov and McKenzie, Proc. N. S. Inst. Sci., *19* (1), 1935: 47 (Nova Scotia); Fowler, Bull. Amer. Mus. nat. Hist., *70* (1), 1936: 31 (Canaries, Madeira, Senegambia, N. Jersey); Breder, Bull. N. Y. zool. Soc., *39* (3), 1936: 161 (Bahamas, N. Jersey); Bull. N. Y. Zool. Soc., *39* (6), 1936: 243 (Nova Scotia); Bigelow and Schroeder, Bull. U.S. Bur. Fish., *48*, 1936: 322 (Gulf of Maine, Nova Scotia); Heilner, Schrenkeisen and La Monte, Field and Stream, Feb. 1936: 27 (size); White, Bull. Amer. Mus. nat. Hist., *74*, 1937: 14, pl. 12, fig. B; pl. 38, fig. F (anat.); Wise, Tigers of the Sea, 1937: 172 (Virginia, capture by angler); Tortonese, Atti Soc. ital. Sci. nat., 77, 1938: 292 (Medit.); Springer, Proc. Fla. Acad. Sci., *3*, 1939: 36 (Florida); Schroeder, Copeia, 1938: 46 (Massachusetts Bay); Copeia, 1939: 48 (Boston Harbor); Springer, Copeia, 1939: 114 (Florida); Schroeder, Copeia, 1940: 139 (Massachusetts Bay); Anonymous, News Letter, Calif. Acad. Sci., June, 1940 (Nova Scotia); Hildebrand, Copeia, 1941: 222 (N. Carolina); Norris, Plagiost. Hypophysis, 1941: pl. 19, fig. 73–75 (brain); Fowler, Arqu. Zool. Estado São Paulo, *3*, 1942: 127 (Brazil); Lunz, Bull. S. C. St. Planning Bd., *14*, 1944: 27 (Fla.); Fowler, Mongr. Acad. nat. Sci. Philad., 7, 1945: 74 (Smith I., Virginia, 242 lb.), 159 (S. Carolina); Bigelow and Schroeder, Guide Comm. Shark Fish., Anglo Amer. Carib. Comm., Wash., 1945: 105, fig. 36 (descr., ill., habits, range).

Great White Shark, Allen, Bull. Boston Soc. nat. Hist., *24*, 1921: 8 (Woods Hole).

2. Pacific Ocean, Indian Ocean, South Africa:

Carcharodon capensis Smith, A., Ill. Zool. S. Afr. Pisces, 1849: pl. 4 (S. Africa).

Squalus carcharias, Kitlitz, Denkwurd. Reise Micronesien, 2, 1858: 188 (Bonin I., not seen); Bennett, Gatherings Nat. Australias, 1860: 26 (Aust.); Jouan, Mem. Soc. Imp. Sci. Nat. Cherbourg, *8* Suppl., 1861: 5 (New Caledonia).

Carcharias leucas Bennet, Proc. zool. Soc. Lond., 27, 1859: 223 (New South Wales); Jouan, Mem. Soc. Imp. Sci. Nat. Cherbourg, *14*, 1868: 229 (N. Zealand).

Carcharodon rondeletii Hector, Col. Mus. Govt. Surv. Dept. N. Z., 1872: 78 (N. Zealand, not seen); Kluntzinger, S. B. Akad. Wiss. Wien, *80* (1), 1880: 426 (Aust.); Macleay, Proc. Linn. Soc. N. S. W., *4*, 1880: 459 (Aust.); Günther, Introd. Stud. Fish., 1880: 320 (Aust., largest recorded specimen); Ramsay, Proc. Linn. Soc. N. S. W., *5*, 1880: 96 (Aust.); Macleay, Proc. Linn. Soc. N. S. W., *6*, 1881: 358 (Aust.); Tenison-Woods, Fish Fish. N. S. W., 1882: 25 (Aust.); McCoy, Prod. Zool. Victoria, Decade 8, 1883: pl. 24 (descr., Aust.); Haswell, Proc. Linn. Soc. N. S. W., *9*, 1885: 83, pl. 1, fig. 1–4 (skelet.); Ogilby, Rep. Comm. Fish. N. S. W., *2A*, 1887: 2 (Aust.); Parker, Proc. zool. Soc. Lond., 1887: 27, pl. 4–8 (anat., embryo, N. Zealand); Philippi, An. Univ. Chile, *71*, 1884: 550, pl. 4, fig. 4 (ill., tooth, Chilean records); Etheridge, Proc. Linn. Soc. N. S. W., (2) *3*, 1888: 159 (Aust.); Ogilby, Cat. Fish. Aust. Mus., *14*, 1888: 5 (Aust.); Proc. Linn. Soc. N. S. W., (2) *3* (4), 1889: 1771 (Aust., abund.); Lucas, Proc. roy. Soc. Vict., 2, 1890: 43 (Aust.); Günther, J. Mus. Godeffroy, *6*, Fish. Sudsee *3*, Heft 9, 1910: 485 (Pacific); Quijada, Bol. Mus. nac. Chile, *5*, 1913: 111 (listed for Chile).

Carcharodon carcharias Ogilby, Handbk. of Sydney, 1898 (Aust., not seen); Jordan and Snyder, Annot. zool. jap., *3* (2–3), 1901: 127 (Japan); Jordan and Fowler, Proc. U.S. nat. Mus., *26*, 1903: 624 (Japan); Waite, Mem. N. S. W. Nat. Club, 2, 1907: 8 (Aust., not seen); Jordan and Evermann, Bull. U.S. Bur.

Fish., *23* (1), 1905: 44 (Hawaiian Is.); Stead, Fish. Aust., 1906: 223 (Aust.); Waite, Rec. Canterbury [N. Zealand] Mus., *1*, 1907: 6 (N. Zealand); Zietz, Trans. roy. Soc. S. Aust., *32*, 1908: 291 (Aust.); Stead, The Lone Hand, Sydney, 1913: 35 (Aust., not seen); Ogilby, Mem. Qd. Mus., *5*, 1916: 74 (Aust.); McCulloch, Aust. Zool., *1* (7), 1919: 223, pl. 17, fig. 23a (Aust.); Waite, Rec. S. Aust. Mus., *2*, 1921: 21 (Aust.); Jordan and Jordan, Mem. Carneg. Mus., *10*, 1922: 5 (Hawaiian Is.); McCulloch, Roy. Soc. N. S. W., Check List Fish., 1922: 8, pl. 2, fig. 23a (Aust.); Waite, Fish S. Aust., 1923: 40 (Aust.); Phillipps, N. Z. J. Sci. Tech., *6*, 1924: 269 (N. Zealand); Barnard, Ann. S. Afr. Mus., *21* (1), 1925: 33 (S. Afr.); McCulloch and Whitley, Mem. Qd. Mus., *8*, 1925: 129 (Aust.); Jordan and Hubbs, Mem. Carneg. Mus., *10*, 1925: 102 (Japan); Fowler, Proc. Acad. nat. Sci. Philad., 77, 1926: 190 (E. Afr.); Herre, Philipp. J. Sci., *26*, 1925: 114 (Philippines); Barnard, Ann. S. Afr. Mus., *21*, 1927: 2, pl. 1, fig. 7 (S. Afr.); Phillipps, Fish. Bull. Wellington, N. Z., *1*, 1927: 9 (synonyms, N. Zealand); Whitley, Aust. Mus. Mag., *3*, 1927: 13 (Aust.); McCulloch and Whitley, Fish. roy. soc. N. S. W., 2nd ed., 1927: 8, pl. 2, fig. 23; also 3rd ed., 1934 (Aust.); Fowler, Mem. Bishop Mus., *10*, 1928: 18 (Pacif.); Roughley, Aust. Mus. Mag., *3*, 1928: 152 (Aust.); McCulloch, Mem. Aust. Mus., *5*, 1929: 15 (Aust.); Chapman, Open Air Stud. Aust., 1929: 20, pl. opp. p. 68 (Aust., not seen); Soldatov and Lindberg, Bull. Pacif. Fish. Res. Sta., *5*, 1930: 12 (part);[40] Fowler, Proc. Pan-Pacif. sci. Congr., *3* (4), 1930: 488 (Pacific); Tanaka, Jap. Fish. Life Colors, 20, 1933 (Japan, not seen); Coppleson, Med. J. Aust., April 1932: 11 (reputation as man-eater); Marchand, Fish. Mar. Biol. Surv. S. Afr., Fish Bull. 2, 1935: 31 (Natal, S. Afr.); Walford, Fish. Bull. Sacramento, *45*, 1935: 38 (California); Barnhart, Mar. Fish. south. Calif., 1936: 10 (off California); Grey, Amer. Angler in Aust., 1937: 48, pl. 34, 35 (Aust., not seen); Tubb, Proc. roy. Soc. Vict., *49* (2), 1937: 422 (Aust.); Roxas and Marten, Dept. Agr. Comm. Manila Tech. Bull., *6*, 1937: 12 (Philippines, not seen); Beebe and Tee-Van, Zoologica, N. Y., *26*, 1941: 98 (general); Fowler, Bull. U.S. nat. Mus., 100 (*13*), 1941: 110 (Indo-Pacific records); Bonham, Copeia, 1942: 264 (coast of Washington, weights).

Mako, Anonymous, Lond. Illustr. News, July 14, 1942: 53 (C. Province, S. Afr., caught by angler, size, photo).

Carcharias maso Morris, Aust. Eng., 1898: 412 (not seen); not *Squalus* (*Carcharias*) *maou* Lesson, Voy. "Coquille," Zool., 2, 1830: 91, pl. 1.

White Pointer, Stead, Giants and Pigmies of the Deep, 1933: 54 (Aust.).

Carcharhinus carcharias Whitley, Mem. Qd. Mus., *10* (4), 1939: 199 (N. Zealand).

Carcharodon albimors Whitley, Aust. Zool., *9* (3), 1939: 240 (Aust.); Fish. Aust., *1*, 1940: 126 (Aust.).

Great White Shark, Kean, J. Amer. med. Ass., July 23, 1944: 846 (fatal attack, Panama Bay); Scheffer and Slipp, Amer. Midl. Nat., *32*, 1944: 399 (8-ft. 2-in. specimen with entire seal in stomach; coast of Washington).

Doubtful References:

Squalus carcharias Forskål, Descr. Anim., 1775: VIII, 20 (Red Sea, descr. insufficient); Forster, in Latham and Davies, Faunula Indica, 1795: 13 (India, name only); Labillardière, Rel. Voy. Rech. de La Pérouse, *1*, 1800: 40, 226, 399 (near Teneriffe, E. Indies, S. W. Aust., name only); Bowditch, Excurs. Gambia (1823), 1825: 233 (Gambia; name only); Forster, Descr. Anim. Obs. Del. Cur. Lichtenstein, 1844: 256 (Tanna, New Hebrides, name only).

Carcharias vorax Owen, Cat. Osteol. Roy. Coll. Surg., *1*, 1853: 94 (brief notice of vertebrae, teeth, 23-ft. spec.).

Carcharodon rondeletti Quijada, Bol. Mus. nac. Chile, *5*, 1913: 1, 11 (Chile, name only).

Not *Squalus carcharias* Pallas, Zoogr., Rosso Asiat., *3*, 1831: 63[41] (the localities, Kamchatka and Bering Sea, make it probable that these records actually refer to some other shark).

Not *Carcharodon atwoodi* Uhler and Lugger, Rep. Comm. Fish. Maryland, 1st ed., 1876: 191; 2nd ed., 1876: 161 (reported "common" in Chesapeake Bay, hence doubtless some other shark).

40. Includes report by Pallas, 1831, of *Squalus carcharius* from Kamchatka and Bering Sea; probably not this species.

41. First printing may have been 1814: see Sherborn, Ibis, (13), 4, 1934: 166.

Family CETORHINIDAE

Characters. Essentially those of Isuridae (p. 109), except that each gill arch bears a great number of long, horny bristle-like rakers directed forward, analogous to those of many bony fishes; the gill openings are very much larger; the teeth are minute, very numerous, and conical with one cusp; the dorso-rostral cartilages are very slender, and the ventro-rostral cartilage broad and blade-like[1] (in the Isuridae all three of the rostral cartilages are rod-like, and about equally stout).

Remarks. The Cetorhinidae have usually been placed among the Isuridae, of which they appear to be an offshoot. However, the presence of horny rakers on their gill arches, a character which makes them unique among modern sharks, suggests to us that they be classed as a distinct family.

Genera. Only one genus, *Cetorhinus* Blainville, 1816.

Genus *Cetorhinus* Blainville, 1816[2]

Basking Sharks

Cetorhinus Blainville, Bull. Soc. philom. Paris, 1816: 121; type species, *Squalus gunnerianus* Blainville, 1810, equals *Squalus maximus*, Gunnerus, 1765.[3]

Generic Synonyms:

Squalus (in part) Gunnerus, K. norske Vidensk.-Selsk. Skr. Trondh., 1765: 33; and subsequent authors; not *Squalus* Linnaeus, 1758.

Selache Cuvier, Règne Anim., *2*, 1817: 129; type species, *Selache maxima* Cuvier, equals *Squalus maximus* Gunnerus, 1765.

Selanche Jaroki, Zoologi, *4*, 1822: 452 (not seen); type species, *S. maximus* Jaroki, equals *Squalus maximus* Gunnerus, 1765.

Selachus Minding, Lehrb. Naturg., 1832: 52 (not seen); type species, *Selachus maximus* Minding, equals *Squalus maximus* Gunnerus, 1765.

Polyprosopus Couch, Brit. Fishes, *1*, 1867: 68; type species, *P. macer* Couch; type locality English Channel.

Cethorhinus Escribano, An. Soc. esp. Hist. nat., *9*, 1909: 340; type species, *C. maximus* Escribano, equals *Squalus maximus* Gunnerus, 1765.

Doubtful Synonyms:

Halsydrus Fleming, Scots. Mag. Edinb. Misc., 1809: 6; type species, *H. pontoppidani* Fleming; type locality, Orkney Islands.[4]

Tetraoras Rafinesque, Carratt. Gen. Nuov. Sicil., 1810: 46; type species, *T. angiona* Rafinesque.

1. The dorso-rostral cartilages have been pictured either as uniting some distance posterior to the point of union between the resultant bar and ventral cartilage (Senna, Arch. ital. Anat. Embriol., *22*, 1925: pl. 9, fig. 1, 2), or as connected with each other by a pair of transverse bars which unite in the median line and extend thence forward as a single member to the point of union with the ventral cartilage (Pavesi, Ann. Mus. Stor. nat. Genoa, *6*, 1874: pl. 2, fig. 1, 2).
2. For reasons why *Cetorhinus* is retained for this genus rather than *Halsydrus* Fleming, 1809, see footnote 4, p. 146.
3. Type designated by Jordan (Genera Fish., 1, 1917: 95) as *C. gunneri* Blainville, 1816, which was a substitution for *Squalus gunnerianus* Blainville, 1810.
4. Whitley (Mem. Qd. Mus., *10*, 1934: 196), followed by Fowler (Bull. U.S. nat. Mus., 100 [*13*], 1940: 112), has replaced the generic name *Cetorhinus* with *Halsydrus* on the ground that the carcass of the Orkney animal, for which the latter was proposed, was actually that of a very large Basking Shark, as is certainly suggested by pub-

Generic Characters. Those of the family.

Range. Temperate belts of North and South Atlantic including the Mediterranean, North and South Pacific and southern Indian Ocean.

Fossil Gill Rakers. Oligocene to Pliocene, Europe.

Species. *Cetorhinus* had long been thought to be monotypic, but Whitley[5] has recently discussed its Australian representative under a name *maccoyi* Barrett,[6] distinct from that of its northern Atlantic representative *maximus.* Comparison of Whitley's photographs of an Australian specimen 25 feet long with a Massachusetts Bay specimen of about the same size, and pictured below, suggests that a longer caudal and perhaps a higher first dorsal may prove diagnostic for the former. If correct, this opens the whole question of the specific relationship of the Basking Sharks of the western and eastern South Atlantic[7] and of the northern and southeastern Pacific to the North Atlantic form. The discontinuity of geographic distribution suggests that actually the genus may include several species instead of one only. But definite decision must await critical comparison of specimens from different seas, or at least of comparable measurements and photographs.

Cetorhinus maximus (Gunnerus), 1765

Basking Shark, Bone Shark

Figures 23, 24

Study Material. Mounted specimens, about 26 feet 6 inches long (New Eng. Mus. Nat. Hist.) and about 14 feet 6 inches, from New Jersey (Amer. Mus. Nat. Hist.); head of a 12-foot specimen from Fire Island, New York; gills and gill rakers of another from same locality (Amer. Mus. Nat. Hist.); excellent photographs of a specimen about 15 feet long taken off Jones Inlet, New York, June 20, 1941.[8]

Distinctive Characters. The combination of lunate caudal, enormously long gill openings, long rakers on the gill arches, very many minute teeth, and nostrils widely separated from mouth, sets *Cetorhinus* apart from all other sharks.

Description. Proportional dimensions in per cent of total length. Female, 4,400 mm. (4,318 mm. between perpendiculars) from about 15 miles S by E of Long Branch, New Jersey.[9]

lished illustrations of its cranium, vertebrae and pelvic skeleton (Barclay, Mem. Werner. Soc., *1*, 1811: 418). But we agree with Norman ("Discovery" Rep., *16* [2], 1937: 7, footnote 2) that nothing would be gained by abandoning a name as old and as generally used as *Cetorhinus,* at least until some modern student establishes, by personal examination of the remains in question (if they are still in existence), that they actually are those of a Basking Shark and not of some other very large species.

5. Fish. Aust., *1*, 1940: 132.

6. Sun-Nature Book, Pt. IV, Water Life, 1933: 13.

7. Norman ("Discovery" Rep., *16*, 1937: 7) had already suggested that the Falkland Island form may be distinct from the northern. For a recent description of the South African form, see Barnard (Ann. S. Afr. Mus., *32* [2], 1937: 43).

8. Received from New York *Herald Tribune.*

9. This specimen is mounted in the American Museum of Natural History, New York, and the above proportions are based on measurements made by E. W. Gudger at the time of its capture. Measurements of body, fins and gills were made on the curvature.

Snout length: to angle of jaw, in straight line 16.1.
Eye: horizontal diameter 1.0.
Gill opening lengths: 1st 25.7; 2nd 23.1; 3rd 20.0; 4th 17.1; 5th 14.6.
First dorsal fin: height 9.8; length of base 9.8.
Caudal fin: upper margin 22.3; lower anterior margin 13.9.
Pectoral fin: length 17.6.
Distance from snout to: 1st dorsal 36.3; 2nd dorsal 65.3; upper caudal 77.6; pectoral 27.7; pelvics 55.6.
Distance from origin to origin of: 1st and 2nd dorsals 29.1; 2nd dorsal and caudal 13.0; pectoral and pelvics 28.5; pectoral and anal 45.0.

Trunk fusiform, stoutest from shoulders to 1st dorsal, tapering rearward to moderately stout caudal peduncle, the latter somewhat flattened dorso-ventrally, with strongly developed lateral keel on either side originating opposite the tip of anal and extending well out on caudal fin. Well developed precaudal pits both above and below, in the form of lunate furrows. Dermal denticles small, but of various sizes, in patches or stripes with

FIGURE 23. *Cetorhinus maximus.* Drawing based on adult female, 26 feet 6 inches long, mounted in New England Mus. Nat. Hist., and on other available information. *A, B* Head of 12-foot specimen from Fire Island, New York (Amer. Mus. Nat. Hist.). *C* Sector of upper jaw of same, about 1.4 x. *D* Lateral view of two teeth, about 4 x. *E* Dermal denticles from back, above origins of pectorals, about 8 x. *F* Left-hand nostril, about natural size.

bare spaces between,[10] erect, close-set, thorn-like, with recurved tips having a median ridge along the anterior face, their bases large and corrugated.

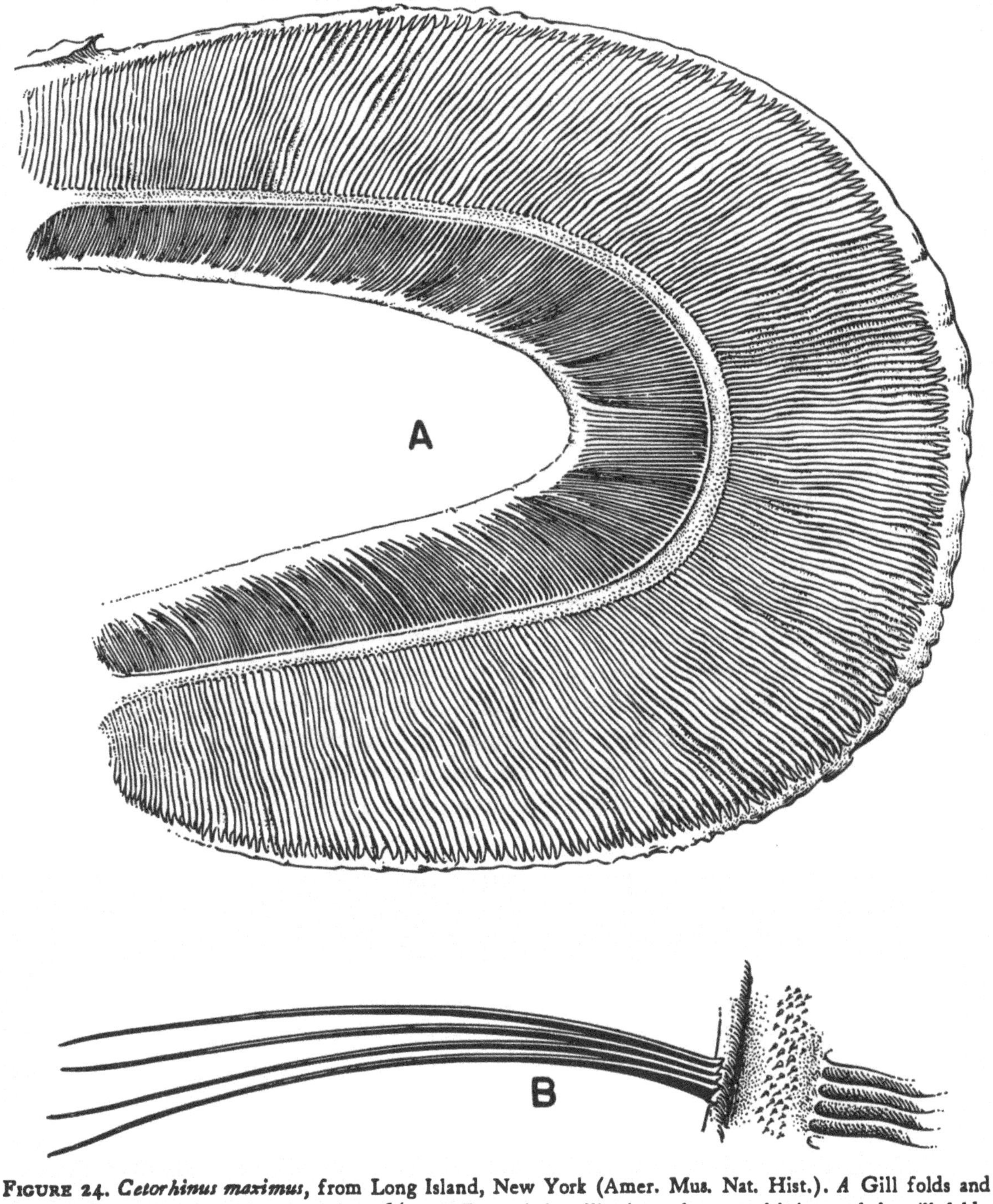

FIGURE 24. *Cetorhinus maximus*, from Long Island, New York (Amer. Mus. Nat. Hist.). *A* Gill folds and gill rakers of one of the gill-arches, about ⅓ x. *B* Four of the gill rakers of same, with bases of the gill folds, about 2 x.

10. Radcliffe (Bull. U.S. Bur. Fish., *34*, 1916: 248) gives an excellent illustration from a North Pacific specimen.

Head slightly compressed laterally opposite mouth (strongly so in small specimens). Snout very short, subconical, with rounded tip in larger specimens but relatively much longer in small ones, forming a subcylindrical proboscis, obliquely truncate in front, terminating dorsally in a sharp point, with many circular pores on its dorsal surface; transition from the juvenile to the adult form takes place at lengths of 12 to 16 feet. Eyes nearly circular without nictitating membrane or subocular folds, their diameters only about ⅛ as great as distance between them opposite, or a little posterior to, front of mouth. Spiracles described as minute, circular, a little posterior to angles of jaws or opposite latter. Gill openings very large, extending from upper sides down onto lower surface of throat, the 1st longest, the 5th shortest, the 1st pair separated below by 6 inches only, the 2nd pair by 9 inches, the 4th pair by 21 inches and 5th pair by 27 inches in a specimen[11] 30 feet 3 inches long. Gill rakers very numerous (about 1,260 on gill studied), flattened basally on the adjacent sides but bristle-like toward the tips, in a continuous series, and directed inward; 1 series on the 1st gill arch, 2 series on the 2nd, 3rd and 4th and only 1 series described for the 5th.[12] Nostrils wide apart at outer edges of snout, small, transverse, considerably nearer to mouth than to tip of snout in young specimens, less so in adults because of decrease in relative length of snout, their anterior margins slightly expanded in subtriangular outline. Mouth very large, occupying most of breadth of head, rounded in adult but varying in young from nearly transverse, with corresponding lateral expansion of sides of head behind the eyes, to broadly V-shaped; these variations probably associated with wide distensibility of mouth and loose articulation at symphyses. A very short labial furrow at corner of mouth on lower jaw, but none on upper.

Teeth minute, being only about 3 mm. high in specimen about 12 feet 9 inches (3,900 mm.) long and about 6 mm. in one of 30 feet; in 4 to 7 functional series, with 100 or more teeth in each row on each side of jaw; those toward center of mouth low and triangular, but those along the sides conical, slightly recurved, somewhat compressed laterally, with a ridge on each side, the basal part striated; a wide space with only scattered teeth at center of upper jaw (106 mm. wide in 12-foot specimen) but not of lower jaw.

First dorsal fin an approximately equilateral triangle, its anterior margin nearly straight, its posterior margin slightly concave or even slightly convex in some cases, its apex subacute, its free rear corner extending only a slight distance beyond the rear termination of its base, the height along anterior margin varying from about 11 to 14% of total length, its origin considerably behind the inner corner of the pectoral when latter is laid back; the midpoint of its base about midway between tip of snout and fork of caudal. Second dorsal's anterior margin only ¼ to ⅕ as high as 1st, with rounded apex, strongly concave rear margin and free rear tip about as long as its base. Caudal ¼ to ⅕ of total length, lunate, its axis steeply raised as in Isuridae, its posterior outline obtusely subangular rather than rounded, with well marked subterminal notch, its lower anterior

11. Storer, Mem. Amer. Acad. Arts Sci., (2) 9, 1867: 229.
12. See White (Bull. Amer. Mus. nat. Hist., 74, 1937: pls. 7, 8) for photographs of gill rakers in position.

margin (lower lobe) about 60–65% as long as upper, each measured from the precaudal pit, its tips subacute. Anal similar to 2nd dorsal, and about as large, its origin under rear part of base of latter. Pelvics about ⅔ as high as 1st dorsal along anterior margin. Claspers described as about 3 feet 3 inches long in 30-foot male. Pectoral with straight or slightly concave distal margin and blunt tip, but broadly rounded inner corners, relatively smaller than in *Isurus*, the length along anterior margin being only about ⅕ of distance from snout to origin of caudal.

Color. Grayish-brown to slaty gray, or nearly black above. The under parts may be either uniformly of the same color as the back, of a paler shade of the same, or grading into white, sometimes with a triangular white patch under the snout and with two pale bands along the ventral surface on either side of the midline or otherwise marked with white, there being a wide variation in this respect.

Size. The Basking Shark rivals, although it does not equal, the Whale Shark (p. 192) in size. It has been credited repeatedly with reaching a maximum length of 40 to 50 feet. For Basking Sharks to reach lengths of 35 to 40 feet is not exceptional, for one of about 45 feet and three of about 40 feet, as well as smaller ones, were taken on the Norwegian coast during the period 1884 to 1905.[13] The six next longest actually measured were 36 feet; 32 feet 2 inches; 32 feet; 31 feet; 30 feet 6 inches; and 30 feet 3 inches. The four largest, for which we find exact measurements for the western Atlantic, were 32 feet 2 inches, 32 feet, 30 feet 3 inches, and 26 feet 6 inches, although others up to 40 feet have been reported without supporting evidence. Similarly, the longest of 21 Basking Sharks landed in Monterey, California, from November to February of 1931, was about 28 feet; the largest ever sold to the particular fishery firm in question was a few inches less than 30 feet.[14]

The smallest free-living specimens of which we find record were of 5 feet 5 inches,[15] 8 feet 4 inches,[16] and about 8 feet 6 inches (2.6 m.),[17] which suggests that Basking Sharks are as a rule at least 5 to 6 feet long at birth. Males mature at a length of perhaps 15 to 20 feet as indicated by the presence of small claspers in specimens up to about 11 feet, with very large ones in specimens of 25 to 26 feet or longer.[18] Similarly, most described specimens of less than 11 to 13 feet have had the immature, proboscis-like form of snout. On the other hand, a 14-foot 3-inch specimen taken recently near New York showed an intermediate state,[19] and all specimens of 20 feet or upward, for which adequate information is at hand, have been of adult conformation in this respect.

We have not succeeded in finding precise weights for the larger sizes in the Atlantic. But since the two Monterey specimens just quoted actually weighed 6,580 pounds at 28 feet and 8,600 pounds at about 30 feet, this no doubt is a fair indication of the weight of

13. Collett, Norges Fiske, *3*, 1905: 83–86.
14. McGinitie, Science, N.S. *73*, 1931: 496.
15. Thompson, Nat. Hist. Ireland, *4*, 1856: 253.
16. Pengelly, Zoologist, (3) *5*, 1881: 337.
17. Nobre, Fauna Marinha Port. Vert., *1*, 1935: 441.
18. Pavesi, Ann. Mus. Stor. nat. Genoa, *12*, 1878: 398, 406.
19. Gudger, J. Morph., *57*, 1935: 96.

the Atlantic specimens, there being no reason to suppose that Atlantic specimens would differ very widely from those of the Pacific. Estimated weights of smaller specimens are: about 6,600 pounds at about 23 feet, 1,000 to 1,800 pounds at 13 to 15 feet, and 800 pounds at 8 feet 4 inches.[20]

Developmental Stages. Developmental stages have not been described, except as noted (p. 152).

Habits. Basking Sharks are sluggish and inoffensive fish. When in coastwise waters they spend much time lying at the surface with backs awash, their dorsal fins standing high above the water with tip of snout and caudal showing; or they swim slowly, with mouth open gathering their diet of plankton. They are also described as sometimes lying on their backs sunning their bellies. They are so little disturbed by boats that it is easy to approach them closely; in fact, excellent moving pictures of them have been taken off Ireland.[21] However, on occasion they are reported as jumping, perhaps in an attempt to shake off remoras or parasites. They often gather in schools of up to 60 or 100 individuals and there are reports of two or three swimming tandem.

In the Gulf of Maine and off the middle Atlantic coast of the United States, as well as in the northern part of their range in the opposite side of the Atlantic, Basking Sharks appear almost exclusively during the warm half of the year,[22] and the early accounts suggest some movement northward during the summer in northern European waters. The winter habitat of the northern species is not known for either side of the Atlantic, although lack of evidence of any increase in abundance to the southward suggests that they simply retire in the fall and winter to deeper water.[23] If so, the scarcity of animal plankton that prevails generally in boreal seas during winter must result in very poor feeding for them, suggesting that they are generally inactive at that time, perhaps lying on or close to bottom.

The only definite information as to breeding habits is the report, more than a century and a half old, that an embryo about one foot long was taken from the mother.[24] It is not known at what season the young are born, for while it has been stated that their habit of schooling is associated with breeding, this seems more likely connected with their pursuit of planktonic food. However, it seems certain that young are produced throughout their entire range, for small ones have been reported both from the north (Ireland, Norway) and from the south (Mediterranean).

The diet of the Basking Shark consists wholly of small planktonic organisms which it sifts out of the water by means of its gill rakers, as do such plankton feeders as some clupeoids, anchovies and whalebone whales. Usually the stomach contents are simply a

20. An estimated weight of about 3,000 pounds for one between 12 and 14 feet long was probably far too high.
21. In the widely popular film, "Men of Arran."
22. A skeleton found on the beach near Provincetown, Massachusetts, in January 1939, may have been there for months.
23. On this, see Lübbert and Ehrenbaum (Handb. Seefisch. Nordeurop., 2, 1936: 281).
24. Pennant, Brit. Zool., 3, 1776: 101.

soupy or gelatinous mass. On several occasions, however, this has been found to consist chiefly of minute crustacea, this being true of the only western Atlantic specimen whose stomach contents has been recorded.[25]

Abundance. The published records show that there is much variation in the number of Basking Sharks in the centers of abundance over periods of years. For example, great schools were seen during the summer of 1776 and for a few succeeding summers off the coast of Wales,[26] but no comparable numbers have ever been reported there subsequently. Similarly, along the Norwegian coast, where Basking Sharks formerly supported an intermittent fishery, a paucity in the first half of the 18th century and again around 1840 alternated with a comparative abundance around 1800 and 1880; since then only occasional specimens have been reported yearly from one Norwegian locality or another.[27] Similar fluctuations are also reported for the western Atlantic, but with less definite evidence (see p. 155).

Basking Sharks Reported as Sea Serpents or Other Monsters. Without entering into the controversy regarding the so-called "sea serpent," we may point out that the Basking Shark has formed the demonstrable basis of sea serpent stories on several occasions; "as the carcase of the shark rots on the shore, or is buffeted against the rocks, the whole of the gristly skeleton of the jaws and gill arches, by far the bulkiest part of the head skeleton, as well as the pectoral and pelvic fins, is soon washed away, leaving only the backbone and the somewhat curiously shaped box-like cranium."[28] As a recent instance of this nature we may point to the wide publicity given by the press and radio to a supposed sea serpent whose identity was based upon the skeleton of a Basking Shark about 25 feet long that was stranded on the beach at the tip of Cape Cod near Provincetown, and which we were able to examine.[29] Also, it has been suggested repeatedly that the dorsal and caudal fins of Basking Sharks, swimming in line at the surface, have been the basis for stories of at least some of the reported sea serpents or other supposed monsters, especially in northern Scandinavian waters.

Relation to Man. The livers of medium-sized to large Basking Sharks will yield anywhere from about 80 to 200 gallons of oil, and occasionally as much as 400 gallons, with a maximum reported yield of 600 gallons.[30] As the oil is considered nearly or quite equal to sperm oil for use in lamps, it was readily saleable up to the time when animal oils were replaced by petroleum products for lighting. For example, the oil of a specimen taken at Provincetown in 1836 or 1837 yielded its captor $103; even as far back as the last part of the eighteenth century a large one in British waters was said to be worth the equivalent of

25. Hussakoff, Copeia, 21, 1915: 25.

26. Pennant, Brit. Zool., 3, 1776: 102.

27. Collett (Norges Fiske, 3, 1905: 83) lists about 25 records for the period 1881–1905. See also Lübbert and Ehrenbaum (Handb. Seefisch. Nordeurop., 2, 1936: 281) for general summary of fluctuations.

28. Norman and Fraser, Giant Fishes, 1937: 21.

29. For detailed account, with photograph, see Schroeder (New Engl. Natural., 2, 1939: 1).

30. Stevenson (Rep. U.S. Comm. Fish. [1902], 1904: 227); many of the older records of yield are expressed in "barrels" of unknown volume.

20 pounds sterling.[31] Their oil and sluggish nature made the Basking Shark the object of intermittent small-boat fisheries with harpoon wherever and whenever they appeared in any numbers, especially in Irish and Norwegian waters and around Iceland. Similarly, the Pacific Basking Shark has supported, and probably still does, a local fishery of small boats, each manned by six or eight men, off the coasts of Peru and Ecuador. Also, considerable numbers are landed in California, where they are utilized for oil and fish meal.

The larger whaling vessels also pursued them in earlier days whenever encountered; for an instance of this in the Gulf of Maine, see p. 155. But it is now more than 100 years since Basking Sharks have been plentiful enough on the western side of the North Atlantic for more than incidental capture. With its large yield one might wish that the liver oil of the Basking Shark had a high vitamin content, but this appears not to be the case.

Range. Once thought to be an Arctic species, and often so characterized, the Basking Shark is now known to be an inhabitant of temperate and boreal waters. In the North Atlantic its range is bounded on the north by a line extending from the eastern side of the Gulf of Maine and Newfoundland to the western and southern coasts of Iceland, the Orkneys, the Faroes and northward along western Norway to the North Cape, with occasional reports of it from the Murman Coast. In general, this line marks the zone of transition from the region of influence of Atlantic waters to those of Arctic waters.[32]

To the southward in the eastern side of the Atlantic it is reported occasionally from the English Channel and the North Sea as far as the Skagerrak and Kattegat (never from the Baltic), along the coasts of France and the Iberian Peninsula, from Madeira, Morocco and the Mediterranean. On the western side it is reported as far as North Carolina. At present its chief centers of abundance appear to be west and south of Iceland, along western Ireland, among the Orkneys, and off southwestern Norway. There is no evidence that it occurs at all in the tropical Atlantic. However, it is represented on both sides of the South Atlantic off South Africa, Argentina and the Falkland Islands, in the South Pacific off Peru and Ecuador, off southern Australia and New Zealand, and in the northern Pacific from California to British Columbia as well as in Japanese and Chinese waters, by a form (or forms) whose precise relationship to the Basking Shark of the North Atlantic is still to be determined (p. 147).

Occurrence in the Western Atlantic. There is no reason to suppose that the Basking Shark ever occurred, other than as a stray, north of about 44° to 45° N. in the western North Atlantic, there being only four positive records of it from the southern part of Newfoundland: one from the outer coast of Nova Scotia, three from the Bay of Fundy and a few from the vicinity of Eastport, Maine, at the mouth of that bay. In colonial days the southern and western parts of the Gulf of Maine appear to have supported a considerable population of them, however, for by old reports many were taken in Massachu-

31. Pennant, Brit. Zool., 3, 1776: 174. An estimate of 80 pounds sterling (Day, Fish. Gt. Brit., 2, 1880–1884: 306) seems too high.

32. There is no recent report of it for any Arctic locality; nor does Jensen (Mindskrif. Japetus Steenstrup, 2 [3], 1914) include it in his survey of the sharks of Greenland.

setts waters, especially off the tip of Cape Cod, in the first half of the eighteenth century, the oil being then in demand for illuminating purposes. But the local stock soon went the way of the Atlantic Right Whale in these same waters, *i.e.*, into the try-pot.

The only positive records of them north of Cape Cod since 1840, of which we have learned, are as follows:

1840, a number seen, and several captured, by a whaler off Cape Elizabeth, Maine.

1847, one killed near Provincetown at the tip of Cape Cod.

1851, a large one reported as about 40 feet long captured at Musquash Harbor, New Brunswick, near the mouth of the Bay of Fundy.

1864, one harpooned but lost in Massachusetts Bay.

1868–1870, several, 25 to 35 feet, killed near Eastport, Maine, at the mouth of the Bay of Fundy.

1876, one stranded in Conception Bay, Newfoundland.

1908, one about 18 feet taken near Provincetown, Mass., in a weir.

1909, one about 22 feet in Provincetown Harbor.

1913, one about 29 feet, Provincetown Harbor.

1925, one about 30 feet off Portland, Maine.[33]

1931, female, 12½ feet, at York Harbor, Maine.

1934 (?) three records from Newfoundland at Petty Harbor, the vicinity of St. John and at Placentia; the last one 32 feet long.

1936, two specimens off Portland, Maine, the first a small one about 20 feet long and 550 pounds dressed, taken about May 1st, the second a large specimen reported to have been about 40 feet, taken August 2nd.

1938, one washed ashore near French Village, Halifax County, Nova Scotia, of which we received a clearly recognizable photograph.

1939, January, a skeleton washed ashore near Provincetown and reported as a Sea Serpent (see p. 153).

Unknown date, a 31-foot specimen taken at Long Point, near Provincetown, Mass.

The hiatus in the foregoing list between 1876 and 1908 probably reflects the fact that fishes generally, and especially sharks, in the Gulf of Maine received little scientific attention during that period. But this large shark is probably no more plentiful now than the paucity of the recent record suggests, for, so great has been the popular interest in sharks of late, and so wide the newspaper publicity given to unusual captures, that any large specimen is almost certain to be reported sooner or later—even if not captured—in such frequented and hard-fished waters as those of the coastwise belt of the Gulf of Maine.

Near Woods Hole, a few miles west of Cape Cod, an incursion by Basking Sharks appears to have taken place in the summer of 1878, when at least twenty were found dead in the local fish traps. However, only occasional specimens have been reported more recently, *e.g.*, one of 26 feet 6 inches (see Study Material, p. 147) taken at Martha's

33. Personal communication from Walter H. Rich.

Vineyard, June 24, 1920, and another of 20 feet 2 inches, stranded in the landlocked waters of a small harbor (Hadley's) on Naushon Island in July 1937. There is nothing in the published record to suggest that the zone of most frequent occurrence ever extended much farther west or south than this along the North American coast, there being occasional records only for Long Island,[34] one or two near New York (one in New York Harbor many years ago) and about six for the coast of New Jersey, with one probable and one positive record for North Carolina.[35] The only reports of Basking Sharks farther south in the western Atlantic are for northern Argentina[36] and the Falkland Islands,[37] which may be distinct from those of the North Atlantic (p. 147).

Synonyms and References:

North and South Atlantic and South Africa:[38]

Squalus maximus Gunnerus, K. norske Vidensk.-Selsk. Skr. Trondh., 1765: 33, pl. 2 (type locality, Trondhjem, Norway); Drontheim Gesellsch. Schr. Leipzig, *3*, 1767: 28, pl. 2 (German translation of the foregoing); Linnaeus, Syst. Nat., 12th ed. 1766: 400 (descr.); Gunnerus, K. norske Gesellsch. Wiss. Skr. Kbh. u. Leipzig, *4*, 1770: 13, pl. 4 (descr., Norway); Olafsen and Povelsen, Reyse en Island, 1772: 988 (Iceland, not seen); Reise durch Island (German translation), *2*, 1774: 204 (South Iceland); Pennant, Brit. Zool., *3*, 1776: 101 (Isle of Arran, embryo, fishery); Olavius, Oecon. Reyse Island, 1780: 80 (Iceland, not seen); Mohr, Forsøg. Island. Naturh., 1786: 60 (Iceland); Bonnaterre, Tabl. Encyc. Meth. Ichthyol., 1788: 10 (descr.); Gmelin, Syst. Nat., *1*, 1789: 1498 (descr.); Bloch and Schneider, Syst. Ichthyol., 1801: 134 (descr.); Olafsen and Povelsen, Voy. en Island, *5*, 1802: 278 (Iceland); Lehmann, Neue Schr. Naturf. Freunde, Berlin, *4*, 1803: 120; Bosc., Nouv. Dict. Hist. nat., *21*, 1803: 185 (diagn.); Latreille, Nouv. Dict. Hist. nat., *24*, 1804: 72 (in table of contents); Turton, Brit. Fauna, 1807: 113 (Gt. Brit.); Home, Philos. Trans., 2, 1809: 206 (descr., anat.); J. Physique, *71*, 1810: 241 (descr., anat.); Philos. Trans., *2* (2), 1813: 227, pl. 6 (descr., anat.); Mitchill, Trans. Lit. Phil. Soc. N. Y., *1*, 1815: 486 (fishery near Provincetown); Couch, Trans. Linn. Soc. Lond., Zool., *14*, 1825: 91 (Cornwall, size); Vrolik, Bijdr. Natuurk. Wetensch. Amsterdam, *1*, 1826: 305 (Holland); Fleming, Hist. Brit. Anim., 1828: 164 (descr., Gt. Brit.); Vrolik, Z. Organ. Physik, *2*, 1828: 490 (Holland); Faber, Fische Islands, 1829: 20 (distrib., habits); Nilsson, Prod. Ichthyol. Scand., 1832: 114 (Scandinavia); Agassiz, L., Poiss. Foss., *3*, 1835–1843: 87, pl. F, fig. 8, 8[a]; Yarrell, Brit. Fish., 1836: 366 (descr., Gt. Brit.); Couch, Cornish Fauna, 1838: 51 (size); Bennett, Narr. Whaling Voy., *2*, 1840: 240 (Gt. Brit.); Owen, Odontogr., 1840–1845: 27 (teeth); Linsley, Amer. J. Sci., *47*, 1844: 77 (Long Island Sound); Gaimard, Voy. Islands et Groenland, Zool. Med., 1851: 163 (Iceland); Schlegel, Natuurl. Hist. Nederland Dieren, 1862: 191, pl. 19, fig. 1 (size); Brito Capello, J. Sci. math. phys. nat. Lisboa, *2*, 1870: 233 (Portugal); Gervais and Gervais, C. R. Acad. Sci. Paris, *82*, 1876: 1237 (descr., Concarneau); J. Zool., *5*, 1876: 319 (descr., ill. of head of juv., gills, rakers, teeth, vertebrae, Concarneau, France); Gervais and Boulart, Poiss., *3*, 1877: 190, pl. 73 (descr.); Buckland, Hist. Brit. Fish., 1881: 215 (Gt. Brit.); Pengelly, Zoologist, (3) *5*, 1891: 337 (Devonshire).

Basking Shark, Pennant, Brit. Zool., *3*, 1769: 38, 342 (descr., distrib., embryo); Shaw, Gen. Zool., *5* (2), 1804: 327, pl. 149, 150 (general); Pennant, Brit. Zool., *3*, 1812: 134 (addit. record); Low, Fauna Orcadensis, 1813: 173 (Orkneys, fishery); Brabazon, Deep Sea Coast Fish. Ireland, 1848: frontispiece, 48 (fishery); Couch, Trans. nat. Hist. Soc. Penzance, 1864: 234 (not seen); Fish. Brit. Isles, 1862: 60, pl. 14 (not seen); *1*, 1867: 60, pl. 14 (descr., habits, Gt. Brit.); Cornish, Zoologist, (2) *5*, 1870:

34. Most recently a 12-foot specimen taken off Fire Island, July 1944 (see Study Material, p. 147).
35. Coles, Proc. biol. Soc. Wash., *28*, 1915: 92; Brimley, J. Elisha Mitchell sci. Soc., *51*, 1935: 311.
36. Lahille, An. Mus. nac. B. Aires, *34*, 1928: 325; Pozzi and Bordale, An. Soc. cient. argent., *120*, 1935: 150.
37. Norman, "Discovery" Rep., *16* (2), 1937: 143.
38. References for the South Atlantic and South Africa are included for convenience.

2253 (descr., meas., Cornwall); Allman, Nature, Lond., *14*, 1876: 368 (gill rakers, food); Harvey, Nature, Lond., *15*, 1877: 273 (Newfoundland); Stevenson, Rep. U.S. Comm. Fish. (1902), 1904: 227 (fishery, yield of oil); Grew, Illustr. London News, Aug. 11, 1945: 166 (general account).

Le Très Grand, Broussonet, Mém. Math. Phys. Acad. Sci. Paris, 1780: 669; Daubenton, Encyc. Method. Hist. nat. Paris, *3*, 1787: 96 (teeth).

Le Squale Très-Grand, Lacépède, Hist. nat. Poiss., 4° ed., *1*, 1798: 165, 209, in Buffon, Hist. Nat. (descr., size); in Sonnini, Hist. Nat. Poiss., *3*, 1801–1802: 400 (general).

Squalus gunnerianus Blainville, J. Physique, 1810: 256, pl. 2, fig. 3 (descr.).

Squalus homianus Blainville, J. Physique, 1810: 257, pl. 2, fig. 1 (descr.).

Squalus pelegrinus Blainville, J. Physique, 1810: 256, pl. 2, fig. 2 (descr.).

Squalus peregrinus Blainville, Bull. Soc. philom. Paris, *2*, 1811: 365 (size); Ann. Mus. Hist. nat. Paris, *18*, 1811: 132 (discuss.).

Squale pelerin Blainville, Ann. Mus. Hist. nat. Paris, *18*, 1811: 88, pl. 6 (detailed descr., ill. of 29-ft. 4-in. male, is the one most often copied).

Cetorhinus gunneri Blainville, Bull. Soc. philom. Paris, 1816: 121 (name).

Cetorhinus homianus Blainville, Bull. Soc. philom. Paris, 1816: 121 (name).

Cetorhinus peregrinus Blainville, Bull. Soc. philom. Paris, 1816: 121 (name).

Cetorhinus shavianus Blainville, Bull. Soc. philom. Paris, 1816: 121 (name); J. Physique, 1816: 264 (not seen).

Selache maxima[39] Cuvier, Règne Anim., *2*, 1817: 129 (general); Règne Anim., 2nd ed., *2*, 1829: 391 (general); Cloquet, Dict. Sci. Nat., *38*, 1825: 308 (general); Bory de St. Vincent, Dict. Class. Hist. Nat., *15*, 1829: 597 (general); Jenyns, Man. Brit. Vert. Anim., 1835: 503 (descr., Gt. Brit.); Richardson, Fauna Boreal. Amer., *3*, 1836: 29; Parnell, Mem. Werner. Soc. Edinb., 7, 1838: 418 (Scot.); Bonaparte, Mém. Soc. neuchâtel. Sci. Nat., *2* (8), 1839: 9 (in synopsis); Owen, Odontogr., 1840–1845: 27 (descr. of teeth of 36-ft. spec.); Müller and Henle, Plagiost., 1841: 71 (descr., distrib.); Cuvier, Règne Anim., ill. ed. Poiss., 1843: 363, pl. 115, fig. 2 (good ill. of teeth); Owen, Cat. Osteol. Roy. Coll. Surg., *1*, 1853: 97 (brief descr., vertebrae, refs.); Traill, Proc. roy. Soc. Edinb., *3*, 1854: 210 (argues "animal of Stronsa" not a Basking Shark); Van der Hoeven, Handb. Dierkunde, 2nd ed., *2*, 1855: 261 (general); Nilsson, Skand. Fauna Fisk., 1855: 720 (Scandinavia); Thompson, Nat. Hist. Ireland, *4*, 1856: 253 (Ireland, fishery); Nardo, Atti Ist. veneto, (3) *5*, 1859–1860: 787 (Medit.); Fitzinger, Bild. Atlas Naturg. Fische, 1864: fig. 173 (ill.); Malmgren, Arch. Naturgesch., *30* (1), 1864: 346 (north. Norway); Duméril, Hist. Nat. Poiss., *1*, 1865: 413, pl. 3, fig. 18 (descr., size, ill., teeth); Bocage and Brito Capello, Poiss. Plagiost., 1866: 14 (Portugal); Malmgren, Oefvers. Vetensk. Akad. Forh. Stockholm, *24*, 1867: 264 (Norway); Brito Capello, J. Sci. math. phys. nat. Lisboa, *2*, 1870: 140 (Portugal); Günther, Cat. Fish. Brit. Mus., *8*, 1870: 394, 518 (descr., distrib., synonyms); Ninni, An. Soc. Nat. Modena, *5*, 1870: 66 (Medit.); Van Beneden, Poiss. Côtes Belg. Parasit., 1870: 70 (parasites, Belgium), also in Mém. Acad. R. Belg. Cl. Sci., *38* (4), 1871: 7; Canestrini, in Cornalia, *et al.*, Fauna d'Ital., 1871–1872: 44 (Medit.); Pavesi, Ann. Mus. Stor. nat. Genova, *6*, 1874: 36, pl. 1–3 (descr.; ill. juv.; anat., Medit.); Collett, Norges Fisk., 1875: 209 (Norway); Trois, Atti Ist. veneto, (5) *1* (6), 1875: 612 (Medit.); Wright, Nature, Lond., *14*, 1876: 313 (gill rakers, diet, value of oil, Ireland); Giglioli, Nature, Lond., *15*, 1877: 273 (name); Hasse, Morph. Jb., Suppl. *4*, 1878: 43, pl. 3, 4, fig. 1 (anat.); Pavesi, Ann. Mus. Stor. nat. Genova, *12*, 1878: 416, pl. 3 (meas., ill. of young male, anat., discus.); Doderlein, Atti Accad. Palermo, *6*, 1878–1879: 30 (Medit.); Stossich, Boll. Soc. adriat. Sci. nat., *5*, 1880: 68 (Adriatic); Turner, J. Anat., Lond., *14*, 1880: 273, pl. 12 (teeth, gill rakers); Day, Fish. Gt. Brit., *2*, 1880–1884: 303, pl. 158, fig. 1 (descr., Gt. Brit.); Doderlein, Man. Ittiol. Medit., *2*, 1881: 70 (Medit.);[40] Moreau, Hist. Nat. Poiss. France, *1*, 1881: 305 (general); Hasse, Naturl. Syst. Elasm. besond. Theil, 1882: 236, pl. 32, fig. 1–5 (vertebrae); Mela, Vert. Fennica, 1882: 365 (Murman coast, not seen); Vieira, Ann. Sci. nat. Porto, *1*, 1894: 137 (Portugal); Brandt, Biol. Zbl., *18*, 1898: 257 (bristles described on snout probably were dermal denticles); Saemundsson, Vidensk. Medd. naturh. Foren. Kbh., 1899:

39. Also variously spelled "*maximus*" or "*maximum*."

40. For additional Mediterranean citations in publications not accessible to us, see Doderlein (above) and Cascia (Bull. Ist. zool. Palermo, *2*, 1935: 173).

420 (Iceland); Facciolà, Revist. Ital. Pesc. Aquic., *15* (4), 1900: 40 (Strait of Messina, meas.); Saemundsson, Vidensk. Medd. naturh. Foren. Kbh., 1903: 51 (Iceland); Carazzi, Zool. Anz., *28*, 1904: 161 (photo of head of juv., meas., Sardinia); Steuer, Verh. zool.-bot. Ges. Wien, *55*, 1905: 275 (gill rakers); Carruccio, Boll. Soc. zool. Ital., (2) 7, 1906: 191 (Medit., not seen); Southwell, Zoologist, (4) *10*, 1906: 355 (a large one, Loch Broom, Scot.); Hendricks, Z. wiss. Zool., *91*, 1908: 427, pl. 18, 19 (anat., histol., gill rakers); S. B. naturh. Ver. preuss. Rheinl. Westf., 1909: 31 (gill rakers); Osorio, Mem. Mus. Bocage Lisbon, *1*, 1909: 49 (food, Portugal); Saemundsson, Skr. Komm. Havunders. Kbh., 1909: 113 (Iceland); Günther, Encycl. Brit., 11th ed., *24*, 1911: 808 (general); Seabra, Poiss. Port., 1911: 194 (Portugal, not seen); Mourgue, Bull. Soc. linn. Provence, *2*, 1913: 53 (descr. of juv., Marseilles); Senna, Monit. zool. ital., *24*, 1913: 229 (Medit., size); Monit. zool. ital., *31*, 1920: 35, pl. 3 (Medit., size, brain); Valle, Congr. Soc. Pesca Mar. Trieste, *28*, 1922: 31 (ill., Adriatic, not seen); Vinciguerra, Ann. Mus. Stor. nat. Genova, 51, 1923: 133 (Medit., ill. head of juv., clasper, stomach contents); Monit. zool. ital., *35*, 1923: 36 (off Genoa); Jenkins, Fish. Brit. Isles, 1925: 314 (general); Saemundsson, Vidensk. Medd. naturh. Foren. Kbh., *84*, 1927: 183, pl. 5 (Iceland, photo, 27-ft. female); Ehrenbaum, in Grimpe and Wagler, Tierwelt N.- u. Ostsee, Lief. 7 12[e], 1927: 16 (general); Faune Ichthyol. Nord Atlant., Cons. Explor. Mer, 1929: plate not numbered (general); Gould, Case for Sea Serpent, 1930: 245, pl. 6 (hist. of "animal of Stronsa"); Monterosso, Atti Accad. gioenia, *18*, 1931: 55, 4 pls. (proportions; photos of juv. and adult, Medit.); Belloc, Rev. des Trav. Pêches Marit., 7 (2), 1934: 140 (ill. after Day; Morocco); Noronha and Sarmento, Peixes Madeira, 1934: 110 (Madeira, not seen); Nobre, Fauna Marinha Port. Vert., *1*, 1935: 440, pl. 58, fig. 183 (Portugal); Otto, Levende Nat., *41* (12), 1937: 373 (spiral valve, vertebrae, North Sea, not seen).

Squalus rostratus Macri, Atti Accad. Sci. fis. mat. Napoli, *1*, 1819: 76, pl. 1, fig. 2 (Medit.).

Squalus isodus Macri, Atti Accad. Sci. fis. mat. Napoli, *1*, 1819: 76, pl. 1, fig. 1, pl. 2 (Medit.).

Selanche maximus Jaroki, Zoologi, *4*, 1822: 452 (not seen).

Squalus elephas Lesueur, J. Acad. nat. Sci. Philad., *2*, 1822: 350, 1 pl. (off N. Jersey, size, descr.); Storer, Mass. Zool. Bot. Surv., Rep. on Fish., Suppl., 1839: 407 (near Provincetown, Massachusetts); Boston J. nat. Hist., *3*, 1841: 270 (Provincetown).

Pelerin tres grand, Cloquet, Dict. Sci. Nat., Atlas Poiss., 1825: pl. 30 (ill.).

Squalus rhinoceras Mitchill, in 1828 newspaper, quoted by DeKay, Zool. N. Y., 4, 1842: 358 (coast of Maine, not seen).

Selache elephas Bory de St. Vincent, Dict. Class. Hist. Nat., *15*, 1829: 597.

Selachus maximus Minding, Lehrb. Naturgesch. Fische, 1832: 52 (not seen); Swainson, Nat. Hist. Fish. Amphib. Rept., *2*, 1839: 314 (general); Storer, Mem. Amer. Acad. Arts Sci., N. S. *2*, 1846: 506 (in synopsis); Yarrell, Brit. Fish., 2nd ed., 1841: 518 (general descr., Gt. Brit.); DeKay, Zool. N. Y., 4, 1842: 357, pl. 63, fig. 208 (name, N. York, descr., size); Hamilton, Brit. Fish., *2*, 1843: 311; also subsequent eds. (Gt. Brit.); Gosse, Nat. Hist. Fish., 1851: 306 (size); Baikie, Zoologist, *11*, 1853: 3846 (Orkneys); Foulis, Proc. Boston Soc. nat. Hist., *4*, 1853: 205 (descr., meas. 40-ft. spec., Bay of Fundy); Storer, Proc. Boston Soc. nat. Hist., *4*, 1853: 206 (name); DeKay, Rep. St. Cab. Nat. Hist. N. Y., *8*, 1858: 64 (N. York); Storer, Mem. Amer. Acad. Arts Sci., N. S. *9* (1), 1867: 229, pl. 39, fig. 3; also as Fishes Mass., 1867: 253, pl. 37, fig. 3 (descr., meas., ill., Massachusetts); Steenstrup, Overs. danske Vidensk.-Selsk. Forh., 1873: 47, 65, pl. 2, and summary in French (descr., ill. of gill rakers); Lütken, Vidensk. Medd. naturh. Foren. Kbh., 1880: 62 (Iceland); Cornish, Zoologist, (3) *9*, 1885: 35 (Cornwall); Gatcombe, Zoologist, (3) *9*, 1885: 266 (Cornwall); Carus, Prod. Faune Medit., *2*, 1889–1893: 507[41] (Medit.); Jungersen, Danish "Ingolf" exped., *2* (1), 1899: 39 (claspers, California); Damiani, Monit. zool. ital., *14*, 1903: 351 (Medit.); Ariola, Atti Soc. ligust. Sci. nat. geogr., *24*, 1913: 14 (Medit.).

Squalus (*Selache*) *maximus* Voigt, in Cuvier, Tierreich, *2*, 1832: 509 (descr.).

Squalus rashleighanus Couch, Cornish Fauna, *1*, 1838: 51 (Cornwall, not seen).

Cetorhinus maximus Gray, List. Fish. Brit. Mus. Chondropt., *1*, 1851: 61; Gill, Proc. Acad. nat. Sci. Philad., Addend., 1861: 60 (name); Abbott, in Cook, Geol. N. J., App. E, 1868: 828 (N. Jersey); Ver-

41. Spelled "*maximum*."

rill, Bull. Essex Inst., *3*, 1871: 6 (Eastport, Maine); Gill, Rep. U.S. Comm. Fish. (1871–72), 1873: 813; Gervais and Gervais, J. Zool., *5*, 1876: pl. 13, fig. 1, 2 (ill., head of juv.); Goode and Bean, Bull. Essex Inst. Salem, *11*, 1879: 29 (Gulf of Maine); Jones, List Fish. N. S., 1879: 9 (Nova Scotia); Proc. N. S. Inst. Sci., 1882: 95 (same as Jones, 1879); Jordan and Gilbert, Bull. U.S. nat. Mus., *16*, 1883: 31, 875 (distrib.); Goode, Fish. Fish. Industr. U.S., 1, 1884: 668, pl. 249 (general); Jordan, Rep. U.S. Comm. Fish. (1885), 1887: 797 (distrib.); Nelson, Rep. State Geol. N. J., 1890: 663 (off N. Jersey, teeth); Dean, Fishes Living and Extinct, 1895: 90 (general); Smitt, Hist. Scand. Fish., *2*, 1895: 1143 (descr., habits, Norway); Jordan and Evermann, Rep. U.S. Comm. Fish. (1895), 1896: 218 (distrib.); Bull. U.S. nat. Mus., 47 (1), 1896: 51 (descr., distrib.); Goode and Bean, Smithson. Contr. Knowl., *30*, 1895: 21; Mem. Harv. Mus. comp. Zool., *22*, 1896: 21 (descr., distrib.); Jordan and Evermann, Bull. U.S. nat. Mus., 47 (4), 1900: pl. 7, fig. 23 (ill.); Knipowich, Annu. Mus. zool. Acad. St. Pétersb., *2*, 1897: 154 (Murman coast); Bean, Rep. For. Comm. N. Y., 1901: 380 (old N. York record); Bull. N. Y. St. Mus., *60*, Zool. *9*, 1903: 42 (general); de Braganza, Result. Invest. Sci. "Amelia," *2*, 1904: 56 (off Portugal); Collett, Norges Fisk., *3*, 1905: 79 (Norway, descr., meas., discus.); Jordan, Guide to Study Fish., *1*, 1905: 540 (general); Fowler, Rep. N. J. Mus. (1905), 1906: 57 (descr., N. Jersey); Proc. Acad. nat. Sci. Philad., *60*, 1908: 55; Kendall, Occ. Pap. Boston Soc. nat. Hist., 7 (8), 1908: 8 (New England records); Berg, Faune Russie, Poiss., Mus. Zool. Acad. Imp. Sci. St. Pétersb., *1*, 1911: 56 (Russia); Roule, Bull. Inst. oceanogr. Monaco, 243, 1912: 10 (Medit.); Garman, Mem. Harv. Mus. comp. Zool., *36*, 1913: 39 (descr.); Halkett, Check List Fish. Canad., 1913: 41 (distrib.); Nichols, Abstr. Linn. Soc. N. Y., 20–23, 1913: 91 (off N. York); Sumner, Osburn and Cole, Bull. U.S. Bur. Fish., *31* (2), 1913: 737 (Woods Hole); Coles, Proc. biol. Soc. Wash., *28*, 1915: 92 (off C. Lookout); Gudger, Science, *42*, 1915: 653 (southern hemisphere records); Hussakoff, Copeia, 21, 1915: 25 (descr. of Long Island specimen); Scharff, Irish Nat., *24*, 1915: 99 (abund. off Ireland); Fowler, Copeia, 30, 1916: 36; Murphy and Nichols, Brooklyn Mus. Quart., *3*, 1916: 156 (general); Nichols and Murphy, Brooklyn Mus. Sci. Bull., *3* (1), 1916: 28 (habits, food, fishery, occur. near N. York); Radcliffe, Bull. U.S. Bur. Fish., *34*, 1916: 248 (denticles), pl. 38, fig. 4 (teeth); Smith, J. Amer. Mus. nat. Hist., *16*, 1916: 342 (size, habits); Fowler, Proc. biol. Soc. Wash., *33*, 1920: 144 (off N. Jersey); Allen, Bull. Boston Soc. nat. Hist., *24*, 1921: 3 (occur. off New England, habits); Fowler, Copeia, 101, 1921: 89 (N. Jersey); Proc. Acad. nat. Sci. Philad., *74*, 1922: 3 (N. Jersey); Huntsman, Contr. Canad. Biol. (1921), *3*, 1922: 56 (Bay of Fundy); Legendre, Bull. Soc. zool. Fr., *48*, 1923: 275 (north. France, size, food); Bull. Soc. zool. Fr., *49*, 1924: 322 (add. records, sizes, oil); Barnard, Ann. S. Afr. Mus., *21* (1), 1925: 34 (descr., S. Afr.); Bigelow and Welsh, Bull. U.S. Bur. Fish., *40* (1), 1925: 41 (descr., habits, Gulf of Maine); Senna, Arch. ital. Anat. Embriol., *22*, 1925: 84, pl. 9, 10 (skull, branchial skelet.); Lahille, An. Mus. nac. B. Aires, *34*, 1928: 325 (Argentina); Barnard, Ann. S. Afr. Mus., *21* (2), 1927: pl. 2, fig. 1, 1a (ill.); Nichols and Breder, Zoologica, N. Y., *9*, 1927: 20 (Woods Hole, Martha's Vineyard, Long Island, descr.); Rey, Faune Iberica Peces, *1*, 1928: 415 (descr., distrib.); Borri, Mem. Soc. tosc. Sci. nat. Proc. Verb., *38*, 1929: 27 (notes, Medit., not seen); Breder, Field Bk. Mar. Fish. Atlant. Coast, 1929: 23 (general); Jordan, Manual of Vert. Anim. NE. U.S., 1929: 13 (general); also earlier eds.; Navarro, Bol. Pesc. Madr., *14*, 1929: 105 (Balearic Is., spelled "*Cethorinus*," not seen); Desbott, Arch. Medit. Malta, *8*, 1930: 5 (Medit., not seen); Fowler, Fish Culturist, *9* (8), 1930: 115 (N. Jersey, not seen); Gudger, Science, 72, 1930: 341 (N. Jersey, general); MacCoy, Bull. Boston Soc. nat. Hist., *61*, 1933: 17 (Maine, meas., photos); Young and Mazet, "Shark, Shark," 1933: 266 (general); Bertolini, R. C. Accad. Lincei., *18*, 1933: 34; Boll. Soc. Biol. sper., *9*, 1934: 1270 (gill rakers); Bigelow and Schroeder, Canad. Atlant. Fauna, biol. Bd. Canad., *12*[e], 1934: 16 (descr., distrib.); Borri, Mem. Soc. tosc. Sci. nat., *44*, 1934: 93 (discus.); Petit, Terre et La Vie Paris, *4*, 1934: 337 (hist., not seen); Daniel, Elasmobranch Fishes, Univ. Calif. Press, 1934: 37, 38, 155, 268 (gill rakers, gills, eye); Piers, Proc. N. S. Inst. Sci., *18* (3), 1934: 203 (Bay of Fundy); Brimley, J. Elisha Mitchell sci. Soc., *51* (2), 1935: 311 (N. Carolina); Cascia, Boll. Ist. zool. Palermo, *2*, 1935: 137 (descr., meas., Medit.);[42] Gudger, J. Morph., *57*, 1935: 95 (anat.); Anonymous,

42. For additional citations for the Mediterranean in publications not accessible to us, see Cascia (1935) and Doderlein (Man. Ittiol. Medit., *2*, 1881: 70).

Newfoundland Dept. Nat. Res., Fish. Res. Lab., Ann. Rep., *2*, 1935: 79, pl. 2 (Newfoundland, photos); Pozzi and Bordale, An. Soc. cient. argent., *120*, 1935: 150 (Argentina); Vladykov and McKenzie, Proc. N. S. Inst. Sci., *19* (1), 1935: 47 (general); Lübbert and Ehrenbaum, Handb. Seefisch. Nordeurop., *2*, 1936: 280 (descr., distrib., fishery); Barnard, Ann. S. Afr. Mus., *32* (2), 1937: 43 (descr., meas., S. Afr.); Norman, 'Discovery' Rep., *16* (2), 1937: 143 (Falkland Is.); Norman and Fraser, Giant Fishes, 1937: 20, fig. 14 (general); White, Bull. Amer. Mus. nat. Hist., *74*, 1937: 64, 79, 114, pl. 4, fig. 5, pl. 7–9, pl. 29, fig. c, pl. 46, fig. b (anat., photo of gill rakers); Tortonese, Atti Soc. ital. Sci. nat., *77*, 1938: 292 (Medit.); Babic, Zool. Anz., *127*, 1939: 39 (anat. notes; ill. of denticles); Schroeder, New Engl. Naturalist, *2*, 1939: 1 (supposed sea serpent stranded on C. Cod, 1939, a Basking Shark); McKenzie, Proc. N. S. Inst. Sci., *20* (2), 1940: 42 (Nova Scotian records; size; weights); Nybelin, Fauna Flora Uppsala, 1940: 236 (Kattegat); Hildebrand, Copeia, 1941: 222 (N. Carolina, old records); Norris, Plagiost. Hypophysis, 1941: 29, pl. 10, fig. 35–38 (brain).

Squalus cetaceous Gray, Cat. Fish. Coll. Descr. by L. T. Gronow, 1854: 6 (Norway).

Polyprosopus rashleighanus Gill, Ann. N. Y. Lyc., 7, 1862: 398 (name); Couch, Hist. Brit. Fish., *1*, 1867: 67, pl. 15, fig. 1 (descr., size, Cornwall).

Polyprosopus macer Couch, Hist. Brit. Fish., *1*, 1867: 68, pl. 15, fig. 2 (descr., young female, Plymouth, England).

No name, Hannover, K. danske Vidensk.-Selsk. Skr., (5) 7, 1868: 491 (descr. and ill. of gill rakers in Copenhagen Mus., not identified).

Cetorhinus blainvillii Brito Capello, J. Sci. math. phys. nat., Lisboa, *2*, 1870: 234, 1 pl. (descr., ill. of small spec., Portugal).

Cetorhinus rostratus Cornish, Zoologist, *5*, 1870: 2259 (Cornwall, descr., meas.).

Selache rostrata Pavesi, Ann. Mus. Stor. nat. Genoa, 1874: 36 (descr. of young, anat., gill rakers, Medit.); Giglioli, Nature, Lond., *15*, 1877: 273 (discuss.); Doderlein, Atti Accad. Palermo, *6*, 1878–1879: 30 (Medit.).

Selachus pennanti Cornish, Zoologist, (3) *9*, 1885: 351 (Cornwall).

Cetorhinus (Selache) maximus Bridge, Camb. nat. Hist., 7, 1904: 453 (general).

Cethorinus (Selache) maximus Escribano, Bol. Soc. esp. Hist. nat., *9*, 1909: 340 (Medit., Morocco).

Halsydrus maximus Fowler, Bull. U.S. nat. Mus., 100 (*13*), 1941: 113 (descr., synonymy, Australasian refs.); Peces Peru, Mus. Hist. Nat. Javier Prado, 1945: 12 (listed Peru).

Doubtful references:

Halsydrus pontoppidani Fleming, Scots. Mag. Edinb. Misc., 1809: 6; Edinb. Encycl., 1817: 713 (Stronsa, Orkney Is.[43]).

Tetroras angiova Rafinesque, Carrat. Gen. Nuov. Sicil., 1810: 11 (Sicily).

Not *Cetorhinus maximus* Poey, An. Soc. esp. Hist. nat., *5*, 1876: 184 (Cuba, no doubt *Rhincodon*).

Family ALOPIIDAE

Characters. Two dorsal fins, the 1st much shorter than caudal, the rear end of its base anterior to origin of pelvics; caudal nearly or quite ½ of total length, not lunate in form, but its lower anterior corner expanded as a definite lobe, its axis raised but little;[1] inner margins of pelvics entirely separate, posterior to cloaca; caudal peduncle not depressed dorso-ventrally, but moderately compressed laterally; a well marked precaudal pit above, and sometimes below; sides of trunk anterior to anal without longitudinal

43. The animal of Stronsa, named *Halsydrus pontoppidani* by Fleming, seems to have been the partly decomposed remains of a large Basking Shark; see footnote 4, p. 146.

1. The enormously elongate caudal is the most striking feature of the family.

dermal ridges; snout short, thick, fleshy, the jaws not greatly protrusible; 3rd to 5th gill openings over origins of pectorals; gill arches without rakers and not interconnected by a sieve of modified denticles; nostrils entirely separate from mouth, without barbels; spiracles present; eyes without nictitating folds or membranes; each jaw with a labial furrow (or furrows) near corner; teeth small, blade-like, with 1 cusp; head and skull normal in shape (not widely expanded); rostral cartilages 3, united at tip; radials of pectoral nearly all borne on mesopterygium and on metapterygium. Development ovoviviparous; the egg case, in early development, soft, thin, oval.[2]

Genera. Only one genus, *Alopias.*

Genus *Alopias* Rafinesque, 1810

Thresher Sharks

Alopias Rafinesque, Carratt. Gen. Nuov. Sicil., 1810: 12; type species, *A. macrourus* Rafinesque, 1810, Sicily, equals *Squalus vulpinus* Bonnaterre, Tabl. Encyc. Meth. Ichthyol., 1788: 9, pl. 85, fig. 349.

Generic Synonyms:

Squalus (in part) Bonnaterre, Tabl. Encyc. Meth. Ichthyol., 1788: 9; also subsequent authors; not *Squalus* Linnaeus, 1758.

Galeus (in part) Rafinesque, Indice Ittiol. Sicil., 1810: 46.

Carcharhinus (in part) Blainville, Bull. Soc. philom. Paris, 1816: 121.[3]

Carcharias (in part) Cuvier, Règne Anim., *2*, 1817: 126; and subsequent authors; not *Carcharias* Rafinesque, 1810.

Alopecias Müller and Henle, Arch. Naturg., (3) *1*, 1837: 397; type, *Carcharias vulpes* Cuvier, 1817, equals *Squalus vulpinus* Bonnaterre, 1788.

Vulpecula Garman, Mem. Harv. Mus. comp. Zool., *36*, 1913: 30; type species, *Vulpecula marina* Valmont, Dict. Hist. Nat., *3*, 1768: 740.[4]

Alopes Vladykov and McKenzie, Proc. N. S. Inst. Sci., *19*, 1935: 46, for *A. vulpes* (wrongly referred by them to Bonnaterre, 1788, instead of to Gmelin, 1789); evident misspelling for *Alopias.*

Generic Characters. Those of the family.

Range. Cosmopolitan in low and mid latitudes of all oceans, including the Mediterranean.

Fossil Teeth. Eocene, Africa; Oligocene to Miocene, Europe.

Species. The species of this genus fall in two sharply defined groups. In one the rear tip of the first dorsal terminates far in front of the origin of the pelvics; in the other the first dorsal overlaps the pelvics.

The first group includes: *pelagicus* Nakamura, which is set apart by its notched and denticulate teeth; the well known *vulpinus* of the Atlantic and eastern Pacific; also two

2. Photograph of eggs from female taken off Florida, contributed by Stewart Springer.

3. See footnote 12, p. 320.

4. By ruling of the International Commission on Zoological Nomenclature (Smithson. misc. Coll., *73* [3], 1925: 27) Garman's revival of the name *Vulpecula* Valmont is not acceptable, because such of the latter's names as were binomial were only accidentally so.

other named species, *caudatus* Phillipps (1932) and *greyi* Whitley (1937). However, it is doubtful whether either of these last two is actually separable from *vulpinus*. The characters which are supposed to distinguish *caudatus*[5] are: tail no longer than trunk, caudal peduncle up to one-half as deep as trunk at first dorsal and the anterior margin of pectoral straight instead of convex. But the first two of these characters apply equally to some Atlantic specimens (see p. 171), perhaps leaving only the shape of the pectoral as diagnostic. The only characters supposedly diagnostic of *greyi* are eyes "modified for backward vision, are situated much further forward" and greenish color.[6] Its author has in fact suggested recently that *greyi* may be merely a color variety of *caudatus*.[7] Until Australasian specimens are actually compared with those of the eastern Pacific and Atlantic, the question whether or not they are specifically distinct must remain open.

The second group, in which the rear tip of the first dorsal reaches as far back as the origins of the pelvics, or even overlaps the latter, includes two well marked species, *superciliosus* Lowe, 1840, of the tropical Atlantic and *profundus* Nakamura, 1935, so far reported only from Formosa. These two differ further from the *vulpinus* group in the enormous size of their eyes (cf. Fig. 25 with 27); this is, in fact, their most arresting feature apart from their elongate tails.

Key to Species

1a. Rear tip of 1st dorsal terminates considerably anterior to origin of pelvics.
- 2a. Teeth with central cusp strongly oblique, the outer margins with 1 or 2 denticles. *pelagicus* Nakamura, 1935.
  Formosa.
- 2b. Teeth with central cusp erect or only slightly oblique; no marginal denticles.
  - 3a. Anterior margin of pectoral convex. *vulpinus* Bonnaterre, 1788, p. 167.
  - 3b. Anterior margin of pectoral nearly straight. *caudatus* Phillipps, 1932.[8]
    New Zealand, Australia.

1b. Rear tip of 1st dorsal extends at least as far as origin of pelvics, or even overlaps the latter.
- 4a. Rear tip of 2nd dorsal terminates considerably anterior to origin of anal; pelvics a little higher vertically than 1st dorsal and a little larger in area; anterior margin of 1st dorsal strongly convex; no lower precaudal pit. *superciliosus* Lowe, 1840, p. 163.
- 4b. Rear tip of 2nd dorsal terminating over base of anal; pelvics less than ½ as high vertically as 1st dorsal and much smaller in area; anterior margin of 1st dorsal only very weakly convex; a precaudal pit below as well as above. *profundus* Nakamura (1935).
  Formosa.

5. Phillipps, N. Z. J. Sci. Tech., *13*, 1932: 226.
6. Whitley, Rec. Aust. Mus., *20*, 1937: 5.
7. Whitley, Fish. Aust., *1*, 1940: 132
8. Including *greyi* Whitley, 1937.

Alopias superciliosus (Lowe), 1840

Big-eyed Thresher

Figures 25, 26

Study Material. Young male, 1,296 mm. in total length; two embryos, 64 mm. and 632 mm. long; jaws of 18-foot specimen (Harv. Mus. Comp. Zool.); all from the north coast of Cuba;[9] also photographs of embryos from Salerno, Florida.[10]

Distinctive Characters. This species is set apart from the Common Thresher, the only Atlantic Shark with which it might be confused, by the following features: its relatively enormous eye, longer snout, the tip of its first dorsal fin overlapping the pelvics, its second dorsal terminating considerably in advance of the anal and only 10 or 11 teeth on a side in each of its jaws (about 20 in the Common Thresher).

Remarks. The original description of *superciliosus* was limited to the statement that it is "at once distinguished from the only other known species of the genus, *Carcharias vulpes*, Cuv., by the enormous eye and its prominent brow."[11] But the size of the eye is so striking a character that the specimens described here can safely be referred to Lowe's old species.

Description. Proportional dimensions in per cent of total length. Male embryo, 632 mm., from Cuba (Harv. Mus. Comp. Zool., No. 36155). Male, 1,296 mm., from Cuba (Harv. Mus. Comp. Zool., No. 36090).

Trunk at origin of pectoral: breadth 5.8, 7.4; height 8.2, 9.1.
Snout length in front of: outer nostrils 4.3, 4.6; mouth 7.4, 6.0.
Eye: horizontal diameter 4.1, 2.8.
Mouth: breadth 4.6, 4.4; height 3.0, 2.6.
Nostrils: distance between inner ends 1.9, 1.8.
Labial furrow length: upper part 1.9, 1.9; lower 0.7, 0.5.
Gill opening lengths: 1st 2.1, 2.9; 2nd 2.5, 2.9; 3rd 2.4, 2.8; 4th 2.1, 2.2; 5th 1.8, 1.9.
First dorsal fin: vertical height 5.2, 5.2; length of base 5.5, 6.3.
Second dorsal fin: vertical height 0.6, 0.7; length of base 0.8, 0.9.
Anal fin: vertical height 0.8, 0.9; length of base 1.1, 1.0.
Caudal fin: upper margin 48.7, 49.1; lower anterior margin 6.3, 6.3.
Pectoral fin: outer margin 20.4, 19.1; inner margin 4.3, 3.8; distal margin 17.2, 18.1.
Distance from snout to: 1st dorsal 33.3, 31.2; 2nd dorsal 45.3, 46.0; upper caudal 51.3, 50.9; pectoral 20.1, 17.0; pelvics 37.2, 36.5; anal 48.2, 47.8.
Interspace between: 1st and 2nd dorsals 8.9, 8.9; 2nd dorsal and caudal 5.4, 4.6; anal and caudal 1.6, 1.9.

9. Contributed by Luis Howell-Rivero.
10. Contributed by Stewart Springer.
11. Lowe, Proc. zool. Soc. Lond., *8*, 1840: 39.

Distance from origin to origin of: pectoral and pelvics 19.2, 20.8; pelvics and anal 11.1, 11.2.

Trunk, opposite 1st dorsal, a little less than ⅕ as high as the length to origin of caudal, thus more slender than in *vulpinus.* Caudal peduncle compressed laterally, without lateral keels or ridges. A well marked precaudal pit above, but none below. Dermal denticles of two kinds, mostly minute, very widely spaced, lanceolate, but expanded anteriorly on either side and spinous rather than scale-like, the blades not definitely marked off from the pedicels; interspersed among these small denticles are much larger ones, in pairs, the one close behind the other, of shapes more easily illustrated than described (Fig. 26 C).

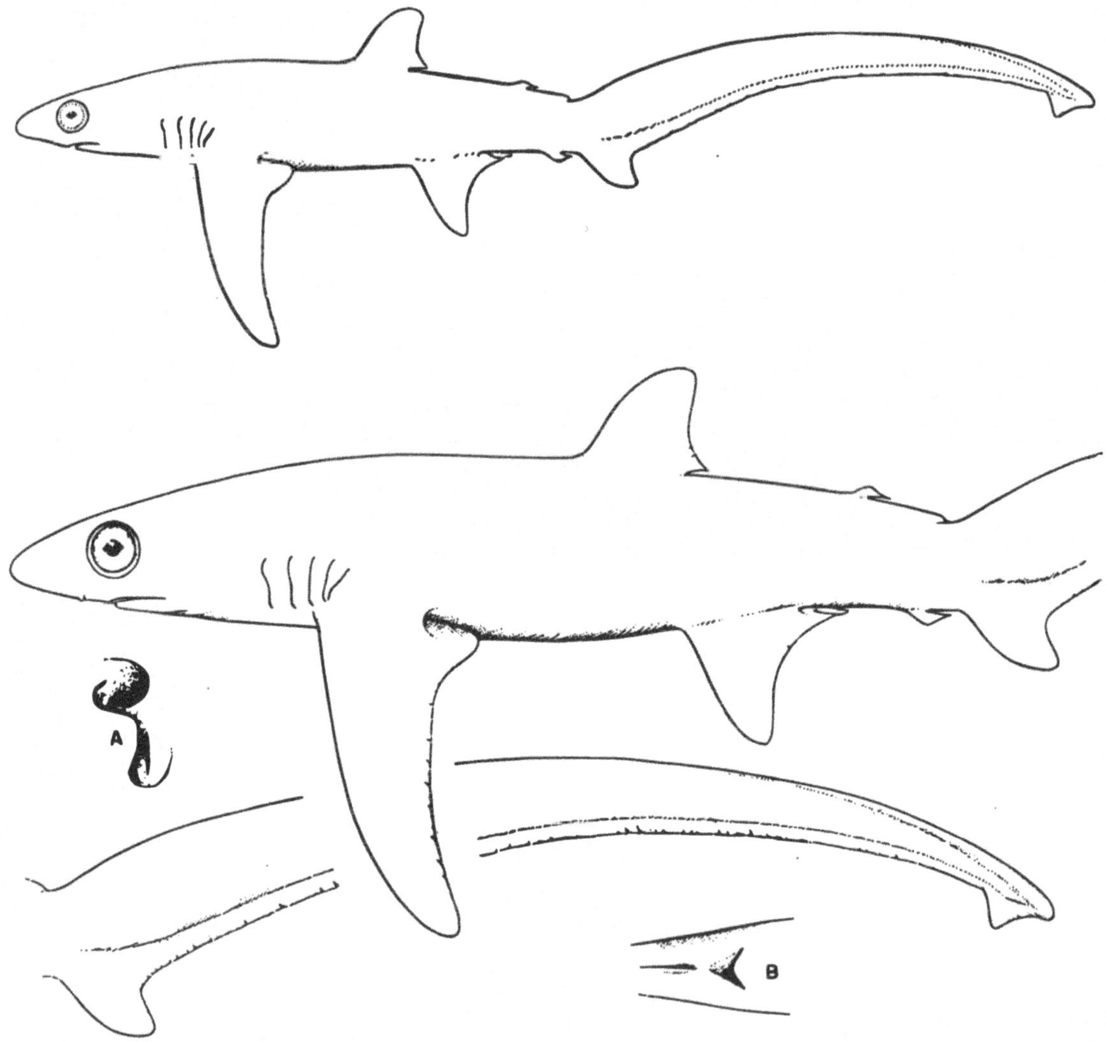

FIGURE 25. *Alopias superciliosus,* young male, 1,296 mm. long, from north coast of Cuba (Harv. Mus. Comp. Zool., No. 36090). *A* Left nostril, about 2.5 x. *B* Caudal peduncle, to show precaudal pit as viewed from above.

FIGURE 26. *Alopias superciliosus.* Same specimen as shown in Fig. 25. *A* Anterior part of head from below, a little more than ½ natural size. *B* Left-hand corner of mouth from below to show labial furrows, about 1.3 x. *C* Dermal denticles, large and small. *D* Lateral view of small dermal denticle. *E* Apical view of same. *F* Lateral view of a pair of large dermal denticles. *C–F*, about 130 x. *G* Left-hand upper and lower teeth, about 0.9 x natural size. *H* Second upper tooth. *I* Sixth upper tooth. *J* Tenth upper tooth. *K* Second lower tooth. *L* Sixth lower tooth. *M* Ninth lower tooth. *H–M* about 1.8 x.

Head about ⅓ as long as trunk to origin of caudal. Snout blunt-conical, its length in front of mouth about ⅓ length of head to origin of pectorals. Distance between nostrils about ⅓ as long as snout in front of mouth. Eye approximately spherical, or a little higher than broad, much larger than in *vulpinus*, its diameter a little more than ½ as long as snout in front of mouth, or between ⅕ and ⅙ as long as head to origin of pectorals. Spiracle a minute pore in embryo, about level with middle of eye, and behind the latter by a distance equal to about ½ the diameter of eye; also visible on one side on young male but not on other, and probably obsolete in adult. First and 2nd gill openings (a little the longest) about as long as diameter of eye, the 5th between ⅔ and ¾ as long as 2nd, and strongly oblique, the 3rd above origin of pectoral. Nostril approximately transverse, about ⅕ as long as horizontal diameter of eye, its inner end only about ⅓ as far from front of mouth as from tip of snout, its anterior margin expanded as a low, subtriangular lobe. Mouth broadly rounded, about ⅔ as high as broad. Two labial furrows above, the outer originating at corners of mouth and overlapping the inner, which originates a little farther forward and extends ⅓ to ¼ of the way toward the symphysis; one labial furrow below, extending forward a short distance along lower jaw and around corner of mouth.

Teeth $\frac{11-11}{10-10}$ in specimen examined, with one subtriangular cusp; similar in the two jaws, the 1st erect, nearly symmetrical, about as broad basally as long, but the others increasingly oblique toward corners of jaws, their outer edges more and more strongly convex, their inner margins increasingly concave with even curvature; the 1st lower tooth a little shorter than 2nd, and the outer 3 (lower jaw) or 4 (upper jaw) progressively smaller, the outermost of all much the smallest.

First dorsal originates about midway between perpendiculars at inner corner of pectoral and at origin of pelvics, its vertical height about ⅓ as great as length of head to origin of pectorals, its anterior margin strongly convex, its apex rounded, its posterior margin weakly concave, its free rear corner only about ⅕ to ⅙ as long as its base, its rear tip overlapping the pelvics for a short distance. Second dorsal only about ⅙ as long at base as 1st dorsal, its apex broadly rounded, its free rear tip very slender, a little longer than its base, ending anterior to origin of anal by a distance nearly as long as base of 2nd dorsal. Interspace between 2nd dorsal and caudal about ⅘ as long as base of 1st dorsal. Caudal a little less than ½ or about 48% of total length, slender, scythe-shaped, much as in *vulpinus* but with terminal sector somewhat broader and more sharply demarked, though without definite notch, its lower anterior margin more strongly convex. Anal similar to the 2nd dorsal in shape. Pelvics about as long at bases as high, about 1.2 as high vertically as 1st dorsal, and a little larger in area, the anterior margins weakly convex, the apices narrowly rounded, the posterior margins strongly and evenly concave. Pectoral about 1.1 times as long as head and a little less than ½ as broad as long, its anterior margin strongly convex toward tip, its apex moderately rounded, the inner margin weakly concave toward tip but deeply so toward inner corner, the latter narrowly rounded.

Color. Dark mouse gray above and hardly paler below, the posterior margins of the 1st dorsal, pectorals and pelvics more or less dusky.

Size. That the Big-eyed Thresher grows as large as the Common Thresher is indicated by the following facts: one of our specimens was from a 12-foot mother, others have been taken from a female of about the same size, and an 18-foot specimen has been taken (teeth pictured in Fig. 26).

Developmental Stages. An embryo of 64 mm., taken from the horny egg capsule, already showing the extremely elongate caudal so characteristic of the adult, still had well developed external gills and a long yolk stalk. One of 632 mm. in total length is evidently ready for birth, since a well marked scar is alone reminiscent of the yolk stalk; this resembles the young male pictured in Fig. 25, except that its eyes are somewhat larger relatively, which is a characteristic common to embryo sharks; the longest gill openings are about ⅔ as long as the diameter of the eye, its snout is blunter, and its pectorals are narrower toward their tips.

Habits. Nothing definite is yet known as to the habits of the Big-eyed Thresher of the Atlantic. Its very large eyes and its coloration (nearly as dark below as above) suggest that it is chiefly a deep-water species like its Formosan relative, *profundus.*[12] But it is not exclusively so, for the specimen pictured in Fig. 25 was near the surface,[13] and perhaps others of the scanty list of captures likewise.

Abundance and Range. The Big-eyed Thresher was first reported from Madeira more than a century ago. It was not seen again until August 1941 when a female of 11 to 12 feet, containing embryos, was taken off Englewood on the west coast of Florida.[14] Very recently, females containing embryos (the young male pictured in Fig. 25) and an 18-foot specimen have been taken off the north coast of Cuba (off Matanzas, and near Havana). We have been informed[15] that the Museum Poey in Havana has a large mounted specimen from Miami, Florida. No doubt the species is widespread in the tropical and subtropical Atlantic.

Synonyms and References:

Alopecias superciliosus Lowe, Proc. zool. Soc. Lond., *8*, 1840: 39 (Madeira).

Alopias (no specific name), Springer, Copeia, *1*, 1943: 54 (off Englewood, Florida, brief descr., embryos, comparison with *vulpinus* and with *profundus*).

Alopias vulpinus (Bonnaterre), 1788

Common Thresher

Figures 27, 28

Study Material. Three alcoholic specimens from Massachusetts, 1,225 to 1,315 mm. long (Harv. Mus. Comp. Zool., No. 486, 706, 1166); mounted specimens, about 9 feet long, from Massachusetts (Harv. Mus. Comp. Zool., No. 926), and 4 feet 5 inches long

12. For description of the latter, see Nakamura (Mem. Fac. Sci. Agric. Taihoku., *14* [1], 1935: 1).
13. It was harpooned.
14. Springer, Copeia, 1943: 54. We have received photographs of one of the embryos.
15. By Luis Howell-Rivero.

from Miami, Florida, in the same collection; fresh specimens not preserved—an adult male of 13 feet taken August 1, 1941 (jaws saved), a female 7 feet 1 inch taken June 15, 1943, immature males of 5 feet 2 inches (1,577 mm.) and 6 feet 10 inches (2,083 mm.); and immature female, 4 feet 10 inches (1,478 mm.) from Woods Hole, Massachusetts; also two small specimens from San Francisco and one from the west coast of South America (Harv. Mus. Comp. Zool., No. 345, 519, 705).

Distinctive Characters. The enormously elongate tail sets the Common Thresher apart at a glance from all other Atlantic sharks, except for its close relative, the Big-eyed Thresher. It is marked off from the latter by its much smaller eye, shorter snout, by the tip of its first dorsal considerably anterior to the origin of its pelvics, by the tip of its second dorsal overlapping the base of its anal, and in having about 20 teeth on each side in each jaw (only about 10 in *superciliosus*).

Description. Proportional dimensions in per cent of total length. Female, 1,225 mm., from Nahant, Mass. (Harv. Mus. Comp. Zool., No. 486). Male, 2,083 mm., from Buzzards Bay, Mass. (field specimen).

Trunk at origin of pectoral: breadth 8.3, 7.3; height 9.6, 9.3.

Snout length in front of: outer nostrils 2.5, 1.7; mouth 3.6, 3.8.

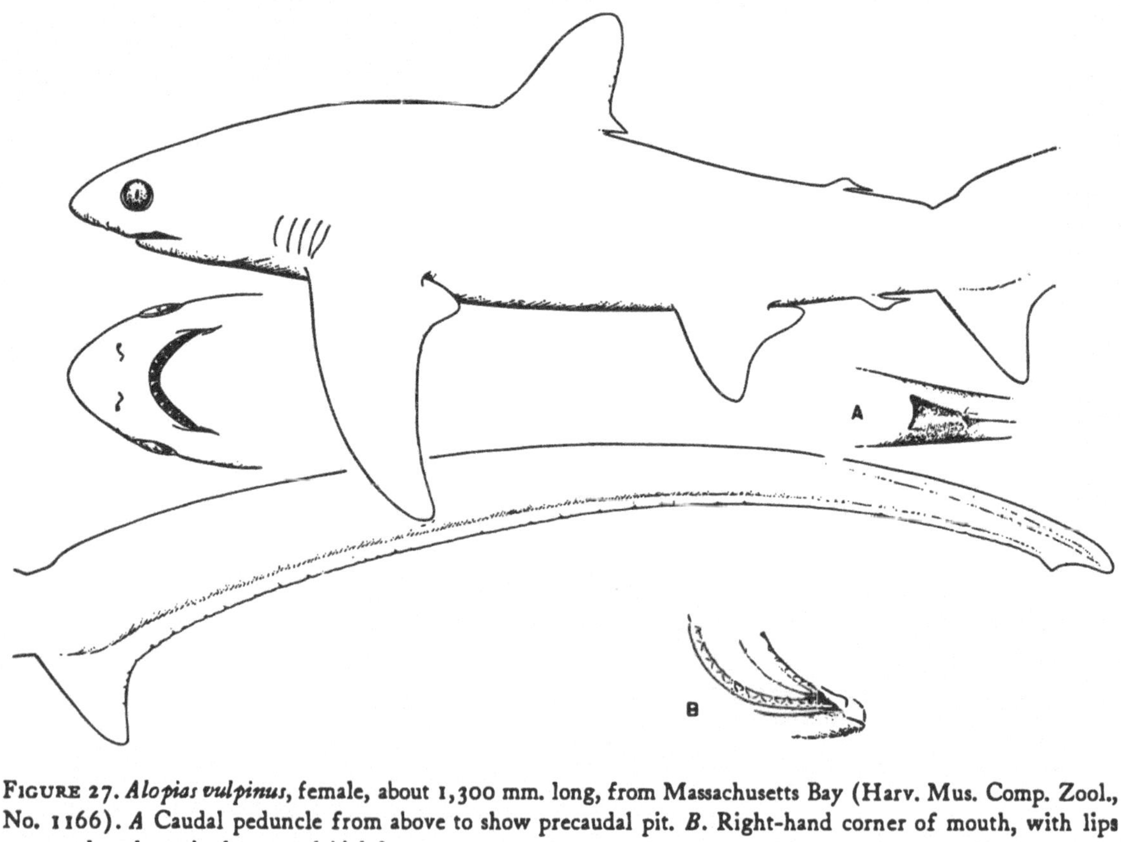

FIGURE 27. *Alopias vulpinus*, female, about 1,300 mm. long, from Massachusetts Bay (Harv. Mus. Comp. Zool., No. 1166). *A* Caudal peduncle from above to show precaudal pit. *B*. Right-hand corner of mouth, with lips separated to show single upper labial furrow.

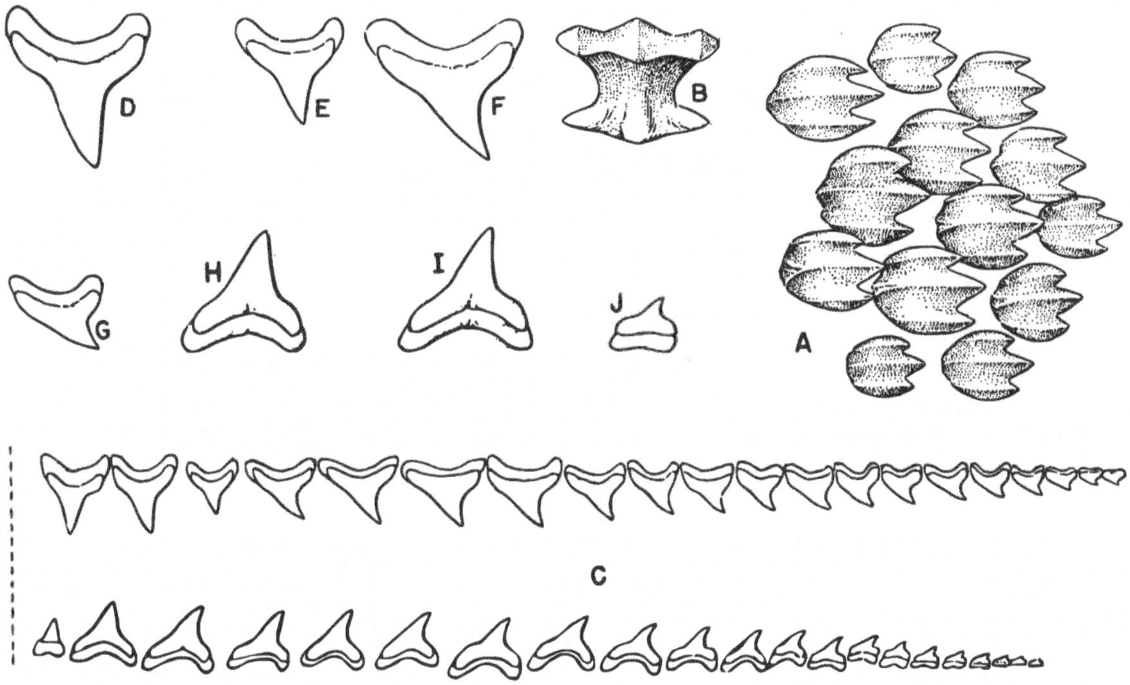

FIGURE 28. *Alopias vulpinus.* *A* Dermal denticles of specimen pictured in Fig. 27, about 50 x. *B* Apical view of dermal denticle, about 100 x. *C* Teeth of a 13-foot Woods Hole, Massachusetts, specimen (Harv. Mus. Comp. Zool., No. 36089), about natural size. *D* Second upper tooth of same. *E* Third upper tooth. *F* Fifth upper tooth. *G* Fifteenth upper tooth. *H* Second lower tooth. *I* Sixth lower tooth. *J* Fifteenth lower tooth. *D-J*, about 2 x.

Eye: horizontal diameter 1.5, 1.2.
Mouth: breadth 4.4, 3.7; height 3.1, 2.1.
Nostrils: distance between inner ends 1.4, 1.2.
Labial furrow length: upper part, 1.7, ——; lower 0.8, 0.6.
Gill opening lengths: 1st 2.0, 1.7; 2nd 2.1, 1.9; 3rd 2.1, 2.0; 4th 1.9, 1.9; 5th 1.8, 1.9.
First dorsal fin: vertical height 6.6, 6.9; length of base 5.6, 6.4.
Second dorsal fin: vertical height 0.6, 0.4; length of base 0.7, 0.8.
Anal fin: vertical height 0.8, 0.7; length of base 0.7, 0.9.
Caudal fin: upper margin 53.0, 53.9; lower anterior margin 5.7, 6.7.
Pectoral fin: outer margin 15.1, 15.8; inner margin 3.3, 2.3; distal margin 11.5, 14.4.
Distance from snout to: 1st dorsal 22.5, 21.3; 2nd dorsal 39.8, 40.5; upper caudal 47.0, 46.0; pectoral 14.0, 13.2; pelvics 32.1, 31.4; anal 40.7, 42.6.
Interspace between: 1st and 2nd dorsals 12.2, 12.6; 2nd dorsal and caudal 6.1, 5.0; anal and caudal 2.7, 2.2.
Distance from origin to origin of: pectoral and pelvics 18.7, 18.5; pelvics and anal 9.1, 11.2.

Trunk stout, somewhat compressed laterally, its dorsal profile strongly convex anterior to 1st dorsal, its depth opposite the latter about ¼ its length to origin of caudal. Caudal peduncle strongly compressed laterally, about 1½ times as high as thick, without longitudinal lateral ridges or keels. A well marked precaudal pit above (Fig. 27 A) but none below. Dermal denticles closely overlapping and very small, being only about 0.2 × 0.21 mm. in a 15-foot specimen, blades horizontal, usually with 3, sometimes with 5, low keels, and as many rather short marginal teeth, the median largest; moderately long pedicels.

Head and snout together subconical, between ⅓ and ¼ as long as trunk to origin of caudal. Snout rounded at tip and very short, its length in front of mouth only about ¼ to ⅕ the length of head. Eye circular, moderately large, its margin considerably anterior to front of mouth, its diameter about ½ as long as snout in front of mouth in small specimens, but only about ⅓ that length in large. Spiracle pore-like, on same level as center of eye and behind latter by a distance about ½ as great as length of snout in front of mouth. Gill openings terminating relatively high up on the sides of neck, noticeably short, the longest only a little longer than diameter of eye in small specimens, but about twice as long in large ones; 1st to 4th evenly spaced, the lower ends of 4th and 5th close together over origin of pectoral. Nostril transverse, considerably nearer to mouth than to tip of snout, its anterior margin expanded in low, subtriangular contour. Mouth broadly rounded, about twice as wide as high. Upper labial furrow reaching about ⅓ of distance to symphysis, the lower furrow only about ½ as long as upper and visible only when mouth is open.

Teeth $\frac{20-20}{21-21}$ in specimen counted; similar in the 2 jaws, blade-like, subtriangular, with single sharp-pointed cusp and smooth edges, the 1st to 3rd uppers and 1st and 2nd lowers nearly symmetrical, but successive teeth increasingly oblique, with their outer margins increasingly deeply concave; the 3rd upper tooth only about ⅔ as high as 1st and 2nd, or as 4th to 10th; 1st lower tooth also very small; 10th or 11th and subsequent teeth in each jaw decreasing successively in size toward corners of mouth, the outermost minute in lower jaw; 1 and sometimes 2 rows functional in front of mouth, 2 rows toward corners.[16]

Origin of 1st dorsal only slightly behind inner corner of pectoral, its rear tip anterior to origin of pelvics by a distance about as great as length of snout in front of mouth, its anterior margin moderately convex, the apex rounded, the posterior margin only slightly concave basally, its free rear tip only about ⅕ as long as its base, its vertical height less than ½ as great as length of pectoral. Second dorsal only about ⅛ as long as 1st dorsal along anterior margin, its origin much nearer to origin of caudal than to rear end of base of 1st dorsal, its apex rounded, its rear tip slender, elongate, nearly or quite twice as long as its base, the rear end of its base about over origin of anal, or a little anterior to latter.

16. According to Moreau (Hist. Nat. Poiss. France, *1*, 1881: 288) the first small tooth in the lower jaw is lost with age; also, in the upper jaw there may be a minute first tooth in small specimens, which is similarly lost with age. But the specimens we have seen lack this small median upper tooth, the first pair of large teeth being close together at the upper symphysis.

Caudal usually a little more than ½ the total length,[17] its upper lobe narrow, scythe-shaped, with a small rounded subterminal prominence but no definite subterminal notch, its lower anterior corner produced as a small but definite subtriangular lobe, the lower anterior margin about ⅛ to ⅑ as long as the upper margin, its axis raised only slightly. Anal similar to 2nd dorsal in size and shape, its origin posterior to rear end of base of 2nd dorsal by a distance as long as its own base, or a little longer. Pelvics about as large in area as 1st dorsal, with weakly convex anterior margins, rounded apices, moderately concave distal margins and subacute inner corners. Claspers of adult males about 4 times as long as pelvic fins and very slender. Pectoral nearly or quite twice as long as height of 1st dorsal along anterior margin, falcate, with very broad base, the anterior margin rather strongly convex in small specimens but tending to become less so in large,[18] the apex broadly rounded, the distal margin deeply and evenly concave, the inner corner subacute, the inner margin only a little more than ½ as long as breadth of base.

Color. Back and upper sides varying between brown, blue slate, slate gray, blue gray and dark lead, even nearly black, often with metallic luster; shading along sides to white below, except that lower surface of snout in front of nostrils, as well as lower surfaces of pectorals, may be of same hue as upper sides; white of lower surface reaching farthest upward on sides from axil of pectoral to opposite rear part of 1st dorsal and again rearward from pelvics; the sides near pectorals, the lower surface from pelvics to caudal, and sometimes the belly may be more or less mottled with gray; iris black or green.

Size. Maximum length 20 feet or more, with lengths of 13 to 16 feet common. Judging from the sizes of females with embryos and of males with large claspers, sexual maturity probably is not attained at a length less than 14 feet. A female of 4,410 mm. (14 feet 6 inches) was found to contain an embryo of 1,550 mm. (5 feet 1 inch).[19] On the other hand, a free-living specimen as small as 46 inches has been reported, while many of 4 to 5 feet have been taken at Woods Hole. One of 149 cm. (about 4 feet, 10½ inches) taken there is described as still showing the umbilical scar,[20] but no trace of it is to be seen on another slightly smaller specimen (4 feet 4 inches) which we have examined from the same locality. The few recorded weights range from about 300 to 320 lbs. at 10 feet, 375 to 400 lbs. at about 13 feet, and about 500 lbs. at 14 feet 5 inches, up to a maximum of perhaps 1,000 lbs.

Developmental Stages. No information is available about the embryo, except that the caudal is about as long, relatively, as in the adult. But the decrease in relative size of the eye with growth after birth suggests that it is even larger in the embryo, as is so commonly the case. Apparently the number of young in any one litter is much smaller than in many other ovoviviparous species, for females have been reported as containing two

17. Ratio of caudal length to trunk length (snout to origin of caudal) ranges from 1.1:1 to 1.3:1 in specimens examined from Massachusetts, San Francisco and the west coast of South America, but only 1.04:1 in a Mediterranean example measured by Tortonese (Atti Soc. ital. Sci. nat., 77, 1938: 293).

18. Owing to the obliquity of their basal insertions, the pectorals often appear as straight-edged, or nearly so, in photographs of large Threshers suspended by the caudal peduncle, as we have observed.

19. Poey, An. Soc. esp. Hist. Nat., 5, 1876: 383.

20. Springer, Copeia, 1943: 55.

or four only. But they are correspondingly large when born, for those in a mother of about 15½ feet (4,700 mm.) measured respectively 1,500 and 1,550 mm. (about 5 feet).[21] Others, perhaps from smaller mothers, are considerably smaller at birth, judging from the small sizes of free-living specimens repeatedly recorded.

Habits. The Thresher is a typically pelagic species, most often seen at least a few miles offshore, but often coming close in to pursue small fish. It is commonly described as usually keeping near the surface. We have seen Threshers jumping on one occasion, these being easily identified by their long tails. But while it is from specimens seen at the surface, or taken in nets set shoal, that the majority of records of its occurrence emanate, it is equally certain that it may descend to moderate depths on occasion, since there is record of at least one specimen captured on hook and line at 35 fathoms in British waters.[22]

It feeds chiefly, if not solely, on whatever smaller schooling fishes may be available; in North American waters most commonly on mackerel, bluefish (*Pomatomus*), shad (*Alosa*) and menhaden (*Brevoortia*), of which it destroys great numbers; no doubt it feeds on herring also, as well as on bonito and squid. In North European waters its diet includes pilchards, garfish, etc. Twenty-seven mackerel have been recorded from a specimen 13½ feet long, and one-half bushel of garfish (*Belone*) from another in Scottish waters. The method by which it captures its prey is highly specialized; in general accounts it has been described repeatedly as using "its whiplike tail to splash the water, while it swims in narrowing circles round a school of fishes, which are thus kept crowded together until the moment of slaughter . . . Sometimes a pair of threshers work together . . ."[23] That it also uses its tail on occasions to stun a prospective victim is proved by eyewitness accounts of one in Irish waters rising and killing a wounded sea bird with a stroke of its tail, then swallowing it,[24] and of another at La Jolla, California,[25] injuring a single small fish by lashing at it repeatedly with its tail. Perhaps it is hardly necessary at this late date to remark that the time-honored tradition that the Thresher leagues with the swordfish to attack whales, which was doubtless based on its being confused with the killer whale (*Orca*), has long since been relegated to the category of myth.

Presumably, young are produced throughout its geographic range, very small specimens having been taken off southern Florida on the one hand and in New England waters on the other. No information is available as to season when the young are born, or when mating takes place.

Abundance and Relation to Man. Of late years the Thresher has not appeared in sufficient abundance anywhere along the Atlantic coast of America to be of any commercial importance one way or the other. However, when it gathers in any numbers in pursuit of small fish, as is said to have happened at times in the past, it has been an annoyance to mackerel fishermen by becoming entangled in their nets. This is a frequent occurrence in British waters where the Thresher is a more familiar species. It is entirely harmless.

21. Vaillant, Bull. Soc. philom. Paris, (7) *10*, 1885: 41.
22. Day, Fish. Gt. Brit., *2*, 1880–1884: 302.
23. Nichols and Murphy, Brooklyn Mus. Sci. Bull., *3*, 1918: 21.
24. Blake-Knox, Zoologist, (2) *1*, 1866: 509.
25. Allen, Science, N. S. *58*, 1923: 31.

Range. Pelagic; in warm temperate and subtropical latitudes; north commonly in the eastern Atlantic to southern Ireland, less regularly to the North Sea and inward as far as the Danish coast and Kattegat; occasionally to the Orkneys and to the Norwegian coast as far north as Lofoten and Trondhjem. Also widespread in the Mediterranean, and recorded from Madeira and Cape of Good Hope. It is known in the west as far north as Nova Scotia and the Gulf of St. Lawrence, south to Brazil and northern Argentina; also, in the eastern Pacific, from Oregon to the Isthmus of Panama and Chile.[26] It is also reported from the Hawaiian Islands, Fanning Island and "Polynesia," Japan, Korea and China, New Zealand and Australia, and from Ceylon, Arabia and Natal. But whether the Thresher (or Threshers) of the *vulpinus* group of the western Pacific, New Zealand and the Indian Ocean are identical with *vulpinus* of the eastern Pacific and Atlantic, or whether more than one species of the group occurs in those regions, is still an open question (p. 162).

Occurrence in the Western Atlantic. The Thresher is reported more frequently and in larger numbers from the offing of southern New England than from anywhere further south along the east coast of the United States. Over the continental shelf off Block Island it has been described as the commonest shark,[27] appearing in May, being most plentiful in June, and remaining until autumn. In the vicinity of Woods Hole, Vineyard Sound and Buzzards Bay, Threshers are taken from time to time in the traps between April and late autumn (see Study Material, p. 168). There is record, in fact, of three fish of 16 feet in one trap in a single morning, and specimens up to 20 feet have been reported locally. Although only two specimens have been recorded in print from Nantucket, Threshers enter the Gulf of Maine in some numbers, at least during some years. Thus, we saw several large ones leaping in Pollock Rip off the southern angle of Cape Cod on August 4, 1913; it is recorded from Provincetown at the tip of Cape Cod and repeatedly from various localities in Massachusetts Bay (*e.g.*, Boston Harbor, Nahant) as well as from various localities along the coast of Maine, in Passamaquoddy Bay,[28] and from the cold waters of the Bay of Fundy (Basin of Minas).[29] From time to time Threshers are taken entangled in the nets off the outer coast of Nova Scotia; seemingly they are not rare on the Scotian Fishing Banks and they have even been reported from the Bay of Chaleur in the Gulf of St. Lawrence, this being the most northerly known record for them on the western side of the Atlantic.

Being a creature of at least moderately warm waters, it is surprising that the Thresher has been reported more frequently and in larger numbers off southern New England than

26. Our own comparison of specimens from San Francisco, California, and the west coast of South America (p. 168) with a considerable series from Massachusetts fails to reveal any significant differences in proportionate dimensions, shape and relative location of fins, length of tail relative to trunk, size of eye, or in shape and number of teeth.
27. Nichols and Murphy, Brooklyn Mus. Sci. Bull., 3, 1916: 20.
28. McKenzie, Proc. N. S. Inst. Sci., 20 (1), 1939: 14.
29. But the old report of Threshers as common there (Knight, Cat. Fishes N. S., 1866: 8) may not have been well founded.

from anywhere farther south along the Atlantic coast of America. No doubt the fact that there is no record of the Thresher for Georges Bank is accidental. But the paucity of reported captures westward and southward from the Block Island–Woods Hole region cannot be explained thus, for so striking is the Thresher in general aspect that any specimen taken is likely to be reported in the daily press, if not in strictly scientific literature. Actually, we have found but one positive record of it for Rhode Island and Connecticut; four for Long Island, New York; one near New York; three in recent years for New Jersey; one for Maryland; two for Cape Lookout, North Carolina (from which it appears that few come inshore close enough along this sector to be picked up in the pound nets). While Threshers have been described as rather numerous at times among the Florida Keys, there are only three or four reports of it along the east coast of Florida, including a small one from Miami in the Museum of Comparative Zoology. One has been reported to us from Biloxi, Mississippi.[30] Nor does it appear to be any more plentiful anywhere farther to the south, where published captures total only three, one being for the Havana region, one for Santa Catherina, Brazil, and one for northern Argentina (Lat. 38° S.). It has not been reported from Bermuda.

In the northern sector of its range the Thresher appears only in spring, summer or autumn, being wholly absent in winter. But nothing is known of its seasonal incidence anywhere else in the western Atlantic.

Synonyms and References:

1. Atlantic:

Sea Fox, Borlase, Nat. Hist. Cornwall, 1758: 265 (Cornwall); Brookes, Nat. Hist., *3*, Fishes, Serpents, 1763: 31 (descr., Medit.); Pennant, Brit. Zool., *3*, 1769: 86, pl. 4 (descr., Gt. Brit.); Couch, Fish. Brit. Isles, *1*, 1867: 37, pl. 7 (descr., Gt. Brit.).

Long Tailed Shark, Pennant, Brit. Zool., *3*, 1776: 110, pl. 14 (descr., Gt. Brit.).

Squalus vulpinus Bonnaterre, Tabl. Encyc. Meth. Ichthyol., 1788: 9, pl. 85, fig. 349 (descr., Medit.).

Squalus vulpes Gmelin, in Linnaeus, Syst. Nat., *1*, 1789: 1496 (descr.); Bloch and Schneider, Syst. Ichthyol., 1801: 127 (descr.); Turton, Brit. Fauna, *1*, 1807: 112 (Gt. Brit.); Risso, Ichthyol. Nice, 1810: 36 (descr., Medit.); Mitchill, Trans. Lit. Phil. Soc. N. Y., *1*, 1815: 482 (descr., Long Island, N. York); Nardo, Prod. Ichthyol. Adriat., 1827: 9 (Medit.); Voigt, in Cuvier, Tierreich, *2*, 1832: 506 (descr.); Jenyns, Manual Brit. Vert. Anim., 1835: 498 (descr.); Couch, Cornish Fauna, 1838: 50 (Cornwall); Buckland, Hist. Brit. Fish., 1881: 218 (size, weight); Gatcombe, Zoologist, (3) *6*, 1882: 434 (Devonshire, England); Zoologist, (3) *9*, 1885: 352 (Sussex, England).

Squale renard, Lacépède, Hist. Nat. Poiss., 4° ed., *1*, 1798: 167, 267, in Buffon, Hist. Nat. (descr.); in Sonnini, Hist. Nat. Poiss., *4*, 1801–1802: 89 (descr., meas.).

Alopias macrourus Rafinesque, Caratt. Gen. Nuov. Sicil., 1810: 12 (Sicily); Indice Ittiol. Sicil., 1810: 45 (Sicily); Swainson, Nat. Hist. Fish. Amphib. Rept., *2*, 1839: 313.

Squalus (Carcharhinus) vulpes Blainville, Bull. Soc. philom. Paris, 1816: 121 (name); in Vieillot, Faune Franc., Poiss., 1825: 94, pl. 14, fig. 1 (descr., meas., French seas).

Carcharias vulpes Cuvier, Règne Anim., *2*, 1817: 126 (descr.); Cloquet, Dict. Sci. Nat., *7*, 1817: 67 (general); Risso, Hist. Nat. Europe Merid., *3*, 1826: 120 (descr., Medit.); Fleming, Hist. Brit. Anim., 1828: 167 (descr., Gt. Brit.); Bory de St. Vincent, Dict. Class. Hist. Nat., *15*, 1829: 597 (general); Yarrell, Brit. Fish., *2*, 1836: 379 (descr., Gt. Brit.); Templeton, Charlesworth's Mag. Nat. Hist., (2) *1*, 1837: 413 (Ireland); Storer, Rep. Fish. Rept. Birds Mass., 1839: 181 (Massachusetts); Boston J. nat. Hist., *2*,

30. Personal communication from Stewart Springer.

1839: 529 (descr., Massachusetts); Yarrell, Brit. Fish., 2nd ed., *2*, 1841: 522 (descr., Gt. Brit.); DeKay, Zool. N. Y., 4, 1842: 348, pl. 61, fig. 199 (descr., near N. York); Cuvier, Règne Anim., Ill. Ed., Poiss., 1843: 361; Linsley, Amer. J. Sci., *47*, 1844: 76 (Connecticut); Guichenot, Explor. Algier. Poiss., 1850: 124 (Algeria); Gosse, Nat. Hist. Fish., 1851: 306 (general); Thompson, Nat. Hist. Ireland, *4*, 1856: 250 (Ireland); DeKay, Rep. St. Cab. Nat. Hist. N. Y. (1855), 1858: 64 (near N. York); Blake-Knox, Zoologist, (2) *1*, 1866: 509 (Ireland); Storer, Mem. Amer. Acad. Arts Sci., N. S. *9*, 1867: 221, pl. 36, fig. 3, also Fishes Mass.: 245, pl. 36, fig. 3 (descr., habits, Massachusetts); Buckland, Hist. Brit. Fish., 1881: 218 (size, weight, Gt. Brit.); Clogg, Zoologist, (3) *5*, 1881: 386 (Cornwall).

Alopecias vulpes Müller and Henle, S. B. Akad. Wiss. Berlin, 1837: 114; Arch. Naturg., (3) *1*, 1837: 397 (name); Couch, Fish. Brit. Isles, *1*, 1867: 37, pl. 7 (descr., Gt. Brit.); Günther, Cat. Fish. Brit. Mus., *8*, 1870: 393, 518 (descr., synonyms, distrib.); Collett, Norges Fisk., 1875: 208 (Norway); Giglioli, Elenc. Pesc. Ital., 1880: 52 (Medit.); Parona, Atti Soc. Nat. Modena, (3) *1*, 1883 (Medit., not seen); Carus, Prod. Faune Medit., *2*, 1889–1893: 507 (Medit.); Oppel, Lehrb. Verg. Mikr. Anat. Wirbelt., *1*, 1897: 51 (histology); Sicher, Atti Accad. gioenia, (4) *11* (5), 1898: 16 (Medit.); Grieg, Norsk Tidskr. Fiskeri, *21*, 1902: 469 (Norway, not seen); Bridge, Camb. nat. Hist., 7, 1904: 452 (general); Werner, Zool. Jb., Syst. Abt., *21*, 1904: 285 (Medit.); Southwell, Zoologist, (4) *10*, 1906: 398 (Norfolk, England); Günther, Encycl. Brit., 11th ed., *24*, 1911: 808 (general); Metzelaar, Trop. Atlant. Visschen, 1919: 190 (Madeira, C. of Good Hope); Borri, Mem. Soc. tosc. Sci. nat., 1934: 93 (Medit.).

Alopias vulpis Swainson, Nat. Hist. Fish. Amphib. Rept., *2*, 1838–1839: 313 (general).

Alopias vulpes Bonaparte, Icon. Faun. Ital., *3* (2), 1835: plate not numbered (descr., ill.); Mém. Soc. neuchâtel. Sci. nat., *2* (8), 1839: 9 (in synopsis); Müller and Henle, Plagiost., 1841: 74, pl. 55 (descr., ill.); Hamilton, Brit. Fish., *2*, 1843: 313; also subsequent eds. (Gt. Brit.); Bonaparte, Cat. Pesc. Europ., 1846: 18 (distrib.); Storer, Mem. Amer. Acad. Arts Sci., N. S. *2*, 1846: 505 (in synopsis); Kneeland, Boston J. nat. Hist., *5*, 1847: 485 (anat.); Nardo, Sinon. Modern. Spec. descr. Pesci St. Chiereghin, 1847: 111 (name); Baikie, Zoologist, *11*, 1853: 3846 (Orkneys); Krøyer, Danmarks Fisk., *3*, 1853: 929, 937 (Denmark); Nardo, Atti Ist. veneto, (3) *5*, 1859–1860: 787 (Medit.); Gill, Proc. Acad. nat. Sci. Philad., Addend., 1861: 60 (name); Duméril, Hist. Nat. Poiss., *1*, 1865: 421 (descr., distrib.); Bocage and Brito Capello, Poiss. Plagiost., 1866: 14 (Portugal); Caruccio, Atti Soc. ital. Sci. nat., *12*, 1869: 568 (Medit.); Brito Capello, J. Sci. math. phys. nat. Lisboa, *2*, 1869: 140; also in separate, Cat. Peixes Port., 1869: 10 (Portugal); Ninni, Ann. Soc. Nat. Modena, *5*, 1870: 66 (Medit.); Canestrini, in Cornelia, *et al.*, Fauna d'Ital., *3*, 1872: 46 (Medit.); Baird, Rep. U.S. Comm. Fish. (1871–1872), 1873: 827 (Woods Hole); Gill, Rep. U.S. Comm. Fish. (1871–1872), 1873: 813 (Cape Cod to Florida); Poey, An. Soc. esp. Hist. Nat., *5*, 1876: 383; Enumerat. Pisc. Cubens., 1876: 187 (size, embryo, Cuba); Gervais and Boulart, Poiss., *3*, 1877: 188, pl. 72 (descr.); Jones, List Fish. N. S., 1879: 9 (Nova Scotia); Doderlein, Man. Ittiol. Medit., *2*, 1881: 52 (Medit.);[31] Moreau, Hist. Nat. Poiss. France, *1*, 1881: 287 (descr., France); Perugia, Elenc. Pesc. Adriat., 1881: 54 (Adriatic); Pascoe, Nature, Lond., *23*, 1881: 35; Hasse, Naturl. Syst. Elasm. besond. Theil, 1882: 221, pl. 29, fig. 14–21 (vertebrae, dermal denticles); Jones, Proc. N. S. Inst. Sci., *5*, 1882: 95 (same as Jones, 1879); Jordan and Gilbert, Bull. U.S. nat. Mus., *16*, 1883: 27 (distrib.); Day, Fish. Gt. Brit., *2*, 1880–1884: 300, pl. 157 (descr., habits, Gt. Brit.); Goode, Fish. Fish. Industr. U.S., 1, 1884: 672 (general); Petersen, Vidensk. Medd. naturh. Foren. Kbh. (1884–1886), 1884: 160 (Kattegat); Vaillant, Bull. Soc. philom. Paris, (7) *10*, 1885: 14 (embryos, sizes); Graeffe, Arb. zool. Inst. Univ. Wien, 7, 1886: 446 (Medit.); Jordan, Rep. U.S. Comm. Fish. (1885), 1887: 797 (distrib.); Garman, Bull. Mus. comp. Zool. Harv., *17*, 1888: 80, pl. 12, 13 (lateral line syst.); Nelson, Rep. St. Geol. N. J., 1890: 662 (common, N. Jersey); Lilljeborg, Sverig. Norg. Fisk., *3* (2), 1891: 627 (Norway, Denmark); Dean, Fishes Living and Extinct, 1895: 89 (general); Smitt, Hist. Scand. Fish., *2*, 1895: 1136 (descr., Scand.); Jordan and Evermann, Rep. U.S. Comm. Fish. (1895), 1896: 217 (distrib.); Bull. U.S. nat. Mus., 47 (1), 1896: 45 (descr., distrib.); Smith, Bull. U.S. Bur. Fish., *17*, 1898: 89 (Woods Hole); Jordan and Evermann, Bull. U.S. nat. Mus., 47 (4), 1900: pl. 6, fig. 20 (ill.); Stirton, Ann. Scot. nat. Hist., 1900: 17 (Scotland); Bean,

31. See Doderlein, above, for additional references for the Mediterranean in publications not accessible to us.

Rep. For. Comm. N. Y. (1900), 1901: 378 (abund. off Martha's Vineyard, and near N. York); Linton, Bull. U.S. Bur. Fish., *19*, 1901: 428 (parasites, Woods Hole); Popta, Ann. Sci. nat., (7) *12*, 1901: 205 (gill arches, Concarneau, France); McIntosh, Ann. Mag. nat. Hist., (7) *10*, 1902: 254 (food, Scotland); de Braganza, Result Invest. Sci. "Amelia," *2*, 1904: 46 (Portugal); Sharp and Fowler, Proc. Acad. nat. Sci. Philad., *56*, 1904: 505 (Nantucket); Bean, Bull. N. Y. St. Mus., *60*, Zool. *9*, 1903: 33 (descr., N. York); Borca, Arch. Zool. exp. gén., (4) *4*, 1905: 204, 207 (France); Collett, Norges Fisk., *3*, 1905: 79 (Norway); Jordan, Guide to Study Fish., *1*, 1905: 536 (general); Fowler, Rep. N. J. Mus. (1905), 1906: 53 (N. Jersey); Fowler, Proc. Acad. nat. Sci. Philad., *58*, 1906: 79 (N. Jersey); Tracy, Rep. R. I. Comm. inl. Fish., 1906: 45 (off Rhode Island); Kendall, Occ. Pap. Boston Soc. nat. Hist., 7 (8), 1908: 5 (New England); Fowler, Proc. Acad. nat. Sci. Philad., *60*, 1908: 54 (Newport, Rhode Island, size); Proc. Acad. nat. Sci. Philad., *61*, 1909: 407 (N. Jersey); Proc. Acad. nat. Sci. Philad., *62*, 1911: 599 (N. Jersey); Tracy, Rep. R. I. Comm. inl. Fish., 1910: 60 (Rhode Island); Roule, Bull. Inst. océanogr. Monaco, 243, 1912: 9 (Medit.); Halkett, Check List Fish. Canad., 1913: 40 (Gulf of St. Lawrence, Nova Scotia, Basin of Minas, Bay of Fundy); Le Danois, Ann. Inst. océanogr. Monaco, *5* (5), 1913: 16 (English Channel); Nichols, Abstr. Linn. Soc. N. Y., 20–23, 1913: 91 (season off N. York); Sumner, Osburn and Cole, Bull. U.S. Bur. Fish., *31* (2), 1913: 736 (Woods Hole, Mass.); Fowler, Proc. Acad. nat. Sci. Philad., *66*, 1914: 342 (N. Jersey); Coles, Proc. biol. Soc. Wash., *28*, 1915: 91 (N. Carolina); Nichols and Murphy, Brooklyn Mus. Sci. Bull., *3* (1), 1916: 20 (descr., habits, distrib.); Lahille, Physis B. Aires, *5*, 1921: 63 (Argentina); Enum. Peces Cartilag. Argent., 1921: 15 (north. Argent.); Townsend, Bull. N. Y. zool. Soc., *26*, 1923: 78 (general); Ribeiro, Mus. nac. Rio de J., Fauna Brasil. Peix., *2* (1), Fasc. 1, 1923: 16, pl. 5 (descr., Brazil); Jenkins, Fish. Brit. Isles, 1925: 313 (Gt. Brit.); Wilson, Proc. U.S. nat. Mus., *64* (17), 1925: 12 (parasites, Woods Hole); Barnard, Ann. S. Afr. Mus., *21* (1), 1925: 34 (S. Afr.); Ehrenbaum, in Grimpe and Wagler, Tierwelt N.- u. Ostsee, 7 (12°), 1927: 14 (general, North Sea); Nichols and Breder, Zoologica, N. Y., *9*, 1927: 18 (general); Lönnberg, Fauna Flora Uppsala, 1927: 258 (Bohusland, Sweden, not seen); Rey, Fauna Iberica Peces, *1*, 1928: 411 (descr., Spain); Breder, Field Bk. Mar. Fish. Atlant. Coast, 1929: 20 (general); Truitt, Bean and Fowler, Bull. Md. Conserv. Dep., 1929: 29 (Maryland); Breder, Amer. Mus. Novit., 382, 1929: 4 (use of tail); Wilson, Proc. U.S. nat. Mus., *60* (15), 1932: 60 (ref.); Pearson, Invest. Rep. U.S. Bur. Fish., (10) *1*, 1932: 18 (off C. Hatteras); Belloc, Rev. des Trav. Pêches Marit., 7 Fasc. 2, 1934: 140 (Morocco, Madeira); Daniel, Elasmobranch Fishes, Univ. Calif. Press, 1934: 38 (Stomodeal denticles); Ehrenbaum, Fauna Ichthyol., Cons. Intern. explor. Mer., 1935: plate not numbered (general); Nobre, Fauna Marinha Port. Vert., *1*, 1935: 439, pl. 63, fig. 198 (descr., Portugal); Pozzi and Bordale, An. Soc. cient. argent., *120*, 1935: 150 (north. Argent.); Lübbert and Ehrenbaum, Handb. Seefisch. Nordeurop., *2*, 1936: 279 (general, N. Europe); Norman and Fraser, Giant Fishes, 1937: 24 (general); Norris, Plagiost. Hypophysis, 1941: 29, pl. 11, fig. 41–43 (brain).

Squalus alopecias Gray, Cat. Fish Coll. Descr. by L. T. Gronow, in Brit. Mus., 1854: 7 (in list).

Alocepias vulpes Yung, Arch. Zool. exp. gén., (3) 7, 1899: 140 (name, evidently misprint for "Alopecias").

Vulpecula marina[32] Garman, Mem. Harv. Mus. comp. Zool., *36*, 1913: 30, pl. 7, fig. 1 (descr.); Radcliffe, Trans. Amer. Fish. Soc., 1914: 37 (N. Carolina); Bull. U.S. Bur. Fish., *34*, 1916: 246, pl. 38, fig. 3 (teeth, N. Carolina); Smith, Amer. Mus. J., *16*, 1916: 347 (habits); Huntsman, Contr. Canad. Biol., (1921), *3*, 1923: 56 (Bay of Fundy); Meek and Hildebrand, Field Mus. Publ., *15*, 1923: 62 (general); Linton, Proc. U.S. nat. Mus., *64* (21), 1924: 13, 35, 56, 82 (parasites); Bigelow and Welsh, Bull. U.S. Bur. Fish., *40* (1), 1925: 32 (descr., habits, Gulf of Maine); Wilson, Bull. U.S. nat. Mus., 158, 1932: 463, 464 (parasites, Woods Hole); Nigrelli, Amer. Mus. Novit., 996, 1938: 7 (parasites).

Alopias vulpinus Fowler, Copeia, 30, 1916: 36 (mid. Atlant., U.S.); Proc. Boston Soc. nat. Hist., *35*, 1917: 110 (Newport, Rhode Island); Proc. biol. Soc. Wash., *33*, 1920: 143 (N. Jersey); Proc. Acad. nat. Sci. Philad., *74*, 1922: 3 (size, N. Jersey); Copeia, 156, 1926: 146 (N. Jersey, small size); Jordan, Manual Vert. Anim. NE. U.S., 1929: 11 (general); Jordan, Evermann and Clark, Rep. U.S. Comm.

32. According to the International Commission on Zoological Nomenclature (Smithson. Misc. Coll., *73* [3], 1925: 27), Garman's revival of the name *Vulpecula marina* Valmont is not acceptable because such of Valmont's (1768) names as were binomial were only accidentally so.

Fish. (1928), 2, 1930: 18 (distrib.); Bigelow and Schroeder, Canad. Atlant. Fauna, biol. Bd. Canad., 12e, 1934: 11 (descr., Gulf of St. Lawrence, Nova Scotia); Marchand, Fish Mar. Biol. Surv. S. Afr., Fish. Bull. 2, 1935: 32 (S. Africa); Howell-Rivero, Proc. Boston Soc. nat. Hist., 41 (4), 1936: 45 (embryos, Cuba); Fowler, Bull. Amer. Mus. nat. Hist., *70* (1), 1936: 43 (descr., distrib.); Bigelow and Schroeder, Bull. U.S. Bur. Fish., *48*, 1936: 322 (name); Tortonese, Atti Soc. ital. Sci. nat., 77, 1938: 293 (Medit.); Springer, Proc. Fla. Acad. Sci., *3*, 1939: 34 (Florida); McKenzie, Proc. N. S. Inst. Sci., *20* (1), 1939: 14 (Bay of Fundy); Hildebrand, Copeia, 1941: 221 (N. Carolina); Fowler, Arqu. Zool. Estado São Paulo, *3*, 1942: 128 (Brazil); Lunz, Bull. S. Carolina St. Planning Bd., 14, 1944: 27 (Florida); Bigelow and Schroeder, Guide Comm. Shark Fish., Anglo Amer. Carib. Comm., Wash., 1945: 107, fig. 37 (descr., ill., habits, range); Fowler, Mongr. Acad. nat. Sci. Philad., 7, 1945: 93 (600-lb. spec., N. Carolina).

Alopias vulpinus Bonnaterre, *A. vulpes* Gmelin, or *Vulpecula marina* Valmont; Jordan, Copeia, 49, 1917: 89 (name).

Alopes vulpes Vladykov and McKenzie, Proc. N. S. Inst. Sci., *19*, 1935: 46 (Nova Scotian Banks and Bay of Fundy).

2. Pacific Coast of America:

Carcharias vulpes Gray, Hist. Chile, Zool., *2*, 1858: 363 (Chile).

Alopias vulpes Ayers, Proc. Calif. Acad. Sci., *3*, 1863: 66 (California); Eigenmann, Proc. U.S. nat. Mus., *15*, 1893: 134 (San Diego, California); Allen, Science, N. S. *58*, 1923: 31 (use of tail, California); Walford, Fish Bull. Sacramento, *45*, 1935: 41 (California).

Alopecias vulpes Perez Canto, Estud. Escual. Chile, 1886: 5 (listed for Chile); Philippi, An. Univ. Chile, *71*, 1887: 551, pl. 5, fig. 1 (descr., ill., Chile); Delfin, Rev. chil. Hist. nat., *4*, 1900: 71 (Chile); Quijada, Boll. Mus. nac. Chile, *5*, 1913: 111 (listed for Chile).

Alopecias barrae Perez Canto, Estud. Escual. Chile, 1886: 6 (descr., Chile); Philippi, An. Univ. Chile, *71*, 1887: 553, pl. 5, fig. 2 (descr., ill., Chile).

Alopecias chilensis Philippi, An. Univ. Chile, *109*, 1901: 310 (Chile).

Alopecias longimana Philippi, An. Univ. Chile, *109*, 1901: 308 (descr., Chile); Quijada, Boll. Mus. nac. Chile, *5*, 1913: 111 (listed for Chile).

Alopias vulpinus Barnhart, Mar. Fish. south. Calif., 1936: 10 (off California); Schultz, Univ. Wash. Publ. Biol., *2* (4), 1936: 131 (Washington, Oregon); Beebe and Tee-Van, Zoologica, N. Y., *26*, 1941: 97 (ref.).

3. Central and Western Pacific, Australasia, Indian Ocean:[33]

Alopecias vulpes Hutton and Hector, Colonial Mus. Geol. Surv. N. Z., 1872: 78 (N. Zealand); Macleay, Proc. Linn. Soc. N. S. W., *6*, 1881: 359 (descr., Aust.); McCoy, Prod. Zool. Victoria, *1* Decade 9, 1884: pl. 88 (descr., meas., ill., Aust.); Boulenger, Proc. zool. Soc. Lond., 1889: 243 (Arabia); Lucas, Proc. roy. Soc. Vict., N. S. *2*, 1890: 43 (Aust.); Sauvage, in Grandidier, Hist. Phys. Nat. Madagascar, *16*, 1891: 511 (Madagascar); Günther, J. Mus. Godeffroy, *6* Heft 16, Fische der Sudsee, *3* Heft 8, 1910: 486 (distrib.); Zugmayer, Abh. bayer. Akad. Wiss., *26* (6), 1913: 8 (Oman).

Alopias vulpes Day, Fish. India, Suppl. 1888: 810 (Ceylon); in Blanford, Fauna Brit. India, *1*, 1889: 28 (Ceylon); Ogilby, Proc. Linn. Soc. N. S. W., (2) *3*, 1889: 1772 (Aust., Tasmania); Gill, Mem. nat. Acad. Sci. Wash., *6*, 1893: 111 (N. Zealand); Jordan and Snyder, Annot. zool. jap., *3*, 1901: 127 (Japan); Jordan and Fowler, Proc. U.S. nat. Mus., *26*, 1903: 619 (Japan); Jordan and Snyder, Proc. U.S. nat. Mus., *27*, 1904: 939 (Hawaiian Is.); Jordan and Evermann, Bull. U.S. Bur. Fish., *23* (1), 1905: 42 (Honolulu); Stead, Fish. Aust., 1906: 232 (Aust.); Zietz, Trans. roy. Soc. S. Aust., *32*, 1908: 291 (Aust.); Thompson, Mar. biol. Rep. Cape Town, *2*, 1914: 147 (S. Afr.); Jordan and Jordan, Mem. Carneg. Mus., *10*, 1922: 4 (Hawaiian Is.); Fish Soc. Japan, Ills. Japan. Aquat. Plants Anim., *1*, 1931: pl. 7 (ill., Japan).

33. If *A. caudatus* Phillipps, including *greyi* Whitley, finally proves identical with *vulpinus* (see discussion, p. 162), the following references are to be included; if it is proved distinct then some of them may still refer to *vulpinus*.

Alopias vulpinus Ogilby, Mem. Qd. Mus., *5*, 1916: 74 (Aust.); McCulloch, Biol. Result. Fish. F. I. S. "Endeavour," *4* (4), 1916: 170 (Aust.); Waite, Rec. S. Aust. Mus., *2*, 1921: 18 (Aust.); Phillipps, N. Z. J. Sci. Tech., *6*, 1924: 265 (N. Zealand); McCulloch and Whitley, Mem. Qd. Mus., *8* (2), 1925: 128 (Aust.); Jordan and Hubbs, Mem. Carneg. Mus., *10*, 1925: 10 (Japan); McCulloch, Fish. N. S. W., 1927: 8, pl. 2 (Aust., not seen); Phillipps, N. Z. Mar. Dept. Fish. Bull., 1, 1927: 8 (N. Zealand); Mori, J. Pan-Pacif. Res. Instn., *3*, 1928: 3 (Korea); Fowler, Mem. Bishop Mus., *10*, 1928: 18 (Hawaiian Is., Fanning I.); Proc. Pan-Pacif. sci. Congr. Java, (4) *3*, 1930: 488 (distrib.); Hongkong Nat., *1*, 1930: 87 (Hong Kong, China, not seen); Phillipps, N. Z. J. Sci. Tech., *13*, 1932: 226 (comp. with *caudatus*); Coppleson, Med. J. Aust., April 15, 1933: 11 (Aust., not dangerous); Fowler, Bull. U.S. nat. Mus., 100 (*13*), 1941: 25 (descr., Indo-Pacif. distrib.).

Alopias caudatus Phillipps, N. Z. J. Sci. Tech., *13*, 1932: 226 (descr., N. Zealand); Whitley, Fish. Aust., *1*, 1940: 129 (N. Zealand, Aust.).

Alopias greyi Whitley, Rec. Aust. Mus., *20* (1), 1937: 5 (diagn., Aust.); Fish. Aust., *1*, 1940: 132 (Aust.).

Family ORECTOLOBIDAE

Carpet Sharks, Nurse Sharks

Characters. Two dorsal fins, the 1st much shorter than the caudal, its origin over or posterior to the pelvics; caudal much less than ½ the total length, not lunate in form, its lower anterior corner not expanded as a definite lobe, its axis but little raised; inner margins of pelvics posterior to cloaca either separated, or united for only a very short distance; caudal peduncle not greatly flattened dorso-ventrally or expanded laterally, without precaudal pits; sides of trunk anterior to anal with or without longitudinal ridges; snout not elongate, nor jaws widely protrusible; 4th and 5th gill openings over base of pectoral; gill arches without rakers and not interconnected by a sieve of modified denticles; nostril connected with mouth by a deep groove, its anterior margin with a well developed fleshy barbel or cirrus;[1] spiracles present; lower eyelid without nictitating fold or membrane, but orbit, in some, with a longitudinal fleshy fold above and below, inside the eyelids, but entirely free from latter; a labial furrow on each jaw near the corner; teeth small, with several cusps; head of normal form (not widely expanded); rostral cartilages either none, 1, or 3, but separate at tip and very small; mesopterygium of pectoral nearly as large as metapterygium and with nearly as many radials; mesopterygia and metapterygia separated by a foramen; heart valves in 2 rows. Development ovoviviparous in some (*Brachaelurus, Orectolobus, Ginglymostoma*), but oviparous in others (*Chiloscyllium, Hemiscyllium, Nebrodes, Stegostoma*), the horny egg capsules of which are attached to algae either by terminal tendrils or by fibrous extensions of the margin.[2]

Genera. Most of the members of this large family of warm-water sharks are inhabitants of the western Pacific, Australian region, or Indian Ocean, including the Red Sea; only one genus (*Ginglymostoma*) occurs in the Atlantic. Many of them live on bottom in shallow water, are brilliantly marked, especially when young (hence the common name

1. This is the most striking characteristic of the family as a whole.
2. For illustrations of the egg case of *Chiloscyllium*, see Southwell and Prashad (Rec. Indian Mus., *16*, 1919: 222, pl. 19, fig. 5) and Whitley (Fish. Aust., *1*, 1940: 39, fig. 28, 4); for *Stegostoma* (Zebra Shark) and *Nebrodes* (Tawny Shark), see Whitley (Fish. Aust., *1*, 1940: 39, fig. 28, 5, 6).

Carpet or Zebra Sharks), or are ornamented with fleshy flaps on the head. The majority are small, but a few grow to a considerable size. Their diet is chiefly bottom dwelling invertebrates and fishes.

Key to Genera

1a. Sides of head fringed with fleshy lobes.

2a. A continuous series of branching dermal lobes around lower jaw close to mouth.
Eucrossorhinus Regan, 1908.
East Indies.

2b. No dermal lobes on lower jaw, or at most only a few small ones below chin.

3a. Back smooth, without papillae or tubercles. *Orectolobus* Bonaparte, 1834.
Eastern Pacific, Australia.

3b. Back with rows of papillae or tubercles. *Sutorectus* Whitley, 1940.
Australia.

1b. Sides of head without fleshy lobes.

4a. Second dorsal originating posterior to origin of anal.

5a. Throat with a pair of thread-like barbels.
Cirrhoscyllium Smith and Radcliffe, 1913.
China Sea.[3]

5b. Throat without barbels. *Parascyllium* Gill, 1862.
Australia, Tasmania.

4b. Second dorsal originating anterior to origin of anal.

6a. Spiracle minute.

7a. Teeth with central cusp largest and several rows functional.
Ginglymostoma Müller and Henle, 1837, p. 180.

7b. Teeth with all cusps about equal, only 1 or 2 rows functional.
Nebrius Rüppell, 1835.
Australasia, Malay Peninsula, Indian Ocean, Red Sea.[4]

6b. Spiracle nearly or quite as large as eye.

8a. Caudal more than 1/3 of total length; 1st dorsal originating in front of pelvics and terminating over latter.
Stegostoma Müller and Henle, 1837.
Western Pacific, Australasia, Indian Ocean.

3. Whitley (Rec. Aust. Mus., *15*, 1927: 289) proposes to substitute a new name, *Zev*, on the ground that *Cirrhoscyllium* Smith and Radcliffe, 1913, was preoccupied by *Cirriscyllium* Ogilby, 1908, which in turn is a synonym of *Brachaelurus* Ogilby, 1907, as pointed out below (footnote 5, p. 180). According to the International Rules on Zoological Nomenclature, however, this substitution is not required; see recommendation under Article 36 (Proc. biol. Soc. Wash., *39*, 1926: 87).

4. Garman's (Mem. Harv. Mus. comp. Zool., *36*, 1913: 56) substitution of *Nebrodes* for *Nebrius* Rüppell, because of the earlier use of *Nebria* by Latreille (1802) for insects and of *Nebris* by Cuvier and Valenciennes (1830) for bony fishes, was not necessary, according to the recommendations under Article 36, International Code of Zoological Nomenclature (Proc. biol. Soc. Wash., *39*, 1926: 87).

8b. Caudal considerably less than ⅓ of total length; 1st dorsal originating over or behind pelvics and terminating behind latter.
  9a. Cloaca much nearer to snout than to tip of caudal.
    10a. Mouth closer to tip of snout than to front of eye; lower labial furrow not crossing chin.
      Hemiscyllium Andrew Smith, 1837.
      Australasia, East Indies, India.
    10b. Mouth closer to front edge of eye than to tip of snout; lower labial furrow continuous across chin.
      Chiloscyllium Müller and Henle, 1837.
      Western Pacific, Australasia, Indian Ocean, Red Sea, South Africa.
  9b. Cloaca as near to tip of caudal as to tip of snout, or nearer.
    11a. Rear end of base of anal terminating close to caudal and its rear tip overlapping latter. *Brachaelurus* Ogilby, 1907.
      Australia.[5]
    11b. Rear end of base of anal separated from caudal by a space twice as long as the base. *Heteroscyllium* Regan, 1908.
      Australia.[5]

Genus *Ginglymostoma* Müller and Henle, 1837

Nurse Sharks

Ginglymostoma Müller and Henle, S. B. Akad. Wiss. Berlin, 1837: 113; also Arch. Naturg., (3) *1*, 1837: 396 (no species given); type species, *Squalus cirratus* Bonnaterre, Tabl. Encyc. Meth. Ichthyol., 1788: 7, American Seas, designated by Hay, U.S. Geol. Surv. Bull., 179, 1902: 310.

Generic Synonyms:

Squalus (in part) Bonnaterre, Tabl. Encyc. Meth. Ichthyol., 1788: 7; Gmelin, in Linnaeus, Syst. Nat., *1*, 1789: 1492; also subsequent authors; not *Squalus* Linnaeus, 1758.

Scyllium (in part) Griffith, in Cuvier, Règne Anim., *10*, 1834: pl. 30; for *S. cirrhosum* Griffith, equals *Squalus cirratus* Bonnaterre, 1788.

Ginglimostoma L. Agassiz, Poiss. Foss., *3*, 1838: 85; equivalent to *Ginglymostoma* Müller and Henle, 1837.

Gingylostoma Springer, Proc. Fla. Acad. Sci., *3*, 1939: 13; evident misspelling for *Ginglymostoma*.

Generic Characters. Sides of head without fleshy lobes; back without papillae; no longitudinal ridges on back or sides of trunk, and no longitudinal fold below eye; eye very small, the orbit with a longitudinal fleshy fold above and below, wholly inside the eyelid, and not connected with latter; spiracle minute, about on a level with eye; 4th and 5th gill

5. The new generic name *Brachaelurus* was proposed by Ogilby (Proc. roy. Soc. Qd., *20*, 1907: 27) for *Chiloscyllium modestum* Günther, 1871 (equals *Squalus waddi* Bloch and Schneider, 1801), in which the anal is close to the caudal; but the name was transferred by him a year later (Proc. roy. Soc. Qd., *21*, 1908: 3) to his new species *colcloughi*, in which the anal is far from the caudal; he proposed a new generic name, *Cirriscyllium*, for *modestum* (Proc. roy. Soc. Qd., *21*, 1908: 4). *Cirriscyllium* Ogilby, 1908, is therefore a synonym of *Brachaelurus* Ogilby, 1908, as pointed out by Regan (Ann. Mag. nat. Hist. [8] *2*, 1908: 455), who proposed *Heteroscyllium* for *colcloughi* to clarify this confusion.

openings very close together; anterior marginal expansions of nostrils reach to mouth, but are wide apart; teeth with several cusps, the central much the largest, and several rows functional; lower labial furrow not continuous across chin; 1st dorsal originates over or slightly posterior to origin of pelvics; 2nd dorsal originates anterior to origin of anal; rear tip of 2nd dorsal not reaching to origin of caudal; caudal ⅓ to ¼ of total length; cloaca about midway between tip of snout and tip of caudal. Development ovoviviparous. Characters otherwise those of the family.

Range. Both sides of tropical and subtropical Atlantic; west coast of Mexico; western tropical Pacific; Malaysia, Indian Ocean, Red Sea.

Fossil Teeth. Upper Cretaceous, West Indies; Upper Cretaceous to Eocene, Europe; Eocene, Africa and North America.

Key to Species

1a. Corners of fins angular. *ferrugineum* Lesson, 1830.
Western tropical Pacific, Malaysia, Indian Ocean, Red Sea.

1b. Corners of fins rounded.

- 2a. Anal much smaller in area than 2nd dorsal; nasal barbels reach to mouth. *cirratum* Bonnaterre, 1788, p. 181.
- 2b. Anal nearly as large in area as 2nd dorsal; nasal barbels reach only about halfway to mouth. *brevicaudatum* Günther, 1866.
  Zanzibar, Seychelles.

Ginglymostoma cirratum (Bonnaterre), 1788

Nurse Shark

Figure 29

Study Material. 14 specimens, 275 to 650 mm. long, from Florida, Cuba, Jamaica, Sombrero I., West Indies, and Panama Bay (Harv. Mus. Comp. Zool.); 2 eggs (about 125 mm. by 63 mm.) and an embryo (125 mm. long with traces of external gills visible) from Key West, Florida (Harv. Mus. Comp. Zool., No. 783 and 819); female, 936 mm. long, from Key West, Florida (U.S. Bur. Fish., No. 13927); several specimens fresh caught, as well as others at liberty, from southern Florida.

Distinctive Characters. The "Nurse" is set apart from all other sharks of the western Atlantic by the presence of a long barbel on the anterior margin of each nostril and of a deep groove connecting the nostril with the mouth, together with the terminal position of the latter. For characters distinguishing it from its several allies in the Indo-Pacific, see the preceding Keys (pp. 179, 181).

Description. Proportional dimensions in per cent of total length. Female, 650 mm., from Cuba (Harv. Mus. Comp. Zool., No. 518).

Trunk at origin of pectoral: breadth 15.7; height 10.9.
Snout length in front of: outer nostrils 1.4; mouth 2.5.
Eye: horizontal diameter 1.1.
Mouth: breadth 6.5; height 0.6.
Nostrils: distance between inner ends 4.8.
Labial furrow length: upper part 2.6; lower 2.6.
Gill opening lengths: 1st 2.0; 2nd 2.5; 3rd 2.8; 4th 3.1; 5th 3.4.
First dorsal fin: vertical height 10.5; length of base 10.0.
Second dorsal fin: vertical height 7.1; length of base 8.6.
Anal fin: vertical height 4.8; length of base 6.1.
Caudal fin: upper margin 30.7; lower anterior margin 9.2.
Pectoral fin: outer margin 16.9; inner margin 6.8; distal margin 13.7.
Distance from snout to: 1st dorsal 42.4; 2nd dorsal 56.9; upper caudal 69.3; pectoral 22.6; pelvics 42.7; anal 61.4.
Interspace between: 1st and 2nd dorsals 5.4; 2nd dorsal and caudal 3.8; anal and caudal 1.7.
Distance from origin to origin of: pectoral and pelvics 23.4; pelvics and anal 18.9.

Trunk very broad anteriorly, its breadth abreast origin of pectorals about 1/6 to 1/7 of total length, tapering rearward. Caudal peduncle strongly compressed laterally, without lateral ridges or precaudal pits. Dermal denticles closely spaced and more or less

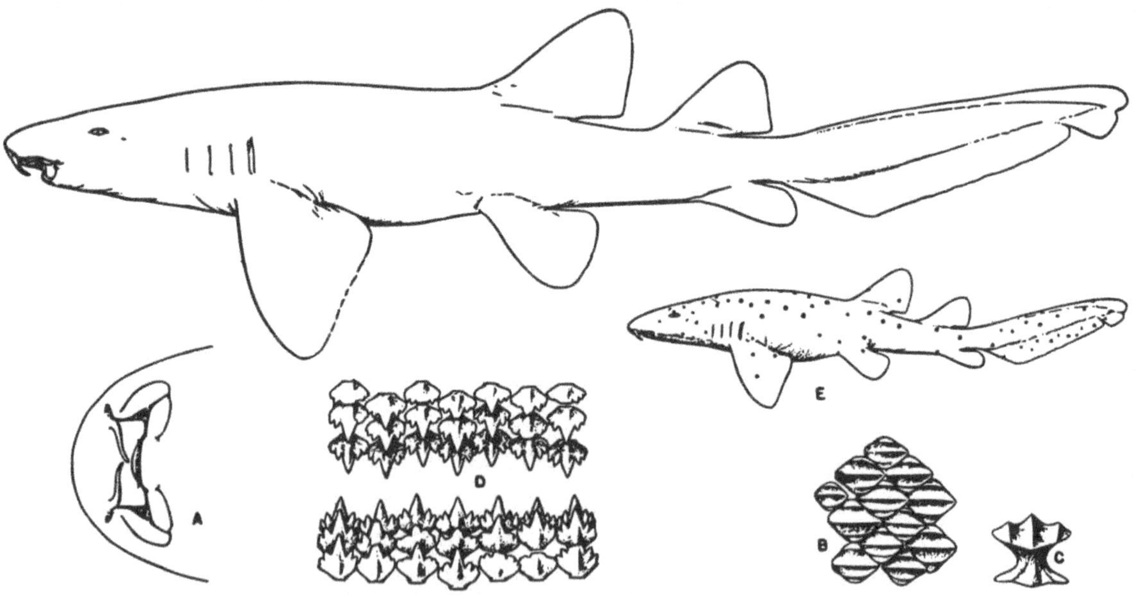

FIGURE 29. *Ginglymostoma cirratum*, female, 650 mm. long, from Cuba (Harv. Mus. Comp. Zool., No. 518). *A* Anterior part of head of same from below, about 0.4 x natural size. *B* Dermal denticles, about 12 x. *C* Apical view of dermal denticle, about 24 x. *D* Upper and lower teeth about 3.3 x. *E* Newborn male, 283 mm. long, from Miami, Florida (Harv. Mus. Comp. Zool., No. 33393).

overlapping,[6] large (about 0.4 × 0.5 mm. in a specimen 650 mm. long), but varying much in size, scale-like, their blades ovate, sharp-pointed or blunted, usually with 3 ridges, the median longest, reaching about halfway to the apex; pedicels high and slender on broad stellate bases.

Head flattened above (more so in males than in females), widest opposite 1st gill opening in males but opposite 5th gill opening in females. Snout broadly rounded and very short, its length in front of mouth only about 1/8 to 1/9 as great as length of head to origin of pectoral. Eye oval, about twice as broad as high, its horizontal diameter only about 1/8 as great as distance from eye to 1st gill slit. Orbital folds as described above for the genus (p. 180). Spiracle a minute slit or pore on a level with lower edge of eye and behind the latter by a distance about 1 to 1½ times the horizontal diameter of eye. Gill openings high on sides and nearly straight, the 3rd over origin of pectoral, the 1st to 4th widely spaced, but 4th and 5th very close together, the margin of the former sometimes overlapping and thus concealing the latter in large specimens, the 5th about 1.7 times as long as the 1st and about 3 times as long as the diameter of eye. Nostril nearly longitudinal, its inner (posterior) end connected with front of mouth by a deep, open groove, its anterior edge outwardly with a tapering, fleshy barbel reaching backward to mouth and also expanded posteriorly as a subrectangular flap that is continuous across front of mouth with that of the opposite nostril, and also with the upper lip. Mouth close to tip of snout, notably small, its breadth a little less than 1/3 as great as length of head, its corners with very deep furrows which form the outline of thick fleshy labial folds on both jaws, the upper extending inward to edge of nostril and the lower a little further.

Teeth $\frac{30\text{ to }36}{28\text{ to }31}$; similar in the 2 jaws, with high triangular central cusp flanked on either side by 1 to 3 smaller cusps (their number increasing with age of tooth) except when worn away; cusps progressively smaller and curving outwardly more toward angles of jaw; 7 to 9 series functional in upper jaw and 8 to 12 series in lower, in medium-sized specimens.

Fins large, with rounded corners. Origin of 1st dorsal over or a little behind origin of pelvics, its vertical height about 2/3 as great as length of pectoral, its anterior margin slightly convex, its rear margin nearly straight, its free rear tip 1/2 to 2/3 as long as its base and reaching rearward considerably past tips of pelvics. Second dorsal similar to 1st in shape, and 2/3 to 4/5 as large in linear dimensions, its origin a little anterior to origin of anal, the distance from its rear tip to origin of caudal about 1/2 to 2/3 as long as its base. Caudal a little less than 1/3 of total length, only a little narrower toward tip than toward base, its axis very little raised, its tip rounded and slightly bilobed in some specimens but scarcely so in others, its lower anterior and lower posterior margins nearly straight, except for the deep subterminal notch, its lower anterior corner obtuse and not expanded as a lobe. Anal less than half as large in area as 2nd dorsal, although nearly as long as at base, its rear margin broadly rounded and slightly overlapping the caudal, its origin under or a little posterior to midpoint of base of 2nd dorsal. Pelvics about 2/3 as large in area as 1st

6. Evidently there is considerable variation in this respect, for Radcliffe (Bull. U.S. Bur. Fish., *34*, 1916: 249) shows them as widely spaced.

dorsal, with broadly rounded corners. Pectoral about 2/3 as broad as long, with broad base, its outer and inner margins moderately convex, its distal margin nearly straight or very slightly concave, its corners broadly rounded; about 1½ times as large in area in males as in females.

Color. Rich yellowish to grayish-brown, darker above than below. Small specimens are usually sparsely and variously marked with small dark spots below as well as above, sometimes with brown crossbars across the snout and through the dorsals, ventrals and anal; adults may or may not retain these markings; also, some young specimens are plain-colored.

Size. The Nurse Shark is small at birth, free-living specimens of only 270 to 290 mm.[7] being recorded, but it grows to a very considerable size, specimens of 7 to 10 or 11 feet being commonly reported, with 11 to 12 feet not unusual. The maximum so far reported is about 14 feet, but maturity may be attained at a comparatively small size, as in the case of a female of only 5 feet that contained well developed embryos.[8] The weight is given as about 330 to 370 pounds at about 8½ feet; 4¼ pounds at 2 feet 3½ inches (692 mm.).

Developmental Stages. Both ovaries may be functional,[9] or only one, with the other atrophied. Mature eggs are very large (reported up to 130 × 180 mm.), blunt-ended, with brownish-black, thin, horny shells. They remain in the hinder parts of the oviducts until the shells break and the young are hatched into the uterus.[10]

Later in development embryos have a short umbilical cord, with very large subspherical or oval yolk sacs; the external gill filaments are retained up to a length of 130 to 140 mm. The length of the nasal barbel and the size of the eye decrease from embryo to adult, but the fins increase in relative size. Females have been described as containing as many as 28 large eggs; a West African specimen (2.43 meters long) has been reported as giving birth to 26 young on capture.[11]

Habits. In its centers of abundance, from Florida to the Caribbean region, this Shark appears chiefly inshore, often in water as shallow as two to ten feet. It is frequently encountered in channels between the mangrove keys. Schools of one to three dozen are sometimes seen on sand flats and over rocky bottom where they are easily approached, the sharks often lying motionless and close to one another, with dorsal fins out of water. Proverbially sluggish in habit, the "Nurse" feeds chiefly on invertebrates — squids, shrimps, crabs, spiny lobsters (*Palinurus*), sea urchins — and small fish. They bite readily on almost any bait. It is common knowledge that they come into very shallow water to breed, and here they are often seen mating. While in the act of copulation the male grasps the

7. See Study Material, p. 181; also Beebe and Tee-Van, Zoologica, N. Y., *10*, 1928: 26.

8. Beebe, Zoologica, N. Y., *26*, 1941: 9.

9. Bell and Nichols, Copeia, 92, 1921: 17.

10. For an account of eggs and early development, see Gudger (Yearb. Carnegie Instn., 1912: 11, 149; Copeia, 98, 1921: 57).

11. Budker (Bull. Mus. Hist. nat. Paris, [2] 7, 1935: 183) substantiates the general report that the number of young in a litter is large.

female with his mouth at the edge of her pectoral fin, these fins in females often being much frayed in consequence.[12] No information is available as to the duration of gestation or the season when young are born.

Relation to Man. Nurse Shark hides are used to some extent for leather, having at present a higher value in the Florida fishery than those of other sharks, but the fins are not in demand. The yield of oil is relatively low.[13] In the West Indies they are sold to some extent in the fish markets, as are most other sharks. On the islands off the southern coast of Brazil the liver oil is said to be in high repute, and the otoliths of this species, as well as those of other sharks, are used by the local fishermen as a diuretic.[14] This shark is perfectly harmless to bathers, and is too sluggish when hooked to be of any interest to sport anglers.

Range. Littoral, on both sides of the tropical and subtropical Atlantic; tropical West Africa and the Cape Verde Islands in the east; southern Brazil to North Carolina and accidentally to Rhode Island in the west; also west coast of America, from the Gulf of California to Panama and Ecuador.

Occurrence in the Western Atlantic. The Nurse Shark is very generally distributed throughout the Caribbean–West Indian region.[15] It is common around Jamaica and Cuba,[16] and in southern Florida waters among the Keys; and it is a year-round resident on the west coast north to Tampa and for some distance up the east coast. It is likewise known from Bermuda. In the warm months it expands its range to the northern coast of the Gulf of Mexico, and occasional Nurse Sharks are taken near Charleston, South Carolina; schools of them sometimes appear in summer off Cape Lookout, North Carolina, and one has been taken in the enclosed waters of Bogue Sound. But this is the boundary to regular migrations in a northerly direction, the only records of them further north being one stray individual for Chesapeake Bay, and one for Rhode Island.

To the southward the Nurse Shark probably occurs all along the northeastern coast of South America, it being known from Maceio in northern Brazil, from Rio de Janeiro, and from South Trinidad Island off southern Brazil (Lat. 20° 30′ S., Long. 29° 22′ W.). There is no report of it farther south. Curiously enough there is but one record of it for the western shores of the Gulf of Mexico–Caribbean region, that being from Colón, Panama. But it is to be expected anywhere there, judging from the generality of its distribution throughout the West Indian region.

Synonyms and References:

Chien de Mer Barbillon, Broussonet, Mem. Math. Phys. Acad. Sci. Paris, 1780: 656.

Gata, Parra, Hist. Nat., 1787: 86, pl. 34, fig. 2 (descr., Cuba).

Squalus cirratus Bonnaterre, Tabl. Encyc. Meth. Ichthyol., 1788: 7 (descr., American seas); Gmelin, in Lin-

12. See Gudger (Yearb. Carneg. Instn., *11*, 1912: 149) for a description.
13. Springer, Proc. Fla. Acad. Sci., *3*, 1939: 14.
14. Nichols and Murphy, Bull. Amer. Mus. nat. Hist., *33*, 1914: 262.
15. Recorded from French and British Guiana, Trinidad, St. Croix, Turks Island, St. Martins and Barbados in the Lesser Antilles, Jamaica, Haiti, Porto Rico, the Bahamas and Cuba.
16. Personal communication from Luis Howell-Rivero.

naeus, Syst. Nat., *1*, 1789: 1492 (descr.); Bloch and Schneider, Syst. Ichthyol., 1801: 128 (descr., spelled *cirrhatus*); Bosc. Nouv. Dict. Hist. Nat., *21*, 1803: 188 (diagn., Pacif.); Latreille, Nouv. Dict. Hist. Nat., *24*, 1804: 72 (in table of contents); Bennett, Zool. J. Lond., *5*, 1835: 87 (refs.); Bancroft, Zool. J. Lond., *5*, 1835: 416 (Jamaica).

Squale barbillon, Lacépède, Hist. Nat. Poiss., 4° ed., *1*, 1798: 166, 245 (diag., descr., Pacif. Amer.); in Sonnini, Hist. Nat. Poiss., *4*, 1801–1802: 54 (descr.).

Squalus punctulatus Lacépède, Hist. Nat. Poiss., 8° ed., *3*, 1800: 153, pl. 4, fig. 3 (Cayenne, descr.); Bloch and Schneider, Syst. Ichthyol., 1801: 549 (descr., ref. to Lacépède, 1800); Lacépède, Hist. Nat. Poiss., 4° ed., *2*, 1800: 120, pl. 4, fig. 3, in Buffon, Hist. Nat.

Squalus punctatus Bloch and Schneider, Syst. Ichthyol., 1801: pl. 134; Bennett, Zool. J. Lond., *5*, 1835: 87 (ref.).

Scyllium cirrhosum Griffith, in Cuvier, Anim. Kingd., Lond., 1834: 10, pl. 30 (ill.).

Squalus argus Bancroft, Zool. J. Lond., *5*, 1835: 82 (West Indies).

Scyllium cirratum Bennett, Zool. J. Lond., *5*, 1835: 87.

Ginglymostoma cirrhosum Bonaparte, Mém. Soc. neuchâtel. Sci. nat., *2* (8), 1839: 11 (in synopsis).

Ginglymostoma cirratum Müller and Henle, Plagiost., 1841: 23 (descr., Cayenne); Duméril, Rep. Poiss. Afr. Occid., 1861: 261 (trop. W. Afr.); Hist. Nat. Poiss., *1*, 1865: 334 (descr., distrib.); Poey, Repert. Fisico-Nat. Cuba, *2*, 1868: 455 (Cuba); Günther, Cat. Fish. Brit. Mus., *8*, 1870: 408 (distrib., descr., synonyms); Poey, Enumerat. Pisc. Cubens., 1876: 204; An. Soc. esp. Hist. Nat., *5*, 1876: 400 (Cuba); Goode, Bull. U.S. nat. Mus., *5*, 1876: 73 (Bermuda); Lugger, Rep. Comm. Fish. Md. (1877), 1877: 90 (Chesapeake Bay); Hasse, Naturl. Syst. Elasm. besond. Theil, 1882: 201, pl. 26, fig. 12–14, 16–20, 22–24 (vertebrae); Rochebrune, Act. Soc. linn. Bordeaux, (4) *6*, 1882: 47; Faune Senegambie, Poiss., *1*, 1883–1885: 24 (Senegambia); Goode and Bean, Proc. U.S. nat. Mus., *5*, 1882: 240 (Gulf of Mexico); Jordan and Gilbert, Bull. U.S. nat. Mus., *16*, 1883: 18 (Gulf of Mexico and W. Indies); Bull. U.S. Fish Comm., *2*, 1883: 105 (Mazatlan, Mexico); Proc. U.S. nat. Mus., *5*, 1883: 371 (Colima, Mexico), 581 (S. Carolina), 620 (Panama); Jordan, Proc. U.S. nat. Mus., *8*, 1885: 363 (Mazatlan, Panama); Rep. U.S. Comm. Fish. (1885), 1887: 794 (distrib.); Vaillant, Bull. Soc. philom. Paris, (8) *6*, 1894: 70 (Gulf of Calif.); Jordan, Stanford Univ. Pub., Univ. Ser., *1*, 1895: 381 (Gulf of California); Jordan and Evermann, Rep. U.S. Comm. Fish. (1895), 1896: 214 (distrib.); Bull. U.S. nat. Mus., 47 (1), 1896: 26 (descr., distrib.); Jordan and Rutter, Proc. Acad. nat. Sci. Philad., *49*, 1897: 91 (Jamaica); Osorio, J. Sci. math. phys. nat. Lisboa, (2) *5* (19), 1898: 200 (Cape Verde); Boulenger, Boll. Mus. Zool. Anat. comp. Torino, *13* (329), 1898: 1 (Ecuador); Jordan and Evermann, Bull. U.S. nat. Mus., 47 (4), 1900: pl. 4, fig. 13; Gilbert, Proc. Wash. Acad. Sci., *2*, 1900: 161 (Maceio, Brazil); Evermann and Kendall, Rep. U.S. Comm. Fish., 1900: 48 (Florida); Gregg, Where to Catch Fish, 1902: 17, fig. 288 (Florida); Evermann and Marsh, Bull. U.S. Fish Comm., *20*, 1902: 60 (W. Indies); Werner, Zool. Jb., Syst. Abt., 21, 1904: 288 (Jamaica); Gilbert and Starks, Mem. Calif. Acad. Sci., *4*, 1904: 5 (Panama Bay); Hargreaves, Fish. Brit. Guiana, 1904: 14 (Brit. Guiana); Jordan and Thompson, Bull. U.S. Bur. Fish., *24*, 1905: 232 (Tortugas, Florida); Bean, B. A., in Shattuck, Bahamas Islands Fish., 1905: 296 (size, no. of eggs, Bahamas); Fowler, Proc. Acad. nat. Sci. Philad., *58*, 1906: 79 (Marquesas Keys, Florida, breeding, large size, color); Garman, Bull. Mus. comp. Zool. Harv., *46*, 1906: 229 (Panama); Bean, T. H., Field Mus. Publ., 7 (2), 1906: 29 (Bermuda); Ribeiro, Arch. Mus. nac. Rio de J., *14*, 1907: 166, 204 (descr., ref. Brazil); Regan, Proc. zool. Soc. Lond., 1908: 350 (descr., distrib., Jalisco, Mexico); Fowler, Proc. Acad. nat. Sci. Philad., *60*, 1908: 53 (St. Martins, W. Indies); Vincent, Sea Fish. Trinidad, 1910: 53 (Trinidad); Rosen, Lund Univ. Arsskr., N. F. 7, *2* (5), 1911: 47; Fysiog. Sallsk. Handl., N. F. *22* (5), 1911: 47 (Bahamas); Roule, Bull. Inst. océanogr. Monaco, 243, 1912: 5 (C. Verde); Gudger, Yearb. Carneg. Instn., *11*, 1912: 148 (breeding habits, eggs, early devel., Tortugas, Florida); Garman, Mem. Harv. Mus. comp. Zool., *36*, 1913: 54, pl. 7, fig. 4–6 (descr.); Gudger, J. Elisha Mitchell sci. Soc., *29*, 1913: 8; Science, *40*, 1914: 386 (Florida Keys, abund., habits); J. Elisha Mitchell sci. Soc., 30, 1914: 64; Nichols and Murphy, Bull. Amer. Mus. nat. Hist., *33*, 1914: 262 (South Trinidad I., Lat. 20° 30′ S., Long. 29° 22′ W.); Fowler, Proc. Acad. nat. Sci. Philad., *67*, 1916: 521 (Trinidad); Coles, Proc. biol. Soc. Wash., *28*, 1915: 89 (C. Lookout, N. Carolina, size); Gudger, Yearb. Carneg. Instn.,

14, 1916: 207 (Tortugas, Florida); Fowler, Copeia, 30, 1916: 36 (mid. Atlant., U.S.); Radcliffe, Bull. U.S. Bur. Fish., *34*, 1916: 248, pl. 39, fig. 1 (teeth, Florida, dermal denticles, N. Carolina); Metzelaar, Trop. Atlant. Visschen, 1919: 5 (W. Indies), 186 (w. coast Afr.); Fowler, Proc. Acad. nat. Sci. Philad., 71, 1919: 133 (St. Martins, W. Indies), 146 (Jamaica); Nichols, Bull. Amer. Mus. nat. Hist., *44*, 1921: 22 (Turks I.); Nichols, J. Amer. Mus. nat. Hist., *21*, 1921: 275 (ill. adult and young, teeth, no. of eggs, food); Bell and Nichols, Copeia, 92, 1921: 17 (N. Carolina, food, size, weight, no. of eggs); Gudger, Copeia, 98, 1921: 57 (eggs, devel., near Key West, Florida); Meek and Hildebrand, Field Mus. Publ., *15* (1), 1923: 29 (descr., Colón, Balboa, Panama); Fowler, Proc. biol. Soc. Wash., *36*, 1923: 27 (Tortugas, Florida); Ribeiro, Arch. Mus. nac. Rio de J., Fauna brasil. Peixes, *2* (1) Fasc. 1, 1923: 23 (descr., Rio de Janeiro); Fowler, Proc. Acad. nat. Sci. Philad., *78*, 1926: 249 (south. Florida); Nichols and Breder, Zoologica, N. Y., *9*, 1927: 11 (probably Rhode Island); Fowler, Proc. Acad. nat. Sci. Philad., *80*, 1928: 451 (embryos, Key West, Florida); Hildebrand and Schroeder, Bull. U.S. Bur. Fish., *43*, 1928: 45 (general, Chesapeake Bay); Beebe and Tee-Van, Zoologica, N. Y., *10* (1), 1928: 26 (size, weight, color, food, Haiti); Breder, Field Bk. Mar. Fish. Atlant. Coast, 1929: 11 (general); Jordan, Manual Vert. Anim. NE. U.S., 1929: 9 (general); Nichols, Ann. N. Y. Acad. Sci., *10* (2), 1929: 181 (Porto Rico); Truitt, Bean and Fowler, Bull. Md. Conserv. Dep., *3*, 1929: 29 (Chesapeake Bay); Jordan, Evermann and Clark, Rep. U.S. Comm. Fish. (1928), 2, 1930: 12 (distrib.); Manter, Yearb. Carneg. Instn., 29, 1930: 340; *ibid.*, 30, 1931: 388 (parasites, Tortugas, Florida); Beebe and Tee-Van, Field Bk. Shore Fish. Bermuda, 1933: 23 (Bermuda); Fowler, Proc. biol. Soc. Wash., *36*, 1933: 27 (Tortugas, Florida); Borodin, Bull. Vanderbilt Oceanogr. (Mar.) Mus., *1* (4), 1934: 107 (Miami, Florida); Brooks, Parasitology, *26* (2), 1934: 264 (parasites, Tortugas, Florida); Breder, Copeia, 1934: 27 (sexual diffs.); Zoologica, N. Y., *18*, 1934: 58–59 (Bahamas); Budker, Bull. Mus. Hist. nat. Paris, (2) 7, 1935: 183 (Dakar, W. Afr.); Burton, Sci. Mon., N.Y., *40*, 1935: 283 (S. Carolina); Fowler, Bull. Amer. Mus. nat. Hist., *70* (1), 1936: 36 (descr., trop. Atlant.); Norman and Fraser, Giant Fishes, 1937: 28, fig. 16 (general); Butsch, J. Barbados Mus. Hist. Soc., 7 (1), 1939: 18 (common at Barbados, not seen); Beebe and Tee-Van, Zoologica, N. Y., *26* (2), 1941: 96 (west coasts Mexico, Costa Rica, Panama); Beebe, Zoologica, N. Y., *26* (1), 1941: 9 (embryos, Bermuda); Hildebrand, Copeia, 1941: 221 (recent N. Carolina records); Longley and Hildebrand, Yearb. Carneg. Instn., *34*, 1941: 11 (Tortugas, Florida); Norris, Plagiost. Hypophysis, 1941: 27, pl. 6, fig. 25 (brain); Fowler, Fish Culturist, *21* (9), 1942: 66 (listed, Cuba); Lunz, Bull. S. C. St. Planning Bd., 14, 1944: 26 (S. Carolina, Florida).

Ginglymostoma fulvum Poey, Memorias, *2*, 1861: 342, pl. 19, fig. 15, 16 (teeth, Havana, Cuba); Repert. Fisico-Nat. Cuba, *2*, 1868: 455, pl. 4, fig. 12–14 (teeth, Cuba); Enumerat. Pisc. Cubens., 1876: 204; An. Soc. esp. Hist. Nat., *5*, 1876: 400 (Cuba, Brazil); Stahl, Fauna Puerto Rico, 1883: 167 (Porto Rico); Jordan, Proc. U.S. nat. Mus., *8*, 1885: 363 (west coast Cent. Amer., doubtful species); Bigelow and Schroeder, Guide Comm. Shark Fish., Anglo Amer. Caribb. Comm., Wash., 1945: 98 (descr., ill., range).

Ginglymostoma caboverdianus Brito Capello, J. Sci. math. phys. nat. Lisboa, *1* (2), 1867: 167 (C. Verde).

Ginglymostoma cirrhatum Jordan, Guide to Study Fish., *1*, 1905: 533 (general).

Ginglymostoma cirrotum[17] Gudger, Yearb. Carneg. Instn., 12, 1914: 176 (abund. near Key West, Florida).

Gingylostoma[18] *cirratum* Springer, Proc. Fla. Acad. Sci., *3*, 1939: 13 (descr., Florida).

Nebrius cirratum Fowler, Arqu. Zool. Estado São Paulo, *3*, 1942: 127 (Brazil); Mongr. Acad. nat. Sci. Philad., 7, 1945: 94 (sizes, weights, N. Carolina).

Family RHINCODONTIDAE

Whale Sharks

Characters. Two dorsal fins, the 1st much shorter than the caudal, its origin in front of origin of pelvics, but its base overlapping the latter; 2nd dorsal and anal much smaller than 1st dorsal; caudal much less than ½ total length, lunate in form, its axis steeply

17. Evident misspelling for *cirratum*.

18. Evident misspelling for *Ginglymostoma*.

raised; caudal peduncle not greatly expanded laterally, with a precaudal pit above but none below; sides of trunk anterior to anal with longitudinal ridges; snout very short and mouth not widely protrusible; gill openings very large, the 4th and 5th over base of pectoral; gill arches connected, one with the next, by numerous transverse cartilaginous bars which support soft spongy masses of tissue developed from clumps of modified denticles, the entire gill apparatus forming a sieve of innumerable minute meshes (1 to 2 × 2 to 3 mm. in specimen 31 ft. 6 in.) through which water is forced when the mouth is closed, the planktonic food thus being retained and swallowed;[1] oesophagus lined with large papillae, covered with denticles; nostril entirely separate from mouth,[2] its anterior margin without barbel; spiracles present; lower eyelid without nictitating fold or membrane; each jaw with a labial fold near the corner; teeth minute, very numerous, many rows functional; head of normal shape (not widely expanded); no rostral cartilages; no foramen between mesopterygium and metapterygium of pectoral[3] (relative number of radials on meta- and mesopterygia not known); heart valves in 2 rows. Development probably ovoviviparous (see p. 192).

Genera. *Rhincodon*, the only known representative of the family, has sometimes been associated with the *Orectolobidae* (*e.g.*, by Regan[4]) on the supposition that its nostrils are connected with the mouth by oronasal grooves. Recently, however, it has been found that this is not the case;[2] and it is so widely separated from the *Orectolobidae* in other respects, especially by its complex gill sieve and its lunate caudal with sharply raised axis, that it clearly represents a distinct family.

Only one genus known, *Rhincodon*.

Genus *Rhincodon* Smith, 1829

Whale Sharks

Rhincodon Smith, Zool. J., *4*, 1829: 443; type species, *Rhincodon typus* Smith. Table Bay, South Africa.

Generic Synonyms:[5]

Rineodon Müller and Henle, Charlesworth's Mag. Nat. Hist., *2*, 1838: 37; no species mentioned.

Rhineodon Swainson, Nat. Hist. Fish. Amphib. Rept., *1*, 1838–1839: 142.

Rhiniodon Swainson, Nat. Hist. Fish. Amphib. Rept., *2*, 1838–1839: 317.

Rhinodon Müller and Henle, Plagiost., 1841: 77; type species, "*Rhinodon typicus* Smith"; evident misquotation for *Rhincodon typus* Smith, 1829.

Rhineodon Gray, List Fish. Brit. Mus., 1, 1857: 66; *Rhineodon typicus* Gray, equals *Rhincodon typus* Smith, 1829.

Micristodus Gill, Proc. Acad. nat. Sci. Philad., 1865: 177; type species, *M. punctatus* Gill. Gulf of California.

1. For photographs of this gill structure, unique among sharks, see White (Bull. Amer. Mus. nat. Hist., *74*, 1937: pl. 9, 10) and Gudger (J. Morph., *68*, 1941: 91–95, fig. 6–8).
2. It has been stated repeatedly that the nostril is connected with the mouth by an oronasal groove (for example, see Garman, Mem. Harv. Mus. comp. Zool., *36*, 1913: 41); it has been found recently that such is not the case (Barnard, Ann. S. Afr. Mus., *30*, 1935: 647, pl. 25).
3. See White (Bull. Amer. Mus. nat. Hist., *61*, 1931: 130) for account of skeletal characters.
4. Proc. zool. Soc. Lond., 1906: 745.
5. Most of the generic synonyms here listed are simply emended spellings of *Rhincodon* Smith, 1829.

Cetorhinus Poey, An. Soc. esp. Hist. nat., *5*, 1876: 184 (380); for *C. maximus* Poey (Cuba), which equals *Rhincodon typus* Smith, 1829; not *Cetorhinus* Blainville, 1816.

Selache Thomas, Cannibals and Convicts West Pacif., 1887: 38; for *S. maxima* Thomas (New Guinea), which equals *Rhincodon typus* Smith, 1829; not *Selache* Cuvier, 1817.

Generic Characters. Those of the family (p. 187).

Range. Tropical belts of all oceans.

Species. One species only, *R. typus* Smith, 1829.

Rhincodon typus Smith, 1829

Whale Shark

Figure 30

Study Material. Excellent mounted specimen, 17 feet 4¾ inches long, from Acapulco, Mexico (Amer. Mus. Nat. Hist.); dried skin and dental plates of 18-foot specimen from Ormond, Florida (U.S. Nat. Mus., No. 27618); also photographs of a newly caught Cuban specimen from Luis Howell-Rivero.

Distinctive Characters. Distinguished from all other sharks by its enormous size, spotted color pattern, lunate tail, very wide gill openings, unique gill apparatus (see p. 188), and mouth at tip of snout.

Description. Proportional dimensions in per cent of total length from mounted specimen; female, 5,302 mm. (17 ft. 4¾ in.) from Acapulco, west coast of Mexico (Amer. Mus. Nat. Hist., New York).

Trunk at origin of pectoral: breadth 18.5; height 14.2.

Snout length in front of: outer nostrils 1.0; mouth 0.5.

Eye: horizontal diameter 0.5.

Mouth: breadth 16.3; height 0.

Nostrils: distance between inner ends 10.8.

Gill opening lengths: 1st 8.8; 2nd 9.3; 3rd 8.9; 4th 8.0; 5th 6.6.

First dorsal fin: vertical height 6.9; length of base 8.5.

Second dorsal fin: vertical height 2.8; length of base 3.2.

Anal fin: vertical height 2.7; length of base 2.7.

Caudal fin: upper margin 26.4; lower anterior margin 11.8.

Pectoral fin: outer margin 15.9; inner margin 4.4; distal margin 14.4.

Distance from snout to: 1st dorsal 39.7; 2nd dorsal 62.1; upper caudal 73.6; pectoral 18.0; pelvics 46.1; anal 63.0.

Interspace between: 1st and 2nd dorsals 13.9; 2nd dorsal and caudal 8.3; anal and caudal 10.0.

Distance from origin to origin of: pectoral and pelvics 28.3; pelvics and anal 16.8.

Trunk moderately stout, each side with 2 prominent dermal ridges originating close together high on the shoulders, the lower one extending backward the whole length of trunk and as a keel out onto anterior part of caudal, the upper one dividing anterior to the

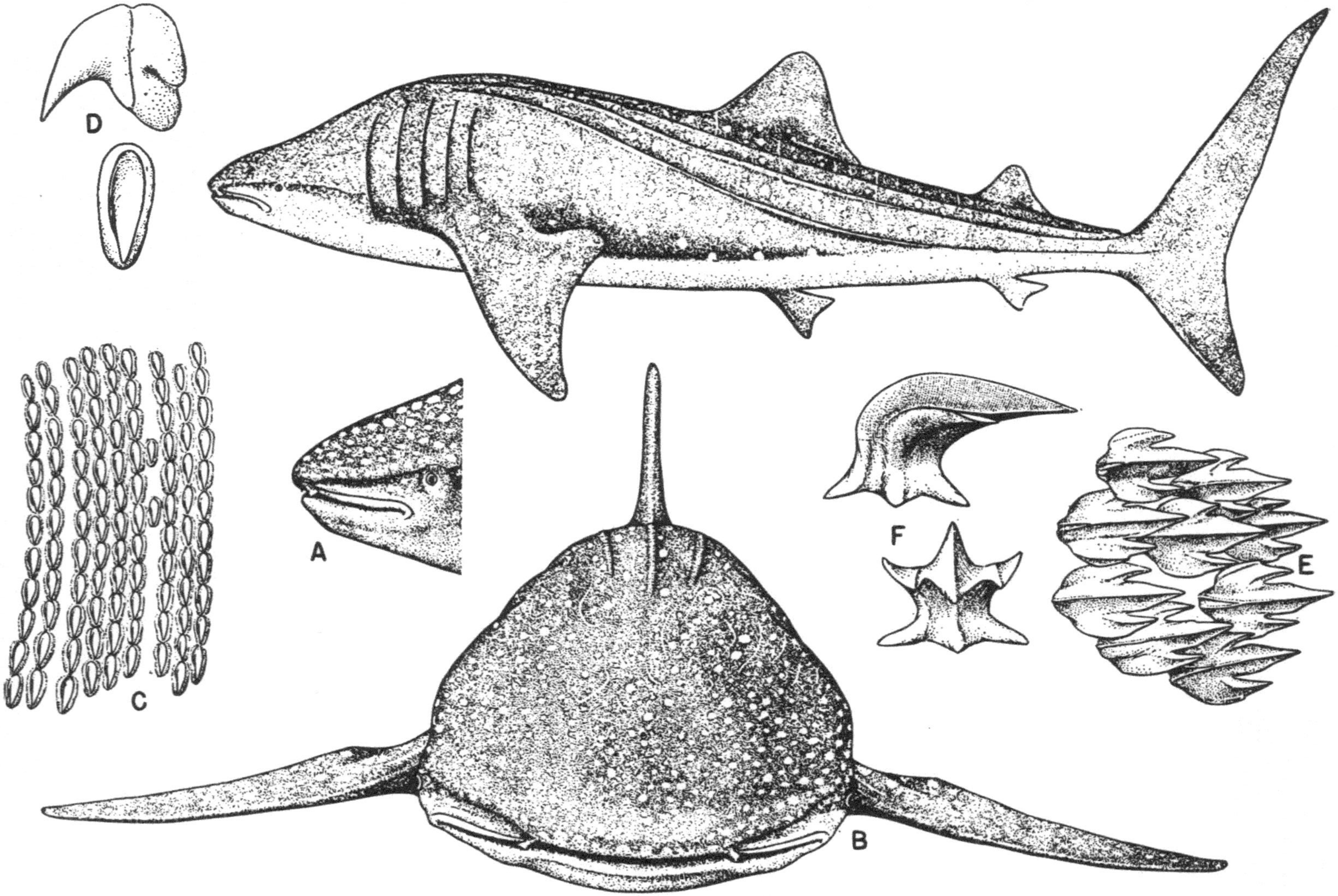

FIGURE 30. *Rhincodon typus*, female, 17 feet 4 inches long, from Acapulco, Mexico (mounted, Amer. Mus. Nat. Hist.). *A* Side view of snout of same to larger scale. *B* Front view of same, same scale as *A*. *C* Segment of upper dentary band of an 18-foot specimen, from Ormond, Florida (U. S. Nat. Mus., No. 27618), about 2.5 x. *D* Front and lateral views of one upper tooth, about 9 x. *E* Dermal denticles of same, about 35 x. *F* Lateral and apical views of dermal denticle, about 75 x.

origin of the 1st dorsal, with both its branches terminating just in front of, under, or well behind, the 2nd dorsal; also a ridge along midline of the back from the rear part of head to 1st dorsal in some specimens, but seemingly not in others. Caudal peduncle with a precaudal pit above but none below. Dermal denticles very small (less than 1 mm. broad in specimen 31 ft. 5 in. long), slightly overlapping, scale-like, their blades varying in shape, but usually with 3 to 5 ridges, the axial ridge very high, their margins with 3 to 7 teeth, the median much the longest, on relatively high pedicels.[6]

Head strongly flattened above, its dorsal profile concave anterior to gill openings and broadly rounded in front. No distinct snout, the mouth being at the anterior margin of the head. Eye minute, its horizontal diameter less than 1/10 as great as length of nostril, its center situated a little posterior to angle of mouth. Spiracle about as large as eye, a short distance behind and above the latter. Gill openings notably large but high up on the sides and widely separated ventrally, the 3rd–4th over origin of pectoral, the 2nd in front of pectoral and extending below it, the 2nd and 3rd longest. Nostrils at front margin of head, widely separated, the space between them about 2/3 as wide as mouth, their anterior margins without barbels, but each expanded as a broad, quadrilateral lobe with rounded corners, overlapping the upper lip, entirely separate from mouth but with outer end continuous with upper labial furrow. Mouth very large, its width nearly as great as breadth of head, transverse, hardly arched. Upper labial furrow extending so far that it joins outer end of nostril,[7] the lower furrow hardly extending beyond corner of mouth.

Teeth similar in the 2 jaws, minute (averaging about 1.5 mm. long in an 18-foot specimen, but about 4.5 mm. in a 31-foot fish), in about 310 rows, with 10 to 15 rows (average about 12 to 13) functional all along the dental band, or a total count of around 3,600 teeth in an 18-foot specimen (Fig. 30) and perhaps still more in larger specimens; the rows vertical toward center of mouth but somewhat oblique toward corners; each tooth with a single sharp cusp curved backward.[8]

First dorsal nearly an equilateral triangle, of moderate size, its vertical height a little less than 1/3 as great as length of head, its anterior margin nearly straight, posterior margin moderately concave, apex rounded, its free rear tip triangular, a little shorter than the base, its origin considerably in front of origins of pelvics with the rear end of its base about over rear ends of bases of pelvics. Second dorsal about 1/3 as large in area as 1st and similar in form, but with its free rear tip a little longer than its base; its origin considerably anterior to that of anal. Caudal noticeably large, its posterior margin broadly concave in subangular outline, its upper lobe about 22% of total length, its lower lobe about 45% as long as upper, the tips of both lobes pointed, the upper without subterminal notch. Anal nearly as large as 2nd dorsal and similar in shape but with shorter free rear tip, its origin

6. See White (Bull. Amer. Mus. nat. Hist., *61*, 1931: 144) for account of the variation in shape of the denticles.
7. This is shown very clearly in Barnard's (Ann. S. Afr. Mus., *30*, 1935: pl. 25) photograph of a newly captured specimen.
8. For excellent photographs of teeth, see Bean (Smithson. misc. Coll., *48*, 1905: pl. 36), and White (Bull. Amer. Mus. nat. Hist., *61*, 1931: pl. 10, 11); for additional counts, see Mowbray (Prelim. Rep. Sci. Cruise "Nourmahal," 1933: 2).

about under rear end of base of 2nd dorsal. Pelvics notably small, being only about as large as anal. Claspers of male of usual galeoid type. Pectoral about ⅙ to ⅐ as long as total length, its distal margin moderately concave, its apex subacute.

Color. Described as varying from dark gray to reddish or greenish brown on back and sides, including upper surface of pectorals, and marked with round white or yellow spots (2 to 3 inches in diameter in 38-foot specimen), these being smallest and most crowded on the head, largest and most scattered rearward; also a variable number of narrow white or yellow transverse stripes; lower parts plain white or yellow; lips, tongue and lining of mouth whitish; lining of oesophagus black.

Size. This is the largest of modern fish-like vertebrates; specimens so far measured have ranged between 6[9] and about 45[9a] feet in length, with 6–34 feet recorded for Cuban specimens, 18–34 feet for Florida examples, and 31.5 feet for one taken at Fire Island, New York. But the Whale Shark is creditably reported to reach lengths of 60 feet or even more. The estimated weight of a 38-foot Whale Shark, taken at Knight's Key, Florida, in June 1912 was 26,594 pounds.[10] The size at which sexual maturity is attained is not known, nor is the size at birth.

Developmental Stages. Sixteen eggs have been counted in a specimen from Ceylon, these being of the "same form as in dog fish."[11] Whether or not these hatch before birth is not definitely known.

Habits. Notwithstanding the extensive literature regarding the Whale Shark that has developed in the past few years, much of which is repetitious, very little is known of its habits, other than that it gathers in schools, often basks at the surface and is so sluggish and so little alert that specimens are rammed by steamers from time to time. It feeds by gulping mouthfuls of small animals, as does the Basking Shark; it then drives out through its branchial sieve the water that it takes with them into its mouth. It has been seen at the surface with open mouth when so employed, swimming or even vertical in the water. Its diet may be either small crustacea, as in the case of a Galapagos specimen where 98 per cent of the stomach contents consisted of such,[12] or perhaps more often small fish. Cuban fishermen, for example, describe it as gorging on schools of anchovies, sardines and albacores, apparently standing vertical below the school while feeding. It is also known to devour small squids when they are available. But the fact that a Philippine specimen had in its stomach 47 buttons, 3 leather belts, 7 leggings and 9 shoes[13] is evidence that it is not very discriminating, if the individual in question actually was a Whale Shark.[14]

Relation to Man. The Whale Shark has been the object of a fishery along the north-

9. Personal communication from Luis Howell-Rivero.
9a. Seychelles Is., Indian Ocean, reported by Wright (Spicil. Biol., Dublin, 1870: 64–65 [not seen]).
10. Field and Stream, Feb. 1936: 27. The length of this specimen was originally given as 45 feet (Garman, Mem. Harv. Mus. comp. Zool., *36*, 1913: 456) but was later found to be only 38 feet (Gudger, Proc. Zool. Soc. Lond., 1934: 881).
11. Southwell, Ceylon Adm. Rep. Mar. Biol., 1912: E. 44, E. 49.
12. Mowbray, Prelim. Rep. Sci. Cruise "Nourmahal," *1*, 1933: 2.
13. Jordan, Science, N. S. *26*, 1915: 463.
14. Gudger (Amer. Nat., *75*, 1941: 550) suggests that it may have been some other shark.

west coast of India, but elsewhere it is of no commercial importance. It is entirely harmless to bathers or small boats, unless by accidental contact.

Range. Pelagic in tropical belts of all oceans. Reliable reports of it are from South Africa (type locality), Red Sea and Straits of Bab-el-Mandeb, Seychelles, west coast of India, Ceylon, Bay of Bengal, many localities in the Malaysian-Papuan region, Philippines (a center of abundance), southeastern Australia, Indo-China, Gulf of Siam, Bonin Islands, Japan, Paumotos, Gulf of California (especially numerous near Cape St. Lucas), west coast of Mexico (taken frequently at Acapulco), Panama-Galapagos region, coast of Peru, in the Indo-Pacific; South Africa, Gulf of Guinea, Brazil, Caribbean–West Indian region, Florida, and casually to New York in the Atlantic. Up to the present time, the most northerly locality for it is about 42° North Latitude (near New York), the most southerly locality 33° 55′ South (Table Bay, South Africa).

Occurrence in the Western Atlantic. Records of the Whale Shark in the western Atlantic are distributed as follows, from south to north: Abrolhas Island, Brazil, Lat. 17° 15′ S. (one); western Caribbean, between Colón and Cartagena (one); Haiti (one); around Cuba;[15] central part of the Gulf of Mexico (single individuals or schools reported on eight occasions); Bahamas (a school in Tongue of the Ocean, and one at Bimini); Gulf Stream between Bahamas and Florida; southern and eastern Florida (five); mouth of Cape Fear River, North Carolina (one); south shore of Long Island, near New York (one). From the foregoing, the Caribbean–West Indian region is evidently the center of population for it on this side of the Atlantic. Occasional captures on the coasts of North Carolina and New York show merely that Whale Sharks, like other tropical animals, occasionally stray far northward beyond their normal range in the warm months. There is one report of it from Bermuda.[15a] The fact that it is sometimes reported in schools prevents estimation of the numbers actually reported up to the present.

Synonyms and References:

Rhincodon typus Smith, Zool. J., *4*, 1829: 443 (Table Bay, S. Afr.); Garman, Mem. Harv. Mus. comp. Zool., *36*, 1913: 42, 456 (descr., large Florida specimen); Herre, Philipp. J. Sci., *26*, 1925: 116 (Philippines); Fowler, Bull. U.S. nat. Mus., 100 (*13*), 1941: 116 (descr., synonyms, distrib.); Fowler, Bol. Mus. Hist. N. Javier Prado, Lima, *5* (17), 1941: 220; Fowler, Fish Culturist, 21 (9), 1942: 66 (listed Cuba); Bigelow and Schroeder, Guide Comm. Shark Fish., Anglo Amer. Caribb. Comm., Wash., 1945: 96 (descr., ill., range).

Rhineodon (no specific name) Swainson, Nat. Hist. Fish. Amphib. Rept., *1*, 1838–1839: 142 (general); *2*: 191, 314, 317[16] (general); Gudger, Nat. Hist. N. Y., *22*, 1922: 246 (a shark sucker, *Remora*, in its mouth); Smith, Science, *62*, 1925: 438 (Siam); Gudger, Nature, Lond., *131*, 1933: 165 (Ceylon); Denison, Bull. Amer. Mus. nat. Hist., *73*, 1930: 477 (skelet.); Gudger, Nature, Lond., *143*, 1939: 79 (feeding, Bahamas); Sci. Mon., N. Y., *48*, 1939: 261 (Haiti, Gulf of Mexico).

Rhinodon typicus Müller and Henle, Plagiost., 1841: 77, pl. 35, fig. 2 (descr.); Smith, Ill. Zool. S. Afr. Pisces, 1849: pl. 26 (descr., meas., color, ill. most often copied); Duméril, Hist. Nat. Poiss., *1*, 1865: 428 (descr.); Günther, Cat. Fish. Brit. Mus., *8*, 1870: 396 (descr.); Guide to Study Fish., 1880: 323 (general); Haly, Ann. Mag. nat. Hist., (5) *12*, 1883: 48 (Ceylon, meas.); Günther, Nature, Lond., *30*,

15. "Caught quite often in open waters, and seen many times feeding" (personal communication from Luis Howell-Rivero); also five published records.

15a. Personal communication from John Tee-Van.

16. Spelling on page 317 is "*Rhiniodon.*"

1884: 365 (Gulf of Panama); Chierchia, Coll. Sci. Nat. viaggio... Vett. Pisani (1882–1885), 1885: 68 (Bay of Panama); Day, Fish. India, suppl., 1888: 811 (Ceylon); Day, in Blandford, Fauna Brit. India, Fishes, *1*, 1889: 29 (general); Thurston, Bull. Madras Govt. Mus., 1894: 36–38, pl. 3 (descr., Ceylon, Madras); Casto de Elera, Cat. Fauna Philipp., 1895: 615 (Philippines); Gill, Science, *15*, 1902: 824 (history); Bean, B. A., Science, *15*, 1902: 353 (Ormond, Florida); Bridge, Camb. Nat. Hist., 7, 1904: 287 (general); Bean, B. A., Smithson. misc. Coll., *48*, 1905: 139, pl. 34–36 (general, photos); Weber-Van Bosse, Ein Jahr am "Siboga," Leipzig, 1905: 220 (Malaysia); Wilson, Proc. U.S. nat. Mus., *33*, 1907: 439 (parasites); Van Kampen, Natuurk. Tidschr. Ned.-Ind., *67*, 1908: 124 (Batavia, Java); Fountain, Rambles Aust. Nat., 1907: 119 (Aust., not seen); Lloyd, Rec. Indian Mus., *2*, 1908: 306 (Bay of Bengal); Regan, Proc. zool. Soc. Lond., 1908: 353 (general); Günther, J. Mus. Godeffroy, *6* Heft 17, Fische der Sudsee, *3* Heft 9, 1910: 486 (size); Smith, Proc. biol. Soc. Wash., *25*, 1911: 97 (Philippines, size); Günther, Encycl. Brit., 11th ed., *24*, 1911: 808 (general); Southwell, Ceylon Adm. Rep. Mar. Biol., 1919–1923: E44, E49 (Ceylon, eggs); Weber, "Siboga" Exped., *57*, 1913: 593 (Malaysia); Jordan, Science, N. S. *41*, 1915: 463 (Philippines, exceptional diet); Smith, J. Amer. Mus. nat. Hist., *16*, 1916: 352 (distrib., habits); Barnard, Ann. S. Afr. Mus., *21* (1), 1925: 37 (descr., S. Afr.); Pillay, J. Bombay nat. Hist. Soc., *33*, 1929: 351 (India, not seen); Tanaka, Jap. Fish. Life Colors, 24, 1933 (Japan, not seen).

Great Basking Shark, Buist, Proc. zool. Soc. Lond., *18*, 1850: 100 (fishery, Karachi, India).

Rhineodon typicus Gray, List Fish. Brit. Mus., 1851: 67 (descr.); Jordan and Fowler, Proc. U.S. nat. Mus., *26*, 1903: 626 (ref.); Jordan, Guide to Study Fish., *1*, 1905: 540 (general); Grey, Fishing Virgin Seas, 1925: 204–216 (Lower California, not seen); Jordan, Evermann and Clark, Rep. U.S. Comm. Fish. (1928), 2, 1930: 18 (distrib.); Gowanloch, Bull. La. Conserv. Dep., 23, 1933: 223, 224, 229 (no Louisiana record); Howell-Rivero, Proc. Boston Soc. nat. Hist., *41*, 1936: 44 (Cuba); Springer, Proc. Fla. Acad. Sci., *3*, 1939: 33 (Florida); Hildebrand, Copeia, 1941: 222 (N. Carolina); Lunz, Bull. S. Carolina St. Planning Bd., 14, 1944: 27 (Florida).

Micristodus punctatus Gill, Proc. Acad. nat. Sci. Philad., 1865: 177 (Gulf of California); Jordan and Gilbert, Bull. U.S. nat. Mus., *16*, 1883: 32 (Gulf of California); Jordan, Rep. U.S. Comm. Fish. (1885), 1887: 798 (in list); Jordan and Evermann, Rep. U.S. Comm. Fish. (1895), 1896: 218 (Gulf of California); Bull. U.S. nat. Mus., 47 (1), 1896: 52 (descr., Lower California); Günther, Encycl. Brit., 11th ed., *24*, 1911: 808 (general); Mowbray, Prelim. Rep. Sci. Cruise "Nourmahal," 1933: 2 (descr., no. of teeth, color, size, stomach content).

Cetorhinus maximus Poey, Enumerat. Pisc. Cubens., 1876: 184; Ann. Soc. esp. Hist. nat., *5*, 1876: 380 (Cuba).

Selache maxima Thomas, Cannibals Convicts West. Pacif., 1887: 38 (New Guinea, not seen).

Rhinodon typus Sauvage, in Grandidier, Hist. Phys. Nat. Madagascar, Poiss., 1891: 511 (Madagascar); Roxas and Martin, Dept. Agric. Comm. Manila, Tech. Bull. 6, 1937: 12 (Philippines, not seen).

Rhinodon pentalineatus Kishinouye, Zool. Anz., *24*, 1901: 694 (labial folds, descr., Japan); Zool. Mag. Tokyo, *15* (172), 1903: 41 (Japan).

Rhineodon typus Gill, Science, *21*, 1905: 790 (Indian fishery, habits); Gudger, Science, *38*, 1913: 270 (Florida); Fowler, Proc. Acad. nat. Sci. Philad., *67*, 1915: 245 (Knights Key, Florida); Gudger, Zoologica, N. Y., *1*, 1915: 349 (Knights Key, Florida); Science, *48*, 1918: 623; Science, *52*, 1920: 192 (general); Science, *56*, 1922: 251 (Florida record); Science, *58*, 1923: 180 (Florida); Nat. Hist. N. Y., *23*, 1923: 62 (Brazil); Beebe, Arcturus Adv., 1926: 414 (Galapagos, not seen); Gudger, Zoologica, N. Y., *30*, 1927: 76 (trop. W. Afr.); Science, *65*, 1927: 211 (Gulf of California); Science, *65*, 1927: 545 (Galapagos); Gudger and Hoffmann, Amer. Mus. Novit., 318, 1928: 1 (Florida Straits); Fowler, Mem. Bishop Mus., *10*, 1928: 18 (ref.); Breder, Field Bk. Mar. Fish. Atlant. Coast, 1929: 23 (general); Gudger and Mowbray, Nat. Hist. N. Y., *40*, 1930: 182 (Florida); Gudger and Hoffmann, Science, *70*, 1930: 639 (near Havana, Cuba); White, Bull. Amer. Mus. nat. Hist., *61*, 1930: 129 (anat., denticles, teeth, skelet.); Gudger, Bull. Amer. Mus. nat. Hist., *61*, 1931: 613 (Florida, size); Gudger and Hoffmann, Sci. Mon. N. Y., *32*, 1931: 33 (near Havana); Gudger, Nature, Lond., *130*, 1932: 169 (Seychelles); Science, *75*, 1932: 412 (5th Florida record); Herre, Science, *75*, 1932: 413 (N.

Borneo); Pearson, Nature, Lond., *131*, 1933: 729 (Ceylon); Gudger, Nature, Lond., *132*, 1933: 569 (Galapagos); Bertolini, R. C. Accad. Lincei, (6) *18*, 1933: 235 (teeth in relation to food); Boll. Soc. Biol. Sper., *9*, 1934: 1271; Barnard, Nature, Lond., *134*, 1934: 66 (C. of Good Hope); Delsman, Nature, Lond., *133*, 1934: 176 (Str. Bab-el-Mandeb); Gudger, Hydrogr. Bull. Wash., 2362, Dec. 12, 1934 (general); Barnard, Ann. S. Afr. Mus., *30*, 1935: 647, pl. 23–25 (descr., meas., discuss., photos, S. Afr.); Herre, Science, *81*, 1935: 253 (N. Borneo); Brimley, J. Elisha Mitchell sci. Soc., *51*, 1935: 160 (C. Fear R., N. Carolina); Marchand, Fish. Mar. Biol. Surv. Union S. Afr., Fish. Bull. 2, 1935: 28 (S. Afr.); Gudger, Nat. Hist. N. Y., *36*, 1935: 128 (general, Florida records); Proc. zool. Soc. Lond., 1935: 863 (records up to end of 1934); Gudger and Smith, Bull. N. Y. zool. Soc., *38*, 1935: 71 (sizes, photos, Acapulco, Mexico); Gudger, Science, *84*, 1936: 227 (Bimini, Bahamas); Nat. Hist. N. Y., *37*, 1936: 159 (Fire Island, N. York); Sci. Mon. N. Y., *42*, 1936: 84 (photos, near Havana, Cuba); Beebe, Bull. N. Y. zool. Soc., *39*, 1936: 241–242 (photos, Gulf of California); Chevey, Note Inst. Oceanogr. Indochine, 28, 1936: 3 (Indo-China); Suvatti, Index Fish. Siam., 1936: 2 (Gulf of Siam); Howell-Rivero, Mem. Soc. cubana Hist. nat., *11* (2), 1937: 109, pl. 14, 15 (descr., meas., photos, Cuba); White, Bull. Amer. Mus. nat. Hist., *74*, 1937: 29, 41, 59, 61, 64, 73, 79, 90, 91, 110, pl. 4, fig. l, pl. 9 to 11, pl. 12, fig. e, pl. 21, pl. 29, fig. m, pl. 36, pl. 37, pl. 46, fig. c (anat., photos of gills); Gudger, Science, *85*, 1937: 314 (general); Nature, Lond., *139*, 1937: 549 (Ceylon); Copeia, 1937: 60 (W. Caribbean); Norman and Fraser, Giant Fishes, 1937: 30, pl. 2 (general); Gudger, Nature, Lond., *141*, 1938: 516 (off Pacific Panama); Copeia, 1938: 172 (rammed by steamers, Red Sea, Gulf of Aden, Ceylon); Calif. Fish Game, *24*, 1938: 420 (Lower California); Beebe, Zaca Venture, N. Y., 1938: 162–170 (Gulf of California, not seen); Gudger, Sci. Mon., N. Y., *50*, 1940: 225 (size, Fire I., N. York); Beebe and Tee-Van, Zoologica, N. Y., *26*, 1941: 97, pl. 1 (photo, Gulf of California); Gudger, J. Morph., *68*, 1941: 81 (feeding, teeth, gill apparatus, photos); J. Elisha Mitchell sci. Soc., 57, 1941: 57 (N. Carolina specimen); Amer. Nat., 75, 1941: 550 (general, photo); Deraniyagala, J. Bombay nat. Hist. Soc., *44* (3), 1944: 427 (photos, Ceylon).

Family SCYLIORHINIDAE[1]

Cat Sharks

Characters. Two (rarely only one) dorsal fins, the 1st much shorter than the caudal, at least ½ of its base posterior to origin of pelvics; caudal much less than ½ of total length, not lunate in form, its lower anterior corner not expanded as a definite lobe, its axis but little raised; caudal peduncle not greatly flattened dorso-ventrally or expanded laterally; sides of trunk anterior to anal without longitudinal ridges; no precaudal pits, at least in most species; inner margins of pelvics more or less united posterior to cloaca; snout not greatly elongate or jaws widely protrusible; 5th, or 4th and 5th, gill openings over origin of pectoral; gill arches without rakers and not interconnected by a sieve of modified denticles; nostril not connected with mouth by a groove, or if so connected, its anterior margin does not bear a well developed fleshy barbel; no nictitating membrane within lower eyelid, but there may be a well developed longitudinal fold below the latter; spiracles present; labial furrows more or less developed; teeth small, numerous, with several cusps, and several rows functional; head of normal shape, not widely expanded laterally; rostral cartilages 3, united at tip; radials of pectoral mostly on metapterygium;

1. We include White's (Bull. Amer. Mus. nat. Hist., *74*, 1937: 107, 108) Halaeluridae and Atelomycteridae under the Scyliorhinidae, the differences in vertebral calcification on which they were based not seeming sharply enough alternative to warrant the rank of families.

mesopterygium much smaller, with few radials; meso- and metapterygia separated by a foramen, or not; vertebral calcifications widely variable in type; heart valves in 2 or 3 rows. Development oviparous so far as known.

The family includes numerous species of small sharks in tropical and temperate latitudes, from both shoal water and deep. Although it embraces two of the most common and best known of the European sharks,[2] the centers of abundance for both genera and species are the western Pacific, Australasian region and Indian Ocean to South Africa. It is represented in the western North Atlantic by only a few little known deep water species.

Genera. Opinions have differed widely as to the number of genera deserving recognition in this family. At the one extreme Garman[3] recognizes eleven, a list to which no less than seven more genera or subgenera have subsequently been added by Fowler[4] and Whitley.[5] At the other extreme Barnard[6] unites in a single genus the ten South African representatives of the family, which would fall in some seven different genera under the contrasting scheme. An intermediate view is taken by Norman, who suggests the recognition of "some four natural groups as genera."[7]

Generic characters in so uniform a family must be based on definitely alternative and easily discernible characters to be of any value to working ichthyologists. For instance, one group of some nine recognizable species is set apart from all other members of the family by the fact that the denticles along the dorsal margin of the anterior part of the caudal are not only enlarged but modified in shape and directed laterally so as to form a definite crest, which is outlined below by a narrow band of naked skin.[8] The members of this group fall into two categories: one with the posterior margin of the nostril widely expanded, the snout short and thick and the body cavity longer; the other with the posterior margin of the nostril expanded little, if at all, the snout long and thin and the body cavity shorter. These characters seem sufficiently alternative for the retention of the genus *Parmaturus* Garman, 1906, for the first group, as distinct from *Galeus* Rafinesque, 1810, for the second. But Whitley's segregation of some members of the latter into a separate subgenus (*Figaro*) because of the presence of a crest on both lower and upper sides of the caudal peduncle seems to us an unnecessarily minute subdivision. Among the other Scyliorhinidae *Pentanchus profundicolus* Smith and Radcliffe, 1912,[9] and another unnamed species[10] are set apart from the rest and from all other galeoid sharks by the fact that they have only one dorsal fin.[11] Among the species that remain after subtraction of the foregoing, the first dorsal of one, *Catulus cephalus* Gilbert, 1891, origi-

2. *Scyliorhinus caniculus* Linnaeus and *S. stellaris* Linnaeus, the so-called Spotted Dogfishes.
3. Mem. Harv. Mus. comp. Zool., *36*, 1913: 68 [Catulidae].
4. Proc. Acad. nat. Sci. Philad., *85*, 1934: 233.
5. Aust. Zool., *9*, 1939: 227.
6. Ann. S. Afr. Mus., *21* (1), 1925: 39.
7. Nature, Lond., *148*, 1941: 7.
8. These fish are commonly called File Tails in California.
9. The type specimen, now in the U.S. National Museum, shows no sign of mutilation. It is further interesting for the fact that its gill openings are of the character pictured in Fig. 38, 39 for *Apristurus profundorum* and *A. riveri*.
10. A Japanese scyliorhinid with only one dorsal fin is briefly described, but without specific name, by Jordan and Hubbs (Mem. Carneg. Mus., *10*, 1925: 100).
11. Jordan and Hubbs (Mem. Carneg. Mus., *10*, 1925: 100) also suggest that the doubtful genus *Caninoa* of

nates considerably in front of the origin of the pelvics, whereas in all the others it originates over or considerably behind the latter.[12] In this respect, and in others also, it is so aberrant that we have recently proposed the new genus *Cephalurus* for it,[13] based on study of a specimen from the original series.

Among the remaining scyliorhinids eight clearly distinct species from various parts of the world are set apart by the facts that labial furrows, well developed on the lower jaw, do not extend around the corner of the mouth or onto the upper jaw, and that the upper lip is expanded to close over the lower near the corner of the mouth. It was for a member of this group (*caniculus* Linnaeus, 1758) that the earliest scyliorhinid genus was proposed (*Scyliorhinus* Blainville, 1816). It is true that in this instance generic diagnosis, based primarily on the morphology of the labial furrows, runs counter to the grouping that might be based on the details of the nostril, and on the relationship of the latter to the mouth, for among the species with well developed lower labial furrows and no upper furrow are some in which the anterior margins of the nostrils reach to the mouth but others in which they fall short of the latter, and a similar range of variation, based on whether or not the nostril is connected to the mouth by a shallow groove, exists among them. But the varietal series are so continuous in these respects that nothing would be gained by abandoning the labial furrows in favor of the nostrils as the primary character. Therefore, it seems logical to use the labial furrow rather than the nostrils as the generic criterion, except for two South African species, *Poroderma pantherinum* Müller and Henle, 1841,[14] and *P. marleyi* Fowler, 1934,[15] in which the anterior margin of the nostril is extended as a long tapering barbel. For these a separate genus seems appropriate. Unfortunately, however, the old name *Poroderma* is not available for them, because its type species[16] lacks the barbel, and is in fact a typical *Scyliorhinus*. But there is no need to coin a new name, Fowler[17] having proposed *Conoporoderma* as a subgenus for the species with barbels.

In some of the members of the family still to be considered the labial furrows extend from the lower jaw around the angle of the mouth onto the upper jaw, while in others they are wholly lacking. The latter category includes the peculiar Swell Sharks, which are able

Nardo (Atti Riun. Sci. Ital., 1841: 312) from the Mediterranean may be *Pentanchus*-like, and Fowler (Bull. U.S. nat. Mus., 100 [*13*], 1941: 26) retains it provisionally among the Scyliorhinidae. But it does not seem likely that the combination of characters credited to it by Nardo (single dorsal fin, but with only 5 gill openings and no spiracle) actually applies to any existing shark. For a history of the case, with suggestions as to what the specimen in question may actually have been, see Fowler, 1941; also Doderlein (Man. Ittiol. Medit., 2, 1881: 82).

12. Another species with first dorsal far forward, classed in this family by Garman (*Proscyllium habereri* Hilgendorf, 1904), is placed among the Triakidae by us.
13. See Bigelow and Schroeder (Copeia, 2, 1941: 73) for discussion and detailed description.
14. Usually credited to Andrew Smith, 1837, but Smith (Proc. zool. Soc. Lond., 1837: 85) listed it only by name; the earliest account of it was by Müller and Henle (Plagiost., 1841: 13).
15. Fowler (Proc. Acad. nat. Sci. Philad., *85*, 1934: 234) has pointed out that it was actually a specimen of this species that he pictured earlier (Proc. Acad. nat. Sci. Philad., 77, 1925: 188) under the name *Scyliorhinus regani*.
16. *Poroderma africanum* A. Smith (Proc. zool. Soc. Lond., 1837: 85) equals *Squalus africanus* Gmelin, 1789.
17. Proc. Acad. nat. Sci. Philad., *85*, 1934: 234.

to inflate themselves with air, and which have widely distensible jaws provided with vertical "accordion" folds in the corners, as well as very broad, flat heads. The majority of recent writers have grouped these in the genus *Cephaloscyllium*. Fowler[18] has also proposed the subgenus *Holohalaelurus* for two other species[19] that agree with the Swell Sharks in lacking labial furrows, but which differ from them in having no ability to inflate and in having less distensible mouths, more slender trunks, shorter body cavities, as well as in different relative sizes and locations of the fins. We propose to raise this subgenus to generic rank.

The remaining species in which there is a well marked furrow around the corner of the mouth reaching out onto both jaws are subdivisible by the relation of nostril to mouth, size of the second dorsal relative to anal fin, and length of the interspace between anal and caudal.

In one rather sharply defined category of some thirteen named species, all from deep water, the nostril is widely separated from the mouth and wholly distinct from the latter, the anal is more than two and one-half times as long as the second dorsal, the interspace between the anal and the caudal is very short or even reduced to a mere notch, there are no folds below the eyes, and the snout is long and fleshy with very prominent mucous pores. Fowler[20] has recently distributed these species among three subgenera, based on the presence or absence of cirri on one or both margins of the nostril. But according to published accounts and to our own examination of three members of the group, there is too much intergradation in this respect for sharp separation. We therefore refer all of them to the genus *Apristurus* Garman, 1913.

In a second category the anterior margins of the nostrils similarly fall considerably short of the mouth and there is a labial furrow around the corner of the latter; but they differ from *Apristurus* by having a much longer interspace between caudal and anal, a considerably smaller anal relative to the second dorsal, and a well marked fold below the eye. Although the twelve named members of this group (*Halaelurus* Gill, 1861) resemble one another so closely that some reduction in the number of species is to be expected eventually, Fowler[21] divides them among two subgenera, *Aulohalaelurus* and *Halaelurus*, according to the lengths of the labial furrows, while Whitley has raised the former to generic rank, besides proposing two new genera, *Juncrus* and *Asymbolus*.[22] But the differences between the several species of this group are so slight that we refer all of them to the old genus *Halaelurus*.

There remain only those species which fall with *Halaelurus* in most respects, except for the anterior margin of the nostril, which more or less overlaps the anterior part of the mouth, and except for a shallow groove which extends either from the nostril to the mouth

18. Proc. Acad. nat. Sci. Philad., *85*, 1934: 235; Bull. U.S. nat. Mus., 100 (*13*), 1941: 41.
19. *Scylliorhinus punctatus* Gilchrist (Mar. biol. Rep. Cape Town, *2*, 1914: 129) and *S. regani* Gilchrist (Mar. biol. Rep. Cape Town, *2* [3], 1923: 45, 46).
20. Bull. U.S. nat. Mus., 100 (*13*), 1941: 53.
21. Bull. U.S. nat. Mus., 100 (*13*), 1941: 41.
22. Aust. Zool., *9*, 1939: 229; *Juncrus* for *Scyllium vincenti* Zietz, 1908, and *Asymbolus* for *Scyllium anale* Ogilby, 1885, both from Australia.

or part way to the latter. Although this group includes only three known species,[23] our own examination of specimens in the collection of the Museum of Comparative Zoology satisfies us that Garman's[24] reference of them to his two new genera, *Haploblepharus* and *Atelomycterus*, was justified by the sharp differences summarized in the following key.

Key to Genera

1a. Only one dorsal fin. *Pentanchus* Smith and Radcliffe, 1912.
Philippines, Japan.

1b. Two dorsal fins.

2a. Origin of 1st dorsal considerably anterior to origin of pelvics; rear contours of dorsal fins straight or concave. *Cephalurus* Bigelow and Schroeder, 1941.
Gulf of California and Revillagigedo Islands, off west coast of Mexico.

2b. Origin of 1st dorsal over, or usually behind, origin of pelvics.

3a. Denticles along dorsal margin of anterior part of caudal enlarged and modified in shape, forming a distinct crest, outlined below by a narrow band of naked skin.

4a. Nostrils far from mouth, the distance from their inner angles to corners of latter about ½ as great as horizontal diameter of eye; posterior margin of nostril not lobed; snout long, thin, its mucous pores not conspicuous. *Galeus* Rafinesque, 1810, p. 214.

4b. Nostrils close to mouth, although entirely separate from latter; distance from inner angle of nostril to corner of mouth not more than ¼ as great as horizontal diameter of eye; posterior margin of nostril with a well developed lobe; snout short and thick, its mucous pores very prominent. *Parmaturus* Garman, 1906.
California, Japan.

3b. Denticles along dorsal margin of anterior part of caudal similar to those lower down, not forming a distinct crest.

5a. Anterior margin of nostril bilobed, the outer lobe in the form of a fleshy barbel reaching to mouth (Fig. 31). *Conoporoderma* Fowler, 1934.
S. Africa, Natal, Mauritius.

5b. Anterior margin of nostril little or not at all bilobed; without well developed barbel.

6a. A well developed labial furrow on lower jaw, but not around corner of mouth or on upper jaw. *Scyliorhinus* Blainville, 1816, p. 202.

6b. Labial furrows either absent, or extending around corner of mouth if present.

23. *Scyllium edwardsii* Voigt (in Cuvier, Tierreich, 2, 1832: 504), S. Afr.; *S. marmoratum* Bennett (Mem. Raffles, 1830: 693), Malaysia, India; and *Atelomycterus macleayi* Whitley (Aust. Zool., 9, 1939: 230), Australia.

24. Mem. Harv. Mus. comp. Zool., 36, 1913: 100, 101.

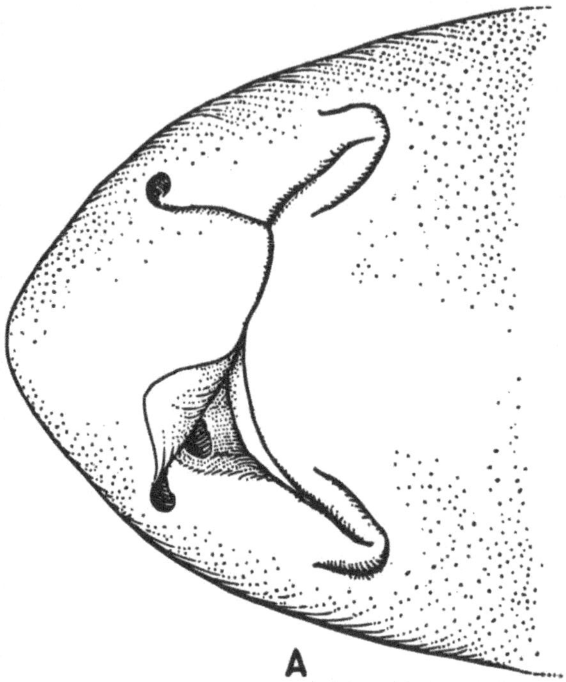

A

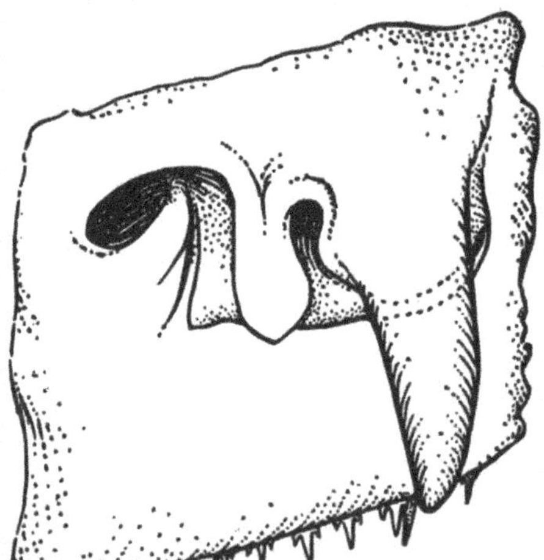

B

Figure 31. *A*, *Haploblepharus edwardsii* (Harv. Mus. Comp. Zool., No. 1028). Showing nasal flap and relationship of nostril to mouth, about 1½ x natural size. *B*, *Conoporoderma pantherinum* (Harv. Mus. Comp. Zool., No. 497). Right-hand nostril and part of upper jaw showing the nasal barbel, about 3 x natural size.

7a. No labial furrow on either jaw, or around corner of mouth.

8a. Mouth broadly distensible, with vertical folds at corners; stomach inflatable with air; anal only about as long as 2nd dorsal, its origin under origin of latter; body sector of trunk to cloaca considerably longer than tail sector.

Cephaloscyllium Gill, 1862.

Eastern Pacific from middle California to Chile; Japan, Australia, Tasmania and New Zealand region; South Africa.

8b. Mouth not distensible, without vertical folds at the corners; stomach not inflatable with air; anal more than 1½ times as long as 2nd dorsal, its origin anterior to origin of latter by a distance equal to at least ½ the length of its base; body sector of trunk to cloaca considerably shorter than tail sector. *Holohalaelurus* Fowler, 1934.

South Africa, Natal.

7b. A labial furrow around corner of mouth and extending forward for a longer or shorter distance on each jaw.

9a. Anterior margin of nostril expanded as a flap, overlapping front edge of mouth; no definitely outlined lower nasal flap; a shallow groove extending at least part way from nostril toward mouth (Fig. 31).

10a. A groove extending from nostril to mouth; anterior flaps of the 2 nostrils not separated by a definite gap opposite symphysis of upper jaw, their outlines nearly straight; anal larger than 2nd dorsal, its base wholly anterior to base of latter; origin of 1st dorsal behind rear end of base of pelvics; fold below eye hardly defined, if at all.

Haploblepharus Garman, 1913.

South Africa.

10b. Nasal grooves not extending to mouth; anterior nasal flaps widely separated, their outlines, as well as outlines of intervening isthmus, forming 3 rounded lobes; anal at least no larger than 2nd dorsal, the rear end of its base under midpoint of latter; origin of 1st dorsal in front of rear end of base of pelvics; a strongly developed fold below eye.

Atelomycterus Garman, 1913.

China, Indo-China, Siam, Malaysia, Philippines, India.

9b. Anterior nasal flaps fall considerably short of mouth; a posterior nasal flap is also present in most cases.

11a. Interspace between anal and caudal at least as long as base of anal; base of anal not more than twice as long as base of 2nd dorsal; folds below eyes strongly developed; mucous pores on snout not conspicuous. *Halaelurus* Gill, 1862.

South Africa; tropical Indian Ocean and Arabian Gulf; India; Australasia; Philippines, China, Formosa, Japan; Chile and Patagonia; Argentina.

11b. Interspace between anal and caudal less than 1/5 as long as base of anal; base of anal more than twice as long as base of 2nd dorsal; no fold below eye; mucous pore system on lower surface of snout very conspicuous. *Apristurus* Garman, 1913, p. 219.

Genus *Scyliorhinus* Blainville, 1816

Scyliorhinus Blainville, Bull. Soc. philom. Paris, 1816: 121; type species, *S. caniculus* Blainville,[25] equals *Squalus caniculus* Linnaeus, 1758.

Generic Synonyms:[26]

Catulus Valmont, Dict. Hist. Nat. Paris, *4*, 1768:[27] 51; type species, *C. major vulgaris* Valmont, equals *Squalus caniculus* Linnaeus, 1758; Andrew Smith, Proc. zool. Soc. Lond., 1838: 85 (in part).

Galeus (in part) Rafinesque, Indice Ittiol. Sicil., 1810: 46; for *G. caniculus* Rafinesque, equals *Squalus caniculus* Linnaeus, 1758.

Scyllium Cuvier, Règne Anim., *2*, 1817: 124; type species, *Squalus caniculus* Linnaeus, 1758, designated by Jordan, Genera Fish., *1*, 1917: 97.

Scylliorhinus Blainville, in Vieillot, Faune Franc., Poiss., 1825: 68; substitute for *Scyliorhinus* Blainville, 1816.

Poroderma A. Smith, Proc. zool. Soc. Lond., 1837: 85; type species, *P. africanum* Smith, equals *Squalus africanus* Gmelin, 1789. South Africa.

Halaelurus Tanaka, Fish. Japan, *1*, 1911: 13, pl. 3, fig. 12; for *H. rudis* Tanaka; not *Halaelurus* Gill, 1862.

Generic Characters. Two dorsal fins; origin of 1st dorsal over or slightly anterior to rear ends of bases of pelvics; denticles along dorsal margin of caudal similar to those lower down, not forming a distinct crest; nasal barbels rudimentary or wholly lacking;

25. While Blainville gave no authorship for this or for any of the several other included species, his subsequent diagnosis (in Vieillot, Faune Franc., 1825: 71) of *canicula* showed that it referred to *Squalus caniculus* Linnaeus, 1758, which was later designated as type of the genus by Gill (Ann. N.Y. Lyc., 7, 1862: 407).

26. For list of fossil genera perhaps synonymous with *Scyliorhinus*, see Fowler (Bull. U.S. nat. Mus., 100 [*13*], 1941: 34).

27. Preoccupied (Kniphof, 1759) for insects and not available even otherwise for sharks; Valmont's names, when binomial, were so only accidentally (see ruling by International Commission on Zoological Nomenclature, Smithson. misc. Coll., *73* [3], 1925: 27); the name, as a shark, must therefore date from Andrew Smith, 1837.

anterior nasal flaps may or may not reach mouth; nostril either entirely separate from mouth or connected with latter by a very shallow groove only; a well developed labial furrow on lower jaw, but none on upper; upper lip expanded to close over lower at corners of mouth; eye with or without a longitudinal fold below lower eyelid; spiracle small, close to corner of eye; anal considerably larger than 2nd dorsal, separated from caudal by a considerable interspace; inner margins of pelvics united posterior to cloaca for a short distance in females and for a longer distance in males; teeth with one large central, and several small lateral, cusps, several series functional; dermal denticles lanceolate, strongly ridged. Egg cases horny, oblong, with long filamentous tendrils at the corners which wind around sea weeds, etc. The eggs are said to be expelled two at a time, and the young to hatch about six months after the eggs are laid.

Range. Both sides of North Atlantic; Mediterranean; South Africa; Natal; Japan; Korea.

Fossil Teeth. Upper Cretaceous to Pliocene, Europe; Upper Cretaceous, western Asia, North America; Eocene, North Africa.

Key to Species

1a. Anterior nasal flaps reach rearward nearly or quite to mouth.
   2a. Anterior nasal flaps joined in the midline, or nearly so.
   caniculus Linnaeus, 1758.
   Eastern North Atlantic, Mediterranean.[28]
   2b. Anterior nasal flaps separated one from the other in the midline by a considerable interspace. *stellaris* Linnaeus, 1758.
   Eastern North Atlantic, Mediterranean.
1b. Anterior nasal flaps separated from mouth by a considerable space.
   3a. Origin of 1st dorsal as close to origin of anal as to rear ends of bases of pelvics; 2nd dorsal as large as 1st. *capensis* Müller and Henle, 1841.
   South Africa, Natal, and perhaps India.[29]
   3b. Origin of 1st dorsal over rear end of bases of pelvics, or at least much closer to them than to origin of anal; 2nd dorsal smaller than 1st.
      4a. Origin of 1st dorsal closer to tip of caudal than to tip of snout by a distance equal to length of latter in front of mouth; color pattern a dark network on paler ground. *retifer* Garman, 1881, p. 207.
      4b. Origin of 1st dorsal as close to tip of snout as to tip of caudal, or a little closer; color pattern spotted or blotched.
         5a. Base of anal considerably longer than base of 1st dorsal.

28. Including *duhamelii* Garman (Mem. Harv. Mus. comp. Zool., *36*, 1913: 73), the type specimens of which (Harv. Mus. Comp. Zool., No. 60, 63) appear to represent a dwarf race of *caniculus*, or perhaps only a color variety.

29. Day, Fish. India, 1878: 724, pl. 190, fig. 1.

6a. Caudal only about as long as from tip of snout to 5th gill opening; color pattern white-spotted on dark ground tint.
torrei Howell-Rivero, 1936, p. 211.

6b. Caudal about as long as from tip of snout to axil of pectoral; color pattern dusky or black-spotted on pale ground tint.
boa Goode and Bean, 1895, p. 204.

5b. Base of anal only as long as base of 1st dorsal, or shorter.
torazame Tanaka, 1908.
Japan.

Scyliorhinus boa Goode and Bean, 1895.

Figure 32

Study Material. Type specimen, a newly-hatched male, 151 mm. long, in poor condition, from Barbados, in 200 fathoms (Harv. Mus. Comp. Zool., No. 1335); newly-hatched male, 87 mm. long, from north coast of Cuba, in 235 fathoms (Harv. Mus. Comp. Zool., No. 36156); half-grown male, 316 mm. long, taken 25 to 30 miles ESE. from

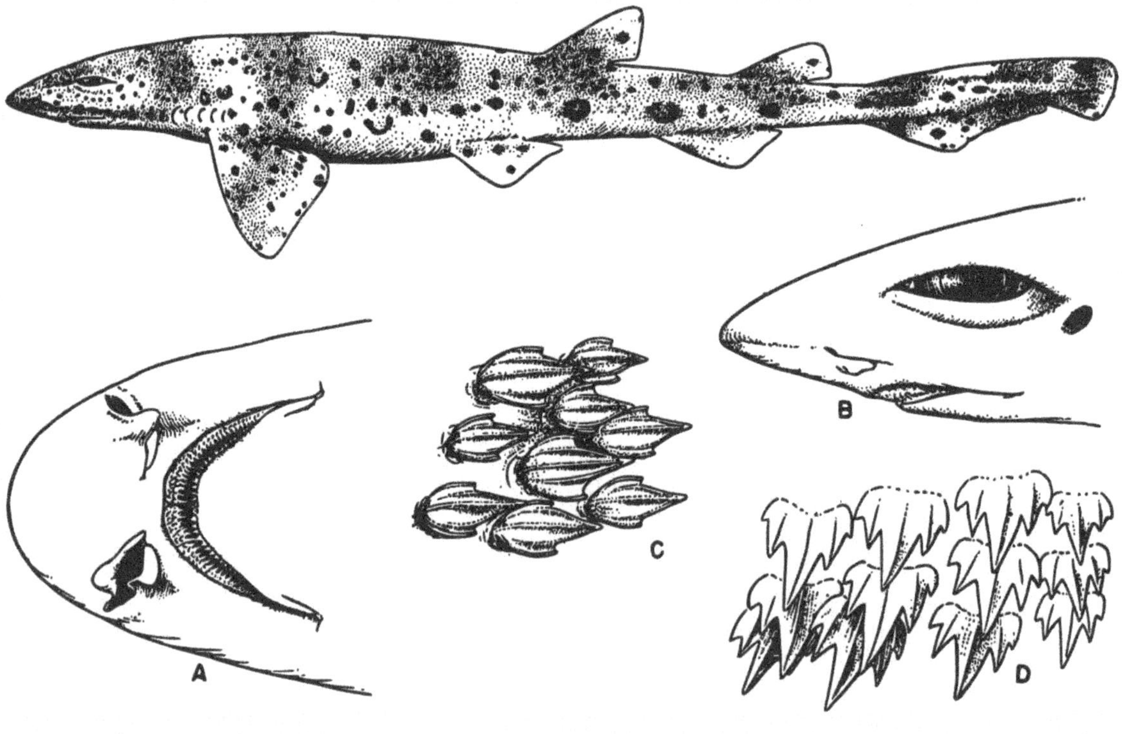

FIGURE 32. *Scyliorhinus boa*, immature male, 316 mm. long, from near Rio de Janeiro, Brazil (Mus. Nac. Rio de Janeiro). *A* Anterior part of head from below to show nostril and labial furrows, about 1.4 x natural size. *B* Side view of anterior part of head, almost 1.5 x natural size. *C* Dermal denticles, about 14 x. *D* Upper teeth from side of jaw, enlarged.

Ilha Rasa near Rio de Janeiro, Brazil, in 80 meters, the type of *S. haeckelii* (Ribeiro), 1907 (Mus. Nac. Rio de Janeiro, No. 494).

Distinctive Characters. Separable from *S. retifer* by its obtusely rounded snout and color pattern, and from *S. torrei*, which it closely resembles, by its relatively longer caudal fin and by its color (see Key, p. 204).

Description. Proportional dimensions in per cent of total length. Male, 151 mm., from Barbados (Harv. Mus. Comp. Zool., type, No. 1335). Male, 316 mm., from Brazil (Mus. Nac. Rio de J., type of *S. haeckelii*, No. 494).

Trunk at origin of pectoral: breadth 10.6, 11.1; height 7.3, 8.9.
Snout length in front of: outer nostrils 3.3, ——; mouth 4.0, 5.1.
Eye: horizontal diameter 3.3, 3.5.
Mouth: breadth 6.3, 6.3; height 2.6, 3.8.
Nostrils: distance between inner ends 2.5, 2.2.
Labial furrow length: lower 1.7, 1.3.
Gill opening lengths: 1st 2.0, 1.6; 2nd 1.5, ——; 3rd 1.5, ——; 4th 1.5, ——; 5th 1.1, 1.2.
First dorsal fin: vertical height 4.6, 5.4; length of base 6.0, 6.3.
Second dorsal fin: vertical height 3.3, 3.2; length of base 4.3, 5.1.
Anal fin: vertical height 3.3, 3.2; length of base 9.4, 9.2.
Caudal fin: upper margin 26.8, 23.4; lower anterior margin 9.0, 8.5.
Pectoral fin: outer margin 10.2, 14.2; inner margin 6.3, 6.3; distal margin 8.0, 9.8.
Distance from snout to: 1st dorsal 43.8, 48.4; 2nd dorsal 59.7, 67.1; upper caudal 73.2, 76.6; pectoral 17.2, 18.7; pelvics 37.9, 39.2; anal 53.0, 60.1.
Interspace between: 1st and 2nd dorsals 10.0, 12.0; 2nd dorsal and caudal 8.0, 6.0; anal and caudal 10.6, 9.2.
Distance from origin to origin of: pectoral and pelvics 23.6, 20.9; pelvics and anal 15.3, 19.7.

Trunk slender, much compressed laterally rearward from pelvics. Dermal denticles rather loosely spaced, much longer than broad, with 3–5 ridges and tridentate margins, the median tooth considerably the largest, their blades erected at an angle of about 40° over trunk as a whole, giving a very rough effect.

Head convex in dorsal profile but flattened below. Snout broadly rounded, its length in front of mouth between ⅛ and ¼ of length of head. Eye narrow, oval, its horizontal diameter nearly as long as snout in front of mouth, the fold below eye well marked when eye is open but hardly distinguishable when it is closed. Spiracle round, very small, posterior to rear corner of eye by a distance about 0.2 times as great as horizontal diameter of latter. Gill openings moderately concave in outline anteriorly, the 1st slightly the longest, about ½ as long as horizontal diameter of eye, the 5th slightly the shortest, the interspaces between them decreasing in breadth rearward, the interspace between 3rd and 4th over origin of pectoral. Nostrils slightly oblique, entirely distinct from mouth and widely separated from each other, the distance between them nearly ½ as great as length of snout in front of mouth, the anterior margin expanded as a rather narrow subtriangular lobe with rounded apex and well marked median crest but falling considerably short of the mouth, the posterior margin also developed as a rounded flap (Fig. 32 A), much as in *S. retifer* and *S. torrei*. Mouth obtusely ovate, about ½ as high as broad. Lower labial furrow slightly less than ⅓ as long as distance from corner of mouth to symphysis of lower jaw.

Teeth $\frac{24-24}{22-22}$;[30] similar in the 2 jaws, usually with 5, occasionally with 3 (or even 7) cusps, the median much the longest, narrow-triangular and sharp-pointed, curving slightly toward corner of mouth in most cases, the anterior surfaces of teeth longitudinally striate; 5 rows (locally only 4) functional in each jaw.

First dorsal brush-shaped, its margins nearly straight, its corners narrowly rounded, its origin slightly behind rear end of bases of pelvics, and a little nearer to snout than to tip of caudal, the interspace between 1st and 2nd dorsals about twice as long as base of 1st dorsal. Second dorsal similar in shape to 1st, and nearly as long at base as latter, but only about ⅔ as high vertically, its origin on a vertical line about halfway between midpoint of base of anal and rear end of latter. Caudal a little less than ¼ of total length, relatively somewhat longer in small specimens than in large,[31] its upper contour nearly straight, its terminal sector transversely truncate, with rounded corners, occupying about ⅓ total length of the fin, lower anterior corner much more than a right angle. Anal a little less than twice as long at base as 2nd dorsal, and longer than 1st dorsal by a distance about as long as horizontal diameter of eye, with nearly straight margins, rounded apex and subacute free rear corner, about ⅓ as long as the base. Pelvics about as large as anal, their anterior margins nearly straight, distal margins weakly concave, corners subangular, their inner edges, in half-grown male, united behind cloaca for a little less than ½ their lengths. Pectoral about 3 times as large as 1st dorsal in area, about 70% as broad as long, the outer and inner margins moderately convex, the distal margin nearly straight, apex narrowly rounded, inner corner more broadly so.

Color. Back and sides pale yellowish brown, marked transversely with seven broad but indistinct dark blotches, one midway of the caudal, one at caudal's origin, one opposite each dorsal fin, and three equally spaced in front of the first dorsal, the most anterior being opposite the origin of the pectorals; also a large number of small dark chocolate-brown spots of varying sizes irregularly spaced, some nearly circular and some in the form of rosettes; one much larger than the others below the first dorsal, with others opposite the origin and rear part of anal; likewise a lunate blotch on each flank about midway between the rear corner of the pectoral and the origin of the pelvics; lower surface very pale yellowish brown, plain except that the lower side of head is faintly mottled; and there are a few dark spots on the pectorals (about twice as many on the one as on the other in the half-

30. It is possible that there was one more series of teeth in each jaw, it being difficult to determine the precise number in the available material.

31. A little longer than from snout to inner corner of pectoral in newly hatched specimen, but only about as long as from snout to axil in a half-grown one.

grown specimen). That the distribution of the dark spots is not the same on the two sides of one specimen, and that there are many more on it and on a newly hatched specimen from Cuba than on another from Barbados, show that their number is not a specific character.

Size. The state of sexual development of the larger specimen, as indicated by the length of its claspers, suggests that this species becomes mature at a length of perhaps two feet. *S. boa* is thus a considerably larger shark than *S. torrei* (p. 213). No females have yet been seen.

Developmental stages. The egg cases have not been identified.

Habits. The depths of capture, listed above, make it likely that this is an inhabitant of moderately deep waters and probably a bottom-dweller. Other than this nothing is known of its habits.

Range. S. boa is positively known only from Brazil, from Cuban waters and from the Barbados (see Study Material, p. 204).

Synonyms and References:

Scylliorhinus boa Goode and Bean, Smithson. Contr. Knowl., *30*, 1895: 17; Mem. Harv. Mus. comp. Zool., *22*, 1896; Spec. Bull. U.S. nat. Mus., *2* (off Barbados); Howell-Rivero, Proc. Boston Soc. nat. Hist., *41*, 1936: 44 (Cuba).

Scylliorhinus retifer Goode and Bean, Smithson. Contr. Knowl., *30*, 1895: pl. 2, fig. 6; Mem. Harv. Mus. comp. Zool., *22*, 1896; Spec. Bull. U.S. nat. Mus., *2* (same); not *Scyllium retiferum* Garman, 1881.

Catulus retifer var. *boa* Ribeiro, Bol. Soc. nac. Agric. Brasil, 1904: 17 (Brazil); not *Scyllium retiferum* Garman, 1881.

Catulus haeckelii Ribeiro, Mem. Mus. nac. Rio de J., *14*, 1907: 163, pl. 8 (Ilha Rasa, near Rio de Janeiro, Brazil, descr.); Fauna brasil. Peixes, *2* (1), Fasc. 1, 1923: 21, pl. 7 (same as Ribeiro, 1907); Fowler, Arqu. Zool. Estado São Paulo, *3*, 1942: 127 (Brazil).

Scyliorhinus retifer (in part) Regan, Ann. Mag. nat. Hist., (8) 1, 1908: 457 (class.).

Catulus boa Garman, Mem. Harv. Mus. comp. Zool., *36*, 1913: 77 (descr.).

Scyliorhinus retifer (Garman), 1881

Chain Dogfish

Figure 33

Study Material. Type specimen, male, 307 mm. long, from off Virginia (Harv. Mus. Comp. Zool., No. 825); a male, 428 mm. long, from off New Jersey (Harv. Mus. Comp. Zool., No. 33932); also two females, 300 and 370 mm., from offing of southern New England, in 50–70 fathoms (Harv. Mus. Comp. Zool.).

Distinctive Characters. This species is most obviously separated from other local species of the genus by its chain-like color pattern, by its wedge-shaped snout, and by the fact that the origin of its first dorsal is closer to the tip of the caudal than to the snout.

Description. Proportional dimensions in per cent of total length. Male, 307 mm., from Lat. 38° N., Long. 73° W. (Harv. Mus. Comp. Zool., type, No. 825). Male, 428 mm., from 110 miles SE. of Atlantic City, N. J. (Harv. Mus. Comp. Zool., No. 33932).

Trunk at origin of pectoral: breadth 10.6, 11.7; height 9.2, 9.6.
Snout length in front of: outer nostrils 3.6, 3.3; mouth 5.2, 5.0.
Eye: horizontal diameter 3.6, 3.7.
Mouth: breadth 6.8, 8.2; height 3.6, 3.3.
Nostrils: distance between inner ends 2.1, 1.9.
Labial furrow length: lower 1.6, 1.6.
Gill opening lengths: 1st 1.5, 2.2; 2nd 1.4, 1.9; 3rd 1.4, 1.9; 4th 1.4, 1.9; 5th 1.1, 1.3.
First dorsal fin: vertical height 6.0, 7.0; length of base 6.5, 6.5.
Second dorsal fin: vertical height 3.7, 4.7; length of base 6.0, 5.1.
Anal fin: vertical height 4.1, 4.2; length of base 8.8, 8.2.
Caudal fin: upper margin 21.9, 19.3; lower anterior margin 12.3, 11.4.
Pectoral fin: outer margin 14.3, 13.3; inner margin 7.5, 7.0; distal margin 8.5, 11.8.
Distance from snout to: 1st dorsal 50.0, 53.0; 2nd dorsal 67.2, 69.8; upper caudal 78.1, 80.7; pectoral 19.2, 21.7; pelvics 42.1, 44.3; anal 61.2, 62.0.
Interspace between: 1st and 2nd dorsals 10.1, 11.8; 2nd dorsal and caudal 6.0, 7.0; anal and caudal 9.1, 8.4.

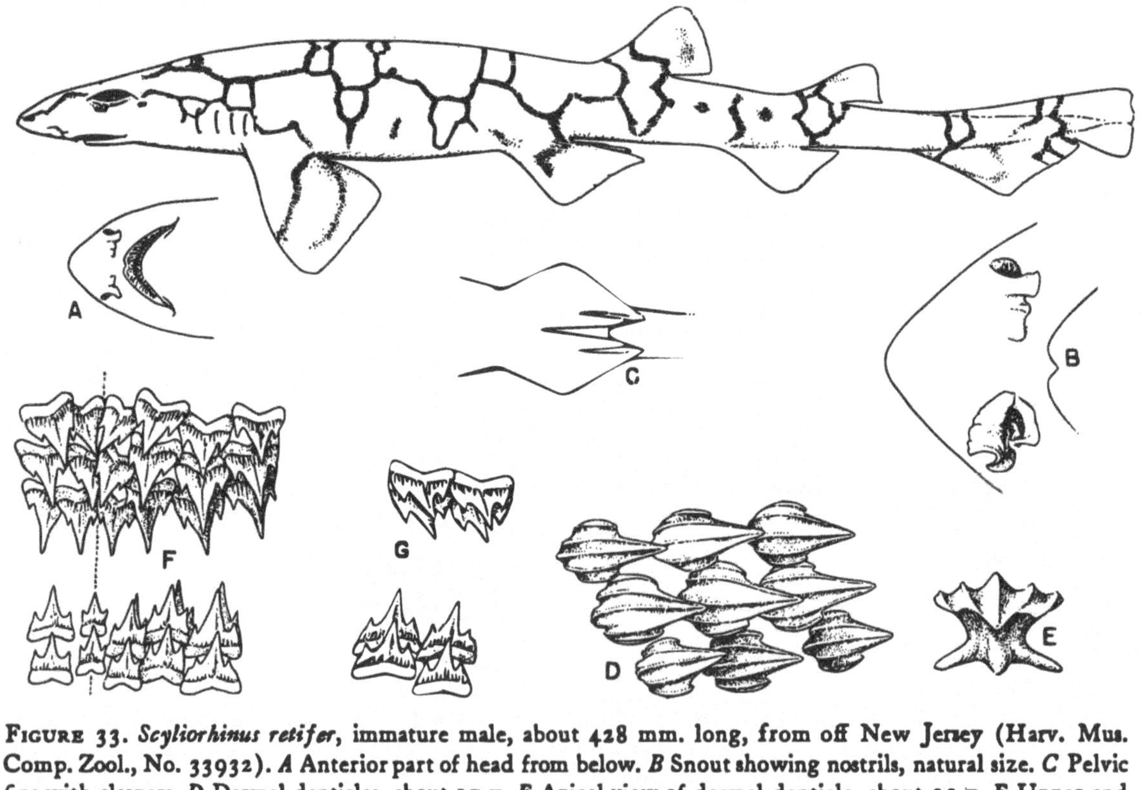

FIGURE 33. *Scyliorhinus retifer*, immature male, about 428 mm. long, from off New Jersey (Harv. Mus. Comp. Zool., No. 33932). *A* Anterior part of head from below. *B* Snout showing nostrils, natural size. *C* Pelvic fins with claspers. *D* Dermal denticles, about 17 x. *E* Apical view of dermal denticle, about 35 x. *F* Upper and lower teeth from center of mouth, about 4 x. *G* Upper and lower teeth from sides of jaws near corners of mouth, about 5 x.

Distance from origin to origin of: pectoral and pelvics 24.8, 22.5; pelvics and anal 19.4, 19.8.

Trunk slender, its breadth at origin of pectorals only about 1/8, its height 1/10, of total length, tapering rearward. Body sector to cloaca about as long as tail sector. Caudal peduncle nearly as broad as deep, oval in cross-section. Dermal denticles narrow, lanceolate, with acute tips, their blades only slightly raised, 3–5 ridged, the axial ridge much the strongest, their posterior margins entire on some denticles but notched between the ridges on others.

Head flattened above. Snout wedge-shaped, but with blunt tip, its length in front of mouth about 1/4 of length of head. Eye moderately narrow, oval, with horizontal diameter twice or more the vertical, its horizontal diameter about 2/3 as long as snout in front of mouth, its anterior edge a little posterior to front of mouth, the longitudinal fold below eye but weakly indicated. Spiracle an oblique slit, about 1/7 as long as horizontal diameter of eye, situated close behind, and a little below, latter. First gill opening the longest, about 1/2 to 2/3 as long as horizontal diameter of eye, the 5th shortest, only about 1/2 as long as 1st, the 4th and 5th over anterior part of pectoral. Nostrils entirely distinct from mouth, nearly transverse, their anterior margins expanded as subtriangular flaps, with strong transverse median crests (no barbel), separated from mouth at nearest point by a distance about 1/4 as great as horizontal diameter of eye, and separated one from the other in the midline by a distance about 1/2 as great as from the median angle of the nostril to the mouth. Mouth ovate, about 1/2 as long as wide; a strongly marked labial furrow at corner of lower jaw extending inward about 1/3 the distance to the symphysis; no furrow on upper jaw but upper lip somewhat expanded at corner of mouth, thus closing over the lower.

Teeth about $\frac{21\text{—}0\text{—}21}{20\text{—}1\text{—}20}$; alike in the 2 jaws, the triangular median cusp flanked near its base on either side by 1 (rarely 2) smaller cusps, the median cusp larger, relative to the laterals, in larger than in smaller specimens; lower jaw, but not upper, with a small median tooth; usually 3 or 4 rows functional.

First dorsal brush-shaped, its origin closer to tip of caudal than to tip of snout by a distance about equal to length of latter in front of mouth, posterior to rear end of bases of pelvics by a distance about 2/3 as long as horizontal diameter of eye, the rear end of its base a little posterior to tips of pelvics; its anterior margin nearly straight, its posterior margin slightly convex, its apex rounded, its free lower margin about as long as its base. Second dorsal about 1/2 as large in area as 1st, its rear margin weakly concave, its free rear corner somewhat more slender than 1st, its origin over rear part of base of anal. Caudal only about 1/5 of total length, with well marked subterminal notch, its terminal sector 1/3 to 1/2 the total length of the fin, brush-shaped, its tip either squarely truncate (Fig. 33) or indented in the midline (type specimen), its lower anterior corner subangular and much more obtuse than a right angle. Anal subtriangular, with nearly straight edges, broadly rounded apex, and moderately acute rear corner, its base about 1/5 to 1/4 longer than that of 2nd dorsal, its origin about midway between perpendiculars at rear end of base of 1st dorsal and at origin of 2nd dorsal. Pelvics a little larger in area than 1st dorsal, subtriangular, with rounded apices and moderately acute rear corners, the inner margins united behind cloaca for a little more than ½ their lengths in immature male. Pectoral about twice as large in area as 1st dorsal and ⅔ to ¾ as broad as long, with rounded corners, slightly convex outer margin and straight distal margin.

Color. The ground tint is dark reddish brown above, yellowish below, with a very characteristic pattern of narrow, sooty black stripes in groups of two crossing the back just behind the pectorals, at the first dorsal, between the first and second dorsals, at the second dorsal, at the anterior end of the caudal, and midway out on the latter; these branching over the sides and out onto the pectorals in a loose net of polygonal meshes which are irregular in size and shape.

Size. The largest specimen so far measured was 17 inches (430 mm.) long, the maximum length probably not being more than 2 to 2½ feet.

Developmental Stages. Horny egg cases, presumably of this species (the only oviparous shark common off the middle Atlantic United States), are 50 to 57 mm. long by 18 to 23 mm. broad, with a long tendril at each corner and brownish amber in color.

Habits. This little shark lives on or close to bottom on the outer part of the Continental Shelf, chiefly at least between about the 40 and 125 fathom contours, all definite records of it having so far been from within this depth range. There is no reason to suppose that it ever strays shoreward into shoal water. Eggs, one with an embryo nearly ready for hatching and others less advanced, have been taken in February off Chesapeake Bay, evidence that the young are produced in late winter or early spring. Nothing more is known of its life history, and nothing of its diet.

Range. All recorded captures of *S. retifer* have been from between the offings of Cape Lookout, North Carolina, and northern New Jersey. Fishermen also report small sharks, probably this species, on the Tilefish grounds at the outer edge of the Continental Shelf off New York. Within this short sector, however, it appears to be very generally distributed in the appropriate depth zone. Its chief center of abundance appears to lie off Virginia (type locality, Lat. 38° 23′ N., Long. 73° 34′ W.), especially in the general offing of Chesapeake Bay, where considerable numbers are taken by the winter trawl fishery from January to March, specimens being brought in daily at times. *S. retifer* has also been reported by name from the Tortugas, Florida,[33] and from some unspecified locality between southern Florida, the Bahamas and Honduras.[34] Re-examination in the first case shows that the shark in question was *Galeus arae* (p. 211); and since the second of these records is by name only, the same may be true of it also.

Synonyms and References:

Scyllium retiferum Garman, Bull. Mus. comp. Zool. Harv., *8*, 1881: 233 (descr., Lat. 38° 23′ N., Long. 73° 34′ W.).

Scylliorhinus retifer Jordan and Gilbert, Bull. U.S. nat. Mus., *16*, 1883: 869 (ref.); Bean, Rep. U.S. Comm. Fish. (1882), 1884: 343 (off Woods Hole); Jordan, Rep. U.S. Comm. Fish. (1885), 1887: 733 (off SE.

33. Longley and Hildebrand, Pap. Tortugas Lab., *34*, 1941: 1.
34. Breder, Bull. Bingham Oceanog. Coll., *1* (1), 1927: 5.

U.S.); Goode and Bean, Smithson. Contr. Knowl., *30*, 1895: 16, 508, pl. 4, fig. 14, 15; Mem. Harv. Mus. comp. Zool., *22*, 1896; Spec. Bull. U.S. nat. Mus., *2* (descr., off SE. U.S.); Jordan, Evermann and Clark, Rep. U.S. Comm. Fish. (1928), 2, 1930: 11 (distrib.).

Scyliorhinus retifer Regan, Ann. Mag. nat. Hist., (8) *1*, 1908: 457 (class.); Fowler, Copeia, 30, 1916: 36 (off mid. Atlant. U.S.); Nichols, Copeia, 1931: 38 (egg cases); Schroeder, Copeia, 1931: 42 (off N. Jersey); Firth, Copeia, 1934: 45 (egg cases, season, off Chesapeake Bay).

Catulus retifer Jordan and Evermann, Rep. U.S. Comm. Fish. (1895), 1896: 214 (Gulf Stream); Bull. U.S. nat. Mus., 47 (1), 1896: 25 (descr., Gulf Stream in deep water off S. Atlant. coast); Smith, Bull. N. C. geol. econ. Surv., *2*, 1907: 31 (off N. Carolina); Gudger, Proc. biol. Soc. Wash., *25*, 1912: 154 (egg cases, perhaps this species, N. Carolina); Garman, Mem. Harv. Mus. comp. Zool., *36*, 1913: 76 (descr.); Radcliffe, Bull. U.S. Bur. Fish., *34*, 1916: 249 (teeth, denticles, off N. Carolina); Breder, Field Bk. Mar. Fishes Atlant. Coast, 1929: 11 (general); Pearson, U.S. Bur. Fish. Invest. Rep., (10) *1*, 1932: 17 (winter trawl fishery, off N. Carolina).

(?) *Catulus retifer* Breder, Bull. Bingham oceanogr. Coll., *1* (1), 1927: 5 (no loc., see p. 210).

Not *Scylliorhinus retifer* Longley and Hildebrand, Pap. Tortugas Lab., *34*, 1941: 1 (this is *Galeus arae*).

Scyliorhinus torrei Howell-Rivero, 1936

Figures 34, 35

Study Material. Type specimen, female, 250 mm. long, off Havana, Cuba (Harv. Mus. Comp. Zool., No. 1457); also 14 others, male and female, 130 to 292 mm. long, collected off the north coast of Cuba by the research ship "Atlantis" in March 1938 and April 1939 at depths of 210 to 250 fathoms (Harv. Mus. Comp. Zool.).

Distinctive Characters. S. torrei is easily separable from *S. retifer* by its very broadly rounded snout and by its color pattern; from *S. boa*, which it closely resembles, by the fact that the caudal is about as long as the distance from the tip of the snout to the origin of pectoral, and by its coloration.

Description. Proportional dimensions in per cent of total length. Female, 250 mm., from Cuba (Harv. Mus. Comp. Zool., type, No. 1457). Male, 292 mm., from Cuba (Harv. Mus. Comp. Zool., No. 36093).

Trunk at origin of pectoral: breadth 10.8, 10.6; height 9.9, 8.2.
Snout length in front of: outer nostrils 3.2, 2.4; mouth 4.4, 3.9.
Eye: horizontal diameter 3.3, 3.1.
Mouth: breadth 7.2, 6.5; height 3.2, 3.4.
Nostrils: distance between inner ends 2.4, 2.1.
Labial furrow length: lower 1.6, 1.5.
Gill opening lengths: 1st 1.8, 2.1; 2nd 1.6, 1.4; 3rd 1.6, 1.4; 4th 1.4, 1.2; 5th 1.0, 1.0.
First dorsal fin: vertical height 5.6, 5.5; length of base 6.8, 6.8.
Second dorsal fin: vertical height 2.6, 2.6; length of base 5.4, 4.3.
Anal fin: vertical height 3.4, 3.1; length of base 9.2, 8.6.
Caudal fin: upper margin 20.0, 21.0; lower anterior margin 9.6, 8.2.
Pectoral fin: outer margin 12.1, 10.3; inner margin 6.8, 6.5; distal margin 8.8, 7.9.

Distance from snout to: 1st dorsal 50.5, 49.4; 2nd dorsal 68.4, 68.3; upper caudal 80.0, 79.0; pectoral 18.8, 18.5; pelvics 41.2, 39.3; anal 60.2, 60.0.
Interspace between: 1st and 2nd dorsal 11.6, 12.3; 2nd dorsal and caudal 7.4, 7.7; anal and caudal 9.6, 10.8.
Distance from origin to origin of: pectoral and pelvics 23.6, 21.7; pelvics and anal 19.2, 20.9.

S. torrei resembles *S. boa* very closely in body form, shape of snout, nostrils and nasal flaps, shape, size and relative position of fins, and in the teeth and dermal denticles. The significant points of difference are as follows: in *torrei* the mouth is slightly the lower-arched, its height being only about 40 per cent of its breadth as against 50 per cent in *boa;* in *torrei* the length of the snout in front of the mouth is slightly less, it being only a little more than half as great as the breadth of the mouth as against about four-fifths in *boa;* the pectorals of *torrei* are only a little larger in area than the first dorsal, whereas in *boa* they are twice as large as the latter; and while the denticles rise steeply from the skin in newly hatched specimens of both *torrei* and *boa,* in larger specimens of the former they lie nearly flat and the surface texture of the skin is smoother compared with the pronounced roughness of *boa.* Also, in the male *torrei* the inner edges of the pelvics are connected to one another and to the ventral surface of the trunk more nearly to their tips than in either *boa* or in *retifer.* However, the most striking difference between the species is in the color

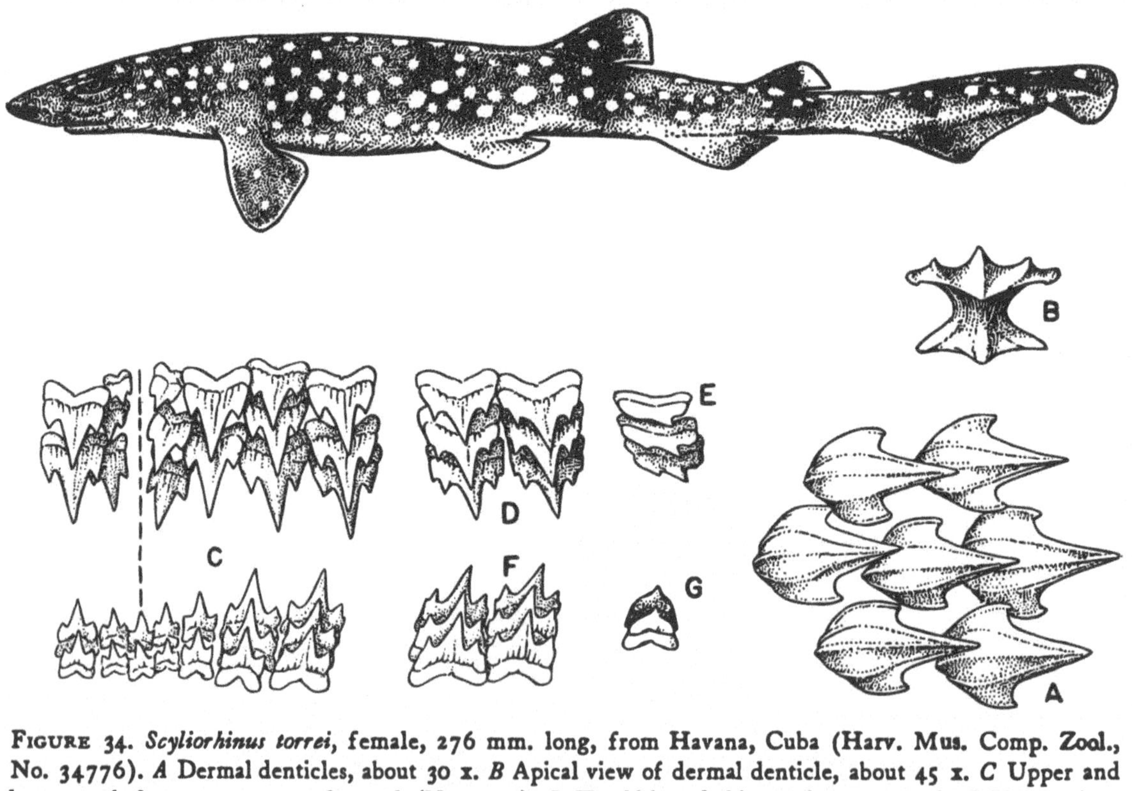

FIGURE 34. *Scyliorhinus torrei,* female, 276 mm. long, from Havana, Cuba (Harv. Mus. Comp. Zool., No. 34776). *A* Dermal denticles, about 30 x. *B* Apical view of dermal denticle, about 45 x. *C* Upper and lower teeth from near center of mouth (Nos. 1–4). *D* Twelfth and thirteenth upper teeth. *E* Upper nineteenth tooth. *F* Lower tenth and eleventh teeth. *G* Lower eighteenth tooth. *C–G,* about 9 x.

pattern. Although the pale brown back and upper sides of *torrei* are transversely marked by a series of indistinct darker blotches, as in *boa,* two of these being on the caudal and one opposite the origin of the pectoral, the finer markings of *torrei* consist of small oval whitish

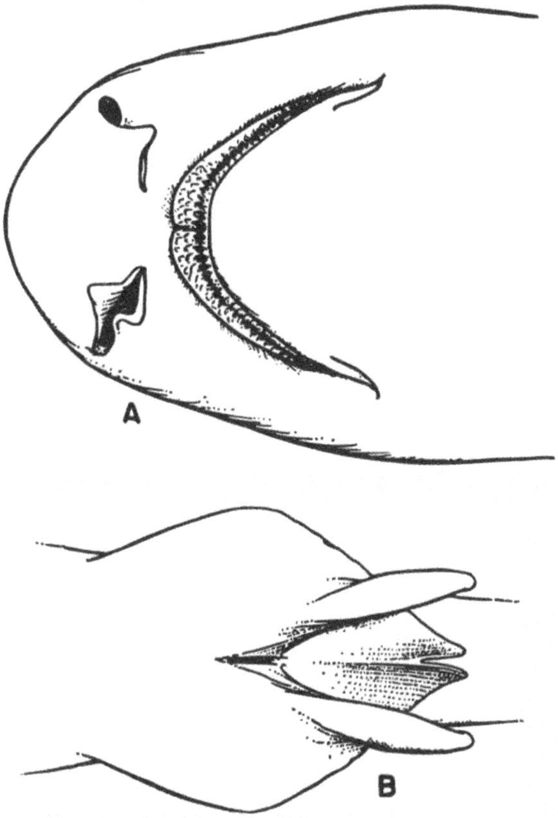

FIGURE 35. *Scyliorhinus torrei.* *A* Anterior part of head of specimen illustrated in Fig. 34, from below, about 1.7 x. *B* Pelvic fins and claspers of male, about 292 mm. long (Harv. Mus. Comp. Zool., No. 36093), about 1.7 x.

spots (in contrast to the dark markings of *boa*) which are rather evenly distributed over the whole back and upper sides. The lower surface is of a very pale shade of the same tint as the upper sides, or nearly white, without evident markings either on the trunk or on the fins.

Size. The male of *torrei* has claspers extending far beyond the tips of the pelvics (suggesting maturity or approaching maturity) at a total length of only about 247 mm., showing that this is a much smaller species than *boa,* perhaps not growing much larger than a maximum of 300 mm. or so.

Developmental Stages. Neither the eggs nor the embryos of *torrei* have yet been seen.

Habits. Nothing is known of its habits.

Range. S. torrei is so far known only off the northern coast of Cuba, but evidently it is common there.

Synonyms and References:

Catulus boae Sanchez-Roig, Revist. Agric. Peces Cubana Commerc. Trabaj., 1931: 17 (Cuba, not seen); not *Scylliorhinus boa* Goode and Bean, 1895.

Scylliorhinus torrei Howell-Rivero, Proc. Boston Soc. nat. Hist., *41*, 1936: 43, pl. 9 (descr., deep water off Havana, Cuba);[35] Fowler, Fish Culturist, *21* (9), 1942: 66, fig. 1 (listed, Cuba); Bigelow and Schroeder, Guide Comm. Shark Fish., Anglo Amer. Caribb. Comm., Wash., 1945: 114 (ill.).

Genus *Galeus* Rafinesque, 1810.

Galeus Rafinesque, Carratt. Gen. Nuov. Sicil., 1810: 13; type species, *G. melastomus* Rafinesque, designated by Fowler, Proc. Acad. nat. Sci. Philad., *60*, 1908: 53.[1]

Generic Synonyms:

Squalus (in part) Gunnerus, Trondh. Gesellsch. Schr. Leipzig, *2*, 1766: 249; not *Squalus* Linnaeus, 1758.

Scyllium (in part) Risso, Ichthyol. Nice, 1810: 30; not *Scyllium* Cuvier, 1817.

Scylliorhinus (in part) Blainville, in Vieillot, Faune Franc., 1825: 68, 75.

Pristiurus[2] Bonaparte, Icon. Faun. Ital., *3*, 1834: 4th p. (not numbered) in description of "*Scyllium canicula*"; type species, *P. melanostomum* Bonaparte, 1834, equals *Galeus melastomus* Rafinesque, 1810.

Pristidurus[3] Bonaparte, Mém. Soc. neuchâtel. Sci. nat., *2* (8), 1839: 11; evident emendation of *Pristiurus* Bonaparte, 1834.

Figaro Whitley, Rec. Aust. Mus., *16*, 1928: 238; type species, *Figaro boardmani* Whitley.

Generic Characters. Two dorsal fins, the 1st originating over rear part of pelvics; denticles along dorsal margin of anterior part of caudal enlarged and modified in shape, forming a distinct crest, bounded below by a narrow band of naked skin on either side; lower margin of caudal peduncle with or without a similar crest of enlarged denticles; nostrils far from mouth and far apart, their anterior margins without barbels, their posterior margins not expanded as flaps; snout long, thin, its mucous pores not conspicuous; labial furrow extending from lower jaw around corner of mouth onto upper jaw, the upper lip not closing outside lower at corner of mouth; upper eyelid not closing outside lower at corner of eye; a longitudinal fold or none below eye; 4th gill opening close in front of pectoral, the 5th over pectoral; teeth alike in the 2 jaws, with long pointed median cusp and 1 to 3 smaller cusps on each side, much as in *Scyliorhinus*, with several rows

35. Howell-Rivero states that the specimen on which he bases his new species, *torrei*, is the same one earlier referred to by Sanchez-Roig as *Catulus boae*.

1. The name *Galeus* was first used by Klein, 1775 (Neuer Schauplatz) and by Valmont (Dict. Hist. Nat., *1*, 1798: 371); but it must date from Rafinesque (Carrat. Gen. Nuov. Sicil., 1810: 13), both Klein's and Valmont's names having been ruled inapplicable by the International Committee on Zoological Nomenclature because such of them as were binomial were so only accidentally (Smithson. misc. Coll., 73 [3], 1925: 27, Opinion 89). In his account of the genus, Rafinesque mentioned only two species, *melastomus* Rafinesque and *uyato* Rafinesque, although he expanded the genus to include seven species in his list of Sicilian fishes published later the same year (Indice Ittiol. Sicil., 1810).

2. The name *Pristiurus* has frequently been credited to Bonaparte, 1831 (Saggio Anim. Vert.: 121). But this first mention of it was nominal only, without diagnosis or reference to any actual species, *i.e.*, it was a *nomen nudem*. For the actual dates of appearance of the individual plates and accompanying text of the Fauna Italica, see Salvadori (Boll. Mus. Zool. Anat. comp. Torino, *3* [48], 1888).

3. The generic name *Pristidurus* was used a year earlier by L. Agassiz (Poiss. Foss., *3*, 1838: 85) with a brief account of the teeth, but without mention of any particular species.

functional; tail sector of trunk considerably longer than body sector; caudal axis raised but slightly, if at all; anal much longer than 2nd dorsal, separated from caudal by a considerable interspace.

Remarks. Sharply diagnostic of this genus as contrasted with all other scyliorhinids are the presence of the caudal crest and the wide separation of the nostrils from the mouth and from each other in combination with the presence of a labial fold on each jaw, the absence of a barbel and the absence or rudimentary state of the posterior nasal flap.

Range. Mediterranean; eastern North Atlantic northward to Norway; Iceland; Madeira; Cuba and southern Florida in the western Atlantic; Japan; Formosa; Australia.

Key to Species

1a. Ventral margin of caudal peduncle, as well as anterior part of dorsal margin of caudal fin, with a conspicuous crest of modified denticles. *boardmani* Whitley, 1928.
Australia.

1b. Denticles along ventral margin of caudal peduncle not modified to form a crest.

- 2a. Tip of anal falls short of a vertical line at rear end of base of 2nd dorsal by a distance about equal to that from eye to spiracle.
  - 3a. Base of 2nd dorsal nearly twice as long as that of 1st dorsal.
    murinus Collett, 1905.
    Iceland.
  - 3b. Base of 1st dorsal only about as long as that of 2nd dorsal.
    eastmani Jordan and Snyder, 1904.
    Japan.
- 2b. Tip of anal extends rearward nearly or quite as far as rear tip of 2nd dorsal.
  - 4a. Trunk plain-colored.
    - 5a. Interspace between anal and caudal about as long as snout in front of mouth. *sauteri*[4] Jordan and Richardson, 1909.
      Formosa.
    - 5b. Interspace between anal and caudal less than ½ as long as snout in front of mouth. *jenseni* Saemundsson, 1922.
      Iceland.
  - 4b. Trunk marked with conspicuous dark stripes, spots or blotches.
    - 6a. Interspace between anal and caudal at least ½ as long as base of anal; base of anal only about twice as long as that of 2nd dorsal.
      arae Nichols, 1927, p. 216.
    - 6b. Interspace between anal and caudal only about ¼ as long as base of anal; base of anal about 3 times as long as that of 2nd dorsal.
      melastomus Rafinesque, 1810.
      Eastern North Atlantic, Mediterranean.

4. Perhaps including *hertwigi* Englehard, 1912, Japan, the description of which is not sufficiently detailed for us to locate it more precisely in this key.

Galeus arae (Nichols), 1927

Figures 36, 37

Study Material. 21 specimens, male and female, 138 to 329 mm. long, taken off the north coast of Cuba, at "Atlantis" stations 2981, 2982, 2985, 2987, 3431, 3437, 3441, and near Tortugas, Florida, in 200 to 345 fathoms (Harv. Mus. Comp. Zool.). Also a specimen from Tortugas, Florida (U.S. Nat. Mus.).

Distinctive Characters. The presence of the caudal crest of large denticles marks *G. arae* off from all other scyliorhinids yet known from the western Atlantic.

Description. Proportional dimensions in per cent of total length. Female, 202 mm., from off Tortugas, Florida (Harv. Mus. Comp. Zool., No. 35250). Male, 324 mm., from Cuba (Harv. Mus. Comp. Zool., No. 36118).

Trunk at origin of pectoral: breadth 8.8, 9.2; height 7.4, 7.3.

Snout length in front of: outer nostrils 4.0, 3.4; mouth 7.6, 7.1.

Eye: horizontal diameter 4.2, 4.3.

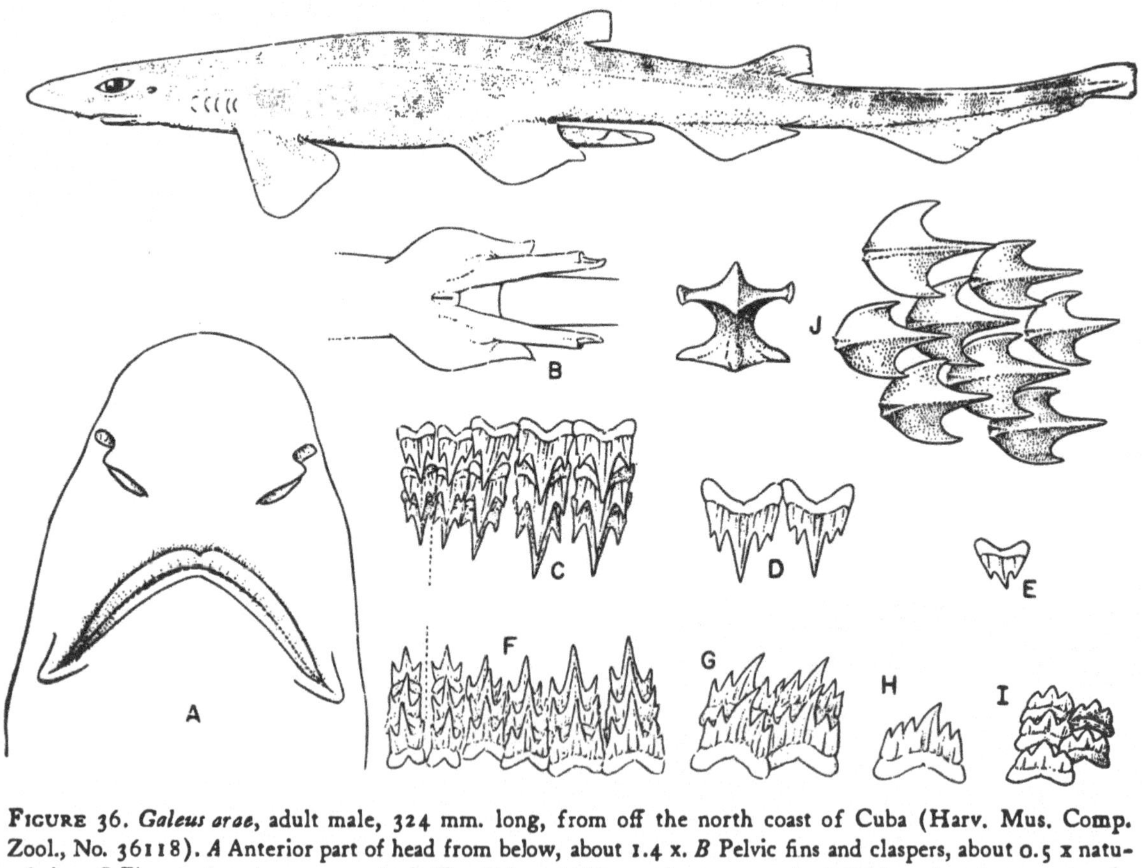

FIGURE 36. *Galeus arae*, adult male, 324 mm. long, from off the north coast of Cuba (Harv. Mus. Comp. Zool., No. 36118). *A* Anterior part of head from below, about 1.4 x. *B* Pelvic fins and claspers, about 0.5 x natural size. *C* First to fourth upper teeth. *D* Twelfth and thirteenth upper teeth. *E* Thirty-fourth upper tooth. *F* First to fifth lower teeeth. *G* Sixteenth and seventeenth lower teeth. *H* Twenty-third lower tooth. *I* Thirtieth and thirty-first lower teeth. *C–I*, about 12 x. *J* Dermal denticles, about 60 x.

Mouth: breadth 8.1, 7.7; height 3.5, 3.3.
Nostrils: distance between inner ends 2.8, 2.8.
Labial furrow length: upper 1.5, 1.7; lower 1.5, 1.8.
Gill opening lengths: 1st 1.8, 1.4; 2nd 1.5, 1.3; 3rd 1.2, 1.2; 4th 1.0, 1.1; 5th 1.0, 1.1.
First dorsal fin: vertical height 3.9, 4.2; length of base 6.4, 5.3.
Second dorsal fin: vertical height 3.7, 4.0; length of base 5.9, 5.2.
Anal fin: vertical height 3.5, 3.5; length of base 13.5, 11.4.
Caudal fin: upper margin 30.2, 29.2; lower anterior margin 11.5, 10.2.
Pectoral fin: outer margin 12.2, 10.8; inner margin 6.5, 5.9; distal margin 9.3, 9.0.
Distance from snout to: 1st dorsal 43.3, 45.7; 2nd dorsal 59.7, 64.0; upper caudal 69.8, 72.8; pectoral 16.8, 19.1; pelvics 37.7, 37.7; anal 51.5, 56.0.
Interspace between: 1st and 2nd dorsals 12.2, 13.0; 2nd dorsal and caudal 3.2, 3.7; anal and caudal 4.0, 4.6.
Distance from origin to origin of: pectoral and pelvics 19.1, 18.6; pelvics and anal 14.7, 19.4.

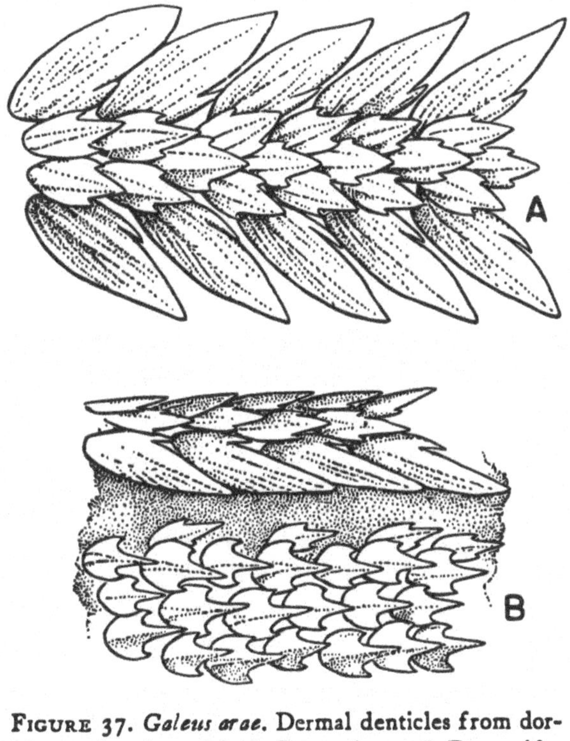

FIGURE 37. *Galeus arae.* Dermal denticles from dorsal margin of caudal. *A* From above. *B* From side, about 17 x.

Trunk slender, its breadth opposite pectorals about 1/10, and its height about 1/12, of total length. Body sector to cloaca considerably shorter than tail sector. Dermal denticles on trunk close-spaced, their blades only slightly raised, with 3 low ridges, their posterior margins with 3 strong teeth, the median much the longest; the 2 or 3 rows along the dorsal margin of the anterior half of the caudal larger, only weakly dentate and without ridges, flanked on either hand by a single row of very much larger blade-like denticles, lanceolate in shape, their inner margins with a deep notch, their tips directed outward and bounded below by a narrow band of naked skin forming a noticeable crest, but grading rearward into denticles of the usual size and shape.

Head strongly flattened above. Snout thin, broadly rounded in front and slightly narrowed opposite nostrils, its length in front of mouth about ⅓ of length of head to origin of pectoral. Eye narrow-oval, its horizontal diameter a little more than ½ as long as snout in front of mouth, with a weakly marked longitudinal fold below it. Spiracle oval, its diameter about ¼ as great as that of eye, behind the latter by a distance about ⅓ as great as the horizontal diameter of eye, and a little below it. Gill openings concave anteriorly in outline, the 4th and 5th closest together, the 1st and 2nd (slightly the longest) about ⅓ as long as horizontal diameter of eye, the 5th (shortest) about ¾ as long as 1st; the 5th above or a little posterior to origin of pectoral. Nostrils oblique, separated one from the other by a distance equal to about ⅖ the length of snout in front of mouth, and separated from mouth by a distance about ½ that great, the anterior margin expanded as a low, subtriangular lobe with rounded tip, the posterior margin not expanded. Mouth obtusely ovate, about ½ as long as broad, with labial furrows extending a short distance inward along both jaws.

Teeth about $\frac{36-36}{35-35}$, with slender median cusp, and a much smaller cusp on each side in central part of mouth, but usually with 2, or even 3, lateral cusps on each side toward corners of mouth; 4 or 5 series functional in front of mouth, with 2 to 3 toward its corners in upper jaw and 3 to 4 series in lower jaw.

Dorsals small, similar in size and shape, quadrate, with weakly convex anterior margins, straight distal margins and subrectangular corners, their bases about as long as snout in front of eye or a little shorter, their free lower margins about ½ as long as their bases or a little less; origin of 1st dorsal over rear ⅓ of bases of pelvics, origin of 2nd dorsal about over midpoint of base of anal. Caudal about ¼ of total length, and noticeably narrow, its axis only very slightly raised, its tip squarely truncate posteriorly, its lower anterior corner much more obtuse than a right angle, the subterminal notch scarcely marked. Interspace between caudal and anal varying from about ½ as long as base of anal to almost as long as latter. Anal about twice as long at base as 2nd dorsal, its rear tip a little anterior to rear tip of latter, with nearly straight margins and rounded apex, its free basal margin very short. Pelvics with broadly rounded apices and tapering, blunted tips, widely divergent in adults but less so in smaller specimens, their inner edges joined and attached to ventral surface of trunk for about ½ their lengths posterior to cloaca, both in males and in females. Pectoral with very broad base, nearly straight margins and broadly rounded corners, about as broad as long, and about 3 times as large in area as 1st dorsal.

Color. Ground tint pale yellowish brown, strikingly marked along sides and back with rows of dark brown blotches and spots of various sizes, forming an especially intricate pattern on top of head; a dark streak from snout to eye; a large and conspicuous blotch extending up onto each dorsal fin, one on the upper half of the caudal near its anterior end, one on its lower part and another across it abreast of the subterminal notch. The precise sizes, shapes and arrangements of the finer markings vary considerably, and they become more or less confluent on the larger specimens, in which the upper surface, anterior to the first dorsal fin, has a clouded rather than a spotted and striped appearance; roof of mouth dusky or sooty, tongue and floor of mouth similar in some specimens, but pale in others, perhaps faded in the preservative.

Size. The largest specimen yet seen is a male of 329 mm. (listed on p. 216). Since the claspers fall considerably short of the tips of the pelvics in a specimen of about 295 mm., but extend far beyond them in another of about 317 mm., maturity is probably attained at about 300 mm.

Developmental Stages. Presumably oviparous, but the eggs have not been seen.

Habits. Knowledge of the habits of *G. arae* is confined to the fact that it is a deep-water species; recorded depths of capture range from 200 down to 345 fathoms.

Range. So far known only off the north coast of Cuba where it is evidently common at suitable depths, off the Tortugas, Florida, and off Miami, Florida (the type locality).

Synonyms and References:

Pristiurus arae Nichols, Amer. Mus. Novitates, 256, 1927: 1, fig. 1 (descr., off Miami Beach, Florida); Jordan, Evermann and Clark, Rep. U.S. Comm. Fish. (1928), 2, 1930: 12 (off Florida).

Scyliorhinus retifer Longley and Hildebrand, Pap. Tortugas Lab., *34*, 1941: 1 (depth, color, Tortugas, Florida).[5]

Genus *Apristurus* Garman, 1913

Apristurus Garman, Mem. Harv. Mus. comp. Zool., *36*, 1913: 96; type species, *Scylliorhinus indicus*, Brauer, Wiss. Ergebn. 'Valdivia,' *15*, 1908: 8, pl. 14, fig. 1 (Indian Ocean and Gulf of Aden), desig. by Jordan, Genera Fish., *4*, 1920: 548.

Generic Synonyms:

Catulus (in part) Gilbert, Proc. U.S. nat. Mus., *14*, 1891: 542; for *C. brunneus* Gilbert (west coast of N. America); not *Catulus* A. Smith, 1837.

Scylliorhinus (in part) Brauer, Wiss. Ergebn. 'Valdivia,' *15*, 1908: 8; and subsequent authors; not *Scyliorhinus* Blainville, 1816.

Pristiurus (in part) Garman, Mem. Harv. Mus. comp. Zool., *36*, 1913: 94; for *Catulus spongiceps* Gilbert, 1905; not *Pristiurus* Bonaparte, 1834.

Scyllium Saemundsson, Vidensk. Medd. naturh. Foren. Kbh., *74*, 1922: 173; type species *S. laurussonii* Saemundsson, monotypic; not *Scyllium* Cuvier, 1817.

Apristurius Schulze, Kükenthal, *et al.*, Nomencl. Anim., *1* (2), 1926: 244; evident misspelling for *Apristurus* Garman, 1913.

Pentanchus (in part) Fowler, Proc. Acad. nat. Sci. Philad., *85*, 1934: 237; not *Pentanchus* Smith and Radcliffe, Proc. U.S. nat. Mus., *41*, 1912: 490.

Parapristurus Fowler, Proc. Acad. nat. Sci. Philad., *85*, 1934: 237; type species, *Catulus spongiceps* Gilbert, 1905.

5. We have examined and identified this specimen, now in the United States National Museum.

Generic Characters. Two dorsal fins, the origin of the 1st considerably posterior to origin of pelvics; dermal denticles along dorsal margin of anterior part of caudal not enlarged or modified as a distinct crest, or bounded below by a band of naked skin on either side; nostrils entirely separate from mouth, their anterior margins without barbels and falling considerably short of mouth; posterior as well as upper margin of nostril expanded as a flap; labial furrow around corner of mouth and on each jaw; interspace between anal and caudal less than 1/5 as long as base of anal; base of anal more than twice as long as base of 2nd dorsal; no fold below eye; mucous pore system on lower surface of snout very conspicuous; gill openings either of the usual conformation, or so deeply concave anteriorly that tips of gill filaments are exposed; 5th gill opening over or behind origin of pectoral; teeth numerous, those in front of mouth with one chief cusp and one or more smaller cusps on each side; several series of teeth functional.

Range. Both sides of North Atlantic, including Iceland; South Africa; west coast of North America from Gulf of California to Puget Sound; Hawaiian Islands; Japan; Philippines and East Indies; Indian Ocean and Gulf of Aden; west coast of South Africa.

Species. These are little known sharks of deep water, the majority of them so far known from very few specimens. The named species of the genus, numbering 13 and from widely separated seas, resemble one another very closely in general appearance, but they appear to be separable by sufficiently precise differences. Fowler has even distributed them among three subgenera,[1] according to the degree of cirrus-like development on the margins of the nostrils, anterior and posterior. According to published accounts, however, and to our own study of three of the species, the differences in this respect are not sharp enough to serve as a basis for generic separation. Nevertheless, the members of the genus do fall into two sharply contrasting categories as regards the gill openings, for while these are of the ordinary type in one group, typified by *A. brunneus* Gilbert from the west coast of North America, they are close together above and below in other species, but so deeply concave anteriorly at the midlevel that the tips of the gill filaments are exposed (p. 227). It is astonishing that attention has not been directed to this earlier, for the gills are clearly pictured thus in *A. atlanticus* Koefoed,[2] as well as in *A. microps* Gilchrist.[3] Furthermore, a re-examination of the specimens in the United States National Museum shows gills of this same type in *profundorum* Goode and Bean, 1895 (p. 222), *verweyi* Fowler, *herklotsi* Fowler, 1934,[4] *spongiceps* Gilbert, 1905, and *platyrhynchus* Tanaka, 1909,[5] although no suggestion of the fact appears in the published accounts or in the illustrations of these species. Under ordinary circumstances a difference so striking would demand the institution of a new genus. In the present case, however, such action does not seem advisable because neither the account nor the illustration of the type

1. Bull. U.S. nat. Mus., 100 (*13*), 1941: 53; *Parapristurus*, *Pentanchus* and *Apristurus*.
2. Koefoed, Rep. Sars N. Atlantic Deep Sea Exped., *4* (1), 1932: 18, pl. 3.
3. Mar. biol. Rep. Cape Town, *2*, 1922: 46, pl. 7, fig. 1.
4. Proc. Acad. nat. Sci. Philad., *85*, 1934: 237, 238.
5. There is a specimen of this species in the United States National Museum, although not the type.

species of *Apristurus*[6] gives any information as to its gill openings, *i.e.*, there is no way of knowing to which subdivision of the old genus *Apristurus* it belongs; nor are the specimens available for study at present, being presumably in Berlin. Therefore, it seems wiser to use *Apristurus* in the more inclusive sense for the time being.[7] Neither can a dependable Key to Species be constructed for the genus as a whole until more complete information is available in other respects regarding *indicus* Brauer, also one of the two supposedly distinct species that have been named from Japan,[8] and *sibogae* Weber[9] from the East Indies.[10]

Key to Atlantic and South African Species

1a. Distance between 1st and 2nd dorsal fins as great as from tip of snout to spiracle.
saldanha Barnard, 1925.
South Africa.

1b. Distance between 1st and 2nd dorsal fins at least no greater than from tip of snout to eye.

- 2a. Interspace between 1st and 2nd dorsals less than 1/2 as long as from tip of snout to eye; eye minute, its diameter only about 1/11 as long as from tip of snout to 5th gill opening. *microps* Gilchrist, 1922.
South Africa.

- 2b. Interspace between 1st and 2nd dorsals nearly or quite as long as from tip of snout to eye; eye larger, its diameter at least 1/8 as long as head to 5th gill slit.
  - 3a. Second dorsal about twice as large in area as 1st; 1st to 3rd gill openings nearly as long as distance between nostrils.
riveri Bigelow and Schroeder, 1944, p. 225.
  - 3b. Second dorsal little if any larger in area than 1st; 1st to 3rd gill openings only about 1/2 as long as distance between nostrils.
    - 4a. Horizontal diameter of eye slightly longer than distance between nostrils; rear ends of bases of pelvics slightly nearer to tip of snout than to tip of caudal; caudal about 1/3 of total length. *atlanticus* Koefoed, 1932.
Eastern North Atlantic.
    - 4b. Horizontal diameter of eye only about 3/5 as long as distance between nostrils; rear ends of bases of pelvics nearer to tip of caudal than to tip of snout; caudal only about 1/4 of total length.
profundorum[11] Goode and Bean, 1895, p. 222.

6. *indicus* Brauer, Wiss. Ergebn. 'Valdivia,' *15*, 1908: 8, pl. 14, fig. 1.
7. For further discussion, see Bigelow and Schroeder (Proc. New Engl. zool. Cl., *23*, 1944: 21).
8. *macrorhynchus* Tanaka, J. Coll. Sci. Tokyo, *27*, 1909: 1.
9. Siboga Exped., *57*, 1913: 595.
10. For comparable illustrations of *spongiceps* Gilbert, 1905, from the Hawaiian Islands, *verweyi* Fowler, 1934, from Borneo, *herklotsi* Fowler, 1934, of the Philippines, and *platyrhynchus* Tanaka, 1909, from Japan, see Fowler (Bull. U.S. nat. Mus., 100 [*13*], 1941: 53–55, 57).
11. Including *laurussonii*, Saemundsson, 1922, Iceland.

Apristurus profundorum (Goode and Bean), 1895

Figure 38

Study Material. Type specimen, mature male, 510 mm. long, taken off Delaware Bay in 816 fathoms (U.S. Nat. Mus., No. 35646); newly hatched male, 145.5 mm., from a nearby locality (U.S. Nat. Mus., No. 83894).

Distinctive Characters. The adult *profundorum* is separated from *riveri* by its considerably smaller second dorsal relative to the first dorsal, shorter snout relative to length of head, smaller eye, much shorter gill openings, relatively broader mouth and much shorter caudal; from *atlanticus* by its relatively smaller eye, shorter caudal, and by the fact that the tips of the pelvics are closer to the tip of the caudal than to the tip of the snout. But it may be difficult to distinguish newly hatched specimens of the three species from one another.

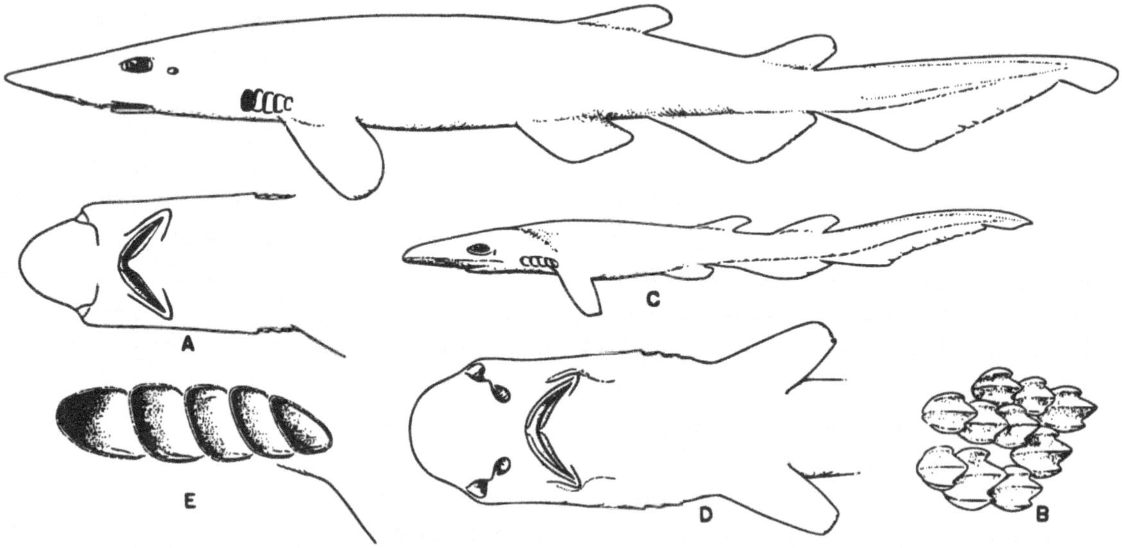

FIGURE 38. *Apristurus profundorum*, female, 510 mm. long, from off Delaware Bay (U. S. Nat. Mus., No. 35646, type). *A* Head of same from below. *B* Dermal denticles of same, about 12 x. *C* Newly hatched male, 146 mm. long, from off Delaware Bay (U. S. Nat. Mus., No. 83894). *D* Head and pectorals of same from below, about 1.1 x. *E* Gill openings of same, about 4 x.

Description. Proportional dimensions in per cent of total length. Male, 510 mm., from Lat. 39° N., Long. 72° W. (U.S. Nat. Mus., type, No. 35646).

Trunk at origin of pectoral: breadth 11.0; height 9.2.

Snout length in front of: mouth 8.9.

Eye: horizontal diameter 2.7.

Mouth: breadth 8.4; height 2.9.

Nostrils: distance between inner ends 4.1.

Labial furrow lengths: upper 2.9; lower 3.5.

Gill opening lengths: 1st 1.8; 5th 1.3.

First dorsal fin: vertical height 3.2; length of base 7.0.
Second dorsal fin: vertical height 3.3; length of base 6.9.
Anal fin: vertical height 4.3; length of base 13.9.
Caudal fin: upper margin 25.0.
Pectoral fin: outer margin 10.6; inner margin 6.4; distal margin 5.1.
Distance from snout to: 1st dorsal 49.5; 2nd dorsal 62.5; upper caudal 75.0; pectoral 24.7; pelvics 43.3; anal 56.6.
Interspace between: 1st and 2nd dorsals 8.2; 2nd dorsal and caudal about 3; anal and caudal 0.0.
Distance from origin to origin of: pectoral and pelvics 19.6; pelvics and anal 12.5.

Trunk slender, highest opposite axil of pectoral, tapering evenly rearward, its height at axil of pectoral (where highest) about 1/7 its length to origin of caudal. Body sector to cloaca a little longer than tail sector. Dermal denticles with 3 ridges and 3 teeth, as in *riveri*, but with the teeth shorter and overlapping more, so that the skin is more concealed.

Head about 1/4 of total length, flattened above, and contracted laterally just anterior to outer ends of nostrils. Snout broadly rounded, its length in front of mouth a little more than 1/3 as great as length of head to origin of pectoral, with a median belt of conspicuous mucous pores in 8 or 9 irregular rows on its ventral surface. Eye oval, its horizontal diameter about 3/5 as great as distance between nostrils, its midpoint opposite corner of mouth. Spiracle oval, its diameter about 1/3 as great as that of eye, and behind latter by a distance about 2/3 as great as diameter of eye. Gill openings much smaller relatively than in *riveri*, the 1st to 3rd (longest) a little less than 1/2 as long as distance between nostrils, or about 2/3 as long as horizontal diameter of eye, the 5th (shortest) about 2/3 as long as 1st, of the same general type as in *riveri* (p. 227), their anterior outlines so deeply concave that the tips of the gill filaments are exposed on all 5 of the interbranchial septa; the 4th and 5th over origin of pectoral. Nostrils moderately oblique, at margins of head, their outer ends about equidistant between tip of snout and center of mouth, the distance between them a little less than 1/2 as great as length of snout in front of mouth, the anterior margins more broadly rounded than in *riveri* (the condition of the specimen is not good enough for description of the inward cirroid extensions of the nostril, if any). Mouth ovate, nearly 3 times as broad as high. Labial furrows very prominent, the upper extending about 1/2 the distance toward the symphysis, and more nearly parallel with the jaw than in *riveri*, the lower a little shorter than the upper.

Teeth about $\frac{25-25}{25-25}$; uppers with long, sharp median cusp, flanked on either side by 2 or 3 smaller cusps; lowers similar to uppers, except with the lateral cusps somewhat larger relative to the median cusp, and more often 3 in number on one or both sides; no median tooth in either jaw; several series functional.

Dorsals similar in form, brush-shaped, with rounded tips and weakly convex anterior margins. Origin of 1st about over midpoint of bases of pelvics, its base a little less than 3/4 as long as snout in front of mouth, its rear tip about over origin of anal. Second dorsal about as long at base as 1st dorsal, and only a very little larger than the latter in area, if at all so, its origin about over midpoint of base of anal. No definite interspace between 2nd dorsal and caudal. Caudal about 1/4 of total length, with rounded tip and weakly marked subterminal notch, its lower anterior corner subangular, its axis not appreciably raised above main axis of trunk. No measurable interspace between lower origin of caudal and rear end of base of anal. Anal a nearly equilateral and very obtuse triangle, with nearly straight edges, slightly rounded corners and very short free tip, its origin about under tip of 1st dorsal, its base about twice as long as that of 2nd dorsal. Pelvics quadrate, with nearly straight edges and blunted corners, apices broadly rounded, the rear corners more narrowly so. Pectoral more than twice as large in area as 1st dorsal, brush-shaped, about as broad at base as at tip, with rounded corners, the outer margin nearly straight, but the distal and inner margins moderately convex.

Color. Uniform grayish brown below as well as above after preservation in alcohol.

Size. The fact that the claspers of the type specimen (510 mm. long) are only moderately developed suggests that this deep-sea shark does not mature until a length of perhaps 550 to 600 mm. is reached.

Developmental Stages. Presumably *A. profundorum* is oviparous, but its eggs have not been identified, although Gudger[12] suggests that certain egg cases found on the coast of North Carolina might be of this parentage. If the very small specimen listed above and illustrated in Fig. 38 C actually belongs to this species and not to *riveri*, as seems probable (from the shortness of its gill openings, its small eyes, as well as from the locality of its capture), *profundorum* more closely resembles *riveri* when newly hatched than later in growth, for the length of its caudal is then as great as in *riveri* (about 1/3 of total length) and its second dorsal considerably larger than its first dorsal. Furthermore, the snout is considerably longer, relatively, in newborn specimens than in adults of either of the two possible parent species, since it occupies considerably more than one-third of the length of the head, and the anal is actually confluent with the lower edge of the caudal. More interesting still is the great breadth of the basal lines of attachment of the pectorals to the lower sides of the trunk (Fig. 38 C).

Habits. Nothing is known positively of its habits, but the depth of capture listed above and its uniformly dark coloration above and below suggest a deep-sea habitat.

Range. *A. profundorum* is definitely known only from the continental slope off Delaware Bay and from the specimen (or specimens) listed above (p. 222). However, if *A. laurussonii* from Iceland is identical with it, as the only published account of *laurussonii* suggests, it is no doubt wide-ranging around the slopes of the northern North Atlantic in the appropriate latitudinal belt.[13]

12. Proc. biol. Soc. Wash., 25, 1912: 154.

13. *Profundorum* has also been reported from British Columbia (Halkett, Check List Fish. Canad., 1913: 117). But probably the shark in question was actually *A. brunneus* Gilbert, which is rather common along the Pacific coast of North America from the Gulf of California northward, in deep water.

Synonyms and References:

Scylliorhinus profundorum Goode and Bean, Smithson. Contr. Knowl., *30*, 1895: 17, pl. 5, fig. 16; Mem. Harv. Mus. comp. Zool., *22*; Spec. Bull. U.S. nat. Mus., *2*, same date and pagination (descr., off Delaware Bay, Lat. 39° 9′ N., Long. 72° 3′ W., 816 fathoms); Jordan and Evermann, Rep. U.S. Comm. Fish. (1895), 1896: 213 (name, N. Atlant.); Bull. U.S. nat. Mus., 47 (1), 1896: 22 (descr.); Bull. U.S. nat. Mus., 47 (4), 1900: pl. 3, fig. 11; Gudger, Proc. biol. Soc. Wash., *25*, 1913: 154 (egg cases perhaps this species).

Apristurus profundorum Garman, Mem. Harv. Mus. comp. Zool., *36*, 1913: 99 (descr.); Fowler, Copeia, 30, 1916: 36; Smith, J. Amer. Mus. nat. Hist., *16*, 1916: 349 (ref. to type specimen); Jordan, Evermann and Clark, Rep. U.S. Comm. Fish. (1928), 2, 1930: 12 (ref. to type specimen).

Not *Scylliorhinus profundorum* Halkett, Check List Fish. Canad., 1913: 117.

Probable synonym:

Scyllium laurussonii Saemundsson, Vidensk Medd. naturh. Foren. Kbh., *74*, 1922: 173, pl. 4, fig. 1 (Iceland).

Apristurus riveri Bigelow and Schroeder, 1944

Figure 39

Study Material. The type specimen, female, 407 mm. long, trawled at "Atlantis" Sta. 2993, off north coast of Cuba, March 15, 1938, in 580 fathoms (Harv. Mus. Comp. Zool., No. 36092).

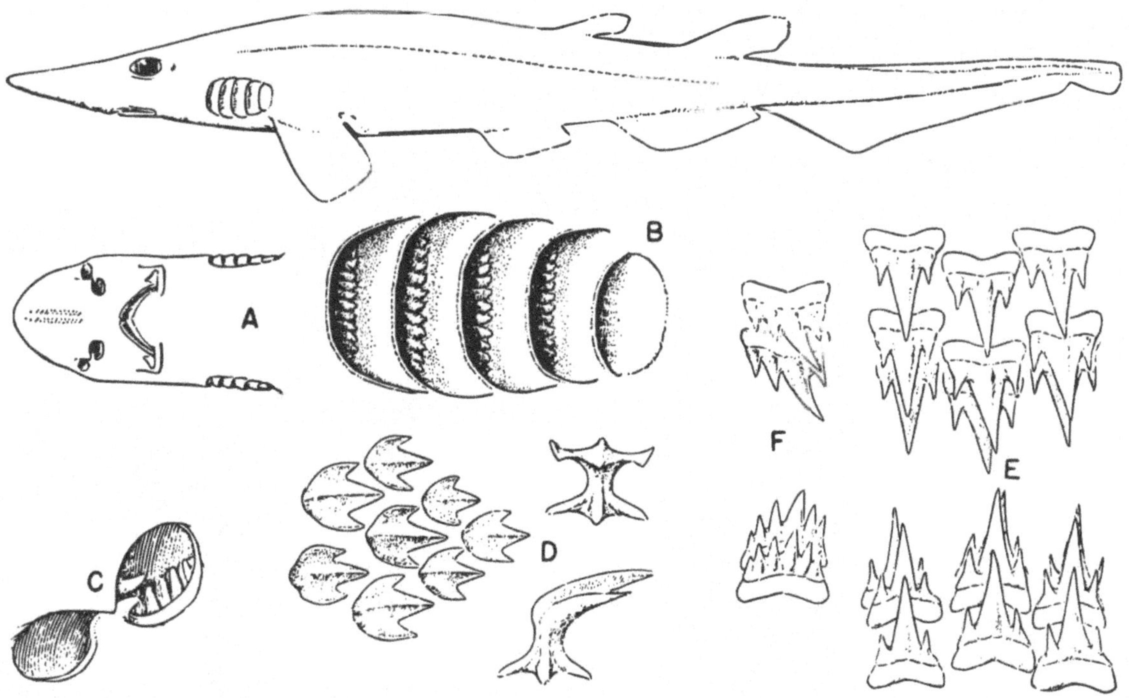

FIGURE 39. *Apristurus riveri*, female, 407 mm. long, from off northern Cuba (Harv. Mus. Comp. Zool., No. 36092, type). *A* Head from below. *B* Gill openings, about 1.8 x. *C* Right-hand nostril, about 3 x. *D* General view of dermal denticles, about 22 x; lateral and apical views, about 43 x. *E* Upper and lower teeth from near center of mouth. *F* Upper and lower teeth from outer parts of jaws, about 16 x.

Distinctive Characters. The adult is separated from *A. profundorum*, *A. laurussonii* and *A. atlanticus*[14] by the considerably greater size of its second dorsal relative to the first, and by the fact that its first to third gill openings are about as long as the distance between the nostrils but only about one-half that relative length in the other three species. It is further separated from *profundorum* (the only local species with which it might be confused) by its relatively larger eye, and by the fact that its caudal occupies about one-third of the total length as compared to only about one-quarter. Young specimens of the different species may be difficult to separate.

Description. Proportional dimensions in per cent of total length. Female, 407 mm., from Cuba (Harv. Mus. Comp. Zool., type, No. 36092).

Trunk at origin of pectoral: breadth 10.1; height 11.8.
Snout length in front of: outer nostrils 5.8; mouth 9.3.
Eye: horizontal diameter 2.7.
Mouth: breadth 6.1; height 2.2.
Nostrils: distance between inner ends 3.9.
Labial furrow lengths: upper 2.1; lower 2.5.
Gill opening lengths: 1st 3.3; 2nd 3.4; 3rd 3.2; 4th 3.0; 5th 2.1.
First dorsal fin: vertical height 2.9; length of base 4.4.
Second dorsal fin: vertical height 4.2; length of base 6.1.
Anal fin: vertical height 3.7; length of base 13.7.
Caudal fin: upper margin 33.0; lower anterior margin 10.1.
Pectoral fin: outer margin 10.1; inner margin 4.9; distal margin 7.7.
Distance from snout to: 1st dorsal 47.8; 2nd dorsal 57.7; upper caudal 67.0; pectoral 23.9; pelvics 40.2; anal 52.2.
Interspace between: 1st and 2nd dorsals 7.0; 2nd dorsal and caudal indefinite; anal and caudal 0.0.
Distance from origin to origin of: pectoral and pelvics 18.5; pelvics and anal 12.5.

Trunk highest and broadest opposite axil of pectoral, narrowing evenly rearward. Caudal peduncle strongly compressed laterally, about ½ as broad as deep. Body sector to cloaca about as long as tail sector. Dermal denticles small in specimen examined, moderately erect, leaf-like, with short pedicels, their blades with weak median crest but tridentate free margin, the median tooth much the longest; the denticles slightly the largest, relatively the narrowest, and the most closely spaced along upper sides of caudal.

Head strongly flattened anteriorly, and contracted laterally just anterior to the outer ends of the nostrils, its dorsal surface noticeably concave, with a triangular belt of about 110 very prominent pores along the midzone anterior to eyes. Snout thin, broadly rounded at tip, its length in front of mouth slightly less than ½ as great as distance from its tip to 5th gill opening, its lower surface with a belt of conspicuous pores in 4 rows along

14. The only other members of the genus yet known from the North Atlantic.

the midzone. Eye oval, its horizontal diameter about twice its vertical height and nearly as long as distance between nostrils, its midpoint about opposite corner of mouth. Spiracle round, about 1/5 as long as horizontal diameter of eye, situated close behind latter. Gill openings with their anterior margins so deeply concave in outline that the tips of the gill filaments on the 1st to 4th arches are exposed, but with the dorsal and ventral ends so close together that the successive margins form an apparent frame around the gill area as a whole; 1st to 4th gill openings about as long as horizontal diameter of eye, or about 4 times as long as from posterior margin of eye to spiracle, the 5th considerably shortest, and close in front of origin of pectoral. Nostrils moderately oblique and far apart, their inner corners about 3 times as far from tip of snout as from symphysis of upper jaw, their outer corners at outer edge of snout, their anterior margins obtusely triangular in outline, the posterior margin of nostril also expanded, as shown in Fig. 39 C. Mouth obtusely ovate, nearly 3 times as broad as high, occupying only about 1/2 the breadth of head. Labial furrows very prominent, forming approximately a right angle at corner of mouth when latter is closed, the upper extending forward about 1/3 the distance toward outer end of nostril, the lower (slightly the longer) reaching only a short distance past corner of mouth.

Teeth, about $\frac{26-26}{25-25}$; uppers with 3 cusps, the median erect and much the longest, except that there are 2 small cusps on one or both sides toward corners of mouth, with the median cusp curved outward; lowers similar to uppers in front of mouth, but usually with 5 cusps along sides of mouth, the median only a little the longest; no tooth at symphysis of either jaw; a very small tooth next to the symphysis in lower; mostly 3 series functional in upper jaw, but 3 to 4 in lower.

First dorsal very small, its base a little longer than horizontal diameter of eye, brush-shaped, with convex anterior margin and rounded tip, its origin over rear ends of bases of pelvics. Second dorsal similar to first but nearly 1 1/2 times as long (at base) and 1 1/2 times as high vertically (correspondingly larger in area), its origin about over midpoint of base of anal. Caudal about 1/3 of total length, with brush-shaped tip and well marked subterminal notch, its lower anterior corner more obtuse than a right angle, its axis only very slightly raised, if at all. No measurable interspace between caudal and anal. Anal with rounded anterior corner and angular rear corner, its base slightly more than twice as long as base of 2nd dorsal. Pelvics a little less than 1/2 as long at base as anal, with rounded anterior and angular rear corners, their inner margins very short and entirely separate one from the other behind the cloaca in female. Interspace between pelvics and anal about 2/3 as long as base of former. Pectoral a little more than 1/3 as long as head, nearly as broad at base as at tip, with slightly convex outer margin, nearly straight distal and inner margins and rounded corners.

Color. Uniform chocolate-brown above and below, in alcohol, the tongue and lining of the mouth blackish.

Size. No information is available as to the length to which this species may grow. The specimen at hand does not contain eggs.

Developmental Stages. Not known; see *profundorum*, p. 224.

Range. Known only from the specimen recorded above, taken off the north coast of Cuba. The depth of capture, combined with the uniformly dark coloration, suggests that this is a deep-water species.

Synonym and Reference:

Apristurus riveri Bigelow and Schroeder, Proc. New Engl. zool. Cl., *23*, 1944: 23, pl. 7 (descr, ill., type loc. off the north coast of Cuba, "Atlantis" Sta. 2993, 580 fathoms).

Family PSEUDOTRIAKIDAE

False Cat Sharks

Characters. Two dorsal fins, the 1st as long as caudal,[1] or longer, the rear end of its base over or a little anterior to origin of pelvics; 2nd dorsal as high as 1st or higher; caudal less than ¼ of total length, not lunate, its lower anterior corner not expanded as a definite lobe, its axis but little raised; caudal peduncle not flattened dorso-ventrally or expanded laterally, without precaudal pits above or below; sides of trunk without longitudinal dermal ridges; snout not greatly elongate; jaws not widely protrusible; gill openings very short, 5th over origin of pectoral; gill arches without rakers and not interconnected by a sieve of modified denticles; nostril entirely separate from mouth, its anterior margin without barbel; spiracles present; lower eyelid without nictitating membrane, but with a well marked longitudinal fold; teeth small, numerous, with larger median and smaller lateral cusps, 6 to 13 series functional; head of normal shape, not widely expanded laterally; rostral cartilages 3, united terminally; radials of pectoral mostly on metapterygium, those on meso- and propterygia fused; mesopterygium and propterygium much smaller than metapterygium.[2] Development ovoviviparous.

Genera. One genus only.

Genus *Pseudotriakis* Brito Capello, 1867

Pseudotriakis Brito Capello, J. Sci. math. phys. nat. Lisboa, *1* (2), 1867: 315, 321; type species, *P. microdon* Brito Capello, Setubal, Portugal.

Generic Characters. Those of the family.

Range. Both sides of North Atlantic, in deep water; Japan.

Species. Only two species known.

1. The very long first dorsal is the most striking character of this family.
2. For detailed account and illustrations of the skeleton, see Jacquet (Bull. Inst. océanogr. Monaco, *36*, 1905).

Key to Species

1a. Caudal about 1/4 of total length; origin of anal only a little posterior to origin of 1st dorsal; distance from tip of snout to angle of mouth only about 1/4 as long as from snout to 5th gill opening; length of snout in front of mouth about 1/3 as great as width of mouth. *acrages* Jordan and Snyder, 1904.
Japan.[3]

1b. Caudal only about 1/5 of total length; origin of anal considerably posterior to origin of 1st dorsal; distance from snout to angle of mouth 1/2 as great as from snout to 5th gill opening or greater; length of snout in front of mouth about 1/2 as great as width of mouth. *microdon* Brito Capello, 1867, p. 229.

Pseudotriakis microdon Brito Capello, 1867

Figure 40

Study Material. None.

Distinctive Characters. Separable from all other Atlantic sharks by the great length of its first dorsal fin.

Description. Proportional dimensions in per cent of total length. Specimen, 2,950 mm., from Amagansett, N. Y. (after Goode and Bean).

Trunk at origin of 1st dorsal: breadth 8.5; height 12.0.
Snout length in front of: mouth 3.0.
Eye: horizontal diameter 2.3.
Mouth: breadth 9.0.
Nostrils: distance between inner ends 4.2.
Gill opening lengths: 1st 2.6.
First dorsal fin: greatest height 3.2; length of base 22.7.
Second dorsal fin: greatest height 5.4; length of base 12.5.
Anal fin: greatest height 4.0; length of base 8.5.
Caudal fin: upper margin 18.0; lower anterior margin 7.7.
Pectoral fin: greatest length 11.2; greatest width 8.0.
Distance from snout to: 1st dorsal 34.0; 2nd dorsal 67.0; upper caudal 82.0; pectoral 20.0; pelvics 56.0; anal 70.7.
Interspace between: 1st and 2nd dorsals 10.5; 2nd dorsal and caudal 3.9.

Male embryo, ready for birth, 850 mm., from Iceland (after measurements by Saemundsson).

Snout length in front of: mouth 5.3.
First dorsal fin: height 4.1; length 21.2.

3. For these and other differences in proportionate measurements, see Jordan and Snyder (Smithson. misc. Coll., *45*, 1904: 233). The original spelling was "*acrales*," but Garman (Mem. Harv. Mus. comp. Zool., *36*, 1913: 105) has pointed out that this was a misprint.

Second dorsal fin: height 6.2; length 11.7.
Anal fin: height 3.8; length 9.0.
Caudal fin: upper margin 19.4.
Pectoral fin: length 10.0.
Distance from snout to: 1st dorsal 32.4; 2nd dorsal 62.3; upper caudal 80.6; pectoral 20.0; pelvics 52.9; anal 65.3.

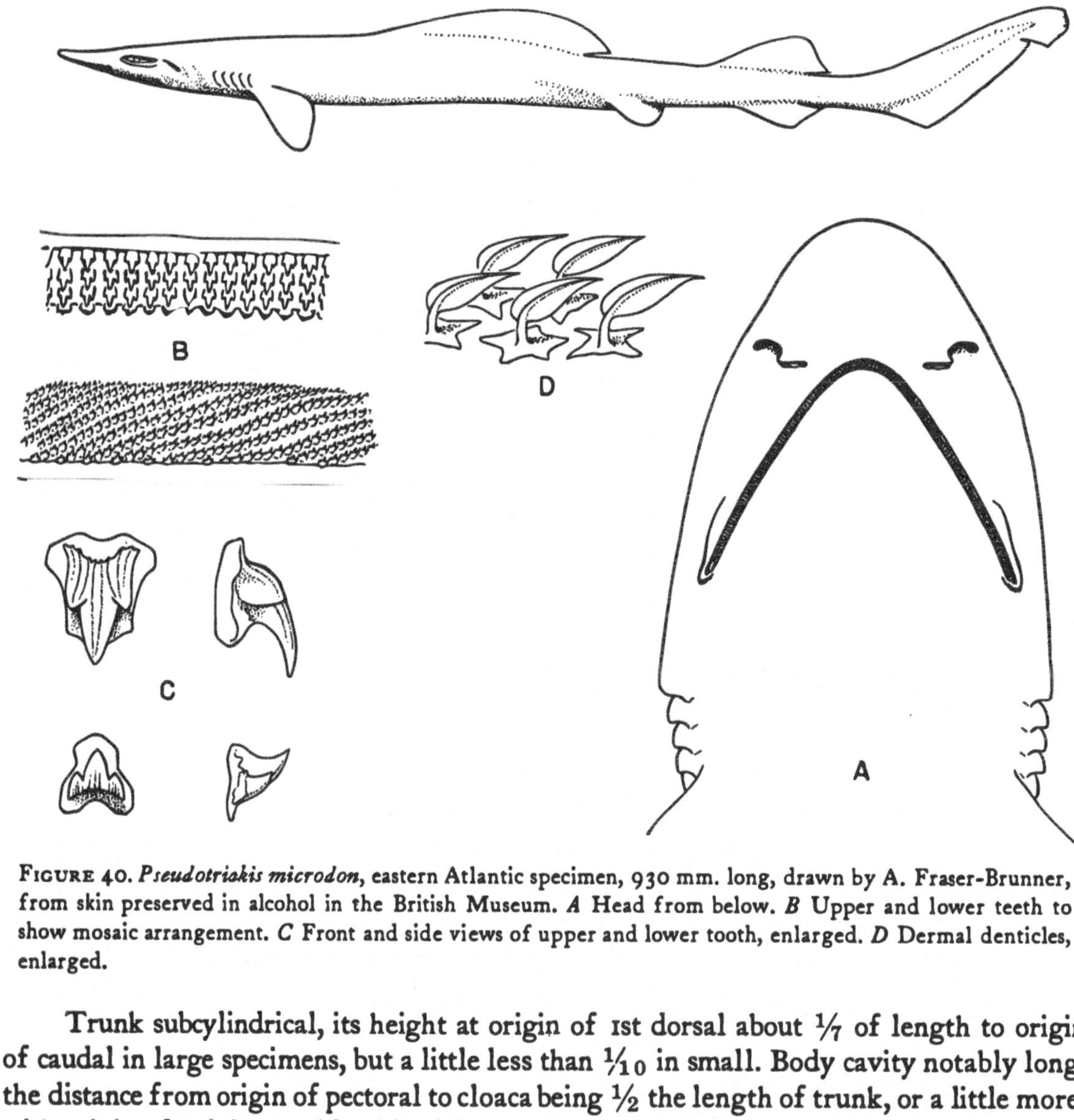

FIGURE 40. *Pseudotriakis microdon*, eastern Atlantic specimen, 930 mm. long, drawn by A. Fraser-Brunner, from skin preserved in alcohol in the British Museum. *A* Head from below. *B* Upper and lower teeth to show mosaic arrangement. *C* Front and side views of upper and lower tooth, enlarged. *D* Dermal denticles, enlarged.

Trunk subcylindrical, its height at origin of 1st dorsal about 1/7 of length to origin of caudal in large specimens, but a little less than 1/10 in small. Body cavity notably long, the distance from origin of pectoral to cloaca being 1/2 the length of trunk, or a little more, with origin of pelvics considerably closer to tip of caudal than to snout. Dermal denticles lanceolate, with 1 to 5 low longitudinal ridges, raised steeply from the skin on short pedicels.

Head about 1/5 of total length and somewhat flattened above. Snout in front of mouth a little less than 1/6 to 1/7 of length of head to 5th gill opening in large specimens, but relatively somewhat longer in small (Fig. 40 A), ovate with rounded tip. Eye oval, with fold below it. Spiracle oval, about as long as diameter of eye or longer, its long axis oblique, situated close behind the eye. Second gill opening a little longer than diameter of eye in large specimens, but a little shorter than eye in small. Nostrils far apart and much closer to mouth than to tip of snout, only slightly oblique, the anterior margins with a low, subtriangular lobe. Mouth rounded in front with nearly straight sides, its height between 2/3 and 3/4 its width. A well marked labial furrow on upper jaw extending 1/3 to 1/4 the distance toward symphysis, the lower furrow very short.

Teeth minute, extremely numerous, arranged in mosaic; smooth-edged; uppers with 3 to 5 triangular cusps, the median much the largest in front part of mouth, but the laterals more nearly equalling it toward corners of mouth; lowers with 3 erect cusps in front of mouth (the median longest) but usually with 4 along its sides, the cusps of the outermost being of nearly equal lengths;[4] about 6 to 13 series functional, more being so in lower jaw than in upper.

First dorsal very sloping, about 1/6 to 1/7 as high as the length at base, relatively somewhat higher in small specimens than in large (see proportional dimensions, p. 229), convex in upper contours, but without definite apex, its free rear margin very short with acute tip, its origin 1/2 to 1/3 as far from axil of pectoral as from origin of pelvics, its base between 1/4 and 1/5 as long as the total length, and about 1 1/3 times as long as caudal in large specimens, but only about as long as caudal in small, the rear end of its base terminating about over origin of pelvics or a little anterior to the latter. Second dorsal a little higher than 1st in small specimens and about 1.7 times that high in large, about twice as long as high, subtriangular, with weakly convex anterior margin, very weakly concave or nearly straight distal margin, broadly rounded apex and very short free rear corner, its origin a little posterior to tips of pelvics, its tip over or a little posterior to tip of anal. Caudal only between 1/5 and 1/6 of total length, slightly less than 1/2 as broad as long, with well marked subterminal notch, its posterior outline nearly straight, its lower anterior corner obtuse, subangular. Anal similar to 2nd dorsal, but only about 2/3 as long at base and 2/3 as high, its origin a little posterior to origin of 2nd dorsal. Pelvics 1/3 to 1/4 as long at base as 1st dorsal. Pectoral noticeably small, about 1.3 times as long as base of anal, a little more than 2/3 as broad as long, with nearly straight distal margin, weakly convex outer margin, rounded corners and broad base.

Color. Described as uniformly dark brownish gray, darkest on posterior margins of pelvics, dorsals, anal and caudal; the embryo is described as slate gray.[5]

Size. This is one of the larger deep-water sharks. Specimens so far measured have ranged from 930 mm. to 2,950 mm. (9 feet 8 inches) in length. The length at which it matures is not known.

4. For recent illustrations of the teeth of a large specimen, see Jacquet (Bull. Inst. océanogr. Monaco, *36*, 1905: pl. 8).

5. Saemundsson (Vidensk. Medd. naturh. Foren. Kbh., *74*, 1922: 197).

Developmental Stages. Knowledge of its development is limited to the fact that a gravid female of nine feet contained two embryos, each about 850 mm. long.

Habits. The fact that most of the captures of specimens for which pertinent information is available have been made at depths ranging between 300 and 1,477 meters (164 and 807 fathoms) shows this to be a deep-water species, a habitat with which its uniformly dark coloration accords. But the Long Island specimen mentioned below was washed ashore on the beach and the New Jersey example was taken in a pound net, which is evidence that it occasionally wanders into shallow water, as various other deep-water fishes do. Nothing more is known of its habits.

Range. Both sides of the North Atlantic; also represented in Japanese waters by a closely allied form (*acrages* Jordan and Snyder, 1904). Apparently this Shark is rare everywhere, for positively identified specimens so far captured number only nine; these are the type specimen and two others from the coast of Portugal; one from the Cape Verde Islands; three from Iceland; one from an unspecified Atlantic locality; and one from Amagansett, Long Island, New York. It is also reported nominally from Madeira, and it is probable that an eight-foot shark taken in a pound net at Manasquam, New Jersey, in July 1936 also belonged to this species. All the numerous references to it in scientific literature are based on the foregoing.

Synonyms and References:

Pseudotriakis microdon Brito Capello, J. Sci. Math. Phys. nat. Lisboa, *1* (2), 1867: 315, 321, pl. 5, fig. 1 (type loc. Setubal, Portugal); *ibid.*, *2*, 1870: 139; Günther, Cat. Fish. Brit. Mus., *8*, 1870: 395 (Portugal); Brito Capello, Cat. Peix. Portugal, 1880: 44 (Portugal); Bean, T. H., Proc. U.S. nat. Mus., *6*, 1883: 147 (Long Island, N. York); Baird, Bull. U.S. Fish Comm., *4*, 1884: 177 (same specimen as Bean, 1883); Jordan, Rep. U.S. Comm. Fish. (1885), 1887: 794 (distrib.); Goode and Bean, Smithson. Contr. Knowl., *30*, 1895: 18, 508, pl. 5, fig. 17; Mem. Harv. Mus. comp. Zool., *22*, 1896 (same ref.); Special Bull. U.S. nat. Mus., *2*, same pagination (descr., ill. of Long Island specimen, meas.); Jordan and Evermann, Rep. U.S. Comm. Fish. (1895), 1896: 214 (distrib.); Bull. U.S. nat. Mus., 47 (1), 1896: 27; Bull. U.S. nat. Mus., 47 (4), 1900: pl. 4, fig. 14 (descr., distrib.); Bean, T. H., Bull. N. Y. St. Mus., 60, Zool. 9, 1903: 18 (Long Island spec.); de Braganza, Result. Invest. Sci. "Amelia," *2*, 1904: 28, 29 (off Portugal, depths); Prince of Monaco, Bull. Inst. océanogr. Monaco, *6*, 1904: 11 (depth); Jacquet, Bull. Inst. océanogr. Monaco, 36, 1905: 1, pl. 1–8 (anat., teeth, depth, C. Verdes); Regan, Ann. Mag. nat. Hist., (8) *1*, 1908: 464 (descr., Atlant.); Richard, Bull. Inst. océanogr. Monaco, 162, 1910: 152, fig. 109 (photo, C. Verdes); Roule, Bull. Inst. océanogr. Monaco, 243, 1912: 6 (C. Verdes); Garman, Mem. Harv. Mus. comp. Zool., *36*, 1913: 104 (descr.); Nichols and Murphy, Brooklyn Mus. Sci. Bull., (1) *3*, 1916: 6 (Long Island record); Fowler, Copeia, 30, 1916: 36; Roule, Result. Camp. sci. Monaco, *52*, 1919: 114 (C. Verdes); Saemundsson, Vidensk. Medd. naturh. Foren. Kbh., *74*, 1922: 197 (meas. embryos, size, depth, Iceland); Noronha, Ann. Carneg. Mus., *16*, 1926: 385 (Madeira); Nichols and Breder, Zoologica, N. Y., *9*, 1927: 11 (the Long Island record); Rey, Fauna Iberica Peces, *1*, 1928: 320 (descr., distrib.); Breder, Field Bk. Mar. Fishes Atlant. Coast, 1929: 12 (general); Jordan, Manual Vert. Anim. NE. U.S., 1929: 8 (general); Fowler, Proc. Pan-Pacif. sci. Congr. (1929), Java, 1930: 489 (part); Jordan, Evermann and Clark, Rep. U.S. Comm. Fish. (1928), 2, 1930: 12 (distrib.); Frade, Cons. Explor. Mer., 1932: plate not numbered (general); Daniel, Elasmobranch Fishes, Univ. Calif. Press, 1934: 54, fig. 59 (skull); Nobre, Fauna Marinha Port. Vert., *1*, 1935: 442 (descr., Portugal); Fowler, Bull. Amer. Mus. nat. Hist., *70* (1), 1936: 42 (general); Proc. Acad. nat. Sci. Philad., *89*, 1937: 303, 304 (specimen in pound net, Manasquam, N. Jersey, July 1936, ident. probable, size); White, Bull. Amer.

Mus. nat. Hist., *74*, 1937: 41, 119 (class.); Fowler, Bull. U.S. nat. Mus., 100 (*13*), 1941: 118 (part, class.).

Family TRIAKIDAE

Smooth Dogfishes

Characters. Two dorsal fins, the 1st much shorter than the caudal, its base terminating anterior to origin of pelvics; caudal less than ¼ of total length, not lunate, its lower anterior corner expanded as a low lobe only, if at all; caudal peduncle not flattened dorso-ventrally or expanded laterally, with or without a precaudal pit above, but none below; sides of trunk, anterior to anal, without longitudinal ridges; inner margins of pelvics entirely separate posterior to cloaca in both sexes; snout not greatly elongate, or jaws widely protrusible; 5th gill opening posterior to origin of pectoral; gill arches without rakers and not interconnected by a sieve of modified denticles; nostril either separate from mouth or connected with latter by a groove, its anterior margin without barbel; spiracles present or absent; lower eyelid with a longitudinal fold externally, but without internal nictitating membrane; jaws with or without labial furrows; teeth small, rounded, or with 3 to 4 distinct cusps; several series functional along entire length of jaw; dermal denticles ovate, lanceolate or tridentate; head of normal shape and not widely expanded laterally. Rostral cartilages 3, united terminally; most of the radials of pectoral on metapterygium; meso- and propterygia much smaller; meso- and metapterygia separated by a foramen, at least in some cases; heart valves in 3 rows. Development either ovoviviparous, or viviparous with a well developed yolk-sac placenta.

Genera. These small sharks, of shoal or moderate depths, are widely distributed in tropical and warm temperate belts of all oceans. They are perfectly harmless. The group is very close to the Carcharhinidae where it is placed as a subfamily by Fowler,[1] although it more resembles the Orectolobidae and the Scyliorhinidae in its dentition. However, the triakids seem sufficiently different both from the typical carcharhinids (with respect to the teeth and the absence of a true nictitating membrane), as well as from the orectolobids and scyliorhinids, to be ranked as a distinct family for convenience, if for no better reason.

The family, as defined above, corresponds to the Galeorhinidae as used by Garman and by many subsequent authors. But the latter name is not appropriate in the present connection, because *Galeorhinus* Blainville, 1816, from which it is derived, is the correct name of the so-called "Topes," a genus of carcharhinid sharks (see footnote 4, p. 264), and not of the Smooth Dogfishes (*Mustelus*) to which Garman[2] and others have applied it. For this reason Bigelow and Schroeder[3] substituted Mustelidae for the sharks now in question, overlooking the fact that this name had been in common use for a family

1. Bull. U.S. nat. Mus., 100 (*13*), 1941: 127.
2. Mem. Harv. Mus. comp. Zool., *36*, 1913: 3, 169.
3. Canad. Atlant. Fauna, biol. Bd. Canad., 12[e], 1934: 6.

of carnivorous mammals (weasels, etc.) for many years previous. To avoid the confusion that would result from the use of this family name for sharks, as well as for mammals, it seems wiser to replace it with "Triakidae" White,[3a] first used by that author in a more restricted sense, but subsequently expanded by her.[3b]

Key to Genera

1a. Nostril connected with mouth by a groove. *Scylliogaleus* Boulenger, 1902.
Natal, southeastern Africa.

1b. Nostril entirely separate from mouth.

2a. Teeth low, rounded, or with only slightly sinuous cutting edge.
Mustelus Link, 1790, p. 240.

2b. Teeth somewhat compressed, with 3 to 5 pointed cusps.

3a. No labial furrow at corner of mouth. *Eridacnis* H. M. Smith, 1913.
Philippines.

3b. A labial furrow at corner of mouth.

4a. Anterior margin of nostril expanded as a well developed and conspicuous barbel. *Fur* Whitley, 1943.
Australia.[4]

4b. Anterior margin of nostril not expanded as a well developed barbel.

5a. Spiracle moderately large, easily seen.
Triakis Müller and Henle, 1838, p. 235.[5]

5b. Spiracle minute or absent.

6a. Caudal peduncle with precaudal pit above; lower anterior corner of caudal expanded as a lobe.
Triaenodon Müller and Henle, 1837.
Tropical Indian Ocean, Red and Arabian Seas; India and Ceylon, Malaysia, Melanesia and Polynesia, Hawaiian Islands, Cocos Island, and Panama.

6b. Caudal peduncle without precaudal pit; lower anterior corner of caudal not expanded as a lobe.
Leptocharias Müller and Henle, 1838.
West Africa.

Remarks. The separation between the only two triakid genera that are known to occur in the Atlantic, *Triakis* and *Mustelus*, is partly bridged by *T. maculata* of the west coast of South America, the teeth of which are only a little more definitely cuspidate than those of *Mustelus dorsalis* of the same geographic province.[6]

3a. Amer. Mus. Novit., 879, 1936: 19. 3b. Bull. Amer. Mus. nat. Hist., 74, 1937: 121.
4. Whitley, Aust. Zool., 10 (2), 1943: 167; Rec. Aust. Mus., 7 (4), 1943: 397; Aust. Zool., 10, 1944: 260, pl. 3.
5. Including *Rhinotriacis* Gill, 1862, *Calliscyllium* Tanaka, 1912, and *Hemitriakis* Herre, 1923, all of which seem clearly referable to the old genus *Triakis* (see p. 235).
6. For discussion, see Bigelow and Schroeder (Proc. Boston Soc. nat. Hist., 41 [8], 1940: 431).

Genus *Triakis* Müller and Henle, 1838

Triakis Müller and Henle, Charlesworth's Mag. Nat. Hist., *2*, 1838: 36; Arch. Naturg., Jahrg. 4, *1*, 1838: 84 (no species mentioned); Plagiost., 1841: 63; type species, *T. scyllium* Müller and Henle. Japan.

Generic Synonyms:

Triacis Günther, Cat. Fish. Brit. Mus., *8*, 1870: 384; emended spelling for *Triakis* Müller and Henle, 1838.
Rhinotriacis Gill, Proc. Acad. nat. Sci. Philad., 1862: 486; type species, *R. henlei* Gill. California.
Calliscyllium Tanaka, Fish. Japan, *10*, 1912: 17; type species, *C. venusta* Tanaka. Japan.
Hemitriakis Herre, Philipp. J. Sci., *23*, 1923: 70; type species, *H. leucoperiptera* Herre. Philippines.

Generic Characters. Nostrils entirely distinct from mouth, their anterior margins without barbel; teeth compressed, with 2, 3 or 4 pointed cusps, the axial longest; a labial furrow on each jaw at corner of mouth; spiracles moderately large, easily seen. Characters otherwise those of the family.

Range. Until recently the genus was known only from the Indian Ocean (including the Red and Arabian Seas), Malaysia, Melanesia and Polynesia, the western North Pacific, and the eastern Pacific, north and south. Within the last few years it has been encountered in Cuban waters (p. 240).

Species. The members of this genus show a wide gradation in the number of dental cusps and in the relative length and acuteness of the median member;[7] the dermal denticles, too, may be either simple-lanceolate, partly or weakly tridentate, or strongly so;[8] and the labial furrows are very prominent in some but very short and inconspicuous in others.[9] The several species also differ in the relative positions of the dorsal and anal fins, the size and shape of the caudal, and in the proximity of nostril to mouth, these being the characters employed in the accompanying Key because they are the most conspicuous.

Key to Species

1a. Origin of 2nd dorsal about over origin of anal or behind it, its tip over or posterior to tip of latter; labial furrows very short.

  2a. Anterior margin of nostril close to mouth; rear tip of 1st dorsal a little anterior to origin of pelvics; tips of pelvics anterior to origin of 2nd dorsal by a distance considerably longer than base of 2nd dorsal. *venusta* Tanaka, 1912.
  Japan.

  2b. Anterior margin of nostril far from mouth; rear tip of 1st dorsal posterior to origin of pelvics; tips of pelvics anterior to origin of 2nd dorsal by a distance only about 1/3 as long as base of 2nd dorsal.
  barbouri Bigelow and Schroeder, 1943, p. 236.

7. Median cusp long and slender in *scyllia*, *semifasciata* and *barbouri*; only moderately so in *venusta*; short and blunter in *maculata* and *henlei*.
8. Denticles simple-lanceolate in *scyllia*, *semifasciata* and *henlei*; weakly tridentate in *venusta*; partly so in *maculata*; strongly tridentate in *barbouri*.
9. Labial furrows long and conspicuous in *scyllia*, *semifasciata*, *henlei* and *maculata*; very short in *barbouri* and *venusta*.

1b. Origin of 2nd dorsal considerably anterior to origin of anal, its tip terminating anterior to tip of latter; labial furrows long, conspicuous.

3a. Rear end of base of 1st dorsal about over origin of pelvics.

maculata Kner and Steindachner, 1867.[10]
Peru.

3b. Rear end of base of 1st dorsal so far in advance of origin of pelvics that its tip is over origin of latter.

4a. Lower anterior corner of caudal expanded as triangular lobe, with pointed tip; pectoral also pointed, with deeply concave distal margin.

leucoperiptera Herre, 1923.
Philippines.

4b. Lower anterior corner of caudal rounded and only slightly expanded; pectoral rounded at tip, its distal margin only weakly concave, or nearly straight.

5a. Origin of 1st dorsal anterior to inner corner of pectoral, its tip anterior to origin of pelvics; plain colored. *henlei* Gill, 1862.
California.

5b. Origin of 1st dorsal over or behind inner corner of pectoral, its tip about over origin of pelvics; back and sides with dark bars, blotches or spots.

6a. Caudal about as long as from tip of snout to axil of pectoral; teeth in sides of jaws similar to those in front, tricuspidate, the median cusp straight although directed somewhat obliquely outward.

scyllia[10a] Müller and Henle, 1841.
Japan, Korea, Formosa, China.[11]

6b. Caudal shorter than from tip of snout to axil of pectoral by a distance about equal to that between nostrils; teeth in sides of jaws noticeably unlike those in front, usually with the inner lateral cusp lacking, and the median cusp strongly curved outward.

semifasciata Girard, 1854.
Oregon to Magdalena Bay, Lower California.

Triakis barbouri Bigelow and Schroeder, 1944

Figure 41

Study Material. About 100 specimens, males and females, 225 to 338 mm. long, including the type, collected off the north coast of Cuba in March 1938 and May 1939 by the research ship "Atlantis" (Harv. Mus. Comp. Zool.).

10. Including *nigromaculata* Evermann and Radcliffe, 1917; *nigromaculata* has sometimes been referred to *Mustelus*, but our own examination of Peruvian specimens inclines us to follow Garman (Mem. Harv. Mus. comp. Zool., *36*, 1913: 167) in retaining it in *Triakis* where it was placed by its discoverers, its teeth being definitely cuspidate, though usually with only two cusps, rarely three (see Bigelow and Schroeder, Proc. Boston Soc. nat. Hist., *41* [8], 1940: 428, pl. 17, fig. M).

10a. The original spelling, *scyllium*, is emended here to accord with the gender of the Greek noun (ἀκίς) on which the generic name *Triakis* was based.

11. Old reports of this species from South Australia probably were not correct; see Whitley (Fish. Aust., *1*, 1940: 115).

Distinctive Characters. The very short labial furrows of *barbouri* are enough to separate it at a glance from all other members of the genus, except from *venusta,* from which it differs in its strongly tridentate dermal denticles, in the fact that the anterior marginal expansion of its nostril is far in advance of the front of the mouth, and in its teeth. It is not likely to be confused with any other Atlantic shark, being the only member of its genus yet known from this ocean.

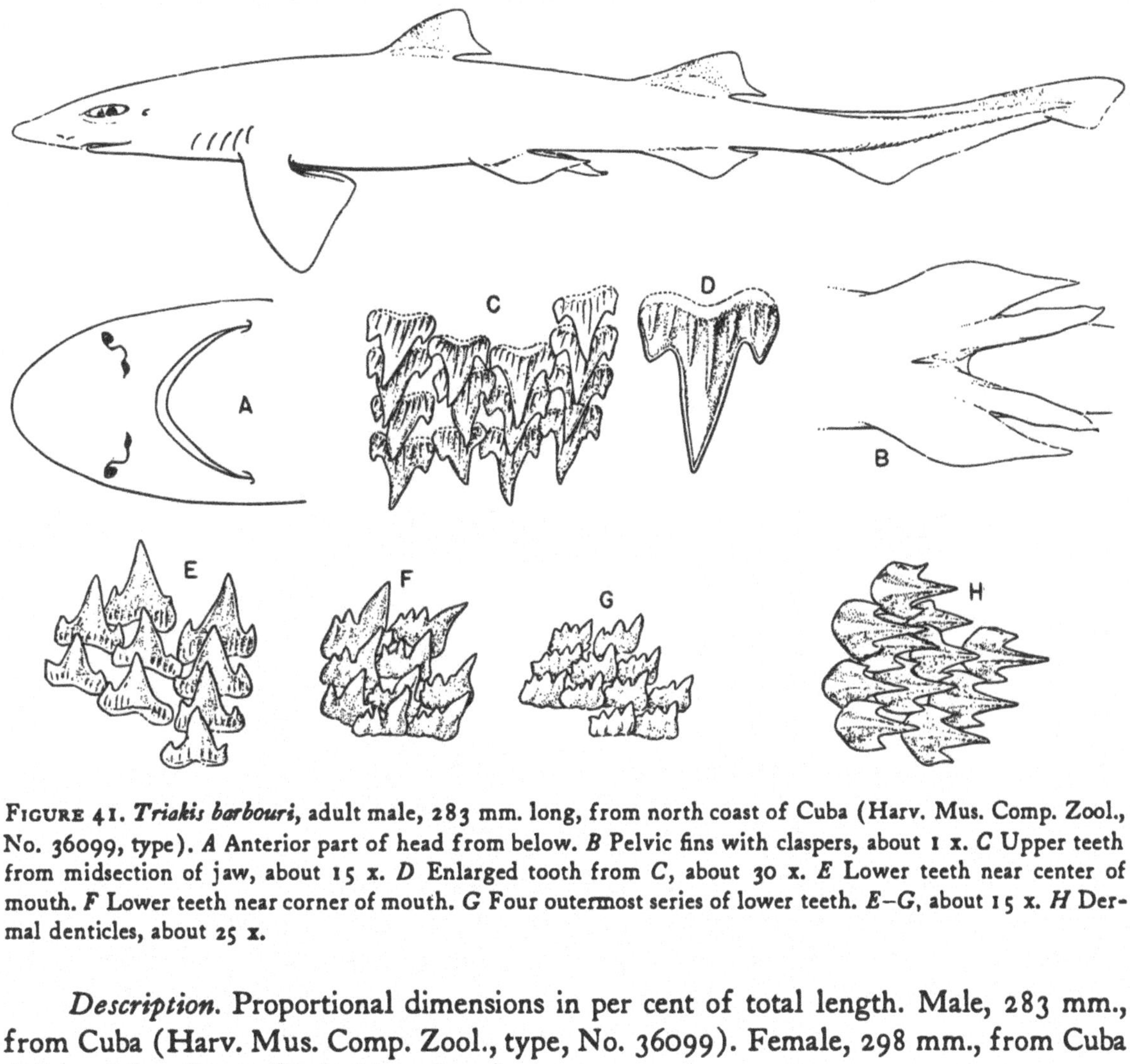

FIGURE 41. *Triakis barbouri,* adult male, 283 mm. long, from north coast of Cuba (Harv. Mus. Comp. Zool., No. 36099, type). *A* Anterior part of head from below. *B* Pelvic fins with claspers, about 1 x. *C* Upper teeth from midsection of jaw, about 15 x. *D* Enlarged tooth from *C*, about 30 x. *E* Lower teeth near center of mouth. *F* Lower teeth near corner of mouth. *G* Four outermost series of lower teeth. *E–G*, about 15 x. *H* Dermal denticles, about 25 x.

Description. Proportional dimensions in per cent of total length. Male, 283 mm., from Cuba (Harv. Mus. Comp. Zool., type, No. 36099). Female, 298 mm., from Cuba (Harv. Mus. Comp. Zool., No. 36151).

Trunk at origin of pectoral: breadth 9.5, 9.5; height 8.8, 8.4.

Snout length in front of: outer nostrils 3.5, 3.5; mouth 5.6, 5.4.

Eye: horizontal diameter 3.9, 4.3.

Mouth: breadth 7.1, 7.2; height 3.5, 3.3.

Nostrils: distance between inner ends 2.9, 2.9.

Labial furrow lengths: upper 0.5, 0.6; lower 0.5, 0.7.

Gill opening lengths: 1st 1.8, 1.7; 2nd 1.8, 1.8; 3rd 1.8, 1.8; 4th 1.8, 1.7; 5th 1.4, 1.2.

First dorsal fin: vertical height 4.9, 5.4; length of base 6.9, 7.4.

Second dorsal fin: vertical height 4.2, 5.2; length of base 8.8, 8.7.

Anal fin: vertical height 2.9, 3.5; length of base 8.1, 9.1.

Caudal fin: upper margin 27.0, 27.8; lower anterior margin 9.9, 8.4.

Pectoral fin: outer margin 13.0, 13.8; inner margin 8.1, 8.5; distal margin 9.5, 10.7.

Distance from snout to: 1st dorsal 31.4, 31.2; 2nd dorsal 56.5, 56.4; upper caudal 73.0, 72.2; pectoral 20.4, 19.4; pelvics 39.2, 42.6; anal 57.8, 56.7.

Interspace between: 1st and 2nd dorsals 16.9, 17.1; 2nd dorsal and caudal 7.6, 7.7; anal and caudal 5.0, 4.7.

Distance from origin to origin of: pectoral and pelvics 22.2, 24.3; pelvics and anal 16.9, 14.6.

Trunk slender, its height abreast of origin of 1st dorsal only about 1/7 as great as its length to origin of caudal, its dorsal profile only slightly arched, its ventral profile nearly straight except in gravid females. Tail sector about 1/6 longer than body sector to cloaca. Caudal peduncle laterally compressed, about twice as high as broad at origin of caudal, without lateral keels or precaudal pit above or below. Dermal denticles close-spaced and usually overlapping, with short pedicels and horizontal blades, the latter regularly tridentate, the median tooth narrow and much the longest; 3 ridges, the median ridge flat-topped distally but rounded proximally.

Head to 5th gill opening about 1/5 of total length, flattened above, its dorsal profile sloping steeply downward. Snout in front of mouth about 1/3 of length of head to origin of pectoral, or a little less, its tip thin and broadly ovate. Eye oval, about 3 times as long as high, and noticeably large, its horizontal diameter about 2/3 as long as snout in front of mouth, the larger specimens with a well marked subocular fold which is separated from the margin of the lower eyelid at both ends when the latter is drawn down, but this is only faintly indicated when the lower eyelid is drawn up over eyeball, or in small specimens even when the eye is wide open. Spiracle oval, its longest diameter only about 1/5 to 1/6 as long as horizontal diameter of eye, situated slightly below level of center of latter and behind it by a distance slightly less than 1/2 an eye's length. Gill openings deeply concave anteriorly, the 4th to 5th over the origin of the pectoral, the 1st, 2nd, 3rd, and 4th about 1/2 as long as the horizontal diameter of the eye, the 5th considerably shorter and situated noticeably higher on the sides than the 1st to 4th; the interspaces between gill openings about 1/2 to 2/3 as wide as lengths of latter. Nostril moderately oblique, its inner corner a little less than 1/3 as far from front of mouth as from tip of snout, its anterior margin expanded as a low, subquadrate lobe. Mouth broadly ovate, occupying about 3/4 of breadth of head, and a little less than 1/2 as high as broad, with a very short labial furrow on each jaw.

Teeth $\frac{62 \text{ to } 62}{60 \text{ to } 62}$, closely crowded, with those of successive rows overlapping; 3 or 4 rows functional in front of upper jaw, but only 2 or 3 rows near corners; 3 to 4 rows functional in front of lower jaw, and 5 to 6 rows toward the angles where the serial arrangement is increasingly oblique. Upper teeth with 3 cusps, the axial erect or slightly curved and much the longest, with irregular longitudinal striae basally, those toward corners of mouth somewhat smaller, with relatively shorter median cusp and broader base than those in front of mouth; lowers similar to uppers toward center of mouth, except that the outer lateral cusp may be minute or lacking, the teeth along sides of jaw usually with 2 or 3 basal cusps on the inner side, but none on the outer side, so that the cusp that is primarily median is at the outer edge of the tooth.

Origin of 1st dorsal a little behind inner corner of pectoral, its base about as long as from center of eye to 1st gill opening, its anterior margin very slightly convex, its apex rounded, its posterior margin slightly concave, the free rear corner a little more than ½ as long as its base, the rear end of latter a little anterior to origin of pelvics. Second dorsal similar to 1st, and about as high, but about 1½ times as long at base, its origin about over that of anal and slightly closer to origin of caudal than to rear tip of 1st dorsal. Caudal ¼ to ⅓ of total length, noticeably narrow, obliquely truncate at tip, with well marked subterminal notch, its rounded, lower anterior corner not expanded to form a distinct lobe. Anal similar to 2nd dorsal in shape, and about as long at base, but only about ⅔ as high. Interspace between anal and caudal a little shorter than base of anal, that between anal and tips of pelvics a little more than ½ as long as base of anal. Pelvics with nearly straight anterior and distal margins but slightly concave inner margins, and broadly rounded outer corners, the inner corners narrowly elongate in both sexes, with pointed tips. Pectoral between 2 and 3 times as large in area as 1st dorsal, about ⅔ as broad as long, with weakly convex outer and inner margins, very weakly concave distal margin and narrowly rounded corners.

Color. Upper parts of trunk, caudal, dorsals and pectorals pale gray; lower surface grayish white. At least in some specimens the anterior edge of the dorsals, as well as that of the caudal midway of its length, is marked with faintly defined smoky blotches. The embryo just before birth shows more extensive blotches in these same general regions, in addition to similar blotches on the upper side of the caudal peduncle, on the back midway between the two dorsals, and on the side just posterior to the pectoral; also, a sooty area, irregular in outline, extending upward obliquely from the region of the gill openings to the dorsal surface of the nape.

Size. Evidently this is one of the smallest of sharks. The fact that a female of only 303 mm. contained two large embryos, and that the claspers in a 338-mm. male are very large and extend rearward as far as the origin of the anal, suggests a maximum length of perhaps not more than about 350 to 400 mm., or about one foot two to four inches.

Developmental Stages. Two embryos, 90 to 100 mm. long, and nearly ready for birth, contained in a female of 303 mm., resemble the parent, except for the facts that the caudal is relatively longer with slightly convex lower margin, that the origin of the 1st dorsal is relatively somewhat farther back, that the dorsals are more broadly rounded, that the tips of the pelvics are relatively much shorter, and that the eyes are relatively much larger (embryonic characters to be expected); the color pattern is not only more pronounced but more extensive, as noted above. There is no placental attachment between embryo and mother.

Habits. The recorded depths of capture of this little Shark range only between 235 and 335 fathoms, but since the trawls that yielded them were hauled up to the surface open, the specimens may have been swimming at some mid-depth and not on bottom. Nothing more is known of its habits.

Range. So far known only off the north coast of Cuba in the offing of the province of Santa Clara, where it must be tolerably plentiful, judging from the number of specimens taken.

Synonym and References:

Triakis barbouri Bigelow and Schroeder, Proc. New Engl. zool. Cl., *23*, 1944: 27, pl. 8 (descr., ill., Cuba); Bigelow and Schroeder, Guide Comm. Shark Fish., Anglo Amer. Caribb. Comm., Wash., 1945: 124 (descr., ill., range).

Genus *Mustelus* Link, 1790[1]

Smooth Dogfishes

Mustelus Link, Mag. Physik Naturg. Gotha, *6* (2), 1790: 31; type species, *Squalus mustelus* Linnaeus, 1758.

Generic Synonyms:

Galeus (in part) Rafinesque, Indice Ittiol. Sicil., 1810: 46 (includes *Squalus mustelus* Linnaeus, 1758); Leach, Mem. Werner. Soc. Edinb., *2*, 1812: 62 (includes *Squalus mustelus* Linnaeus, 1758); not *Galeus* Rafinesque, Carratt. Gen. Nuov. Sicil., 1810: 13; see p. 214.

Galeorhinus (in part) Blainville, Bull. Soc. philom. Paris, 1816: 121.

Myrmillo Gistel, Naturg. Tier., 1848: 4; substitution for *Mustelus* Link, 1790; not seen.

Pleuracromylon Gill, Proc. Acad. nat. Sci. Philad., 1864: 148; type species, *Mustelus laevis* Risso, 1826.

Rhinotriakis (in part) Jordan and Gilbert, Bull. U.S. nat. Mus., *16*, 1883: 60; includes *Mustelus laevis* Risso, 1826.

Galeus Jordan and Evermann, 1896, Bull. U.S. nat. Mus., 47 (1), 1896: 29; for *Mustelus dorsalis* and *M. californicus* Gill, 1864; not *Galeus* Rafinesque, Carratt. Gen. Nuov. Sicil., 1810: 13.

Cynias Gill, Proc. U.S. nat. Mus., *26*, 1903: 960; type species, *Squalus canis* Mitchill, 1815.

Cynois Fowler, Proc. Acad. nat. Sci. Philad., *60*, 1908: 55; evident misprint for *Cynias*.

Galeorhinus Garman, Mem. Harv. Mus. comp. Zool., *36*, 1913: 169; same as *Mustelus* Link, 1790, substituted because of the diversity of application of the latter; not *Galeorhinus* Blainville, 1816, the type species of which is *Squalus galeus* Linnaeus, 1758, desig. by Gill, Ann. N. Y. Lyc., 7, 1862: 402.

1. The International Commission on Zoological Nomenclature has ruled that *Mustelus* Link, 1790, is applicable to sharks, in spite of the fact that *Mustela* was earlier used by Linnaeus, 1758, for weasels. See Opinion 93 (Smithson. misc. Coll., *73* [4], 1929: 8; and Science, N. S. *65*, 1927: 300).

Generic Characters. Nostrils entirely separate from mouth; teeth numerous, alike in the 2 jaws, low, rounded, or with somewhat sinuous cutting edge, in mosaic arrangement, several rows functional; spiracles present, of at least moderate size; corners of mouth with a strongly marked labial furrow on each jaw; caudal with well marked subterminal notch, its lower anterior corner expanded as a short lobe in some species but not in others; dorsals, anal and pelvics with their posterior corners considerably produced and their posterior margins moderately concave; origin of anal below 2nd dorsal; rear end of base of 1st dorsal considerably anterior to origin of pelvics; in some species there is a placental attachment between mother and embryo, but in others this is lacking. Characters otherwise those of the family.

Remarks. The members of the genus are separated from others of this family by their low rounded teeth, arranged in mosaic.

Range. Widely distributed in coastal waters in the warm and warm-temperate belts of the Atlantic and Indo-Pacific, north and south; Gulf of Maine to northern Argentina in the western Atlantic; Shetland Islands, North Sea and mouth of the Baltic to tropical West Africa (Senegambia, Cameroons) in the eastern Atlantic, including the Mediterranean, Madeira and the Canaries; also South Africa; western Indian Ocean (Natal); Red and Arabian Seas; Australia, Tasmania and New Zealand; Indo-China, China, Korea and Japan; west coasts of North, Central, and South America from northern California to northern Chile.

Fossil Teeth. Oligocene to Pliocene, Europe.

Species. Although the members of this genus resemble one another very closely in general appearance, certainly two, and probably three, recognizably distinct species occur in the eastern Atlantic, four in the western Atlantic, four along the western coasts of America; also at least two or three and perhaps more in the western Pacific, in the Australian–New Zealand region and in the Indian Ocean with its tributary seas. We have already published a comparison of the western Atlantic species with those of the eastern Atlantic and eastern Pacific.[3] The next step would be to compare these with their western Pacific–Australian–Indian Ocean relatives. However, since we lack sufficient material for this, the following key is limited to the Atlantic and to the eastern side of the Pacific.

Key to Atlantic and Eastern Pacific Species

1a. Lower anterior corner of caudal expanded as a pointed lobe directed rearward (Fig. 43 D).

  2a. Upper labial furrow as long as lower, or longer; denticles on shoulders loosely spaced. *norrisi* Springer, 1939, p. 254.

  2b. Upper labial fold shorter than lower; denticles on shoulders closely imbricate. *lunulatus* Jordan and Gilbert, 1882.
  Southern California to Colombia.

3. Bigelow and Schroeder, Bull. Boston Soc. nat. Hist., *41* (8), 1940: 417.

1b. Lower anterior corner of caudal broadly rounded.

3a. Midpoint of base of 1st dorsal about midway between origin of pelvics and inner corner of pectoral when latter is laid back.

4a. Teeth noticeably asymmetrical, their outer margins deeply concave, their tips moderately pointed; dermal denticles strongly ridged, usually to their tips. *punctulatus* Risso, 1826.
Mediterranean.

4b. Teeth nearly symmetrical, their outer margins only weakly concave, their tips broadly rounded; dermal denticles ridged only at their bases, their tips smooth, transparent. *californicus* Gill, 1864.
Northern California to Cape San Lucas and Gulf of California.

3b. Midpoint of base of 1st dorsal much closer to inner corner of pectoral when this is laid back than to origin of pelvics.

5a. Teeth symmetrical, evenly rounded in both jaws.

6a. Lower anterior corner of caudal not expanded as a lobe; snout in front of mouth about 4 times as long as horizontal diameter of eye. *fasciatus* Garman, 1913.
Southern Brazil, Uruguay, northern Argentina.

6b. Lower anterior corner of caudal expanded as a low, rounded lobe; snout in front of mouth only 2 to 3 times as long as horizontal diameter of eye.

7a. Horizontal diameter of eye considerably longer than distance between nostrils, and about ½ as long as snout in front of mouth; denticles on shoulders strongly ridged, usually to their tips, and opaque. *asterias* Cloquet, 1819.
Eastern Atlantic, Mediterranean.

7b. Horizontal diameter of eye slightly shorter than distance between nostrils, and only ⅓ as long as snout in front of mouth; denticles on shoulders ridged only near their bases and so transparent that the pedicels are visible from without. *mento*[4] Cope, 1877, p. 259.

5b. Teeth noticeably asymmetrical, their margins more or less concave.

8a. Sides in adults conspicuously marked with white spots. *asterias* Cloquet, 1819.
Eastern Atlantic, Mediterranean.

8b. Sides in adults plain colored.

9a. Lower anterior corner of caudal not appreciably expanded even in adult. *schmitti* Springer, 1938.
Southern Brazil, Uruguay and northern Argentina.

4. Including *abbotti* Evermann and Radcliffe, 1917, and *edulis* Perez Canto, 1886.

9b. Lower anterior corner of caudal noticeably expanded as a rounded lobe from shortly after birth.

10a. Teeth high-crowned, their outer margins deeply concave even in adults, and sometimes the inner as well.

dorsalis Gill, 1864.
Pacific Panama, Peru.

10b. Teeth low-crowned, their tips bluntly rounded, their margins only slightly concave in adults.

11a. Inner margin of pectoral 60 to 63% as long as outer; distance between nostrils only about 1.3 times as long as upper labial furrow; dermal denticles on shoulders ridged only near their bases, their tips smooth.

mustelus Linnaeus, 1758.
Eastern Atlantic, Mediterranean.

11b. Inner margin of pectoral only about 48 to 54% as long as outer; distance between nostrils at least 1.3 to 1.7 times as long as upper labial furrow; denticles usually strongly ridged to their tips.

canis Mitchill, 1815, p. 244.

Key to the Western Atlantic Species[5]

1a. Head to origin of pectoral considerably longer than interspace between 1st and 2nd dorsals. *fasciatus* Garman, 1913, p. 256.

1b. Head to origin of pectoral not longer than interspace between 1st and 2nd dorsals.

2a. Lower anterior corner of caudal sharp-pointed and directed rearward (Fig. 43 D). *norrisi* Springer, 1939, p. 254.

2b. Lower anterior corner of caudal broadly rounded.

3a. Horizontal diameter of eye only about 60% as long as 3rd gill opening; teeth symmetrical, evenly convex in outline (Fig. 42 J).

mento Cope, 1877, p. 259.

3b. Horizontal diameter of eye about as long as 3rd gill opening; teeth somewhat asymmetrical, their cutting edges more or less concave.

4a. Lower anterior corner of caudal not appreciably expanded, even in adult (Fig. 42 F); distance between nostrils only about 1/3 as great as breadth of mouth. *schmitti* Springer, 1939, p. 261.

4b. Lower anterior corner of caudal noticeably expanded as a rounded lobe from shortly after birth (Fig. 42); distance between nostrils considerably more than 1/2 as great as breadth of mouth.

canis Mitchill, 1815, p. 244.

5. This simplified Key is offered for assistance in the identification of species from the American Atlantic coast.

Mustelus canis (Mitchill), 1815

Smooth Dogfish

Figures 42 A–E, 43 C

Study Material. Numerous specimens from southern New England and New York, newborn to adult, living, fresh-caught, and preserved; also preserved specimens from South Carolina; Galveston, Texas; Havana, Cuba; Jamaica; Trinidad (embryo); Bermuda; and Brazil, including three embryos from Rio de Janeiro.

Distinctive Characters. M. canis is very closely allied to *M. mustelus* of the eastern Atlantic, but is separable from the latter by the narrower pectoral fin, relatively greater distance between the nostrils and the much more strongly sculptured denticles. Among western Atlantic species of the genus it is distinguished from *schmitti* in that the lower anterior corner of its caudal forms a lobe and that the distance between its nostrils is relatively greater, that is, approximately half as great as the breadth of the mouth (only about a third as great in *schmitti*); from *norrisi* in that its lower caudal lobe is rounded rather than pointed and that the midpoint of its first dorsal is as close to the axil of the pec-

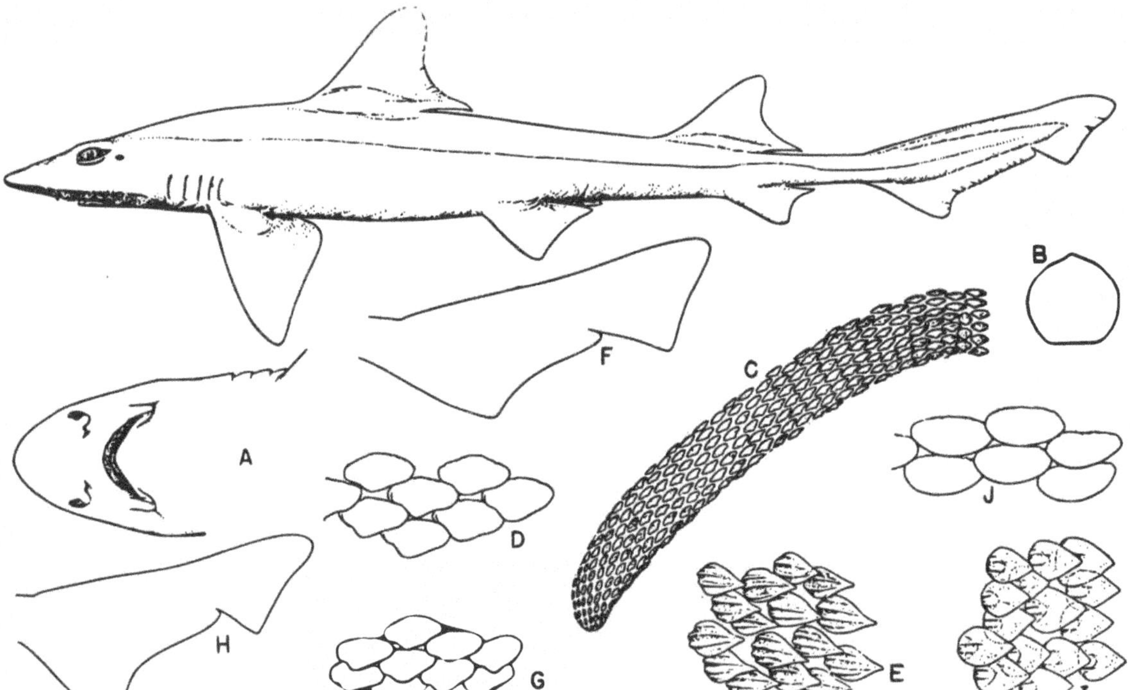

FIGURE 42. *Mustelus canis*, male, 781 mm. long, from Woods Hole, Massachusetts (Harv. Mus. Comp. Zool., No. 35245). *A* Head to pectorals. *B* Cross section of trunk, midway between the two dorsal fins, to show mid-dorsal ridge. *C* Dentition of right-hand side of upper jaw, about 2 x. *D* Teeth of another specimen of about the same size, about 7 x. *E* Dermal denticles of a female, about 678 mm. long, from Woods Hole, Massachusetts, about 22 x. *F* Caudal fin of *Mustelus schmitti*, 578 mm. long, from Muldonado, Uruguay (Harv. Mus. Comp. Zool., No. 530), for comparison with *M. canis*. *G* Upper teeth of same, about 7 x. *H* Caudal fin of *M. mento*, 1,024 mm., from Payta, Peru. *I* Denticles of same, about 22 x. *J* Upper teeth of same, about 7 x.

toral as to the origin of the pelvics; from *fasciatus* by the shape of the teeth (cf. Fig. 42 D with 43 A), by its lobed lower caudal, and by its plain coloration; from *mento*[6] by a much larger eye relative to the gill openings (the horizontal diameter of its eye being about as long as the third gill opening but only about 60% as long in *mento*), by the distance between its nostrils which is at least approximately half as great as the width of the mouth (considerably less than ½ that great in *mento*), by a relatively smaller pectoral with inner margin only about 1.3 times as long as snout (1.8 times as long in *mento*), by its more strongly sculptured and more opaque denticles (cf. Fig. 42 E with 42 I) and by the shapes of its teeth (cf. Fig. 42 D with 42 J). Adults of *canis* are plain colored also, whereas those of *mento* often are white-spotted (p. 260).

Description. Proportional dimensions in per cent of total length. Male, 781 mm., from Buzzards Bay, Mass. (Harv. Mus. Comp. Zool., No. 35245). Female, 1,231 mm., same locality.

Trunk at origin of pectoral: breadth 10.0, 10.3; height 9.9, 10.1.
Snout length in front of: outer nostrils 3.8, 3.2; mouth 6.4, 5.4.
Eye: horizontal diameter 2.8, 2.3.
Mouth: breadth 5.5, 5.5; height 2.7, 2.3.
Nostrils: distance between inner ends 2.9, 2.6.
Labial furrow length: upper 2.0, 1.9; lower 1.5, 1.4.
Gill opening lengths: 1st 2.2, 2.7; 2nd 2.4, 2.8; 3rd 2.6, 3.2; 4th 2.6, 3.2; 5th 2.0, 2.4.
First dorsal fin: vertical height 9.5, 8.3; length of base 11.1, 12.3.
Second dorsal fin: vertical height 6.8, 6.7; length of base 8.9, 10.6.
Anal fin: vertical height 3.6, 4.2; length of base 5.4, 5.5.
Caudal fin: upper margin 21.2, 18.8; lower anterior margin 8.7, 7.0.
Pectoral fin: outer margin 13.6, 13.9; inner margin 7.4, 6.7; distal margin 12.4, 12.5.
Distance from snout to: 1st dorsal 26.8, 28.6; 2nd dorsal 59.7, 61.7; upper caudal 78.8, 81.2; pectoral 18.1, 19.1; pelvics 41.5, 47.6; anal 64.3, 68.0.
Interspace between: 1st and 2nd dorsals 22.0, 22.0; 2nd dorsal and caudal 10.5, 9.5; anal and caudal 6.8, 6.7.
Distance from origin to origin of: pectoral and pelvics 24.3, 29.2; pelvics and anal 22.4, 19.8.

Trunk slender and tapering rearward, its height at origin of 1st dorsal about ⅐ its length to origin of caudal, the midline of the back with a low but sharp-edged and unmistakable dermal ridge running rearward from about opposite the 1st or 2nd gill opening nearly to the origin of the caudal. Trunk sector to cloaca a little shorter than tail sector. Caudal peduncle subcylindrical, without lateral ridges or precaudal pit either above or below. Dermal denticles irregularly spaced, sometimes overlapping but sometimes with

6. *Mento* may occur in Argentine waters, see p. 260.

areas of skin visible between, the blades nearly horizontal, on short pedicels, lanceolate, strongly sculptured with 2 to 6 longitudinal ridges, the median pair flanking the axis of the blade and usually extending out to its extreme tip, the basal part so opaque that the pedicel is not visible from without, even in fresh specimens, their margins usually entire but sometimes weakly notched between the tips of the ridges, and often irregularly worn.

Head flattened above, its dorsal outline nearly straight, sloping to thin-tipped snout. Snout broadly ovate at tip, its length in front of mouth a little greater than 1/3 the length of head to origin of pectoral. Eye oval, its horizontal diameter slightly shorter than distance between nostrils, the subocular fold separate at both ends from the margin of the lower eyelid in small specimens, but joining the latter at the anterior corner of the eye by the time a total length of about 700 mm. is reached; in very large specimens it becomes continuous with the margin of the upper lid at both corners. Spiracle oval, about 1/8 to 1/7 as long as horizontal diameter of eye, about on a level with center of latter and behind it by a distance 1/2 as long as horizontal diameter of eye. Third and fourth gill openings slightly the longest, about 1 1/3 times as long as horizontal diameter of eye, the 5th considerably the shortest, their outlines nearly straight or weakly concave anteriorly, the 5th the most so, the 4th gill opening above origin of pectoral. Nostril about as long as horizontal diameter of eye, oblique, its inner corner about 1/3 as far from front of mouth as from tip of snout, its anterior margin expanded as a well developed subpentagonal lobe with blunt tip. Mouth occupying between 1/2 and 2/3 of breadth of head, ovate, about twice as broad as high. A strongly marked labial furrow on each jaw, the upper usually considerably the longer in northern specimens, but sometimes only about as long as the lower, or even slightly shorter, in West Indian and South American races.[7]

Teeth $\frac{74}{80}$ in specimen counted, usually 5 to 7 rows functional, the cutting edges with bluntly rounded apices directed somewhat outward (*i.e.*, asymmetrical), their margins slightly concave (the outer usually the more deeply so) or sometimes even notched, especially in small specimens.

First and second dorsals similar in shape, with very slightly convex or nearly straight anterior margins, narrowly rounded apices, deeply concave rear margins and acute rear corners, the free lower edges about 1/3 as long as the bases. Origin of 1st dorsal about over midpoint of inner margin of pectoral, the midpoint of its base about as close to axil of pectoral as to origin of pelvics. Second dorsal nearly or quite as long as 1st at base, but only about 2/3 to 4/5 as high, its origin at a perpendicular about midway between tips of pelvics and origin of anal, its rear tip a little anterior to rear tip of anal. Caudal about 1/5 of total length, with truncate tip, the terminal sector noticeably large or a little more than 1/3 of total length of caudal, with well marked subterminal notch, the lower anterior contour expanded as a low but well marked lobe, with broadly rounded apex. Anal only about 2/3 as long at base as 2nd dorsal, and about 1/2 as high, with less deeply concave posterior margin and shorter free rear corner, its origin about under midpoint of base of 2nd dorsal.

7. See Bigelow and Schroeder, Proc. Boston Soc. nat. Hist., *41* (8), 1940: 422.

Pelvics with nearly straight anterior and weakly concave posterior margins, narrowly rounded apices and subacute tips, their origins considerably closer to origin of anal than to origin of pectoral. Pectoral ⅔ as broad as long, or a little more, and a little larger in area than 1st dorsal, with moderately convex outer and inner margins, nearly straight distal margin, narrowly rounded apex and broadly rounded inner corner.

Color. Adults in life are plain grayish olive, slaty-gray or brown above, without any definite darker markings; yellowish or grayish white below with the margins of the fins paler. In embryos, however, and in young specimens up to a length of 400 mm. or so, the upper part of the first dorsal is more or less widely edged with dusky gray, the apex of the second dorsal sooty-edged or tipped, its posterior margin white; the caudal with a sooty blotch above at tip, white-edged below. The strength of these dark markings is variable, however, and by the time a length of 600 to 700 mm. is reached they are only faintly discernible, or have wholly faded.

Color Changes. This species is capable of changing shade—to a degree unusual among sharks—by expansion or contraction of its melanophores, from dark gray above on a dark background to a pale, translucent pearly tint when lying or swimming over a pale sand bottom. It has been found by experiment that it requires about two days[8] for it to attain the maximum paleness.

Size. Smooth Dogfish range from about 340 mm. to about 390 mm. in length (average about 360 mm.) at birth. The majority of mature females with young are between 1,000 and 1,300 mm. (about 3 feet 3 inches to 4 feet 4 inches) long. The maximum length is about five feet.

Remarks. There is some evidence that in the tropical part of its range, where the stocks of *canis* appear to be resident rather than migratory as they are in the north, local populations may differ slightly from the typical form in their proportionate dimensions, especially in the relative lengths of the labial folds, in the outlines of the fins and in the sculpturing of the denticles.

Developmental Stages. It has long been known that this species, like its close relative *M. mustelus* of the eastern Atlantic, is truly viviparous, the embryo deriving its nourishment from its mother by a highly organized yolk-sac placenta.[9] The number of young in a litter usually ranges between 10 and 20 (average about 16 in a large number of gravid females recently examined at Woods Hole), but litters as small as four have been reported.

Habits. This is an inhabitant of the continental shelf and is not pelagic. During its summer stay on the coast in the northeastern part of its range (see p. 249), it is most commonly taken in comparatively shoal water of 10 fathoms or less, often coming into enclosed bays and harbors, or even into fresh water on occasion.[10] Large numbers of them

8. For accounts of these experiments, some of which we have witnessed, see Parker and Porter (Biol. Bull. Wood's Hole, *66*, 1934: 30–37) and Parker, G. H. (Color Changes of Animals, Univ. Penn. Press, Sect. 2, 1936: 12–20).
9. For account of this structure, see Fowler (Science, *30*, 1909: 815) and Ranzi (Pubbl. Staz. zool. Napoli, *13*, 1934: 387) for the European *M. mustelus* with list of earlier references.
10. Definitely reported in fresh water in the North East River, Maryland, by Gunter (Amer. Midl. Nat., *28*, 1942: 316).

are taken in pound nets, as well as on hook and line close to bottom or actually on it. However, we are told by a well informed fisherman that a few are also caught along the outer edge of the continental shelf on the so-called "Tilefish Grounds," in the offings of New York and of southern New England, in depths of 80 to 90 fathoms during June and September; and the species is described to us as a midwater form around Cuba.[11]

Its food consists chiefly of the larger crustacea, with crabs of one species or another ranking first in most localities. It also preys largely on lobsters, of which it is perhaps the most destructive enemy off the southern New England coast, where lobsters have been found to form up to 16 per cent of its food. Smooth Dogfish also feed on squid, especially in spring, and on whatever small fish may be available, such as menhaden (*Brevoortia*), tautog (*Tautoga*), puffers (*Spheroides*), sticklebacks (*Gasterosteus*), scup (*Stenotomus*) and sculpins (*Myoxocephalus*). It has been estimated that in Buzzards Bay 10,000 Smooth Dogfish might annually devour over 60,000 lobsters, and perhaps ⅕ million crabs, as well as possibly 70,000 fish of one kind or another. They also feed on mollusks to a lesser extent, both univalve and bivalve, as well as on worms (*Nereis*). And they swallow considerable quantities of eel grass (*Zostera*), although probably only incidental to the capture of their animal food.[12] They are also scavengers when occasion offers; off Havana, for example, they have often been taken with garbage (chicken-heads, etc.) in their stomachs.[13] Experiments have shown that food is found chiefly by the sense of smell,[14] although they also have fairly keen vision for nearby objects. A crab, for instance, is found as quickly when hidden as if lying in the open. In captivity, and no doubt at liberty also, they constantly search the bottom for food. When a crab is found it is shaken to and fro and devoured. It has also been observed in the aquarium that they never molest active fish, but soon devour any sick or injured specimens, suggesting that it is only the smaller fishes that they normally capture in any considerable numbers.

The fact that every female of 1,000 mm. or longer taken at Woods Hole in the first half of July during a recent investigation had either ovulated or was in the process of so doing shows this to be the mating season. Corresponding to this, many kept in captivity became pregnant during the late summer. The period of gestation appears to be about ten months, *i.e.*, the Smooth Dogfish carries her young during the winter migration.[15] Off southern New England the young are born between early May and mid-July, when newborn specimens are often caught in the pound nets.

Such facts as are known regarding its winter and summer ranges show that the north and south migrations of the northern stock of this species are chiefly thermal in nature. Thus the temperature of its winter home ranges from about 6° to 7° C. (43°–45° F.) up to

11. Personal communication from Luis Howell-Rivero.
12. For lists of stomach contents and estimates of destructiveness, see Field (Rep. U.S. Comm. Fish. [1906], Spec. Pap. *6*, 1907: 12, 15).
13. Personal communication from Luis Howell-Rivero.
14. Sheldon, J. exp. Zool., *10*, 1911: 51; Parker and Sheldon, Bull. U.S. Bur. Fish., *32*, 1913: 33; Parker, Bull. U.S. Bur. Fish., *33*, 1914: 61.
15. Unpublished studies by F. L. Hisaw and A. Abramowitz.

about 10° to 15° C. (50°–59° F.); it does not appear on the coast of the Middle Atlantic United States or southern New England until the bottom water has warmed to at least 6° to 7° C. or higher; it departs in autumn when the temperature falls below about 10° to 12° C. At the opposite extreme, specimens kept in the tanks at Woods Hole show no ill effects at the highest summer readings which are usually up to about 22° to 23° C. or 72° to 73° F., butt here is some evidence of withdrawal locally when the water warms above about 70° F. The Caribbean populations of the species are inhabitants of tropical temperatures.

Between North Carolina and Cape Cod the Smooth Dogfish moves north and south regularly with the seasons, wintering chiefly within the sector between the southern half of North Carolina and the offing of Chesapeake Bay. In the southern part of this range many are caught in beach seines at times during the cold months. We have found odd specimens on the flats in Pamlico Sound in early winter, chilled to death by the sudden onset of freezing weather. Further north, however, along the southern Virginia coast, where water temperatures are lower, winter records of them are chiefly from the offshore fishing banks, in depths of 30 to 60 fathoms, where they are abundant enough to be a nuisance.

The northward migration is progressive from early spring on, considerable numbers lingering in North Carolina waters until June, with occasional specimens present there into July, although none are seen thereafter. They appear by May along the peninsula that separates Chesapeake Bay from the Atlantic,[16] and they arrive nearly simultaneously at about this same season all along the coasts of New Jersey, New York and southern New England, the average date of their vernal arrival at the entrance to Long Island Sound for a fifteen-year period being May 10, the earliest record being May 2. As a rule they also appear in appreciable numbers at Woods Hole some time in May. However, the date of their vernal arrival varies locally from year to year. Near New York, for example, they usually are not plentiful until June. They summer all along the coast from Delaware Bay to Cape Cod in such numbers that every fisherman is familiar with them, and they are also present to some extent along the outer part of the continental shelf off southern New England (p. 248). But the return movement of "Smooth Dogs" southward from southern New England may commence as early as July, a decrease in their numbers often having been reported there after June, with a corresponding increase in their numbers from summer to September near New York and offshore on the outer edge of the continental shelf. But no mass movement southward takes place until late October or November, when they withdraw almost simultaneously from the coast line as a whole to as far south as Chesapeake Bay, though stragglers (most often small specimens) may linger in the vicinity of New York, and no doubt elsewhere, as late as the second week in December during some years.[17] Presumably they also withdraw from the outer edge of the continental shelf off

16. For some reason, odd specimens only have been taken within the Bay.

17. Latest date for the mouth of Long Island Sound, Dec. 13.

New York at about the same time, for none are taken there in winter according to reports of fishermen; in fact, there is no record of their capture between mid-December and early May anywhere to the northward of the offing of Chesapeake Bay.

Numerical Abundance. Along southern New England and on the mid-Atlantic coast of the United States the Smooth Dogfish is the second most numerous shark, although falling far short of the Spiny Dogfish (*Squalus acanthius*, p. 466). Old accounts report occasional catches of as many as 100 at a time in pound nets during their periods of abundance, with 10 to 20 on a hand line not exceptional in a few hours' fishing, though catches of 5 or 6 in this way are more usual. More precise information is that 373 specimens were brought in to the Laboratory of the U. S. Bureau of Fisheries at Woods Hole during an investigation of the food of the species, in the summer of 1903, from pound net catches varying from 1 to 41, and averaging about 7. It is also described as the most common local shark in Uruguayan waters at the opposite extreme of its geographic range (p. 251). While the populations of the intervening regions (Caribbean and Gulf of Mexico) have attracted very little attention, they may be numerous there also, for Smooth Dogfish are common around Cuba[18] and have been so described around Bermuda.[19]

Relation to Man. In spite of its abundance the Smooth Dogfish is of no commercial importance except for classroom study in schools, for which purpose considerable numbers are preserved yearly. Many are caught incidentally by anglers, for they bite freely when fish or squid are used for bait, and they usually take the hook more freely by night than by day, as so many sharks do. But few anglers consider them game fish.

Range. Western Atlantic; abundant northward to Cape Cod during part of year, occasionally to Massachusetts Bay, and as a stray to Passamaquoddy Bay at the mouth of the Bay of Fundy; southward to Texas, Cuba, the Caribbean region, central Brazil (Rio de Janeiro) and Uruguay; also Bermuda. Present indications are that several more or less isolated populations of Smooth Dogfish exist, their areas of distribution being separated one from the next by wide gaps between which little or no intermigration occurs. The best known of these populations is found along the coasts of the Middle Atlantic United States. To the northward the Smooth Dogfish occurs regularly as far as Cape Cod, but only as a stray in the southwestern part of the Gulf of Maine (odd specimens reported from time to time for different localities in Massachusetts Bay), while only a single specimen has been reported from farther north, *i.e.*, from St. Andrews on Passamaquoddy Bay at the mouth of the Bay of Fundy. What the barrier may be to a more general dispersal of them into the Gulf of Maine is not clear. Since they may appear on the coast of southern New England when the temperature has not yet risen above 7° to 8° C., and since they are most plentiful there during June when the water is still only 13° to 15° C., it is unlikely that their failure to pass Cape Cod more regularly or to reach Georges Bank at all is the result of temperature. Nor is there any other obvious explanation, for it seems hardly

18. Personal communication from Luis Howell-Rivero. 19. Goode, Bull. U.S. nat. Mus., 5, 1876: 73.

likely that the change in the specific composition of the crab fauna (on which they largely subsist) from the waters west of Cape Cod to those to the east can be responsible.

Present indications also are that this particular population is bounded equally sharply to the southward, for while "Smooth Dogs" are common in season off the coasts of southern Virginia[20] and North Carolina, at least as far as Cape Lookout, there are only three reports of the species for South Carolina (including one specimen in the collection of the Harvard Museum of Comparative Zoology), and no positive record or rumor of its presence on the east coast of Florida.[21]

In contrast with the considerable amount of information that has accumulated about this species along the east coast of the United States, little is known of it in the Gulf of Mexico and the Caribbean, except that it occurs on the coast of Texas,[22] around Cuba and Jamaica in some numbers (p. 250), at Curaçao,[23] and at Trinidad.[24] Whether the Cuban center of population receives any recruits in winter from the northern stock is not known. Positive knowledge of the distribution farther south is even more scant, for while it has been reported repeatedly by name from southern Brazil, Uruguay and northern Argentina, there is no knowing how many of these records actually may have referred to the newly described *schmitti* (p. 261).[25] But *M. canis* does occur in southern Brazil, as proved by the fact that the collection in the Harvard Museum of Comparative Zoology includes several small specimens from Rio de Janeiro, as well as 31 embryos and the head of a large female from an indeterminate Brazilian locality that we cannot distinguish from *canis*. And we judge from an excellent illustration[26] that a *Mustelus*, said to be the most common Uruguayan shark,[27] is likewise identical with the North American *canis*. Comparison also of a specimen from Bermuda with extensive series from North America, West Indies and Brazil failed to reveal differences sufficient to warrant specific separation.[28] The coastwise nature of this species makes it likely that the Bermudian population has long been entirely isolated.

Synonyms and References:[29]

Squalus canis Mitchill, Trans. Lit. Phil. Soc. N. Y., *1*, 1815: 486 (type loc., N. York).

Mustelus canis DeKay, Zool. N. Y., 4, 1842: 355, pl. 64, fig. 209 (descr., N. York); Linsley, Amer. J. Sci., *47*, 1844: 77 (Connecticut); Storer, Mem. Amer. Acad. Arts Sci., *2*, 1846: 505 (in synopsis); Baird,

20. Only occasional specimens are taken in Chesapeake Bay.
21. *Canis* has been recorded from southern Florida, but it is probable that these reports actually referred to the newly discovered *M. norrisi*, for which that is the center of abundance (p. 255).
22. See Study Material, p. 244. It has also been reported as common in fresh water in Louisiana (Fowler, Proc. biol. Soc. Wash., *46*, 1933: 57), but perhaps not on good evidence.
23. Measurements given by Metzelaar (Trop. Atlant. Visschen, 1919: 5) identify this specimen as *canis*, not *norrisi*.
24. We have examined the embryo reported by Fowler (Proc. Acad. nat. Sci. Philad., *67*, 1916: 521) from Trinidad.
25. For discussion of this species, see Bigelow and Schroeder (Proc. Boston Soc. nat. Hist., *41* [8], 1940: 423).
26. Devincenzi and Barattini, An. Mus. Hist. nat. Montevideo, Suppl. Atlas, (2) *2*, 1926: pl. 2, fig. 2.
27. Devincenzi, An. Mus. Hist. nat. Montevideo, (2) *1*, 1920: 12.
28. Bigelow and Schroeder, Proc. Boston Soc. nat. Hist., *41*, 1940: 417.
29. The many studies of the physiology of *Mustelus canis* are omitted here, unless pertinent to knowledge either of its geographic distribution or of its habits.

Rep. Smithson. Instn. (1854), 1855: 337 extra (abund., N. Jersey); DeKay, Rep. St. Cab. nat. Hist. N. Y. (1855), 8, 1858: 64 (N. York); Gill, Proc. Acad. nat. Sci. Philad., Addend., 1861: 59 (ref.); Proc. Acad. nat. Sci. Philad., 1863: 333 (Massachusetts); Putnam, Bull. Mus. comp. Zool. Harv., *1*, 1863: 10 (ref.); Gill, Proc. Acad. nat. Sci. Philad., 1864: 263 (class.); Duméril, Hist. Nat. Poiss., *1*, 1865: 402 (ref., N. York); Storer, Mem. Amer. Acad. Arts Sci., *9*, 1867: 227, pl. 37, fig. 2; also Fishes Mass., 1867: 251, pl. 37, fig. 2 (descr., Massachusetts); Poey, Repert. Fisico.-Nat. Cuba, *2*, 1868: 453 (Cuba); Günther, Cat. Fish. Brit. Mus., *8*, 1870: 386, 518 (ref.); Gill, Rep. U.S. Comm. Fish. (1871–72), 1873: 813 (C. Cod to C. Hatteras); Verrill, Rep. U.S. Comm. Fish. (1871–72), 1873: 521 (Massachusetts, food); Goode, Bull. U.S. nat. Mus., *5*, 1876: 73 (common, Bermuda); Poey, Enumerat. Pisc. Cubens., 1876: 201; An. Soc. esp. Hist. nat., *5*, 1876: 397 (Cuba); Goode, Amer. J. Sci., *14*, 1877: 293 (Bermuda); Goode and Bean, Bull. Essex Inst. Salem, *11*, 1879: 30 (Massachusetts Bay); Bean, T. H., Proc. U.S. nat. Mus., *3*, 1881: 115 (N. Jersey to Massachusetts); Jordan and Gilbert, Proc. U.S. nat. Mus., *5*, 1882: 581 (S. Carolina); Bull. U.S. nat. Mus., 16, 1883: 870 (comp. with Europ. species); Bean, T. H., Rep. U.S. Comm. Fish. (1882), 1884: 343 (off Woods Hole); Bull. U.S. Fish Comm. (1887), 7, 1889: 152 (N. Jersey); Nelson, Rep. St. Geol. N. J., *2* (2), 1890: 660 (common, N. Jersey); Jordan and Evermann, Rep. U.S. Comm. Fish. (1895), 1896: 214, part (C. Cod, Cuba); Bull. U.S. nat. Mus., 47 (1), 1896: 29, part (C. Cod, Cuba); Bean, T. H., Bull. Amer. Mus. nat. Hist., *9*, 1897: 329 (near N. York); Linton, Proc. U.S. nat. Mus., *20*, 1897: 423, 433 (parasites, Woods Hole); Smith, Bull. U.S. Fish Comm., *17*, 1898: 88 (food, abund. near Woods Hole); Linton, Bull. U.S. Bur. Fish., *19*, 1901: 270 (parasites, Woods Hole); Bean, Rep. For. Comm. N. Y., 1901: 376 (abund., season, near N. York); Bull. N. Y. St. Mus., *60*, Zool. 9, 1903: 23 (general); Sharp and Fowler, Proc. Acad. nat. Sci. Philad., *56*, 1904: 505 (Nantucket); Bean, Field Mus. Publ., Zool., (7) *2*, 1906: 29 (Bermuda); Tracy, Rep. R. I. Comm. inl. Fish., 1906: 49 (Rhode Island); Field, Rep. U.S. Comm. Fish. (1906), Spec. Pap. 6, 1907: 10 (food, Woods Hole); Linton, Bull. U.S. Bur. Fish., *27*, 1907: 125, 410–414, 425 (parasites, Woods Hole); Wilson, Proc. U.S. nat. Mus., *31*, 1907: 707, 711 (parasites); Proc. U.S. nat. Mus., *33*, 1907: 360, 422, 423 (parasites); Smith, Bull. N. C. geol. econ. Surv., *2*, 1907: 32 (N. Carolina); Kendall, Occ. Pap. Boston Soc. nat. Hist., 7 (8), 1908: 3 (New England); Parker, Bull. U.S. Bur. Fish., *29*, 1910: 43 (senses, movements); Linton, Bull. U.S. Bur. Fish., *28*, 1910: 120 (parasites, Woods Hole); Tracy, Rep. R. I. Comm. inl. Fish., 1910: 58 (Rhode Island); Sheldon, J. exp. Zool., *10*, 1911: 54 (sense of smell, feeding); Parker and Sheldon, Bull. U.S. Bur. Fish., *32*, 1913: 33–36 (food found by smell); Fowler, Proc. Acad. nat. Sci. Philad., *65*, 1913: 62 (Virginia); Nichols, Abstr. Linn. Soc. N. Y., 20–23, 1913: 91 (season off N. York); Sumner, Osburn and Cole, Bull. U.S. Bur. Fish., *31*, 1913: 734 (Woods Hole); Parker, Bull. U.S. Bur. Fish., *33*, 1914: 61 (food, found by smell); Fowler, Proc. Acad. nat. Sci. Philad., *66*, 1914: 353 (N. Jersey); Coles, Proc. biol. Soc. Wash., *28*, 1915: 89 (N. Carolina); Fowler, Proc. Acad. nat. Sci. Philad., *67*, 1916: 521 (Trinidad, W. Indies); Wiegmann and Nichols, Copeia, 23, 1915: 43 (near N. York); Nichols and Murphy, Brooklyn Mus. Sci. Bull., *3* (1), 1916: 7 (Long Island, N. York); Fowler, Copeia, 27, 1916: 12 (N. Jersey); Jordan, Copeia, 49, 1917: 87 (name); Latham, Copeia, 41, 1917: 17 (mouth Long Island Sd.); Copeia, 57, 1918: 53 (no. of embryos); Copeia, 71, 1919: 53 (dates of arrival and departure, mouth Long Island Sd.); Fowler, Copeia, 68, 1919: 13 (N. Jersey); Wilson, Proc. U.S. nat. Mus., *55*, 1920: 596 (parasites, N. Carolina); Fowler, Proc. biol. Soc. Wash., *33*, 1920: 144 (N. Jersey); Latham, Copeia, 87, 1920: 91 (mouth Long Island Sd.); Devincenzi, An. Mus. Hist. nat. Montevideo, (2) *1*, 1920: 12 (Uruguay); Breder, Copeia, 101, 1921: 85 (food); Fowler, Proc. Acad. nat. Sci. Philad., *74*, 1922: 3 (N. Jersey); Breder, Copeia, 127, 1924: 26 (near N. York); Mowbray, Bull. N. Y. zool. Soc., *28*, 1925: 75 (near N. York); Fowler, Copeia, 143, 1925: 41, 42 (N. Jersey, Delaware Bay); Devincenzi and Barattini, An. Mus. Hist. nat. Montevideo, (2) *2*, 1926: Suppl. Atlas, pl. 2, fig. 2 (Uruguay); Coles, Copeia, 151, 1926: 105 (N. Carolina); Fowler, Copeia, 156, 1926: 146 (N. Jersey); Copeia, 165, 1927: 89, 90 (Maryland, Virginia, Delaware); Nichols and Breder, Zoologica, N. Y., *9*, 1927: 13 (general); Truitt, Bean and Fowler, Bull. Md. Conserv. Dep., 3, 1929: 29 (Maryland, Chesapeake Bay); Breder, Field Bk. Mar. Fish. Atlant. Coast, 1929: 14 (general); Fowler, Proc. Acad. nat. Sci. Philad., *80*, 1929: 607 (N. Jersey); McCallum, Proc. U.S. nat. Mus., *79* (26), 1931: 2 (parasites); Breder, Copeia, 1932: 31 (Block Island); Lundstrom and Bard, Biol. Bull.

Wood's Hole, *62*, 1932: 1 (color changes); Pearson, Invest. Rep., U.S. Bur. Fish., (10) *1*, 1932: 19 (winter trawl fishery off North Carolina); Fowler, Proc. biol. Soc. Wash., *46*, 1933: 57 (Calcasieu R., Louisiana, fresh water, but we question identification); Bigelow and Schroeder, Canad. Atlant. Fauna, biol. Bd. Canad., *12*[e], 1934: 6 (descr., distrib.); Burton, Sci. Mon. N. Y., *40*, 1935: 283 (S. Carolina); Parker and Porter, Biol. Bull. Wood's Hole, *66*, 1934: 30 (color changes); Fowler, Bull. Amer. Mus. nat. Hist., 70 (1), 1936: 68 (in part); Parker, Biol. Bull. Wood's Hole, *70*, 1936: 6, pl. 1 (color changes of young); Color Changes of Animals, Univ. Penn. Press, 1936: 12 (color changes); Smith, H. W., Biol. Rev., *11*, 1936: 64 (ref. to report of it in fresh water by Fowler, 1933); Fowler, Proc. Acad. nat. Sci. Philad., *89*, 1937: 303 (N. Jersey); Hubbs, Occ. Pap. Mus. Zool. Univ. Mich., 374, 1938: 1, 13 (correct name); Parker, Proc. Amer. Acad. Arts Sci., *72* (7), 1938: 269 (color changes in young); Tortonese, Atti Soc. ital. Sci. nat., 77, 1938: 303 (part, Brazil); Springer, Proc. Fla. Acad. Sci., *3*, 1939: 14 (had no Florida record); Breder, Bull. N. Y. zool. Soc., *41*, 1938: 28 (N. York Harbor); Springer, Proc. U.S. nat. Mus., *86*, 1939: 461 (comp. with other species); Bigelow and Schroeder, Proc. Boston Soc. Nat. Hist., *41*, 1940: 417, pl. 14, fig. A, B, pl. 15, fig. A, B, C, pl. 17, fig. A, B (compar. with other species); Gunter, Amer. Midl. Nat., *28*, 1942: 316 (in fresh water, Maryland); Fowler, Arqu. Zool. Estado São Paulo, *3*, 1942: 129 (Brazil); Fish Culturist, *21* (9), 1942: 66 (listed); Lunz, Bull. S. C. St. Planning Bd., 14, 1944: 26 (S. Carolina, Florida); Merriman and Warfel, Trans. N. Amer. Wild Life Conf., 1944: 234 (Block Island, Rhode Island); Bigelow and Schroeder, Guide Comm. Shark Fish., Anglo Amer. Caribb. Comm., Wash., 1945: 113 (descr., ill., range).

Mustelus laevis Günther, Cat. Fish. Brit. Mus., *8*, 1870: 385 (in part); Metzelaar, Trop. Atlant. Visschen, 1919: 5 (meas., Curaçao, W. Indies); not *M. laevis* Risso, 1826.

Mustelus hinnulus Jordan and Gilbert, Bull. U.S. nat. Mus., 16, 1883: 19 (part, C. Cod); not *Squalus* (*Galeorhinus*) *hinnulus* Blainville, in Vieillot, Faune Franc., 1825: 83.

Galeus canis Jordan, Rep. U.S. Comm. Fish. (1885), 1887: 795; Moore, Bull. U.S. Fish Comm., *12*, 1894: 358 (abund., N. Jersey); not *Galeus canis* Müller and Henle, 1841.

Cynias canis Gill, Proc. U.S. nat. Mus., *26*, 1903: 960 (name); Jordan, Manual Vert. Anim. NE. U.S., 1929: 9 (general); Jordan, Evermann and Clark, Rep. U.S. Comm. Fish. (1928), 2, 1930: 13 (part); Beebe and Tee-Van, Field Bk. Shore Fish. Bermuda, 1933: 24 (Bermuda).

Cynais canis Fowler, Rep. N. J. Mus. (1905), 1906: 60 (N. Jersey); Fowler, Rep. N. J. Mus. (1906), 1907: 254 (N. Jersey); Proc. Acad. nat. Sci. Philad., *60*, 1908: 55 (N. Jersey, Rhode Island, Nantucket); Proc. Acad. nat. Sci. Philad., *61*, 1909: 407 (N. Jersey).[30]

Mustelus mustelus Fowler, Science, *30*, 1909: 815 (embryo with placenta, N. Jersey); Proc. Acad. nat. Sci. Philad., *64*, 1912: 57 (Virginia); Copeia, 30, 1916: 36 (Mid. Atlant. Coast); Copeia, 31, 1916: 41 (N. Jersey); Proc. Acad. nat. Sci. Philad., *69*, 1917: 109 (N. Jersey); Proc. Boston Soc. nat. Hist., *35*, 1917: 110 (Rhode Island, Nantucket); Copeia, 39, 1917: 4 (Argentina, name); Occ. Pap. Mus. Zool. Univ. Mich., *56*, 1918: 15, pl. 2 (ill., descr., embryos with placenta, N. Jersey); Proc. Acad. nat. Sci. Philad., *71*, 1920: 292 (N. Jersey); Hildebrand and Schroeder, Bull. U.S. Bur. Fish., *43*, 1928: 47 (Chesapeake Bay); Jordan, Evermann and Clark, Rep. U.S. Comm. Fish. (1928), 2, 1930: 13 (part); Beebe and Tee-Van, Zoologica, N. Y., *13*, 1932: 119 (Bermuda); Field Bk. Shore Fish. Bermuda, 1933: 24 (Bermuda); Bigelow and Schroeder, Bull. U.S. Bur. Fish., *48*, 1936: 321 (off N. Carolina, winter).

Galeorhinus laevis Garman, Mem. Harv. Mus. comp. Zool., *36*, 1913: 176 (part) pl. 4, fig. 6–9;[31] Radcliffe, Trans. Amer. Fish. Soc., 1914: 38 (N. Carolina); Bull. U.S. Bur. Fish., *34*, 1916: 267, pl. 42, fig. 3 (denticles, teeth, N. Carolina); Huntsman, Contr. Canad. Biol. (1921), 3, 1922: 56 (St. Andrews, Passamaquoddy Bay); Meek and Hildebrand, Field Mus. Publ., Zool., *15*, 1923: 34 (part); Bigelow and Welsh, Bull. U.S. Bur. Fish., *40*, 1925: 25 (descr., Gulf of Maine); White, Bull. Amer. Mus. nat. Hist., 74, 1937: 64, 92, 123, pl. 6, fig. C, pl. 13, fig. P, pl. 41, fig. A (class., anat.); not *Squalus* (*Galeorhinus*) *laevis* Blainville, 1825.

30. The locality suggests that a nominal record of *C. canis* from the Florida Keys (Fowler, Proc. Acad. nat. Sci. Philad., *58*, 1906: 79) probably referred to the newly described *Mustelus norrisi*; see Synonyms, p. 256.

31. Garman's description was based on a Mediterranean specimen of the European *M. mustelus*, but his illustrations on a *canis* from Long Island, New York.

"*Mustelus asterias* (Valmont) or *Cynias canis* Mitchill," Jordan, Copeia, 49, 1917: 87 (name).

Doubtful South American References:

Galeus canis[32] Berg, An. Mus. nac. B. Aires, (2) *4*, 1895: 7 (Argentina, might equally refer to *schmitti*); Lahille, Rev. Mus. paul., *6*, 1895: 276 (Argentina, might equally refer to *schmitti*).

Mustelus vulgaris Bassett-Smith, Proc. zool. Soc. Lond., 1899: 468 (Rio de Janeiro, might refer equally to *schmitti*).

Mustelus canis Eigenmann, Rep. Princeton Exped. Patagonia (1896–1899), *3* (2), 1910: 377 (La Plata R., might equally refer to *schmitti*); Gilbert, Proc. Wash. Acad. Sci., *2*, 1900: 161 (Maceió, Brazil, might equally refer to *schmitti*); Schreiner and Ribeiro, Arch. Mus. nac. Rio de J., *12*, 1903: 78 (Rio de Janeiro, might equally apply to *schmitti*); Evermann and Kendall, Proc. U.S. nat. Mus., *31*, 1906: 68 (Rio de la Plata, might equally refer to *schmitti*); Puyo, Bull. Soc. Hist. nat. Toulouse, *70*, 1936: 89 (French Guiana, name only; might equally refer to *norrisi*).

Mustelus laevis Schreiner and Ribeiro, Arch. Mus. nac., Rio de J., *12*, 1903: 78 (Rio de Janeiro, might equally refer to *schmitti*).

Cynias canis Ribeiro, Arch. Mus. nac., Rio de J., *14*, 1907: 161, 203 (Brazil, might equally apply to *schmitti*); Fauna brasil. Peixes, Mus. nac. Rio de J., *2* (1), Fasc. 1, 1923: 19, pl. 1 (Brazil, might equally refer to *schmitti*).

Mustelus mustelus Fowler, Copeia, 39, 1917: 4 (Argentina, might equally apply to *schmitti*); Proc. Acad. nat. Sci. Philad., *78*, 1926: 261 (Argentina, might equally refer to *schmitti*).

Mustelus asterias Lahille, Enum. Peces Cartilag. Argent., 1921: 13 (Argentina, name only); probably not *asterias* Cloquet, 1819, of the eastern North Atlantic.

Mustelus norrisi Springer, 1939

Florida Dogfish

Figure 43 D–F

Study Material. Three males, 677 to 825 mm. long, taken near Englewood, western Florida (Harv. Mus. Comp. Zool., No. 35222, 35223, 35224); one male, 623 mm., from the Florida Keys.

Distinctive Characters. Among Atlantic members of the genus, the most distinctive feature of *M. norrisi* is the expansion of the lower anterior corner of its caudal as a sharp-pointed lobe directed rearward (Fig. 43 D).

Description. Proportional dimensions in per cent of total length. Male, 671 mm., from Englewood, Florida (Harv. Mus. Comp. Zool., No. 35223). Male, 753 mm., from same locality (Harv. Mus. Comp. Zool., paratype, No. 35222).

Trunk at origin of pectoral: breadth 9.1, 8.5; height 8.2, 8.0.

Snout length in front of: outer nostrils 3.4, 3.2; mouth 6.1, 5.9.

Eye: horizontal diameter 2.6, 2.6.

Mouth: breadth 4.4, 4.2; height 3.1, 3.2.

Nostrils: distance between inner ends 2.5, 2.5.

Labial furrow length: upper 1.3, 1.2; lower 1.1, 1.1.

Gill opening lengths: 1st 1.9, 1.6; 2nd 2.1, 1.7; 3rd 2.1, 1.9; 4th 2.1, 1.9; 5th 1.7, 1.5.

32. *Galeus canis* Lahille (Physis, B. Aires, *5*, 1921: 63), Marini (Physis, B. Aires, *9*, 1929: 452), and Pozzi and Bordale (An. Soc. cient. argent., *120*, 1935: 151) reported for Argentina appears to refer not to a *Mustelus* but to the genus *Galeorhinus*; see p. 264.

First dorsal fin: vertical height 7.9, 8.1; length of base 10.4, 10.9.
Second dorsal fin: vertical height 6.1, 5.8; length of base 8.2, 7.9.
Anal fin: vertical height 3.0, 3.5; length of base 5.9, 5.4.
Caudal fin: upper margin 18.4, 17.8; lower anterior margin 8.6, 8.2.
Pectoral fin: outer margin 12.5, 12.6; inner margin 6.1, 5.8; distal margin 8.5, 9.1.
Distance from snout to: 1st dorsal 28.6, 27.5; 2nd dorsal 61.2, 63.2; upper caudal 81.6, 82.2; pectoral 17.9, 17.2; pelvics 41.9, 42.4; anal 65.2, 67.7.
Interspace between: 1st and 2nd dorsals 23.4, 25.2; 2nd dorsal and caudal 12.8, 11.1; anal and caudal 8.3, 8.0.
Distance from origin to origin of: pectoral and pelvics 24.4, 26.1; pelvics and anal 23.8, 25.6.

M. norrisi resembles *canis* so closely that the points of difference alone need be noted. These are: lower anterior corner of caudal acute instead of rounded, directed rearward, forming a distinct but short lobe (Fig. 43 D); midpoint of base of 1st dorsal nearer to origin of pelvics than to axil of pectoral by a distance about equal to horizontal diameter of eye, instead of as near to axil of pectoral as to origin of pelvics, or nearer; body cavity relatively shorter, with origin of pelvics about midway between origins of pectorals and of anal, and under rear tip of 1st dorsal, instead of being considerably nearer to origin of anal than to origin of pectorals and behind rear tip of 1st dorsal; pectorals with relatively shorter inner and distal margins, the latter more deeply concave; mouth narrower; fins generally smaller; labial folds not only shorter but the upper and lower folds of more nearly equal length than is usual in typical *canis*, in which the upper is in most cases considerably the longer; space between nostrils relatively narrower; trunk more slender, with its dorsal outline less highly arched. The teeth also are higher-crowned in general, their margins more deeply concave or even notched (Fig. 43 F); dermal denticles, however, not distinguishable from those of typical *canis*.

Color. No information is available as to its color in life; preserved specimens are gray above, paler gray or dirty white below and without definite markings.

Size. Norrisi reaches a corresponding stage in development at a size somewhat smaller than does *canis*, the subocular fold being continuous anteriorly with the edge of the upper eyelid, and the claspers of large size in males as small as about 620 mm. in length. And females also may mature at a size no greater than this, one of 825 mm. having been found to contain embryos nearly ready for birth.

Developmental Stages. It is not yet known whether or not the embryo develops a placental connection with the mother, as in *canis* and in *mustelus* (p. 247).

Habits. Nothing is known of its diet or of its breeding habits.

Range. This Dogfish is known up to the present time only from the Florida Keys and from the west coast of southern Florida (Englewood and Naples, where large numbers have been taken in mackerel nets). It has been taken only in winter, suggesting that it has a center of abundance in moderate depths and comes into shallow water only when the temperature of the latter is near the seasonal minimum.[33]

Synonyms and References:

Mustelus norrisi Springer, Proc. U.S. nat. Mus., *86*, 1939: 462 (descr., type loc. off Englewood, Florida, also vicinity Key West, Florida); Proc. Fla. Acad. Sci., *3*, 1939: 15 (same loc. as the preceding, comp. with other species); Bigelow and Schroeder, Proc. Boston Soc. nat. Hist., *41*, 1940: 417, pl. 14, fig. A, pl. 15, fig. F, pl. 17, fig. C (meas., comp. with other species); Lunz, Bull. S. Carolina St. Planning Bd., 14, 1944: 26 (Florida); Bigelow and Schroeder, Guide Comm. Shark Fish., Anglo Amer. Caribb. Comm., Wash., 1945: 110 (descr., ill., range).

?*Mustelus canis* Evermann and Kendall, Rep. U.S. Comm. Fish. (1899), 1900: 48 (Key West, Florida).[34]

?*Cynais canis* Fowler, Proc. Acad. nat. Sci. Philad., *58*, 1906: 79 (Key West, Florida).[34]

Genus *Mustelus*, Addendum

Under this heading we include accounts of two more species of the genus that occur in the coastal waters of Uruguay and southern Brazil; likewise of a third that has been reported from northern Argentina.

Mustelus fasciatus (Garman), 1913

Striped Dogfish

Figure 43 Upper, A, B, C

Study Material. Female and male, 484 and 613 mm. long (the types), Rio Grande do Sul, Brazil (Harv. Mus. Comp. Zool., No. 154 and 315).

Distinctive Characters. Fresh specimens of *fasciatus* are separable at a glance from *canis*, *norrisi* and *schmitti* by their dark striped color pattern, and further, by a much longer snout relative to the size of the eye,[35] by the teeth, which are more nearly symmetrical, their cutting edges evenly convex (Fig. 43 A), and by the distal margin of the pectoral, which is considerably longer relative to the other margins of the fin. A further distinction between *fasciatus* on the one hand and *canis* and *norrisi* on the other is that the lower anterior corner of its caudal is not expanded as a definite lobe. The most obvious distinctions between it and *mento*, which may also be dark-striped when young, are that the head (to pectoral) is considerably longer than the interspace between the first and second dorsals in *fasciatus* but shorter than the interspace in *mento*, and that the caudal of *fasciatus* lacks a definite lower lobe (cf. Fig. 43 upper with 42 H).

Description. Proportional dimensions in per cent of total length. Female, 484 mm., from Rio Grande do Sul, Brazil (Harv. Mus. Comp. Zool., No. 315). Male, 613 mm., same locality (Harv. Mus. Comp. Zool., No. 154).

33. Springer, Proc. Fla. Acad. Sci., *3*, 1939: 5.
34. These nominal records are referred tentatively to this species because of the locality.
35. Snout in front of mouth about 4.5 as long as horizontal diameter of eye in *fasciatus*, but only about 2.5 to 3 as long in either of these other three species.

Trunk at origin of pectoral: breadth 11.6, 12.0; height 8.1, 8.5.
Snout length in front of: outer nostrils 5.7, 5.9; mouth 8.2, 8.0.
Eye: horizontal diameter 2.1, 2.4.
Mouth: breadth 7.2, 7.4; height 3.3, 3.7.
Nostrils: distance between inner ends 3.1, 3.3.
Labial furrow lengths: upper 2.3, 2.1; lower 2.0, 1.8.
Gill opening lengths: 1st 3.0, 2.5; 2nd 3.0, 2.5; 3rd 3.0, 2.5; 4th 2.9, 2.3; 5th 2.5, 2.0.
First dorsal fin: vertical height 8.5, 8.6; length of base 14.6, 14.0.
Second dorsal fin: vertical height 6.0, 6.9; length of base 10.7, 10.4.
Anal fin: vertical height 3.0, 3.6; length of base 6.4, 6.8.
Caudal fin: upper margin 21.1, 21.6; lower anterior margin 7.9, 9.3.
Pectoral fin: outer margin 15.1, 14.7; inner margin 8.7, 9.0; distal margin 11.4, 12.9.
Distance from snout to: 1st dorsal 28.5, 28.5; 2nd dorsal 59.8, 59.5; upper caudal

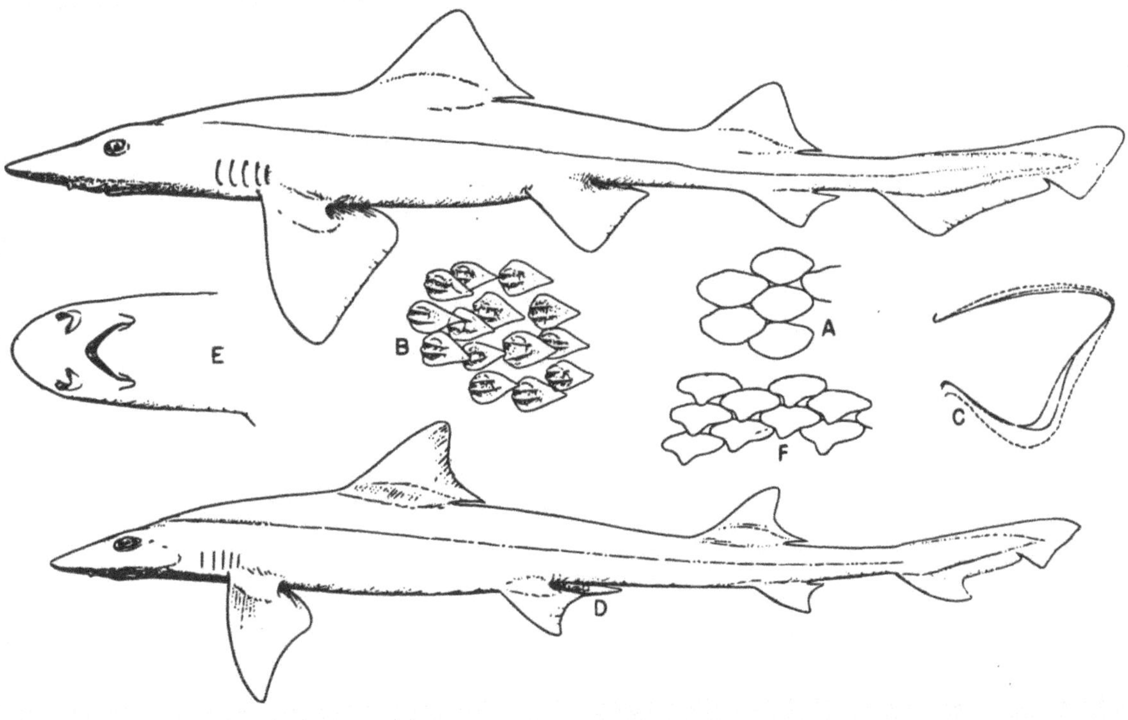

FIGURE 43. *Mustelus fasciatus*, male, 607 mm. long, from Rio Grande do Sul, Brazil (Harv. Mus. Comp. Zool., No. 154, type). *A* Upper teeth of same, about 6 x. *B* Dermal denticles of same, about 24 x. *C* Tracings of pectoral fins adjusted to equal lengths along outer margin: dotted line, *M. canis*, 768 mm. long, from Woods Hole, Massachusetts; solid line, *M. schmitti*, 578 mm. long, from Muldonado, Uruguay; broken line, *M. fasciatus*, same specimen as in *A*, *B*. *D Mustelus norrisi*, adult male, 643 mm. long, from Florida Keys (Harv. Mus. Comp. Zool., No. 442). *E* Head of same, from below. *F* Upper teeth of same, about 7 x.

78.9, 78.4; pectoral 22.7, 22.0; pelvics 49.0, 45.4; anal 68.5, 65.7.

Interspace between: 1st and 2nd dorsals 17.7, 17.1; 2nd dorsal and caudal 8.7, 8.4; anal and caudal 5.6, 4.6.

Distance from origin to origin of: pectoral and pelvics 26.9, 24.5; pelvics and anal 19.2, 20.4.

General form much as in *canis*. Dermal denticles more loosely spaced, with noticeably weaker sculpture, usually with only 2 ridges, and these as a rule confined to the anterior basal half of the blades, the latter so transparent that the outlines of the pedicels are visible from outside.

Head more flattened above than in *canis* and relatively much longer, its length to origin of pectoral about equal to distance from rear base of 1st dorsal to midbase of 2nd dorsal. Snout in front of mouth a little more than ½ as long as pectoral (considerably shorter than that in other western Atlantic species), more narrowly ovate than in *canis*. Eye relatively small, its horizontal diameter only about ¼ to ⅕ as long as snout in front of mouth. Spiracle a little less than ½ as long as horizontal diameter of eye. Third gill opening a very little longer than horizontal diameter of eye, the 4th about over origin of pectoral. Anterior margin of nostril with a well marked rounded lobe. Mouth a little less than ½, or about 45%, as high as broad. Upper labial furrow about 70% as long as distance between nostrils, the lower furrow about ¾ as long as upper.

Teeth nearly symmetrical, with evenly convex cutting edges.

First dorsal about as long at base as along anterior margin, the posterior margin only very weakly concave (much less so than in *canis* or *norrisi*), its origin a little posterior to axil of pectoral, the midpoint of base about midway between origin of pelvics and axil of pectoral. Interspace between 1st and 2nd dorsals about as long as from snout to 1st gill opening. Second dorsal between ⅔ and ¾ as long as 1st at base and about ¾ as high, its origin about midway between tips of pelvics and origin of anal. Caudal with lower anterior corner considerably more obtuse than a right angle, not expanded as a definite lobe. Anal about ⅔ as long at base as 2nd dorsal, its origin about under midpoint of base of latter. Pectoral a little less than ⅔, or about 60%, as long as head, with nearly straight distal margin and very broadly rounded inner corner, the distal margin only a little shorter (about 75–90%) than outer margin.

Color. Described[36] as "back brown with narrow transverse bands of darker, separated by spaces of about equal width: one or a pair crossing the orbits, one across the spiracles, one on the nape, four between the nape and the dorsal, four or five on the base of the dorsal, six between the dorsals, three on the base of the second dorsal, and two or three between it and the caudal. A yellow spot in front of each eye above each orbit. Fins dark with narrow edgings lighter. Lower surfaces whitish. On a larger individual pairs of bands are more or less confluent and all are more indefinite, indicating a probable loss of the

36. Garman (Mem. Harv. Mus. comp. Zool., *36*, 1913: 173), perhaps from field notes from the "Thayer" Expedition on which the specimens were collected.

bands later in life." After many years in alcohol the dark bands have wholly faded, leaving the specimens, which were collected in 1858 or 1859, mouse-gray above and a paler shade of the same tint below.

Size. The fact that the claspers are only about half as long as the inner margins of the pelvics in a male of 613 mm. suggests that maturity is not reached until at a length of perhaps 900 to 1,000 mm., or at about the same size as in *canis* (p. 247).

Developmental Stages. Not known.

Habits. Nothing known.

Range. M. fasciatus is so far known only from Rio Grande do Sul, southern Brazil (the type locality), and from the vicinity of Montevideo, Uruguay, and from Argentina, Lat. 35° 30′ S.[37]

Synonyms and References:

Galeorhinus fasciatus Garman, Mem. Harv. Mus. comp. Zool., *36*, 1913: 172 (type descr., south. Brazil).

Mustelus striatus Devincenzi, An. Mus. Hist. nat. Montevideo, (2) *1*, 1920: 122, pl. 12 (descr., good photos; vicinity of Montevideo, Uruguay); Devincenzi and Barattini, An. Mus. Hist. nat. Montevideo, (2) *2*, Suppl. Album Ictiol., 1926: pl. 2, fig. 3 (the dark cross bars indicated on ill.); Pozzi and Bordale, An. Soc. cient. argent., *120*, 1935: 151 (Lat. 35° 30′ S.).

Mustelus fasciatus Tortonese, Atti Soc. ital. Sci. nat., 77, 1938: 305 (descr., meas., ill., Rio Grande do Sul, Brazil); Springer, Proc. U.S. nat. Mus., *86*, 1939: 467 (in Key to species of *Mustelus*); Bigelow and Schroeder, Proc. Boston Soc. nat. Hist., *41* (8), 1940: 417, 418, Table column J; pl. 14, fig. B, pl. 15, fig. D, pl. 17, fig. E, pl. 18, fig. E (meas., discus., proport. dimensions, ill. of fins, teeth, denticles); Fowler, Arqu. Zool. Estado São Paulo, *3*, 1942: 129 (listed for Brazil).

Mustelus mento Cope, 1877

Figure 42 H–J

Study Material. Female, 1,024 mm. long, from Payta, Peru (Harv. Mus. Comp. Zool., No. 35246).

Distinctive Characters. See following Description.

Description. Proportional dimensions in per cent of total length. Female, 1,024 mm., from Payta, Peru (Harv. Mus. Comp. Zool., No. 35246).

Trunk at origin of pectoral: breadth 10.5; height 9.5.
Snout length in front of: outer nostrils 3.8; mouth 5.6.
Eye: horizontal diameter 2.0.
Mouth: breadth 6.0; height 2.5.
Nostrils: distance between inner ends 2.5.
Labial furrow lengths: upper 2.4; lower 2.0.
Gill opening lengths: 1st 2.9; 2nd 2.9; 3rd 2.9; 4th 2.9; 5th 2.2.
First dorsal fin: vertical height 10.2; length of base 13.7.
Second dorsal fin: vertical height 7.0; length of base 10.7.
Anal fin: vertical height 4.0; length of base 6.8.

37. Pozzi and Bordale, An. Soc. cient. argent., *120*, 1935: 151.

Caudal fin: upper margin 18.9; lower anterior margin 8.6.

Pectoral fin: outer margin 16.0; inner margin 11.4; distal margin 14.2.

Distance from snout to: 1st dorsal 27.1; 2nd dorsal 61.0; upper caudal 81.1; pectoral 17.9; pelvics 45.1; anal 66.8.

Interspace between: 1st and 2nd dorsals 21.1; 2nd dorsal and caudal 9.5; anal and caudal 6.0.

Distance from origin to origin of: pectoral and pelvics 27.8; pelvics and anal 22.0.

Since it is doubtful (see below) whether this species actually occurs in the Atlantic, its most distinctive features alone need be mentioned as an aid toward its identification. It falls with *canis, norrisi* and *schmitti* among the western Atlantic species in the relative shortness of its head; but it is separable at a glance from *norrisi* by the shape of its caudal (cf. Fig. 42 H with 43 D); from *schmitti* by a much smaller eye relative to the gill openings (only about 60% as long as the 3rd gill opening in *mento,* but about as long as the 3rd gill opening in *canis* and *schmitti*), and by the shape of its caudal (cf. Fig. 42 F with 42 H); from *canis* similarly by a small eye, by the distance between its nostrils which is considerably less than ½ as great as the breadth of its mouth (approximately one-half as great as that in *canis*) and by a relatively larger pectoral (inner margin about 1.8 times as long as snout in *mento* and only about 1.3 times that long in *canis*). It differs from *schmitti, norrisi* and *canis* in the fact that its teeth are symmetrical and with evenly convex cutting edges. The most striking differences between *mento* and *fasciatus* are its much shorter head and a caudal that has a well defined lower lobe (cf. Fig. 42 H with 43 upper). At least some of the adults of *mento* resemble *asterias* of the eastern Atlantic and Mediterranean in being conspicuously marked with many small white spots, but *mento* differs from *asterias* in the features stated in the Key, p. 242.

Color. Adults are often conspicuously marked with many small white spots, but sometimes they are plain colored,[38] whereas young specimens may be marked with dark colored bars.

Range. Coasts of Peru and Chile; perhaps Argentina.

Remarks. This species is included because a white-spotted *Mustelus,*[39] said to be common in northern Argentina, seems more likely (on geographic grounds) to be *mento* than the European *asterias,* under which name it was reported.

Synonyms and References:

Mustelus mento Cope, Proc. Amer. phil. Soc., *17*, 1877: 47 (descr., Peru) · Bigelow and Schroeder, Proc. Boston Soc. Nat. Hist., *41*, 1940: 429, Table column K, pl. 14, fig. C, pl. 16, fig. D, pl. 17, fig. I, pl. 19, fig. C (discus.; meas.; ills. fins, teeth, denticles); Fowler, Peces Peru, Mus. Hist. Jav. Prado, 1945: 12 (listed, Callao, Peru).

Mustelus edulis Perez Canto, Estud. Escual. Chile, 1886: 4 (descr., Chile); Philippi, An. Univ. Chile, *71*, 1887: 547, pl. 6, fig. 5 (descr., meas., color, ill., Chile).

38. If our reference of *edulis* Perez Canto, 1886, and *abbotti* Evermann and Radcliffe, 1917, to the synonymy of *mento* be correct. For discussion, see Bigelow and Schroeder (Proc. Boston Soc. nat. Hist., *41* [8], 1940: 429).

39. Lahille, An. Mus. nac. B. Aires, *34*, 1928: 310.

Galeus mento, Quijada, Bol. Mus. nac. Chile, *5*, 1913: 107 (listed for Chile).
Mustelus abbotti Evermann and Radcliffe, Bull. U.S. nat. Mus., 95, 1917: 6 (descr., meas., color, ill., Peru).

Possible South Atlantic References:

Mustelus asterias Lahille, Physis, B. Aires, *5*, 1921: 63 (name only, in list for Argentina); Enum. Peces Cartilag. Argent., 1921: 13 (brief account of teeth, Argentina); An. Mus. nac. B. Aires, *34*, 1928: 310 (Argentina, said to be common); Pozzi and Bordale, An. Soc. cient. argent., *120*, 1935: 151 (Argentina, Lat. 35°–42° S.; name only).

Mustelus schmitti Springer, 1939

Figures 42 F, G; 43 C

Study Material. Four specimens, 542 to 742 mm., from Rio Grande do Sul, Brazil, and from Uruguay (U.S. Nat. Mus. and Harv. Mus. Comp. Zool.).

Distinctive Characters. See following Description.

Description. Proportional dimensions in per cent of total length. Male, 571 mm., from Rio Grande do Sul, Brazil (Harv. Mus. Comp. Zool., No. 35316). Female, 596 mm., from Maldonado, Uruguay (Harv. Mus. Comp. Zool., No. 530).

Trunk at origin of pectoral: breadth 10.0, 9.7; height 9.5, 8.4.
Snout length in front of: outer nostrils 4.6, 3.5; mouth 6.5, 5.4.
Eye: horizontal diameter 3.0, 2.5.
Mouth: breadth 5.6, 6.4; height 3.3, 2.8.
Nostrils: distance between inner ends 2.8, 2.2.
Labial furrow lengths: upper 1.9, 1.9; lower 1.2, 1.5.
Gill opening lengths: 1st 2.1, 1.8; 2nd 2.3, 1.9; 3rd 2.3, 2.0; 4th 2.4, 2.0; 5th 2.0, 1.9.
First dorsal fin: vertical height 8.2, ——; length of base 12.6, 12.9.
Second dorsal fin: vertical height 5.8, 6.4; length of base 10.2, 10.9.
Anal fin: vertical height 3.5, 3.4; length of base 7.0, 7.4.
Caudal fin: upper margin 19.1, 19.5; lower anterior margin 8.9, 9.1.
Pectoral fin: outer margin 14.9, 14.9; inner margin 8.1, 8.9; distal margin 11.1, 11.7.
Distance from snout to: 1st dorsal 28.4, 28.5; 2nd dorsal 61.5, 58.8; upper caudal 80.9, 80.5; pectoral 20.2, 16.8; pelvics, 44.6, 42.8; anal 65.7, 65.1.
Interspace between: 1st and 2nd dorsals 21.9, 21.7; 2nd dorsal and caudal 10.3, 11.7; anal and caudal 6.3, 6.1.
Distance from origin to origin of: pectoral and pelvics 24.2, 26.7; pelvics and anal 21.2, 21.2.

M. schmitti so closely resembles *canis* in general appearance, in the size, shape, and location of fins, and in the teeth and denticles, that the points of difference alone need be mentioned.[40] The most striking difference between *schmitti* and *canis* lies in the outline of

40. For discussion, see Bigelow and Schroeder (Proc. Boston Soc. nat. Hist., *41*, 1940: 420).

the caudal fin; in *schmitti* this lacks any definitely outlined lower lobe, which (added to a proportionately longer terminal sector) gives the fin an aspect quite different from that of *canis* (Fig. 42). The distance between the nostrils is only about ½–⅓ as great as the breadth of the mouth in *schmitti*, but averaging more than one-half that great in *canis*. The inner margin of the pectoral is longer relatively in *schmitti* (Fig. 43 C). The upper labial furrow averages somewhat longer (see Proportional Dimensions) as does the interspace between the pelvics and the anal. The denticles, which are otherwise similar, are so transparent in *schmitti* that the pedicels are visible from the outside, which is seldom the case in *canis*. The teeth are not distinguishable from those of *canis* (cf. Fig. 42 G with 42 D).

Color. Preserved specimens are mouse-gray above and of a pale shade of the same tint below (much as in preserved *canis*) without any conspicuous markings.

Size. Seemingly this is a somewhat smaller species than *canis*, for males may mature at a length of only 600 mm.

Developmental Stages. Not known.

Habits. No information is available.

Range. So far known only from Rio Grande do Sul, in southern Brazil, from Uruguay (including Maldonado), and from Buenos Aires in northern Argentina.

Synonyms and References:

Mustelus schmitti Springer, Proc. U.S. nat. Mus., *86*, 1939: 465 (descr. of type, meas., size at maturity); Uruguay, and Buenos Aires, Argentina.

Mustelus schmidti[41] Bigelow and Schroeder, Proc. Boston Soc. nat. Hist., *41* (8), 1940: 420, column I; pl. 14, fig. B, 15, fig. E, 17, fig. D (meas., discus., ill., of pectoral, caudal and teeth). For other doubtful references, see under *M. canis*, p. 254.

Family CARCHARHINIDAE

Characters. Two dorsal fins, the 1st much shorter than the caudal, its base wholly anterior to origin of pelvics; caudal much less than ½ of total length, with well marked subterminal notch, not lunate, but its lower anterior corner expanded as a definite lobe, its axis raised but little; caudal peduncle not strongly flattened dorso-ventrally or widely expanded laterally; precaudal pits more or less strongly marked; sides of trunk, anterior to anal, without longitudinal ridges; inner margins of pelvics entirely separate, posterior to cloaca, in both sexes; jaws not widely protrusible; 5th gill opening over or posterior to origin of pectoral; gill arches without rakers, not interconnected by a sieve of modified denticles; nostrils entirely separate from mouth, their anterior margins without barbel; spiracles present or absent; eyes with well developed nictitating membrane within lower lid, its upper edge continuous with edge of lower eyelid anteriorly, but enclosed far within the latter posteriorly; labial furrows more or less developed, on one or both jaws; teeth blade-like, with only 1 cusp, only 1 row functional along sides of jaws, or 2, depending on the stage in their replacement; head of normal shape, not widely expanded laterally;

41. Misprint for "schmitti."

rostral cartilages 3, united anteriorly; radials of pectoral mostly on metapterygium; meso- and propterygia much smaller; meso- and metapterygia separated by a foramen, at least in some cases; heart valves in 3 rows. Development either ovoviviparous, or viviparous with well developed yolk-sac placenta.

Genera. This is not only the largest family of sharks, but the majority of modern sharks fall within it. All its western Atlantic members are inhabitants of the tropical or warm temperate belts, only entering the boreal zone in summer with the vernal expansion of high water temperatures, if at all. Some of them are cosmopolitan in the appropriate thermal belt, but others are confined to comparatively narrow areas of distribution. The majority of species are harmless, but a few bear evil reputations as dangerous to bathers.

Key to Genera

1a. Anal nearly 4 times as long at base as 2nd dorsal. *Physodon* Müller and Henle, 1841.
India, China, Australia.

1b. Anal at base less than 3 times as long as 2nd dorsal.
- 2a. Spiracles present, from large to minute.
  - 3a. Second dorsal originates behind rear end of base of anal. *Loxodon* Müller and Henle, 1841.
    Red Sea, Mauritius.
  - 3b. Second dorsal originates over or in front of midpoint of base of anal.
    - 4a. Midpoint of base of 1st dorsal considerably nearer to origin of pelvics than to axil of pectoral. *Thalassorhinus* Müller and Henle, 1841.[1]
      Eastern North Atlantic, Mediterranean.
    - 4b. Midpoint of base of 1st dorsal at least as near to axil of pectoral as to origin of pelvics, or nearer.
      - 5a. Caudal peduncle with a low longitudinal dermal ridge on each side; upper labial furrow as long as snout in front of mouth. *Galeocerdo* Müller and Henle, 1837, p. 265.
      - 5b. Caudal peduncle without longitudinal dermal ridges; upper labial furrow not more than 1/3 as long as snout in front of mouth.
        - 6a. Inner margins of upper teeth regularly serrate nearly to tips, but without basal denticles; their bases very deeply incised in the midline. *Hemipristis* L. Agassiz, 1843.
          Red Sea.[2]

1. The genus seems not to have been reported since 1881, when Moreau (Hist. Nat. Poiss. France, *1*: 319) described a specimen from Cette on the French coast of the Mediterranean. Watch should be kept for it, for it is likely to be confused with *Prionace*, from which it differs chiefly in having spiracles.

2. We follow several previous authors in referring the *Dirrhizodon elongatus* of Klutzinger (Verh. zool.-bot. Ges. Wien, *21*, 1871: 664) to *Hemipristis* on the strength of Probst's (Wurt. Jahresh., *34*, 1878: 141) statement that the teeth of the single known specimen resemble very closely the fossil shark's teeth that have been described under that name from the Upper Cretaceous to Miocene of North America, Upper Cretaceous to Pliocene of Europe, Eocene and Miocene of Africa, Miocene of Asia and South America, and Tertiary of the West Indies.

6b. Inner margins of upper teeth either smooth, or with 1 to several basal denticles (but not serrate); their bases not deeply incised in the midline.

7a. Precaudal pits well developed, both above and below.

8a. Lower teeth erect, smooth-edged all along jaw.

Negogaleus[3] Whitley, 1931.
India, Philippines, Australia, East Indies, Indo-China.

8b. Lower teeth along sides of jaw strongly oblique, their outer margins notched and denticulate.

Paragaleus Budker, 1935, p. 275.

7b. No precaudal pit above or below.

Galeorhinus Blainville, 1816.
Eastern Atlantic, including Mediterranean; southern Scandinavia to tropical West Africa; South Africa; Uruguay and Argentina; west coast of America (Chile, Peru, Lower California, California); Central Pacific; Japan, Formosa, East Indies, Australia, Tasmania, New Zealand.[4]

2b. Spiracles lacking.

9a. Midpoint of base of 1st dorsal considerably nearer to origin of pelvics than to axil of pectoral. *Prionace* Cantor, 1849, p. 280.

9b. Midpoint of base of 1st dorsal at least as near to axil of pectoral as to origin of pelvics, or nearer.

10a. Cusps of upper teeth smooth-edged, as well as those of lower.

11a. Second dorsal at least 3/4 as long at base as 1st, its posterior margin deeply concave. *Negaprion* Whitley, 1939, p. 308.

3. Proposed by Whitley (Aust. Zool., *6*, 1931: 334) to replace *Hemigaleus* Bleeker, 1852, the latter being preoccupied by Jourdain, 1837, for mammals.

4. Recorded nominally from Argentina and from Uruguay as *Galeorhinus galeus* (Berg, An. Mus. nac. B. Aires, [2] *1*, 1895: 7; Devincenzi, An. Mus. Hist. nat. Montevideo, [2] *1*, 1920: 119), and as *Galeus canis* (Lahille, Physis B. Aires, *5*, 1921: 63; Enum. Peces Cartilag. Argent., 1921: 13; An. Mus. nac. B. Aires, *34*, 1928: 310; Marini, Physis B. Aires, *10*, 1929: 452; Pozzi and Bordale, An. Soc. cient. argent., 120, 1935: 150).

None of these citations include any description of the South American specimens. But the illustration of one from Uruguay by Devincenzi and Barattini (An. Mus. Hist. nat. Montevideo, Suppl. Album Ictiol., 1926: pl. 1, fig. 3) resembles the common Tope of Europe (*G. galeus*) so closely in general appearance, especially in the very characteristic shape of the caudal, that the Uruguayan form must be regarded as identical with it, at least until specimens can be compared critically. This appears to be true also of the Oil Shark of the eastern side of the Pacific, described as *Galeorhinus zyopterus* Jordan and Gilbert, from California (Bull. U.S. nat. Mus., 16, 1883: 871); also as *Galeus chilensis* Perez Canto (Estud. Escual., Chile, 1886: 3), and as *G. molinae* Philippi (An. Univ. Chile, *71*, 1887: 544, pl. 4, fig. 1) from Chile.

That this species does not occur on the western side of the North Atlantic is one of the puzzling features in the geographic distribution of sharks.

11b. Second dorsal less than ½ as long at base as 1st and much smaller in area, its posterior margin only weakly concave, or nearly straight.

12a. Bases of upper teeth, as well as of lowers, smooth-edged.

13a. Teeth slender, symmetrical, erect in both jaws; longest gill opening nearly or quite ½ as long as base of 1st dorsal.[5] *Aprionodon* Gill, 1861, p. 303.

13b. Teeth in sides of jaws oblique, their outer edges notched; longest gill opening only about ¼ as long as base of 1st dorsal.

14a. Teeth with swollen rounded bases.
Protozygaena Whitley, 1940.
Australia.

14b. Teeth not swollen at base.
Scoliodon Müller and Henle, 1837, p. 292.

12b. Bases of upper teeth with serrate or denticulate edges.
Hypoprion Müller and Henle, 1841, p. 315.

10b. Margins of cusps of upper teeth regularly serrate; lowers either serrate or smooth. *Carcharhinus* Blainville, 1816, p. 320.

Genus *Galeocerdo* Müller and Henle, 1837

Galeocerdo Müller and Henle, Arch. Naturg., (3) *1*, 1837: 397; type species, *Squalus arcticus* Faber, 1829, Iceland, equals *Squalus cuvier* Lesueur, 1822, Australia.

Generic Synonyms:

Squalus (in part) Lesueur, J. Acad. nat. Sci. Philad., *2*, 1822: 351; Faber, Fische Islands, 1829: 17; not *Squalus* Linnaeus, 1758.

Galeus L. Agassiz, Poiss. Foss., *3*, 1835: pl. E, fig. 5, 6; Poiss. Foss., *3*, 1838: 91; type species, *G. cepedianus* L. Agassiz, East Indies; not *Galeus* Rafinesque, 1810.

Prionodon (subgenus in part) Bleeker, Verh. batavia. Genoot., *24*, Plagiost., 1852: 37; for *P. fasciatus*, Java; not *Prionodon* Müller and Henle, 1841.

Boreogaleus Gill, Ann. N. Y. Lyc., *7*, 1861: 402, 411; type species, *Squalus arcticus* Faber, 1829.

Isurus Townsend, Bull. N. Y. zool. Soc., *34* (6), 1931: 168, photograph of *Galeocerdo* mislabelled "*Isurus tigrinus*," evidently by mistake.

Generic Characters. Anal only a little longer at base than 2nd dorsal; spiracles present, small; 2nd dorsal originates over or a little in front of origin of anal; 2nd dorsal only a little more than ½ as long at base as 1st dorsal, and considerably less than ½ as large in area; midpoint of base of 1st dorsal only about ⅓ as far from axil of pectoral as from origin of pelvics; caudal peduncle with a low longitudinal dermal ridge on each side;

5. The gill openings are as long as this in the most recent illustration of *A. brevipinna* Müller and Henle, 1841 (Whitley, Fish. Aust., *1*, 1940: 108, as "*Longmania brevipinna*"), although pictured as considerably shorter in the original illustration of that species (Müller and Henle, Plagiost., 1841: pl. 9).

upper labial furrow about as long as snout in front of mouth; a well marked precaudal pit below as well as above; a low dermal ridge along midline of back between dorsal fins; caudal with pointed tip and lower lobe; teeth alike in the 2 jaws, large, few in number, with coarsely serrate edges, convex inwardly, their outer margins deeply notched; longest gill opening about ⅓ as long as base of 1st dorsal, the 4th over origin of pectoral. Development ovoviviparous. Characters otherwise those of the family.

Range. Cosmopolitan in tropical and subtropical seas.

Fossil Teeth. Upper Cretaceous to Miocene, North America; Eocene to Miocene, Africa; Eocene to Pliocene, Europe; Miocene, Asia, South America, West Indies.

Species. It is probable that all published references to the genus, from all parts of the world, concern one or another race of a single wide-ranging species, the common Tiger Shark of tropical seas; although the Australasian *Galeocerdo* has been considered a distinct species by some authors,[6] there is nothing in the published accounts to suggest any clear distinction between it and the *Galeocerdo* of the tropical Atlantic.

Galeocerdo cuvier (Lesueur), 1822

Tiger Shark, Leopard Shark

Figure 44

Study Material. Two young females, 1,368 and 1,380 mm. (about 4 feet 6 inches) long, and a young male of 1,245 mm. (about 4 feet 1 inch), from southern New England; very small female, 585 mm. long (about 23 inches), with well marked umbilical scar, hence either late embryo or newborn, from Cuba; also jaws of larger specimens from various localities.

Distinctive Characters. There is no danger of confusing the "Tiger" with any other shark, so diagnostic are its teeth, combined with the very short snout, very long upper labial furrows and sharp-pointed tail.

Description. Proportional dimensions in per cent of total length. Male, 1,245 mm., from Rhode Island (Harv. Mus. Comp. Zool., No. 35145).

Trunk at origin of pectoral: breadth 10.1; height 10.9.
Snout length in front of: outer nostrils 1.9; mouth 4.2.
Eye: horizontal diameter 2.1.
Mouth: breadth 8.4; height 5.0.
Nostrils: distance between inner ends 4.3.
Labial furrow lengths: upper 4.3, lower 1.9.
Gill opening lengths: 1st 2.4; 2nd 2.5; 3rd 2.5; 4th 2.6; 5th 2.3.
First dorsal fin: vertical height 8.3; length of base 8.4.

6. Most recently by Whitley (Fish. Aust., *1*, 1940: 113) as *G. rayneri* MacDonald and Barron, 1868. If the Australian form should finally prove to be distinct from the Atlantic, its correct name is *cuvier* Lesueur, 1822, type locality "New Holland," the name then in use for eastern Australia.

Second dorsal fin: vertical height 2.7; length of base 4.7.
Anal fin: vertical height 3.4; length of base 5.0.
Caudal fin: upper margin 30.2; lower anterior margin 12.6.
Pectoral fin: outer margin 14.2; inner margin 5.4; distal margin 12.1.
Distance from snout to: 1st dorsal 26.2; 2nd dorsal 56.8; upper caudal 69.7; pectoral 19.8; pelvics 44.1; anal 57.5.
Interspace between: 1st and 2nd dorsals 23.0; 2nd dorsal and caudal 9.6; anal and caudal 7.3.
Distance from origin to origin of: pectoral and pelvics 26.0; pelvics and anal 12.4.

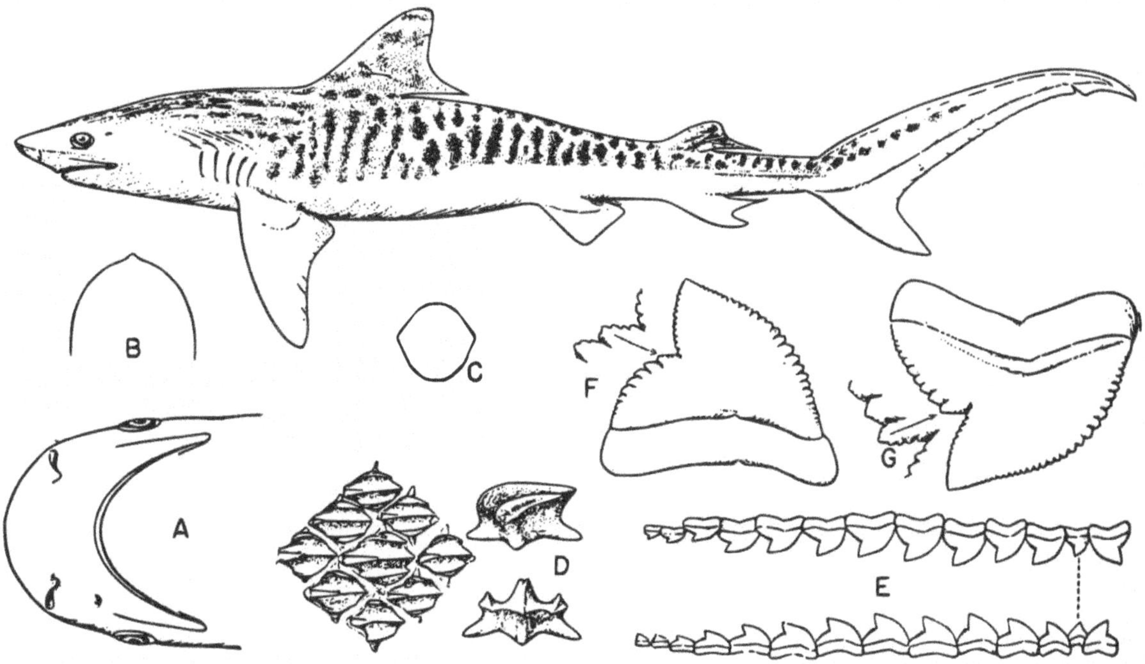

FIGURE 44. *Galeocerdo cuvier*, young male, 1,245 mm. long, from Newport, Rhode Island (Harv. Mus. Comp. Zool., No. 35145). *A* Anterior part of head of same. *B* Cross section of upper part of trunk opposite origin of pelvics showing mid-dorsal ridge. *C* Cross section of caudal peduncle to show lateral ridges. *D* General view of dermal denticles, about 28 x; lateral and apical views, about 56 x. *E* Teeth of larger specimen (U.S. Nat. Mus., No. 110900), about 0.2 natural size. *F* Third lower tooth, enlarged. *G* Fifth upper tooth of same, enlarged.

Trunk stoutest opposite 1st dorsal and tapering evenly rearward, the midline of back with a low dermal ridge extending rearward from a short way anterior to rear tip of 1st dorsal about 5/6 of distance to 2nd dorsal, where it gives place to a narrow, ill-defined furrow that reaches to the latter.[7] Caudal peduncle slender, oval in cross-section, with a low rounded ridge along each side at the midlevel, from opposite rear end of 2nd dorsal to a little beyond origin of caudal. Precaudal pits in the form of obtusely subangular furrows, the upper considerably the more distinct and larger. Dermal denticles large, variously

7. The combination of ridge and furrow in this region appears to be unique.

spaced but usually not overlapping, their blades nearly horizontal, longer than broad, usually with 3 ridges, the median very high and divided anteriorly, the lateral margins upturned, the posterior margin with a short, broad median tooth, usually flanked by a pair of very small teeth, but sometimes by one only; pedicels very short; basal plates very broad, distinctly 4-rayed.

Head flattened above, nearly or quite as broad opposite front of mouth as at origin of pectorals. Snout very broadly rounded, noticeably short, its length in front of mouth a little less than ¼ of length of head to origin of pectorals. Eye broadly oval, its horizontal diameter about ½ as long as snout in front of mouth, its center a little anterior to mid-height of mouth. Spiracle a narrow but easily visible slit, ⅓ to ¼ as long as horizontal diameter of eye and behind the latter by a distance a little greater than ½ the diameter of eye. Gill openings about evenly spaced, the 2nd, 3rd, and 4th (slightly the longest) a little more than ½ as long as snout in front of mouth, or about 1.2 times as long as diameter of eye, the 1st and 5th the shortest, the 4th over origin of pectoral. Nostril nearly transverse, its inner corner about equidistant between tip of snout and front of mouth, its anterior margin expanded at the inner end as a low triangular lobe with rounded apex. Mouth broadly ovate, a little more than ½ as high as broad, occupying between ⅘ and ⅚ of breadth of head, lateral (not inferior) for most of its length. Upper labial furrow as long as snout in front of mouth or a little longer, extending forward to a point about opposite anterior edge of eye. Lower labial furrow. a little less than ½ as long as upper, approximately parallel to lower jaw.

Teeth $\frac{9-9}{9-9}$ to $\frac{11-1-12}{12-1-12}$ in specimens examined; very large in front and sides of jaws (up to an inch or more high in large specimens) but decreasing in size toward corners with the outermost very small, similar in the 2 jaws, about ½ to ⅔ as high as broad, their inner outlines convex, their tips directed obliquely outward, their outer margins with a deep primary notch; both edges serrated, very finely so near the tip which may be worn smooth, but more coarsely so toward the base, especially on the outer margin basal to the notch, where the primary serrations are themselves finely serrate secondarily along their edges. There may or may not be a small symmetrical tooth at the symphysis of either jaw (Fig. 44 E); if this is lacking the tooth next to the symphysis in the upper jaw is usually considerably smaller than the 2nd and subsequent teeth, on either one side or on both, though similar to them in shape.

Anterior margin of 1st dorsal about ½ as long as from snout to axil of pectoral, its origin over or a little posterior to the latter, its anterior edge very slightly convex, its posterior margin deeply concave, its apex narrowly rounded, the free rear tip slender, about ¾ as long as base. Second dorsal about ½ as long as 1st at base, but a little less than ⅓ as high vertically, its anterior edge more sloping, its free rear tip more narrowly acuminate and relatively longer (about as long as base), its origin a little anterior to origin of anal. Caudal a little less than ⅓ of total length, with very slender pointed tip and deep subterminal notch, its lower anterior corner expanded as a narrow sharp-pointed lobe, a little more than 1/3 as long on its anterior margin as the upper lobe. Anal about as long as 2nd dorsal at base, and slightly higher vertically, but with posterior margin much more deeply concave and free rear tip a little shorter relatively, its rear tip a little posterior to rear tip of 2nd dorsal. Pelvics with nearly straight edges and narrowly rounded corners. Pectoral about 1/2 as long as head, or a little longer than 1st dorsal and a little larger in area, about 1/2 as broad as long, with moderately convex outer margin, moderately and evenly concave inner margin, and narrowly rounded corners.

Color. Gray or grayish brown, darker above than on sides and belly; small specimens up to about five or six feet long are more or less prominently marked on back with darker brown spots, often fusing irregularly into oblique or transverse bars on the sides and fins, sometimes surrounded with pale reticulations; but these markings fade with growth, leaving the larger specimens only faintly marked on the caudal peduncle, or even plain-colored in some cases.

Size. This is one of the sharks to which a gigantic size (up to 30 feet in length) has been accredited. However, the majority of specimens that are taken in its centers of abundance are less than 12 to 13 feet long.[7a] The longest of which we find positive record within recent years in the western Atlantic have been a Cuban specimen of about 18 feet,[8] and one of 15 feet 2 inches from South Carolina.[9] The weight at different lengths varies with fatness, and with the stage of development of the embryos in gravid females. Australian specimens are reported as weighing 710 to 825 pounds at 11 to 12 feet, 850 to 1,324 pounds at 12 to 13 feet, and 1,028 to 1,395 pounds at 13 to 14 feet;[10] recorded weights from the Pacific coast of Central America are 37 pounds at 5 feet 4 inches (1,625 mm.), 366 pounds at 10 feet 1 inch (3,073 mm.), 505 pounds at 10 feet 6 inches (3,200 mm.), and 780 pounds at 12 feet 9 inches. One 1,368 mm. long from Woods Hole, Mass., weighed 25¾ pounds, fresh. And there is no reason to suppose that the weights of larger Atlantic specimens would be different at equal lengths from Pacific examples, although they have been previously estimated as somewhat less.[11]

Although they may grow very large, Tiger Sharks are comparatively small at birth corresponding to the large numbers produced at one time, free-living specimens having been recorded as small as 18 to 19 inches.

Developmental Stages. Development is ovoviviparous; the embryos have no placental connection with the mother. The broods are very large, gravid females having been reported repeatedly as containing as many as 30 to 50 embryos, some more nearly ready for birth than others; recently we have received an account of an 18-foot Cuban specimen

7a. Stewart Springer informs us that none of the many measured by him in Florida waters were as long as 14 feet.
8. Personal communication from Luis Howell-Rivero.
9. Burton, Copeia, 1941: 40.
10. Whitley, Fish. Aust., *1*, 1940: 113.
11. 450 to 636 pounds at 11 to 12 feet (Bell and Nichols, Copeia, 92, 1921: 17; Nichols and Breder, Zoologica, N.Y., *9*, 1927: 15). The following weights are also mentioned, without locality; 58.8 pounds at 5 feet 2 inches, 168.4 pounds at 6 feet, 718.3 pounds at 10 feet 8 inches (Schultz, J. Mammal., *19*, 1938: 484).

containing 82 young.[12] On the other hand, litters as small as 10 to 14 have been recorded.[13]

Habits. The "Tiger" is found indifferently far out on the high seas and in coastwise waters. In tropical and subtropical seas they have been seen pursuing sting-rays on the flats in only a few feet of water and even in harbors; they are caught from the shore; and it is not unusual for them to enter enclosed sounds and river mouths in Florida and North Carolina. Most of the few records of them further north are from pound nets set out from the land in a few fathoms of water only. Except when aroused by the scent of food or other stimuli, the "Tiger" is rather sluggish; when stimulated, however, it is one of the most vigorous and strong swimming of sharks. In Florida waters, and presumably throughout its normal range, its young may be born at any time of year.

Although this is perhaps the commonest large shark in the tropics, little more is known about its life history, except for its diet, it being proverbially one of the most voracious of sharks. It is also one of the most omnivorous, for its diet ranges from objects as small as crabs and the smaller migrating land birds that have fallen into the sea to others as big as the larger sea turtles, other sharks, and sea lions.[14] "The large, coarsely serrated teeth are extremely efficient cutting instruments. . . . Bites on large objects are made by a rolling motion with both jaws cutting much in the manner of a saw";[14a] and a Tiger Shark has no difficulty in cutting through the shell of a sea turtle. The recorded list of its stomach contents includes crabs (half a bushel of them were taken from a 13-foot specimen in Florida), gastropods (*Buccinum, Lunatia*), spiny lobsters (*Palinurus*), horseshoe crabs (*Limulus*), squid, a wide variety of fishes, among them sharks smaller than themselves (a case in point is a specimen taken off Morehead City, North Carolina, which contained a *Carcharhinus limbatus*), skates, and even sting-rays, which they devour regardless of the poisonous spines, these often being imbedded in their jaws or elsewhere in their bodies. It is a common habit of this species to bite great chunks from other sharks, often of its own kind or of any other species which may be entangled in nets. The stomach contents of 34 specimens, netted off North Carolina, contained crabs, *Limulus*, sharks (small and large, entire and in pieces), large amounts of mackerel and unidentified small fish, sea turtles (entire and in pieces), bones and feathers of sea fowl, pieces of shark and porpoise that had seemingly been bitten from the nets, and garbage (sheep-bones, etc.).[15] As further evidence of its voracity we may quote an instance in which a large one, rearing head out of water, tore out the throat and belly from another shark that had been hoisted up to a boom.[16] The "Tiger" is also known as a scavenger, feeding on any kind of carrion, for example, parts of sheep, dead dogs, beef bones, remains of poultry, and even on such unappetizing objects as lumps of coal, tin cans or empty sacks. There is a recorded case in

12. Personal communication from Luis Howell-Rivero.

13. Whitley, Fish. Aust., *1*, 1940: 113. For an account of early embryonic development, see Sarangdhar (J. Bombay nat. Hist. Soc., *44* [1], 1943: 105).

14. See Beebe (Galapagos Worlds End, 1924: 201) for an eye-witness account of a Tiger Shark devouring a young sea lion.

14a. Springer, Proc. Fla. Acad. Sci., *3*, 1939: 16.

15. Bell and Nichols, Copeia, *92*, 1921: 18–19.

16. Radcliffe, Bull. U.S. Bur. Fish., *34*, 1916: 263.

Australia in which one, after capture, vomited the entire arm of a man who had been murdered at sea and his body dismembered.[17] "Tigers" also join the company of various other sharks that are soon attracted to the carcasses of dead horses or cattle in tropical harbors in the vicinity of slaughter houses.

Relation to Man. The "Tiger" is of considerable commercial value wherever a shark fishery is actively carried on in warm seas, as in southern Florida at present and until recently among the Virgin Islands, for it not only probably forms the largest single item in the catches, but yields excellent leather which is used for many purposes. Its yield of liver oil is also higher than that of many other tropical sharks. It likewise affords some sport to anglers, for it bites readily, provided the bait is large and strong-smelling. On the other hand, "these sharks are very destructive to gill nets, biting out great holes to take a single fish, and swimming back and forth through the nets as they feed on the gilled fish."[18] Worse yet, Tiger Sharks, when they come into shallow water, may be a danger to bathers; in the West Indies they are said to be considered the most dangerous of sharks. Some of the many shark fatalities that are well attested in medical journals for Australian waters are also credited to this species, although perhaps not on conclusive evidence. A recent instance is recorded from Malwan, south of Bombay, India.[19]

Range. Widespread in the tropical and subtropical belts of all the oceans, inshore and offshore alike.

Occurrence in the Eastern Atlantic. In the Eastern Atlantic, positive records for the Tiger Shark are comparatively few in number, *i.e.*, for the Canaries, tropical West Africa (Senegambia), western South Africa, and accidentally for Iceland.[20]

Early writers repeatedly credited it to northern Scandinavian waters and to the Faroes, an error springing from the fact that Faber's account of his "*arcticus*" was based on a combination of the latter with *Isurus nasus,* the common Porbeagle of boreal waters. Actually there is no positive record of the Tiger Shark for North Europe, other than the one for Iceland. It has never been reported from the Mediterranean, but no doubt it is much more plentiful along the tropical coast of West Africa and around the off-lying islands than the paucity of published records would suggest.

Occurrence in the Western Atlantic. This is one of the more numerous, if not the most abundant, of the larger sharks in the appropriate thermal zone of the western Atlantic. As with various other tropical species, its center of abundance appears to be the Caribbean–West Indian–South Florida region. Among the West Indies there is a published record of it at Trinidad, Porto Rico, the Virgin Islands, Cuba, between Turks Island and the Barbados, and near Nassau in the Bahamas, where it is so plentiful that 31 "Tigers"

17. For account of this happening and the subsequent investigations, see Whitley (Fish. Aust., *1*, 1940: 34).
18. Springer, Proc. Fla. Acad. Sci., *3*, 1939: 16.
19. Sarangdhar, J. Bombay nat. Hist. Soc., *44* (1), 1943: 104.
20. The identity of this specimen is attested by the account of its teeth by Faber (Fische Islands, 1829: 17) and more recently by Krøyer (Danmarks Fiske, *3*, 1852–1853: 933).

up to 16 feet in length have recently been reported among one catch of 51 sharks of all kinds.[21] No doubt it is equally common among the Antilles generally, and around Cuba. In southern Florida waters it is present among the Keys and on both the Atlantic and the Gulf of Mexico coasts throughout the year. Curiously enough, we have found no records of it for the Atlantic coasts of Central America and only one vague report for the inner Gulf of Mexico. But it has been encountered recently in July in the northern side of the Gulf off Biloxi, Mississippi.[22] And the poverty of the printed record, rather than any local scarcity, probably explains the lack of reports of it along Central America.

The Tiger Shark is only a summer visitor to the Atlantic coast of the United States north of Florida. Although there is only one definite record of it for South Carolina, considerable numbers must pass by there, at least in some years, for they have been reported repeatedly along North Carolina, sometimes in schools, even entering the enclosed sounds and river mouths at times. Only odd specimens have been reported from the sector thence northward past New York, *i.e.*, in Chesapeake Bay (once), Delaware Bay (once), New Jersey (about four times), Long Island, New York (once), and Newport, Rhode Island (once, Fig. 44). But like many other tropical fishes, Tiger Sharks appear more often in the Woods Hole region, where one to three are taken in the pound nets almost every summer, more often small but sometimes large. However, this is the northeastern limit to their occurrence inshore,[23] though odd specimens may be expected to stray much farther in this direction offshore in the tropical waters of the Gulf Stream; the often quoted Icelandic specimen may well have journeyed by that route.

To the southward the Tiger Shark is known from southern Brazil and Uruguay. Probably it occurs commonly all along the northeastern and northern coasts of South America, although it is not yet recorded there in scientific literature. It is also taken or seen from time to time around Bermuda.

Synonyms and References:

1. Atlantic:

Canis carcharias[24] Duhamel, Traité Gén. Pêches, *4* (2) Sect. 9, 1782: 297 (in part), pl. 19, fig. 3 (teeth, not fig. 1–3).

Squale (no spec. name) Lacépède, Hist. Nat. Poiss., 4° ed., *1*, 1798: pl. 8, fig. 2, in Buffon, Hist. Nat. (jaws).

Squalus cuvier Lesueur, J. Acad. nat. Sci. Philad., *2*, 1822: 351 (Aust.).

Carcharhinus lamia Blainville, in Vieillot, Faune Franc., 1825: 88 (teeth, ident. by ref. to Duhamel, Traité Gén. Pêches, *4* [2] Sect. 9, 1782: 298, pl. 19).

Squalus arcticus Faber, Fisches Islands, 1829: 17 (teeth, Iceland, confused with *Isurus nasus*); Nilsson, Prod. Ichthyol. Skand., 1832: 115 (Iceland).

Galeus (no spec. name) Agassiz, L., Poiss. Foss., *3*, 1835: pl. E, fig. 5, 6 (teeth); Owen, Odontogr., 1840–1845: pl. 28, fig. 9 (teeth; shows a sting-ray's spine imbedded in jaw).

Galeus maculatus Ranzani, Nov. Comment. Acad. Sci. Inst. Bonon., 1840: 7, pl. 1 (descr., Brazil).

Galeocerdo arcticus Müller and Henle, Arch. Naturg., (3) *1*, 1837: 398 (name); Plagiost., 1841: 60, pl. 64 (descr., distrib. probably confused with that of *Lamna nasus*); Bonaparte, Mém. Soc. neu-

21. Wise, Nat. Hist. N. Y., 38, 1936: 311.

22. Personal communication from Stewart Springer.

23. Doubtfully reported from Provincetown at the tip of Cape Cod.

24. Duhamel's names, if binomial, are only accidentally so.

châtel. Sci. nat., *2* (8), 1839: 10 (in synopsis); Cat. Pesc. Europ., 1846: 19 (Boreal Ocean); Gray, List Fish. Brit. Mus., 1851: 54 (north. seas); Krøyer, Danmarks Fiske, *3*, 1852: 933 (Iceland); Nilsson, Skand. Fauna, *4*, Fisk., 1855: 717 (ref.); Duméril, Hist. Nat. Poiss., *1*, 1865: 394 (descr., spec. without loc.); Günther, Cat. Fish. Brit. Mus., *8*, 1870: 377 (descr., "Arctic seas" by ref.); Hasse, Naturl. Syst. Elasm. besond. Theil, 1882: 259, pl. 37, fig. 19–24, 27 (vertebrae); Günther, Encycl. Brit., 11th ed., *24*, 1911: 807 (general); Garman, Mem. Harv. Mus. comp. Zool., *36*, 1913: 148 (descr.); Radcliffe, Trans. Amer. Fish. Soc., 1914: 39 (N. Carolina); Fowler, Proc. Acad. nat. Sci. Philad., *67*, 1916: 521 (Trinidad, W. I.); Fowler, Copeia, 30, 1916: 36 (mid-Atlant. coast U.S.); Radcliffe, Bull. U.S. Bur. Fish., *34*, 1916: 261, pl. 42, fig. 1, 2 (teeth, denticles, N. Carolina); Smith, J. Amer. Mus. nat. Hist., *16*, 1916: 348 (Atlant. coast U.S.); Meek and Hildebrand, Field Mus. Publ., Zool., *15*, 1923: 56 (general); Bigelow and Welsh, Bull. U.S. Bur. Fish., *40* (1), 1925: 27 (New England); Fowler, Proc. Acad. nat. Sci. Philad., *80*, 1929: 608 (C. May, N. Jersey); Breder, Field Bk. Mar. Fish. Atlant. Coast, 1929: 15 (general); Jordan, Evermann and Clark, Rep. U.S. Comm. Fish. (1928), 2, 1930: 14 (distrib.); Bellon and Mateu, Notas Inst. esp. Oceanogr., (2) *53*, 1931: 17 (descr., meas., teeth, Canaries); Beebe and Tee-Van, Zoologica, N. Y., *13*, 1932: 119 (Bermuda); Wilson, Bull. U.S. nat. Mus., 158, 1932: 454, 464 (parasites, Woods Hole); Beebe and Tee-Van, Shore Fish. Bermuda, 1933: 25 (Bermuda); Gowanloch, Bull. La. Conserv. Dept., *23*, 1933: 222 (Gulf of Mexico); von Bonde, J. comp. Neurol., 1933: pl. 2, fig. 2, 3 (brain); Bigelow and Schroeder, Canad. Atlant. Fauna, biol. Bd. Canad., *12e*, 1934: 9 (Woods Hole region); Budker, Bull. Mus. Hist. nat. Paris, (2) 7, 1935: 186 (Dakar, W. Afr.); Marchand, Mar. biol. Rep. Cape Town, *2*, 1935: 38 (west. S. Afr., Natal); Fowler, Bull. Amer. Mus. nat. Hist., *70* (1), 1936: 56 (W. Afr.); Cadenat, Rev. des. Trav. Pêches Marit., *10* (4), 1937: 429 (off Dakar); Norman and Fraser, Giant Fishes, 1937: 42 (general); White, Bull. Amer. Mus. nat. Hist., *74*, 1937: 124 (in Key); Nigrelli, Amer. Mus. Novit., 996, 1938: 10 (parasites); Devincenzi, Ann. Mus. Hist. nat. Montevideo, (2) *4*, 1939: 3 (Uruguay); Springer, Proc. Fla. Acad. Sci., 3, 1939: 16 (Florida); Hildebrand, Copeia, 1941: 221 (N. Carolina); Norris, Plagiost. Hypophysis, 1941: pl. 1, fig. 1 (brain); Fowler, Arqu. Zool. Estado São Paulo, *3*, 1942: 129 (Brazil); Fish Culturist, *21* (9), 1942: 66 (sizes, Cuba); Lunz, Bull. S. C. St. Planning Bd., 14, 1944: 26 (S. Carolina, Florida); Fowler, Monogr. Acad. nat. Sci. Philad., 7, 1945: 263 (Long Key, Florida).

Squalus carcharias Blake, Dent. Format Struct., Baltimore, 1848: tab. 6, fig. 5 (teeth, not seen); not *S. carcharias* Linnaeus, 1758.

Boreogaleus arcticus Gill, Ann. N. Y. Lyc., 7, 1861: 402, 411 (name).

Galeocerdo tigrinus Gill, Proc. Acad. nat. Sci. Philad., 1864: 263 (jaws, east coast U.S., W. Indies, Calif., W. Pacif.); Günther, Cat. Fish. Brit. Mus., *8*, 1870: 378 (descr., Japan, E. Indies, Atlant.); Verrill, Rep. U.S. Comm. Fish. (1871–72), 1873: 521 (Massachusetts, food); Lütken, Vidensk. Medd. naturh. Foren. Kbh., 1875: 37 (with a shark sucker, *Echeneis remora*); Goode, Proc. U.S. nat. Mus., *2*, 1879: 121 (Florida); Hasse, Naturl. Syst. Elasm. besond. Theil., 1882: 259; Rochebrune, Act. Soc. linn. Bordeaux, (4) *6*, 1882: 43; Faune Senegambie, Poiss., *1*, 1883–1885: 20 (Senegambia); Jordan and Gilbert, Bull. U.S. nat. Mus., 16, 1883: 21 (C. Cod to Indian Oc.); Nelson, Rep. St. Geol. N.J., *2* (2), 1890: 660 (N. Jersey); Jordan and Evermann, Rep. U.S. Comm. Fish. (1895), 1896: 215 (distrib.); Bull. U.S. nat. Mus., 47 (1), 1896: 32 (descr., distrib.); Linton, Proc. U.S. nat. Mus., *20*, 1897: 424, 448 (parasites, Woods Hole); Smith, Bull. U.S. Fish Comm., *17*, 1898: 88 (Woods Hole region); Evermann and Kendall, Rep. U.S. Comm. Fish. (1899), 1900: 48 (St. John's and Indian Rivers, Florida); Bean, Rep. For. Comm. N. Y., 1901: 377 (Vineyard Sound); Linton, Bull. U.S. Bur. Fish., *19*, 1901: 270, 425 (food, Woods Hole); Bean, Bull. N. Y. St. Mus., *60*, Zool. *9*, 1903: 24 (Martha's Vineyard, size); Fowler, Science, N. S. 24, 1906: 596 (N. Jersey); Proc. Acad. nat. Sci. Philad., *58*, 1906: 79 (Florida Keys); Rep. N. J. Mus. (1906), 1907: 255, pl. 71 (ill., N. Jersey records); Sullivan, Bull. U.S. Bur. Fish., *27*, 1908: 14 (digestion); Kendall, Occ. Pap. Boston Soc. Nat. Hist., 7 (8), 1908: 3 (New Engl. locs.); Gudger, Science, N. S. *37*, 1913: 993 (Key West, Florida); Yearb. Carneg. Instn., *12*, 1913: 177 (Tortugas, Florida); Sumner, Osburn and Cole, Bull. U.S. Bur. Fish., 31 (2), 1913: 735 (Woods Hole region); Coles, Proc. biol. Soc. Wash., *28*, 1915: 89 (N. Carolina); Gudger, Science, N. S. *41*, 1915: 437 (Key West, Florida); Yearb. Carneg. Instn., *13*, 1915: 203 (Tortugas, Florida); Nichols, Bull. Amer. Mus. nat. Hist., *34*, 1915: 141 (Porto Rico); Wiegmann and Nichols, Copeia, 23, 1915: 43

(Long Island, N. York); Gudger, Yearb. Carneg. Instn., *14*, 1916: 207 (Tortugas, Florida); Nichols and Murphy, Brooklyn Mus. Sci. Bull., *3* (1), 1916: 8 (Long Island, N. York, Woods Hole region, food); Coles, Copeia, 69, 1919: 38 (size, meas., food, C. Lookout, N. Carolina); Metzelaar, Trop. Atlant. Visschen, 1919: 188 (trop. W. Afr.); Bell and Nichols, Copeia, 92, 1921: 18 (N. Carolina, size, food); Nichols, Nat. Hist. N. Y., *21*, 1921: 273 (food); Gudger, Nat. Hist. N. Y., *22*, 1922: 245 (ref.); Nichols and Breder, Zoologica, N. Y., *9*, 1927: 14 (general); Jordan, Manual Vert. Anim. NE. U.S., 1929: 9 (general); Truitt, Bean and Fowler, Bull. Md. Conserv. Dept., 3, 1929: 28 (Chesapeake Bay); Nichols, Sci. Surv. Porto Rico, N. Y. Acad. Sci., *10* (2), 1929: 182 (food, Porto Rico); Gudger, Sci. Mon. N. Y., *34*, 1932: 411 (sting-rays eaten); Young and Mazet, "Shark, Shark," 1933: 268 (general); Wise, Tigers of the Sea, 1937: 175 (general); Gudger, Bull. Amer. Mus. nat. Hist., *73*, 1937: 273 (tooth abnormalities); Burton, Sci. Mon. N. Y., *40*, 1935: 283 (S. Carolina); Devincenzi, An. Mus. Hist. nat. Montevideo, (2) *4*, 1939: 3 (jaws, Uruguay); Longley and Hildebrand, Pap. Tortugas Lab., *34*, 1942: 2 (color, Tortugas, Florida).

Galeocerdo maculatus Poey, Repert. Fisico-Nat. Cuba, *2*, 1868: 453 (Cuba); An. Soc. esp. Hist. nat., *5*, 1876: 397, pl. 14, fig. 7; Enumerat. Pisc. Cubens., 1876: 201, pl. 9, fig. 7 (teeth, Cuba); Jordan, Rep. U.S. Comm. Fish. (1885), 1887: 795 (distrib.); Henshall, Bull. U.S. Fish Comm., *9*, 1891: 383 (Florida); Linton, Bull. U.S. Fish Comm., *13*, 1894: 103 (parasite); Ribeiro, Arch. Mus. nac. Rio de J., *14*, 1907; 149, 199, pl. 2 (Brazil, refs.); Fauna brasil. Peixes, *2* (1) Fasc. 1, 1923: 7, pl. 1 (Brazil).

Galeus tigrinus Fowler, Rep. N. J. Mus. (1905), 1906: 61 (N. Jersey).

Isurus tigrinus Townsend, Bull. N. Y. zool. Soc., *34*, 1931: 168 (photo, Porto Rico).

Galeocerdo cuvier Bigelow and Schroeder, Guide Comm. Shark Fish., Anglo Amer. Caribb. Comm., Wash., 1945: 118 (descr., ill., range).

2. West Coast of North and South America:

Galeocerdo tigrinus Gilbert, Bull. U.S. Fish Comm., *2*, 1883: 112 (Pacif. coast Guatemala); Jordan and Gilbert, Bull. U.S. Fish Comm., *2*, 1883: 105 (Mazatlan, Mexico); Jordan, Stanf. Univ. Pub., *1*, 1895: 382 (W. coast Mexico and Guatemala); Pellegrin, Bull. Mus. Hist. nat. Paris, 7, 1901: 161, 166 (danger to man, Gulf of Calif.); Gilbert and Starks, Mem. Calif. Acad. Sci., *4*, 1904: 9 (Pacif., Panama); Snodgrass and Heller, Proc. Wash. Acad. Sci., *6*, 1905: 342 (Galapagos); Fowler, Proc. Acad. Nat. Sci. Philad., *60*, 1908: 61 (Gulf of Calif.).

Galeocerdo arcticus Meek and Hildebrand, Field Mus. Publ. Zool., *15* (1), 1923: 56 (Pacif. coast, Panama); Beebe, Arcturus Adv., 1926: 247, 435 (Cocos Is., not seen); Breder, Bull. Bingham Oceanogr. Coll., *2* (1), 1928: 3 (W. coast Cent. Amer.); Walford, Fish Bull. Sacramento, *45*, 1935: 35 (Calif.); Barnhart, Mar. Fishes south. Calif., 1936: 9 (off Calif.); Schmitt, List Fish. President. Cruise, 1938: privately printed (Cocos, Clipperton, Galapagos Is.); Beebe and Tee-Van, Zoologica, N. Y., *26* (2), 1941: 113 (food, coast Central Amer., Galapagos, Cocos I.).

Galeocerdo maculatus Jordan and Bollman, Proc. U.S. nat. Mus., *12*, 1889: 179 (Panama).

Galeocerdo rayneri Philippi, An. Univ. Chile, *109*, 1901: 304 (Ecuador).

3. Central and Western Pacific, Australasian Region, Indian Ocean:

Squalus cuvier Lesueur, J. Acad. nat. Sci. Philad., *2*, 1822: 351 (descr., Aust.).

Galeus cepedianus L. Agassiz, Poiss. Foss., *3*, 1835: pl. E, fig. 5, 6; 1838: 91 (teeth, East Indies).

Galeocerdo tigrinus Müller and Henle, Plagiost., 1841: 59, pl. 23 (small size, India); Gray, List. Fish. Brit. Mus., 1851: 54 (Indian seas); Bleeker, Verh. batavia. Genoot., *25*, 1853: 80 (ref.); Blyth, J. Asiat. Soc. Beng., 29, 1860: 36 (India); Gill, Ann. N. Y. Lyc., 7, 1862: 402 (class.); Proc. Acad. nat. Sci. Philad., 1864: 263 (jaws, E. coast. U.S., West Indies, Calif., W. Pacific); Duméril, Hist. Nat. Poiss., *1*, 1865: 393 (descr., India); Günther, Cat. Fish. Brit. Mus., *8*, 1870: 378 (descr., Japan, E. Indies, Atlant.); Klunzinger, Verh. zool.-bot. Ges. Wien, *21*, 1871: 663 (Red Sea); Day, Fish. India, 1878: 718 (Red Sea, India, Japan); Brusina, Glasnik Naravosi Drutzva, *3*, 1888: 224 (Red Sea); Ogilby, Cat. Fish. Aust. Mus., *1*, 1888: 3 (Solomon Is.); Day, in Blandford, Fauna Brit. India, *1*, 1889: 21 (India); Boulenger, Proc. zool. Soc. Lond., 1889: 243 (Muscat, Arabia); Sauvage, in Grandidier, Hist. Phys. Nat. Madagascar, Poiss., 1891: 510 (ref.); Jordan and Snyder, Annot. zool. Jap., *3*, 1901: 39, 128 (Japan); Jordan and Fowler, Proc. U.S. nat. Mus., *26*, 1903: 612 (Japan); Fowler, J. Acad. nat. Sci. Philad., (2) *12*,

1904: 499 (Padang, Sumatra); Jordan and Snyder, Proc. U.S. nat. Mus., *27*, 1904: 940 (Hawaiian Is.); Jordan and Evermann, Bull. U.S. Bur. Fish., *23* (1), 1905: 36 (Hawaiian Is.); Günther, J. Mus. Godeffroy, *6*, 17, Fische der Sudsee *3*, 1910: 483 (Hawaii); Southwell, Ceylon Adm. Rep. Mar. Biol., 1912–1913: E 46, 49 (Ceylon); Zugmayer, Abh. bayer. Akad. Wiss., Math. Phys. Kl., *26* (6), 1913: 8 (Arabian Sea); Pearson, Ceylon Adm. Rep. Mar. Biol., 1915–1918: f 12 (Ceylon); Jordan and Jordan, Mem. Carneg. Mus., *10*, 1927: 4 (Hawaiian Is.); Sarangdhar, J. Bombay nat. Hist. Soc., *44* (1), 1943: 103, 3 pls. (food, breeding, embryos, attacks on man, utilization, Bombay coast).

Carcharias (Prionodon) fasciatus Bleeker, Verh. batavia. Genoot., *24*, 1852: 37 (Batavia); Günther, Cat. Fish. Brit. Mus., *8*, 1870: 368 (Java).

Galeocerdo rayneri MacDonald and Barron, Proc. zool. Soc. Lond., 1868: 368, pl. 32 (excellent ill., Aust., Lord Howe I.); Günther, Cat. Fish. Brit. Mus., 8, 1870: 377 (India); Day, Fish. India, *4*, 1878: 718, pl. 187, fig. 3 (India, Aust.); Ramsay, Proc. Linn. Soc. N. S. W., *5*, 1880: 95 (Aust.); Macleay, Proc. Linn. Soc. N. S. W., *6*, 1881: 353 (Aust.); Ogilby, Catal. Fish. N. S. W., 1886: 1 (Solomon Is.); Proc. Linn. Soc. N. S. W., (2) *3*, 1889: 1768 (Aust., Solomon Is.); Day, in Blandford, Fauna Brit. India, *1*, 1889: 20 (India); Kent, Naturalist in Aust., 1897: 193 (Aust.); Ogilby, Handbk. of Sydney, 1898: 117 (Aust.); Stead, Fish. Aust., 1906: 232 (Aust.); Zeitz, Trans. roy. Soc. S. Aust., *32*, 1908: 290 (Aust.); Weber, "Siboga" Exped., *57*, Fische, 1913: 591 (E. Indies); Robinson, Natal Fish. Rep., 1919: 50 (Natal); Pillay, J. Bombay nat. Hist. Soc., *33*, 1929: 350 (India); Whitley, Mem. Qd. Mus., *10* (4), 1934: 199 (ref.); Aust. Zool., *8*, 1937: 216 (Aust.); Fish. Aust., *1*, 1940: 113 (descr., size, weight, Aust.).

Galeocerdo obtusus Klunzinger, Verh. zool.-bot. Ges. Wien, *21*, 1871: 664 (Red Sea).

Galeocerdo fasciatus van Kampen, Bull. Dép. Agric. Ind. néerl., *8*, 1907: 9 (East Indies).

Galeocerdo hemprichii Klunzinger, Verh. zool.-bot. Ges. Wien, *21*, 1871: 663 (name only, and ref. to unpublished [?] ill. by Ehrenberg); Hilgendorf, in Hemprich and Ehrenberg, Symbol Physic., 1899: 8, pl. 5 (ill., Red Sea, not seen).

Galeocerdo arcticus Ogilby, Mem. Qd. Mus., *5*, 1916: 78 (Aust.); Barnard, Ann. S. Afr. Mus., *21* (1), 1925: 27 (Natal); McCulloch and Whitley, Mem. Qd. Mus., *8* (2), 1925: 128 (ref.); Fowler and Ball, Bull. Bishop Mus., *26*, 1925: 4 (N. central Pacif. Is.); Griffin, Trans. Proc. N. Z. Inst., *58*, 1927: 137 (N. Zealand); McCulloch, Fish. N. S. W., 2nd ed., 1927: 6, pl. 1, fig. 11a (Aust.); Fowler, Mem. Bishop Mus., *10*, 1928: 19 (Honolulu, Laysan, Nihoa); Phillipps, N. Z. J. Sci. Tech., *10*, 1928: 226 (N. Zealand); Tirant, Serv. Oceanogr. Pêches Indo-Chine, Note 6, 1929: 62 (Indo-China); Fowler, Proc. Pan.-Pacif. sci. Congr., Java, (4) *3*, 1930: 490 (Indo-Pac. distrib.); Mem. Bishop Mus., *11* (5), 1931: 314 (ref.); Chevey, Inst. Oceanogr. Indo-chine, Note 19, 1932: 6 (Indo-China); Giltay, Mém. Mus. Hist. nat. Belg., Hors Ser., *5* (3), 1933: 12 (East Indies); Fowler, Mem. Bishop Mus., *11* (6), 1934: 385 (Solomon Is.); Marchand, Mar. biol. Rep., Cape Town, *2*, 1935: 38 (west. S. Afr., Natal); Fowler, List Fish. Malaya, 1938: 10 (ref.).

Galeocerdo cuvier Whitley, Mem. Qd. Mus., *10* (4), 1934: 199 (ref., Aust.); Fowler, Bull. U.S. nat. Mus., 100 (*13*), 1941: 186 (refs., loc., W. Pacif., Indian Oc.).

Genus *Paragaleus* Budker, 1935

Paragaleus Budker, Bull. Mus. Hist. nat. Paris, (2) 7, 1935: 107; type species, *P. gruveli* Budker, 1935; tropical West Africa, near Dakar.

Generic Synonyms:

Hemigaleus Garman, Bull. Mus. comp. Zool. Harv., *46*, 1896: 203; for *H. pectoralis* Garman, southern New England.

Hemigaleus (in part) Garman, Mem. Harv. Mus. comp. Zool., *36*, 1913: 149 (in part); not *Hemigaleus* Bleeker, Verh. batavia. Genoot., *24*, Plagiost., 1852: 45; not *Negogaleus*[1] Whitley, Aust. Zool., *6*, 1931: 334.

1. Proposed by Whitley to replace *Hemigaleus* Bleeker, 1852, the latter being preoccupied; see footnote 3, p. 264.

Generic Characters. Anal a little shorter at base than 2nd dorsal; spiracle present and easily detected though small; 2nd dorsal originates over or a little anterior to origin of anal; midpoint of base of 1st dorsal nearer to axil of pectoral than to origin of pelvics; caudal peduncle without lateral ridges; well marked precaudal pits, below as well as above; a well marked labial furrow around corner of mouth and on each jaw, the upper less than ⅓ as long as snout in front of mouth; teeth with smooth-edged cusps, the uppers in sides of jaws oblique, notched outwardly, with 3 to 5 strong denticles toward the base; lower teeth slender, erect, without basal denticles in front of jaw, but increasingly oblique toward its corners, and with 3 to 5 strong denticles on the outer side toward the base, as in the uppers; anterior margin of nostrils expanded as a narrow triangular lobe; gill openings of moderate length, the 4th over origin of pectoral; axis of caudal only very slightly raised, its lower anterior corner expanded as a definite lobe with pointed tip. Characters otherwise those of the family.

Remarks. This genus is separated from *Negogaleus* Whitley by the fact that the lower teeth in the sides of the jaw are oblique, notched, with their bases strongly denticulate on the outer sides. See Key, p. 264.

Range. So far known only from tropical West Africa, and from the coast of southern New England.

Species. Two species known,[2] very closely allied to each other but apparently separable by the shapes of the snout and mouth.[3]

Key to Species

1a. Snout broadly rounded (Fig. 45 A); mouth about 2⅔ times as broad as high.
pectoralis Garman, 1913, p. 276.

1b. Snout subrectangular, with narrowly rounded tip; mouth only about 2⅓ times as broad as high.
gruveli Budker, 1935.
Tropical West Africa.

Paragaleus pectoralis (Garman), 1913

Figures 45, 46

Study Material. The type specimen, a female, 651 mm. long (Harv. Mus. Comp. Zool., No. 847).

Distinctive Characters. This species is characterized, among carcharhinids having spiracles, by the shortness of the anal fin relative to the second dorsal, by the position of the first dorsal far forward, and by the comparatively long snout and very characteristic teeth.

2. Only one specimen of each yet seen.
3. Budker's (Bull. Mus. Hist. nat. Paris, [2] 7, 1935: 110) measurements of the mouth do not agree with his illustration; the present Key is based on the former.

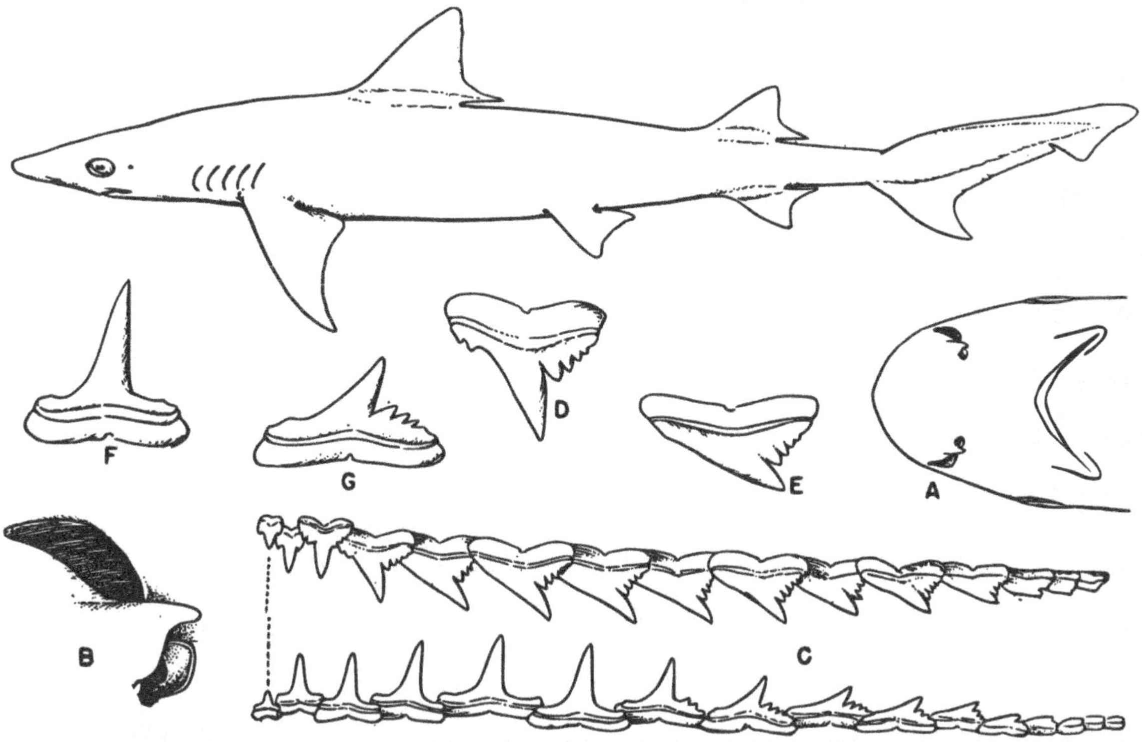

FIGURE 45. *Paragaleus pectoralis*, female, 651 mm. long, from off southern New England (Harv. Mus. Comp. Zool., No. 847). *A* Anterior part of head from below, about 0.5 x. *B* Left-hand nostril, about 2.2 x. *C* Upper and lower teeth, about 4.5 x. *D* Third upper tooth. *E* Tenth upper tooth. *F* Third lower tooth. *G* Seventh lower tooth. *D–G*, about 9 x.

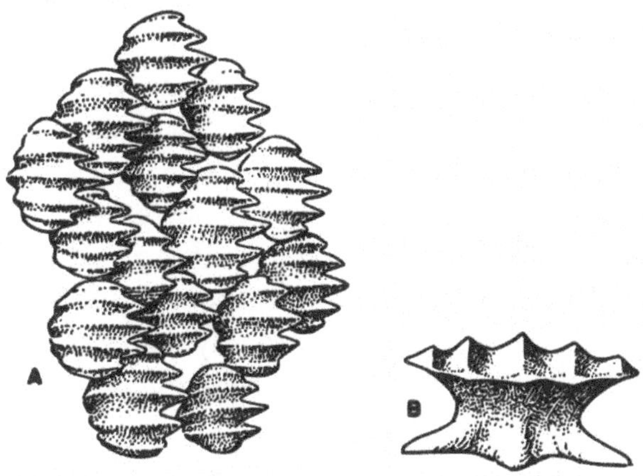

FIGURE 46. *Paragaleus pectoralis*, pictured in Fig. 45. *A* Dermal denticles, about 42 x. *B* Apical view of dermal denticle, about 84 x.

Description. Proportional dimensions in per cent of total length. Female, 651 mm., from the New England coast (Harv. Mus. Comp. Zool., type, No. 847).

Trunk at origin of pectoral: breadth 9.1; height 10.8.
Snout length in front of: outer nostrils 3.4; mouth 7.7.
Eye: horizontal diameter 2.3.
Mouth: breadth 6.5; height 2.5.
Nostrils: distance between inner ends 3.5.
Labial furrow length: upper 2.2; lower 1.6.
Gill opening lengths: 1st 2.2; 2nd 2.2; 3rd 2.2; 4th 2.3; 5th 2.3.
First dorsal fin: vertical height 9.1; length of base 9.8.
Second dorsal fin: vertical height 5.0; length of base 7.2.
Anal fin: vertical height 3.8; length of base 5.8.
Caudal fin: upper margin 23.5; lower anterior margin 10.2.
Pectoral fin: outer margin 14.8; inner margin 5.7; distal margin 11.7.
Distance from snout to: 1st dorsal 27.6; 2nd dorsal 57.7; upper caudal 76.5; pectoral 19.2; pelvics 47.2; anal 61.0.
Interspace between: 1st and 2nd dorsals 21.5; 2nd dorsal and caudal 10.1; anal and caudal 7.6.
Distance from origin to origin of: pectoral and pelvics 26.9; pelvics and anal 14.8.

Trunk slender, without mid-dorsal ridge, its height at origin of 1st dorsal (where highest) a little more than 1/7 of its length to origin of caudal, tapering evenly rearward. Body sector from snout to cloaca about as long as tail sector. Caudal peduncle slender, without lateral ridges, but with a well marked precaudal pit below as well as above. Dermal denticles moderately closely spaced, partially overlapping, their blades on short pedicels, nearly horizontal, with 5 longitudinal ridges, the margins with as many rather blunt teeth, of moderate length, the median a little longest, and the outermost very small.

Head about 1/5 of total length, its dorsal profile moderately arched posteriorly but flattened above anteriorly. Snout moderately thick, broadly rounded, its length in front of mouth about 1/3 length of head. Eye broad-oval, its midpoint about opposite front of mouth, its horizontal diameter a little less than 1/3 as long as snout in front of mouth. Spiracle a small horizontal slit, about 1/7 as long as horizontal diameter of eye, on a level with center of latter and behind it by a distance equal to about 1/2 the horizontal diameter of eye. Gill openings all about equal in length, about as long as horizontal diameter of eye, the 4th over origin of pectoral. Nostril strongly oblique, its inner end a little nearer to front of mouth than to tip of snout, its anterior margin expanded as a prominent subtriangular lobe with sinuous inner margin and blunt tip. Mouth obtusely ovate, about 2 1/2 times as broad as high, occupying about 2/3 of breadth of head. Labial furrows strongly developed, around corners of mouth, the upper extending about halfway toward the symphysis, the lower about 1/2 as long as upper.

Teeth $\frac{12 \text{ or } 13-3-12 \text{ or } 13}{14 \text{ or } 15-3-14 \text{ or } 15}$, not serrate; uppers subtriangular with broad bases, the 3 at symphysis small, symmetrical, erect, the next 9 to 12 increasingly oblique with nearly straight inner margins but outer margins deeply notched, with 3 to 4 strong denticles near base; those toward corners of upper jaw decreasing successively in size, broader relative to height, and with cusps and denticles less prominent, the outermost 2 or 3 low, evenly rounded, and the outermost of all minute; first 6 lower teeth erect, with slender cusps and broad bases, without denticles, the next 5 to 6 increasingly oblique, their bases denticulate on the outer side as in upper teeth, the cusps decreasing in relative length in successive teeth, the outermost 5 lower teeth low and evenly rounded, the outermost of all hemispherical, minute; 1 to 3 rows functional in upper jaw at symphysis, 1 row along sides of jaw, and 2 to 3 rows at corners of mouth; 2 to 4 rows functional at symphysis of lower jaw, 1 to 2 rows along sides of jaw, and 2 to 3 rows near corners.

Origin of 1st dorsal about opposite inner corner of pectoral, the midpoint of its base only about ⅔ as far from axil of pectoral as from origin of pelvics, its anterior margin only very slightly convex, apex subacute, rear margin deeply concave, free rear tip slender and about ⅓ as long as the base, its vertical height about as great as distance from eye to 2nd gill slit. Second dorsal similar to 1st, but only a little more than ⅔ as long at base and not more than ½ as large in area, its origin a little anterior to origin of anal. Caudal with narrowly rounded tip, well marked subterminal notch, its terminal sector nearly ⅓ the length of fin, the lower anterior corner forming an arcuate sharp-tipped lobe directed rearward, about 40% as long as upper lobe, each measured from its respective precaudal pit. Anal similar to 2nd dorsal, but only a little more than ¾ as long at base. Pelvics (in female) a little smaller than anal in area, with nearly straight anterior margins, moderately concave distal margins, narrowly rounded apices and subacute tips. Pectoral about 70% as long as head, only about as large in area as 1st dorsal, and very characteristic in shape,[4] with moderately convex outer margin which is increasingly so toward tip, deeply concave distal margin, nearly straight inner margin, and narrowly pointed tip.

Color. Described as grayish brown in life, paler below, the fins dark with pale hinder margins; after many years in alcohol the type is mouse-gray above and of a paler shade of the same below.

Size. The fact that a female of the closely allied West African species, 1,380 mm. (about 54 inches) long, contained embryos, suggests that this Shark does not reach a large size.

Developmental Stages. It is not known whether or not a placental connection is developed between embryo and mother; the embryos have not been described.

Habits. Nothing is known of its habits, but its teeth suggest a diet of fish or squid.

Range. So far known only from the type specimen taken off the coast of southern New England. All that is known of its origin is that Garman[5] obtained it, apparently in a

4. Hence the specific name *pectoralis.*
5. Garman, Bull. Mus. comp. Zool. Harv., *46,* 1906: 203.

fresh condition, from a public aquarium known as the "Aquarial Gardens," the exhibits for which came from "off the coasts of Massachusetts and Rhode Island."[6]

Remarks. We refer this species to *Paragaleus* Budker, 1935, rather than to *Hemigaleus* Bleeker, 1852 (in which genus Garman placed it), because of the conformation of its lower teeth; Garman's description of the latter as having "erect narrow cusps on broad bases, without denticles"[6] applies only to those in the front of the mouth, and not to those along the sides of the jaw as noted above.

Synonyms and References:

Hemigaleus pectoralis Garman, Bull. Mus. comp. Zool. Harv., *46*, 1906: 203 (descr., off the coast of Massachusetts or Rhode Island); Mem. Harv. Mus. comp. Zool., *36*, 1913: 150, pl. 4, fig. 1–5, pl. 50, fig. 9, pl. 52, fig. 2, pl. 56, fig. 4 (descr., ill. of type spec.); White, Bull. Amer. Mus. nat. Hist., *74*, 1937: 124, pl. 13, fig. g (class., tooth).

Genus *Prionace* Cantor, 1849

Prionace Cantor, J. Asiat. Soc. Beng., *18*, 1849: 1381; proposed in substitution for *Prionodon* Müller and Henle, 1841, preoccupied by Horsfield, 1823, for fossil mammals; type species, *Squalus glaucus* Linnaeus, 1758, designated by Jordan, Genera Fish., *2*, 1919: 242.

Generic Synonyms:

Squalus (in part) Linnaeus, Syst. Nat., *1*, 1758: 235.
Galeus Valmont, Dict. Hist. Nat., *1*, 1768: 371;[1] type species, *G. glaucus* Valmont.
Carcharias (in part) Rafinesque, Indice Ittiol. Sicil., 1810: 45; Cuvier, Règne Anim., *2*, 1817: 126.
Carcharhinus (in part) Blainville, Bull. Soc. philom. Paris, 1816: 121; and many subsequent authors.
Prionodon (in part) Müller and Henle, Plagiost., 1841: 35.[2]
Cynocephalus Gill, An. N. Y. Lyc., 7, 1862: 401;[3] type species, *Squalus glaucus* Linnaeus, 1758.
Galeus Garman, Mem. Harv. Mus. comp. Zool., *36*, 1913: 145;[4] not *Galeus* Rafinesque, 1810 (see discussion, footnote 1, p. 214).
Glyphis Fowler, Mem. Bishop Mus., *10*, 1928: 19; type species, *Squalus glaucus* Linnaeus, 1758; Bull. U.S. nat. Mus., 100 (*13*), 1941: 178; not *Glyphis* L. Agassiz, Poiss. Foss., *3*, 1843: 243.[5]
Carcharhinus Whitley, Fish. Aust., *1*, 1940: 106, 107 (restricted to *Squalus glaucus* Linnaeus, 1758, and to *Prionace mackei* Phillipps, 1935, which appears to be identical with the latter).

6. Mem. Harv. Mus. comp. Zool., *36*, 1913: 150, 151.

---

1. By Opinion 89 of the International Commission on Zoological Nomenclature (Smithson. Misc. Coll., *73* [3], 1925: 27), Valmont's names are not available, because they were not properly binomial.
2. Preoccupied by Horsfield, 1823, for fossil mammals.
3. This pre-Linnaean name, first proposed by Klein (Pisc. Natural., Gedoni, 1742), was revived by Gill to replace *Prionace* Cantor, 1849.
4. Garman's revival of *Galeus* Valmont, 1768, is not acceptable, according to the International Commission on Zoological Nomenclature; see footnote 1, p. 214.
5. The fossil shark's teeth, to which L. Agassiz gave the name *Glyphis*, and which he illustrated (Poiss. Foss., *3*, 1838: pl. 36, fig. 10–13), are not at all suggestive of the corresponding teeth of *Prionace*, being cylindrical near the base and with cutting edge confined to the lanceolate, laterally expanded tip. But they do resemble closely the anterior lower teeth of *Carcharias* (*Prionodon*) *glyphis* Müller and Henle, 1841 (see footnote 4, p. 321).

Generic Characters. Base of anal only about as long as base of 2nd dorsal; midpoint of 1st dorsal considerably nearer to origin of pelvics than to axil of pectoral; 2nd dorsal only about ½ as long at base as 1st dorsal and much smaller in area; spiracles lacking; caudal peduncle without longitudinal ridges, but with well marked precaudal pits both above and below; midline of back, between dorsal fins, smooth, without longitudinal ridge; a very short labial furrow at corner of mouth and on upper jaw, but none on lower; upper teeth subtriangular, oblique, with inner margins strongly convex and outer margins deeply concave; lower teeth more slender, erect; uppers with finely serrate margins, lowers serrate or smooth. Development viviparous, with yolk-sac placenta. Characters otherwise those of the family.

Remarks. Prionace is very closely allied to *Carcharhinus* but separable from it by the location of the first dorsal fin relative to the pelvics and pectorals.

Range. Cosmopolitan in tropical and warm temperate latitudes of all oceans, including the Mediterranean.

Species. The representatives of this genus generally had been considered as belonging to a single wide ranging species until recently, when Phillipps[6] separated its New Zealand representative as a new species *mackei*. According to Phillipps, *mackei* is distinguishable from the Atlantic *glauca* by a shorter head (20% of total length as against 25%), by a shorter snout relative to its head, by pelvics larger than the anal, and by a straight instead of concave distal margin of the pectoral. But these supposed differences are not consistent when tested against Atlantic specimens. Among seven fish from Massachusetts Bay, for example, ranging in length from about three to ten feet, the length of head from snout to pectoral origin ranged from 20 to 24 per cent of the total length, it being 22 per cent in a Japanese example of about 5 feet 6 inches (1,675 mm.).[7] The pelvics are also somewhat larger in area than the anal in five Massachusetts Bay specimens of which we have measurements, just as is the case in the New Zealand form. Neither does a comparison of the outline of the pectorals of the Massachusetts Bay, Japanese and Australian specimens reveal any consistent difference. The teeth, also, of an Australian specimen, as pictured by Whitley,[8] are indistinguishable from those of the Japanese and Atlantic specimens that we have examined (Fig. 47, 48). In short, we find no justification for retaining *mackei* as a distinct species.[9]

6. N. Z. J. Sci. Tech., *16*, 1935: 238.
7. This is the specimen on which Garman (Mem. Harv. Mus. comp. Zool., *36*, 1913: 145) based his description of *glaucus*.
8. Fish. Aust., *1*, 1940: 95, fig. 88, 8.
9. Phillipps (N. Z. J. Sci. Tech., *16*, 1935: 238) further states that the origin of the first dorsal in his *mackei* is midway between tip of snout and rear tip of second dorsal. But his photograph of the type specimen shows it as midway between tip of snout and midlength of caudal peduncle, as it is in one of the larger Massachusetts Bay specimens, as well as in our Japanese example; on the other hand Whitley (Fish. Aust., *1*, 1940: fig. 104) pictures it as about midway between snout and origin of caudal both for Australian and for New Zealand specimens. It is evident, then, that no geographic separation exists in this respect either.

Prionace glauca (Linnaeus), 1758

Great Blue Shark

Figures 47, 48

Study Material. Twenty freshly caught specimens (2 females and 18 males), about 5 to 11 feet long, from various localities in the Gulf of Maine, Georges Bank and from the offing of southern New England (jaws preserved); four preserved specimens, 539 to 2,160 mm. long, from Georges Bank and southern New England; Japanese specimen 1,675 mm. (about 5 feet 6 inches); also several other large specimens caught off the New England coast but not measured, and many seen at liberty.

FIGURE 47. *Prionace glauca*, male, about 2,175 mm. long, from Martha's Vineyard, Massachusetts (Harv. Mus. Comp. Zool., No. 36035). *A* Head from below, about ⅕ natural size. *B* Left nostril, about natural size. *C* Dermal denticles, about 25 x. *D* Lateral and apical views of dermal denticle, about 25 x. *E* Left-hand upper and lower teeth, about ¾ natural size. *F* Third upper tooth. *G* Ninth upper tooth. *H* Third lower tooth. *I* Eighth lower tooth. *F–I*, about 1.5 x.

Distinctive Characters. The Blue Shark is easily distinguished from other West Atlantic Sharks of its family by the combination of very long pointed snout, long falcate pectorals, first dorsal fin set far back, teeth, and brilliant blue upper parts.

Description. Proportional dimensions in per cent of total length. Male, 2,175 mm., from Martha's Vineyard, Mass. (Harv. Mus. Comp. Zool., No. 36035). Male, 2,442 mm., from off Nauset, Cape Cod, Mass. (field specimen).

Trunk at origin of pectoral: breadth 9.1, 10.8; height 8.8, 9.8.
Snout length in front of: outer nostrils 3.8, ——; mouth 8.0, 7.8.
Eye: horizontal diameter 1.4, 1.4.
Mouth: breadth 5.4, 5.4; height 3.7, 3.6.
Nostrils: distance between inner ends 3.5, 3.3.
Labial furrow lengths: upper 0.3, ——.
Gill opening lengths: 1st 1.9, 2.4; 2nd 2.0, 2.9; 3rd 2.3, 3.1; 4th 2.3, 2.9; 5th 1.8, 2.3.
First dorsal fin: vertical height 7.2, 7.2; length of base 7.3, 7.7.
Second dorsal fin: vertical height 3.1, 2.7; length of base 4.4, 4.2.
Anal fin: vertical height 3.5, 3.1; length of base 3.6, 4.0.
Caudal fin: upper margin 25.5, 25.8; lower anterior margin 12.8, 12.1.
Pectoral fin: outer margin 23.4, 21.8; inner margin 4.1, 4.1; distal margin 20.2, 18.0.
Distance from snout to: 1st dorsal 35.5, 35.7; 2nd dorsal 63.7, 62.7; upper caudal

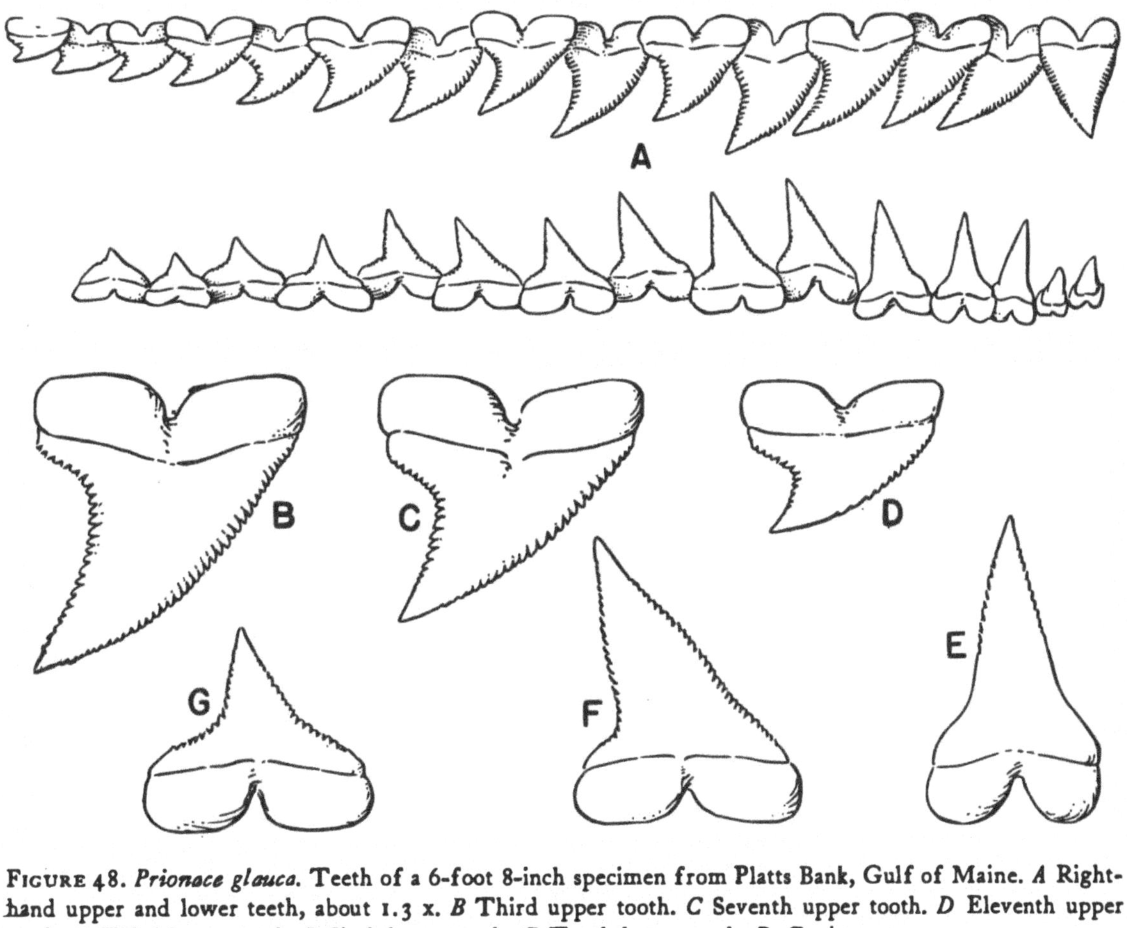

FIGURE 48. *Prionace glauca.* Teeth of a 6-foot 8-inch specimen from Platts Bank, Gulf of Maine. *A* Right-hand upper and lower teeth, about 1.3 x. *B* Third upper tooth. *C* Seventh upper tooth. *D* Eleventh upper tooth. *E* Third lower tooth. *F* Sixth lower tooth. *G* Tenth lower tooth. *B–G*, about 3 x.

74.5, 74.2; pectoral 21.5, 20.4; pelvics 49.5, 49.7; anal 63.7, 63.0.
Interspace between: 1st and 2nd dorsals 21.4, 20.0; 2nd dorsal and caudal 6.8, 7.3; anal and caudal 6.2, 7.9.
Distance from origin to origin of: pectoral and pelvics 28.2, 29.2; pelvics and anal 14.4, 13.2.

Trunk very slender, its height at origin of 1st dorsal (where highest) only about 1/6 to 1/7 of its length to origin of caudal, without mid-dorsal ridge. Body sector from snout to cloaca a little longer than tail sector. Caudal peduncle a little deeper than thick, without lateral ridges but slightly rhomboid in cross-section. Precaudal pits subrectangular. Dermal denticles close-spaced, usually overlapping, and so small that the skin is smooth to the touch, their blades horizontal, as broad as long or broader, usually with 3, and occasionally with 4 or 5, ridges, the apical margins weakly toothed to correspond; pedicels short and stout.

Head noticeably long, its length to origin of pectoral averaging about 1/5 of total length. Snout conical, with narrowly rounded tip, noticeably long, its length in front of mouth about 2/5 of length of head to origin of pectoral in large specimens and relatively a little longer in small specimens. Eye broadly oval or nearly circular, with well developed nictitating membrane, its midpoint about opposite front of mouth, its horizontal diameter between 1/4 and 1/5 as long as snout in front of mouth. Gill openings noticeably short, the 3rd longest, a little shorter than horizontal diameter of eye in small specimens, but about twice as long as eye in large ones, the 1st a little longer than 5th, the 4th over or very slightly posterior to origin of pectorals.[10] Nostril oblique, its inner end a little nearer to front of mouth than to tip of snout, its anterior margin only slightly expanded as a low, inconspicuous subtriangular lobe with rounded tip. Mouth evenly rounded, a little less than 2/3 as high as broad,[11] occupying about 2/3 of breadth of lower surface of head. A deep pit at corner of jaw, concealed when mouth is closed, but subtriangular when open, and extending for a very short distance at approximately a right angle onto the upper jaw but not onto the lower.[12]

Teeth $\frac{14\text{—}0 \text{ or } 1\text{—}14}{13 \text{ to } 15\text{—}1 \text{ to } 4\text{—}13 \text{ to } 15}$ in jaws examined; uppers so closely spaced that successive teeth overlap basally, subtriangular, slightly longer than broad, oblique, their outer margins deeply concave and inner margins convex, with edges serrate; usually one tooth at symphysis much smaller than those flanking it, but similar in form, its point directed toward the right in some specimens, toward the left in others (this tooth is lacking occasionally); next 4 or 5 teeth largest and about equal, the subsequent teeth successively smaller toward the angle of the mouth, with the outermost very small; lower teeth erect, much more slender than uppers toward center of mouth, but increasing in relative breadth and decreasing in length toward angles of mouth, their margins usually very finely serrate,

10. In a Japanese specimen, the third is similarly longest, and the fifth is slightly shorter than the first.
11. 56 to 68 per cent in specimens examined.
12. Much as in *Carcharhinus*.

but an occasional tooth smooth-edged and others partially so; an irregular group of 2 to 4 teeth at the lower symphysis, much smaller than those on either side and with relatively narrower bases; 1 to 2 rows of teeth functional in front and 1 in the sides of mouth in upper jaw; 1 to 3 rows functional in front and 1 row laterally in lower jaw.

Anterior margin of first dorsal about as long as snout in front of mouth, its origin posterior to inner corner of pectoral by a distance about ⅔ as long as its anterior margin and about midway between tip of snout and precaudal pit, the midpoint of its base at a vertical a little less than ¾ (about 70%) as far from origin of pelvics as from axil of pectoral; its anterior margin nearly straight, apex rounded, its posterior margin deeply concave toward the base, its free rear corner moderately acute, sharp-tipped, about ⅔ as long as its base. Second dorsal about ½ as long at base as 1st, and less than ⅓ as large in area, its posterior margin less deeply concave, but its free rear corner more slenderly pointed and about as long as the base, its origin about over or very little posterior to origin of anal. Caudal about ¼ of total length or a little more, its axis moderately raised, its terminal sector slender with pointed tip, the subterminal notch strongly marked, its lower anterior corner expanded as a blunt-tipped lobe, about ½ as long as the upper margin of fin. Anal a little larger than 2nd dorsal, its anterior margin convex, apex rounded, its posterior margin very deeply concave, its free rear corner acutely pointed and about ¾ as long as its base. Pelvics only about as large in area as anal, or a little larger, with nearly straight anterior and inner margins, slightly concave distal margins and narrowly rounded corners. Pectoral noticeably long, being about as long as head to 5th gill opening in medium-sized and large specimens, but relatively somewhat shorter in small ones,[12a] only a little more than ⅓ as broad as long, tapering toward tip, its anterior margin moderately convex (more strongly so in small specimens), the inner margin moderately concave proximally, the apex very narrowly rounded, the inner corner more broadly so.

Color. Living and freshly caught specimens are dark indigo blue along the back, shading to a clear bright blue[13] along the sides, and to snow white below; the tips of the pectorals are usually dusky and the anal partly so. But the beautiful blue of the back and sides darkens to a slaty or sooty gray soon after death.

Size. The Blue Shark is reputed to reach a length of 20 feet and commonly 15 feet. Actually, however, about 12 feet 7 inches (3.83 m.) is the longest of which we have found positive record; an 11-foot specimen is the longest we have handled. Embryos as large as 350 to 450 mm. have been recorded, and free living specimens as small as 21 to 36 inches (539 mm.; 661 mm.; 910 mm.; see Study Material, p. 282). The sizes of the females in which young have been found suggest that this Shark does not mature until a length of at least seven to eight feet is reached. Corresponding to their slender build, Blue Sharks are less heavy, length for length, than the more stout-bodied species; probably the following weights at different lengths, collected from various sources, are representative: 6 to 7

12a. 15.4 to 17.1 per cent of total length in three specimens, 539 to 910 mm. long.
13. "Sailor Blue" in Ridgway, Color Standards and Color Nomenclature, 1912: pl. 21.

feet, 65 to 70 pounds; 7 to 8 feet, 100 to 114 pounds; about 9 feet, 164 pounds.[14] Although we have handled many, we have weighed none.

Developmental Stages. The Blue Shark is viviparous, its embryo having a well developed yolk-sac placenta attached to the uterine wall of the mother.[15] The number of young in a litter is large, 28 to 54 having been reported in the Mediterranean from females of 8 feet 3 inches to 9 feet 4 inches.[16]

Habits. This is a pelagic species, encountered indifferently far out at sea and in continental waters, its wanderings no doubt directed chiefly by the search for food, although it may drift with ocean currents. It is frequently seen at the surface, swimming lazily with first dorsal fin and tip of caudal out of water, or basking in the sun. There is no reason to suppose that it ever descends to any great depth. Many are seen in coastal waters as well as offshore, and in some regions, near Woods Hole for example, it often comes close enough to the land to be caught in pound nets, as many other sharks often are. In our experience it is rather sluggish when not disturbed, but it swims powerfully and swiftly when in pursuit of prey. Normally it feeds on the smaller fishes that may be available locally, and on a variety of cephalopods. In northern waters herring and mackerel, and in European seas sardines, appear to be the chief items in its diet, as well as Spiny Dogfish (*Squalus acanthias*). No doubt it also consumes large quantities of bottom fish on the fishing banks. For example, we have repeatedly had Blue Sharks pick up cod, haddock and American pollock (*Pollachius virens*) that had been returned to the water on Georges Bank during the cod-tagging cruises of the United States Bureau of Fisheries.

In warmer seas they are also known to feed on anchovies and flyingfish, and occasionally on a sea bird that is resting on the water. We find no record of their preying on larger animals while the latter are alive. They sometimes follow sailing vessels in warm seas for days or even weeks picking up offal. And their habit of gathering when a Sperm Whale has been killed, probably by tracing the blood-scent, has long been proverbial among whalemen, one often struggling up on the carcass to "cling there until a descending blubber-spade had put an end to all its ambitions," to quote from an eye-witness account. "If the cutting in of the whale was at any time deferred . . . the sharks . . . would then attack the carcass, and, thrusting their heads partly above the surface, would bite large mouthfuls out of the blubber. . . . A blue shark horribly mutilated by repeated thrusts of a whaleman's blubber-spade, was seen to return immediately to the whale on which it had been feeding and to continue ravenously. . . ."[17] A recent report of one that came to eat scraps thrown to it from a boat, even after it had been transfixed by a harpoon, similarly illustrates its indifference to injury.[18]

14. From Roule, Result. Camp. sci. Monaco, *52*, 1919: 114; Holcombe, Modern Sea Angling, 1921: 144; Schultz (J. Mammal., *19*, 1938: 484, "*Prionace*") gives a weight of 433.6 kg. (about 950 pounds), but this is so far out of line with other recorded weights that some other stouter-bodied shark was doubtless intended.
15. For a recent anatomical account of the placenta, with references, see Calzoni (Pubbl. Staz. zool. Napoli, *15*, 1936: 109).
16. For numbers and sizes of embryos, see Lo Bianco (Mitt. zool. Sta. Neapel, *19*, 1909: 666).
17. Nichols and Murphy, Brooklyn Mus. Sci. Bull., *3* (1), 1916: 11.
18. Piers, Proc. N. S. Inst. Sci., *18*, 1934: 202.

It is not known whether there is a circumscribed breeding season or whether young are produced at all times of the year, which seems more likely, this being a warm-water species. Available information as to its young stages is summarized under Developmental Stages (p. 286).

Relation to Man. The Blue Shark is of no commercial value, nor has it been in the past, but it takes a large bait readily, and a few are caught for sport by anglers.[19] Our own experience, often repeated, has been that a "Blue" puts up little resistance when hooked on a heavy hand line until drawn in nearly to the ship's side, but then it threshes about violently as it is being hoisted aboard. But by anglers' accounts a large one hooked on rod and reel may resist strongly, making long rushes for a considerable time. While most often hooked on natural bait, it will sometimes take an artificial lure, as in the case of one five feet long recently caught on a feather jig tipped with pork rind, off Boone Island, Maine. In spite of its razor-sharp teeth the Blue Shark has always been held in contempt by whalemen who are the most familiar with it. There is no well authenticated record of its attacking swimmers, notwithstanding sailors' yarns to the contrary.

Range. Cosmopolitan, in the tropical, subtropical and warm-temperate belts of all the oceans (including the Mediterranean).

Occurrence in the Atlantic. This is no doubt the most plentiful of the larger oceanic sharks of the Atlantic[20] and it is the one with which we are the most familiar; around it most of the sailors' superstitions about sharks have centered. In the eastern side of the Atlantic it has been reported for so many localities and has been described so often as common that there is adequate evidence that it is practically universal off the coasts of west tropical Africa (Senegambia, Morocco), around the off-lying island groups (Cape Verdes, Canaries, Azores), and throughout the Mediterranean. It is also common, at least in summer, offshore along the Atlantic coasts of the Iberian Peninsula and France, although not often coming close to land. During the warm months it appears regularly off the south and west coasts of England north to Scotland in numbers sufficient for fishermen to be familiar with it, although it is seen less often on the French coast of the Channel, where we find only two records, both for Cherbourg. It penetrates the North Sea eastward to the Skagerrak, occasionally entering the western Baltic, and stray specimens are met with as far north as the Orkneys and southern Norway. Southward, in the eastern Atlantic, it is recorded for the west coast of South Africa.

Old time reports by sperm whalers, who were very familiar with the Blue Shark for reasons given above, show that it is generally, although very irregularly, distributed over the midbelt of the Atlantic. Its latitudinal range is as wide in the western side as it is in the eastern, *i.e.*, from the offing of the Rio de La Plata in the south to Nova Scotian waters (regularly) and to the Banks of Newfoundland (occasionally) in the north. Its

19. For readable accounts of rod and line fishing for Blue Sharks, see Wise (Tigers of the Sea, 1937: 67) and Holcombe (Modern Sea Angling, 1921: 152).

20. Recent authors (Nichols and Murphy, Brooklyn Mus. Sci. Bull., 3 [1], 1916: 10) write of seeing "hundreds, even thousands" of them during a sperm-whaling voyage in the tropical Atlantic.

coastwise distribution in the west is in strong contrast to that in the eastern side of the Atlantic, for while in the latter area it is most often encountered in the tropical-subtropical belt, in the former there are but two published inshore records of it for the entire West Indian region (St. Thomas and Cuba), with one for Florida (Miami) and none for the Gulf of Mexico or Caribbean littoral. But it occurs more commonly there, offshore, than this meagre record would suggest, for it is occasionally caught and often seen out in the open sea around Cuba,[21] while recently (September 1945) one about 12 feet long was taken 600 miles ESE. of Bermuda by the research vessel "Atlantis."

Neither is there any record for it on the coast between southern Florida and Chesapeake Bay; and stray specimens only have been reported from the coast of Maryland, New Jersey (two records), or from the vicinity of New York (two records); this is sufficient evidence that Blue Sharks rarely come inshore anywhere along this extensive sector of the coast. But they are much more common as summer visitors farther to the east and north. For example, 28 were counted 4 to 10 miles off Block Island on August 22, 1943, in an hour's run, with the number seen during the day estimated as 150 to 200.[22] There are many records of specimens taken in the traps close to land at Woods Hole, and it is a well known shark at Nantucket and on the off-lying shoals. Blue Sharks swimming at the surface are a familiar sight to fishermen in summer on Georges Bank, as we can bear witness. It was formerly regarded as a stray only, but it is now known to be a rather regular summer visitor in the Gulf of Maine, where it appears occasionally in July but more commonly in August and September, at least as far northward as Platts Bank, where three were caught and others were in sight of the vessel at nearly all times during the day on September 3, 1925. Two have been reported recently to us as taken on the Maine coast a few miles east of Casco Bay.[23] Many have been seen also within Cape Cod and Massachusetts Bay, and to our own knowledge several have been taken there in recent summers, even close to Boston Harbor.[24]

Still farther to the northward the Blue Shark is quite common in some summers along the Nova Scotian coast as far as Cape Breton, both inshore and on the offshore banks. For example, near Halifax in 1920 it was first reported on August 15, was most plentiful during the last week of that month, and was last reported on October 10th. It has been recorded also as a stray on the Grand Banks of Newfoundland. However, it is strictly a summer visitor to the coasts of the northeastern United States and Canada; none have been reported there later than mid-October. The great majority of those taken or seen there are of medium or large size. Moreover, it appears that few, if any, females take part in this yearly incursion, for all except two regarding which we have pertinent information have been males.

21. Personal communication from Luis Howell-Rivero.

22. Three were harpooned and one, about 21 inches long, was brought into Woods Hole and identified.

23. Personal communication from Walter H. Rich.

24. Eighteen were reported to us from Massachusetts during the summer of 1935; measurements and photographs of several of them were contributed by J. R. Lowes, Jr., a shark angler of wide experience.

Information as to the occurrence of the Blue Shark in coastal waters in the southwestern Atlantic is limited to records for Brazil and the offing of the Rio de La Plata.

Synonyms:

References for Atlantic, South Africa, West Coast of America:[25]

Squalus glaucus Linnaeus, Syst. Nat., *1*, 1758: 235 (Europ. Oc.); Syst. Nat., 12th ed., 1766: 401; Watson, Philos. Trans., *68* (1), 1779: 789, pl. 12, 2 fig. (descr., ill., Devonshire, Eng.); Bloch, Naturg. Fisch. Dtsch., *3*, 1784: 78, pl. 86 (North Sea, Baltic); Bonnaterre, Tabl. Encyc. Meth. Ichthyol., 1788: 9, pl. 7, fig. 22 (general); Gmelin, in Linnaeus, Syst. Nat., *1*, 1789: 1496 (descr.); Walbaum, P. Artedi Genera Pisc. Emend. Ichthyol., 1792: 513 (descr.); Bloch and Schneider, Syst. Ichthyol., 1801: 131 (descr.); Latreille, Nouv. Dict. Hist. Nat., *24*, 1804: 72 (in table of contents); Turton, Gen. Syst. Nature, *1*, 1806: 919 (general); Risso, Ichthyol. Nice, 1810: 26 (descr., Medit.); Retzius, Obs. Anat. Chondropt. Lundae, 1819: 3, pl., fig. 1 (anat.); Martens, Reise Vened., 2, 1824: 408 (Medit., not seen); Nardo, Prod. Ittiol. Adriat., 1827: 9 (Adriatic); Voigt, in Cuvier, Tierreich, 2, 1832: 506 (descr.); Couch, Cornish Fauna, 1838: 50 (Cornwall); Swainson, Nat. Hist. Fish. Amphib. Rept., 2, 1839: 312 (general); Bonaparte, Icon. Faun. Ital., *3*, Fasc. 37: 1840: pl. not numbered, fig. 1 (descr., Medit.); Cat. Pesc. Europ., 1846: 18 (Medit.); Nardo, Pesc. Publ. com. Nuovo Venez., 1853: 15 (Medit.); Machado, Peces Cadiz, 1857: 9 (near Cadiz); Gemmellaro, Atti Acad. Gioenia, (2) *19* (3), 1864: 123 (Medit.); Ninni, Ann. Soc. Nat. Modena, *5*, 1870: 66 (Medit.); Buckland, Hist. Brit. Fish., 1881: 212 (general, Gt. Brit.); Tarel, Act. Soc. linn. Bordeaux, *86*, Proc. Verb., 1934: 113 (Arcachon, France).

Squalus adscensionis Osbeck, Voy. China E. Indies, 1771: 78 (not seen, quoted from Bloch, 1784).

Le Squale Glauque, Lacépède, Hist. Nat. Poiss., *1*, 4° ed., 1798: 165, 213, pl. 9, fig. 1, in Buffon, Hist. Nat. (general); in Sonnini, Hist. Nat. Poiss., *3*, 1802: 405 (color, general).

Blue Shark, Yarrell, Brit. Fish., 2, 1836: 381 (Brit. loc. records).

Carcharias glaucus Rafinesque, Indice Ittiol. Sicil., 1810: 45 (Sicily); Cloquet, Dict. Sci. Nat., 7, 1817: 68 (general); Cuvier, Règne Anim., 2, 1817: 126 (general); Fleming, J., Hist. Brit. Anim., 1828: 167 (Cornwall); Bory de St. Vincent, Dict. Class. Hist. Nat., *15*, 1829: 597 (general); Agassiz, L., Poiss. Foss., *3*, 1835–1837: 12, pl. F, fig. 1 (teeth); Owen, Odontogr., 1840–1845: 27 (teeth); Cuvier, Règne Anim., ill. ed., Poiss., 1843: 361; Hamilton, Brit. Fish., 2, 1843: 303 (Gt. Brit.); Busch, Selach. Ganoid Encephalo, 1848: pl. 3, fig. 4–6 (brain); Gosse, Nat. Hist. Fish., 1851: 306 (general); Baikie, Zoologist, *11*, 1853: 3846 (Orkney I.); Owen, Cat. Osteol. Roy. Coll. Surg., *1*, 1853: 93, 94 (vertebrae); Gray, Hist. Chile Zool., 2, 1854: 364 (Chile, not seen); Van der Hoeven, Handb. Dierkunde, (2) 2, 1855: 262 (general); Thompson, W., Nat. Hist. Ireland, *4*, 1856: 250 (Ireland); Brito Capello, J. Sci. math. phys. nat. Lisboa, 2, 1870: 142 (Portugal); Miklucho-Maclay, Beitr. Vergl. Neurol. Wirbelt., 1, 1870: 23, pl. 4, fig. 10–12 (brain); McIntosh, Ann. Mag. nat. Hist., (4) *14*, 1874: 424 (w. coast of Scot.); Reguis, Ess. Hist. Nat. Provence, 1 (1), 1877: 47 (Medit.); Günther, Guide to Study Fish., 1880: 317, fig. 112 (general); Brito Capello, Cat. Peix. Portugal, 1880: 46 (Portugal); Moreau, Hist. Nat. Poiss. France, *1*, 1881: 329 (France, Medit.); Perugia, Elenc. Pesc. Adriat., 1881: 52 (Adriatic); Day, Fish. Gt. Brit., 2, 1880–1884: 289, pl. 152 (descr., Gt. Brit.); Jordan and Gilbert, Bull. U.S. nat. Mus., 16, 1883: 872 (name); Jouan, Mem. Soc. Imp. Sci. nat. Cherbourg, *24*, 1884: 313 (Cherbourg, France); Günther, "Challenger" Rep., *31*, 1889: 5 (S. Atlant., Japan); Bellotti, Atti Soc. ital. Sci. nat., *33*, 1891: 112; Carus, Prod. Fauna Medit., 2, 1889–1893: 512 (Medit., locs. and refs.); Smitt, Hist. Scand. Fish., 2, 1895: 1130, pl. 50, fig. 3 (descr., refs., Scand. locs.); Murray, "Challenger" Rep., Summary, 2, 1895: 892, 893, 1207 (S. of Japan; west. S. Atlant. off Rio de la Plata); Nobre, Ann. Sci. nat. Porto., 2, 1895: 225 (Portugal); Osorio, J. Sci. math. phys. nat. Lisboa, (2) 5, 1898: 200 (St. Thomé I.); Evans, Ann. Scot. nat. Hist., 7, 1898: 239 (Firth of Forth, Scot.); Sicher, Atti Accad. gioenia, (4) *11*, 1898: 14 (Medit.); Jouan, Mem. Soc. Sci. nat. Cherbourg, *31*, 1900: 222 (Cherbourg); Steinhardt, Arch. Naturg., Jahrg. *69* (*1*),

25. For references for the central and western Pacific, Australasian region and Indian Ocean, see Fowler (Bull. U.S. nat. Mus., 100 [*13*], 1941: 178).

1903: pl. 1, fig. 13–15 (oral and pharyngeal denticles); Richard, Bull. Inst. océanogr. Monaco, *19*, 1904: 29 (off Portugal); Borca, Arch. Zool. exp. gén., (4) *4*, 1905: 204, 208 (Roscoff, France); Imms, Proc. zool. Soc. Lond., 1905: 43 (denticles); Lo Bianco, Mitt. zool. Sta. Neapel, *19*, 1909: 666 (embryos, food, Medit.); Le Danois, Ann. Inst. océanogr. Monaco, *5* (5), 1913: 20 (Eng. Channel); Jenkins, Fish. Brit. Isles, 1925: 307, pl. 118 (Gt. Brit.); Ehrenbaum, in Grimpe and Wagler, Tierwelt N.- u. Ostsee, Lief 7 (12e), 1927: 12 (general, North Sea, Baltic); Wilson, Proc. U.S. nat. Mus., *60* (5), 1932: 60, 440 (parasites); Ehrenbaum, Cons. explor. Mer Faune Ichthyol. N. Atlant., 1933: plate not numbered (N. Europ. distrib.); Belloc, Rev. des Trav. Pêches Marit., 7 (2), 1934: 131 (Morocco, Rio de Oro, Senegal, Madeira, Azores, Canaries, Gorringe Bank); Ranzi, Pubbl. Staz. Zool. Napoli, *13*, 1934: 351, 398 (uterine fluid and refs. to placenta); Nobre, Fauna Marinha Port. Vert., *1*, 1935: 421 (off Portugal); Calzoni, Pubbl. Staz. Zool. Napoli, *15*, 1936: 169 (develop. placenta, Medit.); Needham, Biochem. Morphogen., 1942: 41 (embryonic weight).

Squalus (Carcharhinus) caeruleus Blainville, in Vieillot, Faune Franc., 1825: 90, pl. 23, fig. 1[26] (descr., Medit.).

Squalus (Carcharhinus) glaucus Blainville, 1825: in Vieillot, Faune Franc., 1830: 92, pl. 23, fig. 2[26] (descr.); Gray, List. Fish. Brit. Mus., 1851: 44 (English coast).

Squalus (Carcharias) glaucus Jenyns, Manual Brit. Vert. Anim., 1835: 499 (Gt. Brit.).

Carcharias (Prionodon) glaucus Müller and Henle, Plagiost., 1841: 36, pl. 11 (descr., Medit., Atlant.); Duméril, Hist. Nat. Poiss., 1865: 353 (descr., distrib.); Günther, Cat. Fish. Brit. Mus., *8*, 1870: 364 (descr., East Atlant., Medit., St. Helena, India); Van Beneden, Poiss. Côtes Belg. Parasit., 1870: 4 (parasites, Belgium); Doderlein, Man. Ittiol. Medit., *2*, 1881: 42 (Medit.);[27] Rochebrune, Act. Soc. linn. Bordeaux, (4) *6*, 1882: 41; Faune Senegambie, Poiss., *1*, 1882–1885: 18 (color, Senegambia); Werner, Zool. Jb., Syst. Abt., *21*, 1904: 283 (Medit.); Metzelaar, Trop. Atlant. Visschen, 1919: 187 (St. Thomas, W. I.); Lampe, Dtsch. Sudpol. Exped., (15) Zool., 7 (2), 1914: 213 (Simonstown, S. Afr.).

Carcharias (Prionodon) hirundinaceus Valenciennes, in Müller and Henle, Plagiost., 1841: 37; Duméril, Hist. Nat. Poiss., *1*, 1865: 354 (descr., Brazil).

Prionodon glaucus Bocage and Brito Capello, Poiss. Plagiost. Port., 1866: 17 (chars., Portugal); Canestrini, in Cornalia, *et al.*, Fauna d'Ital., *19*, 1872: 47 (Medit.); Gegenbaur, Unters. Vergl. Anat. Wirbelt., *3*, 1872: pl. 2, fig. 4, pl. 7, fig. 4, pl. 18, fig. 5 (skull, skelet.); Graeffe, Arb. zool. Inst. Univ. Wien, *28*, 1886: 446 (Medit.); de Braganza, Result. Invest. Sci. "Amelia," *2*, 1904: 44, 45 (off Portugal); Borri, Mem. Soc. tosc. Sci. nat., *44*, 1934: 89 (Medit.).

Carcharinus glaucus Jordan and Gilbert, Bull. U.S. nat. Mus., 16, 1883: 22, 60 (descr., range); Jordan, Rep. U.S. Comm. Fish. (1885), 1887: 795 (distrib.); Eigenmann, Proc. U.S. nat. Mus., *15*, 1893: 133 (S. California); Barnard, Ann. S. Afr. Mus., *21* (1), 1925: 26 (Agulhas Bank); Norman and Fraser, Giant Fishes, 1937: 36 (general).

Carcharias pugae Perez Canto, Estud. Escual., Chile, 1886: 2 (Chile); Philippi, An. Univ. Chile, *71*, 1887: 541, pl. 1, fig. 2 (Valparaiso).

Prionace glauca Jordan and Evermann, Rep. U.S. Comm. Fish. (1895), 1896: 215 (distrib.); Bull. U.S. nat. Mus., 47 (1), 1896: 33 (descr., California); Linton, Proc. U.S. nat. Mus., 20, 1897: 423, 424, 439, 443, 447, 449 (parasites, Woods Hole record); Smith, H. M., Bull. U.S. Fish Comm., *17*, 1898: 88 (Woods Hole record); Bean, Rep. For. Comm. N. Y., 1901: 377 (Woods Hole); Bull. N. Y. St. Mus., 60, Zool. *9*, 1903: 25 (descr.); Sharp and Fowler, Proc. Acad. nat. Sci. Philad., *56*, 1904: 505 (Nantucket); Collett, Norges Fiske, *3*, 1905: 76 (south. Norway); Cornish, Contr. Canad. Biol. (1902–1905), 1907: 81 (off C. Breton, Nova Scotia); Ribeiro, Arch. Mus. nat. Rio de J., *14*, 1907: 51, 199 (Brazil); Starks and Morris, Univ. Calif. Publ. Zool., *3*, 1907: 167 (San Diego, California); Kendall, Occ. Pap. Boston Soc. nat. Hist., 7 (8), 1908: 4 (Woods Hole, Nantucket); Fowler, Proc. Acad. nat. Sci. Philad., *60*, 1908: 62 (Italy); Roule, Bull. Inst. océanogr. Monaco, 243, 1912: 6 (loc. records, E. Atlant. and Medit.); Halkett, Check List Fish. Canad., 1913: 40 (Banks of Newfound-

26. We cannot find whether or not this plate was ever published.
27. See Doderlein, 1881, for additional references for the Mediterranean in publications not accessible to us.

land, California); Sumner, Osburn and Cole, Bull. U.S. Bur. Fish., *31* (2), 1913: 735 (Woods Hole region); Nichols and Murphy, Brooklyn Mus. Sci. Bull., *3* (1), 1916: 9 (habits, N. York and south. New Engl.); Osburn and Nichols, Bull. Amer. Mus. nat. Hist., *35*, 1916: 141 (lower California); Ribeiro, Fauna brasil. Peixes, *2* (1) Fasc. 1, Mus. nac. Rio de J., 1923: 9, pl. 1 (descr., Brazil); Nichols and Breder, Zoologica, N. Y., *9*, 1927: 15 (near N. York); Rey, Fauna Iberica Peces, *1*, 1928: 336 (descr., Spain); Jordan, Manual Vert. Anim. NE. U.S., 1929: 9 (general); Breder, Field Bk. Mar. Fish. Atlant. Coast, 1929: 15 (general); Jordan, Evermann and Clark, Rep. U.S. Comm. Fish. (1928), 2, 1930: 17 (distrib.); Schroeder, Bull. Boston Soc. nat. Hist., *58*, 1931: 3 (Gulf of Maine, Georges Bank); Bellon and Mateu, Notas Inst. esp. Oceanogr., (2) 53, 1931: 16 (Canaries, Iberian Penin.); Gudger, Sci. Mon. N. Y., *34*, 1932: 417 (food); Beebe and Tee-Van, Zoologica, N. Y., *13*, 1932: 119 (Bermuda); Field Bk. Shore Fish. Bermuda, 1933: 26 (Bermuda); Gowanloch, Bull. La. Conserv. Dep., 23, 1933: 222 (name only); Young and Mazet, "Shark Shark," 1933: 267 (general); Bigelow and Schroeder, Canad. Atlant. Fauna, biol. Bd. Canad., 12e, 1934: 8 (descr., distrib.); Piers, Proc. N. S. Inst. Sci., *18*, 1934: 202 (off Nova Scotia); Nichols and La Monte, Amer. Mus. Novit., 901, 1935: 6 (Nova Scotia); Vladykov, Proc. N. S. Inst. Sci., *19*, 1935: 7 (Nova Scotia); Vladykov and McKenzie, Proc. N. S. Inst. Sci., *19*, 1935: 45 (Nova Scotia); Walford, Fish Bull. Sacramento, 45, 1935: 34 (Pacif. coast U.S., Gulf of California to Washington); Barnhart, Mar. Fish. south. Calif., 1936: 9 (off California); Bigelow and Schroeder, Bull. U.S. Bur. Fish., *48*, 1936: 32 (small sizes, Gulf of Maine, Grand Banks); Schultz, Publ. Biol., Univ. Wash., Seattle, *2* (4), 1936: 131 (Washington, Oregon); Howell-Rivero, Mem. Soc. cubana Hist. Nat., *11*, 1937: 285 (Cuba); Wise, Tigers of the Sea, 1937: 68, 176 (general); Tortonese, Atti Soc. ital. Sci. nat., 77, 1938: 300 (Medit., Japan); Springer, Proc. Fla. Acad. Sci., *3*, 1939: 17 (Florida); Beebe and Tee-Van, Zoologica, N. Y., *26*, 1941: 104 (Gulf of California); Norris, Plagiost. Hypophysis, 1941: 30, pl. 62, fig. 30–32 (brain); Lunz, Bull. S. Carolina St. Planning Bd., 14, 1944: 26 (Florida); Bigelow and Schroeder, Guide Comm. Shark Fish., Anglo Amer. Caribb. Comm., Wash., 1945: 121, fig. 43 (descr., habits, range, ill.). Not *Prionace glauca* Townsend, Bull. N. Y. zool. Soc., *34* (6), 1931: upper fig. on p. 170, which is an isurid.

Galeus glaucus[28] Garman, Mem. Harv. Mus. comp. Zool., *36*, 1913: 145, pl. 3, fig. 1–3 (descr., Mass., Japan); Nichols, Abstr. Linn. Soc. N. Y., 20–23, 1913: 91 (off N. York); Roule, Result. Camp. sci. Monaco, *52*, 1919: 114 (E. Atlant.); Linton, Proc. U.S. nat. Mus., *64*, 1924: (21), 7, 38, 39, 44 (parasites, Woods Hole); Bigelow and Welsh, Bull. U.S. Bur. Fish., 40 (1), 1925: 28 (Gulf of Maine, Nova Scotia); Bigelow and Schroeder, Bull. Mus. comp. Zool. Harv., *68*, 1927: 244 (Gulf of Maine); Firth, Bull. Boston Soc. nat. Hist., *61*, 1931: 9 (off C. Cod); Wilson, Bull. U.S. nat. Mus., 152, 1932: 440, 453 (parasites, Woods Hole); Wise, Tigers of the Sea, 1937: 176 (general); White, Bull. Amer. Mus. nat. Hist., *74*, 1937: 125, pl. 2, fig. g; pl. 6, fig. h; pl. 13, fig. b; pl. 19, fig. a; pl. 26, fig. a; pl. 38, fig. d, e; pl. 50, fig. a (class., anat.).

Glyphis glaucus Fowler, Copeia, 30, 1916: 36 (mid. Atlant. coast); Proc. Acad. nat. Sci. Philad., *78*, 1926: 276 (Chile); Mem. Bishop Mus., *10*, 1928: 19 (Oceania); Proc. Acad. nat. Sci. Philad., *80*, 1929: 608 (N. Jersey); Bull. Amer. Mus. nat. Hist., *70* (1), 1936: 56 (descr., W. Afr.); Bull. U.S. nat. Mus., 100 (*13*), 1941: 178 (descr., refs. for west. Pacif., Australasia, Indian Ocean, Atlant.); Arqu. Zool. Estado São Paulo, *3*, 1942: 128 (listed, Brazil).

Eulamia glaucus Marchand, Mar. biol. Rep. Cape Town, *2*, 1935: 36 (S. Afr.).

Prionace mackei Phillipps, N. Z. J. Sci. Tech., *16*, 1935: 238 (descr., New Zealand).[28a]

Glyphis mackei Fowler, Bull. U.S. nat. Mus., 100 (*13*), 1941: 181 (based on Phillipps, 1935).

Carcharhinus mackei Whitley, Fish. Aust., *1*, 1940: 106 (descr., New Zealand, Tasmania, Australia).

28. The name *Galeus glaucus* was first proposed by Valmont (Dict. Hist. Nat., 1, 1768: 371) and next by Duhamel (Traité Gén. Pêches, 3 [9], 1777: 298, pl. 19, fig. 6), but by ruling of the International Commission on Zoological Nomenclature neither Valmont's nor Duhamel's names are to be taken into consideration (see footnote 1, p. 214), for if binomial they are so only accidentally.

28a. See footnote 25, p. 289.

Probable References:

Carcharias obscurus Storer, Proc. Boston. Soc. nat. Hist., *2*, 1839: 256 (teeth, Massachusetts Bay); Mass. Zool. Bot. Surv., Rept., Fish., 1839: 184 (same as preceding); Mem. Amer. Acad. Arts Sci., N. S. *9*, 1867: 219, pl. 36, fig. 2; Fishes Mass., 1867: 243, pl. 36, fig. 2 (descr., ill., Provincetown, Massachusetts, not *Squalus obscurus* Lesueur, 1818).

Carcharias gracilis Philippi, An. Univ. Chile, *71*, 1887: 539, pl. 2, fig. 1 (Chile).[29]

Genus *Scoliodon* Müller and Henle, 1837[1]

Scoliodon Müller and Henle, S. B. Akad. Wiss. Berlin, 1837: 114; Arch. Naturg., (3) *1*, 1837: 397, without type or included species; Plagiost. 1841: 28; type species, *Carcharias (Scoliodon) laticaudus* Müller and Henle, India, designated by Gill, Ann. N. Y. Lyc., 7, 1862: 401; equals *Carcharias sorrakowah* Cuvier, 1817.

Generic Synonyms:

Squalus (in part) Richardson, Fauna Boreal. Amer., *3*, 1836: 289; and subsequent authors; not *Squalus* Linnaeus, 1758.

Carcharias (in part) Cuvier, Règne Anim., *2*, 1817: 388; and subsequent authors; not *Carcharias* Rafinesque, 1810.

Cynocephalus Bleeker, Verh. Akad. Wet. Amst., *18*, 1879: 2; type species, *Carcharias* (*Scoliodon*) *macrorhynchus* Bleeker, 1858, E. Indies; not *Cynocephalus* Gill, 1861, which equals *Prionace* (see footnote 3, p. 280).

Carcharinus (in part) Adams and Kendall, Bull. U.S. Bur. Fish, *9*, 1891: 292, 308; and subsequent authors; not *Carcharhinus* Blainville, 1816.

Rhizoprion Ogilby, Mem. Qd. Mus., *3*, 1915: 132; type species *Carcharias* (*Scoliodon*) *crenidens* Klunzinger, 1880, Australia, equals *Carcharias palasorrah* Cuvier, Règne Anim., *2*, 1829: 338; preoccupied by Jourdan, 1861, for fossil mammals (Cetacea).

Rhizoprionodon Whitley, Aust. Zool., *5*, 1929: 354; substitution for *Rhizoprion* Ogilby, 1915, preoccupied by Jourdan, 1861, for fossil mammals (Cetacea).

Doubtful Synonym:

Lamna (in part) Storer, Mem. Amer. Acad. Arts Sci., N. S. *2*, 1846: 504 (Storer's *L. punctata* appears to have been a combination of *Lamna nasus* and *Squalus punctatus* Mitchill, 1815, which has sometimes been thought to have been *Scoliodon terrae-novae* Richardson, 1836; see Synonyms, p. 303); not *Lamna* Cuvier, 1817.

Generic Characters. No spiracles; anal not more than about twice as long at base as 2nd dorsal; 2nd dorsal not more than 40% as long at base as 1st dorsal; midpoint of base of 1st dorsal about equidistant between origin of pelvics and axil of pectoral (sometimes a little nearer one than the other); labial furrow either confined to corner of mouth or extending out onto one jaw or both; caudal peduncle with a triangular precaudal pit below as well as above, but without lateral longitudinal ridges; gill openings short, the length of the longest only about equal to the diameter of eye; anterior margin of nostril with a

29. The forward position of the first dorsal fin in Philippi's (1887) illustration of his *gracilis* makes it doubtful whether or not it is identical with *glaucus*, which it otherwise resembles.

1. The fossil genus *Alopiopsis* Lioy (Atti Soc. ital. Sci. nat., *8*, 1865: 398, pl. 4) is included in the synonymy of *Scoliodon* by Fowler (Bull. U.S. nat. Mus., 100 [*13*], 1941: 131). But its teeth, as pictured, are quite different from those of *Scoliodon*.

small lobe; teeth alike in the 2 jaws, erect, narrow-cusped in front, but broad and strongly oblique along sides of jaws, their outer margins deeply notched, their edges smooth or slightly wavy at base, their bases not swollen. Characters otherwise those of the family.

Remarks. These are small, warm-water sharks, seldom if ever encountered far from land. In some localities they are the most abundant sharks. All are fish-eaters, so far as known, and are entirely harmless.

Range. Coastal waters in tropical and warm-temperate seas; Morocco to Cameroon; North Carolina (accidentally to Bay of Fundy) to Uruguay in the Atlantic; Mexico to Panama in the eastern Pacific; China and Japan to Australia in the western Pacific; Indian Ocean (including Red Sea and Arabian Gulf) south to Natal.

Fossil Teeth. Eocene to Miocene, Europe.

Species. Of the dozen or so named forms that fall in *Scoliodon,* as defined here, all but one (possibly two) are Indo-Pacific and represent not more than eight good species at most. And it is likely that critical comparison of collections from different seas would result in a further reduction, because *terrae-novae* of the Atlantic, the only species of which a large series has been examined, shows considerable variation (see discussion, p. 299); hence, others may also. Furthermore, the differences that now seem to be diagnostic are so inconspicuous, and the several supposed species all resemble one another so closely in general appearance, that identification of individual specimens calls for close examination if they happen to be from regions where more than one kind is to be expected. For this reason it is not yet possible to define the ranges of any of the Indo-Pacific species in detail.

Tentative Key to Species[2]

1a. Origin of 2nd dorsal posterior to base of anal; labial furrow confined to corner of mouth, not extending inward along either jaw. *dumerilii* Bleeker, 1856.
East Indies, southern China.

1b. Origin of 2nd dorsal over rear part of base of anal; labial furrow extends inward from corner of mouth for some distance along one jaw or both.

- 2a. Lower labial furrow considerably longer than upper, which is very short; base of anal about twice as long as base of 2nd dorsal. *sorrakowah* Cuvier, 1829.
India, Malaysian region, China, Japan.
- 2b. Upper labial furrow at least as long as lower, if latter is present; base of anal less than twice as long as base of 2nd dorsal.
  - 3a. Origin of 2nd dorsal definitely anterior to rear end of base of anal.
    - 4a. A short labial furrow on upper jaw directed outward at right angles to the jaw; none on lower jaw. *jordani* Ogilby, 1908.
Australia.

2. *Carcharias palasorrah* of Cuvier (Règne Anim., 2, 1829: 388), commonly referred to *Scoliodon,* falls in *Hypoprion* as here defined, its teeth being conspicuously serrate or denticulate at the base on the outer side.

4b. A labial furrow on lower jaw as well as on upper.

5a. Origin of 1st dorsal over inner corner of pectoral when latter is laid back, or a little anterior to it.

6a. Lower labial furrow nearly or quite as long as upper; distance from tip of 2nd dorsal to upper precaudal pit only 2/3 to 3/4 as long as horizontal diameter of eye. *vagatus* Garman, 1913.
Zanzibar.

6b. Upper labial furrow considerably longer than lower; distance from tip of 2nd dorsal to upper precaudal pit longer than horizontal diameter of eye.

7a. Upper labial furrow only about 1/2 to 3/4 as long as horizontal diameter of eye.
terrae-novae Richardson, 1836, p. 295.[3]

7b. Upper labial furrow as long as horizontal diameter of eye, or a little longer.

8a. Distance from tip of 2nd dorsal to upper precaudal pit only about as long as horizontal diameter of eye; upper labial furrow about 1 1/3 times as long as horizontal diameter of eye; lower furrow only about 1/2 as long as upper. *longurio* Jordan and Gilbert, 1882.
Eastern Pacific, Mexico to Panama.

8b. Distance from tip of 2nd dorsal to upper precaudal pit about 1 1/2 times as long as horizontal diameter of eye; upper labial furrow only about as long as diameter of eye; lower furrow about 2/3 as long as upper.
intermedius Garman, 1913.
Philippines, East Indies.[4]

5b. Origin of 1st dorsal a little posterior to inner corner of pectoral when latter is laid back. *longmani* Ogilby, 1912.
Australia.[4]

3b. Origin of 2nd dorsal over rear end of base of anal.
walbeehmi Bleeker, 1856.
Indian Ocean south to Natal, Malaysian region, southern China, Formosa, Japan.[4]

3. Including *lalandii* Müller and Henle, 1841.

4. *Intermedius*, *longmani* and *walbeehmi* resemble one another so closely that it is doubtful whether they actually represent more than one rather variable species. We have studied the type specimens of *intermedius*. The specimens recorded by Garman (Mem. Harv. Mus. comp. Zool., *36*, 1913: 112) as *walbeehmi* show the diagnostic characters of *sorrakowah*, with a specimen of which (identified by Garman) we have compared them. We have also examined the type of *vagatus*, specimens of *longurio* and the extensive series of *terrae-novae* listed on p. 295.

Scoliodon terrae-novae (Richardson), 1836

Sharp-nosed Shark

Figures 49, 50

Study Material. 115 specimens, 175 to 930 mm. long, from Uruguay, Brazil, Venezuela, Cuba, the Bahamas, Texas, Alabama, Florida, South and North Carolina, and one from Grand Manan Island at the mouth of the Bay of Fundy (Harv. Mus. Comp. Zool., U.S. Nat. Mus. and Bingham Oceanogr. Coll.); also two, about 316 mm. long, from Ashantee, tropical West Africa (U.S. Nat. Mus., No. 42212, 42247).

Distinctive Characters. S. terrae-novae is easily separable from such of the other West Atlantic members of its family as lack spiracles and have the 1st dorsal far forward, by the presence of well marked labial furrows around the corners of the mouth and inward along both jaws, and by its teeth.

Description. Broad snout type. Proportional dimensions in per cent of total length. Male, 522 mm., from Rio de Janeiro, Brazil (Harv. Mus. Comp. Zool., No. 160). Female, 783 mm., from the Bahamas (Harv. Mus. Comp. Zool., No. 1144).

Trunk at origin of pectoral: breadth 9.4, 10.6; height 9.0, 11.1.

Snout length in front of: outer nostrils 4.8, 4.0; mouth 7.7, 7.5.

Eye: horizontal diameter 2.4, 2.2.

Mouth: breadth 7.3, 7.2; height 4.4, 5.3.

Nostrils: distance between inner ends 4.9, 5.4.

Labial furrow lengths: upper 1.8, 1.8; lower 1.6, 1.4.

Gill opening lengths: 1st 1.9, 1.9; 2nd 2.3, 2.3; 3rd 2.5, 2.4; 4th 2.3, 2.4; 5th 1.9, 1.8.

First dorsal fin: vertical height 8.0, 9.1; length of base 8.6, 9.0.

Second dorsal fin: vertical height 2.1, 2.6; length of base 2.9, 3.5.

Anal fin: vertical height 2.5, 3.2; length of base 4.5, 5.3.

Caudal fin: upper margin 27.8, 25.6; lower anterior margin 10.7, 11.2.

Pectoral fin: outer margin 13.0, 14.0; inner margin 4.8, 5.1; distal margin 9.1, 11.4.

Distance from snout to: 1st dorsal 29.8, 31.7; 2nd dorsal 59.8, 63.3; upper caudal 72.2, 74.4; pectoral 20.5, 20.6; pelvics 44.2, 46.3; anal 57.0, 60.5.

Interspace between: 1st and 2nd dorsals 21.7, 24.9; 2nd dorsal and caudal 8.1, 7.8; anal and caudal 8.6, 9.1.

Distance from origin to origin of: pectoral and pelvics 24.4, 24.7; pelvics and anal 13.4, 13.3.

Narrow snout type. Proportional dimensions in per cent of total length. Male, 581 mm., from Rio de Janeiro, Brazil (Harv. Mus. Comp. Zool., No. 412). Female, 608 mm., from Rio de Janeiro, Brazil (Harv. Mus. Comp. Zool., No. 91).

Trunk at origin of pectoral: breadth 9.2, 10.8; height 9.6, 10.3.

Snout length in front of: outer nostrils 4.8, 5.5; mouth 7.8, 8.6.

Eye: horizontal diameter 2.2, 2.3.

Mouth: breadth 6.2, 6.7; height 5.2, 5.1.
Nostrils: distance between inner ends 4.6, 4.6.
Labial furrow lengths: upper 1.8, 2.2; lower 1.5, 1.5.
Gill opening lengths: 1st 1.9, 2.1; 2nd 2.1, 2.2; 3rd 2.2, 2.3; 4th 2.2, 2.3; 5th 1.9, 2.1.
First dorsal fin: vertical height 7.8, 8.4; length of base 9.1, 8.1.
Second dorsal fin: vertical height 2.1, 2.2; length of base 2.7, 3.0.
Anal fin: vertical height 2.4, 2.8; length of base 4.7, 5.0.
Caudal fin: upper margin 24.4, 25.5; lower anterior margin 9.7, 11.0.
Pectoral fin: outer margin 12.6, 13.5; inner margin 5.0, 5.7; distal margin 9.3, 10.0.
Distance from snout to: 1st dorsal 31.2, 32.7; 2nd dorsal 64.9, 64.0; upper caudal 75.6, 74.5; pectoral 22.2, 22.0; pelvics 46.5, 47.2; anal 61.5, 61.0.
Interspace between: 1st and 2nd dorsal 24.6, 25.8; 2nd dorsal and caudal 7.7, 8.4; anal and caudal 8.1, 8.8.
Distance from origin to origin of: pectoral and pelvics 25.2, 28.3; pelvics and anal 15.1, 15.8.

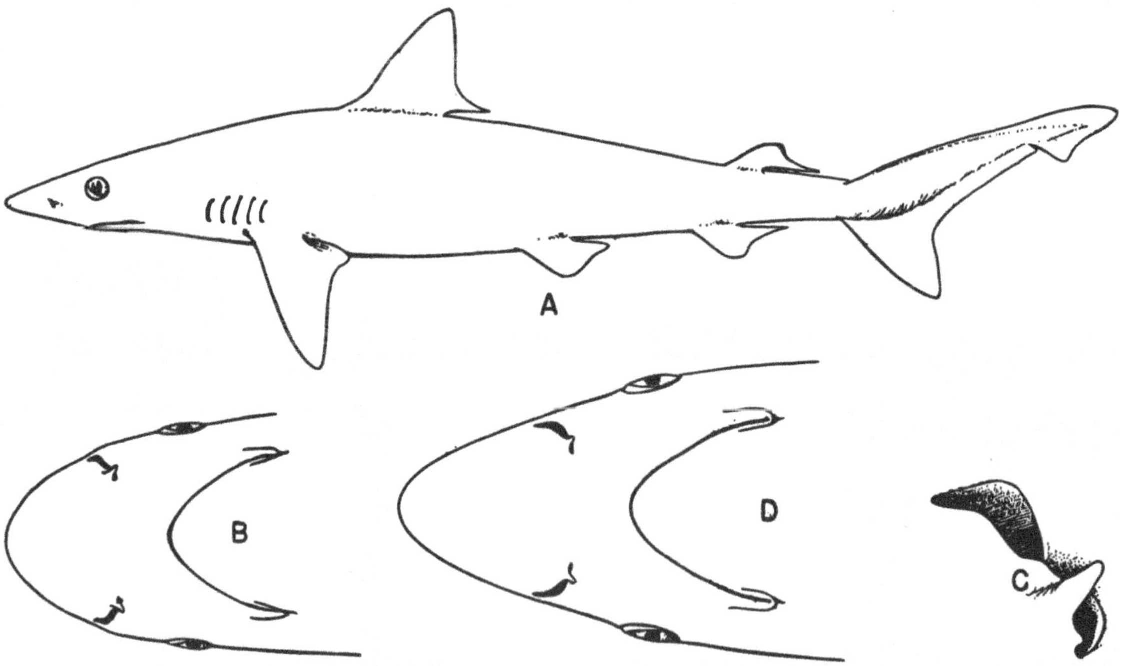

FIGURE 49. *Scoliodon terrae-novae. A* Female, about 783 mm. long, from the Bahamas (Harv. Mus. Comp. Zool., No. 1144). *B* Anterior part of head of same from below, about 0.4 x. *C* Left-hand nostril, about 2.2 x. *D* Head of another specimen with relatively longer snout, 608 mm. long, from Rio de Janeiro, Brazil, about 0.55 x (Harv. Mus. Comp. Zool., No. 91). See discussion p. 299.

Trunk rather slender, its height at origin of 1st dorsal (where highest) about 1/6 of length to origin of caudal. No mid-dorsal ridge. Body sector to cloaca about as long as tail sector. Caudal peduncle about 2/3 to 3/4 as thick as deep. Upper and lower precaudal pits well developed as triangular furrows, the upper the larger. Dermal denticles very small (aver. 0.17 x 0.17 mm. in specimen 610 mm. long), close-spaced, usually overlapping, their blades about as broad as long, usually with 5, but sometimes with only 3, low keels, their posterior margins with as many teeth, the median somewhat the longest, on short pedicels.

Head (to 5th gill opening) a little less than 1/4 of total length, its dorsal profile only slightly convex. Snout flattened above and rather thin toward tip, varying in shape from broadly to more narrowly ovate, its length, in front of a line connecting inner corners of nostrils, also varying from a little shorter than the distance between the inner ends of the latter to nearly 1 1/2 times that long.[5] Eye nearly circular, its anterior edge a little posterior to front of mouth, or nearly opposite latter, its diameter nearly or quite 1/2 as long as distance between inner corners of nostrils. Gill openings evenly spaced, the 3rd (slightly longest) a very little longer than diameter of eye, the 5th slightly the shortest, the 4th above origin of pectoral. Nostril strongly oblique, its inner corner varying with length of snout from a little less to a little more than 1/2 as far from front of mouth as from tip

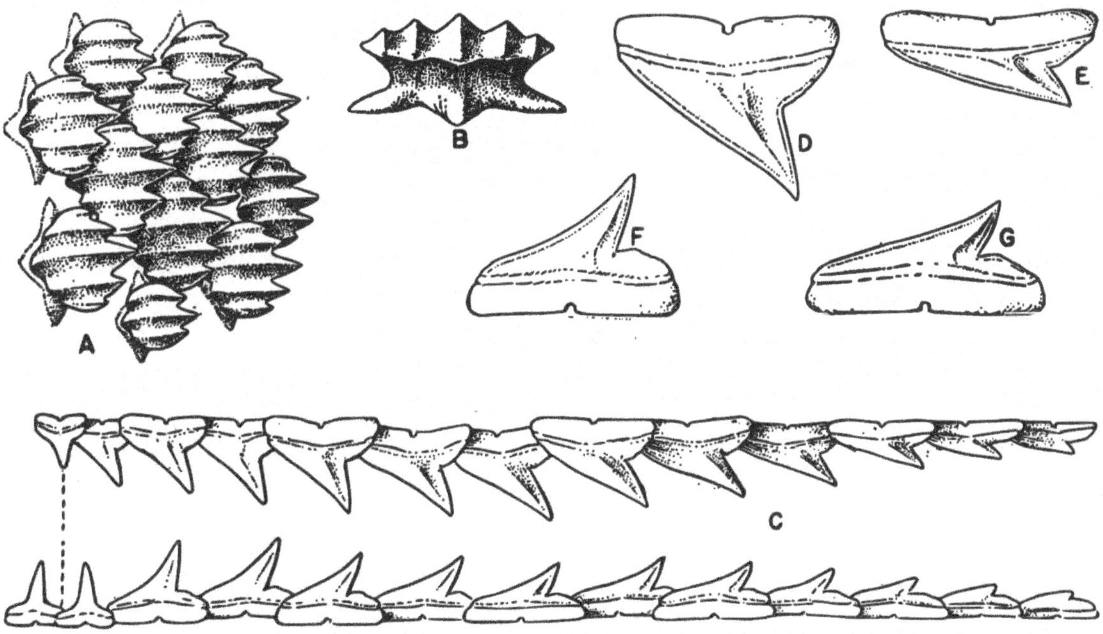

FIGURE 50. *Scoliodon terrae-novae*, illustrated in Fig. 49. *A* Dermal denticles, about 22 x. *B* Apical view of dermal denticle, about 70 x. *C* Upper and lower teeth, left-hand side, about 3 x. *D* Fourth upper tooth. *E* Tenth upper tooth. *F* Fourth lower tooth. *G* Eighth lower tooth. *D-G*, about 6 x.

5. For further comments on this variation, as regards the relationship of the supposedly long, narrow-mouthed form *lalandii* to the shorter, broader-mouted *terrae-novae*, see remarks, p. 299.

of snout, its anterior margin with a short, blunt-tipped, finger-like lobe near the inner corner. Mouth ovate, about 2/3 to 3/4 (60 to 83%) as high as broad, occupying about 2/3 of breadth of head. Labial furrow extending around corner of mouth and onto each jaw, the upper furrow averaging a little more than 3/4 as long as the diameter of eye,[6] the lower averaging about 2/3 as long as upper.

Teeth usually $\frac{12-1-12}{12-12}$, similar in the 2 jaws, except that the lowers are a little smaller than the uppers, with triangular cusp; median upper tooth and 1st lower tooth usually erect, symmetrical and smaller than those on either side, but those along sides of jaws increasingly oblique toward corners of mouth; inner margins slightly concave, outer margins deeply notched about midway toward base, with the basal sector strongly convex, the edges smooth, or at most slightly wavy, basally, on the outer side; the 10th to 12th successively smaller than 2nd to 9th.

First dorsal originates about over inner corner of pectoral when latter is laid back or a little anterior to it, the midpoint of its base varying from a little nearer to axil of pectoral than to origin of pelvics to a little nearer to the latter than to the former, its anterior margin only weakly convex, apex subacute, its posterior margin moderately concave basally, its free rear corner slender and a little more than 1/3 as long as its base, its rear tip over, or a little anterior to, origin of pelvics. Second dorsal on an average only about 1/3 as long at base as 1st, relatively lower, its apex broadly rounded, its free rear tip much more slender and elongate, being about 1 1/2 times as long as its base, its origin about over, or a little posterior to, midpoint of base of anal; the distance from rear tip of 2nd dorsal to precaudal pit about 1 1/4 to 1 1/3 times as long as diameter of eye. Caudal about 1/4 of total length, with moderately raised axis, narrowly rounded tip and deep subterminal notch, its terminal sector about 1/3 the length of fin, its lower anterior corner forming a definite triangular lobe with subacute tip, a little more than 1/3 as long as the upper, each measured from the respective precaudal pit. Anal similar to 2nd dorsal in shape, but averaging about 1 1/2 times as long at base, hence considerably larger in area. Pelvics a little longer at base than anal, with nearly straight anterior margins, weakly concave distal margins, rounded apices and subacute tips. Pectoral only about as long as length of 1st dorsal along its outer margin and smaller than the latter in area,[7] a little more than 1/2 as broad as long, its anterior margin weakly convex, distal margin nearly straight toward tip but moderately concave toward inner corner, the latter subacute, the apex more rounded.

Color. Brownish to olive-gray above, white below and along rear margins of pectorals; dorsals and caudals more or less dark-edged, the 2nd dorsal and lower lobe of caudal the most widely so, especially in small specimens.

Size. Commonly these sharks are about 26 to 30 inches long when mature, rarely growing much larger than 36 inches, the greatest length definitely recorded for a West

6. Average 82 per cent; extremes 64 and 110 per cent.

7. In most of the Carcharhinidae the pectoral is considerably longer than the 1st dorsal, and larger in area.

Atlantic specimen being only about 36½ inches or 930 mm. (Harv. Mus. Comp. Zool., No. 702, from Rio de Janeiro, a male with large claspers).[8]

Developmental Stages. The eggs, in early development, are enclosed in thin yellow shells with pointed ends and are imbedded in crypt-like depressions in the walls of the maternal uteri. A preliminary account[9] suggests that a placental connection later develops between yolk-sac and mother, *i.e.*, that the shark is truly viviparous, as are its close relatives, *S. sorrakowah* and *S. walbeehmi* of the Indian Ocean.[10]

Size. Newborn specimens are usually about 275 to 400 mm. long.[11] It appears that some males may mature when only perhaps 600 mm. long, for we have seen one of 650 mm. with claspers 61 mm. long (Rio de Janeiro); but in another of 660 mm. from the same locality they were only 52 mm. long, while in two others of about the same size (642 and 650 mm.) from Florida they were 33 mm. and 30 mm. long respectively. A 20-inch specimen from Haiti weighed three pounds.

Remarks. Opinions have differed as to whether or not the form with the longer and more pointed snout deserves recognition as a distinct species (*lalandii* Müller and Henle, 1841; see Synonyms, p. 301). Examination of the extensive series above (p. 295) shows that an unbroken gradation occurs from those with longer, narrower snouts to those with shorter and broader snouts. Since we have not been able to draw any sharp line between them in this or in any other respect, the two extremes are included here under the one specific name. But the situation still remains somewhat obscure, for while the broader-snouted specimens appear to be the more common throughout the latitudinal range of the combined species, north to south, the range of the narrower-snouted members appears to be definitely restricted to warmer waters, being recorded only for the southern part of the Gulf of Mexico, West Indies (Martinique, Guadeloupe), Pernambuco and Rio de Janeiro. This raises the interesting question whether the two forms may not represent two species which were originally distinct and with distinct ranges, but which have so hybridized (their ranges having overlapped) that it is not possible to distinguish between them now.[12]

8. A reputed length of 2,135 mm. (Fowler, Bull. Amer. Mus. nat. Hist., *70* [1], 1936: 45) is so much larger than the usual run of adults as to suggest an error.

9. Gudger, Science, N. S. *41*, 1915: 439.

10. For accounts of the placental cord in these, see Southwell and Prashad (Rec. Indian Mus., *16*, 1919: 223, pl. 17, fig. 1, 2, 4, 7, 8 [*walbeehmi*], and 225, pl. 17, fig. 6, 9, 10 [*sorrakowah*]); see also Thillayampalam (Indian zool. Memoir 2, Lucknow, *Scoliodon*, 1928: 107, fig. 93 [*sorrakowah*]).

11. A series of eleven newly born specimens from Texas, with traces of the umbilical scar still visible, range from 280 to 407 mm. in length.

12. In twelve specimens with the broadly-rounded snout, 410 to 930 mm. in total length, measurements are: distance from tip of snout to outer corner of nostril 67 to 76% (average 71%) of the distance between outer nostrils; width of head at outermost part of nostril 95 to 103% (average 99%) of length of snout in front of mouth; shortest distance from inner end of nostril to mouth 42 to 52% (average 45%) of length of snout in front of mouth; distance between nostrils 65 to 73% (average 69%) of length of snout in front of mouth.

In seven specimens with narrowly rounded snout, 544 to 660 mm. in total length, the measurements are: distance from tip of snout to outer end of nostril 77 to 87% (average 83%) of distance between outer ends of nostrils; width of head to outer end of nostril 78 to 90% (average 84%) of length of snout in front of mouth; shortest distance from inner end of nostril to mouth 33 to 38% (average 35%) of length of snout in front of mouth; distance between nostrils 51 to 59% (average 56%) of length of snout in front of mouth.

Habits. This little shark is often taken along the beach, even in the surf, as well as in harbors and partially enclosed sounds and estuaries. In fact, so far as we are aware it has never been reported more than a mile or two out from the land or from water more than a few fathoms deep. It occurs in brackish water in Mississippi (Pascagoula River) and even in tidal fresh water elsewhere.

It feeds chiefly on small fish that may be available locally; in North Carolina waters, for example, its stomach is often full of menhaden (*Brevoortia*); parrotfish have been found in its stomach in Haitian waters. It is also known to eat shrimps and mollusks, and it bites readily on almost any bait.

It is probable that young are born chiefly in late spring and summer in the northern sector of its range, for newborn specimens still showing traces of the umbilical scar have been reported from Florida in July, when they are common in Texan waters also;[13] they are abundant off the mouth of the Mississippi in August, and in June and July in North Carolina waters, where gravid females containing both eggs and late-term embryos are reported in August. All that is known of its breeding in more tropical waters is that newly born specimens have been reported from Haiti in early April, and that pregnant females with as many as twelve embryos are taken around Cuba.[14]

Relation to Other Species. It closely resembles *S. longurio* Jordan and Gilbert of the Pacific coasts of Mexico and Panama, but it is separable from the latter by the facts that its upper labial furrow is definitely shorter than the diameter of the eye (as long, or longer in *longurio*) and that it has only 25 rows of teeth in the upper jaw (27 to 29 rows in *longurio*).

Relation to Man. The only commercial value of this little Shark is that some are sold in fish markets in the West Indies and perhaps in South America. On the other hand, its habit of taking the bait intended for better fish makes it a great nuisance to the fishermen at times and places where it is numerous.

Range. Both sides of the tropical and subtropical Atlantic; Morocco to Cameroon and the Cape Verde Islands in the east; Uruguay to North Carolina in the west, and north accidentally to the mouth of the Bay of Fundy.

Occurrence in the Western Atlantic. The chief center of abundance of this Shark appears to lie in the West Indian–Caribbean region and in the Gulf of Mexico, whence it has been recorded at many localities[15] as plentiful. For example, considerable numbers are caught by the Louisiana shrimp-trawlers, and it is present throughout the year around southwestern Florida and among the Keys. However, to the northward it is chiefly a summer visitor only, present in abundance off the mouth of the Mississippi from June until September, and the commonest summer shark along South Carolina and the southern part

13. One small collection of sharks taken in the vicinity of Galveston, Texas, in July included eleven *S. terrae-novae*, ranging in size from 280 to 407 mm., all with a more or less conspicuous umbilical scar.

14. Personal communication from Luis Howell-Rivero.

15. Venezuela, Yucatán, Colón, Curaçao, Trinidad, Martinique, Guadeloupe, Puerto Cabello, Saba, St. Croix, Jamaica, Cuba, Haiti, Bahamas, Florida Keys, western and northwestern Florida, Mississippi, Louisiana, Texas.

of the coast of North Carolina. In some years (*e.g.*, in 1891) it is taken in some numbers even in winter as far north as Cape Lookout. But it has been recorded only once at the mouth of Chesapeake Bay and not at all within the Bay; and it reaches New Jersey and the vicinity of New York only rarely, four specimens being reliably recorded. Occasional specimens do wander even farther to the northward at rare intervals, for several were taken near Woods Hole in the summer of 1916, while the collection of the Harvard Museum of Comparative Zoology contains a specimen taken at Grand Manan Island at the mouth of the Bay of Fundy in 1857 by A. E. Verrill. Early reports of it from Newfoundland are unfounded.[18]

To the southward it occurs in abundance along the coast of Brazil as far as Rio de Janeiro and Rio Grande do Sul, and the collection of the Harvard Museum of Comparative Zoology contains one taken many years ago at Maldonado, Uruguay. But apparently the estuary of the Rio de La Plata marks the southern limit to its usual range in that direction, for it has not been recorded from Argentina. Neither is it known at Bermuda.

Synonyms and References:

Squalus (Carcharias) terrae-novae Richardson, Fauna Boreal. Amer., *3*, 1836: 289 (locality given as "Newfoundland" for specimen received from Audubon, but probably either Florida or South or North Carolina; see footnote 16, p. 301).

Carcharias (Scoliodon) lalandii Müller and Henle, Plagiost., 1841: 30 (descr., Rio de Janeiro, Martinique, Guadeloupe); Duméril, Hist. Nat. Poiss., 1865: 346 (descr., Brazil).

Lamna terrae-novae (in part)[17] Storer, Mem. Amer. Acad. Arts Sci., N. S. *2*, 1846: 504 (in synopsis).

Scoliodon lalandii Castelnau, Anim. Nouv. Rares Amer. Sud, 1855: 100 (Brazil); Hasse, Naturl. Syst. Elasm. besond. Theil, 1882: 268, pl. 39, fig. 1–4 (vertebrae); Garman, Mem. Harv. Mus. comp. Zool., *36*, 1913: 113 (descr.); Fowler, Proc. Acad. nat. Sci. Philad., *69*, 1917: 128 (Colón); Meek and Hildebrand, Field Mus. Publ. Zool., *15*, 1923: 53, pl. 2, fig. 2 (descr., discus., Colón); Jordan, Evermann and Clark, Rep. U.S. Comm. Fish. (1928), 2, 1930: 15 (distrib.); White, Bull. Amer. Mus. nat. Hist., *74*, 1937: 129 (in Key); Fowler, Arqu. Zool. Estado São Paulo, *3*, 1942: 128 (Brazil).

Scoliodon terrae-novae Gill, Proc. Acad. nat. Sci. Philad., Addend., 1861: 59 (class., name); Rep. U.S. Comm. Fish. (1871–1872), 1873: 813 (range); Jordan, Proc. U.S. nat. Mus., *5*, 1882: 245 (Pensacola, Florida); Goode and Bean, Proc. U.S. nat. Mus., *5*, 1883: 240 (Gulf of Mexico); Jordan and Gilbert, Bull. U.S. nat. Mus., 16, 1883: 24 (descr., distrib.); Proc. U.S. nat. Mus., *5*, 1883: 581 (abund., S. Carolina); Nelson, Rep. St. Geol. N. Jersey, *2* (2), 1890: 661 (diagn., N. Jersey distrib.); Bean, B. A., Proc. U.S. nat. Mus., *14*, 1891: 94 (C. Charles, Virginia); Jordan and Evermann, Rep. U.S. Comm. Fish. (1895), 1896: 217 (C. Cod to Brazil); Bull. U.S. nat. Mus., 47 (1), 1896: 43 (descr., C. Cod to Brazil); Jordan and Rutter, Proc. Acad. nat. Sci. Philad., 1897: 91 (Jamaica); Evermann and Bean, Rep. U.S. Comm. Fish. (1896), 1898: 239 (Indian R., Florida); Gilbert, Proc. Wash. Acad. Sci., *2*, 1900: 161 (Brazil); Evermann and Kendall, Rep. U.S. Comm. Fish. (1899), 1900: 48 (Florida); Bean, T. H., Rep. For. Comm. N. Y., 1901: 378 (distrib.); Rep. N. York St. Mus., *60*, Zool. 9, 1903: 29 (descr., general); Jordan and Thompson, Bull. U.S. Bur. Fish., *24*, 1905: 232 (Tortugas, Florida); Linton, Bull. U.S. Bur. Fish., *24*, 1905: 342 (food, parasites); Cole and Barbour, Bull. Mus. comp. Zool. Harv., *50*, 1906: 155 (Yucatán); Fowler, Rep. N. Jersey Mus. (1905), 1906: 63 (N. Jersey); Rep. N. J. Mus. (1906), 1907: pl. 74 (ill.); Ribeiro, Arch. Mus. nac. Rio de J., *14*, 1907: 155,

16. "This species, with others belonging to the Florida fauna, is said by Richardson to have been brought from Newfoundland by Audubon. They doubtless came from some locality in Florida or Carolina" (Jordan and Evermann, Bull. U.S. nat. Mus., 47 [1], 1896: 43; footnote).

17. This was a combination of *Lamna caudata* Dekay (equals *Carcharhinus milberti*, see p. 376) with *Scoliodon terrae-novae* Richardson.

200 (descr., refs., Brazil); Linton, Bull. U.S. Bur. Fish., *26*, 1907: 122, 125 (parasites); Smith, Bull. N.C. geol. econ. Surv., *2*, 1907: 34 (N. Carolina); Wilson, Proc. U.S. nat. Mus., *33*, 1908: 340, 360, 423, 431, 626 (parasites); Fowler, Proc. Acad. nat. Sci. Philad., *60*, 1908: 66 (Florida); Gudger, Amer. Nat., *44*, 1910: 399 (N. Carolina); Linton, Bull. U.S. Bur. Fish., *28*, 1910: 1201, pl. 2 (parasites); Garman, Mem. Harv. Mus. comp. Zool., *36*, 1913: 115, pl. 2, fig. 1–4 (descr.); Radcliffe, Copeia, 26, 1916: 2 (Woods Hole); Coles, Proc. biol. Soc. Wash., *28*, 1915: 90 (N. Carolina); Gudger, Science, N. S. *41*, 1915: 439 (eggs, uterine attachment of young, Tortugas, Florida); Nichols and Murphy, Brooklyn Mus. Sci. Bull., *3* (1), 1916: 18 (distrib.); Radcliffe, Bull. U.S. Bur. Fish., *34*, 1916: 250, pl. 39, fig. 2 (denticles, teeth, N. Carolina); Fowler, Copeia, 30, 1916: 36 (E. coast U.S.); Proc. Acad. nat. Sci. Philad., *71*, 1919: 146 (Jamaica); Wilson, Proc. U.S. nat. Mus., *55*, 1920: 592 (parasites); Fowler, Proc. biol. Soc. Wash., *33*, 1920: 144 (N. Jersey); Proc. Acad. nat. Sci. Philad., *72*, 1921: 394 (Pensacola, Florida); Wilson, Proc. U.S. nat. Mus., *60* (5), 1922: 60 (parasites); Meek and Hildebrand, Field Mus. Publ. Zool., *15*, 1923: 55, pl. 2, fig. 3 (descr., discus., Colón); Ribeiro, Fauna brasil. Peixes, Mus. nac. Rio de J., *2* (1), Fasc. 1, 1923: 12, pl. 4 (Brazil); Linton, Proc. U.S. nat. Mus., *64* (21), 1924: 7, 35, 36, 46, 47, 48 (parasites, Woods Hole, N. Carolina); Chabanaud and Monod, Bull. Etud. Hist. Sci. Afr. Occid. Franc. (1926), 1927: 229 (trop. W. Afr., not seen); Monod, Faune Colon-Franc., 1927: 646 (not seen); Beebe and Tee-Van, Zoologica, N. Y., *10*, 1928: 27 (Haiti); Fowler, Proc. Acad. nat. Sci. Philad., *80*, 1928: 456 (Haiti); Hildebrand and Schroeder, Bull. U.S. Bur. Fish., *43* (1), 1928: 49 (descr., C. Charles, Virginia); Breder, Field Bk. Mar. Fish. Atlant. Coast, 1929: 18 (general); Jordan, Manual Vert. Anim. NE. U.S., 1929: 10 (general); Jordan, Evermann and Clark, Rep. U.S. Comm. Fish. (1928), 2, 1930: 14 (in check list); Wilson, Bull. U.S. nat. Mus., 158, 1932: 439 (parasites); Fowler, Proc. biol. Soc. Wash., *36*, 1933: 27 (Bayport, Louisiana); Borri, Atti Soc. tosc. Sci. nat., *44*, 1934: 90 (Bahia, Brazil); Burton, Sci. Mon. N. Y., *40*, 1935: 283 (abund., S. Carolina); Budker, Bull. Mus. Hist. nat. Paris, (2) 7, 1935: 184 (Dakar, W. Afr.); Fowler, Bull. Amer. Mus. nat. Hist., *70* (1), 1936: 45 (trop. W. Afr.); Bere, Amer. Midl. Nat., *17*, 1936: 604 (Florida); White, Bull. Amer. Mus. nat. Hist., *74*, 1937: 129, pl. 13, fig. C (anat., in Key); Springer, Proc. Fla. Acad. Sci., *3*, 1939: 18 (Florida); Tortonese, Atti Soc. ital. Sci. nat., 77, 1938: 298 (Rio de Janeiro); Longley and Hildebrand, Pap. Tortugas Lab., *34*, 1941: 2 (Tortugas, Florida); Gunter, Amer. Midl. Nat., *28*, 1942: 316 (in brackish and tidal fresh water, Mississippi and Maryland); Fowler, Arqu. Zool. Estad. São Paulo, *3*, 1942: 128 (listed for Brazil); Lunz, Bull. S. C. St. Planning Bd., *14*, 1944: 26 (S. Carolina, Florida); Bigelow and Schroeder, Guide Comm. Shark Fish., Anglo Amer. Caribb. Comm., Wash., 1945: 128, fig. 46 (descr., habits, range, ill.); Fowler, Monogr. Acad. nat. Sci. Philad., 7, 1945: 94, 263 (Beaufort, N. Carolina, south. Florida locs.), 159 (Charleston, S. Carolina).

Carcharias (*Scoliodon*) *terrae-novae* Duméril, Hist. Nat. Poiss., 1865: 346 (descr., Newfoundland, after Richardson, 1836); Günther, Cat. Fish. Brit. Mus., *8*, 1870: 360 (W. Indies, Bahia, Brazil); Werner, Zool. Jb., Syst. Abt., *21*, 1904: 283 (Jamaica [?], N. Carolina); Lampe, Dtsch. Sudpol. Exped., *15*, Zool. 7, 1914: 204, 213 (C. Verde Is.); Metzelaar, Trop. Atlant. Visschen, 1919: 186 (trop. W. Afr.).

Carcharias (*Scoliodon*) *acutus* Steindachner, S. B. Akad. Wiss. Wien, *61* (1), 1870: 575 (Senegambia);[18] Denkschr. Akad. Wiss. Wien, *44*, 1882: 51 (Senegambia); Osorio, Mem. Mus. Bocage Lisbon, 1, 1909: 77 (C. Verde Is.); Metzelaar, Trop. Atlant. Visschen, 1919: 186 (Senegambia); not *Carcharias acutus* Rüppell, 1835, which is a synonym of *Hypoprion palasorrah* (Cuvier); see footnote 2, p. 293.

Carcharhinus (*Scoliodon*) *terrae-novae* Jordan, Rep. U.S. Comm. Fish. (1885), 1887: 796 (distrib.).

Carcharhinus terrae-novae Adams and Kendall, Bull. U.S. Fish Comm., *9*, 1891: 292, 308 (W. Florida); Henshall, Bull. U.S. Fish Comm., *9*, 1891: 384 (SW. Florida); Lönnberg, Ofvers. Förhand. Vet Akad. Stockholm, *51* (3), 1894: 111 (Florida Keys); Wilson, Amer. Nat., *34*, 1900: 355 (listed, North Carolina).

Carcharias terrae-novae Jordan and Gilbert, Proc. U. S. nat. Mus., *5*, 1883: 245 (Gulf of Mexico); Duerden, J. Inst. Jamaica, 6, 1899: 614 (Jamaica); Schreiner and Ribeiro, Arch. Mus. nac. Rio de J., *12*, 1903: 79 (Brazil); von Ihering, Rev. Mus. paul., *2*, 1897: 34 (Brazil).

Carcharias eumeces Pietschmann, Jb. nassau. Ver. Naturk., 1913: 172 (descr., Cameroon, trop. W. Afr.).

18. Steindachner's excellent and detailed description leaves no doubt that his West African specimens were actually *terrae-novae*.

Scoliodon eumeces Fowler, Proc. U.S. nat. Mus., *56*, 1919: 249 (trop. W. Afr.).
Scoliodon acutus Fowler, Bull. Amer. Mus. nat. Hist., *70* (1), 1936: 46; not *Carcharias acutus* Rüppell, 1835.

Probable Synonyms and References:

Squalus porosus Poey, Memorias, *2*, 1860: 339, 452, pl. 19, fig. 11, 12 (teeth, Cuba).[19]

Scoliodon porosus Poey, Repert. Fisico-Nat. Cuba, *2*, 1868, 452; An. Soc. esp. Hist. nat., *5*, 1876: 396; Enumerat. Pisc. Cubens., 1876: 200 (Cuba); Fowler, Fish Culturist, 21 (9), 1942: 66 (size of mature female, and embryos, Cuba); Jordan, Evermann and Clark, Rep. U.S. Comm. Fish. (1928), 2, 1930: 15 (Cuba); Fowler, Proc. Acad. nat. Sci. Philad., *83*, 1931: 391 (Trinidad).

Doubtful References:

Squalus punctatus Mitchill, Trans. Lit. Phil. Soc. N.Y., 1815: 483 (near New York); not *Squalus punctatus* Bloch and Schneider, 1801.

Lamna punctata (in part) Storer, Mem. Amer. Acad. Arts Sci., N. S. *2*, 1846: 504 (this appears to be a combination of *Squalus punctatus* Mitchill with *Lamna nasus*).

Lamna punctata Linsley, Amer. J. Sci., *47*, 1844: 76 (name only, Connecticut).

Aprionodon punctatus Gill, Proc. Acad. nat. Sci. Philad., Addend., 1861: 59 (in synopsis).

Carcharias porosus Goeldi, Bol. Mus. Paraense, *2*, 1898: 488 (Brazil).

Carcharias (*Scoliodon*) *walbenii* Osorio, J. Sci. math. phys. nat. Lisboa, (2) *5*, 1898: 200 (C. Verde, St. Thomé); Metzelaar, Trop. Atlant. Visschen, 1919: 186 (C. Verde); not *Carcharias* (*Scoliodon*) *walbeehmi* Bleeker, 1856.

Carcharias longurio Engelhardt, Zool. Anz., 39, 1912: 648 (St. Thomas, W. Indies); not *Carcharias* (*Scoliodon*) *longurio* Jordan and Gilbert, 1888.

Probably not *Scoliodon porosus* Garman, Mem. Harv. Mus. comp. Zool., *36*, 1913: 112 (descr., Cuba).[19a]

Genus *Aprionodon* Gill, 1861

Aprionodon Gill, Proc. Acad. nat. Sci. Philad., Addend., 1861: 59; Ann. N. Y. Lyc., 7, 1862: 400, 401, 411; substitution for *Aprion* Müller and Henle, 1841, preoccupied by Cuvier and Valenciennes, 1830, for bony fishes, and by Audinet-Serville 1839, for insects; type species, *Aprionodon punctatus* Gill, equals *Carcharias* (*Aprion*) *isodon* Müller and Henle, 1841.[1]

Generic Synonyms:

Carcharias (in part) Rüppell, Neue Wirbelt. Abyssinia, Fische, 1835: 65; and subsequent authors; not *Carcharias* Rafinesque, 1810.

Aprion Müller and Henle, Plagiost., 1841: 31; type species, *Carcharias* (*Aprion*) *brevipinna* Müller and Henle, Java; not *Aprion* Cuvier and Valenciennes, 1830, for bony fishes.

Squalus (in part) Gray, List Fish. Brit. Mus., 1851: 41; not *Squalus* Linnaeus, 1758.

Carcharhinus (in part) Jordan, Rep. U.S. Comm. Fish. (1885), 1887: 796; not *Carcharhinus* Blainville, 1816.

Longmania Whitley, Aust. Zool., *9*, 1939: 231; type species, *Carcharias* (*Aprion*) *brevipinna* Müller and Henle, 1841.

19. *Porosus* is classed as a probable synonym of *terrae-novae* on the strength of Poey's descriptions of it and his illustrations of its teeth. Although a photograph of an unpublished drawing by him of the lower side of its head fails to show any labial furrows on the lower jaw, this may have been an oversight.

19a. Probably a *Carcharhinus*, not a *Scoliodon*, because described as with weakly serrated teeth and wider gill openings. The specimen on which he based his account is no longer in existence.

1. Of the two species included by Gill (1861), *Carcharias* (*Aprion*) *isodon* Müller and Henle must necessarily be taken as the type and not *Squalus punctatus* Mitchill, 1815, which was designated by Jordan (Genera Fish., 3, 1919: 303), because *Aprionodon* was obviously a substitution for *Aprion*. Furthermore, the name *Squalus punctatus*, having been long antedated by Bloch and Schneider (1801), would not be available in the present connection.

Generic Characters. Carcharhinidae with anal less than twice as long at base as 2nd dorsal; without spiracles; caudal peduncle without lateral ridges, but with a precaudal pit below as well as above; midpoint of base of 1st dorsal as near to axil of pectoral as to origin of pelvics, or nearer; 2nd dorsal only about ½ as long at base as 1st and much smaller in area; teeth slender and symmetrical in both jaws, their bases as well as their cusps with smooth edges; gill openings notably large, the longest nearly or quite ½ as long as base of 1st dorsal and more than twice as long as horizontal diameter of eye (see footnote 5, p. 265), the 5th being over origin of pectoral; anterior margin of nostril only slightly expanded; labial furrow around corner of mouth but extending inward for only a very short distance onto either jaw, if at all. Characters otherwise those of the family.

Range. Tropical and subtropical. Senegambia in the eastern Atlantic; North Carolina (perhaps New York) soutn to Cuba and Texas in the western Atlantic; tropical Indian Ocean; Red Sea and Arabian Gulf; India; East Indies and Australia, Indo-China, Japan, Micronesia.

Species. This genus of small sharks, about which little is known, is represented by one species in the Atlantic and by one in the Indo-Pacific.[2]

Key to Species[3]

1a. Snout in front of mouth only about 2/7 (about 29%) as long as head, or ⅔ as long as from eye to 1st gill opening; distance between nostrils a little more than ⅔ as long as snout; pectoral about ½ as long as head. *isodon* Müller and Henle, 1841, p. 304.

1b. Snout in front of mouth about ⅖ (40%) as long as head, and about as long as from eye to 1st gill opening; distance between nostrils ½ as long as snout; pectoral about ⅔ as long as head. *brevipinna* Müller and Henle, 1841.
Arabia, East Indies, Australia, Japan.

Aprionodon isodon (Müller and Henle), 1841

Figure 51

Study Material. Two females, 460 and 504 mm., from off Biloxi, Mississippi, and from Texas (U.S. Nat. Mus.); 4 young males, 500 to 567 mm., from Texas (Harv. Mus. Comp. Zool.).[4]

Distinctive Characters. Easily recognizable among local Carcharhinidae by its slender, symmetrical, smooth-edged teeth, very long gill openings and a 2nd dorsal fin that is much smaller than its 1st dorsal.

2. Another Indo-Pacific species, *Carcharias acutidens* Rüppell, 1835, has usually been referred to this genus, but is placed in *Negaprion* (p. 309) according to generic definitions adopted here.

3. *Carcharias fronto* Jordan and Gilbert (Proc. U.S. nat. Mus., *5*, 1882: 102) is also referred to this genus by Beebe and Tee-Van (Zoologica, N. Y., *28*, 1941: 105), but it is placed here in *Negaprion* because of the large size of its second dorsal fin; for discussion, see footnote 1, p. 309.

4. Contributed by J. L. Baughman.

Description. Proportional dimensions in per cent of total length. Female, 504 mm., from Galveston, Texas (U.S. Nat. Mus., No. 118457). Male, 560 mm., same locality (Harv. Mus. Comp. Zool., No. 35831).

Trunk at origin of pectoral: breadth 11.3, 10.7; height 11.3, 12.1.
Snout length in front of: outer nostrils 4.1, 3.8; mouth 6.8, 7.1.
Eye: horizontal diameter 1.9, 1.8.
Mouth: breadth 9.1, 8.7; height 5.2, 5.6.
Nostrils: distance between inner ends 5.5, 5.3.
Labial furrow lengths: upper 0.9, 0.8; lower 0.8, 0.7.
Gill opening lengths: 1st 4.4, 5.4; 2nd 4.7, 5.7; 3rd 4.8, 5.7; 4th 4.8, 5.6; 5th 4.3, 4.7.
First dorsal fin: vertical height 9.4, 9.0; length of base 10.0, 9.6.
Second dorsal fin: vertical height 2.8, 2.9; length of base 4.8, 4.8.
Anal fin: vertical height 3.4, 3.4; length of base 5.2, 5.6.
Caudal fin: upper margin 28.1, 28.2; lower anterior margin 11.3, 11.8.

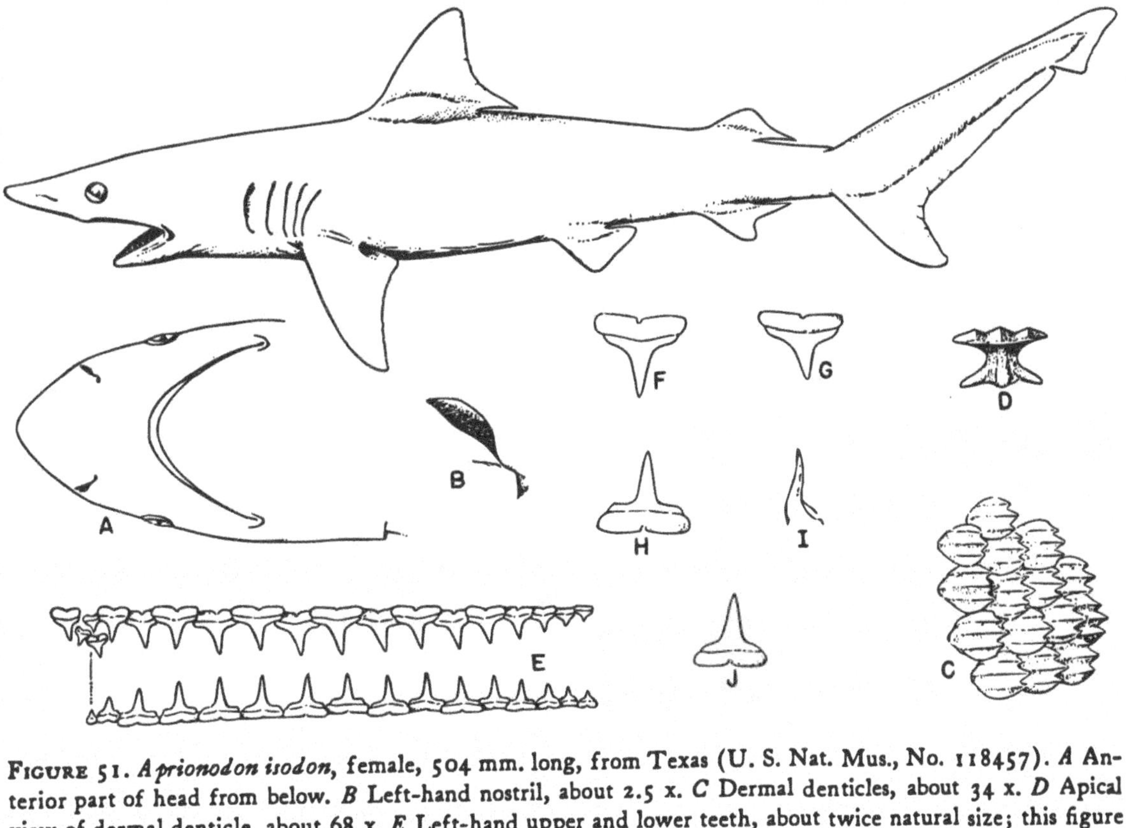

FIGURE 51. *Aprionodon isodon*, female, 504 mm. long, from Texas (U. S. Nat. Mus., No. 118457). *A* Anterior part of head from below. *B* Left-hand nostril, about 2.5 x. *C* Dermal denticles, about 34 x. *D* Apical view of dermal denticle, about 68 x. *E* Left-hand upper and lower teeth, about twice natural size; this figure is inverted by error. *F* Fourth upper tooth. *G* Tenth upper tooth. *H, I* Fourth lower tooth. *J* Sixth lower tooth. *F–J*, about 4 x.

Pectoral fin: outer margin 14.0, 15.5; inner margin 4.9, 5.3; distal margin 10.7, 11.4.

Distance from snout to: 1st dorsal 30.3, 30.7; 2nd dorsal 60.7, 60.7; upper caudal 71.9, 71.8; pectoral 25.0, 24.0; pelvics 48.0, 46.6; anal 59.5, 59.7.

Interspace between: 1st and 2nd dorsals 19.3, 19.6; 2nd dorsal and caudal 6.6, 7.5; anal and caudal 6.0, 6.4.

Distance from origin to origin of: pectoral and pelvics 23.4, 24.1; pelvics and anal 12.5, 13.4.

Trunk moderately slender, its height at 1st dorsal a little less than 1/5 its length to caudal pit. No mid-dorsal ridge. Caudal peduncle only slightly compressed, without lateral ridges. Precaudal pits subtriangular. Body sector to cloaca a little longer than tail sector. Dermal denticles small, closely overlapping, a little broader than long, their blades horizontal, broadly oval, with 3 low ridges and as many short teeth, the median only a little the longest; pedicels slender.

Head about 1/4 of total length, moderately flattened above and a little broader opposite corners of mouth than in region of gill openings. Snout wedge-shaped, its tip narrowly rounded, its length in front of mouth about 1/4 of length of head. Eye approximately circular, its anterior edge about opposite front of mouth, its horizontal diameter about 1/4 as long as snout in front of mouth. Gill openings 3 and 4 (slightly the longest) about 2/3 as long as snout in front of mouth, 2½ times as long as diameter of eye and about as long as distance between nostrils, the 1st and 5th slightly the shortest; the spaces between successive gill openings about equally broad at upper ends, but those between 3rd and 4th, and 4th and 5th much narrower at lower end, the 5th opening moderately oblique and over origin of pectoral, the 4th close in front of latter. Nostril strongly oblique, its outer corner at margin of snout, its inner corner nearer to mouth than to tip of snout by a distance about equal to horizontal diameter of eye, its anterior margin only very slightly expanded in subtriangular outline. Mouth broadly rounded in front, about ½ as high as broad, occupying about 4/5 of breadth of head. A well marked labial furrow around corner of mouth and extending inward for a very short distance on each jaw, the lower usually concealed when mouth is closed.

Teeth $\frac{12\text{ to }15—1—14\text{ to }15}{13\text{ to }14—3—13\text{ to }14}$; similar in the 2 jaws, smooth-edged, symmetrical, with sharp, slender, erect median cusp without lateral denticles, on broad bases; 1 small tooth at symphysis in upper jaw and 3 in lower, the median minute, as are the outermost 2 teeth in each jaw.

Origin of 1st dorsal a little posterior to axil of pectoral, its anterior margin slightly convex, its posterior margin deeply concave, its apex rounded, its free rear corner moderately slender and a little less than ½ as long as its base, its anterior margin about ½ as long as head. Second dorsal about ½ as long at base as 1st dorsal, but only 1/4 as high vertically, its free rear corner nearly as long as its base, its origin a little posterior to origin of anal. Caudal between 1/3 and 1/4 (about 28%) of total length, its axis raised at an angle of about 30°, its upper margin nearly straight, its tip slender and narrowly rounded, its terminal sector only about ¼ the length of fin, the subterminal notch well marked, its lower anterior corner a narrow-tipped lobe, about 40% as long as upper and with convex anterior margin. Anal a little longer at base than 2nd dorsal, but about as large in area and similar in shape except that its posterior margin is much more deeply concave. Pelvics about as long at base as 2nd dorsal, with nearly straight edges, their apices broadly rounded and their tips narrowly so, their origin posterior to rear tip of 1st dorsal by a distance about equal to diameter of eye. Pectoral only a little more than ½ (about 56%) as long as head, and little, if any, longer than anterior margin of 1st dorsal, a little more than ½ as broad as long, the outer margin moderately convex, distal margin moderately and evenly concave, apex and inner corner narrowly rounded, or subacute.

Color. Slate-blue above and on upper surface of pectorals, shading through grayish white on lower sides to pure white below; pelvics and anal white.

Size. The few specimens reported so far have ranged between 500 and 747 mm. (20 to 30 inches) in length for the western Atlantic, but up to 1.2 meters (about 4 feet) off West Africa. The maximum size may be considerably greater, for a male of 747 mm. was immature.

Developmental Stages. Not known.

Habits. The teeth suggest that this is a fish-eater, like others of its family. All recorded specimens have been taken close inshore. Nothing definite is known of its habits or diet.

Range. Both sides of the Atlantic; Senegambia, West Africa, in the east; Cuba, Texas, off Biloxi, Mississippi, Southwest Florida, South and North Carolina, Virginia and New York in the west.[5] It is described as common in Senegambian waters, and several have been reported from southwestern Florida, from Biloxi on the north shore of the Gulf of Mexico and from Texas (see Study Material, p. 304). However, the more northerly records are for single individuals only. The above facts suggest that it is a tropical species which occasionally strays northward along the east coast of the United States in summer, as do so many other fishes of warm-water origin.

Synonyms and References:

Carcharias (Aprion) isodon Müller and Henle, Plagiost., 1841: 32 (descr., no locality given for type specimen in Paris Museum; but received from Milbert, hence probably New York).

Aprionodon punctatus Gill, Proc. Acad. nat. Sci. Philad., Addend., 1861: 55; and later eds. (name); Ann. N. Y. Lyc., 7, 1862: 401 (name); Proc. Acad. nat. Sci. Philad., 1864: 262 (descr., probably N. York because type spec. from Milbert); Jordan and Gilbert, Bull. U.S. nat. Mus., 16, 1883: 24 (Atlant.); not *Squalus punctatus* Mitchill, 1815, which was probably *Scoliodon terrae-novae*; see p. 292.

Aprionodon isodon Gill, Ann. N. Y. Lyc., 7, 1862: 411 (name); Poey, An. Soc. esp. Hist. nat., 5, 1876: 396; Enumerat. Pisc. Cubens., 1876: 200 (teeth, Cuba); Jordan and Gilbert, Bull. U.S. nat. Mus., *16*, 1883: 874 (discus.); Jordan and Evermann, Rep. U.S. Comm. Fish. (1895), 1896: 217 (Atlant.); Bull. U.S. nat. Mus., 47 (1), 1896: 42 (descr., N. York, Virginia, Cuba); Bean, T. H., Rep. For. Comm. N. Y.,

5. No locality was given by Müller and Henle, 1841, for the type specimen, which is in the Paris Museum, but Duméril (Hist. Nat. Poiss., 1865: 349) states that it was from the coast of New York state.

1901: 377; Bull. N. Y. St. Mus., 60, Zool. *9*, 1903: 28 (descr., N. York, Virginia, Cuba); Garman, Mem. Harv. Mus. comp. Zool., *36*, 1913: 119 (descr.); Fowler, Copeia, 30, 1916: 36 (mid-Atlant., U.S.); Radcliffe, Bull. U.S. Bur. Fish., *34*, 1916: 252, pl. 39, fig. 3 (meas., teeth, N. Carolina); Nichols and Murphy, Brooklyn Mus. Sci. Bull., *3* (1), 1916: 17 (N. York, Virginia, Cuba); Breder, Field Bk. Mar. Fish. Atlant. Coast, 1929: 18 (general); Jordan, Manual Vert. Anim. NE. U.S., 1929: 10 (general); Jordan, Evermann and Clark, Rep. U.S. Comm. Fish. (1928), *2*, 1930: 15 (N. York, Virginia, Cuba); Fowler, Bull. Amer. Mus. nat. Hist., 70 (1), 1936: 46 (W. Afr.); Bere, Amer. Midl. Nat., *17*, 1936: 604 (W. Florida); White, Bull. Amer. Mus. nat. Hist., *74*, 1937: 129 (in Key); Springer, Proc. Fla. Acad. Sci., *3*, 1939: 29 (Florida, off Mississippi); Whitley, Aust. Zool., *9* (3), 1939: 231 (comparison with *brevipinna* Müller and Henle, 1841); Burton, Copeia, 1940: 140 (S. Carolina, size); Hildebrand, Copeia, 1941: 221 (N. Carolina, old record); Norris, Plagiost. Hypophysis, 1941: pl. 18, fig. 72 (brain); Lunz, Bull. S. Carolina St. Planning Bd., *14*, 1944: 27 (Florida); Bigelow and Schroeder, Guide Comm. Shark Fish., Anglo Amer. Caribb. Comm., Wash., 1945: 126, fig. 45 (descr., range, ill.); Fowler, Monogr. Acad. nat. Sci. Philad., 7, 1945: 159 (Morris I., S. Carolina).

Carcharias (*Aprionodon*) *isodon* Duméril, Hist. Nat. Poiss., *1*, 1865: 349 (descr. of type specimen, coast of New York); Rochebrune, Act. Soc. linn. Bordeaux, (4) *6*, 1882: 41; Faune Senegambie, Poiss., *1*, 1883–1885: 18 (Senegambia, size); Metzelaar, Trop. Atlant. Visschen, 1919: 187 (Senegambia).

Carcharias (*Aprionodon*) *punctatus* Günther, Cat. Fish. Brit. Mus., *8*, 1870: 361 (N. York, not *Squalus punctatus* Mitchill, 1815, which probably was *Scoliodon terrae-novae*; see p. 292).

Carcharhinus (*Aprionodon*) *isodon* Jordan, Rep. U.S. Comm. Fish. (1885), 1887: 796 (W. Indies, accidental on U.S. coast).

Genus *Negaprion* Whitley, 1940

Negaprion Whitley, Fish. Aust., *1*, 1940: 111; type species, *Aprionodon acutidens queenslandicus* Whitley, Aust. Zool., *9*, 1939: 233, Queensland.

Generic Synonyms:

Hypoprion (in part) Poey, Repert. Fisico-Nat. Cuba, 1868: 451; and subsequent authors; not *Hypoprion* Müller and Henle, 1841.

Carcharias (in part) Jordan, Bull. U.S. Fish Comm., *4*, 1884: 79; Proc. U.S. nat. Mus., 7, 1887: 104; not *Carcharias* Rafinesque, 1810.

Carcharinus (in part) Henshall, Bull. U.S. Fish Comm., *9*, 1891: 383; not *Carcharhinus* Blainville, 1816.

Aprionodon Whitley, Aust. Zool., *9*, 1939: 233, for *A. acutidens queenslandicus* Whitley; not *Aprionodon* Gill, 1861.

Generic Characters. Carcharhinidae with anal not longer at base than 2nd dorsal; without spiracles; midpoint of base of 1st dorsal at least as near to axil of pectoral as to origin of pelvics; 2nd dorsal at least ¾ as long at base as 1st dorsal; caudal peduncle without lateral ridges; a precaudal pit above but none below; back without mid-dorsal ridge; gill openings relatively large, the longest at least ½ as long as snout in front of mouth and more than ⅓ as long as base of 1st dorsal; a labial furrow at corner of mouth and extending outward a very short distance on upper jaw, but none on lower; teeth erect, symmetrical in front of mouth but increasingly oblique toward corners of latter, their cusps smooth-edged, their bases smooth, wavy, or even indistinctly serrate. Characters otherwise those of the family.

Range. Western Atlantic in tropical and subtropical belt; tropical Indian Ocean;

Red Sea and Gulf of Arabia; India; Indo-China; North Australia and Queensland; Micronesia.

Species. Medium-sized tropical sharks of littoral waters; one species so far known from the Atlantic; four from the Indo-Pacific.

Key to Species

1a. Snout obtusely wedge-shaped.

2a. Posterior margins of pectorals and pelvics deeply concave.

queenslandicus Whitley, 1939.
Queensland, Australia.

2b. Posterior margins of pectorals and pelvics only very weakly concave.

fronto Jordan and Gilbert, 1882.
Pacific coasts of Mexico and Costa Rica.[1]

1b. Snout broadly and evenly rounded.

3a. Bases of teeth, as well as cusps, smooth-edged. *odontaspis* Fowler, 1908.
Indian Ocean.[2]

3b. Edges of bases of upper teeth at least wavy, irregularly serrate, or denticulate.

4a. Bases of upper teeth with wavy or irregularly serrate edges, those of lower teeth smooth; distance between outer ends of nostrils only about 3/4 as great as breadth of mouth. *brevirostris* Poey, 1868, p. 310.

4b. Bases of some of the teeth, upper or lower, with one strong denticle on the outer side;[3] distance between outer ends of nostrils equal to breadth of mouth.

acutidens Rüppell, 1835.
Tropical Indian Ocean, including Red Sea (type locality) and Arabian Gulf, India and Indo-China, Torres Strait, Micronesia; perhaps Philippines.[4]

1. Beebe and Tee-Van (Zoologica, N. Y., *26*, 1941: 105) have pointed out that the two specimens on which Jordan and Gilbert's (Proc. U.S. nat. Mus., *5*, 1882: 102) original account of *fronto* was based represented two different species: one with narrow-cusped, broad-based teeth and with the second dorsal nearly as large as the first; the other with small second dorsal and serrate teeth. The second of these was obviously a *Carcharhinus*, perhaps *azureus* Gilbert and Starks, 1904, but the first, designated by Beebe and Tee-Van as the type of the species *fronto*, falls in *Negaprion* as defined here, for Beebe and Tee-Van (Zoologica, N. Y., *26*, 1941: 106) found the teeth of another specimen to be smooth-edged, except where "nicked by some external agency," this last explaining Jordan and Gilbert's original account of them as appearing minutely serrulated under a lens.
2. Fowler (Bull. U.S. nat. Mus., 100 [*13*], 1941: 194) recently has relegated this species to the synonymy of *Triaenodon obesus* Rüppell, 1835. But in his original account of it (Proc. Acad. nat. Sci. Philad., *60*, 1908: 63) he stated that the teeth are not only slender, erect and smooth-edged, but without basal cusps, and he so pictures them, whereas those of *Triaenodon* have one or two lateral cusps on each side of the longer median cusp, this being a family characteristic.
3. According to Müller and Henle (Plagiost., 1841: 33) it was the lower teeth that were denticulate at the base in the specimen (probably the type) that they examined and for which they gave measurements. However, if the *Aprionodon sitakaiensis* of Herre, 1934 (Herre, Philippine Exped. Fish., 1931: 11), is identical with *acutidens*, as it appears to be, the upper teeth may be so armed.
4. That is, if *sitakaiensis* Herre, 1934, is identical with *acutidens*; see footnote 3, p. 309.

Negaprion brevirostris (Poey), 1868

Lemon Shark

Figures 52, 53

Study Material. Six specimens, male and female, 610 to 893 mm. (about 24 to 35 inches), from coasts of Texas, Louisiana and Florida (U.S. Nat. Mus. and Amer. Mus. Nat. Hist.); also jaws of a 6-foot 10-inch female from Bay of Florida.

Distinctive Characters. Made easily recognizable among western Atlantic Carcharhinidae by a second dorsal that is nearly as large as the first, with a very broadly rounded snout, and by its characteristic teeth.

Description. Proportional dimensions in per cent of total length. Female, 610 mm.,

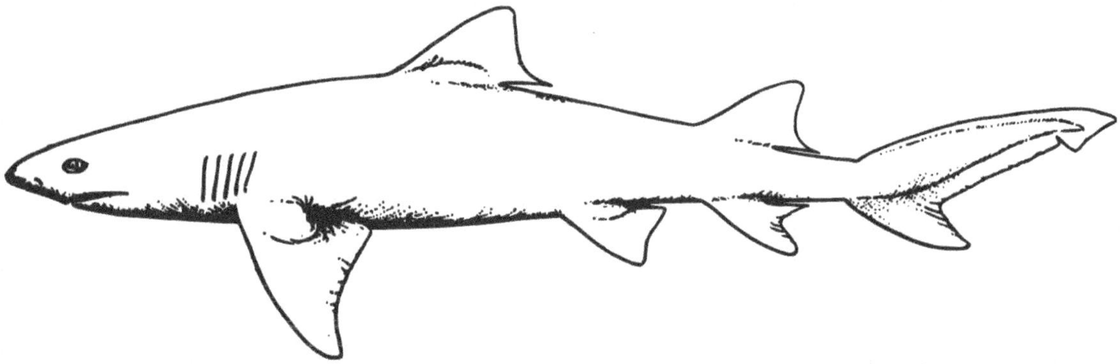

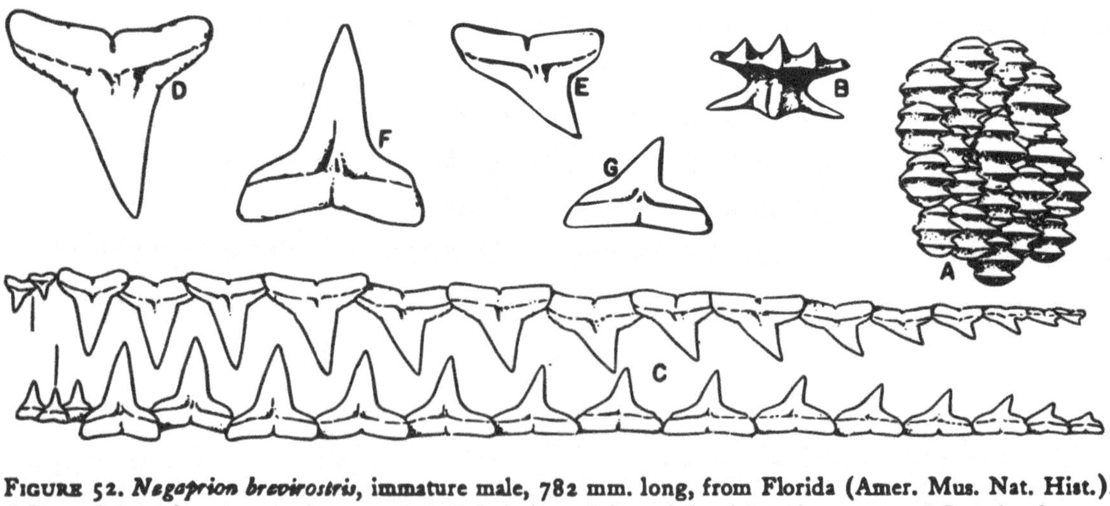

FIGURE 52. *Negaprion brevirostris*, immature male, 782 mm. long, from Florida (Amer. Mus. Nat. Hist.). *A* Dermal denticles of same, about 26 x. *B* Apical view of dermal denticles, about 52 x. *C* Left-hand upper and lower teeth of a 6-foot 10-inch female from the Bay of Florida (Amer. Mus. Nat. Hist., No. 5735), about 1.1 x. *D* Fourth upper tooth. *E* Tenth upper tooth. *F* Second lower tooth. *G* Tenth lower tooth. *D–G*, about 2.2 x.

from Florida (U.S. Nat. Mus., No. 104332). Male, 782 mm., from Florida (Amer. Mus. Nat. Hist.).

Trunk at origin of pectoral: breadth 11.5, 12.8; height 10.8, 11.0.
Snout length in front of: outer nostrils 2.9, 2.9; mouth 5.7, 5.3.
Eye: horizontal diameter 2.1, 1.8.
Mouth: breadth 8.5, 9.3; height 5.1, 5.0.
Nostrils: distance between inner ends 5.6, 5.6.
Gill opening lengths: 1st 3.1, 4.0; 2nd 3.2, 4.1; 3rd 3.4, 4.1; 4th 3.4, 4.0; 5th 3.3, 3.7.
First dorsal fin: vertical height 6.6, 6.9; length of base 10.0, 10.2.
Second dorsal fin: vertical height 5.4, 6.0; length of base 8.2, 7.8.
Anal fin: vertical height 4.3, 5.4; length of base 6.6, 6.4.
Caudal fin: upper margin 23.6, 24.0; lower anterior margin 12.0, 12.0.
Pectoral fin: outer margin 15.7, 16.2; inner margin 6.6, 6.6; distal margin 11.5, 13.4.
Distance from snout to: 1st dorsal 35.1, 34.1; 2nd dorsal 61.5, 61.0; upper caudal 76.4, 76.0; pectoral 21.9, 21.5; pelvics 50.3, 48.0; anal 61.3, 61.0.
Interspace between: 1st and 2nd dorsals 16.7, 18.1; 2nd dorsal and caudal 6.9, 7.0; anal and caudal 6.6, 6.4.
Distance from origin to origin of: pectoral and pelvics 27.0, 27.0; pelvics and anal 13.4, 14.4.

Trunk moderately stout, tapering only slightly rearward, without mid-dorsal ridge. Body sector from snout to cloaca about 1¼ times as long as tail sector. Caudal peduncle

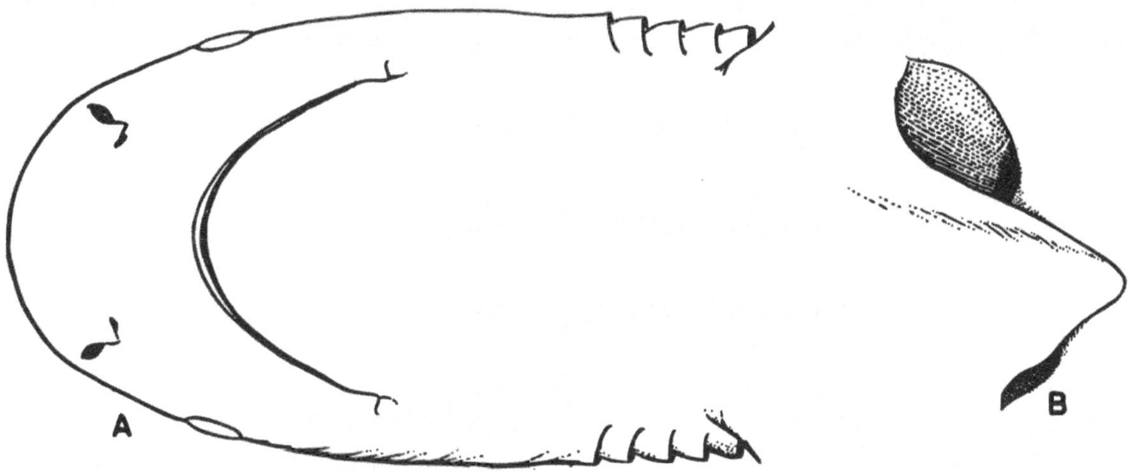

FIGURE 53. *Negaprion brevirostris. A* Head of specimen pictured in Fig. 52, from below, about ⅗ natural size. *B* Left-hand nostril, about 4.5 x.

only slightly compressed laterally, about 4/5 as high as thick. Upper precaudal pit strongly marked as a subtriangular depression, its concavity rearward, but no lower precaudal pit. Dermal denticles comparatively large (average about 0.4 x 0.6 mm. in 2,490-mm. specimen), mostly overlapping, with 3 or 5 ridges, the median ridge and the pair next to it high, sharp-topped and separated by deep furrows, the posterior margins with prominent teeth opposite the 3 primary ridges, the median a little the longest, with or without small teeth opposite the outermost pair of ridges on such of the denticles as have the latter; pedicels moderately broad, as are the basal plates.

Head moderately flattened above, its length to 5th gill opening a little less than 1/4 of total length, its breadth opposite corners of mouth a little less than 2/3 its length. Snout very broadly and evenly rounded, its length in front of a line connecting outer ends of nostrils only about 1/2 as great as distance between inner ends of latter, its length in front of mouth equal to distance between nostrils, or to about 1/4 the length of the head. Eye oval, noticeably small, its horizontal diameter a little less than 1/3 as long as distance between nostrils, and only about 1/2 to 2/3 as long as 1st gill opening. Gill openings all of very nearly equal lengths, about 1 1/2 to 2 times as long as diameter of eye, evenly spaced, the 1st about perpendicular, but the 3rd to 5th increasingly oblique, the 4th above origin of pectoral. Nostril moderately oblique, its inner end a little nearer to front of mouth than to tip of snout, its anterior margin expanded as a triangular lobe about as long as broad, the distance between inner ends of nostrils about 1/2 to 2/3 as great as breadth of mouth and about 3/4 as great between their outer ends. Mouth broadly rounded and moderately arched, its height slightly more than 1/2 as great as its breadth (51 to 60% in specimens examined), occupying about 3/4 of breadth of head. Upper labial furrow extending outward nearly at right angles to upper jaw as a deep groove for a distance about 1/3 as long as horizontal diameter of eye; no furrow on lower jaw.

Teeth $\frac{15\text{—}1 \text{ to } 3\text{—}15}{13 \text{ or } 14\text{—}3\text{—}13 \text{ or } 14}$, with very sharp cutting edges; upper teeth with narrow triangular cusps and broad bases, symmetrical and erect in central part of jaw but increasingly oblique toward its corners, the outer margins increasingly notched with the outermost deeply so, the edges of cusps smooth but edges of basal sectors with moderately fine, irregularly rounded serrations, except for the smooth small teeth at the symphysis and near the corners of the jaw; lower teeth similar in general to uppers, except somewhat more slender and more erect, the bases, as well as cusps, smooth-edged, except near the corners of the jaw where they are somewhat wavy; 1 to 3 minute, smooth-edged teeth at symphysis in upper jaw and 3 in lower; outermost 3 or 4 teeth in each jaw very small; 1 row functional, or 2 rows in places.

First dorsal low relative to its length, its anterior margin about 1.25 times as long as its base, its vertical height slightly less than 1/2 as great as length of pectoral, its origin a little posterior to corner of pectoral, its anterior margin nearly straight, apex narrowly rounded, posterior margin nearly straight toward apex but moderately concave basally, its free rear corner about 1/2 as long as the base. Second dorsal similar to 1st dorsal in shape, nearly as high, and about ¾ to ⅝ as long at base, its origin over or slightly anterior to origin of anal, its free rear corner a little longer than ½ the base (55%). Caudal a little less than ¼ of total length, its axis raised at an angle of about 15 to 18°, its upper margin weakly convex, tip subacute, the terminal sector a little less than ¼ the length of fin, the lower anterior corner expanded as a definite lobe about ½ as long as the upper, with moderately convex anterior edge narrowing to a subacute tip. Anal slightly but evidently smaller than 2nd dorsal, its anterior margin a little more convex, its apex more broadly rounded and its posterior margin much more deeply concave, its free rear corner a little more than ⅔ as long as base (69 to 70%). Pelvics with weakly concave rear margins, narrowly rounded distal corners and subacute tips, their anterior margins about as long as anterior margin of anal. Pectoral a little less than ¾ (about 71%) as long as head, about ⅔ as broad as long, with noticeably long base (as long as, or a little longer than, inner margin), the outer margin moderately convex, distal margin moderately concave and corners very narrowly rounded.

Color. Usually yellowish brown above, but sometimes dark brown or dark bluish gray; lower sides more or less tinged with yellow, or with greenish olive; lower surface either white, pale yellowish or in some cases grayish olive, like the back; anal usually yellowish, edged with gray; other fins grayish, either with or without dark edges; margins of gill openings white, shading to dark gray; inside of mouth white, at least in some specimens.

Size. Matures at about 7 to 7½ feet and grows to a maximum length of about 11 feet. One of 9 feet 6 inches is said to have weighed only 265 pounds.

Developmental Stages. It is not known whether or not the embryo develops a yolk-sac placenta.

Habits. Enough information has now accumulated to show that this, like *Ginglymostoma*, is strictly an inshore species, common around docks (*e.g.*, at Key West, Florida), in salt-water creeks (*e.g.*, around southern Florida) and in enclosed sounds as along the coast of North Carolina. It has even been reported from within the mouth of the Amazon River and from fresh water elsewhere in Brazil. But it is not known from Bermuda, nor is there any positive record that it appears elsewhere more than a very short distance out from land. Around southwestern Florida it evidently breeds in spring and summer, for newborn specimens with umbilical scars still open (624 to 630 mm.) are taken in shallow inlets from May to September. The only direct information available as to its diet is that cowfish (*Lactophrys*) were found in the stomach of one, and a sting-ray's spine was imbedded in the jaw of another. But this, with the fact that it readily takes a hook baited with fish, makes it likely that it feeds indiscriminately on whatever fish may be available locally, as its teeth would suggest.

Relation to Man. Around southern Florida it has some value commercially, its hides, fins and liver oil being of good quality. On the other hand, it has been suspected of attacks on bathers in South Carolina waters, whether justly or not.

Range. Littoral, in the western Atlantic, northern Brazil to North Carolina, and accidentally to New Jersey; also reported from tropical West Africa.[5]

Occurrence in the Western Atlantic. This is one of the more plentiful of the larger sharks along the Florida Keys and on the southern and southwestern coast of Florida, where it constitutes a considerable portion of the catches of the shark fishery. And it is common along the west coast of Florida, at least as far north as Tampa and Pensacola. In all probability its center of abundance covers the West Indian–Caribbean region as a whole, and the southern part of the Gulf of Mexico, although definite records of it there are confined to the Bahamas,[6] Cuba, Jamaica,[7] and the Atlantic coast of Panama.

To the northward its presence has been established recently off Mississippi in July,[8] and it ranges in the summer not uncommonly as far as South Carolina and the southern part of North Carolina. But it appears that few pass the latitude of Cape Hatteras, unless perhaps they enter the warm enclosed waters of Pamlico Sound, for the only record of it further north is of a single large specimen from Beach Haven, New Jersey, in July 1919. Present indications are that its range is equally circumscribed in the opposite direction, the only South American records for it being of a very small specimen from Para in northern Brazil, and another from fresh water of some unspecified Brazilian locality.

Synonyms and References:

Hypoprion brevirostris Poey, Repert. Fisico-Nat. Cuba, *2*, 1868: 451, pl. 4, fig. 5, 6, 20 (descr., teeth, Cuba); An. Soc. esp. Hist. nat., *5*, 1876: 394; Enumerat. Pisc. Cubens., 1876: 198 (Cuba); Goode and Bean, Proc. U.S. nat. Mus., *2*, 1879: 156 (W. Florida); Jordan and Gilbert, Proc. U.S. nat. Mus., *5*, 1882: 581 (descr., color, S. Carolina); Bull. U.S. nat. Mus., 16, 1883: 61 (W. Indies, Gulf coast of U.S.); Goode and Bean, Proc. U.S. nat. Mus., *5*, 1883: 240 (Gulf of Mexico); Jordan and Evermann, Rep. U.S. Comm. Fish. (1895), 1896: 217 (W. Indies to S. Carolina); Bull. U.S. nat. Mus., 47 (1), 1896: 41 (descr., W. Indies to S. Carolina); Bull. U.S. nat. Mus., 47 (4), 1900: pl. 5, fig. 18 (ill.); Evermann and Kendall, Rep. U.S. Comm. Fish. (1899), 1900: 48 (Florida); Bean, B. A., in Shattuck, Bahama Islands, Fish., 1905: 296 (Bahamas); Rosen, Lunds Univ. Årsberätt., (7) *2* (5), 1911: 47 (Watlings I.); Garman, Mem. Mus. comp. Zool. Harv., *37*, 1913: 120 (descr., Pensacola, Florida); Gudger, Science, N. S. *37*, 1913: 993 (Key West, Florida); Starks, Stanford Univ. Publ., Univ. Ser., 1913: 5 (Para, Brazil); Gudger, Yearb. Carneg. Instn. (1913), *12*, 1914: 177 (Tortugas, Florida); Yearb. Carneg. Instn. (1914), *13*, 1915: 204 (Tortugas, Florida); Science, N. S. *41*, 1915: 437 (anat., Key West, Florida); Coles, Proc. biol. Soc. Wash., *28*, 1915: 90 (N. Carolina); Radcliffe, Bull. U.S. Bur. Fish., *34*, 1916: 253, pl. 39, fig. 1 (meas., descr., N. Carolina); Nichols, Bull. Amer. Mus. nat. Hist., *37*, 1917: 875 (Florida); Fowler, Proc. Acad. nat. Sci. Philad., *72*, 1921: 386 (size, weight, Beach Haven, N. Jersey); Meek and Hildebrand, Field Mus. Publ. Zool., *15*, 1923: 50 (descr., Atlant.); Borodin, Bull. Vanderbilt Oceanogr. (Mar.) Mus., *1* (1), 1928: 5 (Florida); Breder, Field Bk. Mar. Fish. Atlant. Coast, 1929: 18 (general); Gudger, Publ. Carneg. Instn., 391, 1929: 200 (food, Tortugas, Florida); Jordan, Evermann and Clark, Rep. U.S. Comm. Fish. (1928), 2, 1930: 17 (distrib.); Gudger, Sci. Mon. N. Y., *34*, 1932: 409 (sting-ray spine in jaw, west coast, Florida); Breder, Zoologica, N. Y., *18*, 1934: 59 (W. Indies, Bahamas); Brooks, Parasitology, *26*, 1934: 260 (Tortugas,

5. Budker, Bull. Mus. Hist. nat. Paris, (2) 7, 1935: 185. The collection of the Harvard Museum of Comparative Zoology also contains a female of 872 mm. received in 1864 and catalogued as from the Kingsmill Islands. But we hesitate to include the West Pacific in the range of the species, in view of the possibility that the specimen may not have come from the stated locality.

6. There is a specimen from Watlings Island in the United States National Museum (No. 38497).

7. Personal communication from Luis Howell-Rivero.

8. Personal communication from Stewart Springer.

Florida, parasites); Budker, Bull. Mus. Hist. nat. Paris, (2) 7, 1935: 185 (Dakar, W. Afr.); Burton, Sci. Mon. N. Y., *40*, 1935: 279 (perhaps attacks on bathers, S. Carolina); White, Bull. Amer. Mus. nat. Hist., *74*, 1937: 128 (in Key); Springer, Proc. Fla. Acad. Sci., *3*, 1939: 28 (breeding season, Florida); Hildebrand, Copeia, 1941: 221 (N. Carolina); Longley and Hildebrand, Pap. Tortugas Lab., *34*, 1941: 2 (Tortugas, Florida); Bomkamp, Contr. biol. Lab. Cath. Univ. Amer., 44, 1942: 4 (chemistry of liver, Salerno, Florida); Fowler, Arqu. Zool. Estad. São Paulo, *3*, 1942: 128 (listed for Brazil); Fish Culturist, *21* (9), 1942: 66 (listed Cuba, but not p. 67, fig. 56); Boos, Contr. biol. Lab. Cath. Univ. Amer., 45, 1943: 10 (chemistry of pancreas, Salerno, Florida), Lunz, Bull. S. C. St. Planning Bd., 14, 1944: 27 (S. Carolina, Florida); Gunter, Publ. Inst. Mar. Sci. Univ. Texas, 1945: 20 (temp., breeding, Texas); Bigelow and Schroeder, Guide Comm. Shark Fish., Anglo Amer. Caribb. Comm., Wash., 1945: 116, fig. 41 (descr., habits, range, ill.); Fowler, Monogr. Acad. nat. Sci., 7, 1945: 96 (Beaufort, N. Carolina), 160 (S. Carolina), 263 (Bahamas).

Carcharias (Hypoprion) brevirostris Günther, Cat. Fish. Brit. Mus., *8*, 1870: 362 (descr., Cuba).

Carcharias brevirostris Jordan and Gilbert, Proc. U.S. nat. Mus., *5*, 1883: 245 (Gulf of Mexico); Jordan, Bull. U.S. Fish Comm., *4*, 1884: 79 (Key West, Florida); Proc. U.S. nat. Mus., 7, 1887: 104 (Key West, Florida); Englehardt, Zool. Anz., *39*, 1912: 648 (Brazil, in fresh water).

Carcharhinus (Hypoprion) brevirostris Jordan, Rep. U.S. Comm. Fish. (1885), 1887: 796 (W. Indies).

Carcharhinus brevirostris Henshall, Bull. U.S. Fish. Comm., *9*, 1891: 383 (Florida, teeth).

Genus *Hypoprion* Müller and Henle, 1841

Hypoprion Müller and Henle, Plagiost., 1841: 34; type species, *Carcharias (Hypoprion) macloti* Müller and Henle, 1841, designated by Jordan and Gilbert, Bull. U.S. nat. Mus., 16, 1883: 61.

Generic Synonyms:

Squalus (in part) Gray, List. Fish. Brit. Mus., 1851: 43; not *Squalus* Linnaeus, 1758.

Hypoprionodon Gill, Ann. N. Y. Lyc., 7, 1862: 399, 401, 409; type species, *Carcharias (Hypoprion) hemiodon* Müller and Henle, 1841.

Carcharias (in part) Günther, Cat. Fish. Brit. Mus., *8*, 1870: 362; not *Carcharias* Rafinesque, 1810.

Generic Characters. Carcharhinidae with anal less than 1½ times as long at base as 2nd dorsal; without spiracles; midpoint of 1st dorsal at least as near to axil of pectoral as to origin of pelvics; 2nd dorsal only about ½ as long at base as 1st dorsal; caudal peduncle without lateral ridges, but with a precaudal pit below as well as above; back with a low mid-dorsal ridge, at least in some species; a labial furrow at corner of mouth and extending for a short distance on upper jaw, but not onto lower; upper teeth strongly oblique and notched outwardly, or erect, their cusps smooth-edged, but their bases with several coarse marginal serrations or low denticles on the outer side, more or less wavy or indistinctly serrate on the inner; the lowers slender, erect, both bases and cusps smooth-edged. Characters otherwise those of the family.

Range. Western Atlantic; Chile; China and Indo-China; East Indies; Philippines; New Guinea; India and tropical Indian Ocean, including Red Sea and Gulf of Arabia.[1]

Species. One species is known in the warm belt of the western Atlantic, and two or perhaps three in the Pacific and Indian Oceans.[2]

1. Also reported for Australia, but incorrectly, according to Whitley (Fish. Aust., *1*, 1940: 107).
2. An additional species, *H. brevirostris* Poey, 1868, has been referred to this genus previously, but it falls in *Negaprion*, according to the generic definitions adopted here.

Key to Species

1a. Snout in front of mouth considerably longer than breadth of mouth.

2a. Tip of 1st dorsal terminates anterior to origin of pelvics by a distance at least as long as base of latter; origin of 2nd dorsal opposite or only a very little posterior to origin of anal. *signatus* Poey, 1868, p. 316.

2b. Tip of 1st dorsal reaches nearly to a perpendicular at origin of pelvics; origin of 2nd dorsal over or posterior to midbase of anal.
macloti Müller and Henle,[3] 1841.
New South Wales, East Indies, India; also probably Chile.[4]

1b. Snout in front of mouth not longer than breadth of mouth.

3a. Base of 2nd dorsal only ⅔ as long as base of anal; posterior margin of 2nd dorsal not concave, but that of anal deeply so; upper teeth oblique, lowers erect.
hemiodon Müller and Henle, 1841.
Arabian Gulf, India, Indo-China, East Indies, Philippines.[5]

3b. Base of 2nd dorsal as long as base of anal; 2nd dorsal of same shape as anal; upper teeth erect, like lowers. *playfairii* Günther, 1870.
Madagascar.

Hypoprion signatus Poey, 1868

Night Shark

Figure 54

Study Material. Immature female, 935 mm. long, from offing of South Carolina, Lat. 33° 37′ 30″ N., Long. 77° 36′ 30″ W., in 14 fathoms on October 20, 1885 (U.S. Nat. Mus., No. 38508); three embryos (two females, one male), 385 to 407 mm. long (Harv. Mus. Comp. Zool., No. 36091); also photograph of adult female taken off north coast of Cuba.[6]

Distinctive Characters. Easily separable from all other local carcharhinids by the combination of very long, pointed snout with smooth-cusped teeth, but the uppers strongly serrate at the base, and with the presence of a low but unmistakable mid-dorsal ridge.

Description. Proportional dimensions in per cent of total length. Female, 935 mm., from Lat. 33° 37′ 30″ N., Long. 77° 36′ 30″ W. (U.S. Nat. Mus., No. 38508).

3. Proportional dimensions calculated from measurements given by Müller and Henle (Plagiost., 1841: 34) for the type specimen.
4. Reported as *Hypoprion ? (Hemigaleus ?) heterodus* and *Hypoprion ? (Hemigaleus ?) isodus* by Philippi (An. Univ. Chile, 71, 1887: 541, 542).
5. According to Whitley (Fish. Aust., 1, 1940: 107), a report of this species from South Australia by Zietz probably was not correct.
6. Contributed by Luis Howell-Rivero.

Trunk at origin of pectoral: breadth 11.6; height 11.6.
Snout length in front of: outer nostrils 6.0; mouth 10.0.
Eye: horizontal diameter 2.2.
Mouth: breadth 7.4; height 5.2.
Nostrils: distance between inner ends 5.2.
Gill opening lengths: 1st 2.2; 2nd 2.3; 3rd 2.5; 4th 2.3; 5th 1.8.
First dorsal fin: vertical height 7.8; length of base 9.7.
Second dorsal fin: vertical height 2.0; length of base 3.5.
Anal fin: vertical height 2.7; length of base 4.0.
Caudal fin: upper margin 26.9; lower anterior margin 11.4.
Pectoral fin: outer margin 16.8; inner margin 5.4; distal margin 13.2.
Distance from snout to: 1st dorsal 34.7; 2nd dorsal 64.3; upper caudal 73.1; pectoral 25.2; pelvics 51.5; anal 63.0.
Interspace between: 1st and 2nd dorsals 23.4; 2nd dorsal and caudal 5.8; anal and caudal 6.3.
Distance from origin to origin of: pectoral and pelvics 29.2; pelvics and anal 12.1.

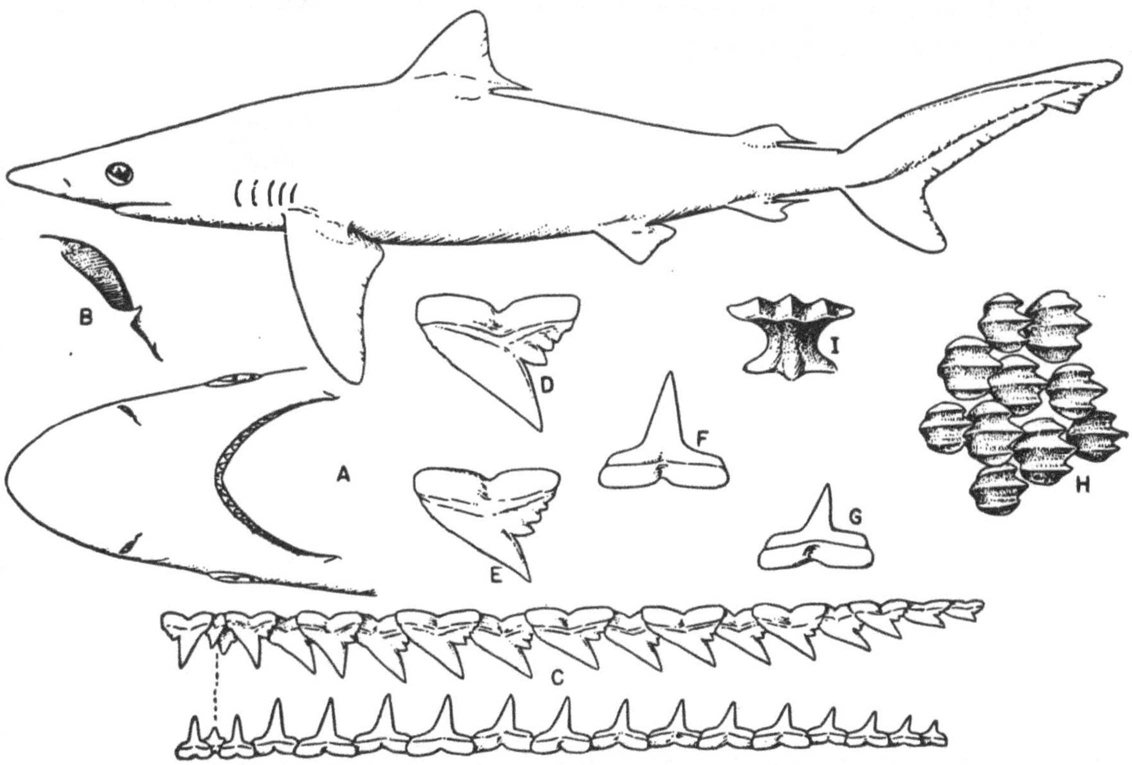

FIGURE 54. *Hypoprion signatus*, female, 935 mm. long, from off South Carolina (U.S. Nat. Mus., No. 38508). *A* Anterior part of head from below, about 0.3 x natural size. *B* Left-hand nostril, about 2.6 x. *C* Left-hand upper and lower teeth, about 1.5 x. *D* Sixth upper tooth. *E* Twelfth upper tooth. *F* Fourth lower tooth. *G* Eleventh lower tooth. *D–G*, about 3.0 x. *H* Dermal denticles, about 30 x. *I* Dermal denticle, apical view, about 60 x.

Trunk comparatively stout, about 1/5 as high at 1st dorsal (where highest) as length to origin of caudal. Caudal peduncle moderately slender, the lower as well as upper precaudal pit strongly marked, subrectangular. Dermal ridge low but unmistakable along midline of back from close behind rear end of base of 1st dorsal to origin of 2nd dorsal.[7] Dorsal profile sloping sharply, in convex contour, from 1st dorsal forward. Dermal denticles rather loosely spaced and overlapping but little, their blades nearly horizontal, about as broad as long, usually with 3 low ridges, the posterior margins usually with 3 teeth, but occasionally 5, the median considerably the largest; pedicel rather slender.

Head noticeably long, forming about 1/3 of trunk to origin of caudal. Snout narrow, ovate at tip and very long, its length in front of mouth a little more than 1/3 of length of head, or about 1.25 times as great as breadth of mouth, and its length anterior to outer ends of nostrils a little more than 1/2 as great as length in front of mouth. Eye nearly circular, its diameter about 1/2 as great as distance between nostrils. Gill openings noticeably small, the 1st to 3rd (slightly the longest) about as long as diameter of eye, the spaces between 1st and 2nd and between 2nd and 3rd of about equal breadth, but those between 3rd and 4th and between 4th and 5th a little narrower, the 4th gill opening over origin of pectoral. Nostril strongly oblique, its inner corner nearer to mouth than to tip of snout by a distance about 1/2 as great as that between them, the anterior margin with a pronounced narrow triangular lobe near the inner end; the distance between nostrils a little greater than 2/3 breadth of mouth and about 1/2 length of snout. Mouth broadly ovate, about 1 1/3 times as broad as high. Upper labial fold about 1/2 as long as nostril.

Teeth $\frac{15\text{—}1 \text{ or } 2\text{—}15}{15\text{—}1\text{—}15}$; those at symphysis small, triangular, the base with or without a blunt denticle on each side; uppers otherwise acute-triangular, increasingly oblique toward corners of mouth, the inner margins nearly straight and smooth-edged on cusps but more or less wavy or irregularly serrate basally, the outer margins strongly notched, smooth toward tips, but with 2 to 4 very prominent serrations or low denticles on basal sector, the distal serration considerably the largest; lowers symmetrical, more slender than uppers, nearly erect, bases as well as cusps with smooth edges.

First dorsal comparatively small, its anterior margin only about as long as snout in front of mouth, and about 1/2 as long as pectoral, its origin about over inner corner of pectoral,[8] its anterior margin weakly convex, its posterior margin strongly concave basally, its apex rounded, its free corner a little less than 1/2 as long as its base. Second dorsal a little less than 1/2 as long as 1st dorsal at base, relatively much lower, and only about 1/5 to 1/8 as great in area, its apex broadly rounded, posterior margin only very weakly concave, its free rear tip very slender and nearly as long as its base, its origin very little posterior to origin of anal. Caudal about 1/4 of total length, with bluntly rounded tip, its terminal sector about 1/4 the length of fin, the lower anterior lobe about 40% as long as upper margin, with rather broadly rounded tip. Interspace between caudal and anal about 1 1/2 times as long as base of anal. Anal about as high and long as 2nd dorsal and with similarly slender

7. In the preserved state this ridge lies at the bottom of a groove of muscular contraction.

8. Its precise point of origin is difficult to determine in the preserved specimen.

free rear tip, but with much more deeply concave posterior margin. Pelvics with nearly straight edges and narrowly rounded corners, about as long as anal along anterior margins, about 1½ times as large in area, their origin considerably nearer to rear end of base of 1st dorsal than to origin of anal. Distance from cloaca to caudal about ⅔ as great as from cloaca to inner corner of pectoral in female, but perhaps somewhat longer, relatively, in male (not seen). Pectoral about ⅔ (about 66%) as long as head, and about ½ as broad as long, with strongly convex outer margin, weakly and evenly concave distal margin, and rather narrowly rounded corners.

Color. Adults in life are bluish gray above, grayish white below, with small black spots scattered over the body; embryos are silvery gray above, dirty white below.[9]

Size. The lengths of recorded specimens, for which the identity is established, are 955 mm. (immature, see above), 2,766 and 2,270 mm. (male and female); thus this Shark grows to at least a moderately large size.

Developmental Stages. The embryos bear long yolk stalks and well developed yolk-sac placentae, spongy in texture, showing that development is viviparous. Females have been taken with as many as 12 embryos.[10]

Habits. The only available information as to its habits is that Cuban fishermen report it as caught well offshore only, on set lines at depths greater than 150 fathoms, and only at night (hence its local name "Tiburon de Noche").

Range. The Night Shark is known only off the north coast of Cuba, where it is so common in the Gulf Stream that 36 specimens have been counted at one time at the shark fishery station at Cojimar,[11] and from the offing of South Carolina (a single specimen, see Study Material, p. 316). It has been recorded also by name from Georgetown, British Guiana, as well as from Key West and the Tortugas, Florida, but without any supporting evidence as to the actual identity of the specimens concerned. The fact that only one specimen has been reported from the east coast of the United States, although *signatus* is easily recognizable, suggests that it wanders northward from its tropical home only as a stray.

Synonyms and References:

Hypoprion signatus Poey, Repert. Fisico.-Nat. Cuba, 1868: 452, pl. 4, fig. 7, 8 (descr., teeth, Cuba); An. Soc. esp. Hist. nat., 5, 1876: 395; Enumerat. Pisc. Cubens., 1875: 199 (Cuba); Jordan and Evermann, Bull. U.S. nat. Mus., 47 (1), 1896: 41 (descr., Cuba); Blosser, Ann. Carneg. Mus., *6*, 1909: 295 (Brit. Guiana, ident. queried); Garman, Mem. Harv. Mus. comp. Zool., *36*, 1913: 122 (descr.); Gudger, Science, N. S. *41*, 1915: 437 (name only, Tortugas, Florida); Yearb. Carneg. Instn. (1914), *13*, 1915: 203 (name only, Key West, Florida); Jordan, Evermann and Clark, Rep. U.S. Comm. Fish. (1928), 2, 1930: 17 (Cuba); White, Bull. Amer. Mus. nat. Hist., *74*, 1937: 128 (in Key); Bigelow and Schroeder, Guide Comm. Shark Fish., Anglo Amer. Caribb. Comm., Wash., 1945: 130, fig. 47 (descr., range, ill.).

Carcharias (Hypoprion) signatus Günther, Cat. Fish. Brit. Mus., 1870: 362.

Hypoprion longirostris Poey, An. Soc. esp. Hist. nat., *5*, 1876: 198, pl. 9, fig. 8, 9.[12]

9. Information contributed by Luis Howell-Rivero.

10. Information contributed by Luis Howell-Rivero.

11. Personal communication from Luis Howell-Rivero.

12. We agree with Garman (Mem. Harv. Mus. comp. Zool., *36*, 1913: 123) that *longirostris*, based on a 2,266 mm. specimen, was a synonym of *signatus* (based on jaws only), there being no essential differences in the teeth; a photograph of unpublished drawings by Poey, showing the anterior part of the head of "*longirostris*," agrees closely with the specimen illustrated here (Fig. 54) in shape and length of snout relative to breadth of mouth.

Genus *Carcharhinus*[1] Blainville, 1816

Carcharhinus Blainville, Bull. Soc. philom. Paris, 1816: 121; J. Phys. Chem. Hist. nat., *83*, 1816: 264; in Vieillot, Faune Franc., 1825: 88; type species, *C. commersonii* Blainville, 1816, 1825.[1a]

Generic Synonyms:

Carcharinus Cloquet, Dict. Sci. Nat., 7, 1817: 77; and subsequent authors; emended spelling for *Carcharhinus* Blainville, 1816.

Carcharias (in part) Cuvier, Règne Anim., *2*, 1817: 125; and subsequent authors; not *Carcharias* Rafinesque, 1810.

Prionodon (in part) Müller and Henle, Plagiost., 1841: 35 (preoccupied by Horsfield, 1823, for mammals).

Galeolamna Owen, Cat. Osteol. Roy. Coll. Surg. London, *1*, 1853: 96, no. 427; type, *G. greyi* Owen. South Australia.[2]

Eulamia Gill, Ann. N. Y. Lyc., 7, 1862: 401 (name), 410 (diagn.); type species, *Carcharias* (*Prionodon*) *milberti* Müller and Henle, 1841.[3]

1. Often spelled *Carcharinus*.

1a. The generic name *Carcharhinus* has long been a "football" in Elasmobranch nomenclature. Briefly, its history is as follows: Blainville, in his original diagnosis of the genus, listed the following species by name only: *commersonii*, *lamia*, *lividus*, *ustus*, *heterodon*, *verus*, *broussonetii*, *glaucus*, *caeruleus*, *megalops*, *heterobranchialis*, *cornubicus*, *monensis* and *vulpes*. This assemblage includes representatives of *Lamna* Cuvier, 1817, as now understood (*cornubicus*, *monensis*); of *Alopias* Rafinesque, 1810 (*vulpes*); of *Prionace* Cantor, 1849 (*glaucus*, *caeruleus*); and of *Carcharodon* L. Agassiz, 1838 (*carcharias*); also one (*lamia*) that by subsequent evidence apparently represented the combination of some unidentifiable member of the genus here named *Carcharhinus*, with the teeth of *Galeocerdo*. The remaining members of the list were nominal only, and have so remained except for *commersonii* which was later stated by Blainville (in Vieillot, Faune Franc., 1825: 90) to have been based by him on the shark pictured by Lacépède, "T. 1, pag. 169, pl. 5, fig. 1." Unfortunately, this reference was erroneous, for pl. 5, fig. 1 pictures a skate and not a shark. However, if it was pl. 8, fig. 1, that was intended (as seems almost certain), *commersonii* seems to have been a member of the genus now under discussion, though neither the illustration in question nor the accompanying measurements of a "requin" suffice for specific identification.

The specific name *commersonii* seems then to have lain in abeyance until 1913, when Garman (Mem. Harv. Mus. comp. Zool., *36*, 1913: 140) revived it. However, his account appears to have been based on a combination of *longimanus* Poey, 1861, with *leucas* Müller and Henle, 1841, and various authors following Garman have subsequently used the name *commersonii* for one or the other of these two species, although there is no apparent reason for identifying *leucas* with Lacépède's illustration, while *longimanus* certainly cannot be so identified.

Bosc (Nouv. Dict. Hist. Nat., *5*, 1816: 277) was the first to designate a type species for *Carcharhinus* in the words "*Squalus carcharias* lui sert de type." But this designation is not valid, because the original assemblage of species listed in the genus by Blainville (Bull. Soc. philom. Paris, 1816: 121 and J. Phys. Chem. Hist. nat., 1816: 264) did not include a "*carcharias*," although he did describe a *Carcharhinus* under that specific name subsequently (in Vieillot, Faune Franc., 1825: 89). The next designation of a type was *commersonii* Blainville, 1816, by Jordan and Gilbert (Bull. U.S. nat. Mus., 16, 1883: 22); and this selection appears to be valid, for while Fowler (Proc. Acad. nat. Sci. Philad., *60*, 1908: 62) has preferred *vulpes* as the type on the ground that *commersonii* was a *nomen nudem*, thus reducing *Carcharhinus* to a synonym of *Alopias* Rafinesque, 1810, *commersonii* cannot properly be discarded, for, as pointed out above, the illustration on which it was based is identifiable with reasonable certainty at least as to genus if not to species.

2. According to Whitley (Aust. Zool., *9*, 1939: 230), the type, which he saw in London, is the jaws of the common "Whaler Shark" of Australia (*Carcharias brachyurus* of Günther, 1870, and *C. macrurus* of Ramsay and Ogilby, 1887) which seems clearly referable to *Carcharhinus* Blainville 1816, as here recognized.

3. In his first mention of the genus, Gill (Ann. N. Y. Lyc., 7, 1862: 401) included only one species, "*Eulamia lamia* Gill," which he stated to be the type, and Jordan and Gilbert (Bull. U.S. nat. Mus., 16, 1883: 60), and later Jordan (Genera Fish., 3, 1919: 306), accepted this designation, changing the authorship of *lamia*, however, from "Gill" to "Risso" in the one case and to "Rafinesque" in the other. But this alteration was not warranted, for "*lamia* Gill" was a *nomen nudem*, no account having ever been published by Gill himself of any shark under that name, or even any indication as to whether he referred to the "*lamia*" of Rafinesque, 1810 (a name substituted for *Squalus carcharias* Linnaeus, 1758, and therefore equivalent to *Carcharodon*), of Risso, 1826, or of Müller and Henle, 1841. Therefore, it is fortunate that the only species included by Gill under his diagnosis of *Eulamia* on a later

Platypodon Gill, Ann. N. Y. Lyc., 7, 1862: 401; type, *Carcharias (Prionodon) menisorrah* Müller and Henle, 1841.

Isogomphodon Gill, Ann. N. Y. Lyc., 7, 1862: 401; type, *Carcharias (Prionodon) oxyrhynchus* Müller and Henle, 1841.

Lamiopsis Gill, Ann. N. Y. Lyc., 7, 1862: 401; type, *Carcharias (Prionodon) temmincki* Müller and Henle, 1841.

Isoplagiodon Gill, Ann. N. Y. Lyc., 7, 1862: 401; type, *Carcharias (Prionodon) sorra* Müller and Henle, 1841.

Gymnorhinus Hilgendorf, in Hemprich and Ehrenberg, Symbol. Phys. Icon. Ined. Pisces, 1899: 8; type, *Carcharias (Prionodon) menisorrah* Müller and Henle, 1841 (not seen); preoccupied by Maximillian, 1841, for birds.

Mapolamia Whitley, Mem. Qd. Mus., *10* (4), 1934: 185, 188; type, *Carcharias melanopterus* Quoy and Gaimard, 1824.

Gillisqualus Whitley, Mem. Qd. Mus., *10* (4), 1934: 185; type, *Carcharias (Prionodon) amblyrhynchus* Bleeker, 1856.

Galeolamnoides Whitley, Mem. Qd. Mus., *10* (4), 1934: 185, 191; type, *Carcharias macrurus* Ramsay and Ogilby, 1887.

Galeolamna Whitley, Aust. Zool., *9*, 1939: 230; revives *Galeolamna* Owen, 1853, which see above.

Ogilamia Whitley, Aust. Zool., *9*, 1939: 231, subgenus; type, *Carcharias stevensi* Ogilby, 1911. Australia.

Uranga Whitley, Proc. Linn. Soc. N. S. W., *68*, 1943: 115; type, *U. nasuta* Whitley. Australia.

Uranganops Whitley, Proc. Linn. Soc. N. S. W., *68*, 1943: 117, subgenus; type, *Galeolamna (Uranganops) fitzroyensis* Whitley. Australia.

Lamnarius Whitley, Proc. Linn. Soc. N. S. W., *68*, 1943: 119, subgenus; type, *Carcharias spenceri* Ogilby, 1910. Australia.

Bogimba Whitley, Proc. Linn. Soc. N. S. W., *68*, 1943: 123, subgenus; type, *Galeolamna (Bogimba) bogimba* Whitley. Australia.

Longmania (in part) Whitley, Aust. Zool., *10* (3), 1944: 257; for *L. calamaria* Whitley, Australia; not *Longmania* Whitley, 1939, which is a synonym of *Aprionodon* (see p. 303).

Doubtful synonym:

Glyphis L. Agassiz, Poiss. Foss., *3*, 1838: pl. 36, fig. 10–13; Poiss. Foss., *3*, 1843: 243; type species, *G. hastalis* L. Agassiz, 1838.[4]

Generic Characters. Carcharhinidae with anal fin little if any longer at base than 2nd dorsal; without spiracles; midpoint of 1st dorsal nearer to axil of pectoral than to origin of pelvics; cusps of upper teeth regularly serrate, those of lowers serrate or smooth; back with or without mid-dorsal ridge; caudal peduncle without lateral ridges, but with well developed upper precaudal pit, the lower varying from well developed to hardly discernible; 1st dorsal subtriangular, its posterior margin more or less deeply concave, its lower posterior angle more or less prolonged as a free corner; 2nd dorsal and anal of approximately equal size and much smaller than 1st dorsal; pelvics quadrilateral, their inner

page of the same publication (Ann. N. Y. Lyc., 7, 1862: 410) was the well known *Carcharias (Prionodon) milberti* Müller and Henle, 1841. This must necessarily stand as the type of the genus. See also footnote 82, p. 368.

4. The fossil shark's teeth to which L. Agassiz gave this name resemble the anterior lower teeth of *Carcharias (Prionodon) glyphis* Müller and Henle, 1841, which falls in *Carcharhinus* as here defined; they are slender, erect, cylindrical near the base, and with cutting edge confined to the lanceolate, laterally-expanded tips. But we doubt the propriety of reviving the name *Glyphis* for any modern shark on the evidence of these two teeth alone. See also footnote 5, p. 280.

corners not elongate; dermal denticles either overlapping or not, their blades with 3 or more ridges, their apical margins toothed or not; axis of caudal raised only moderately; upper labial furrow very short, at an obtuse angle with the jaw; no lower labial furrow. Development viviparous, with well developed yolk-sac placenta in the few cases where it is known (see pp. 359, 394). Characters otherwise those of the family.

Range. Tropical and warm-temperate belts of all oceans, including the Mediterranean, both inshore and on the high seas; also landlocked in Lake Nicaragua in fresh water.

Fossil Teeth, closely resembling those of *Carcharhinus* (perhaps including *Hypoprion* and *Aprionodon*), have been described under various names from: Eocene, Africa; Eocene to Pliocene, Europe and North America; Oligocene to Miocene, South America; and Miocene, West Indies.

Species. Carcharhinus includes a much larger number of species than any other genus of modern sharks and many of the most familiar of the larger sharks of warm seas. Its members cover a wide range as regards teeth, the relative sizes and shapes of fins, and to a lesser degree the relative positions of the latter. But the extremes are connected by such a continuous series of intermediate stages in all these respects that attempts to subdivide the genus have not been easy.[5] In fact, the only alternative character which might form a sharp-cut basis for such subdivision, from the standpoint of specific identification, is the presence or absence of a mid-dorsal ridge. But the use of this would entail the generic separation of species that closely resemble one another in other respects, and the union of others that do not, which seems too high a price to pay for reducing the length of the generic Key, which would be the only advantage gained.

The genus has received much less attention than it deserves, no doubt due to the fact that most of the species are rather large, with consequent paucity of specimens in collections. Many of the species resemble one another so closely in general appearance that little or no dependence can be placed on published reports of occurrence unless accompanied by some indication as to fins, teeth, etc. Therefore, we are very fortunate in having been able to study specimens of all 13 species now known to occur in the western Atlantic; finding that while some of them look much alike on cursory examination, they are separable by characters so precise and so little variable that specific identification is not difficult, although attention to detail is required.

The genus is as universally distributed in the warmer belt of the Pacific and Indian Oceans as it is in the Atlantic, and some of its Indo-Pacific representatives are evidently very close to some of the Atlantic species, if not identical with them. However, to attempt to revise the genus as a whole would be idle without access to adequate material of at least a majority of the supposed Indo-Pacific species, which we have not had. The following Key is therefore limited to the western Atlantic.

5. Whitley, in a series of papers, has recently broken the genus down into no less than nine genera and subgenera (see Generic Synonyms, p. 321). But the characters on which these are based seem to us more appropriate for the definition of species within the genus.

Key to Western Atlantic Species

1a. Length of snout, anterior to a line connecting outer corners of nostrils, about 1.5 times as great as distance between nostrils.
oxyrhynchus Müller and Henle, 1841, p. 391.

1b. Length of snout anterior to a line connecting outer corners of nostrils less than distance between nostrils.

2a. Origin of 2nd dorsal over or behind midpoint of base of anal.
porosus Ranzani, 1839, p. 394.

2b. Origin of 2nd dorsal over origin of anal, or anterior to it.

3a. Midline of back between 1st and 2nd dorsal fins with a low but distinct dermal ridge.

4a. Free rear corner of 2nd dorsal more than twice as long as vertical height of the fin and notably slender (Figs. 56 B, 59 B).

5a. Anterior margin of pectoral nearly as long (about 92%) as from tip of snout to origin of pectorals, and longer than from snout to 1st gill opening.
floridanus Bigelow and Schroeder, 1943, p. 333.

5b. Anterior margin of pectoral not more than ⅔ as long as from tip of snout to origin of pectorals, and shorter than from snout to 1st gill opening by a distance as great as that between nostrils.
falciformis Müller and Henle, 1841, p. 329.

4b. Free rear corner of 2nd dorsal considerably less than twice as long as vertical height of fin, and not notably slender.

6a. Apex of 1st dorsal very broadly rounded; tip of anal reaches nearly to origin of caudal. *longimanus* Poey, 1861, p. 354.

6b. Apex of 1st dorsal subangular, or very narrowly rounded; tip of anal separated from origin of caudal by a distance at least as long as diameter of eye.

7a. Origin of 1st dorsal about over inner corner of pectoral, its vertical height less than distance from eye to 1st gill opening; dermal denticles regularly overlapping, with strongly marked marginal teeth.

8a. Distance from tips of pelvics to origin of anal longer than base of anal; 5th gill opening at least 1.5 times as long as horizontal diameter of eye; vertical height of 1st dorsal only a little more than length of snout in front of mouth; anterior margin of nostril not lobed.
obscurus Lesueur, 1818, p. 382.

8b. Distance from tips of pelvics to origin of anal less than ¾ as long as base of anal; 5th gill opening only about as long as horizontal diameter of eye; vertical height of 1st dorsal about 1 ⅓ times as long as snout in front of mouth; anterior margin of nostril with low, subtriangular lobe.

springeri Bigelow and Schroeder, 1944, p. 404.

7b. Origin of 1st dorsal over axil of pectoral, its vertical height (after birth) at least as great as distance from eye to 3rd gill opening; dermal denticles loosely spaced, without well marked teeth. *milberti* Müller and Henle, 1841, p. 368.

3b. Midline of back between dorsals smooth, without dermal ridge.

9a. Apex of 1st dorsal very broadly rounded; tip of anal reaches nearly to origin of caudal. *longimanus* Poey, 1861, p. 354.[5a]

9b. Apex of 1st dorsal subangular or narrowly rounded; tip of anal separated from origin of caudal by a distance at least as long as diameter of eye.

10a. Snout, in front of line connecting outer ends of nostrils, less than ½ as long as distance between inner ends of nostrils.

11a. Anterior margin of eye opposite or a little anterior to front of mouth; 1st gill opening not more than ½ as long as distance between nostrils.

leucas Müller and Henle, 1841, p. 337.

11b. Anterior margin of eye a little posterior to front of mouth; 1st gill opening nearly ⅔ as long as distance between nostrils.

nicaraguensis Gill and Bransford, 1877, p. 378.

10b. Snout, in front of a line connecting outer ends of nostrils, at least ⅔ as long as distance between inner ends of nostrils.

12a. Upper teeth strongly asymmetrical, their outer margins deeply concave in subangular contour (notched).

acronotus Poey, 1860, p. 325.

12b. Upper teeth along inner half of jaw erect, nearly symmetrical, their outer margins not much more concave than the inner, if at all.

13a. First gill opening not more than 1.5 times as long as horizontal diameter of eye; lower precaudal pit only faintly indicated; fins without conspicuous black markings; anterior margin of nostril conspicuously lobed (Fig. 76 F).

remotus Duméril, 1865, p. 400.

5a. *Longimanus* is included under alternative 3b, as well as under 3a, because of uncertainty as to whether or not the mid-dorsal ridge, evident in embryos, persists after birth.

13b. First gill opening nearly twice as long as horizontal diameter of eye, or longer; lower precaudal pit strongly marked; fins conspicuously tipped with black; anterior margin of nostril only slightly expanded (Figs. 63 G, 67 B).

14a. Origin of 1st dorsal about over midpoint of inner margin of pectoral; 1st gill opening less than 2.5 times as long as horizontal diameter of eye; horizontal diameter about ⅕ as long as snout in front of mouth; edges of lower teeth regularly though very finely serrate.
limbatus Müller and Henle, 1841, p. 346.

14b. Origin of 1st dorsal over or a little posterior to inner corner of pectoral; 1st gill opening more than 4 times as long as horizontal diameter of eye; horizontal diameter of eye only about ⅐ to ⅛ as long as snout in front of mouth; margins of lower teeth smooth.
maculipinnis Poey, 1866, p. 364.

Carcharhinus acronotus (Poey), 1861

Black-nosed Shark

Figure 55

Study Material. Two males, about 485 mm. long, either embryo or newborn with umbilical scar still showing, from Rio de Janeiro; 2 male embryos, 358 and 371 mm., from Cuba; male embryo, 480 mm., about ready for birth, from Brazil (Harv. Mus. Comp. Zool., No. 723); immature female, 637 mm., from Englewood, Florida (U.S. Nat. Mus., No. 104331); also photographs of Poey's unpublished drawings of this species.

Distinctive Characters. Easily recognized, among those of the smooth-backed members of the genus in which the second dorsal originates above the origin of the anal, by the relatively long snout, combined with the fact that the upper teeth are noticeably asymmetrical with deeply notched outer edges and much more coarsely serrate bases than tips.

Description. Proportional dimensions in per cent of total length. Male, 485 mm., from Rio de Janeiro (Harv. Mus. Comp. Zool., No. 701). Female, 637 mm., from Englewood, Florida (U.S. Nat. Mus., No. 104331).

Trunk at origin of pectoral: breadth 10.7, 11.0; height 11.5, 11.0.
Snout length in front of: outer nostrils 3.7, 3.6; mouth 8.0, 8.0.
Eye: horizontal diameter 2.5, 1.9.
Mouth: breadth 7.2, 7.2; height 4.5, 4.2.
Nostrils: distance between inner ends 4.7, 4.7.

Labial furrow length: upper 0.8, 0.7.

Gill opening lengths: 1st 2.5, 2.9; 2nd 2.7, 3.0; 3rd 2.7, 3.2; 4th 2.5, 2.9; 5th 2.3, 2.5.

First dorsal fin: vertical height 8.2, 9.9; length of base 9.5, 9.7.

Second dorsal fin: vertical height 3.0, 2.9; length of base 3.7, 4.1.

Anal fin: vertical height 3.9, 3.8; length of base 4.5, 5.0.

Caudal fin: upper margin 28.3, 27.4; lower anterior margin 11.7, 11.6.

Pectoral fin: outer margin 15.6, 14.8; inner margin 6.0, 5.6; distal margin 11.1, 12.1.

Distance from snout to: 1st dorsal 29.7, 32.4; 2nd dorsal 60.8, 61.3; upper caudal 71.7, 72.6; pectoral 21.4, 21.9; pelvics 45.6, 47.2; anal 60.2, 60.6.

Interspace between: 1st and 2nd dorsals 22.4, 21.1; 2nd dorsal and caudal 7.2, 7.4; anal and caudal 6.5, 7.2.

Distance from origin to origin of: pectoral and pelvics 24.5, 27.0; pelvics and anal 15.1, 13.8.

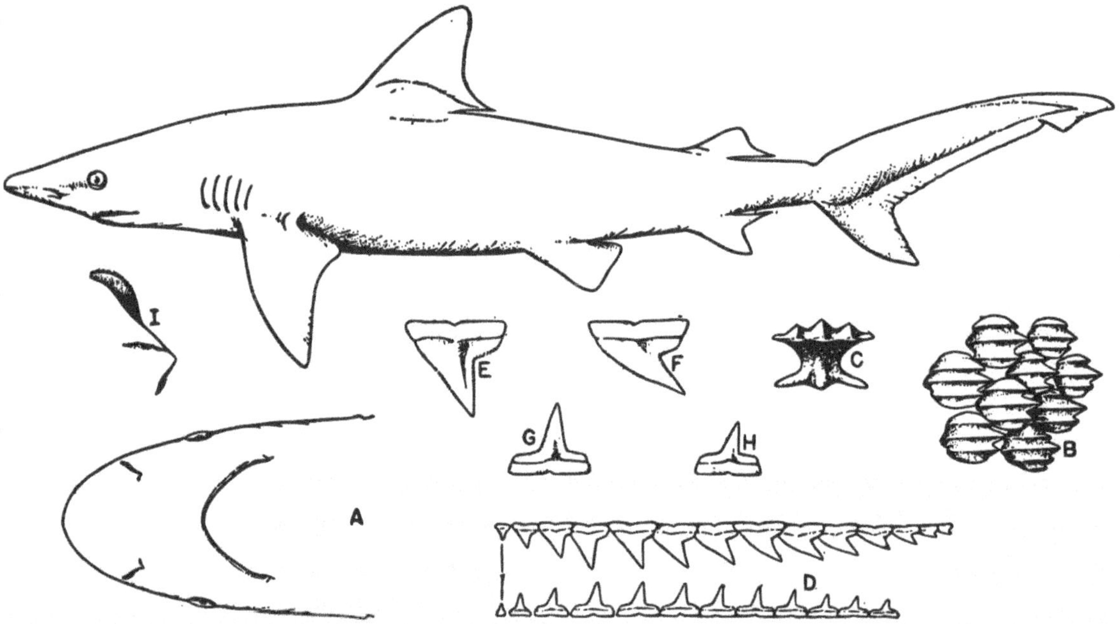

FIGURE 55. *Carcharhinus acronotus*, immature female, 637 mm. long, from Englewood, Florida (U.S. Nat. Mus., No. 104331). *A* Anterior part of head from below, about 0.4 x. *B* Dermal denticles, about 23 x. *C* Apical view of dermal denticle, about 46 x. *D* Left-hand upper and lower teeth, about 1.5 x. *E* Third upper tooth. *F* Eighth upper tooth. *G* Third lower tooth. *H* Eighth lower tooth. *E–H*, about 3 x. *I* Left-hand nostril, about 2 x.

Trunk moderately stout, its height at origin of 1st dorsal (where highest) about ⅕ its length to origin of caudal, back smooth, without mid-dorsal ridge. Body sector to cloaca a little shorter than tail sector. Caudal peduncle only slightly compressed laterally. Lower precaudal pit strongly marked, similar to upper. Dermal denticles mostly overlapping, their leaf-like blades nearly horizontal with usually 3 (sometimes 5) low keels separated by shallow valleys, their edges with 3 or 5 sharp teeth, the median considerably the longest.

Head a little more than 1/5 of total length. Snout rather thin-tipped, narrow-ovoid, its length in front of a line connecting outer ends of nostrils 2/3 to 3/4 as great as distance between inner ends of latter, its length in front of mouth about 1 1/2 times as great as from front of latter to origin of pectoral. Eye approximately circular, its diameter about 1/4 as great as length of snout in front of mouth, or about 2/3 as long as 1st gill opening, its midpoint about opposite front of mouth. First gill opening about 2/3 times as long as diameter of eye, the 3rd slightly longest, 5th a little the shortest, evenly spaced, the 4th above origin of pectoral. Nostrils strongly oblique, the inner corners nearer to front of mouth than to tip of snout by a distance equal to about 1/2 the diameter of eye, the anterior margins slightly sinuous and expanded near inner end as a prominent triangular lobe with narrowly rounded tip. Mouth broad-ovoid, about 1.7 times as broad as high in young specimens, but 1.5 times in larger.

Teeth $\frac{12\text{ or }13\text{—}1\text{ or }2\text{—}12\text{ or }13}{11\text{—}1\text{—}11}$; uppers broadly triangular, their edges serrate, most coarsely so on outer side of base, their inner margins straight near center of mouth but weakly convex along outer part of jaw, the outer margins deeply notched and increasingly oblique along the jaw, the outermost teeth very strongly so; lowers with slender cusps and broad bases, erect in sides of jaw as well as in front, more finely serrate than uppers; one or two very small erect teeth at symphysis in upper jaw, and one at symphysis in lower; outermost 2 or 3 teeth also very small in upper jaw, and with very short cusps in lower.

Origin of 1st dorsal over or slightly behind inner corner of pectoral, its anterior margin moderately convex in young but only slightly so in larger specimens, the rear margin nearly straight toward apex but concave toward base, the apex narrowly rounded in young but subacute in adult, the free rear corner only moderately slender, about 1/3 as long as base; its vertical height a little greater than length of snout in front of mouth and about 57% as great as length of pectoral. Second dorsal a little more than 1/3 as long at base as 1st dorsal, its origin about opposite that of anal, its anterior margin convex in young but nearly straight in adult, its posterior margin weakly concave, apex broadly rounded in young but more narrowly so in adult, its free rear corner about as long as base. Caudal a little more than 1/4 (27 to 28%) of total length, with slender, subacute tip, the lower anterior lobe (expanded lower anterior corner) 40 to 45% as long as upper margin, each measured from its respective precaudal pit, with moderately convex lower anterior margin (more so in young than in adult), and subacute tip (but rounded in newborn). Anal a little longer at base than 2nd dorsal, its anterior margin more convex, its posterior margin much more deeply concave, its apex subacute, its free rear corner about as long as its base, with tip extending slightly farther rearward than tip of 2nd dorsal and anterior to lower precaudal pit by a distance 2/3 as long as base of anal or about twice as long as diameter of eye. Distance between origin of anal and tips of pelvics a little longer than base of anal. Pelvics about 1.3 times as long at base as anal, their margins nearly straight. Pectoral about ⅔ (about 67%) as long as head, and about 1.7 times as long as broad, its outer margin strongly convex in young but decreasingly so with growth, the distal margin concave for its entire length, most deeply so in young, the inner margin weakly convex, tip subacute, inner corner rounded.

Color. Described as variable in life; some specimens cream-colored or yellowish gray above, and of a paler shade of the same, or white, below; others uniform brown below as well as above; the fins are without markings but the tip of the snout is dusky, this nose-spot being strongest in the young, becoming diffuse or even obscure in adults.

Size. This is one of the smaller members of the genus, maturing while still no longer than three feet four inches to four feet six inches, and perhaps seldom exceeding a length of five to six feet. To judge from our Study Material (p. 325) the usual length at birth is not far from 450 mm.

Developmental Stages. It is not known whether or not there is a yolk-sac placenta in this species.

Habits. Nothing definite is known of its diet or habits beyond the fact that females, with embryos (usually three to six) nearly ready for birth, have been taken off southwestern Florida from January to April. It is often found in the stomachs of larger sharks.

Relation to Man. It is not numerous enough anywhere to be of any commercial importance, or of interest to anglers.

Range. Western tropical and subtropical Atlantic, Rio de Janeiro to North Carolina. The southwestern coast of Florida is the only region where this species has been reported in any numbers. It is also known from Rio de Janeiro, from Cuba, from the north coast of the Gulf of Mexico (Biloxi, Mississippi, one specimen), and from North Carolina, where stray specimens have been taken from time to time. This suggests that its center of abundance lies in the Florida–West Indies region, probably including the southern part of the Caribbean generally and northern coasts of South America to Brazil.

Synonyms and References:

Squalus acronotus Poey, Memorias, *2*, 1860: 335, pl. 19, fig. 3, 4 (descr., teeth, Cuba).

Platypodon acronotus Poey, Repert. Fisico.-Nat. Cuba, *2*, 1868: 450 (Cuba); An. Soc. esp. Hist. nat., *5*, 1876: 389; Enumerat. Pisc. Cubens., 1876: 193 (Cuba).

Carcharias (Prionodon) acronotus Günther, Cat. Fish. Brit. Mus., *8*, 1870: 369 (Cuba).

Carcharhinus (Platypodon) acronotus Jordan and Evermann, Rep. U.S. Comm. Fish. (1895), 1896: 216 (Cuba); Bull. U.S. nat. Mus., 47 (1), 1896: 36 (descr., Cuba).

Carcharinus acronotus Garman, Mem. Harv. Mus. comp. Zool., *36*, 1913: 136 (descr., Cuba); Gudger, J. Elisha Mitchell sci. Soc., *28*, 1913: 158 (N. Carolina); Coles, Proc. biol. Soc. Wash., *28*, 1915: 90 (N. Carolina); Radcliffe, Bull. U.S. Bur. Fish., *34*, 1916: 259, pl. 41, fig. 1, 2 (denticles, teeth, N. Carolina); Smith, J. Amer. Mus. nat. Hist., *16*, 1916: 348 (Cuba, N. Carolina); Nichols, Bull. Amer. Mus. nat. Hist., *37*, 1917: 875, pl. 111, fig. 2 (south. Florida); Breder, Field Bk. Mar. Fish. Atlant. Coast, 1929: 16 (Florida, W. Indies); White, Bull. Amer. Mus. nat. Hist., *74*, 1937: 127, pl. 50, fig. d (in Key, cartilages of claspers); Springer, Proc. Fla. Acad. Sci., *3*, 1939: 21 (Florida, embryos, color); Hildebrand, Copeia, 1941: 221 (N. Carolina); Lunz, Bull. S. Carolina St. Planning Bd., *14*, 1944: 27 (Florida).

Carcharias acronotus Jordan, Manual Vert. Anim. NE. U.S., 1929: 10 (general); Jordan, Evermann and Clark, Rep. U.S. Comm. Fish. (1928), 2, 1930: 16 (Cuba).
Eulamia acronotus Fowler, Monogr. Acad. nat. Sci. Philad., 7, 1945: 95 (N. Carolina, but 8-ft. specimen of 250 pounds is perhaps not this species).

Carcharhinus falciformis (Müller and Henle), 1841

Figures 56, 57

Study Material. Five immature specimens, males and females, 782 to 1,065 mm. long, from offing of Delaware Bay and of Cape Fear, North Carolina, and from the Bahamas (U.S. Nat. Mus. and Harv. Mus. Comp. Zool.);[6] jaws, measurements of embryos, and photographs of latter from Florida Keys;[7] a skin, without locality (Harv. Mus. Comp. Zool., No. 1384); also photographs of Poey's unpublished drawings of this species.

Distinctive Characters. Among the ridge-backed members of the genus, *falciformis* resembles *floridanus* in the very elongate free tips of the second dorsal and anal, and in the teeth, but it has a much shorter pectoral, a larger eye and a more broadly ovate snout.

Description. Proportional dimensions in per cent of total length. Female, 785 mm., from Lat. 33° 38′ N., Long. 77° 36′ W. (U.S. Nat. Mus., No. 38510). Male, 848 mm., from Lat. 38° 37′ N., Long. 73° 11′ W. (U.S. Nat. Mus., No. 35643).

Trunk at origin of pectoral: breadth 11.2, 11.0; height 11.4, 9.6.
Snout length in front of: outer nostrils 4.5, 4.4; mouth 8.7, 8.0.
Eye: horizontal diameter 1.9, 1.8.
Mouth: breadth 7.9, 7.6; height 5.4, 5.1.
Nostrils: distance between inner ends 5.8, 5.7.
Labial furrow lengths: upper 0.6, 0.5.
Gill opening lengths: 1st 2.2, 2.1; 2nd 2.3, 2.4; 3rd 2.3, 2.5; 4th 2.2, 2.4; 5th 1.8, 1.8.
First dorsal fin: vertical height, 7.3, 7.5; length of base 8.5, 8.6.
Second dorsal fin: vertical height 2.0, 1.9; length of base 3.3, 3.4.
Anal fin: vertical height 3.0, 2.9; length of base 3.5, 3.5.
Caudal fin: upper margin 27.7, 28.2; lower anterior margin 10.9, 11.6.
Pectoral fin: outer margin 15.3, 14.7; inner margin 4.2, 4.2; distal margin 12.2, 11.8.
Distance from snout to: 1st dorsal 33.9, 32.7; 2nd dorsal 63.2, 60.3; upper caudal 72.3, 71.8; pectoral 24.5, 23.0; pelvics 50.3, 47.7; anal 63.0, 60.7.
Interspace between: 1st and 2nd dorsals 22.5, 21.6; 2nd dorsal and caudal 6.6, 7.0; anal and caudal 6.2, 6.1.
Distance from origin to origin of: pectoral and pelvics 26.8, 26.2; pelvics and anal, 13.4, 14.8.

6. Lat. 38° N., Long. 73° 11′ W., Sept. 12, 1884, off Delaware Bay, male (U.S. Nat. Mus., No. 35643); Lat. 33° 38′ N., Long. 77° 36′ W., off Cape Fear, North Carolina, Oct. 20, 1885, "Albatross" Sta. 2617 and 2623, 1 male, 2 females (U.S. Nat. Mus., No. 38509, 38510); female, Bahamas.
7. Contributed by Stewart Springer.

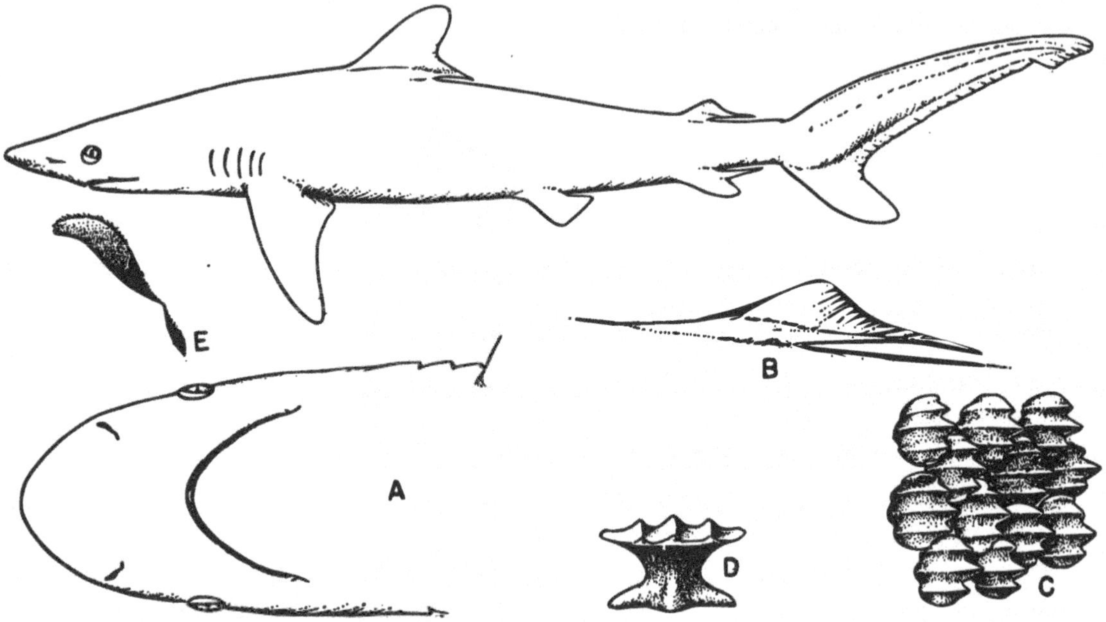

FIGURE 56. *Carcharhinus falciformis*, immature male, 848 mm. long, off Delaware Bay (U. S. Nat. Mus., No. 35643). *A* Head from below, about ½ natural size. *B* Second dorsal fin, about 1 x. *C* Dermal denticles, about 36 x. *D* Apical view of dermal denticle, about 72 x. *E* Left-hand nostril, about 2.5 x.

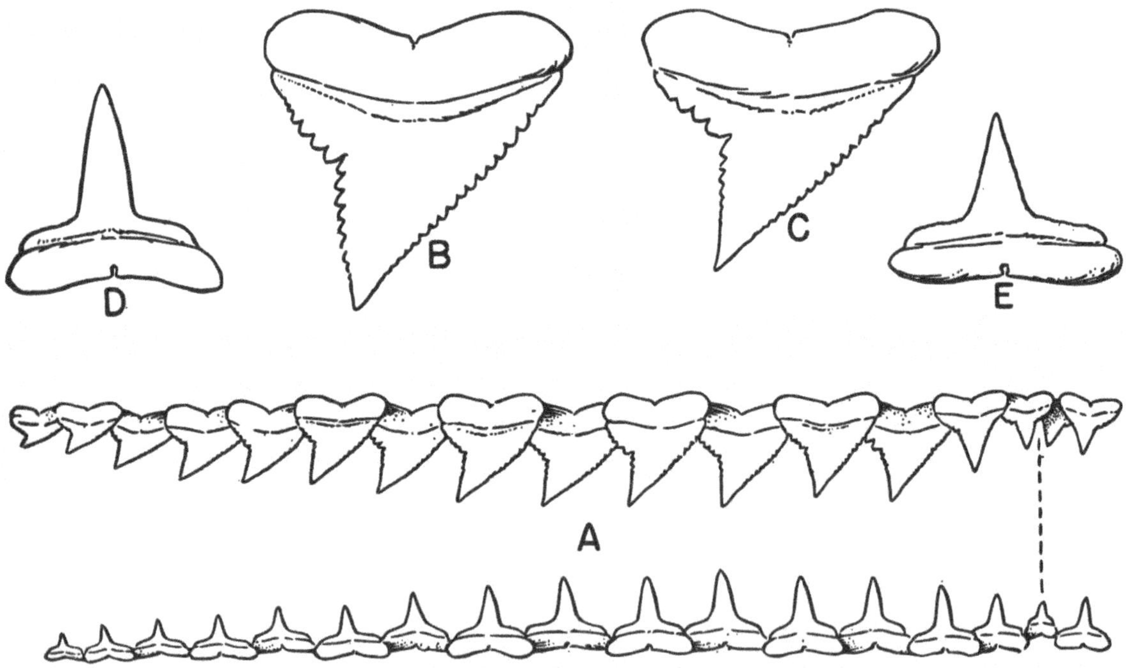

FIGURE 57. *Carcharhinus falciformis*, illustrated in Fig. 56. *A* Right-hand upper and lower teeth, about 2.5 x. *B* Fifth upper tooth. *C* Ninth upper tooth. *D* Second lower tooth. *E* Tenth lower tooth. *B-E*, about 7.5 x.

Trunk slender, its height at origin of 1st dorsal about 1/6 its length to origin of caudal. Midline of back with a low but unmistakable dermal ridge from close behind tip of first dorsal to a point about as far in front of origin of 2nd dorsal as length of base of latter.[8] Upper precaudal pit subrectangular in outline, the lower pit more obtusely angular and only faintly marked. Dermal denticles small, regularly overlapping, with 3 or sometimes 5 low ridges, the apical margin with a corresponding number of teeth, the axial only a little larger than the pairs flanking it.

Head a little less than 1/4 (22 to 24%) of total length, nearly as wide at eyes as at origin of pectorals, and about 3/4 as wide opposite outer ends of nostrils as at eyes. Snout rather thin-tipped, broadly rounded, its length in front of a line connecting outer ends of nostrils about 70% as great as distance between inner ends of latter, its length in front of mouth about 1/3 of length of head. Eye nearly circular, its diameter about 1/3 as long as distance between inner ends of nostrils. First gill opening about 1 1/5 times as long as diameter of eye, the 3rd slightly longest, the 5th about 5/8 as long as 1st, the 4th and 5th more narrowly spaced than 1st to 4th, the space between 3rd and 4th above origin of pectoral. Nostril strongly oblique, its inner end nearer to mouth than to tip of snout by a distance about equal to diameter of eye, the anterior margin only very slightly expanded toward inner end, without definite lobe. Mouth ovate, about 70% as high as broad.

Teeth $\frac{14\text{—}1 \text{ or } 2\text{—}14}{14\text{—}1\text{—}14}$;[9] upper teeth broadly triangular, the 1st tooth erect and nearly symmetrical, with both margins slightly concave and strongly serrate, especially toward base, the 2nd and successive teeth increasingly oblique with inner margins slightly convex and outer margins notched more and more deeply toward angles of jaws, their inner edges serrate, their outer margins similarly so on cusps and even more coarsely so on basal sectors where the 1st one or two serrations are the most prominent; lower teeth with narrow, lanceolate cusps on broad bases, symmetrical and erect along entire jaw, their edges smooth, or at most slightly wavy or irregularly serrate toward tips; 1 or 2 very small symmetrical teeth with notched margins at symphysis in upper jaw, and 1 minute, slender, symmetrical tooth in lower; outermost teeth in each jaw also very small.

Vertical height of 1st dorsal about as great as length of snout in front of mouth, or about 1/2 length of pectoral, its origin a little posterior to inner corner of pectoral, its anterior margin weakly convex, the apex moderately to broadly rounded, the posterior margin convex distally but very deeply concave proximally, the free rear corner only moderately slender, a little less than 1/2 as long as base. Second dorsal only about 1/3 as long at base as 1st dorsal, its origin about over or a little anterior to that of anal,[10] its anterior margin nearly straight, posterior margin moderately concave, its apex rounded, the free rear corner very slender and noticeably elongate, being more than twice as long as the vertical height, and a little longer than the base, extending nearly 2/3 of the distance from rear end of base toward precaudal pit. Caudal between 1/4 and 1/3 of total length, its upper margin

8. In preserved specimens this ridge may lie along a deep longitudinal furrow of muscular contraction.

9. Fourteen teeth could be counted in the side of each jaw in one specimen, with perhaps another very small one at the corner.

10. It is difficult to determine its point of origin exactly in the specimens seen.

moderately convex, its terminal sector only about ⅕ of total length of fin, slender, with rounded tip and weakly concave lower margin, its lower lobe (expanded anterior corner) about ⅖ (42%) as long as upper lobe with convex lower anterior margin and narrowly rounded tip; general posterior re-entrant contour, included by the two lobes, more broadly rounded than in *C. floridanus* (cf. Fig. 56 with 58). Anal about as long at base as 2nd dorsal, or a little longer, with similarly slender free rear tip, about 1½ times as long as the vertical height and about as long as the base, but with much more deeply incised rear margins and more broadly rounded apex. Distance from origin of anal to tips of pelvics about 2½ times as long as base of anal. Pelvics a little longer at base than anal, with nearly straight edges, their origins about midway between origins of 1st and 2nd dorsals. Pectoral only about ⅔ as long as head (about as long as head in *floridanus*) and only about twice as long as vertical height of 1st dorsal, a little more than twice as long as broad, the outer margin weakly convex toward tip, distal margin moderately concave, inner margin only weakly convex, inner corner and apex narrowly rounded.

Color. Described as dark gray above, grayish white below; those we have seen after preservation are mouse gray above and a paler shade of the same tint below.

Size. The claspers of the males listed above have not yet reached the tips of the pelvic fins, and a female of 7½ feet has been found to contain embryos.[11] These facts suggest that *falciformis* matures at a length of perhaps six feet, but it is said to attain a length of about 10 feet (3,050 mm.).[12]

Developmental Stages. It is not yet known whether this is an ovoviviparous or a viviparous species; its embryos have been reported only once.[13]

Habits. While several *falciformis* have been taken on set lines along the reef off Metacumbe Key in southeastern Florida during summer in about 60 feet of water, none have ever been reported in the passages between the Keys, in spite of the great amount of angling that is done from the bridges that span the latter; similarly, one of the three records of it farther to the north was from the outer edge of the continental shelf, the other two being from about 17 to 20 miles out from the nearest land, although taken in shoal water of 14 to 15 fathoms. Thus it appears that this is an offshore species, not to be expected close to the beach unless as a stray. Nothing further is known of its habits, and nothing of its diet.[14]

Relation to Man. Falciformis is not caught anywhere in numbers large enough to make it of commercial importance.

Range. Both sides of the Atlantic, in waters of high temperature; Gorée, West Africa, on the one side, West Indies to the offing of Delaware Bay on the other. So far known in the western Atlantic from: Trinidad (nominal record only); Porto Rico; Haiti; Bermuda; Cuba; Bahamas; east coast of Florida (Salerno), where half-grown specimens as

11. Personal communication from Stewart Springer.
12. 3,050 mm. (Fowler, Bull. Amer. Mus. nat. Hist., *70* (1), 1936: 49).
13. Personal communication from Stewart Springer.
14. The head of a menhaden (*Brevoortia*), found in the stomach of one of the present series taken off Cape Fear, North Carolina, may have been a bait taken from a hook before this Shark was caught.

well as adults with embryos have been taken often enough in summer to mark it as common outside the reefs; two stations 17 to 20 miles off Cape Fear, North Carolina, in October 1885; and the outer edge of the continental shelf off Delaware Bay, in September 1884. The geographic distribution of these localities, together with the fact that the more northerly captures in September and October were at stations where the temperature at the surface (where it is probable that the specimens were caught) was 74° to 75° F., shows this to be a tropical-subtropical species, occasionally straying northward along the coast of the United States during the late summer and early autumn. It is to be expected throughout the Caribbean region generally, and at least as far to the south as northern Brazil, if not farther.

Synonyms and References:

Carcharias (Prionodon) falciformis Müller and Henle, Plagiost., 1841: 47 (descr., Cuba); Duméril, Hist. Nat. Poiss., *1*, 1865: 374 (descr. of type, Cuba); Günther, Cat. Fish. Brit. Mus., *8*, 1870: 363, footnote (Cuba); Steindachner, Denkschr. Akad. Wiss. Wien., *45* (1), 1882: 14 (Gorée, W. Africa); Metzelaar, Trop. Atlant. Visschen, 1919: 187 (both coasts of Atlantic).

Prionodon falciformis Guichenot, in Sagra, Hist. Cuba, 1855: 248, pl. 5, fig. 3 (ill., Cuba, not seen); Poey, Repert. Fisico.-Nat. Cuba, *2*, 1868: 172 (discus.).

Squalus tiburo Poey, Memorias, *2*, 1860: 331, 334 (descr., Cuba); Repert. Fisico.-Nat. Cuba, *2*, 1868: 172[15] (Cuba); not *Squalus tiburo* Linnaeus, 1758.

Prionodon tiburo Poey, Memorias, *2*, 1860: pl. 19, fig. 1, 2 (teeth, Cuba).

Platypodon tiburo Poey, Repert. Fisico.-Nat. Cuba, *2*, 1868: 448, pl. 4, fig. 18 (teeth, Cuba).

Platypodon falciformis Poey, Repert. Fisico.-Nat. Cuba, *2*, 1868: 449; Ann. Soc. esp. Hist. nat., *5*, 1876: 387; Enumerat. Pisc. Cubens., 1876: 191 (discus., refs. to Müller and Henle, 1841, and to Duméril, 1865).

Carcharhinus (Platypodon) falciformis Jordan and Evermann, Rep. U.S. Comm. Fish. (1895), 1896: 216 (Cuba); Bull. U.S. nat. Mus., 47 (1), 1896: 36 (descr., Cuba).

Carcharhinus falciformis Evermann and Marsh, Bull. U.S. Bur. Fish., *20* (1), 1902: 62 (meas., Porto Rico; one of two specimens, the other being *C. floridanus* Bigelow, Schroeder and Springer, 1943: 71); Vincent, Sea Fish. Trinidad, 1910: 53 (name only, Trinidad); Garman, Mem. Harv. Mus. comp. Zool., *36*, 1913: 129 (descr., Cuba, West Indies); Beebe and Tee-Van, Zoologica, N. Y., *10* (6), 1928: 28 (Haiti, meas.); Nichols, Sci. Surv. Porto Rico, N. Y. Acad. Sci., *10* (2), 1929: 183 (Porto Rico); White, Bull. Amer. Mus. nat. Hist., *74*, 1937: 125 (in Key).

Carcharias falciformis Jordan, Evermann and Clark, Rep. U.S. Comm. Fish. (1928), 2, 1930: 16 (Cuba and neighboring waters); Beebe and Tee-Van, Zoologica, N. Y., *13*, 1932: 119 (Bermuda); Field Bk. Shore Fish. Bermuda, 1933: 27 (Bermuda).

Eulamia falciformis Fowler, Bull. Amer. Mus. nat. Hist., *70* (1), 1936: 49 (trop. W. Afr., max. size); Bigelow and Schroeder, Guide Comm. Shark Fish., Anglo Amer. Caribb. Comm., Wash., 1945: 78, fig. 23 (descr., range, ill.).

Carcharhinus floridanus Bigelow, Schroeder and Springer, 1943

Silky Shark

Figures 58, 59

Study Material. Type, female about 8 feet (2,328 mm.) long, taken off Fort Pierce, Florida, in about 100 feet of water on November 2, 1942 (Harv. Mus. Comp.

15. Poey concluded that *tiburo* was a synonym of *falciformis*, but later (ibid., p. 448) questioned this. However, there appears to be nothing to differentiate it from *falciformis*.

Zool., No. 35807); also jaws, fins, photographs and measurements of a female about 8 feet long caught at the surface from the "Atlantis" in Cochinas Bay, Cuba, on April 5, 1939 (Harv. Mus. Comp. Zool., No. 35515).

Distinctive Characters. Among the ridge-backed members of the genus, with which it falls, *floridanus* most closely resembles *falciformis* in the elongate free rear tips of the

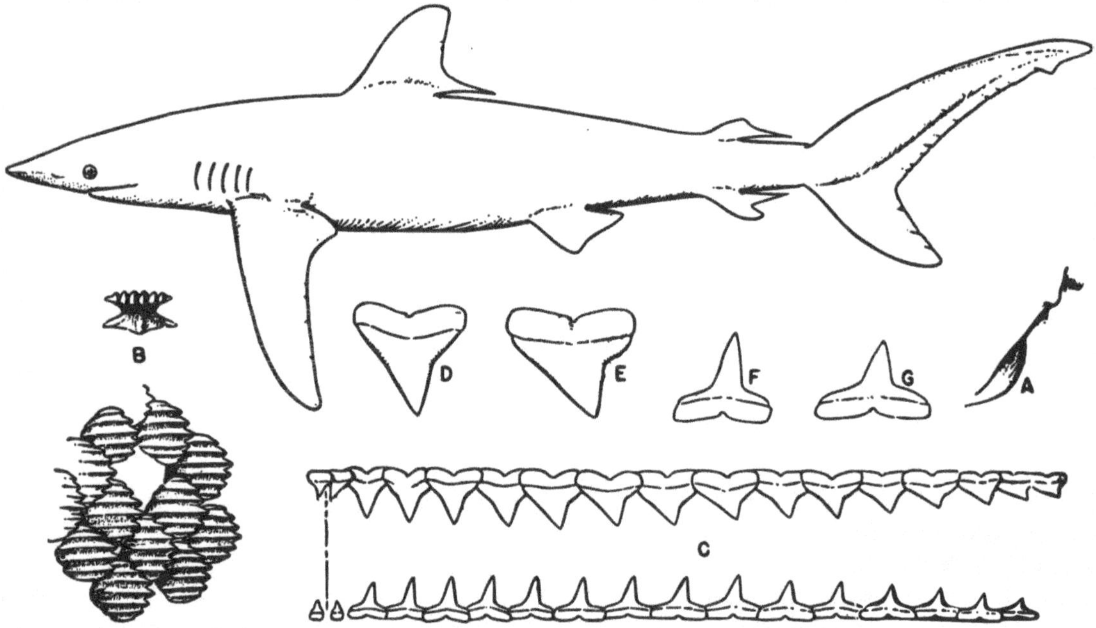

FIGURE 58. *Carcharhinus floridanus*, adult female, about eight feet long, from Fort Pierce, Florida (Harv. Mus. Comp. Zool., No. 35807). *A* Right-hand nostril, about 1 x. *B* Dermal denticles, about 26 x. *C* Left-hand teeth, upper and lower, about ⅔ natural size. *D* Fourth upper tooth. *E* Sixth upper tooth. *F* Fourth lower tooth. *G* Eleventh lower tooth. *D–G*, about 1.3 x.

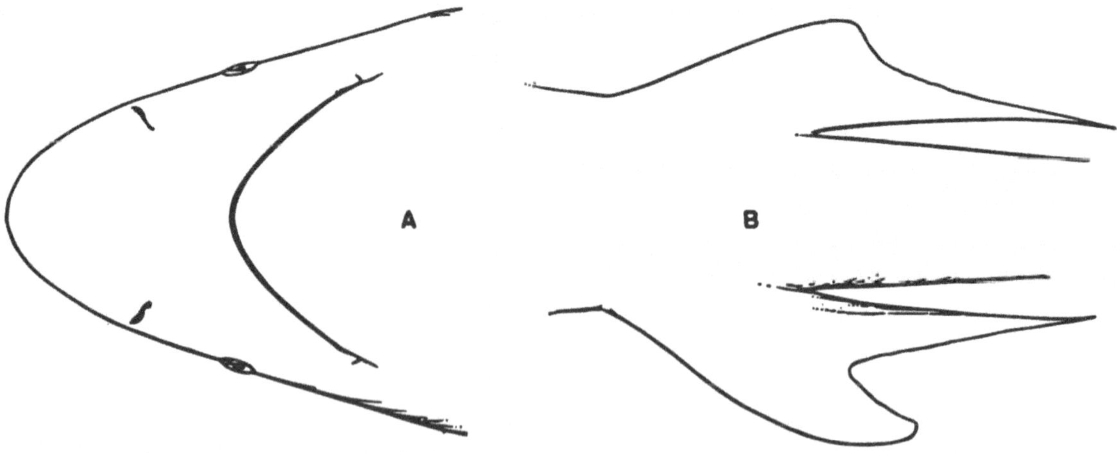

FIGURE 59. *Carcharhinus floridanus*, pictured in Fig. 58. *A* Anterior part of head from below, about ⅙ x. *B* Second dorsal and anal fins, about ⅓ natural size.

second dorsal and anal, and in its teeth. But it is easily separable from *falciformis* by its much longer pectorals, much smaller eye, and the more narrowly ovate snout (cf. Fig. 56 with 58).

Description. Proportional dimensions in per cent of total length. Female, 2,328 mm., from Fort Pierce, Florida (Harv. Mus. Comp. Zool., type, No. 35807).

Snout length in front of: outer nostrils 4.0; mouth 7.2.

Eye: horizontal diameter 1.2.

Mouth: breadth 8.3; height 4.4.

Nostrils: distance between inner ends 5.5.

Labial furrow length: upper 0.5.

Gill opening lengths: 1st 2.9; 2nd 3.0; 3rd 3.3; 4th 3.1; 5th 2.6.

First dorsal fin: vertical height 8.6; length of base 8.7.

Second dorsal fin: vertical height 2.2; length of base 2.8.

Anal fin: vertical height 3.6; length of base 3.2.

Caudal fin: upper margin 26.5; lower anterior margin 13.6.

Pectoral fin: outer margin 20.7; inner margin 5.0; distal margin 16.7.

Distance from snout to: 1st dorsal 30.0; 2nd dorsal 63.7; upper caudal 73.5; pectoral 20.7; pelvics 48.5; anal 63.7.

Interspace between: 1st and 2nd dorsals 25.8; 2nd dorsal and caudal 7.1; anal and caudal 5.8.

Distance from origin to origin of: pectoral and pelvics 27.8; pelvics and anal 15.7.

Trunk rather slender and tapering evenly both anteriorly and posteriorly, its height at origin of 1st dorsal only about ⅙ its length to origin of caudal. Body sector to cloaca considerably longer than tail sector. A low but unmistakable dermal ridge along midline of back extending part way between 1st and 2nd dorsal fins. Lower precaudal pit, as well as upper, well marked. Dermal denticles so small and flat that the skin feels silky, close set, regularly overlapping, those on trunk usually with 7 ridges but occasionally with only 6, the marginal teeth short and rather blunt, those on head minute, less strongly sculptured and broadly oval.

Head about ⅕ of total length. Snout flattened above, its tip narrow-ovate, its length in front of a line between outer ends of nostrils a little less than ¾ (71%) as great as distance between inner ends of latter, the length in front of mouth about ⅓ of length of head. Eyes round and noticeably small, the horizontal diameter only about ¼ to ⅕ as long as distance between nostrils, or ⅙ the length of snout in front of mouth. Gill openings evenly spaced, of medium length, the first a little more than twice as long as diameter of eye, the 3rd a little the longest and 5th a little the shortest, the 3rd, or space between 3rd and 4th, above origin of pectoral. Nostril oblique, its inner end nearer to upper jaw than to tip of snout by a distance about twice as great as diameter of eye, the anterior margin without definite lobe but only somewhat sinuous in outline. Mouth broad-ovate, its length slightly less than ½ its breadth.

Teeth $\frac{15 \text{ or } 16\text{—}2 \text{ or } 3\text{—}15 \text{ or } 16}{15\text{—}2 \text{ or } 3\text{—}15}$; uppers triangular, increasingly oblique toward corners of mouth, their outer edges more or less notched, with some toward center of jaw notched on both sides, the edges serrate, very finely toward tips and somewhat more coarsely toward bases, but very much worn;[16] lower teeth smaller, more slender and more nearly erect, symmetrical, with narrow cusps and broad bases, their margins smooth toward bases and slightly wavy toward tips but not definitely serrate; 2 or 3 minute teeth at symphysis in both jaws.

Origin of 1st dorsal over, or slightly posterior to, inner corner of pectoral, its vertical height about ⅓ as great as length of head to 5th gill opening, its apex rounded, its rear margin deeply concave basally, its free rear tip very slender and about ⅔ as long as its base. Origin of 2nd dorsal about over origin of anal, its vertical height only about ¼ to ⅕ as great as that of 1st dorsal, its posterior margin only weakly concave, its free rear tip very slender and greatly elongate, it being about 2½ times as long as the vertical height of the fin, or 1½ times as long as its base, and extending about ¾ the distance from rear end of base toward precaudal pit. Caudal a little more than ¼ of total length, its terminal sector a little less than ⅕ total length of fin, slender, with concave lower margin and narrowly rounded tip, the lower lobe (expanded lower anterior corner) about ½ as long as upper, with subacute tip and moderately convex anterior margin, the re-entrant corner included between the two lobes subrectangular (rounded in *falciformis*). Anal a little longer at base than 2nd dorsal, with much more deeply incised posterior and more broadly rounded posterior margins, its free rear tip slender, a little longer than its base, about 1½ times its vertical height, and terminating about under tip of 2nd dorsal. Distance from origin of anal to tips of pelvics nearly or quite twice as long as base of anal. Pelvics about 1½ times as long at base as anal, with weakly concave distal margins, their origin about midway between origins of 1st and 2nd dorsals. Pectoral nearly or quite as long as head or about 2½ times as long as vertical height of 1st dorsal, a little less than ½ as broad as long, the outer margin moderately convex, distal margin deeply concave basally, the apex and inner corner narrowly rounded.

Color. The type specimen was shiny black above when fresh caught, dirty white below; the tips of the pectorals somewhat dusky below; the Cuban specimen was dark gray above in life.

Size. This is one of the larger members of the genus; all the specimens so far reported have been between about eight and ten feet long.

Developmental Stages. Embryos of this species have been reported but not described.

Habits. Little is known of the habits of this newly described species except that off Salerno, Florida, it is usually taken only where lines are set at a depth of 100 feet or more. While one large individual was taken on the reef near Metacumbe Key in about 45 feet of water, it was so thin as to suggest that its excursion into relatively shoal water was abnormal.

16. Those of the Cochinos Bay specimen are more definitely notched than those of the type specimen, and with coarser basal serrations.

Relation to Man. Such specimens as are caught in the Florida fishery (see below) are used for leather, etc., like other large sharks.

Range. It is known to the present time from the south coast of Cuba (see Study Material, p. 333) and the north coast near Havana,[17] Porto Rico, and southeastern Florida. Hence it probably occurs generally throughout the tropical belt of the western Atlantic. Evidently it is plentiful locally, for it is taken so frequently in the shark fishery that is now carried on from Salerno, Florida, that it has been given the vernacular name "Silky Shark," appropriate because of the small size of its dermal denticles. As many as 60 adults of nine to ten feet were caught there in a single day during the winter of 1943, making it dominant in the total catch of sharks of all sorts for the time being. It is less numerous there in summer.[18] That a shark so common, so large and so easily recognized should have continued unknown for so long casts an unflattering light on scientific knowledge of the group to which it belongs.

Synonyms and References:

Carcharhinus falciformis (in part) Evermann and Marsh, Bull. U.S. Fish Comm., *20* (1), 1902: 62 (meas., Porto Rico, one of two specimens, the other being correctly identified as *C. falciformis* Müller and Henle, 1841).

Carcharinus floridanus Bigelow, Schroeder and Springer, Proc. New Engl. zool. Cl., *22*, 1943: 71, pl. 13 (ill., descr., Salerno, Florida, type loc., and south coast of Cuba).

Eulamia floridanus Bigelow and Schroeder, Guide Comm. Shark Fish., Anglo Amer. Caribb. Comm., Wash., 1945: 76, fig. 22 (descr., range, ill.).

Carcharhinus leucas (Müller and Henle), 1841[19]

Cub Shark, Bull Shark, Ground Shark

Figures 60, 61

Study Material. Female, 924 mm. long, from Florida (Amer. Mus. Nat. Hist.); skin of a female, about 1,137 mm., at Miraflores Locks, Panama Canal (U.S. Bur. Fish., No. 13961); head of specimen, about 6½ feet, from Bahamas (from Daniel Merriman); embryo, 435 mm. long, from Cuba (Harv. Mus. Comp. Zool., No. 722); tracings of fins of an adult male, 2,310 mm., from Metacumbe, Florida (from Stewart Springer); male embryos, 490 mm. (Harv. Mus. Comp. Zool.) and 625 mm. (U.S. Nat. Mus., No. 108,456), from Englewood, Florida; female, 692 mm. long, either embryo or newborn to judge from the umbilical scar, also female, 920 mm., from Lake Yzabal, Guatemala (U.S. Nat. Mus.).

Distinctive Characters. Leucas is separable from all other Atlantic carcharhinids except *Negaprion brevirostris, Carcharhinus longimanus* and *C. nicaraguensis* by its ex-

17. Personal communication from Luis Howell-Rivero. 18. Personal communication from Stewart Springer.

19. The account of *leucas* by Müller and Henle (Plagiost., 1841: 42) agrees in detail with the present species, except that the dried specimens on which it was based were described as white above as well as below. But their paleness (probably from fading) was perhaps exaggerated, for the same specimens (presumably) were described later as whitish gray (Duméril, Hist. Nat. Poiss. *1*, 1865: 358).

tremely short, very broadly rounded snout and smooth back (without dorsal ridge). Its second dorsal is much smaller than the first and its teeth are regularly serrate on the cusps, features which mark it off from *N. brevirostris;* from *C. longimanus* it is separated by the shape of its dorsal with subacute apex, by its relatively shorter and broader pectorals, and by the long interspace between the tip of its anal and its caudal, but relatively much

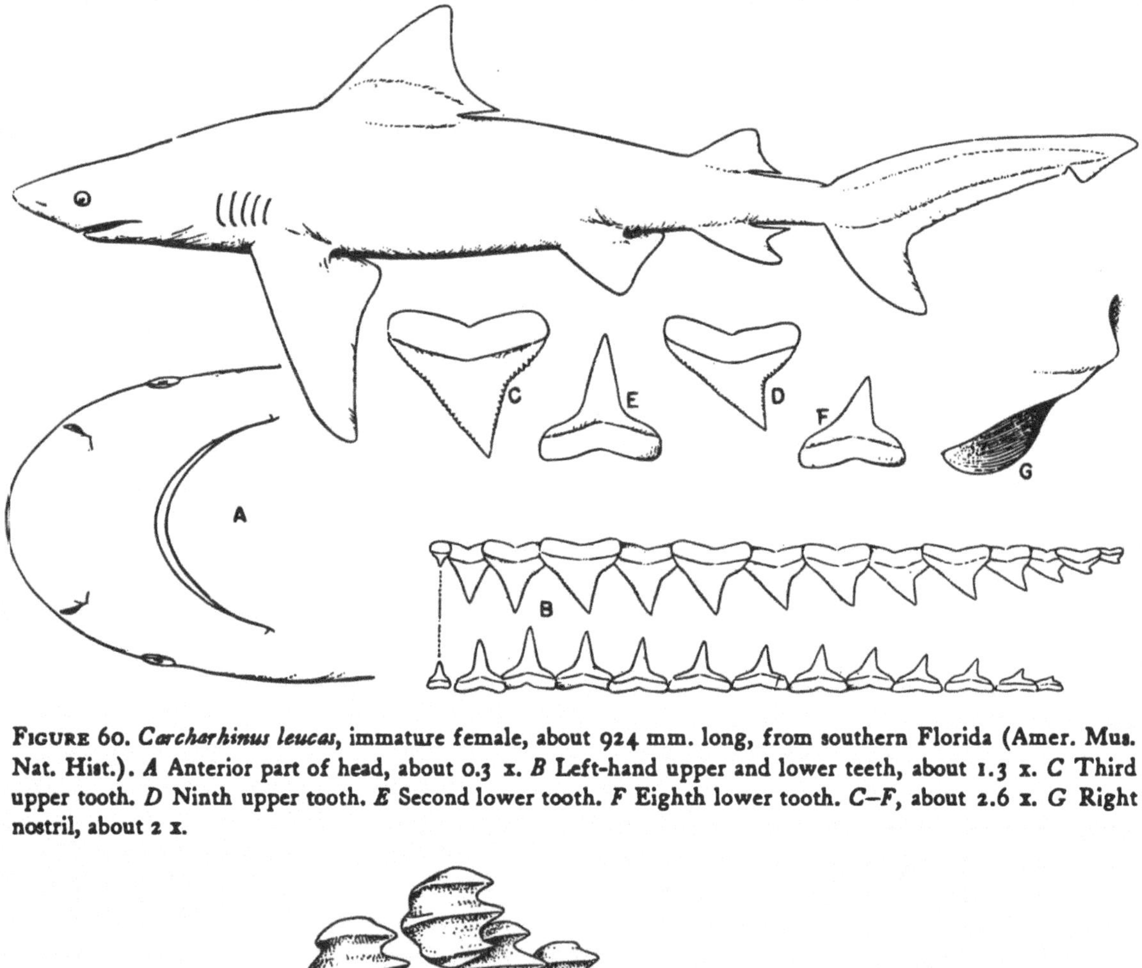

FIGURE 60. *Carcharhinus leucas*, immature female, about 924 mm. long, from southern Florida (Amer. Mus. Nat. Hist.). *A* Anterior part of head, about 0.3 x. *B* Left-hand upper and lower teeth, about 1.3 x. *C* Third upper tooth. *D* Ninth upper tooth. *E* Second lower tooth. *F* Eighth lower tooth. *C–F*, about 2.6 x. *G* Right nostril, about 2 x.

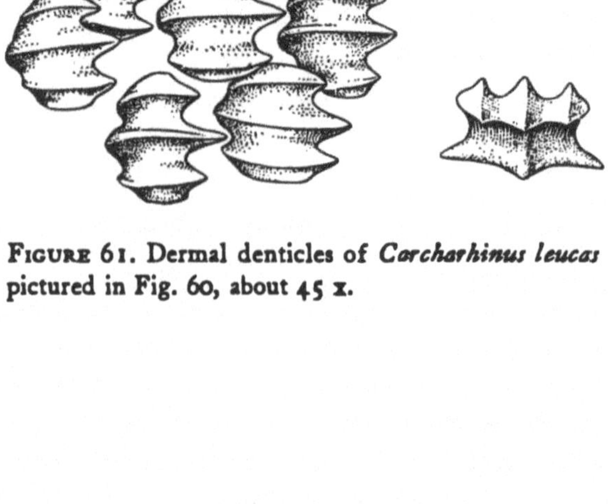

FIGURE 61. Dermal denticles of *Carcharhinus leucas* pictured in Fig. 60, about 45 x.

shorter distance from origin of anal to tips of pelvics; *C. nicaraguensis* appears to be a landlocked variant of it (see discussion, p. 381).

Description. Proportional dimensions in per cent of total length. Female, 924 mm., from Florida (Amer. Mus. Nat. Hist.).

Trunk at origin of pectoral: breadth 12.8; height 12.9.
Snout length in front of: outer nostrils 2.0; mouth 6.6.
Eye: horizontal diameter 1.5.
Mouth: breadth 9.8; height 4.9.
Nostrils: distance between inner ends 6.7.
Gill opening lengths: 1st 2.9; 2nd 3.2; 3rd 3.2; 4th 2.9; 5th 2.4.
First dorsal fin: vertical height 9.6; length of base 13.4.
Second dorsal fin: vertical height 4.1; length of base 5.5.
Anal fin: vertical height 4.9; length of base 5.3.
Caudal fin: upper margin 28.3; lower anterior margin 12.3.
Pectoral fin: outer margin 20.6; inner margin 6.6; distal margin 17.6.
Distance from snout to: 1st dorsal 27.0; 2nd dorsal 60.0; upper caudal 71.7; pectoral 21.4; pelvics 48.3; anal 60.0.
Interspace between: 1st and 2nd dorsal 21.0; 2nd dorsal and caudal 7.0; anal and caudal 5.5.
Distance from origin to origin of: pectoral and pelvics 28.4; pelvics and anal 11.3.

Trunk moderately stout, its height at origin of 1st dorsal about 1/5 of its length to precaudal pits, its breadth at origin of pectorals about equally great, its dorsal profile more convex than the ventral. Midline of back smooth, without dermal ridge, either in embryo or in adult. Anterior outline of upper caudal pit obtusely subangular in smaller specimens, the lower caudal pit less well marked than upper but similar in outline. Dermal denticles moderately raised so that skin feels slightly rough when stroked toward head, loosely spaced and overlapping but little, as broad as long, or a little broader, usually with 3 low ridges in smaller specimens, but perhaps more often 5 in larger, with 3 prominent teeth (the axial considerably the largest) separated by rounded notches; if there are 5 teeth, the outermost pair are very small; pedicels very short and broad.

Head very wide, its breadth as great opposite corners of mouth as at origin of pectoral or slightly greater and only a little narrower at eyes. Snout thick, very broadly rounded and notably short, its length in front of a line connecting outer ends of nostrils only between 1/3 and 1/2 (30 to 42%) as great as distance between inner ends of latter in specimens seen, its length in front of mouth a little less than 1/3 of length of head to origin of pectoral. Eye approximately circular, small, its diameter a little less than 1/5 as great as distance between inner ends of nostrils. Longest gill openings (2nd and 3rd) a little more than twice as long as diameter of eye or about half as long as distance between inner ends of nostrils, the 5th only a little more than 2/3 that long; all about evenly spaced, the 4th about over origin of pectoral. Nostril strongly oblique, its inner end nearly or quite as close to tip of snout as to mouth and about equidistant between rear edge of eye and tip of snout; its anterior margin slightly expanded in obtusely triangular outline with rounded corner. Mouth broadly ovate, its height about ½ its breadth. Upper labial furrow very short even for this genus, and at approximately a right angle with the jaw.

Teeth $\frac{12 \text{ or } 13—1—12 \text{ or } 13}{12 \text{ or } 13—1—12 \text{ or } 13}$; uppers broadly triangular, both edges regularly and moderately coarsely serrate from base to tip, the first 2 teeth erect and nearly symmetrical, with both margins slightly concave, but 3rd and subsequent teeth with only very slightly concave or nearly straight inner margins, and outer margins increasingly concave toward corners of mouth, the outermost 3 or 4 teeth rather definitely notched; lower teeth with narrow triangular cusps on broad bases, nearly erect in front and sides of mouth and only slightly oblique toward its corners, the cusps successively shorter from 8th or 9th to 12th or 13th, both edges serrate, but more finely so than on uppers. One small symmetrical tooth at symphysis in each jaw.

First dorsal noticeably large, its origin about over or a little anterior to axil of pectoral, its vertical height about ⅖ (40–45%) as great as distance from tip of snout to origin of pectorals, its base nearly as long as its anterior margin, its anterior margin nearly straight, its apex subacute or very narrowly rounded, its posterior outline moderately concave (a little the more so basally), its free rear corner relatively obtuse and only about ¼ as long as its base, the midpoint of latter only a little more than ⅛ as far from axil of pectoral as from origin of pelvics. Second dorsal about ⅖ (41%) as long as 1st at base, relatively lower, its origin over or a little anterior to that of anal,[20] its anterior outline nearly straight, apex rounded, posterior outline only weakly concave, its free rear corner broad and about ½–⅔ as long as its base, its rear tip definitely though only a little anterior to that of anal and separated from origin of caudal by a distance about ¾ as long as its base. Caudal with weakly convex upper margin and subacute tip, the narrow-triangular terminal sector about ¼ the total length of fin, the lower lobe (expanded lower anterior corner) a little less than ½ (about 44%) as long as upper, with moderately convex anterior outline, nearly straight posterior outline and narrowly rounded tip; the re-entrant corner, included by the 2 lobes, broadly rounded. Distance from origin of caudal to tip of anal about ½ as long as base of anal (a convenient field mark separating this species from *longimanus*). Anal with base and free rear corner about as long as those of 2nd dorsal, but anterior margin about 1⅓ times as long and more convex, and posterior edge much more deeply incised. Distance from origin of anal to tips of pelvics only about ½ to ⅔ as long as base of former (longer than base of anal in *longimanus*). Pelvics with nearly straight anterior margins and slightly concave distal margins, their bases a little longer than base of anal. Pectoral nearly as long (about 87%) as head to origin of pectoral, about ½ as broad as long, the outer margin nearly straight toward base but

20. The origin of the second dorsal is about over that of the anal in the female studied, but a little anterior to it in male embryos and in a tracing of an adult male contributed by Stewart Springer.

moderately convex toward tip, the distal margin only weakly concave, the inner corner moderately rounded, the tip more narrowly so.

Color. Described in life as white below, gray above, varying from very pale to much darker, apparently as the result of environmental conditions, for those living over white sand bottom may be very pale. Adults show no conspicuous fin markings. But in embryos the tip and lower edge of the caudal and the margin of the second dorsal are sooty, and perhaps the tips of the other fins likewise, in some cases, and these fin markings may persist for a considerable time after birth, at least in some instances, for a 32-inch specimen, apparently of this species, has been described as with second dorsal and anal dusky-tipped, and the caudal wholly so.[21]

Size. Mature at a length of about seven feet, this species certainly grows to 10 feet and perhaps somewhat longer, but reports of specimens longer than 12 feet may have referred to *C. longimanus,* with which it has often been confused. Specimens of 8 to 8½ feet have been reported as weighing 250 to 375 pounds; the weight of a male of 10 feet caught off North Carolina is given as about 400 pounds.[22] It has been suggested that the three-foot Florida specimen illustrated in Fig. 60, taken in winter, was a yearling,[23] and it is probable that the young are born there in spring. Also, the advanced stage of development of the embryos listed above indicates a length of perhaps 650 to 700 mm. as usual at birth.

Developmental Stages. Presumably development is viviparous, but the presence of a yolk-sac placenta has not been definitely recorded for this species, so far as we are aware. Embryos have relatively stouter bodies and blunter heads than their parents, but they do not differ nearly so much from the adults in the shapes and relative sizes of the fins as do those of *longimanus;* like the adults, they are smooth-backed. Five or six appear to be the usual number of young in a litter.

Habits. This is a heavy, slow-swimming species, most common inshore in shoal water, perhaps never very far from land except by accident. They are most often caught around docks, at the entrances to the passages between islands, in estuaries and in harbors. They often run up rivers for considerable distances, and it seems that they do not hesitate to enter fresh water. Thus the series we have studied includes one from the Panama Canal at Miraflores Locks, besides others from Lake Yzabal, Guatemala, a body of water that is said to vary between fresh and brackish, and a 55-pound specimen has been reported, at least by name, as having been caught in the Atchafalaya River, Louisiana, 160 miles from the sea.[24] We have also received a photograph of a shark four or five feet long that appears to be of this species (unless possibly of the landlocked form *nicaraguensis,* which cannot be determined from the photograph), taken 180 miles up the Patuca River, northeastern Honduras.[25] *C. leucas* is, in fact, the only Shark that is known to have permanently adapted

21. Jordan and Gilbert, Proc. U.S. nat. Mus., 5, 1882: 243.
22. Bell and Nichols, Copeia, 92, 1921: 17.
23. Nichols, Bull. Amer. Mus. nat. Hist., 37, 1917: 874.
24. Gunter, Copeia, 1938: 69.
25. This capture has already been reported, without attempt at specific identification (Strong, Explor. Smithson. Instn. [1933], 1934: 46, 47, fig. 55).

itself to fresh water and developed a local race (see under *C. nicaraguensis*, p. 381). On the other hand, it rarely shows itself at the surface, as the more pelagic members of the genus commonly do, unless lured up by the scent of food, such as floating offal. We have never heard of one jumping, whether at liberty or after being hooked.

Under natural conditions its diet is perhaps no less varied than that of the Tiger Shark (p. 270). Thus the stomach contents of a series taken off North Carolina in summer have been reported as including crabs, smaller sharks that had been taken in the same net, the fin of a porpoise which was probably obtained in the same way, shad (*Alosa*) and mackerel. That they had eaten the last-named proves that they have a greater ability to capture fast-swimming fishes than their rather sluggish habits would suggest. Also found in the stomach were large pieces of devil-rays (*Mobula*), which, when fitted together, showed the victim to have been bitten into five parts. It also attacks sting-rays in Florida waters, as evidenced by a spine imbedded in one's mouth. It is notorious also as a scavenger, devouring any kind of offal, and it bites readily on almost any large bait of fish or meat.

Females with embryos nearly ready for birth have been taken in Florida in October, January and February, which suggests that the young are born there in late winter and early spring, but beyond this nothing is known of the breeding habits.

Relation to Man. The only commercial importance of this species is that it contributes something to the general catch of sharks around Florida. But no data are available to show its importance relative to other species. Like various other sharks it has been named a man-eater, but we think it unlikely that this reputation is deserved, for otherwise shark fatalities probably would be far more frequent than they actually are in Florida and the West Indies, where it is one of the more common of the larger sharks.

Range. In the western Atlantic from southern Brazil to North Carolina and occasionally north to the vicinity of New York. It is not yet certain whether *leucas* occurs at all in the eastern Atlantic. It is true that sharks have been recorded under that name from tropical West Africa and from Algeria, but the first of these reports was by name only,[26] while the brief description accompanying the second[27] would apply equally well to any other *Carcharhinus* with moderately blunt snout, for example, *longimanus*, *milberti* or *obscurus*. The *commersonii* described by Rey[28] from the coast of Spain was doubtless an *obscurus*, for it had a mid-dorsal ridge and agreed with that species in other respects as well. Nor is any *Carcharhinus* other than *longimanus* included in the most recent survey of the fishes of Portugal.[29]

Occurrence in the Western Atlantic. Published accounts are in line with word-of-mouth reports to the effect that *leucas* is a tropical species, expanding its range northward during the warm months and perhaps southward as well. But it is not yet possible to

26. Rochebrune, Act. Soc. linn. Bordeaux, (4) *6*, 1882: 42; Metzelaar, Trop. Atlant. Visschen, 1919: 187.
27. Guichenot, Explor. Algér., *3*, Poiss., 1850: 124; a specimen sent to Paris (Duméril, Hist. Nat. Poiss., *1*, 1865: 358).
28. Fauna Iberica Peces, *1*, 1928: 342, 343.
29. Nobre, Fauna Marinha Portugal, *1*, 1935.

write of it in more than the most general terms because of the uncertainty in many cases as to whether published records referred to it, to *longimanus* or to *obscurus*, or to a combination of these three.[30] Locality records for it that can be accepted as reasonably supported by description or other information locate its center of abundance in the West Indian–Caribbean region, no doubt including the southern part of the Gulf of Mexico; it is reported from Cayenne (probably), French Guiana, the Antilles[31] (see Study Material, p. 337), Cuba, Porto Rico (probably), the Bahamas, Colón (probably), from the Panama Canal, from Lake Yzabal, Guatemala; and from many localities on both coasts of southern Florida, where it is certainly one of the more common of the larger sharks inshore and one with which commercial fishermen and anglers are far more familiar than might be expected from the meager scientific records. In fact, we would hazard the guess that more of this species are caught there from wharves, etc., than of any other large shark.

To the southward it is no doubt distributed generally along the South American coast as far as southern Brazil, being positively known from Para and Rio de Janeiro. Available information also suggests that this is the most numerous shark in shoal water around Bermuda. In southern Florida it is resident throughout the year, but perhaps not north of Cape Romaine, for it is not seen at Englewood (Lat. about 27° N.) in December, January or February although well known there during the warmer months. To the northward in the Gulf of Mexico it has been described similarly as the commonest large shark in summer on the Texas coast near Galveston; it has been encountered off Biloxi, on the north shore of the Gulf in July,[32] and has even been reported in fresh water in Louisiana, as noted above (p. 341). Also, it has been described, nominally at least, as moderately common near Charleston, South Carolina. But seemingly it does not range much farther north than this with any frequency, for while eleven large ones were reported as caught near Morehead City in July and August of 1930, only three specimens had been definitely reported for the North Carolina coast up to 1916,[33] although a rather productive shark fishery had been carried on there for some years previously. Moreover, some or all of these may have been *longimanus*, *milberti* or *obscurus*. Evidently it occurs only as a stray along the sector thence northward as far as New York, where the only report ostensibly referring to it is of one New Jersey specimen.[34] Perhaps it visits the stretch of coast east of New York oftener, for it has been described as uncommon along Long Island,[35] but there is no way of knowing how many of the captures on which this characterization was based were *leucas*, or *obscurus*, while the single published report of its presence farther east is based only on a photograph of a specimen about five feet long, supposedly of this species, taken at Woods Hole.[36]

30. On this, see p. 361 under *longimanus*, and p. 388 under *obscurus*.

31. The *Eulamia platyodon* reported from Jamaica by Fowler (Proc. Acad. nat. Sci. Philad., *71*, 1919: 146) appears to have been some other species, for its teeth were described as narrow and the lowers as "entire."

32. Personal communication from Stewart Springer.

33. Radcliffe, Bull. U.S. Bur. Fish., *34*, 1916: 261.

34. Fowler (Proc. biol. Soc. Wash., *33*, 1920: 144, footnote 2) states that a shark earlier reported from the Delaware River by him as *obscurus* (Rep. N. J. Mus. [1907], 1908: 12) was actually *commersonii*, i.e., *leucas*.

35. Nichols, Bull. Amer. Mus. nat. Hist., *37*, 1917: 873; Copeia, 140, 1925: 21.

36. Nichols and Breder, Zoologica, N. Y., *9*, 1927: 16.

Synonyms and References:[37]

Carcharias (Prionodon) leucas Müller and Henle, Plagiost., 1841: 42 (descr., Antilles); Duméril, Hist. Nat. Poiss., *1*, 1865: 358 (descr., Antilles, La Trinité, S. Coast Cuba; but spec. from Algeria probably not this species; see discuss., p. 342);[38] Günther, Cat. Fish. Brit. Mus., *8*, 1870: 368 (discuss.).

Carcharias (Prionodon) leucos (in part) Guichenot, Explor. Sci. Algér., *3*, Poiss., 1850: 124 (ref. to Cayenne and Porto Rico probably *leucas*, but not his Algerian spec.; see footnote 27, p. 342.

Squalus platyodon Poey, Memorias, *2*, 1861: 336 (descr., Cuba).

Prionodon platyodon Poey, Memorias, *2*, 1861: pl. 19, fig. 5, 6 (teeth, Cuba).

Squalus obtusus Poey, Memorias, *2*, 1861: 337 (descr., Cuba).

Prionodon obtusus Poey, Memorias, *2*, 1861: pl. 19, fig. 7, 8 (teeth, Cuba).

Eulamia obtusa Poey, Repert. Fisico-nat. Cuba, *2*, 1868: 447, pl. 4, fig. 16 (discuss., dermal denticles, Cuba); An. Soc. esp. Hist. nat., *5*, 1876: 385; Enumerat. Pisc. Cubens., 1876: 189 (Cuba).[39]

Carcharias platyodon Jordan and Gilbert, Proc. U.S. nat. Mus., *5*, 1882: 243 (descr., abund., Galveston, Texas); Bull. U.S. nat. Mus., 16, 1883: 872 (descr., Cuba to Texas); Evermann and Kendall, Bull. U.S. Fish Comm., *12*, 1894: 95 (in list for Texas); Jordan and Dickerson, Proc. U.S. nat. Mus., *34*, 1908: 11 (Tampico, Mexico); Gunter, Amer. Midl. Nat., *28*, 1927: 316 (Atchafalaya R., Louisiana, in fresh water); Jordan, Evermann and Clark, Rep. U.S. Comm. Fish. (1928), 2, 1930: 15 (Cuba to Texas); Beebe and Tee-Van, Field Bk. Shore Fish. Bermuda, 1933: 26 (descr., abund., Bermuda).

Carcharias lamia Jordan, Proc. U.S. nat. Mus., 7, 1884: 104 (ident. seemingly established by ill. of embryo by Jordan and Evermann, Bull. U.S. nat. Mus. 47 [4], 1900: pl. 5, fig. 17; abund., embryos, S. Florida).

Carcharhinus (Eulamia) platyodon Jordan, Rep. U.S. Comm. Fish. (1885), 1887: 795 (ident. by ref., W. Indies, S. Atlant. and Gulf coasts of U.S.).

Carcharhinus (Carcharhinus) platyodon Jordan and Evermann, Rep. U.S. Comm. Fish. (1895), 1896: 216 (ident. by ref., Cuba to Texas); Bull. U.S. nat. Mus., 47 (1), 1896: 39 (descr., size, Cuba to Texas).

Carcharhinus lamia Jordan and Evermann, Bull. U.S. nat. Mus., 47 (4), 1900: pl. 5, fig. 17 (embryo, probably Florida); Evermann and Kendall, Rep. U.S. Comm. Fish. (1899), 1900: 48 (ident. by included refs., S. Florida); Smith, J. Amer. Mus. nat. Hist., *16*, 1916: 346 (ident. probable from context, southeast coast U.S.); Nichols and Murphy, Brooklyn Mus. Quart., *3*, 1916: 152 (name only, but ident. probable because of loc., Key West, Florida); not *Carcharias (Prionodon) lamia*, Müller and Henle, 1841.

Eulamia longimanus Fowler, Proc. Acad. nat. Sci. Philad., *60*, 1908: 65 (spec., no data, probably *leucas* because of meas.; jaws, Florida).

Carcharhinus obscurus Fowler, Rep. N. J. Mus. (1907), 1908: 124 (spec. Delaware R. brackish water; later stated by Fowler, Proc. biol. Soc. Wash., *33*, 1920: 144, to have been "*commersonii*," i.e., *leucas*).

Carcharhinus platyodon Verrill, Trans. Acad. Arts Sci., *11*, 1903: 55 (Bermuda, ident. by Garman); Barbour, Bull. Mus. comp. Zool. Harv., *46*, 1905: 111 (Bermuda, ident. by Garman); Garman, Mem. Harv. Mus. comp. Zool., *36*, 1913: 126, in part (synonymy, but embryos described are *longimanus*; see p. 354); Starks, Stanford Univ. Publ., Univ. Ser., 1913: 4 (descr., 29-inch spec., Para, Brazil); Breder, Field Bk. Mar. Fish. Atlant. Coast, 1929: 17 (general); White, Bull. Amer. Mus. nat. Hist., *74*, 1934: 126 (in Key); Gunter, Copeia, 1938: 69 (fresh water in Louisiana); Springer, Proc. Fla. Acad. Sci., *3*, 1939: 19 (descr., size, discuss., south. Florida); Bomkamp, Contr. biol. Lab. Cath. Univ. Amer., 44, 1942: 4 (chemistry of liver, Florida); Boos, Contr. biol. Lab. Cath. Univ. Amer., 45, 1943: 10 (chemistry of pancreas, Florida); Lunz, Bull. S. Carolina St. Planning Bd., *14*, 1944: 26 (Florida).

Carcharinus commersonii (in part) Garman, Mem. Harv. Mus. comp. Zool., *36*, 1913: 140 (*leucas* Müller and Henle incl. in synonymy, but refers otherwise to *longimanus*); Radcliffe, Bull. U.S. Bur. Fish., *34*, 1916: 260, pl. 41, fig. 3, 4 (denticles, teeth, N. Carolina, but meas. are of a *longimanus* taken north of the

37. This species has so often been confused with others that we confine the following list to such citations as include reasonably conclusive evidence as to specific identity.

38. Spelled "*leucos*."

39. Photographs of unpublished drawings by Poey of his *obtusa* appear clearly referable to *C. leucas*.

Bahamas); Meek and Hildebrand, Field Mus. Publ. Zool., *15* (1), 1923: 43 (*leucas* Müller and Henle, 1841, incl. in synonymy, but specs. descr. from Colón are apparently *limbatus*); Fowler, Bull. Amer. Mus. nat. Hist., *70* (1), 1936: 53 (descr. of Florida specimen but not the synonymy).

Carcharhinus commersonii Nichols, Bull. Amer. Mus. nat. Hist., *37*, 1917: 873, pl. 91, fig. 1, 3, 4 (descr., photo of spec. ill. here, in Fig. 60; size, no. and size of embryos, winter, S. Florida); Nat. Hist. N. Y., *21*, 1921: 275 (food); Bell and Nichols, Copeia, 92, 1921: 17 (food, size, number, N. Carolina); Nichols, Copeia, 140, 1925: 21 (descr., compar. with *milberti* and *obscurus*, a straggler, Long Island, N. York and Woods Hole); Nichols and Breder, Zoologica, N. Y., *9*, 1927: 16 (uncommon, Long Island and near Woods Hole, but evidence perhaps not conclusive, see p. 343); Breder, Field Bk. Mar. Fish. Atlant. Coast, 1929: 17 (general); Townsend, Bull. N. Y. zool. Soc., *34* (6), 1931: 172 (Florida, good photo); Gudger, Sci. Mon. N. Y., *34*, 1932: 417 (ident. by included ref. to Bell and Nichols, 1921, food); White, Bull. Amer. Mus. nat. Hist., *74*, 1934: 127, pl. 42E (in Key, ill. of heart valves, ident. not established); Lunz, Bull. S. C. St. Planning Bd., *14*, 1944: 27 (S. Carolina).

Eulamia commersonii Fowler, Proc. biol. Soc. Wash., *33*, 1920: 144 (footnote 2 states that his earlier N. Jersey record was *commersonii*); Proc. biol. Soc. Wash., *36*, 1923: 27 (name, Florida).

Carcharias melanopterus Ribeiro, Fauna brasil., Peixes, *2* (1) Fasc. 1, 1923: 12, pl. 2 (descr., photo of 785 mm. spec. evidently *leucas*, not *melanopterus* Quoy and Gaimard, 1824, of the Indo-Pacific).

Carcharias commersonii (in part) Jordan, Evermann and Clark, Rep. U.S. Comm. Fish. (1928), 2, 1930: 15 (*leucas* combined with *longimanus* in synon.); Beebe and Tee-Van, Field Bk. Shore Fish. Bermuda, 1933: 28 (ill., apparently *leucas*, but text and photograph of snout appear to be *longimanus*).

Eulamia platyodon Fowler, Arqu. Zool. Estado São Paulo, *3*, 1942: 128 (listed, Brazil); Fish Culturist, *21* (9), 1942: 66 (listed, Cuba).

Eulamia melanoptera Fowler, Arqu. Zool. Estado São Paulo, *3*, 1942: 128 (listed, Brazil).

Eulamia leucas Bigelow and Schroeder, Guide Comm. Shark Fish., Anglo Amer. Caribb. Comm., Wash., 1945: 86, fig. 27 (descr., habits, range, ill.).

References, Presumably to *leucas*, But Not Definitely Identifiable:

Carcharhinus platyodon Henshall, Bull. U.S. Fish Comm., *9*, 1891: 383 (S. Florida); Evermann and Kendall, Rep. U.S. Comm. Fish. (1899), 1900: 48 (Florida); Bean, Field Mus. Publ. Zool., 7 (2), 1906: 29 (Bermuda); Linton, Proc. U.S. nat. Mus., *33*, 1907: 86; Bull. U.S. Bur. Fish., *26*, 1907: 122; Engelhardt, Zool. Anz., *39*, 1913: 648 (S. Atlantic); Linton, Proc. U.S. nat. Mus., *64* (21), 1924: 5, 48 (parasites); von Bonde, J. comp. Neurol., 1933: pl. 2, fig. 2, 3 (brain); Beebe, Proc. Wash. Acad. Sci., *19*, 1933: 184 (Bermuda).

Carcharhinus obscurus (probably in part) Thorne, Copeia, 35, 1916: 69; Nichols, Copeia, 35, 1916: 72–73; Thorne, Bull. N. Y. zool. Soc., *31*, 1928: 114 (*leucas* likely represented among *obscurus* reported from Long Island, N. York).

Carcharinus lamia Nichols, Copeia, 53, 1918: 13 (abund., Florida).

Carcharhinus commersonii Linton, Proc. U.S. nat. Mus., *64* (21), 1924: 6, 7, 12, 33, 38, 49, 54, 84, 87 (parasites); Burton, Sci. Mon. N. Y., *40*, 1935: 283 (Charleston, S. Carolina).

Carcharias commersonii Marini, Physis B. Aires, *9*, 1929: 452 (Argent.); Bere, Amer. Midl. Nat., *17*, 1936: 593 (parasites).

Carcharias carcharias (in part) Jordan, Manual Vert. Anim. NE. U.S., 1929: 10 (account perhaps partly based on *leucas*).

Eastern Atlantic References, Nominally to *leucas*, But Probably Not This Species (see footnote 37, p. 344):

Carcharias (*Prionodon*) *leucos* Guichenot, Explor. Algér., *3*, Poiss., 1850: 124 (spec. from Algeria); Duméril, Hist. Nat. Poiss., 1, 1865: 358 (ref. to Algerian spec. from Guichenot); Doderlein, Man. Ittiol. Medit., *2*, 1881: 44 (credited to Medit., on basis of Algerian spec. reported as *leucos* by Guichenot, 1850); Rochebrune, Act. Soc. linn. Bordeaux, (4) *6*, 1882: 19; Faune Senegambie, Poiss., *1*, 1883–1885: 19 (name only, C. Verde, Senegambia).

? *Carcharias leucos* Carus, Prod. Fauna Medit., *2*, 1889–1893: 513 (name, Algiers).

Red Sea Reference, Probably Not *leucas*:

Carcharhinus commersonii Gudger, Bull. Amer. Mus. nat. Hist., *73*, 1937: 264 (photos of teeth, prob. from Gulf of Aden).

Not *Carcharias leucas* Bennett, Proc. zool. Soc. Lond., 1859: 223 (N. Zealand).[40]

Not *Carcharinus platyodon* Garman, Mem. Harv. Mus. comp. Zool., *36*, 1913: pl. 3, fig. 1–4 (this is a *milberti*, mislabelled).

Not *Eulamia platyodon* Fowler, Proc. Acad. nat. Sci. Philad., *71*, 1919: 146 (a Jamaica spec. not *leucas* because teeth narrow and the lowers entire; perhaps *maculipinnis*).

Not *Carcharhinus commersonii* Rey, Fauna Iberica Peces, *1*, 1928: 342 (this is *obscurus*, see p. 342).

Carcharhinus limbatus (Müller and Henle), 1841

Small Black-tipped Shark

Figures 62, 63

Study Material. Female, 990 mm. (about 3 feet 3 inches) long, from Florida (Amer. Mus. Nat. Hist.); also 14 specimens, 460 to 621 mm. long, from Galveston, Texas;[41] one, 585 mm., from Pine Island, Florida; 3 embryos, 351 to 398 mm., from Captiva Key, Florida (Harv. Mus. Comp. Zool.).

Distinctive Characters. Among smooth-backed members of the genus in the Atlantic, *limbatus* falls with *maculipinnis* (p. 364) in the conspicuously black-tipped fins. But it is easily distinguishable from the latter by its relatively larger eyes, its shorter gill openings and upper labial furrow, and by the more obtuse free rear tip of its second dorsal; also, the edges of its lower teeth, as well as those of the uppers, are regularly, though very finely, serrate, and its first dorsal usually originates somewhat farther forward relative to the pectorals.

Description. Proportional dimensions in per cent of total length. Male, 621 mm., from Galveston, Texas (Harv. Mus. Comp. Zool., No. 35842). Female, 990 mm., from Florida (Amer. Mus. Nat. Hist.).

Trunk at origin of pectoral: breadth 11.4, 10.4; height 12.6, 12.3.

Snout length in front of: outer nostrils 3.5, 3.3; mouth 7.9, 7.9.

Eye: horizontal diameter 1.5, 1.5.

Mouth: breadth 9.0, 8.9; height 4.8, 4.7.

Nostrils: distance between inner ends 5.4, 5.2.

Labial furrow length: upper 0.8, 0.8.

Gill opening lengths: 1st 4.3, 3.6; 2nd 4.5, 3.9; 3rd 4.7, 4.1; 4th 4.2, 3.8; 5th 3.6, 2.9.

40. As *leucas* was preoccupied, the name *brachyurus* was substituted by Günther (Cat. Fish. Brit. Mus., *8*, 1870: 369) for this Indo-Pacific species, which differs from the Atlantic *leucas* in a more pointed snout, and in that its first dorsal originates considerably farther back. It is classed as a synonym of *lamia*, i.e., *longimanus*, by Fowler (Bull. U.S. nat. Mus., 100 [*13*], 1941: 171), but differs equally sharply from the latter in its angular first dorsal, in its much shorter pectoral, and in that the tip of its anal is far in advance of the origin of its caudal. For photograph, see Whitley (Fish. Aust., *1*, 1941: 102, fig. 97).

41. Contributed by J. L. Baughman.

First dorsal fin: vertical height 10.6, 11.3; length of base 11.3, 11.6.
Second dorsal fin: vertical height 3.4, 3.2; length of base 4.5, 5.0.
Anal fin: vertical height 3.7, 3.7; length of base 5.3, 5.4.
Caudal fin: upper margin 28.3, 27.6; lower anterior margin 12.7, 12.5.
Pectoral fin: outer margin 17.7, 17.7; inner margin 5.8, 5.4; distal margin 13.4, 14.8.
Distance from snout to: 1st dorsal 30.9, 31.6; 2nd dorsal 61.1, 62.0; upper caudal 71.7, 72.4; pectoral 22.1, 25.4; pelvics 48.2, 49.0; anal 59.2, 61.1.
Interspace between: 1st and 2nd dorsals 20.5, 20.9; 2nd dorsal and caudal 6.5, 6.3; anal and caudal 5.9, 6.1.
Distance from origin to origin of: pectoral and pelvics 26.1, 25.0; pelvics and anal 11.6, 11.9.

Trunk moderately slender, its dorsal profile only slightly arched, its height at origin of 1st dorsal about 1/5–1/6 its length to origin of caudal. No mid-dorsal ridge. Upper precaudal pit rounded, the lower a subrectangular furrow and less strongly marked than upper. Caudal peduncle about 1 1/5 times as thick as deep. Dermal denticles closely spaced and usually overlapping, the skin visible only here and there between them, the smaller

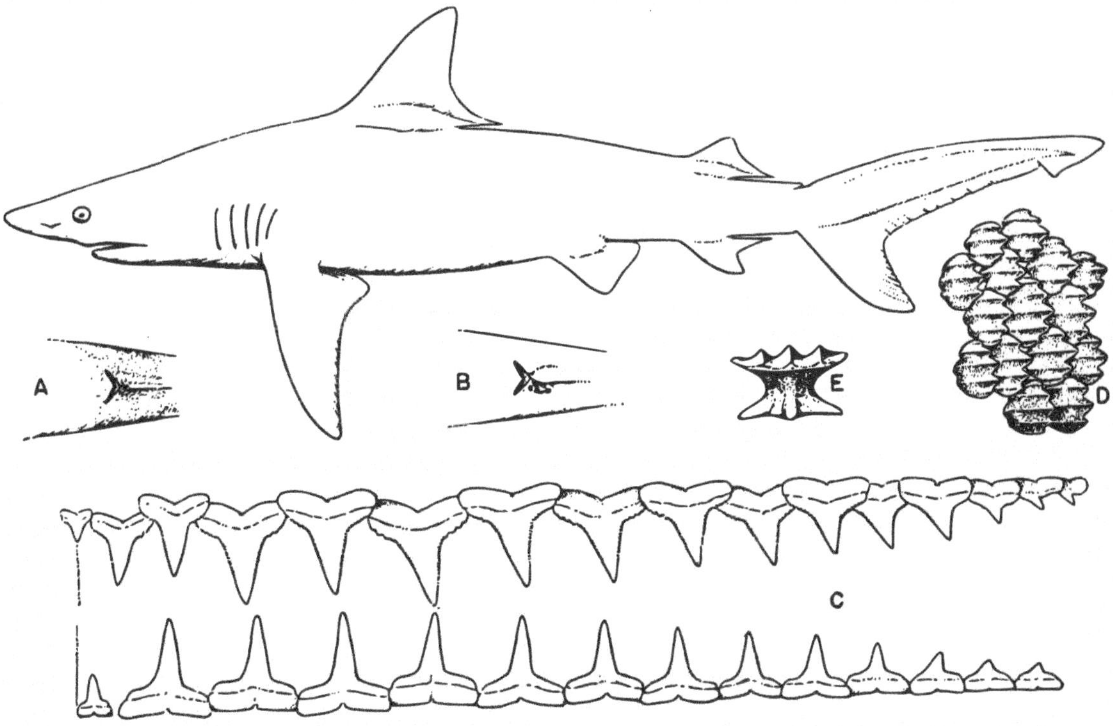

FIGURE 62. *Carcharhinus limbatus*, female, about 990 mm. long, from southern Florida (Amer. Mus. Nat. Hist.). *A*, *B* Caudal peduncle, from above and below respectively, to show precaudal pits. *C* Left-hand upper and lower teeth, about 1.8 x. *D* Dermal denticles, about 30 x. *E* Apical view of dermal denticle, about 60 x.

ones with 3 to 5 low ridges and the larger with 5 or even 7, the marginal teeth relatively low and broad, mostly worn down in specimens seen, except in the cases of newly formed denticles; pedicels moderately slender, on large basal plates.

Head about 1/4 of total length. Snout moderately thick, its lateral outlines converging gradually from opposite front of mouth to opposite ends of nostrils and then

Figure 63. *Carcharhinus limbatus*, pictured in Fig. 62. *A* Lower view of head. *B* Pelvic fins. *C* Fourth upper tooth. *D* Tenth upper tooth. *E* Fourth lower tooth. *F* Ninth lower tooth. *C–F*, about 5.4 x. *G* Right-hand nostril, about 3 x.

more abruptly; its tip rounded (relatively somewhat shorter and broader anteriorly than in *maculipinnis*), its length in front of a line connecting outer ends of nostrils nearly 2/3 (63–65%) as great as distance between inner ends of latter, and its length in front of mouth a little more than 1 1/2 times as great as that, or about 1/3 as great as length of head. Eye approximately circular, its vertical diameter about 0.9 the horizontal diameter, the latter about 1/5 (17 to 20%) as long as snout in front of mouth, or a little less in large specimens. Gill openings relatively somewhat shorter than in *maculipinnis*, the 1st only about 2/3 as long as distance between inner ends of nostrils and 1.9 to 2.8 times as long as horizontal diameter of eye, the 3rd a little the longest, the 5th oblique and a little more than 3/4 as long as 1st, the 4th over origin of pectoral. Nostril strongly oblique, its inner end closer to mouth than to tip of snout by a distance about equal to diameter of eye, the anterior margin slightly expanded near the inner end as an inconspicuous, broadly-triangular lobe with rounded tip. Mouth ovate, moderately arched, its breadth 1.6 to 2 times its height. Upper labial furrow only about half as long as nostril (thus shorter, relatively, than in *C. maculipinnis*), at an angle of about 45° with the jaw.

Teeth $\frac{\text{14 or 15—1 to 3—14 or 15}}{\text{13 to 15—1 or 2—13 to 15}}$; uppers nearly symmetrical with narrow triangular cusps on broad bases, with edges finely serrate on cusps and more coarsely so on bases, those toward center of mouth erect, their margins nearly straight, but those along sides slightly oblique; lowers with slender erect cusps, recurved slightly forward near tips, and broad bases, the cusps very finely serrate, but bases less clearly so; 1 to 3 minute teeth at symphysis in upper jaw, and 1 or 2 in lower; outermost 2 or 3 teeth in each jaw also very small.

Origin of 1st dorsal about over midpoint of inner margin of pectoral, its position somewhat variable, its anterior margin moderately convex in small specimens but only very slightly so in larger, its apex very narrowly rounded, the posterior margin nearly straight toward apex but deeply concave toward base, its free rear tip noticeably slender but only about ⅓ as long as base, its vertical height about equal to distance from corner of mouth to 5th gill slit or slightly more than ½ (62 to 64%) the length of pectoral, the midpoint of its base only about ½ as far from axil of pectoral as from origin of pelvics. Second dorsal between ⅓ and ½ (38 to 41%) as long at base as 1st, its origin over or very slightly behind origin of anal, its anterior margin nearly straight, its rear margin slightly and evenly concave, its apex narrowly rounded or subacute, its free rear tip relatively broad and only about as long as the base or slightly shorter. Caudal between ¼ and ⅓ (26 to 29%) of total length, the terminal sector between ¼ and ⅕ the length of the upper margin, narrow-triangular with slightly sinuous lower contour and narrowly rounded tip, the lower lobe (expanded lower anterior corner) a little less than ½ as long as upper, its anterior margin moderately convex, its tip narrowly rounded or subacute, the re-entrant corner (included by the two lobes) subrectangular and only slightly rounded. Anal about as long at base as 2nd dorsal or a little longer, but its anterior margin somewhat more convex, and its rear margin much more deeply concave, its apex subacute, its free rear tip slightly but definitely shorter than its base. Distance from origin of anal to tip of pelvics a little shorter than base of anal. Pelvics with weakly concave distal margins, their anterior margins slightly convex in small specimens but nearly straight in larger ones, with bases about ½ as long as base of 1st dorsal. Pectoral a little less than ¾ (70 to 72%) as long as head, about ½ as broad as long but with base only about ¼ as long as outer margin, distal margin nearly straight toward tip but deeply concave toward base, apex narrowly rounded and inner corner only a little more broadly so.

Color. Dark gray, dusky bronze, or ashy blue above; pure white or yellowish white below, with a more or less pronounced band of the dark upper tint extending rearward along each side to about over the origin of the pelvic, including above it a forward extension of the pale or white of the lower side, which narrows forward. This dark band varies widely in strength from specimen to specimen and gradually fades after preservation in formalin or alcohol. The pectorals are conspicuously tipped with black; the dorsals, anal and lower lobe of caudal are also black-tipped in young, but fade more or less with growth.

Size. This Shark may mature at a length no greater than about four to five feet; adults perhaps average 5½ to 6½ feet, and few grow longer than seven or possibly eight feet in length. It is, however, relatively much larger at birth than many other sharks. Embryos 21 to 23 inches (nearly ready for birth) have been taken from females only four to five feet long, and embryos 23 to 26 inches (580 to 660 mm.) from a Pacific specimen of about 6 feet 9 inches (2,070 mm.). A specimen of about 28 inches (705 mm.) weighed about 4.2 pounds (1.9 kilo); one of about 47.7 inches (1,210 mm.) about 19.6 pounds (8.9 kilo); one of about 54.2 inches (1,375 mm.) about 32 pounds (14.5 kilo); one of about 61.3 inches about 42.2 pounds (19.1 kilo); and one of about 65.8 inches about 68.3 pounds (31 kilo).[42]

Developmental Stages. Embryos of this species have been reported repeatedly, but we find no statement as to whether it is viviparous or ovoviviparous, although the large size which the embryos reach before birth suggests the former; nor do the specimens we have studied (see Study Material, p. 346) clarify the matter. Three to nine embryos have been recorded, but the usual number is four to six, with the sexes in approximately equal numbers.

Habits. This is a very active, swift-swimming shark, often seen in schools at the surface, frequently leaping clear of the water, sometimes revolving as much as three times around the horizontal axis in the air before falling back into the sea.[43] It is encountered indifferently near shore and out at sea. It feeds on fishes, its teeth suggesting chiefly the smaller species, and probably on squid; menhaden (*Brevoortia*) and butterfish (*Poronotus triacanthus*) have been found in the stomachs of North Carolina specimens, and 24 California sardines (*Sardinops caerulea*) in an eastern Pacific example. Like sundry other sharks, it devours sting-rays on occasion and their spines have been found imbedded in its jaws. On the other hand, smallish Black-tipped Sharks have been found in the stomachs of larger sharks of other species.

Embryos in gravid females taken off Biloxi, Mississippi, in midsummer were small;[44] but larger embryos nearly ready for birth were taken around southern Florida in April, suggesting that the period of gestation is nearly a year, and that the young are born chiefly in late spring. Other than the foregoing, nothing is recorded of its habits, notwithstanding that it has been known to science for more than a century.

Relation to Man. Such of the larger specimens as are taken in shark fisheries in Florida or elsewhere are put to the same uses as other species. Thanks to the readiness with which it takes a hook baited with fish (still or trolled) many are so caught; in fact, it is often caught by anglers trolling for tarpon in Florida waters. But reports differ as to

42. Florida specimens measured and weighed by Stewart Springer.
43. Recent observation off Biloxi, Mississippi, by Stewart Springer.
44. About 125 mm.; personal communication from Stewart Springer.

its game qualities; we have caught none on light tackle. Some anglers describe it as putting up a fast fight when hooked and often jumping clear of the water, but according to others it is more stubborn and "mulish" than active in its resistance.

Range. Pelagic in tropical and subtropical seas; southern Brazil to North Carolina and occasionally to New York and southern New England in the western Atlantic; Madeira, Cape Verde Islands and tropical West Africa (Dakar, and Kribi in Cameroon) in the eastern Atlantic; also eastern tropical Pacific, from Lower California to Peru.[45] A shark (or sharks) is also reported under this same name from Cochin China, India, Red Sea, Seychelles, Madagascar and Natal. But its actual relationship to *limbatus* of the eastern Pacific and Atlantic cannot be determined until specimens from the different ocean areas have been compared critically.

Occurrence in the Western Atlantic. This can be outlined only within broad limits, partly because of the uncertainty in some cases as to whether published records actually refer to *limbatus* or to *maculipinnis*, and partly because reliable records for it are very irregularly distributed. It is certainly one of the commoner, if not the commonest, pelagic shark around the Bahamas and southern Florida; also along the coasts of Mississippi and Louisiana, where many are caught by shrimp fishermen, and of Texas[46] at least in the warm season. In all probability it is equally widespread and locally common throughout the West Indian–Caribbean region in general and in the southern part of the Gulf of Mexico; but published records of it there are confined to Haiti, Porto Rico (where it is said to be one of the commoner sharks), Turks Island, Martinique, Antilles in general and Surinam. Southward it is common along the coast of Brazil at least as far as Rio de Janeiro (reported also from Bahia and from Ilha de Victoria); but it has not been reported farther south. No doubt it is present in the truly tropical belt throughout the year, and it is common around southern Florida throughout spring, summer and autumn.[47] During the warm months many visit the coast of South Carolina, and a few are taken in some summers along the southern half of North Carolina. To the north of Cape Hatteras, however, it occurs on the coast only as a stray, there being only about six reliable reports of it for the vicinity of New York and Long Island. But at least twenty small ones were taken in pound nets on the eastern shore of Buzzards Bay, near Woods Hole, during the summer of 1878, and one other in the summer of 1916, which shows that it reaches the southern New England coast in unusual numbers at rare intervals. Nor is it unusual to see Black-tipped Sharks in the warm oceanic waters off this sector of the continental shelf in summer, drifting north in the Gulf Stream, probably never to return to their tropical home.

45. There appears to be nothing in the accounts of *aethlorus* Jordan and Gilbert, 1882, from the west coast of Mexico and Lower California to separate it from *limbatus* as Garman (Mem. Harv. Mus. comp. Zool., *36*, 1913: 127) has already concluded; also Meek and Hildebrand (Field Mus. Publ. Zool., *15* [1], 1923: 43), after examining the type specimen. Neither has our own examination of the type specimen of *natator* Meek and Hildebrand, 1923, from Panama revealed significant differences from the Atlantic specimens of *limbatus* (listed above) whether in shape of snout, shape and relative positions of fins, teeth, or color.

46. Personal communication from J. L. Baughman; see Study Material, p. 346.

47. Not reported there in December, January or February.

Synonyms and References:

Atlantic, Eastern Pacific:[48]

Carcharias (Prionodon) limbatus Müller and Henle, Plagiost., 1841: 49, pl. 19, fig. 9 (descr., no loc.); Duméril, Hist. Nat. Poiss., 1865: 375 (descr., Martinique [type], Bahia); Günther, Cat. Fish. Brit. Mus., *8*, 1870: 373 (descr., C. Verdes, Seychelles, Indian Oc., Guatemala); Rochebrune, Act. Soc. linn. Bordeaux, (4) *6*, 1882: 42; Faune Senegambie, Poiss., 1883–1885: 19 (C. Verde); Metzelaar, Trop. Atlant. Visschen, 1919: 188 (nominal, both sides of Atlantic).

Prionodon cucuri Castelnau, Anim. Nouv. Rares Amer. Sud., 1855: 99[49] (Bahia, Brazil).

Lamiopsis limbatus Gill, Ann. N. Y. Lyc., 7, 1862: 410 (name only).

Carcharias (Prionodon) mülleri Steindachner, S. B. Akad. Wiss. Wien, *56*, 1867: 356 (Antilles).

Carcharias maculipinna Günther, Trans. zool. Soc. Lond., *6*, 1868: 490 (teeth, meas., Chiapam, Pacific Guatemala); not *Isogomphodon maculipinnis* Poey, 1865; see p. 368.

Platypodon perezii Poey, An. Soc. esp. Hist. nat., *5*, 1876: 390, pl. 14, fig. 2, 3; Enumerat. Pisc. Cubens., 1876: 194, pl. 9, fig. 2, 3 (descr., color, size, teeth, Cuba).[50]

Carcharias aethlorus Jordan and Gilbert, Proc. U.S. nat. Mus., *5*, 1882: 104 (descr., Mazatlan, Mexico); Jordan and Gilbert, Bull. U.S. Fish Comm., *2*, 1882: 105, 109 (Mazatlan, Panama); Gilbert and Starks, Mem. Calif. Acad. Sci., *4*, 1904: 9, 207 (descr., meas., Panama, Gulf of California); Jordan, Evermann and Clark, Rep. U.S. Comm. Fish. (1928), 2, 1930: 17 (Mazatlan to Panama).

Isogomphodon limbatus Jordan and Gilbert, Bull. U.S. nat. Mus., 16, 1883: 23 (tropical seas and Woods Hole region); Nelson, Rep. St. Geol. N. J., *2* (2), 1890: 661 (in N. Jersey list); Springer, Proc. Fla. Acad. Sci., *3*, 1939: 26 (descr., habits, Florida); Norris, Plagiost. Hypophysis, 1941: 20, 28, pl. 3, fig. 10, 10a (brain); Lunz, Bull. S. C. St. Planning Bd., 14, 1944: 27 (S. Carolina, Florida); Gunter, Publ. Inst. Mar. Sci. Univ. Texas, *1*, 1945: 20 (temp., breeding, Texas).

Isogomphodon maculipinnis Goode, Fish. Fish. Industr. U.S., *1*, 1884: 673 (Woods Hole, Massachusetts); not *Isogomphodon maculipinnis* Poey, 1865; see p. 368.

Carcharhinus aethlorus Jordan, Proc. U.S. nat. Mus., *8*, 1885: 363 (refs.); Rep. U.S. Comm. Fish. (1885), 1887: 795 (east. trop. Pacif.); Proc. Calif. Acad. Sci., (2) *5*, 1895: 383 (discuss.); Garman, Bull. Mus. comp. Zool. Harv., *46*, 1906: 229 (Panama).

Eulamia limbata Henshall, Bull. U.S. Fish Comm., *9*, 1891: 372 (color, west coast Florida); Fowler, Bull. Amer. Mus. nat. Hist., *70* (1), 1936: 48 (descr.); Arqu. Zool. Estado São Paulo, *3*, 1942: 128; Fish Culturist, *21* (9), 1942: 66 (listed, Cuba); Peces Peru, Mus. Hist. Nat. Javier Prado, 1945: 12 (listed, Peru); Bigelow and Schroeder, Guide Comm. Shark Fish., Anglo Amer. Caribb. Comm., Wash., 1945: 88, fig. 28 (descr., habits, range, ill.).

Carcharhinus (Isogomphodon) aethlorus Jordan and Evermann, Rep. U.S. Comm. Fish. (1895), 1896: 216 (Mazatlan to Panama); Bull. U.S. nat. Mus., 47 (1), 1896: 40 (descr., Mazatlan to Panama).

Carcharhinus (Isogomphodon) limbatus Jordan and Evermann, Rep. U.S. Comm. Fish. (1895), 1896: 216 (distrib.); Bull. U.S. nat. Mus., 47 (1), 1896: 40 (descr., distrib.).

Carcharhinus (Platypodon) perezi Jordan and Evermann, Rep. U.S. Comm. Fish. (1895), 1896: 216 (name, Cuba); Bull. U.S. nat. Mus., 47 (1), 1896: 36 (descr., Cuba).

Carcharhinus limbatus Smith, Bull. U.S. Fish Comm., *17*, 1898: 88 (Woods Hole region); Evermann and Kendall, Rep. U.S. Comm. Fish. (1899), 1900: 48 (Florida); Pellegrin, Bull. Mus. Hist. nat. Paris, 7,

48. References for "*limbatus*" for the western Pacific and Indian Oceans are omitted because of uncertainty whether the shark (or sharks) to which they refer are identical with *limbatus* of the Atlantic and eastern Pacific.

49. The name *cucuri* dates back to Marcgrave (in Piso de Medicina brasiliense, et G. Marcgravius Hist. rerum natur. Brasiliae, 1648: 164). But Castelnau's revival of it for the shark in question to replace *limbatus*, Müller and Henle, was not a fortunate one, even apart from nomenclatural grounds, for Marcgrave's original account "pinnae omnes cinereae, exceptis parvulibus in inferiori corpore, quae albae" does not suggest this species.

50. *Perezii* was classed by Garman (Mem. Harv. Mus. comp. Zool., *36*, 1913: 138) as a synonym of *remotus*. Actually, however, its color as well as the original illustrations of its teeth more nearly resemble *limbatus*, its pectorals being described as with black tips and its sides with a dark band. A photograph of Poey's unpublished drawing of an adult female shows nothing to separate it from *limbatus*.

1901: 161 (Gulf of California); Evermann and Marsh, Bull. U.S. Fish Comm., *20*, 1902: 62 (descr., Porto Rico); Kendall, Occ. Pap. Boston Soc. nat. Hist., 7 (8), 1908: 5 (Woods Hole region); Garman, Mem. Harv. Mus. comp. Zool., *36*, 1913: 127 (descr., Rio de Janeiro); Gudger, J. Elisha Mitchell Sci. Soc., *28*, 1913: 158 (N. Carolina); Nichols, Abstr. Proc. Linn. Soc. N. Y., 20–23, 1913: 91 (N. York); Sumner, Osburn and Cole, Bull. U.S. Bur. Fish., *31* (2), 1913: 736 (Woods Hole region); Coles, Proc. biol. Soc. Wash., *28*, 1915: 90 (N. Carolina); Nichols and Murphy, Brooklyn Mus. Sci. Bull., *3* (1), 1916: 17 (trop. seas, Atlantic coast to Buzzards Bay, Massachusetts); Nichols, Copeia, 35, 1916: 73 (name); Thorne, Copeia, 35, 1916: 69 (Long Island, N. York); Helmuth, Copeia, 36, 1916: 80 (Long Island, N. York); Radcliffe, Bull. U.S. Bur. Fish., *34*, 1916: 255, pl. 40, fig. 3, 4 (teeth, denticles, food, N. Carolina); Smith, J. Amer. Mus. nat. Hist., *16*, 1916: 348 (Woods Hole spec.); Nichols, Bull. Amer. Mus. nat. Hist., *37*, 1917: 874 (size, embryos, Florida); Nichols and Mowbray, Copeia, 48, 1917: 78 (Porto Rico); Nichols, Copeia, 53, 1918: 13 (Long Island, N. York); Bell and Nichols, Copeia, 92, 1921: 18 (N. Carolina); Nichols, Bull. Amer. Mus. nat. Hist., *44*, 1921: 22 (Turks I.); Nichols and Murphy, Bull. Amer. Mus. nat. Hist., *46*, 1922: 504 (jaw, Peru); Meek and Hildebrand, Field Mus. Publ. Zool., *15*, 1923: 41 (descr., Pacif. Panama); Linton, Proc. U.S. nat. Mus., *64* (21), 1924: 12, 33, 46, 49, 87 (parasites, Woods Hole); Breder, Bull. N. Y. zool. Soc., *28*, 1925: 187 (C. Hatteras, embryos); Nichols and Breder, Zoologica, N. Y., *9*, 1927: 17 (general); Thorne, Bull. N. Y. zool. Soc., *31*, 1928: 114 (Great South Bay, N. York); Beebe and Tee-Van, Zoologica, N. Y., *10*, 1928: 28 (Haiti); Breder, Field Bk. Mar. Fish. Atlant. Coast, 1929: 17 (general); Nichols, Sci. Surv. Porto Rico, N. Y. Acad. Sci., *10* (2), 1929: 183 (Porto Rico); Gowanloch, Bull. La. Conserv. Dep., *21*, 1932: 89 (Louisiana); Budker, Bull. Mus. Hist. nat. Paris, (2) 7, 1935: 183 (Dakar, W. Afr.); Burton, Sci. Mon. N. Y., *40*, 1935: 283 (Charleston, S. Carolina); Cadenat, Rev. des Trav. Pêches Marit., *10* (4), 1937: 430 (Dakar); White, Bull. Amer. Mus. nat. Hist., *74*, 1937: 127, pl. 6, fig. g, pl. 13, fig. a (in Key, denticles, teeth); Breder, Bull. N. Y. Zool. Soc., *41*, 1938: 28 (N. York Harbor); Hildebrand, Copeia, 1941: 221 (N. Carolina).

Carcharias limbatus Jordan and Gilbert, Proc. U.S. nat. Mus., *5*, 1883: 245 (Gulf of Mexico); Schreiner and Ribeiro, Arch. Mus. nac. Rio de J., *12*, 1903: 79 (Rio de Janeiro); Ribeiro, Arch. Mus. nac. Rio de J., *14*, 1907: 153, 200, pl. 3 (descr., photo, W. Atl. refs.); Rev. Mus. paul., *10*, 1918: 707 (Ilha de Victoria); Fauna brasil. Peixes, *2* (2) Fasc. 1, 1923: 10 (same as Ribeiro, 1907); Jordan, Manual Vert. Anim. NE. U.S., 1929: 10 (general); Jordan, Evermann and Clark, Rep. U.S. Comm. Fish. (1928), 2, 1930: 16 (distrib.); Gowanloch, Bull. La. Conserv. Dep., *23*, 1933: 83, 220, 236 (Louisiana); Bere, Amer. Midl. Nat., *17*, 1936: 589, 593, 604 (Florida, ident.?); Gunter, La. Conserv. Rev., *5* (4), 1936: 45 (Louisiana).

Carcharhinus natator Meek and Hildebrand, Field Mus. Publ. Zool., *15* (1), 1923: 41, pl. 1, fig. 1 (descr., ill., Panama City, Panama).

Carcharias natator Jordan, Evermann and Clark, Rep. U.S. Comm. Fish. (1928), 2, 1930: 17 (Panama City).

Prionodon limbatus Borri, Mem. Soc. tosc. Sci. nat., *44*, 1934: 89 (no loc.).

Eulamia aethlorus Beebe and Tee-Van, Zoologica, N. Y., *26*, 1941: 106 (descr., size, weight, parasites, food, embryos, Pacific coasts, Mexico, Guatemala, Panama).

Probable References:

Carcharias (Prionodon) melanopterus Rochebrune, Act. Soc. linn. Bordeaux, (4) *6*, 1882: 42 (ident. probable because of fins descr. as conspicuously black-tipped; trop. W. Africa); not *C. melanopterus* Quoy and Gaimard, 1824.

Carcharhinus commersonii Meek and Hildebrand, Field Mus. Publ. Zool., *15* (1), 1923: 43 (synonymy, refs. to *longimanus* and *leucas*, but Colón specs. apparently *limbatus* because of meas. and color).

Doubtful Reference:

Carcharias microps Lowe, Proc. zool. Soc. Lond., 1840: 38; Trans. zool. Soc. Lond., 3, 1843: 18 (not identif. by brief notice, Madeira).

Carcharhinus longimanus (Poey), 1861[51]

White-tipped Shark

Figures 64, 65

Study Material. Adult male, about 6 feet 9 inches long, taken off Guantanamo, Cuba, in April 1939; adult female of about the same length, taken off Santa Clara Province on the north coast of Cuba in May 1939, the latter specimen with 6 female embryos, 550 to 580 mm. long and nearly ready for birth;[52] jaws from Guadeloupe and Santa Cruz, West Indies, specimens; jaws from six specimens, with pieces of skin from two of these, up to 6½ to 7 feet long by calculation, caught from the research ship "Atlantis" in the Caribbean (precise localities not recorded) in February 1934; also the following embryos: a male, said to have been one of 9 embryos, from north of the Bahamas in Lat. 28° 30′ N., Long. 77° 35′ W. (Harv. Mus. Comp. Zool., No. 35249); two males, 395 and 460 mm., from Guadeloupe, West Indies (Harv. Mus. Comp. Zool., No. 756[52a]); female of about 580 mm., from off Havana, Cuba (Harv. Mus. Comp. Zool., No. 33439); four others, female and male, about 515 to 525 mm. long, from north of the Bahamas (U.S. Nat. Mus., No. 118548, 118549); also photographs of unpublished drawings by Poey.

Distinctive Characters. C. longimanus is set apart from all other Atlantic members of the genus by the very broadly rounded apex of its first dorsal fin, the convexity of the posterior outline of the lower caudal lobe, its very short snout in front of the nostrils (*leucas* alone resembles it in this respect) and by the fact that the rear tip of the anal reaches nearly to the lower precaudal pit (see also comparison with *C. leucas*, p. 338).

Description. Proportional dimensions in per cent of total length. Female, 2,070

51. In early shark literature this species was almost inextricably entangled with the man-eater now universally known as *Carcharodon*, while more recently it has been confused frequently with *Carcharhinus leucas*. Thus it appears to have been combined with *Carcharodon* by Linnaeus, 1758, under the name *Squalus carcharias*, and clearly was so combined by Risso (Ichthyol. Nice, 1810: 25). But this specific name is not available for it, because *S. carcharias* Linnaeus is now universally accepted as the type of *Carcharodon*. In 1826 Risso (Hist. Nat. Europ. Merid., *3*: 119) redescribed it as *Carcharias lamia*, this time omitting such of the features as obviously referred to *Carcharodon* in his earlier account. It was as *Carcharias* (*Prionodon*) *lamia* that Müller and Henle (Plagiost., 1841: 37, pl. 12) published what has continued to be the best account and illustration of it up to the present time. But the name *Carcharias lamia* had been used previously by Rafinesque (Indice Ittiol. Sicil., 1810: 44) as a substitute for *Squalus carcharias* Linnaeus, the White Shark; hence it is a synonym of the latter, according to the rules of zoological nomenclature as generally accepted, and cannot be used for any other shark. To replace it for the species here under discussion we must therefore turn to the next oldest name under which the latter has been cited, *i.e.*, to *longimanus* Poey, 1861. Fortunately there can be no doubt as to the identity of the shark so named by him; his account and the photographs of his unpublished drawings specify the short, broad snout, the rounded first dorsal, the very long pectorals, the close proximity of the tip of the anal to the origin of the caudal, and the teeth of the specimen pictured here reproduce almost exactly the one figured by him.

52. Measurements and photographs were taken of the latter; the jaws and fins, with the embryos, are in the Harvard Museum of Comparative Zoology.

52a. Basis of Garman's (1913) account of *C. platyodon*.

mm., from Cuba (Harv. Mus. Comp. Zool., No. 35516). Male, 2,075 mm., from Cuba (Harv. Mus. Comp. Zool., No. 35517).

Snout length in front of: outer nostrils 2.5, 2.5; mouth 6.5, 6.1.

Eye: horizontal diameter 1.1, 1.2.

Mouth: breadth 9.8, 9.3; height 4.6, 5.0.

Nostrils: distance between inner ends 5.8, 5.8.

Gill opening lengths: 1st 4.0, 3.2; 5th 3.5, 2.8.

First dorsal fin: vertical height 12.2, 13.2; length of base 11.0, 10.4.

Second dorsal fin: vertical height 4.3, 4.0; length of base 3.7, 3.7.

Anal fin: vertical height 5.4, 5.5; length of base 3.7, 4.0.

Caudal fin: upper margin 28.2, 28.3; lower anterior margin 13.5, 14.1.

Pectoral fin: outer margin 23.4, 22.0; inner margin 5.3, 5.2; distal margin 19.9, 19.6.

Distance from snout to: 1st dorsal 30.8, 31.5; 2nd dorsal 63.0, 62.5; upper caudal 71.8, 71.7; pectoral 22.8, 23.3; pelvics 54.7, 49.6.

Interspace between: 1st and 2nd dorsals 21.4, 20.5; base of 2nd dorsal and caudal 5.8, 6.0; anal and caudal 4.1, 4.3.

Distance from origin to origin of: pelvics to caudal pit 21.2, 21.4.

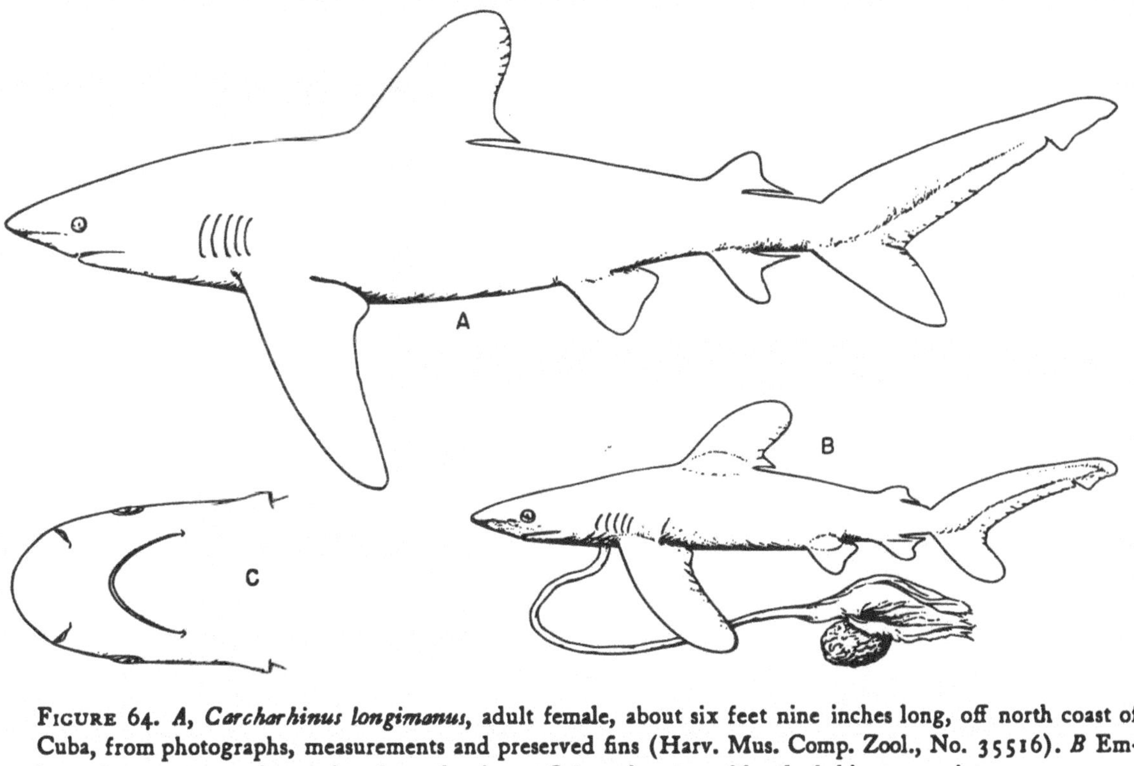

FIGURE 64. *A, Carcharhinus longimanus*, adult female, about six feet nine inches long, off north coast of Cuba, from photographs, measurements and preserved fins (Harv. Mus. Comp. Zool., No. 35516). *B* Embryo about 575 mm. long taken from the above. *C* Anterior part of head of this same embryo.

Proportional dimensions in per cent of total length. Female embryo, 580 mm. (Harv. Mus. Comp. Zool., No. 35518, taken from No. 35516).

Trunk at origin of pectoral: breadth 11.7; height 12.9.
Snout length in front of: outer nostrils 2.7; mouth 7.3.
Eye: horizontal diameter 2.2.
Mouth: breadth 8.6; height 5.9.
Nostrils: distance between inner ends 5.9.
Gill opening lengths: 1st 2.9; 2nd 3.4; 3rd 3.5; 4th 3.4; 5th 2.6.
First dorsal fin: vertical height 12.0; length of base 12.5.

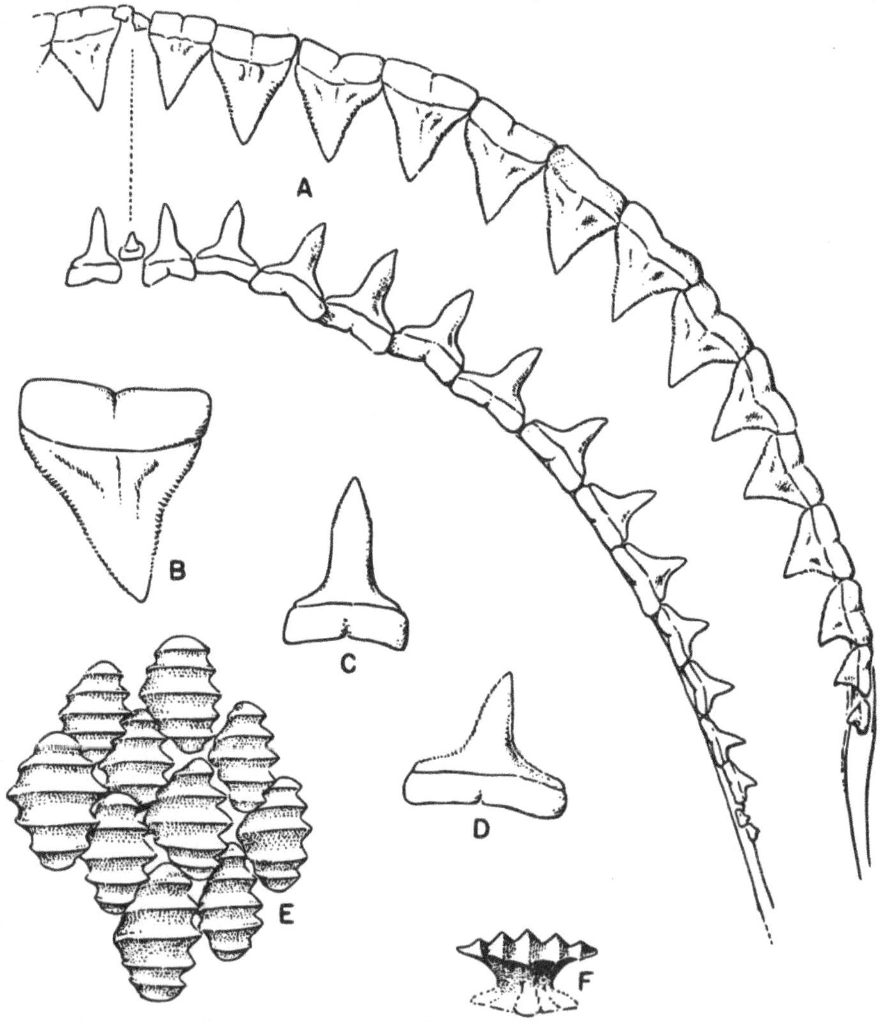

FIGURE 65. *Carcharhinus longimanus*, adult illustrated in Fig. 64. *A* Upper and lower teeth, left-hand side, about natural size. *B* Sixth upper tooth, about 2 x. *C* Second lower tooth, about 2 x. *D* Eighth lower tooth, about 2 x. *E* Dermal denticles, about 35 x. *F* Apical view of dermal denticle.

Second dorsal fin: vertical height 3.1; length of base 4.2.
Anal fin: vertical height 3.6; length of base 4.6.
Caudal fin: upper margin 30.3; lower anterior margin 12.4.
Pectoral fin: outer margin 23.8; inner margin 6.4; distal margin 17.6.
Distance from snout to: 1st dorsal 33.0; 2nd dorsal 61.0; upper caudal 69.7; pectoral 22.0; pelvics 49.2; anal 62.0.
Interspace between: 1st and 2nd dorsals 17.2; 2nd dorsal and caudal 5.3; base of anal and caudal 3.6.
Distance from origin to origin of: pectoral and pelvics 26.7; pelvics and anal 12.4.

Trunk moderately stout, its height at origin of 1st dorsal a little more than 1/4 its length to origin of caudal. Midline of back with a low dermal ridge occupying the middle 2/3 of the space between 1st and 2nd dorsals in embryos;[53] whether or not this ridge persists throughout life, or how regularly, is not yet known.[54] Caudal peduncle moderately flattened, upper precaudal pit well marked, subrectangular, the lower only weakly defined. Dermal denticles so nearly flat that skin is smooth to the touch, overlapping only very little, the skin exposed here and there, their blades broader than long, usually with 5 (occasionally 6 or 7) low, sharp-edged ridges, the posterior margins very broadly ovate, usually with 5 very short teeth (the median only a little the largest), or in some cases merely sinuous in the corresponding radii, depending on the position on the body and on the degree of wear; pedicels rather slender.

Head about 1/5 of total length, its dorsal profile moderately and evenly convex, about 3/4 as wide at outer ends of nostrils as at eyes. Snout thick-tipped, very broadly rounded in front and noticeably short, its length in front of a line connecting outer ends of nostrils a little less than 1/2 as great as distance between inner ends of latter,[55] its length in front of mouth between 1/3 and 1/4 as long as head to origin of pectoral (about 26 to 28%) in adult, but relatively somewhat longer in embryo. Eye approximately circular, its anterior edge about opposite front of mouth, noticeably small and increasingly so with growth, its horizontal width decreasing from about 40% as great as distance between inner ends of nostrils in embryos to only about 20% as great as that in large specimens. Gill openings evenly spaced, the 1st 2 to 3 times as long as diameter of eye in adult, but only a little longer than eye in embryo, the 5th slightly the shortest, the 3rd slightly the longest, the 4th above origin of pectoral. Nostril moderately oblique, its inner end nearer to mouth than to tip of snout by a distance a little shorter than diameter of eye, its inner margin slightly expanded in obtusely subangular outline near inner end. Mouth about twice as broad as high.

Teeth $\frac{14 \text{ or } 15\text{—}1 \text{ or } 2\text{—}14 \text{ or } 15}{13 \text{ to } 15\text{—}1\text{—}13 \text{ to } 15}$; uppers broadly triangular, the 1st and 2nd nearly symmetrical, but subsequent teeth increasingly oblique, the inner margins weakly convex

53. All the embryos listed above show this ridge more or less clearly.
54. We did not appreciate the importance of this character at the time when we examined the adults listed above in the fresh state.
55. About 50 per cent that great as calculated in adults, 42 to 43 per cent by direct measurement in embryos.

toward tip but weakly concave toward base, the outer margins moderately concave (most so toward base), the outermost 3 or 4 teeth being definitely notched outwardly; both margins strongly and evenly serrate from base nearly to tip; lower teeth erect on broad bases with lanceolate cusps narrowing rather abruptly toward the tip, the apical part of cusps very finely serrate but bases smooth except toward corners of mouth, where occasional teeth show more or less serration along the basal expansions as well as on the cusp; 1 or 2 minute teeth at symphysis in upper jaw, 1 in lower; outermost teeth in each jaw very small.

First dorsal noticeably large, its vertical height a little more than ½ as great as distance from tip of snout to 5th gill opening, its origin opposite inner corner of pectoral in embryo but slightly posterior to it in adult specimens, its anterior margin weakly convex in adult but strongly so in embryo, its apex very broadly rounded, its posterior margin convex near apex but deeply concave toward base (much more so in young specimens), its free rear corner about ⅓ as long as the base, the midpoint of its base considerably nearer to axil of pectoral than to origin of pelvics. Second dorsal about ⅓ as long at base as 1st and slightly less than ⅓ as high, but with rear corner much more elongate, relatively, and a little longer than the base, its origin over or slightly before origin of anal. Caudal a little less than ⅓ (28%) of total length, its terminal sector about ¼ the length of fin, moderately slender, the tip rounded in embryo but subacute in adult, the lower lobe (expanded lower anterior corner) nearly or quite ½ as long as upper (relatively somewhat shorter in embryo), its tip very broadly rounded in embryo but altering to subacute with growth, its posterior margin evenly convex, the re-entrant corner (included between the 2 lobes) narrowly rounded in adult but broadly so in embryo. Distance from lower precaudal pit to tip of anal only about ¼ as long as base of anal. Anal about as long at base as 2nd dorsal, with broadly rounded apex, but about 1.3 times as high vertically, and with much more deeply incised rear outline, its free rear tip about as long as its base. Distance from origin of anal to tips of pelvics only about as long as base of anal. Pelvics about as long as anal along anterior margins. Pectoral as long as, or a little longer than, head, or slightly less than ¼ the total length,[56] about 2.3 times as long as broad, its anterior margin moderately convex (increasingly so toward tip) in adult and very strongly so in embryo, its distal margin moderately and increasingly concave toward inner corner, the inner corner moderately rounded, the tip similarly rounded in adult, but much more broadly so in embryo.

Color. Varying from light gray or pale brown to slaty-blue above and yellowish or dirty white below. In the two fresh adults we have seen, the pelvics and the lower surfaces of the pectorals were spotted with gray, the tips of the dorsals being grayish white and similarly spotted. But in some cases these fins, as well as the caudal lobes and the pectorals,

56. Pectoral a little longer relatively in one of the two adults measured and a little shorter in the other. Among seven embryos (five of them from one brood), the ratio of length of pectoral to length of head (snout to pectoral origin) is 1:1.2 at 395 mm. (male); 1:1 at 460 mm. (male); and from about 1:0.9 to about 1:1 in 5 females of 555 to 580 mm.; this range of variation shows a small increase in the relative length of the fin with growth. In a Mediterranean embryo of 420 mm., reported by Moreau (Poiss. France, Suppl., 1891: 7), the ratio was about 1:1.5.

are white tipped. In embryos, however, the dorsals, pectorals, pelvics and lower lobe of caudal are more or less conspicuously tipped with sooty gray or black.

Size. The stage of development of the embryos listed above (up to about 580 mm. long) suggests a length of perhaps 650 to 700 mm. at birth. Maturity probably is not reached at less than about six feet. The longest for which we find exact measurement was 3.5 meters (about 11½ feet). The maximum size is said to be 12 to 13 feet, but we think it likely that at least some may grow considerably longer, for larger sharks, apparently of this species because of the rounded shape of the first dorsal fins, have been described recently to us as seen at the surface over the continental slope in the offing of Woods Hole from the research vessel "Atlantis."

Developmental Stages. Development is viviparous; the embryos which we have collected[57] have a long umbilical cord, about 410 mm. in length, terminating in a well developed yolk-sac placenta by which they were attached to the uterine wall of the mother. *C. longimanus* is also peculiar among carcharhinids for the very considerable changes in the shape of its fins with growth, as illustrated by drawings of the embryo and mother shown in Fig. 64, and emphasized above in the description. It is interesting that in one case all members of a litter of embryos were of the same sex (female), whereas in another case both sexes were represented (see Study Material, p. 354).

Remarks. C. longimanus very commonly has been confused with *leucas,* which it resembles in general form of trunk, very broad head, very short and broadly rounded snout, low-arched mouth, relative positions of fins, and teeth; consequently the synonymy of the two species is almost hopelessly confused. Actually, however, the two species are separable at a glance by the shape of the first dorsal fin (strongly rounded in *longimanus,* but subangular in *leucas*); also the tip of the anal reaches nearly to the precaudal pit in *longimanus* but falls considerably short of it in *leucas,* and the pectoral is much longer, relatively, in the former than in the latter. The outline of the lower lobe of the caudal is also distinctive, being convex posteriorly and nearly straight anteriorly in adult *longimanus* with the reverse in *leucas,* while the margins of the dermal denticles are much more strongly toothed in *leucas* than in *longimanus.* The differences in the shapes of the fins, especially the pectoral, are even more striking in the case of embryos than of adults.

Habits. Astonishingly little is known of the habits of *longimanus,* considering that it is one of the members of its genus that has been recognized the longest. Apparently it is more strictly pelagic than any other members of *Carcharhinus* in the western Atlantic, and more strictly tropical there than most of them are. We have not found a single report of one caught from the beach or taken in a pound net anywhere along the coast of the United States that can be referred with certainty to this particular species.[58] And while "*Carcharias*

57. See Study Material, p. 354.

58. Jordan, it is true (Bull. U.S. nat. Mus., 7, 1884: 104), has characterized a large shark under the name *Carcharias lamia* as very common around wharves and off the Keys of southern Florida, adding that one of 5 or 6 embryos which were taken from a 7½-foot female was kept. But the subsequent illustration, probably of this specimen (Jordan and Evermann, Bull. U.S. nat. Mus., 47 [4], 1900: pl. 5, fig. 17), is not of a *longimanus* but apparently of a *leucas.*

(*Prionodon*) *lamia*" has been said to run up into fresh water in Senegal,[59] there is no way of knowing whether this report (by name only) actually referred to *longimanus* or to some other carcharhinid, the latter being more likely. In the Mediterranean it is often mentioned as entering the tuna nets close to land and as being common offshore.

It is not possible as yet to relate its geographical distribution to physical factors in a satisfactory way. Although it has never been reported reliably in the western Atlantic in temperatures lower than about 21 to 22° C. or 70 to 71.5° F. (Bahamas, March 1914), it is described in the Mediterranean around Sicily as caught most often in winter,[60] *i.e.*, when the water is only about 13 to 15° C. (about 55.5 to 59° F.). Hence, its failure to visit the coasts of the eastern United States in summer (see below) is less likely to be due to unfavorable temperature than to low salinity, for it occurs chiefly where the water is more saline than 35.5 per mille, or even more than 36 per mille. If such is the case, it is exceptional among sharks.

In the Mediterranean it is said to destroy large numbers of fish; we know from personal experience that it takes a large bait readily, and it is so well armed that it would not be astonishing if it preyed on large as well as small fishes, or on sea turtles. But no precise information is available as to its diet.

Of its breeding habits it is only known that a female, caught off the north coast of Cuba in May (see Study Material, p. 354), contained six embryos.

Relation to Man. This species has never been of commercial importance anywhere. On the other hand, it has been accused vaguely of being a man-eater, but we do not know on what evidence.

Range. Tropical and subtropical Atlantic. In the east it is well known in the Mediterranean and along the Iberian peninsula, and is reported by name from off Cape Verde,[61] where it is certainly to be expected, and from Senegal;[61] but it seems likely that its reported occurrence in the latter region refers in part to some other shark (see discussion of its reputed presence in fresh water there, p. 360). In the west its normal zone of occurrence is from Uruguay and southern Brazil to the more northerly waters of the West Indies, and thence northward in the Gulf Stream, perhaps to the offing of southern New England. The species "*lamia*" has also been reported from various localities in the Pacific and Indian Oceans, Australia and Red Sea.[62] All but one of these reports are by name only, however, hence they afford no clue regarding the actual species of *Carcharhinus*; and even that one[63] seems actually to have referred to *C. brachyurus* (Günther), 1870, the "Whaler"

59. Steindachner, S. B. Akad. Wiss. Wien, *61* (1), 1870: 576.
60. Doderlein, Man. Ittiol. Medit., *2*, 1881: 41.
61. Steindachner, S. B. Akad. Wiss. Wien, *61* (1), 1870: 576 (Senegal, in fresh water); Denkschr. Akad. Wiss. Wien, *44*, 1882: 51 (Senegambia); Rochebrune, Act. Soc. linn. Bordeaux, (4) *6*, 1882: 19 (Cape Verde, Senegal).
62. For a list of these records, see Fowler (Bull. U.S. nat. Mus., 100 [*13*], 1941: 169) in synonymy of *Eulamia lamia*.
63. Phillipps, N. Z. J. Sci. Tech., *6*, 1924: 260, fig. 3; we may point out that his illustration (fig. 3) is apparently

of New Zealand seas, a species only remotely resembling the Atlantic *longimanus*. Twelve out of 13 other western Pacific–Indian Ocean species included by Fowler[64] in the synonymy of his *Eulamia lamia* equally fail to show the combination of characters most distinctive of *longimanus*, at least if the published accounts of them are to be relied upon. The several species of *Carcharhinus* that occur along the Pacific coast of Central America are also clearly separable from *longimanus* by one character or another.[65] Although *insularum* Snyder, 1904,[66] from the Hawaiian Islands, does resemble *longimanus* in the roundness of its first dorsal, in the close proximity of the tip of its anal to the origin of the caudal, and in its teeth (particularly in the serration of the lowers), its pectoral appears to be definitely much shorter than that of *longimanus*.[67]

Occurrence in the Western Atlantic. Definite information as to the actual frequency of occurrence of this species in the western Atlantic is astonishingly scant, partly because it is so seldom encountered in continental waters there, but equally because it has been confused so often with *C. leucas*, and perhaps with *C. obscurus* also. The only reports of it there that include evidence as to their actual identity are for: Uruguay; the Island of South Trinidad (Lat. 20° 30′ S., Long. 29° 23′ W.) off southern Brazil; northern Brazil south to Rio de Janeiro; Santa Cruz, Dominica and Guadeloupe, West Indies; off the north and south coasts of Cuba; three stations off Florida north of the Bahama Bank (*i.e.*, seaward of the Gulf Stream); and the Caribbean in general. But the wide distribution of these localities proves it to be generally distributed in the western side of the tropical Atlantic. And the fact that we counted 28 and caught one on an occasion from the research ship "Atlantis" off the north coast of Cuba in May of 1939 is in line with earlier characterizations of it as abundant in the Caribbean–West Indian region, and with reports to us of "White-finned" sharks being seen there very often. The scanty information available suggests that it is also common offshore around Bermuda but not inshore.

We find no reliable record of it for the coast of Florida[68] or for anywhere else on the east coast of the United States farther north, which is in agreement with its oceanic nature. But a school[69] of large sharks, apparently of this species, was encountered at the surface over the continental slope by the research vessel "Atlantis" on one occasion in June 1941 in the offing of southern New England (about Lat. 39° 30′ N., Long. 70° 30′ W., see p. 359), suggesting that the transition-band between oceanic and continental waters is its normal boundary in the western Atlantic north of tropical latitudes.

copied from the one by Jordan and Evermann (Bull. U.S. nat. Mus., 47 [4], 1900: pl. 5, fig. 17) of an embryo, probably of *C. leucas*; see also footnote 58, p. 359.

64. Bull. U.S. nat. Mus., 100 (*13*), 1941: 169.
65. For the most recent survey of these, see Beebe and Tee-Van, Zoologica, N. Y., *26*, 1941: 106.
66. Bull. U.S. Bur. Fish, *22*, 1904: 513, pl. 1, fig. 1, *Carcharias insularum*.
67. Pectoral only 80 per cent of length of head to origin of pectoral in adult *insularum*, and 83 per cent of length of head in embryo, whereas in *longimanus* it is about as long as the head or longer.
68. It is not included in the most recent survey of the sharks of Florida (Springer, Proc. Fla. Acad. Sci., *3*, 1939: 9–41); an earlier characterization of it by Jordan (Bull. U.S. nat. Mus., 7, 1884: 104) as plentiful there seems actually to have referred to *C. leucas*; see footnote 58, p. 359; jaws from two Florida localities were reported by Fowler (Proc. Acad. nat. Sci. Philad., *60*, 1908: 65) as only "probably" this species.
69. Described as "several hundred."

Synonyms and References:[70]

Squalus carcharias (in part) Linnaeus, Syst. Nat., *1*, 1758: 235; *1*, 1776: 400 (this appears to be a combination of the present species with *Carcharodon*, of which *S. carcharias* Linnaeus is now universally considered the type; for discussion, see Jordan, Copeia, 166, 1928: 4); Walbaum, P. Artedi Genera Pisc. Emend. Ichthyol., *3*, 1792: 514 (after Linnaeus); Bloch and Schneider, Syst. Ichthyol., 1801: 132 (after Linnaeus); Risso, Ichthyol. Nice, 1810: 25 (descr., a combination of this species with *Carcharodon*, Medit.); Gray, Cat. Fish. Brit. Mus. descr. by L. T. Gronow, 1854: 5 (combined in descr. with *Carcharodon*, Medit., Atlant.).

Squalus sp. (in part) Gronow, Zoophyl., *1*, 1763: 32 (incl. in synon., not seen).

Le Requin, Cloquet, Dict. Sci. Nat., Atlas. Poiss., 1816–1830: pl. 26 (clearly recognizable ill.); not Le Requin, Cloquet (Dict. Sci. Nat., 7, 1817: 69, which is *Carcharodon*).

Carcharias lamia Risso, Hist. Nat. Europ. Merid., *3*, 1826: 119 (descr., Medit.); Bory de St. Vincent, Dict. Class. Hist. Nat., 15, 1829: 598 (name only, by ref. to Risso, 1826); Jordan and Gilbert, Bull. U.S. nat. Mus., 16, 1883: 873 (by ref., coasts of Europe, not *Carcharinus lamia*, p. 23, which was the tooth of a *Carcharodon*); Carus, Prod. Fauna Medit., *2*, 1889–93: 512 (by refs., Medit.); Ribeiro, Arch. Mus. nac. Rio de J., *14*, 1907: 154, pl. 4 (descr., photo of jaws, north. Brazil south to Rio de Janeiro); Fauna brazil. Peixes, *2* (1) Fasc. 1, 1923: 11 (same as preceding); Nobre, Fauna Marinha Port. Vert., *1*, 1935: 422 (descr., off Portugal).

Carcharias (*Prionodon*) *lamia* Müller and Henle, Plagiost., 1841: 37, pl. 12 (descr., excellent ill., common in Medit. and Ocean); Duméril, Hist. Nat. Poiss., *1*, 1865: 356, in part (good descr., but specs. from India, Aust. and C. of Good Hope probably not this species); Günther, Cat. Fish. Brit. Mus., *8*, 1870: 372 (descr., Medit., Atlant.); Doderlein, Man. Ittiol. Medit., *2*, 1881: 40 (good descr., season, voracity, damage to nets, Medit.).

Squalus longimanus Poey, Memorias, *2*, 1861: 338 (excellent descr., Cuba).

Prionodon longimanus Poey, Memorias, *2*, 1861: pl. 19, fig. 9, 10 (excellent ill. of teeth, Cuba).

Eulamia lamia Gill, Ann. N. Y. Lyc., 7, 1862: 401 (name only, with Gill as authority, see footnote 3, p. 320); Poey, An. Soc. esp. Hist. nat., *5*, 1876: 384; Enumerat. Pisc. Cubens., 1876: 188 (size, ref. to earlier descr., Cuba); Fowler, Bull. U.S. nat. Mus., 100 (*13*), 1941: 170 (Atlant. refs. in synonymy at least in part, but not Pacif. refs.); Arqu. Zool. Estado São Paulo, *3*, 1942: 128 (Brazil).

Prionodon lamia Bocage and Brito Capello, Poiss. Plagiost. Port., 1866: 18 (brief descr., size, off Portugal).

Eulamia longimana Poey, Repert. Fisico-nat. Cuba, *2*, 1868: 448 (descr., one 3.5 meters long, Cuba); Bigelow and Schroeder, Guide Comm. Shark Fish., Anglo Amer. Caribb. Comm., Wash., 1945: 80, fig. 24 (descr., habits, range, ill.).[70a]

Prionodon lamia Canestrini, in Cornelia, *et al.*, Fauna d'Ital., 3, Pesci, 1871–1872: 48 (ident. by refs., size, Medit.).

Carcharias (*Prionodon*) *obtusus* Garman, Bull. Harv. Mus. comp. Zool., *8*, 1881: 232, at least in part (ident. by descr. of long pectorals, rounded dorsal, white-tipped fins, number of embryos; Cuba, Santa Cruz, Guadeloupe, Dominica).

Carcharias obtusirostris Moreau, Hist. Nat. Poiss. France, *1*, 1881: 332 (descr. at least in part, but perhaps combined with some other *Carcharhinus*, Medit.); Poiss. Franc., Suppl., 1891: 7 (meas. of embryo 420 mm., Medit.).

Carcharhinus (*Eulamia*) *lamia* Jordan, Rep. U.S. Comm. Fish. (1885), 1887: 796 (name by ref., W. Europ.).

Carcharhinus (*Carcharhinus*) *lamia* Jordan and Evermann, Bull. U.S. nat. Mus., 47 (1), 1896: 38 (descr., probably in part, and by refs., but apparently confused with *C. leucas*).

Carcharinus commersonii (in part) Garman, Mem. Harv. Mus. comp. Zool., *36*, 1913: 140 (descr., Atlant. refs. in part, but not Pacif.); Radcliffe, Bull. U.S. Bur. Fish., *34*, 1916: 260 (meas. of specs. taken north of Bahama Bank, but not fig. of denticles, or N. Carolina records which probably were *leucas*); Meek and Hildebrand, Field Mus. Publ. Zool., *15* (1), 1923: 43 (Atlant. refs., in part only; not descr., which is of

70. This species has been confused so commonly with others that we limit the following list to such citations as contain evidence that they did in fact refer to this particular species, at least in part, and not to some other. For a list of references to *Carcharhinus* that cannot be allocated with certainty to any particular species, see p. 363.

70a. Spelled *longimanus*.

a *Carcharhinus* with short pectorals); Jordan, Evermann and Clark, Rep. U.S. Comm. Fish. (1928), *2*, 1930: 15 (synonymy in part); White, Bull. Amer. Mus. nat. Hist., *74*, 1934: 127, but not pl. 42E (included in synonymy, but ill. of heart valves probably *leucas*).

Carcharinus platyodon (in part) Garman, Mem. Harv. Mus. comp. Zool., *36*, 1913: 126 (descr. of embryo, in Harv. Mus. Comp. Zool., Guadeloupe, W. I., see Study Material, p. 354; but not synonymy, and not pl. 3, fig. 4–6, which is a mislabeled *milberti*); Devincenzi, An. Mus. Hist. nat. Montevideo, (2) *4*, 1939: 5 (jaws, teeth, apparently this species, Uruguay).

Carcharhinus commersonii Jordan, Copeia, 49, 1917: 87 (name only, substituted for *lamia*); Norman and Fraser, Giant Fishes, 1937: 36 (distrib., based on old accounts).

Carcharhinus sp. Nichols and Murphy, Bull. Amer. Mus. nat. Hist., *33*, 1914: 262 (descr. of embryo, South Trinidad I., S. Atlant., Lat. 20° 30′ S., Long. 29° 22′ W.).

Carcharias commersonii (in part) Beebe and Tee-Van, Field Bk. Shore Fish. Bermuda, 1933: 28 (includes both this species and *leucas*, Bermuda).

Doubtful References:

Squale requin Lacépède, Hist. Nat. Poiss., 4° ed., *1*, 1798: 169, but not pl. 8, fig. 1;[71] in Soninni, Hist. Nat. Poiss., *3*, 1802: 332 (a rewrite of Lacépède's account, but the illustrations, pl. 6, are of the teeth and jaws of *Carcharodon*, combined with a general view so poor that it might be interpreted as representing either the latter or one of the larger carcharhinids).

Carcharhinus lamia Tortonese, Atti Soc. ital. Sci. nat., 77, 1828: 298 (discus., Medit.).

Carcharias carcharias (in part) Jordan, Manual Vert. Anim. NE. U.S., 1929: 10 (class., descr., apparently based partly on *longimanus* and partly on *leucas*, S. Atlant., U.S.).

References, Ostensibly *longimanus*, But Which Cannot be Definitely Identified as Any Particular Species of the Genus:

Carcharias (*Prionodon*) *lamia* Steindachner, S. B. Akad. Wiss. Wien, *61* (1), 1870: 576; Denkschr. Akad. Wiss. Wien, 44 (1), 1882: 51 (Senegambia, name only).

Carcharinus lamia Evermann and Kendall, Rep. U.S. Comm. Fish. (1899), 1900: 48 (south. Florida); Bean, Field Mus. Publ. Zool., 7 (2), 1906: 29 (Bermuda); Fowler, Proc. Acad. nat. Sci. Philad., *58*, 1906: 80 (Florida Keys); Gudger, Proc. biol. Soc. Wash., *26*, 1913: 97 (C. Lookout, N. Carolina); Coles, Proc. biol. Soc. Wash., *28*, 1915: 90 (Carolina); Nichols, Copeia, 53, 1918: 13 (Florida); Borri, Mem. Soc. tosc. Sci. nat., *44*, 1934: 89 (Medit.); Norman and Fraser, Giant Fishes, 1937: 36 (Medit., trop. Atlant.).

Prionodon lamia de Braganza, Result. Invest. Sci. "Amelia" Ichthyol., *2*, 1904: 44, 45 (Portugal); Richard, Bull. Inst. océanogr. Monaco, *19*, 1904: 11 (Canaries).

Carcharias lamia Jordan and Thompson, Bull. U.S. Bur. Fish., *24*, 1905: 232 (Tortugas, Florida); Ribeiro, Rev. Mus. paul., *10*, 1918: 707 (Santos, Brazil); Marini, Physis B. Aires, *9*, 1929: 452 (Argentina); Pozzi and Bordale, An. Soc. cient. argent., *120*, 1935: 150 (Argentina).

Eulamia longimanus Fowler, Proc Acad. nat. Sci. Philad., *60*, 1908: 65 (Florida).

Eulamia commersonii Fowler, Proc. Acad. nat. Sci. Philad., *69*, 1917: 127 [Colón, provisional ident. by ref. to Müller and Henle's (Plagiost., 1841: 37, pl. 12) account of *Carcharias* (*Prionodon*) *lamia*].

Eulamia commersonii (in part) Fowler, Bull. Amer. Mus. nat. Hist., *70*, 1936: 53, pl. 1 [trop. W. Afr. by ref. to *Carcharias* (*Prionodon*) *lamia* Steindachner, 1870, 1882, but descr. is of a Florida spec., probably *leucas*].

Eulamia lamia Fowler, Monogr. Acad. nat. Sci. Philad., 1945: 95, 263 (size, Marco, Florida, off N. Carolina).

Not *Carcharias lamia* Rafinesque, 1810; see footnote 51, p. 354.

Not *Carcharhinus lamia* Blainville, in Vieillot, Faune Franc., 1825: 88, pl. 22, fig. 1[72] (jaws, ident. by ref. to Duhamel, Traité gén. Pêches, *4* [9], pl. 19, as *Galeocerdo*).

71. Much of the confusion in shark nomenclature comes from uncertainty as to actual identity of Lacépède's Squale requin. The measurements on page 184 might apply to *longimanus*; but the account of the first dorsal, and its representation in the illustration as only a little rounded at apex, do not. Most of the description is devoted to a general account of the larger carcharhinids, in general, perhaps combined with *Carcharodon*.

72. We cannot find whether or not this plate was ever published.

Not *Carcharias* (*Prionodon*) *lamia* Putnam, Bull. Essex Inst., *6*, 1874: 72 (this was a *Carcharodon*).
Not *Prionodon lamia* Goode and Bean, Bull. Essex Inst., *11*, 1879: 30 (the specimen of *Carcharodon* reported by Putnam, 1874).
Not *Eulamia lamia* Jordan and Gilbert, Bull. U.S. nat. Mus., 16, 1883: 60 (descr. of San Diego, Calif., spec. actually *lamiella*, Jordan and Gilbert, 1882).
Not *Carcharias lamia* Jordan, Proc. U.S. nat. Mus., 7, 1884: 104 (almost certainly refers to *leucas*; see p. 344).
Not *Carcharhinus lamia* Jordan and Evermann, Bull. U.S. nat. Mus., 47 (4), 1900: pl. 5, fig. 17 (embryo, evidently of *leucas*, see p. 344).
Not *Carcharias lamia* Lahille, An. Mus. nac. B. Aires, *34*, 1928: 305 (appears to be *remotus*, see p. 403).

Carcharhinus maculipinnis (Poey), 1865

Large Black-tipped Shark

Figures 66, 67

Study Material. Female, about 6 feet 6 inches (1,975 mm.) long, from Salerno, Florida (Harv. Mus. Comp. Zool., No. 35772) and jaws of two males, about 6 feet 4 inches (1,915 mm.) long, from Englewood, Florida (U.S. Nat. Mus., No. 109957, 110306).

Distinctive Characters. Among the smooth-backed *Carcharhinus* of the Atlantic, *maculipinnis* is most like *limbatus* in general appearance, fins, teeth and color. But it is separated from *limbatus* by its noticeably smaller eyes, its relatively longer upper labial furrows and gill openings, the noticeably more slender free rear tip of its second dorsal and the perfectly smooth edges of its lower teeth (finely serrate in *limbatus*). Also, its first dorsal usually originates somewhat farther rearward, although there may be some variation in this respect. Its dermal denticles are distinctive, also (cf. Fig. 66 A with 62 D).

Description. Proportional dimensions in per cent of total length. Female, 1,975 mm., from Salerno, Florida (Harv. Mus. Comp. Zool., No. 35772).

Snout length in front of: outer nostrils 4.8; mouth 8.0.
Eye: horizontal diameter 0.9.
Mouth: breadth 9.1; height 4.9.
Nostrils: distance between inner ends 5.3.
Labial furrow length: upper 1.1.
Gill opening lengths: 1st 4.1; 2nd 4.4; 3rd 4.8; 4th 4.7; 5th 3.5.
First dorsal fin: vertical height 9.0; length of base 9.7.
Second dorsal fin: vertical height 2.6; length of base 3.7.
Anal fin: vertical height 2.7; length of base 4.2.
Caudal fin: upper margin 26.5; lower anterior margin 12.2.
Pectoral fin: outer margin 17.2; inner margin 4.8; distal margin 15.0.
Distance from snout to: 1st dorsal 32.5; 2nd dorsal 65.2; upper caudal 73.5; pectoral 23.9; pelvics 52.0; anal 65.2.

Interspace between: 1st and 2nd dorsals 21.7; 2nd dorsal and caudal 5.3; anal and caudal 5.4.

Distance from origin to origin of: pectoral and pelvics 27.7; pelvics and anal 13.5.

Trunk moderately slender, its height at 1st dorsal (where highest) between 1/5 and 1/6 its length to precaudal pits, tapering both anteriorly and posteriorly, the midline of

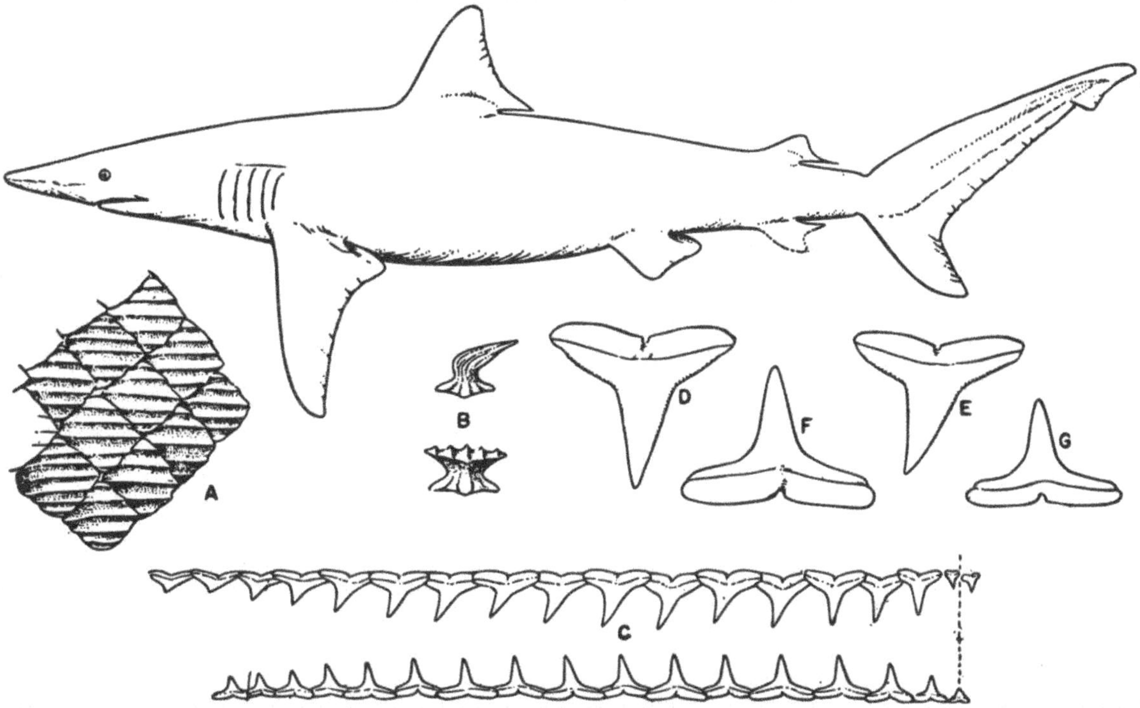

FIGURE 66. *Carcharhinus maculipinnis*, female, about six feet six inches (1,975 mm.) long, from Salerno, Florida (Harv. Mus. Comp. Zool., No. 35772). *A* Dermal denticles, about 18 x. *B* Side and apical views of dermal denticle, about 18 x. *C* Upper and lower teeth, right-hand side, about 2/3 natural size. *D* Fourth upper tooth. *E* Twelfth upper tooth. *F* Fourth lower tooth. *G* Twelfth lower tooth. *D–G*, about 2 x.

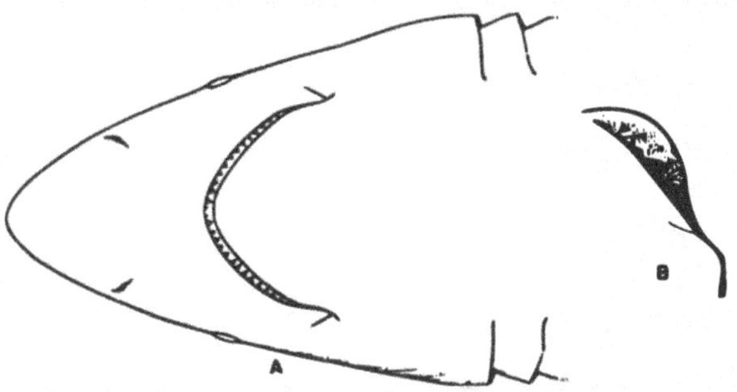

FIGURE 67. *Carcharhinus maculipinnis*, illustrated in Fig. 66. *A* Anterior part of head from below. *B* Left-hand nostril, about 1.2 x.

back evenly rounded, without dermal ridge. Precaudal pits semilunar in outline, the upper the deeper, but the lower also well marked. Dermal denticles so close-spaced and evenly overlapping that the skin is wholly concealed, but their blades raised enough from skin to feel slightly rough when stroked from rear to front, broader than long, with mostly 7 (rarely 5) low ridges, their posterior margins broadly ovate or subangular, with 7 very short teeth (shorter than in *limbatus*) or even entire in some cases; pedicels very short.

Head a little less than 1/4 of total length and flattened above anterior to gill region. Snout rather thin-tipped, ovate, with narrowly rounded tip (relatively somewhat longer and more pointed than in *limbatus*), its length in front of a line connecting outer ends of nostrils about 3/4 as great as distance between inner ends of latter, and length in front of mouth about 1/3 as great as length of head. Eye approximately circular, and noticeably small, its diameter only about 1/9 (11%) as great as length of snout in front of mouth. Gill openings noticeably large, the 3rd (longest) nearly as long as distance between nostrils or about 5 times as long as diameter of eye (only about 2.5 times as long as diameter of eye in *limbatus*), the 5th a little less than 3/4 (73%) as long as 3rd, the 4th over origin of pectoral. Nostril strongly oblique, its inner end nearer to mouth than to tip of snout by a distance a little more than twice as great as diameter of eye, its anterior margin sinuous, with low, rounded expansion near inner end. Mouth ovate, about 1/2 as high as broad, occupying only about 2/3 of breadth of head. Upper labial furrow about as long as nostril, thus considerably longer relatively than in *limbatus*.

Teeth $\frac{16 \text{ or } 17—2 \text{ or } 3—16 \text{ or } 17}{16—1—15 \text{ or } 16}$ in specimen examined; uppers with narrow triangular cusps on broad bases, smaller and relatively narrower than in *limbatus* (cf. Fig. 66 C–E with 62 C and 63 C, D), the first 3 nearly symmetrical and erect, but 4th and subsequent teeth slightly oblique, with outer margin more deeply concave than the inner toward base, and decreasing in length toward corner of jaw, the outermost 3 very low, the edges regularly though finely serrate, except that the tips are smooth-edged; lowers with much more slender cusps than uppers, on very broad bases, their tips not recurved forward as they are in *limbatus*, very slightly oblique along whole length of jaw, the 1st smaller than 3rd, and 12th and 13th to 16th again successively smaller, the edges perfectly smooth on base as well as on cusp; 2 minute teeth at symphysis in upper jaw and 1 in lower.

Origin of 1st dorsal a little posterior to inner corner of pectoral, its apex rather narrowly rounded, its free rear corner relatively obtuse and only about 1/3 as long as its base, the base about 4/5 as long as anterior margin, its vertical height a little less than 1/2 as long as head or about 1/2 as long as pectoral, the midpoint of base about 2/3 as far from axil of pectoral as from origin of pelvics. Second dorsal a little less than 1/2 as long at base as 1st dorsal, its origin about over origin of anal, its free rear tip about as long as the base, and noticeably more slender than that of 1st. Caudal about 1/4 of total length, the terminal sector a little less than 1/4 of the fin, slender, with narrowly rounded tip, the lower lobe (expanded lower anterior corner) a little less than 1/2 as long as upper, with moderately convex lower anterior margin and very narrowly rounded or subacute tip, the re-entrant corner between the two lobes rather broadly rounded.[74] Anal slightly longer at base than 2nd dorsal, its outline presumably similar to that of *limbatus* but damaged in the specimen we have seen, its free rear corner more obtuse and only a little more than ½ as long as the base. Distance from origin of anal to tips of pelvics about 1½ times as long as base of anal. Pelvics a little longer than 2nd dorsal along anterior margin. Pectoral a little less than ¾ (72%) as long as head, and about ½ as broad as long, similar in form to that of *limbatus*, with very narrowly rounded or subacute apex and inner corner.

Color. Varying shades of gray above, the colors said to be more intense in life than in *limbatus;* white or whitish below, resembling *limbatus* in having a narrowing band of the darker tint of the upper parts extending rearward to about over the origin of the pelvics; above this there is a corresponding extension forward of a narrowing band of white (or whitish) from the lower sides to a point ⅔ of the distance forward from the origin of the pelvics toward the axil of the pectoral;[75] lower lobe of caudal, apex of 2nd dorsal and lower surfaces of tips of pectorals broadly and conspicuously tipped with black, with apex of 1st dorsal narrowly so, much as in *limbatus.*

Size. The few specimens of both sexes for which the sizes have been recorded have ranged from five feet eight inches to about eight feet in length; it apparently grows larger than *limbatus.*[76]

Developmental Stages. We have a photograph of a female six feet three inches from Florida with ten young.[77]

Habits. This shark has been seen in schools and leaping at the surface, in which habit it resembles *limbatus,* but nothing definite is known of its diet, other than that in Florida waters it follows shrimp trawlers to pick up the discarded fish; nothing is known of its life in other respects.

Range. So far reported only from Cuba (the type locality), from Porto Rico by name only, and from both coasts of southern Florida. Although more than three-fourths of a century has passed since *maculipinnis* was first described, it generally has been confused with *limbatus* and until very recently[78] reported under that name, if at all; consequently the published records afford no information as to the details of its occurrence, other than that it is common in winter off southeastern Florida, also off northeastern Florida (Jacksonville) in the spring,[79] and that the local shark fishermen are familiar enough with it to have recognized it as distinct from *limbatus.* But whether or not it ranges northward in summer like *limbatus* is not yet known.

Relation to Man. This, like various other species, forms part of the catch of the

74. Slight apparent differences in the shape of the caudal between *maculipinnis* and *limbatus* may represent individual variation.
75. These contrasting lighter and darker bands are still visible in the preserved specimen, although the latter is now much discolored in the preservative.
76. Females of eight feet were recently reported to us off Salerno, Florida, by Stewart Springer.
77. From Stewart Springer.
78. Springer (Proc. Fla. Acad. Sci., 3, 1939: 27) was the first to show that *maculipinnis* is not a synonym of *limbatus* and that it is a distinct and easily recognizable species.
79. Personal communication from Stewart Springer.

Florida shark fishery. But no information is available as to its percentage in the total. Recently an attack was reported on a bather at Mayport near Jacksonville by a 5½- to 6½-foot shark, which, judging by circumstantial evidence, seems to have been *maculipinnis.*[80]

Synonyms and References:

Isogomphodon maculipinnis Poey, Repert. Fisico-nat. Cuba, *1*, 1865: 191, pl. 4, fig. 2, 3; *2*, 1866–1868: 245, 450, pl. 2, fig. 1–3 (descr., teeth, Cuba); Synop. Pisc. Cubens., 1868: 450 (diagn., Cuba); Springer, Proc. Fla. Acad. Sci., *3*, 1939: 27 (descr., size, color, Florida); Lunz, Bull. S. Carolina St. Planning Bd., *14*, 1944: 27 (Florida).

Carcharias limbatus (in part) Günther, Cat. Fish. Brit. Mus., *8*, 1870: 373 (incl. in synon.); Ribeiro, Arch. Mus. nac. Rio de J., 14, 1907: 200; Jordan, Evermann and Clark, Rep. U.S. Comm. Fish. (1928), 2, 1930: 16 (included in synonymy of *limbatus*).

Platypodon (?) *maculipinnis* Poey, Enumerat. Pisc. Cubens., 1876: 197, pl. 9, fig. 6 (ill., denticles, Cuba); An. Soc. esp. Hist. nat., *5*, 1876: 393, pl. 14, fig. 6 (Cuba).

Carcharhinus limbatus (in part) Jordan and Evermann, Bull. U.S. nat. Mus., 47 (1), 1896: 40; Garman, Mem. Harv. Mus. comp. Zool., *36*, 1913: 127; Meek and Hildebrand, Field Mus. Publ. Zool., *15*, 1923: 41 (included in synonymy of *limbatus*).

Eulamia maculipinnis Bigelow and Schroeder, Guide Comm. Shark Fish., Anglo Amer. Caribb. Comm., Wash., 1945: 90, fig. 29 (descr., range, ill.).

Doubtful References:

Platypodon maculipinnis Stahl, Fauna Puerto Rico, 1883: 167 (Porto Rico, name only).

Eulamia platyodon Fowler, Proc. Acad. nat. Sci. Philad., *71*, 1919: 146 (Jamaica spec. perhaps *maculipinnis* because teeth narrow and the lowers "entire").

Not *Carcharias maculipinnis* Günther, Trans. zool. Soc. Lond., *6*, 1867: 490.[81]

Carcharhinus milberti (Müller and Henle), 1841[82]

Brown Shark, Sand-bar Shark

Figures 68, 69

Study Material. Male, 1,496 mm., taken at Woods Hole in August 1941 (Harv. Mus. Comp. Zool., No. 35370); male, 1,400 mm., taken at Woods Hole in August 1944 (not preserved); male, 2,000 mm. (about 6 feet 7 inches), from Vineyard Sound near Woods Hole, taken Sept. 18, 1943; jaws (Harv. Mus. Comp. Zool., No. 36032); head and skin of male, about 800 mm., from Somer's Point, New Jersey (Harv. Mus. Comp. Zool., No. 147); newborn specimens, 562 mm. and 580 mm. (female and male), from Woods Hole (U.S. Bur. Fish.); young male, 760 mm., from Chesapeake Bay, 540-mm. specimen, from off Grand Terre, Texas, and one young male, 747 mm., from Virginia

80. Stewart Springer has supplied us with details of this occurrence, mentioned on page 70 also.
81. See p. 352.
82. If it is finally proved that the Mediterranean form is identical with the American, the name *plumbeus* Nardo, 1827, must be used for the combined species in place of *milberti*; see discussion, p. 374.

Beach, Virginia (U.S. Nat. Mus., No. 104969, 13540, 119698); male embryo, 402 mm. long, from Englewood, Florida (Harv. Mus. Comp. Zool., No. 35361); head and skin of female embryo, about 600 mm., from Cuba (Harv. Mus. Comp. Zool., No. 715);

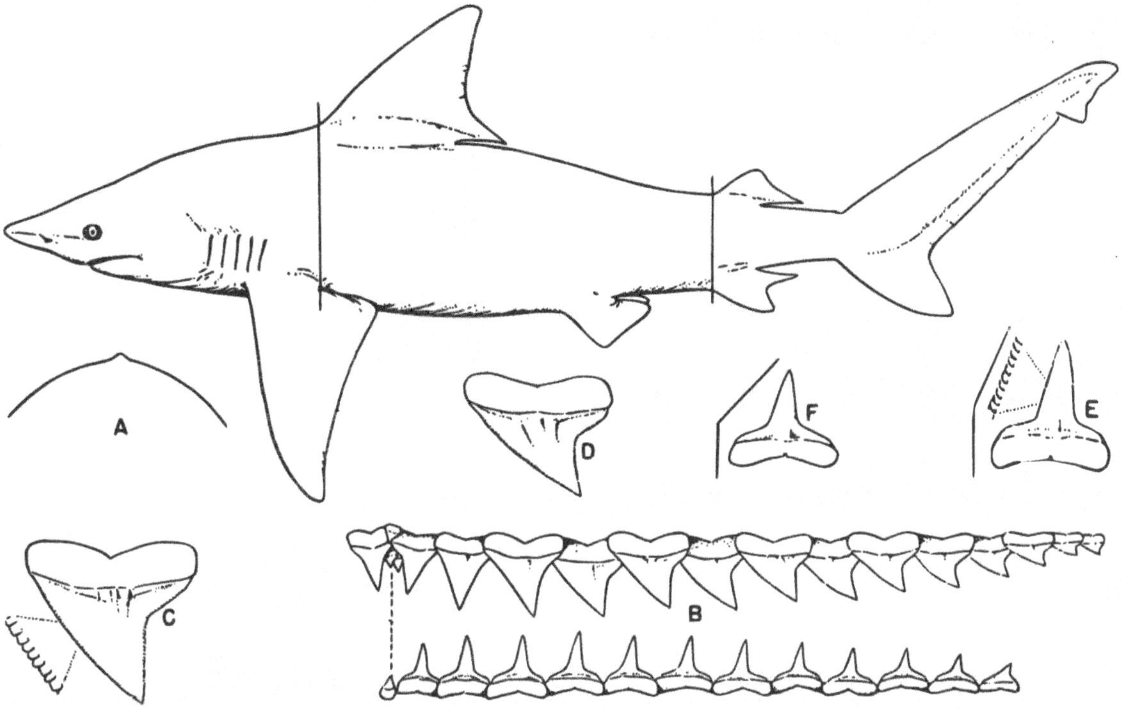

FIGURE 68. *Carcharhinus milberti*, female, 1,496 mm. long, from Woods Hole, Massachusetts (Harv. Mus. Comp. Zool., No. 35370). *A* Cross section of upper part of trunk, midway between the two dorsal fins, to show the mid-dorsal ridge. *B* Left-hand upper and lower teeth, about 1 x. *C* Fourth upper tooth. *D* Ninth upper tooth. *E* Third lower tooth. *F* Eighth lower tooth. *C–F*, about 1.8 x.

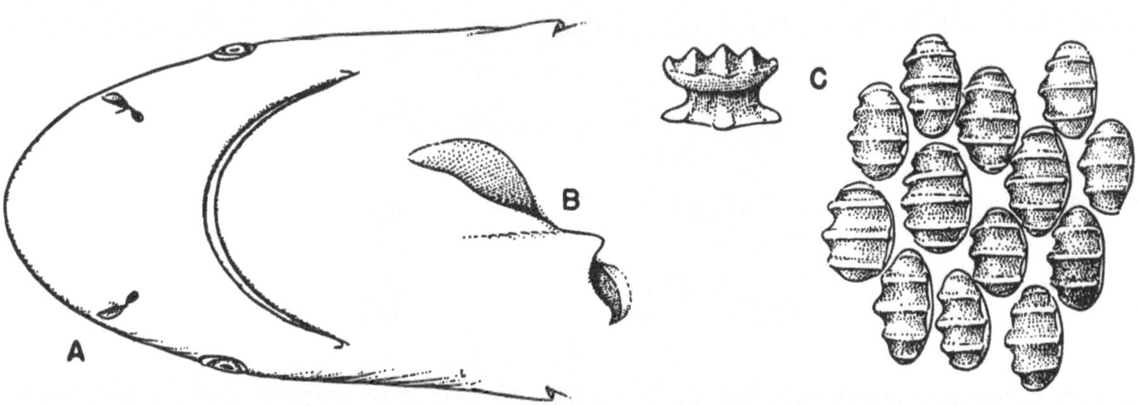

FIGURE 69. *Carcharhinus milberti*, pictured in Fig. 68. *A* Anterior part of head from below. *B* Left-hand nostril, about 1.4 x. *C* Dermal denticles, about 36 x.

measurements by Stewart Springer of an 80-inch specimen taken at Woods Hole in Aug. 1942, and of 3 females, 2,125 to 2,220 mm., from Englewood, Florida; also a full-term embryo, 540 mm., taken in upper Chesapeake Bay (U.S. Nat. Mus., No. 83597).

Distinctive Characters. Among the ridge-backed subdivision of the genus, *milberti* after birth is made easily recognizable by the large size of the first dorsal and by its position far forward (originating over the axil of the pectoral), and by the free rear corner of the second dorsal which is only about as long as the base. The wide spacing of the dermal denticles and their free edges without definite teeth are diagnostic also.

Description. Proportional dimensions in per cent of total length. Male, 747 mm., from Virginia Beach (U.S. Nat. Mus., No. 104969). Male, 1,496 mm., from Buzzards Bay, Mass. (Harv. Mus. Comp. Zool., No. 35370).

Trunk at origin of pectoral: breadth 12.0, —; height 12.9, —.
Snout length in front of: outer nostrils 4.1, 3.3; mouth 8.2, 7.0.
Eye: horizontal diameter 2.1, 1.5.
Mouth: breadth 9.5, 8.5; height 4.7, 4.1.
Nostrils: distance between inner ends 5.9, 5.5.
Labial furrow length: upper —, 0.5.
Gill opening lengths: 1st 3.0, 2.7; 2nd —, 3.0; 3rd —, 3.2; 4th —, 3.0; 5th 2.5, 2.4.
First dorsal fin: vertical height 10.9, 12.4; length of base 13.3, 11.3.
Second dorsal fin: vertical height 3.1, 2.9; length of base 6.0, 4.1.
Anal fin: vertical height 3.5, 3.9; length of base 5.7, 5.0.
Caudal fin: upper margin 26.7, 27.8; lower anterior margin 12.0, 12.1.
Pectoral fin: outer margin 18.1, 20.4; inner margin 6.3, 6.5; distal margin 14.4, 17.7.
Distance from snout to: 1st dorsal 28.3, 28.7; 2nd dorsal 60.3, 61.3; upper caudal 73.3, 72.2; pectoral 24.7, 20.7; pelvics 49.1, 47.1; anal 61.5, 61.0.
Interspace between: 1st and 2nd dorsals 20.7, 22.5; 2nd dorsal and caudal 7.4, 7.1; anal and caudal 7.4, 7.4.
Distance from origin to origin of: pectorals and pelvics 27.4, 27.1; pelvics and anal 13.3, 14.4.

Trunk comparatively stout, its height at 1st dorsal a little more than ⅕ as great as length to origin of caudal, with dorsal profile rather strongly arched. Midline of back with a low but unmistakable dermal ridge from about under rear tip of 1st dorsal to a point about as far in front of origin of 2nd dorsal as length of base of latter; this also discernible in embryos, though less prominent. Caudal peduncle ⅔ to ¾ as thick as deep. Upper precaudal pit strongly marked, obtuse-ovate in outline, the lower pit much smaller than upper and hardly visible in small specimens. Dermal denticles widely spaced, seldom if ever overlapping, the skin exposed between them, blades thick, strongly convex antero-posteriorly, usually with 5 high, moderately sharp ridges separated by round-bottomed valleys, their apical margins not definitely toothed but at most slightly sinuous opposite the ridges; pedicels very short.[83]

Head about 1/4 of total length or a little less, and very broad forward, its breadth being nearly as great at eyes as at 1st gill opening. Snout broad-ovate, relatively more obtuse in large specimens than in small, its length in front of a line connecting outer ends of nostrils a little more than 1/2 as great as distance between inner ends of latter, and length in front of mouth about 1/3 of length of head to 5th gill opening. Eye approximately circular, noticeably small in adult, its diameter varying in medium-sized and large specimens from a little more than 1/5 to a little more than 1/4 as great as distance between nostrils, but relatively larger in young, its diameter being about as long as the 3rd gill opening in late-term embryos, about 1/2 to 2/3 that long in newborn and in adults; its anterior edge about opposite front of mouth. Gill openings nearly evenly spaced, their outlines nearly straight or slightly sinuous, the 3rd (very slightly the longest) a little more than 1/2 as long as distance between nostrils, the 5th slightly the shortest, the 4th above origin of pectoral. Nostril strongly oblique, its inner end nearer to mouth than to tip of snout by a distance a little greater than its own length, its anterior margin expanded in obtusely angular outline, subacute at apex, and weakly crested opposite latter. Mouth very broadly ovate or nearly arcuate, its height approximately 1/2 its breadth (this proportion varies somewhat from specimen to specimen). Upper labial furrow about 1/4 as long as diameter of eye.

Teeth $\frac{14\text{ to }16-2-14\text{ to }16}{12\text{ to }15-1-12\text{ to }15}$;[84] uppers broadly triangular, their margins regularly but finely serrate from tip to base, the first 2 erect and nearly symmetrical, but subsequent teeth increasingly oblique, with inner margins slightly convex and outer margins either evenly concave or slightly subangular in contour, the 9th or 10th and subsequent teeth decreasing successively in size and height relative to breadth; lower teeth erect, symmetrical, with narrow triangular cusps on broadly expanded bases, the edges of cusps more finely serrate than those of uppers, the bases smooth-edged, or at most slightly irregular.

First dorsal origin about over axil of pectoral, its vertical height increasing relatively with growth from a little greater than distance from eye to 1st gill opening in embryo to about as great as distance from eye to 2nd gill opening when newborn and to about as great as from eye to 3rd gill opening in specimens of medium size; apex also more broadly rounded in embryo but very narrowly so in adult, the free rear corner a little more than 1/3 as long as base, the midpoint of base only about 2/5 (40%) as far from axil of pectoral as from origin of pelvics. Second dorsal about 1/3 as long at base as 1st and relatively much lower, its origin about over origin of anal, its posterior margin only weakly concave, its free rear corner moderately tapering and only about as long as the base. Caudal a little more than 1/4 of total length, the terminal sector between 1/4 and 1/5 the length of fin, the

83. Previous accounts (Radcliffe, Bull. U.S. Bur. Fish., *34*, 1916: 257; Springer, Proc. Fla. Acad. Sci., *3*, 1939: 23) locate this loose spacing only along the upper sides; but it extends over the trunk generally, below as well as above, in the specimens we have examined.

84. In one Woods Hole specimen (Harv. Mus. Comp. Zool., No. 36031) the tooth count is $\frac{14-1\text{ or }2-14}{10-1-10}$, but the spacing of the lower teeth near the corners of the jaws is irregular, suggesting some abnormality.

tip narrowly rounded in adults but more broadly rounded in young, the lower lobe (expanded lower anterior corner) about 2/5 (40 to 44%) as long as upper and with subacute tip, the re-entrant corner, included by the two lobes, well rounded. Distance from origin of caudal to tip of anal nearly or quite as long as base of anal. Anal about as long at base as 2nd dorsal, and about as large in area, but with subacute apex and much more deeply concave posterior margin, its free rear corner about 4/5 as long as base. Distance from origin of anal to tips of pelvics a little longer than base of anal. Pelvics a little longer at base than anal. Pectoral nearly as long (87 to 97%) as head to origin of pectoral in large specimens but relatively somewhat shorter (about 73%) in small, a little less than twice as long as broad, the outer margin only weakly convex, distal margin only weakly concave, the corners very narrowly rounded in adult but somewhat more broadly so in young.

Color. Varying from slate-gray to brownish gray or brown above, perhaps depending on color of the environment; a pale tint of the same hue, or whitish, below; fins without any conspicuous markings. When the shark is newly caught some of the dermal denticles may be brilliant blue, at least in Florida specimens.

Size. The usual size at birth is said to be about 22 inches (weight about 2½ pounds). Maturity is attained at about six feet and it appears that very few reach as great a length as eight feet; 7 feet 10 inches is the longest of which we find unquestionable record.[85] If one about three feet long, taken near New York on June 9, was born the previous summer or early autumn, as seems probable,[86] *milberti* may be expected to grow by about 10 inches during the first winter, for the newborn young of 25 to 26 inches have been taken in that general region in September. But nothing whatever is known of the subsequent rate of growth.

Usual weights of adults are about 100 pounds at six feet, 125 to 130 pounds at about seven feet, 200 pounds at about 7 feet 8 inches.

Developmental Stages. Presumably *milberti* is viviparous, but whether the young develop placental connection with the mother has not yet been definitely established. Embryos resemble the adults in general; 6 to 13 are recorded in a litter, the usual number being 8 to 12, with the two sexes about equally represented.

Habits. Although this is undoubtedly the most abundant member of its genus in season along the middle Atlantic coast of the United States, and the one most often seen, knowledge of its habits is scant. Certainly it is littoral rather than pelagic, for considerable numbers enter shallow bays, harbors and river mouths; hence, they are often harpooned or taken in pound nets, and it is said to be the only large shark that regularly visits the small bays on the north shore of Long Island, New York, as it and others also do the much more

85. Two specimens of *milberti* of 8 feet 6 inches have been reported from New Jersey, but by name only (Fowler, Proc. Acad. Nat. Sci. Philad., 72, 1921: 386); an old report (Baird, Rep. Smithson. Instn. [1854], 1855: 352) that the largest in New Jersey waters are of about nine feet may not have been based on actual measurements; and the report of a North Carolina *milberti* of 9 feet 2 inches (Smith, Bull. N. C. geol. econ. Surv., 2, 1907: 35) actually may have referred to some other shark, so far as the brief account goes; perhaps to the larger-growing *obscurus*.

86. Nichols and Breder, Zoologica, N. Y., 9, 1927: 16.

extensive enclosed waters along the south shore of Long Island. Similarly, its representative in the eastern Atlantic is said to enter the larger of the Venetian canals and the neighboring lagoons adjacent to the Mediterranean.[87] We have looked down from a beach side bluff on a middle-sized one swimming so close to the tide line that its dorsal fin was necessarily exposed. Correspondingly, we have never heard of one taken more than a short distance out from the land. But it appears never to enter fresh water. Although it is an inshore species, as a rule it is only when crossing some shoal that the Brown Shark shows itself at the surface.

It feeds chiefly on fish and on mollusks and crustacea also, its diet depending on what may be available for it locally. Near New York, for example, flounders (*Pseudopleuronectes*) are reported as predominant in its stomach contents, with an occasional eel and crab; others taken at Woods Hole have been found to have fed on amphipod crustacea, as well as on the bivalve mollusk *Yoldia*, which is plentiful in 2 to 10 fathoms in the general vicinity. Still others off the east coast of Florida had eaten *Octopus* chiefly, and also small fish and crabs.[88] Skates and Dogfish are listed for it, and it is able to capture fast-swimming fishes also, for bonito (*Sarda*), weakfish (*Cynoscion*), mackerel, menhaden (*Brevoortia*) and pinfish (*Lagodon*) are included in its known diet. But there is no reason to suppose it ever attacks larger prey.

In the bays of Long Island, New York, its young are born from June until August. It seems to be chiefly for that purpose that it enters those very shoal waters, for a great majority of the adults taken there are females, males being very few in number and reported only for August. A large proportion of the adult females that visit the shallow bays of Long Island carried embryos nearly ready for birth. Newborn specimens have often been taken there in summer and early autumn, as well as in Chesapeake Bay in September, but on the other hand, no young ones have been reported from Florida, although some of the large females taken there carry embryos. These facts are evidence that the young are produced chiefly in the northern part of its range. It seems that this applies equally to such *milberti* as visit the Gulf of Mexico, for the only Texas record is for a newborn specimen taken in early August.

Relation to Extralimital Species. Milberti is so closely allied to *azureus* Gilbert and Starks, 1903, of Ecuador and the Pacific coast of Central America that the latter has been classed by some as probably identical with it.[89] But the most recent illustration of *azureus*[90] shows it to be easily separated from *milberti*, the origin of its second dorsal being considerably anterior to that of the anal, instead of about over the latter, to mention only the most obvious difference between the two.

Relation to Man. Milberti is reported as the most abundant of the commercially valuable sharks taken off southeastern Florida. Some are also caught by anglers, and a few

87. Nardo, Sopra due Sp. Pesci Nuov. Venet., 1853: 15.
88. Personal communication from Stewart Springer.
89. Garman, Mem. Harv. Mus. comp. Zool., *36*, 1913: 133; Meek and Hildebrand, Field Mus. Publ. Zool., *15* (1), 1923: 38.
90. Beebe and Tee-Van, Zoologica, N. Y., *26*, 1941: 109, fig. 18.

sportsmen have thought it worth special pursuit with the harpoon.[91] We have never heard any rumor of its molesting bathers, nor would this be likely, judging from the nature of its prey.

Range. Western Atlantic, from southern New England to southern Florida, Louisiana and southern Brazil. It has been reported also from various localities in the Mediterranean, southern coast of Spain (but whether inside or outside the Straits of Gibraltar is not stated), the Canaries and Cape Verdes. But we still await actual comparison of specimens from the two sides of the ocean.[92]

Occurrence in the Western Atlantic. Owing to a long-standing tendency to call any *Carcharhinus* encountered on the middle Atlantic coast of the United States a Dusky Shark, *milberti* has been reported so often as *obscurus* that little dependence can be placed on published locality records for it unless these are accompanied by some evidence of actual identity more convincing than the mere name. However, information is at hand to show that it is a regular seasonal migrant along the east coast of the United States. Thus it is present along both coasts of southern Florida from December to March, being perhaps the most abundant of the species caught commercially at Salerno on the east coast at that season,[93] and it visits the coasts of New Jersey and New York regularly only in summer though in such numbers that it has been repeatedly described as "abundant" and is to be seen most any summer day in the bays of New Jersey.[94] Recent reports of 305 sharks being harpooned in Great South Bay, Long Island, during the summers of 1911 to 1927 (almost all being *milberti*), of 46 being taken in one summer, and of 14 being harpooned there in one day (August 11, 1906),[95] give a more precise indication of the actual numbers concerned. It is rather common off Rhode Island also during the warm months, occasionally entering Narragansett Bay. Likewise it visits the Buzzards Bay–Vineyard Sound–Nantucket Sound region yearly, but so much less abundantly that the number taken near Woods Hole in an average summer probably does not exceed six or seven. And Cape Cod so sharply marks the usual limit to its northerly dispersal that there is no reliable record of it for the Gulf of Maine,[96] for the fishing banks off its mouth or for Nova Scotia. In the vicinity of New York its season of maximum abundance (mostly females as noted above) is from mid-June to mid-September, the latest for Sandy Hook Bay being October 19; extremes reported for it at Block Island are May and November.

91. For a vivid account of harpooning *milberti* in the bays of Long Island, New York, see Nichols and Murphy (Brooklyn Mus. Sci. Bull., *3* [1], 1916: 16).

92. The view is generally held that the *milberti* of the two sides of the Atlantic are one species. Although one from southern Spain, described by Rey (Fauna Iberica, Peces, *1*, 1928: 346), agreed with American specimens as to fins and teeth, the account does not state whether or not it had a mid-dorsal ridge; its denticles, too, were more closely spaced, and their margins more definitely dentate than in those we have examined.

93. Knowledge that *milberti* occurs around southern Florida dates only from the recent development of the local fishery; the only previous Florida record for it, and that by name only, was for the Indian River (Goode, Proc. U.S. nat. Mus., *2*, 1879: 121).

94. Baird, Rep. Smithson. Inst. (1854), *9*, 1855: 352.

95. Rockwell, Brooklyn Mus. Quart., *3*, 1916: 160–167; Thorne, Bull. N. Y. zool. Soc., *31*, 1928: 114.

96. An early statement that it ranges as far as New Hampshire (DeKay, Zool. N. Y., 4, 1842: 350) seems not to have had any factual basis.

Apparently its migrations between its southern wintering and northern summering grounds on the Atlantic coast of the United States are comparatively direct, for it has never been reported in so great abundance anywhere along the intervening sector, although it is occasionally taken on the coast of Delaware, and in some numbers in autumn in Chesapeake Bay.[97] Although our Study Material includes one October specimen from the Virginia coast a few miles south of the entrance to the Bay,[98] only occasional specimens, whose identities are well attested, have been taken off North Carolina. Although it must pass and repass the South Carolina coast twice yearly in its seasonal migrations, the only report of it there is by name only.[99]

The status of *milberti* in the Gulf of Mexico and to the southward continues doubtful. The only record of it for the Gulf is a Texan specimen (see Study Material, p. 368); so far as we know the only records of it from the Caribbean or West Indies are represented by a pair of jaws from the coast of Nicaragua,[100] the head and skin of a Cuban specimen listed above, and a nominal report from Yucatán, with no further clue to identity.[101] Nor is it likely that a shark frequenting shoal inshore waters so regularly would have been overlooked throughout the whole of such an extensive area if it occurred there in numbers at all rivaling those that visit the Atlantic coasts of the United States. Hence, a very small (650 mm.) specimen taken near Rio de Janeiro[102] seems more likely to have been a stray visitor from the north[102a] than a representative of a local southern Brazilian center of population; but it is still possible that it was a representative of such a population. It is not known from Bermuda.

Synonyms and References:[103]

1. Western Atlantic:

Carcharias (*Prionodon*) *milberti* Müller and Henle, Plagiost., 1841: 38, pl. 19, fig. not numbered (descr., ill., teeth, N. York); Duméril, Hist. Nat. Poiss., *1*, 1865: 360 (descr., N. York, Medit.); Günther, Cat. Fish. Brit. Mus., *8*, 1870: 363, footnote 5 (ref. only).

Carcharias ceruleus DeKay, Zool. N. Y., 4, 1842: 349, pl. 61, fig. 200 (descr., ill., N. York to N. Hampshire); Baird, Rep. Smithson. Instn. (1854), *9*, 1855: 352, extra (N. Jersey, ident. probable because of local abund.); Ann. Rep. St. Cab. nat. Hist. N. Y. (1855), 1858: 64 (N. York); Jordan and Gilbert, Bull. U.S. nat. Mus., 16, 1883: 873 (ident. probable from brief descr., New England to Florida); Proc. U.S. nat. Mus., *5*, 1883: 245 (Gulf of Mexico); Nelson, Rep. St. Geol. N. J., *2* (2), 1890: 660 (N. Jersey loc.).

Lamna caudata DeKay, Zool. N. Y., 4, 1842: 354, pl. 62, fig. 205 (descr., recognizable ill., Rhode Island); Linsley, Amer. J. Sci., *47*, 1844: 77 (Rhode Island, N. York); DeKay, Rep. St. Cab. nat. Hist. N. Y. (1855), 1858: 64 (name only).

97. For numbers caught there, see Hildebrand and Schroeder (Bull. U.S. Bur. Fish., *43* [1], 1928: 48).

98. An earlier record of "*milberti*" for Virginia (Linton, Bull. U.S. Bur. Fish., *24*, 1905: 341) was based on so large a specimen (longer than nine feet) that it may have referred to some other *Carcharhinus*.

99. Reported by Gunter (Amer. Midl. Nat., *28* [2], 1942: 28) as entering the Cooper's River and other rivers near Charleston, South Carolina.

100. Fowler, Proc. Acad. nat. Sci. Philad., *75*, 1923: 24.

101. Bean, Bull. U.S. Bur. Fish., *8*, 1890: 206.

102. Ribeiro, Fauna brasil. Peixes, *2* (1) Fasc. 1, 1923: 51, pl. 3; identity established by the excellent illustrations.

102a. Or its mother may have been the stray visitor, since it cannot have been born long previous to its capture.

103. Citations for the western and eastern Atlantic are listed separately, since it is still an open question as to whether the Mediterranean form is identical with the American.

Squalus (*Carcharinus*) *milberti* (in part) Gray, List. Fish. Brit. Mus., 1851: 45 (refs., but specimen from India some other species).

Squalus (*Carcharinus*) *caeruleus* Gray, List. Fish. Brit. Mus., *1*, 1851: 44 (ref.).

Squalus (*Carcharinus*) *caudata* Gray, List. Fish. Brit. Mus., *1*, 1851: 44 (ref.).

Squalus coeruleus Gill, Proc. Acad. nat. Sci. Philad., Addend., 1861: 59 (incl. in list); Abbott, Rep. N. J. geol. econ. Surv., 1868: 828 (N. Jersey, not seen).

Squalus milberti Gill, Proc. Acad. nat. Sci. Philad., Addend., 1861: 59 (in list), and subsequent eds.

Eulamia milberti Gill, Ann. Lyc. nat. Hist. N. Y., 7, 1862: 410 (name); Proc. Acad. Nat. Sci. Philad., 1864: 262 (name, discuss.); Rep. U.S. Comm. Fish. (1871–72), 1873: 813; Fowler, Proc. Acad. nat. Sci. Philad., *60*, 1908: 62 (N. Jersey); Science, N. S. *30*, 1909: 815 (N. Jersey); Proc. Acad. nat. Sci. Philad., *63*, 1911: 5 (Maryland); Copeia, 30, 1916: 36 (in list); Copeia, 31, 1916: 41 (N. Jersey); Proc. Acad. nat. Sci. Philad., *69*, 1917: 122 (Northeast R., Maryland); Proc. Acad. nat. Sci. Philad., *71*, 1920: 292 (N. Jersey); Proc. biol. Soc. Wash., *32*, 1919: 72 (Delaware); Proc. biol. Soc. Wash., *33*, 1920: 144 (N. Jersey); Proc. Acad. nat. Sci. Philad., *72*, 1921: 386 (meas., weights, ident. perhaps doubtful because of large size, N. Jersey); Proc. Acad. nat. Sci. Philad., *74*, 1922: 3 (N. Jersey); Proc. Acad. nat. Sci. Philad., *75*, 1923: 24 (jaws, Caribbean coast of Nicaragua); Monogr. Acad. nat. Sci. Philad., 7, 1945: 95, 263 (Beaufort, N. Carolina; Florida); Bigelow and Schroeder, Guide Comm. Shark Fish., Anglo Amer. Caribb. Comm., Wash., 1945: 84, fig. 26 (descr., habits, range, ill.).

Eulamia coerulea Gill, Proc. Acad. nat. Sci. Philad., 1863: 333 (Massachusetts).

Carcharias milberti Ribeiro, Fauna brasil. Peixes, *2* (1) Fasc. 1, 1923: Append., 50a, pl. 3 (descr., excellent ill., Rio de Janeiro); Jordan, Manual Vert. Anim. NE. U.S., 1929: 10 (general); Jordan, Evermann and Clark, Rep. U.S. Comm. Fish. (1928), 2, 1930: 16 (C. Cod to Florida, Medit.).

Carcharias (*Eulamia*) *coeruleus* Jordan and Gilbert, Proc. U.S. nat. Mus., *5*, 1883: 245 (ref., no loc.).

Carcharinus[103a] *milberti* Jordan and Gilbert, Bull. U.S. nat. Mus., 16, 1883: 22 (ref., C. Cod to Medit.); Smith, Bull. U.S. Fish Comm., 17, 1898: 88 (Woods Hole); Bean, Rep. For. Comm. N. Y., 1902: 377 (Long Island, N. York); Bull. N. Y. St. Mus., *60*, Zool. 9, 1903: 26 (descr., near N. York); Fowler, Rep. N. J. Mus. (1905), 1906: 62 (descr., N. Jersey); Tracy, Rep. R. I. Comm. inl. Fish., 1906: 45 (ident. probable because of loc., Rhode Island); Smith, N. C. geol. econ. Surv., *2*, 1907: 34 (N. Carolina, but doubtful if 9 ft. 2 in. spec. was this species because so large); Fowler, Rep. N. J. Mus. (1906), 1907: pl. 73 (ill.); Kendall, Occ. Pap. Boston Soc. nat. Hist., 7 (8), 1908: 4 (south. New Eng.); Tracy, Rep. R. I. Comm. inl. Fish., 1910: 59 (ident. probable because of local abund., Rhode Island); Garman, Mem. Harv. Mus. comp. Zool., 36, 1913: 133 (descr.); Sumner, Osburn and Cole, Bull. U.S. Bur. Fish., *31* (2), 1913: 735 (Woods Hole); Fowler, Copeia, 2, 1914: 2 (in list, Maryland); Copeia, 13, 1914: [2] (N. Jersey); Nichols, Copeia, 35, 1916: 72–73 (lengths, weights, Long Island, N. York); Nichols and Murphy, Brooklyn Mus. Sci. Bull., *3* (1), 1916: 14, pl. 1 (descr., breeding, habits, food, season, abund., Long Island, N. York); Brooklyn Mus. Quart., *3*, 1916: 151 (N. York); Radcliffe, Bull. U.S. Bur. Fish., *34*, 1916: 257, pl. 39, fig. 4 (descr., denticles, teeth, N. Carolina); Rockwell, Brooklyn Mus. Quart., *3*, 1916: 162 (abund., season, harpoon fishery, Long Island, N. York); Smith, Nat. Hist. N. Y., *16*, 1916: 348 (Atlant. coast, U.S.); Thorne, Copeia, 35, 1916: 69–71 (abund., season, size, weight, sex, embryos, near N. York); Nichols, Bull. Amer. Mus. nat. Hist., *37*, 1917: 874 (Woods Hole specs.); Copeia, 53, 1918: 13 (states that majority of the sharks taken around Long Island, New York, are female *milberti*); Meek and Hildebrand, Field Mus. Publ. Zool., *15* (1), 1923: 38 (descr., discuss., concludes the Pacific *azureus* Jordan and Gilbert, 1882, probably identical); Breder, Copeia, 138, 1925: 2 (Sandy Hook Bay, N. York); Nichols, Copeia, 140, 1925: 21 (Woods Hole); Breder, Copeia, 153, 1926: 122 (Sandy Hook Bay); Nichols and Breder, Zoologica, N. Y., *9*, 1927: 15 (local dist., life hist.); Thorne, Bull. N. Y. zool. Soc., *31*, 1928: 114 (Long Island, N. York); Hildebrand and Schroeder, Bull. U.S. Bur. Fish., *43* (1), 1928: 48 (descr., weights, lengths, numbers, Chesapeake Bay); Breder, Field Bk. Mar. Fish. Atlant. Coast, 1929: 16 (general, size at birth); Truitt, Bean and Fowler, Bull. Md. Conserv. Dep., *3*, 1929: 29 (Maryland, from previous records); White, Bull. Amer. Mus. nat. Hist., *74*, 1937: 126 (in Key), pl. 31, fig. a, pl. 42, fig. a, pl. 50, fig. b (spiral valve, heart valves, cartilages

103a. Also often spelled "*Carcharhinus*."

of claspers, New York spec.); Breder, Bull. N. Y. zool. Soc., 41, 1938: 28 (N. York Harbor, in list); Springer, Proc. Fla. Acad. Sci., *3*, 1939: 22 (descr. denticles, abund., season, S. Florida); Lucey, Contr. biol. Lab. Cath. Univ. Amer., 43, 1942: 9 (chemistry of pancreas, Florida); Bomkamp, Contr. biol. Lab. Cath. Univ. Amer., 44, 1942: 4 (chemistry of liver, Florida); Boos, Contr. biol. Lab. Cath. Univ. Amer., 45, 1943: 10 (chemistry of pancreas, Florida); Lunz, Bull. S. C. St. Planning Bd., 14, 1944: 27 (S. Carolina, Florida); Springer, Copeia, 1946: 174 (food, off Salerno, Florida).

Carcharias plumbeus Jordan and Gilbert, Bull. U.S. nat. Mus., 16, 1883: 872 (near N. York and Medit., by ref.).

Carcharhinus (Eulamia) caudatus Jordan, Rep. U.S. Comm. Fish. (1885), 1887: 796 (name, northeast. U.S.).

Carcharhinus (Carcharhinus) milberti Jordan and Evermann, Rep. U.S. Comm. Fish. (1895), 1896: 216 (C. Cod to Florida); Bull. U.S. nat. Mus., 47 (1), 1896: 37 (compar. with other sp., C. Cod to Florida).

Carcharhinus obscurus (in part) Bean, Bull. N. Y. St. Mus., *60*, Zool. 9, 1903: 25 (abund., N. Jersey Bays, probably refers to ***milberti***, but not the descr.).

Carcharhinus platyodon Garman, Mem. Harv. Mus. comp. Zool., *36*, 1913: pl. 3, fig. 4–6 (good ill., mislabeled); not ***Squalus platyodon*** Poey, 1861.

Eulamia plumbeus Fowler, Bull. Amer. Mus. nat. Hist., *70* (1), 1936: 51 (descr., ill. of American specs.; states ***plumbeus*** Nardo has priority over ***milberti***); Proc. Acad. nat. Sci. Philad., *89*, 1937: 304 (size, N. Jersey); Fish Culturist, *21* (9), 1942: 66 (listed, Cuba); Arqu. Zool. Estado São Paulo, *3*, 1942: 128 (listed, Brazil).

Doubtful references:

Eulamia milberti Verrill and Smith, Rep. U.S. Comm. Fish. (1871–1872), 1873: 521 (Woods Hole); Goode, Proc. U.S. nat. Mus., *2*, 1879: 121 (name only, Indian River, Florida, no clue to actual identity).

Carcharias obscurus Bean, T. H., Bull. U.S. Fish Comm., 7, 1888: 152 (probably ***milberti***, because of reported abund., N. Jersey).

Carcharias coeruleus Bean, T. H., Bull. U.S. Fish Comm., *8*, 1890: 206 (name only, Yucatán).

Carcharhinus milberti Evermann and Bean, Rep. U.S. Comm. Fish. (1896), 1898: 239 (nominal, Indian R., Florida, by ref. to earlier report by Goode, 1879); Evermann and Kendall, Rep. U.S. Comm. Fish. (1898), 1900: 48 (nominal, Indian R., Florida); Linton, Bull. U.S. Bur. Fish., *19*, 1901: 426 (name only, food, parasites, Woods Hole); Bull. U.S. Bur. Fish., *24*, 1905: 341 (ident. doubtful because of large size, Virginia); Wilson, Proc. U.S. nat. Mus., *33*, 1907: 360 (name only, parasites, Woods Hole); Linton, Bull. U.S. Fish Comm., *28*, 1910: 1200 (name only, parasites, no loc.); Proc. U.S. nat. Mus., *64* (21), 1924: 3, 5, 6, 12, 33, 36, 38, 40, 46, 56, 87, 90, 92 (name only, parasites, Woods Hole); Wilson, Proc. U.S. nat. Mus., *64* (17), 1924: 11, 12 (name only, parasites, Woods Hole); MacCallum, Proc. U.S. nat. Mus., *79* (26), 1931: 7 (name only, parasites, no loc.); Wilson, Bull. U.S. nat. Mus., 158, 1932: 427, 434, 440, 453, 464, 524 (name only, parasites, Woods Hole region); Nigrelli, Amer. Mus. Novit., 996, 1938: 10 (name only, parasites, no loc.); Gunter, Amer. Midl. Nat., *28* (2), 1942: 316 (name only, rivers of S. Carolina but not into fresh water).

Carcharinus milberti Norris, Plagiost. Hypophysis, 1941: 28.

Carcharinus sp. (probably ***C. milberti***), Norris, Plagiost. Hypophysis, 1941: pl. 9, fig. 33, 34 (brain).

2. Eastern Atlantic:

Squalus plumbeus Nardo, in Oken's Isis, *20* (6), 1827: 483; also sep., as Prod. Ittiol. Adriat., 1827: 9 (brief diagn., Adriatic);[104] Sinon. Modern. Spec. descr. Pesci. St. Chiereghin, 1847: 111 (name, Medit.).

Squalus caecchia Nardo, Prod. Faun. Mar. Venet. Estuario, 1847: 38 (near Venice, not seen).

Squalus, sp. n., Molin, Revist. I. R. Acad. Padova, Trim. 3–4, 1853: 381 (ref. to ***Squalus milberti*** by Nardo, 1853; not seen).

104. Nardo's statement that his ***plumbeus*** "convenit perfecti ***Squalus glaucus*** Bloch si colorem excipetur et forman rostri quae in exemplari nostro rotunda est" would apply equally to any round-snouted ***Carcharhinus***. But the fact that Nardo (Pesc. Publ. com. Nuov. Venez., 1853: 15) later referred his ***plumbeus*** to ***milberti*** justifies Fowler's (Bull. Amer. Mus. nat. Hist., *70* [1], 1936: 50) substitution of ***plumbeus*** for ***milberti*** as the correct specific name, at least of the European form and of the American as well if the two prove to be identical.

Squalus milberti Bonaparte, Cat. Pesc. Europ., 1846: 18 (incl. in list); Nardo, Pesc. Publ. com. Nuov. Venez., 1853: 15 (brief diagn., in lagoons and canals, Venice); Ninni, Ann. Soc. Nat. Modena, *5*, 1870: 66 (nominal, Medit. near Naples).

Prionodon milberti Canestrini, in Cornalia, *et al.*, Fauna d'Ital., 1872: 48 (descr., ref., Adriatic); Doderlein, Prosp. Metod. Pesc. Sicil., *1*; Atti Acad. Palermo, N. S. *6*, 1878–1879: 30 (in Adriatic list); Graeffe, Arb. zool. Inst. Univ. Wien, *28*, 1886: 446 (Medit.).

Carcharias caecchia Ninni, Revist. Critica Pesc. Adriat. descr. St. Chiereghin, 1872: 6–7 (name, Adriatic, not seen).

Carcharias (*Prionodon*) *milberti* Doderlein, Man. Ittiol. Medit., *2*, 1881: 44 (descr., refs., Medit.);[105] Bellotti, Atti Soc. ital. Sci. nat., *33*, 1891: 112 (discuss.).

Carcharias milberti Moreau, Hist. Nat. Poiss. France, *1*, 1881: 335 (ref., Medit.); Perugia, Elenc. Pesc. Adriat., 1881: 52 (in list); Carus, Prod. Faune Medit., *2*, 1889–1893: 512 (locs., Medit.); Moreau, Poiss. France, Suppl., 1891: 2 (descr. of types from N. York, and meas. Medit. embryo, but ill. of teeth of some other species, if correctly drawn); Sicher, Atti Accad. gioenia, (4) *11* (5), 1898: 14 (Medit.); Pietschmann, Ann. naturh. (Mus.) Hofmus. Wien, *21*, 1906: 133 (descr., discuss., south. Medit.).

Prionace milberti Roule, Bull. Inst. Océanogr. Monaco, 243, 1912: 7 (nominal, C. Verde).

Carcharhinus milberti Rey, Fauna Iberica Peces, *1*, 1928: 346 (descr., S. coast of Spain); Bellon and Mateu, Notas Inst. esp. Oceanogr., (2) *53*, 1931: 15 (nominal, Canaries); Borri, Mem. Soc. tosc. Sci. nat., *44*, 1934: 88 (Medit.).

Carcharhinus plumbeus Tortonese, Atti Soc. ital. Sci. nat., 77, 1938: 299 (discuss. specs. from Medit.).

Carcharhinus nicaraguensis (Gill and Bransford), 1877

Lake Nicaragua Shark

Figure 70

Study Material. Immature male, about 1,511 mm. long, from San Carlos, Lake Nicaragua, caught by Maj. C. M. Duke, U.S.A. (Harv. Mus. Comp. Zool., No. 35896); photographs of two females, fresh-caught, by President Don Anastasio Somoza of Nicaragua and Capt. W. B. Bunker; three males of about 1,330 mm., 1,568 mm. and 1,710 mm., from Lake Nicaragua (U.S. Nat. Mus., No. 120371, 120372, 120373).

Distinctive Characters. Nicaraguensis very closely resembles *leucas,* of which it is a landlocked representative, but is perhaps separable from *leucas* by the following characters. The anterior margin of the eye is posterior to the front of the mouth by a distance equal to half its own diameter in *nicaraguensis* (a little anterior to front of mouth in *leucas*); the gill openings are relatively somewhat longer in *nicaraguensis,* the third being nearly ⅔ as long as the distance between the nostrils (in *leucas* the third is a little less than half that long); the free tip of the second dorsal is about two-thirds as long as its base in *nicaraguensis* (only about half that long in *leucas*).

Description. Proportional dimensions in per cent of total length. Male, 1,511 mm., from Lake Nicaragua (Harv. Mus. Comp. Zool., No. 35896). Male, 1,568 mm., from Lake Nicaragua (U.S. Nat. Mus., No. 120372).

Trunk at origin of pectoral: breadth 14.0, 15.3; height 14.0, 15.3.
Snout length in front of: outer nostrils 2.3, 3.1; mouth 6.0, 6.9.
Eye: horizontal diameter 1.1, 1.0.

105. See Doderlein for additional Mediterranean citations in publications not accessible to us.

Mouth: breadth 9.5, 10.8; height 5.0, 4.9.
Nostrils: distance between inner ends 6.3, 6.9.
Labial furrow length: upper 0.3, 0.3.
Gill opening lengths: 1st 4.1, 3.3; 2nd 4.2, 3.6; 3rd 3.8, 4.0; 4th 3.7, 3.7; 5th 2.7, 3.1.
First dorsal fin: vertical height 12.7, 12.3; length of base 13.2, 12.2.
Second dorsal fin: vertical height 4.9, 4.7; length of base 5.9, 6.4.
Anal fin: vertical height 5.3, 5.4; length of base 6.0, 5.3.
Caudal fin: upper margin 27.1, 29.6; lower anterior margin 13.2, 13.6.
Pectoral fin: outer margin 21.9, 23.6; inner margin 6.4, 6.5; distal margin 19.6, 20.4.
Distance from snout to: 1st dorsal 27.6, 27.9; 2nd dorsal 61.0, 56.6; upper caudal 72.9, 70.4; pectoral 22.3, 20.4; pelvics 47.7, 44.9; anal 61.4, 58.1.
Interspace between: 1st and 2nd dorsals 21.8, 19.9; 2nd dorsal and caudal 7.0, 7.9; anal and caudal 5.8, 5.9.
Distance from origin to origin of: pectoral and pelvics 28.6, 28.0; pelvics and anal 14.1, 12.2.

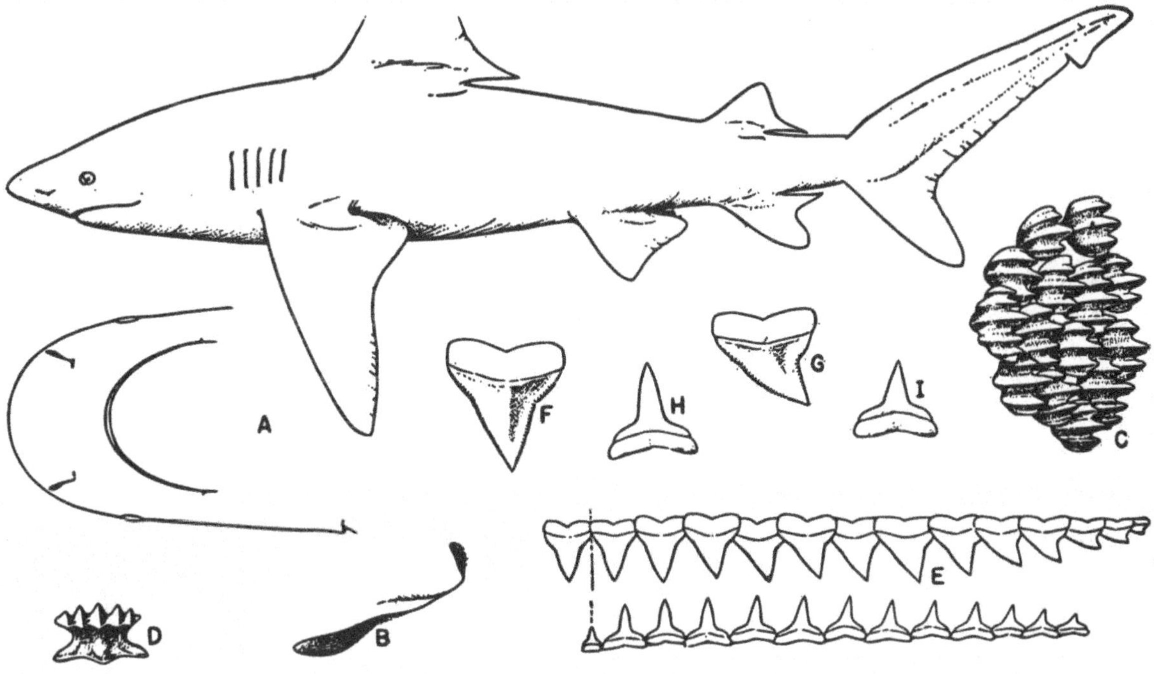

FIGURE 70. *Carcharhinus nicaraguensis*, immature male, 1,511 mm. long (Harv. Mus. Comp. Zool., No. 35896). *A* Anterior part of head from below, about 1/7 natural size. *B* Right-hand nostril, about 0.8 x. *C* Dermal denticles, about 17 x. *D* Apical view of dermal denticle, about 17 x. *E* Upper and lower teeth, left-hand side, about 0.6 x natural size. *F* Third upper tooth. *G* Ninth upper tooth. *H* Second lower tooth. *I* Eighth lower tooth. *F–I*, about 1.2 x.

Trunk moderately stout, as in *leucas*, its height at 1st dorsal about 1/5 its length to origin of caudal. Back without trace of median dermal ridge. Upper precaudal pit strongly marked, semi-lunar, the lower smaller but still apparent. Dermal denticles overlapping in varying degrees, with skin more or less exposed between their blades, thick, strongly arched and rising steeply, about as broad as long, usually with 3–5 high, sharp-topped ridges, 3, 4, or 5 blunt teeth, the median much the longest; pedicels very broad and stout as in *leucas*.

Head strongly flattened above and very wide anteriorly, its breadth about 7/8 as great at eyes as at origin of pectoral. Snout very broadly rounded and shorter than in any other local *Carcharhinus*, its length in front of a line connecting outer ends of nostrils only a little more than 1/3 as great as distance between inner ends of latter, the length in front of mouth about 1/4 to 1/3 (27 to 33%) that of head to origin of pectoral. Eye noticeably small as in *leucas*, its diameter about 1/5 as great as distance between nostrils. Nostril strongly oblique, its inner end only a little more than 1/2 as far from mouth as from tip of snout, its anterior margin expanded toward inner end in subrectangular outline, much as in *leucas*. Gill openings about evenly spaced, the 1st to 3rd between 1/2 and 2/3 as long as distance between nostrils or about 3 1/2 times as long as diameter of eye, the 5th about 2/3 as long as 1st, the 3rd over origin of pectoral. Mouth broadly rounded, its height about 1/2 its breadth. Upper labial fold so short as to be easily overlooked.

Teeth $\frac{\text{12 or 13—0 or 1—12 or 13}}{\text{12—1—12}}$, shaped much as in *leucas*; uppers broadly triangular, their edges moderately serrate from tip to base, the 1st and 2nd teeth symmetrical, erect, their edges slightly concave, but successive teeth increasingly oblique along the jaw with outer edges increasingly concave in rounded or subangular contour, the 9th to 13th with inner edge convex, and successively smaller, the outermost very small; lower teeth erect, symmetrical, with narrow triangular cusps on broadly expanded bases, the cusps finely serrate and bases partially so, the 9th to 12th successively smaller, with relatively lower cusp; a very small symmetrical tooth at symphysis on lower jaw, 1 present or not in upper.[106]

First dorsal of about same size relatively as in *leucas*, with a similar subacute apex, but somewhat more erect in specimens seen, and with a relatively somewhat longer free rear corner (about 1/3 as long as the base), its origin a little anterior to axil of pectoral. Second dorsal about 1/2 as long at base as 1st, as in *leucas*, and of the same general form except that the free rear corner is about 2/3 as long as base (only about 1/2 that long in *leucas*), and the rear margin somewhat more deeply concave, its origin slightly but definitely anterior to origin of anal. Caudal about 28% of total length, of same shape as in *leucas*. Distance from lower origin of caudal to tip of anal about 1/2 as long as base of latter. Anal about as long at base as 2nd dorsal, and nearly 1 1/2 times as long on anterior margin, its shape as in *leucas*. Distance from origin of anal to tips of pelvics about 1/2 as long as base of former. Pelvics a little longer along anterior margins than 2nd dorsal and a little shorter

106. This tooth is recorded for the type specimen (Gill and Bransford, Proc. Acad. nat. Sci. Philad., 1877: 190) but is lacking in one which we have studied.

than anal, as in *leucas*. Pectoral very nearly as long as head and thus a little longer than in *leucas*, but of same general shape, a little more than ½ as broad as long, the outer margin nearly straight except near tip, the distal margin a little more deeply concave toward base than in *leucas*.

Color. Dark mouse-gray above after preservation in salt, grayish to yellowish white below, with lower surfaces of pectorals dusky at tips; photographs at time of capture show the lower surface a clearer white, the lower pectoral tips and tip of lower caudal dusky or nearly black. It is said that the bellies of large specimens may have a reddish bronze tinge.

Size. The fact that the specimens we have seen are immature, although up to 5 feet long, is in line with information reaching us from Nicaragua that the average size of those caught around San Carlos is 6 to 6½ feet. We have a definite report of one of 8 feet, and they are rumored to reach 10 feet. One of 4 feet is reported as weighing about 50 pounds. Specimens of 62 and 67 inches (1,568 mm. and 1,710 mm.) in the United States National Museum weighed 73 pounds and 98 pounds respectively when caught.

Developmental Stages. Embryos have not been seen.

Remarks. *Nicaraguensis* was classed by Garman,[107] and more recently by Meek and Hildebrand,[108] as a synonym of *milberti*, in spite of the fact that firsthand accounts had credited it with a much shorter snout. Actually it is so close to *leucas* that it is undoubtedly an offshoot of the latter. But the several small differences, enumerated above, seem sufficient for retention of a separate name for it, especially since it is the only shark that is known to have adapted itself permanently to life in fresh water. Nor is it astonishing that this should have happened, for *leucas* has been reported in fresh water far up rivers elsewhere (p. 341).

Habits. Very little is known of the habits of this fresh-water shark, except that it comes commonly into very shoal water although it is seldom actually seen at the surface and that it bites very readily on bait of meat or fish. Presumably it feeds on fish, but no precise information is available as to its diet.[109] Nothing is known about its breeding habits.

Relation to Man. The fins are valued locally for food, and the livers are sold for their vitamin content.

It is reputedly a danger to bathers, as well as to any dog that may venture into the lake. And published accounts of its ferocity appear to be well founded, for a correspondent in whom we have full confidence[110] reports that he has not only seen an attack on a youth swimming at San Carlos but has heard of actual fatalities at other localities around the lake. Very recently the press has reported attacks on bathers and fishermen at Granada, where one of the victims lost an arm, while another lost his right leg and had his left leg injured.[111]

Range. Known definitely only from Lake Nicaragua, its tributaries and outlet.[112]

107. Mem. Harv. Mus. comp. Zool., *36*, 1913: 133.
108. Field Mus. Publ. Zool., *15* (1), 1923: 38.
109. For an account of the fishes of Lake Nicaragua, see Meek (Field Mus. Publ. Zool., 7 [4], 1907: 95–132).
110. Capt. W. B. Bunker.
111. In the *Diario Nuevo*, San Salvador, for April 24, 1944.
112. Jordan, Evermann and Clark (Rep. U.S. Comm. Fish. [1928], 2, 1930: 16) include "Bay of Panama" in its

It would be interesting to know whether a shark of this general type, and of which we received a photograph, recently taken in the Patuca River, Honduras, in fresh water was this same landlocked species, or whether it was simply a *leucas* that had run far upstream, which we think more likely (see discussion, p. 341).

Recent contributors who have caught sharks in Lake Nicaragua[113] report them as plentiful there as well as in the upper reaches of the San Juan River, by which the lake discharges into the Caribbean. It is not known how far downstream its range extends in the San Juan, but since there are heavy rapids in the river it seems likely that the sharks from the lake and upper river are entirely landlocked. According to local reports it also runs at least 40 miles up the Rio Frio, which empties into the lake close to where the latter discharges via the San Juan River. Rumors of its presence in Lake Managua appear to lack foundation. The fact that the specimen shown in Fig. 70 was caught from the pier after only 15 minutes fishing is some indication of the abundance of these sharks, at San Carlos at any rate.[114]

Synonyms and References:

Eulamia nicaraguensis Gill and Bransford, Proc. Acad. nat. Sci. Philad., 1877: 190 (descr., Lake Nicaragua); Eigenmann, Proc. U.S. nat. Mus., *16*, 1893: 54 (name only, San Juan R., Nicaragua); Marden, Nat. Geogr. Mag., *86*, 1944: 183 (fatalities, capture of specs., excellent photos, L. Nicaragua).

Carcharias nicaraguensis Lütken, Vidensk. Medd. naturh. Foren. Kbh., 1879–1880: 65 (ill., meas., descr., San Juan R., Nicaragua); Regan, Biol. Cent. Amer. Pisces, 1906–1908: 183 (size, Lake Nicaragua and San Juan R.); Jordan, Evermann and Clark, Rep. U.S. Comm. Fish. (1928), 2, 1930: 16 (Lake Nicaragua, San Juan R., Panama Bay, but see footnote 112, p. 380).

Carcharhinus nicaraguensis Jordan, Proc. U.S. nat. Mus., *9*, 1887: 556 (name only in list, Cent. Amer.); Meek, Field Mus. Publ. Zool., 7, 1907: 103 (local abund., Lake Nicaragua and San Juan R.); Eigenmann, Rep. Princeton Exped. Patagonia (1896–1899), *3* (4), 1910: 377 (listed); Smith, H. W., Biol. Rev., *11*, 1936: 64 (fresh water); Norman and Fraser, Giant Fishes, 1937: 36 (confined to fresh water).

Carcharhinus (*Carcharhinus*) *nicaraguensis* Jordan and Evermann, Rep. U.S. Comm. Fish. (1895), 1896: 216 (Lake Nicaragua and San Juan R.); Bull. U.S. nat. Mus., 47 (1), 1896: 39 (descr., Lake Nicaragua and San Juan R.).

Carcharinus milberti (in part) Garman, Mem. Harv. Mus. comp. Zool., *36*, 1913: 133 (referred to synonymy of *milberti*); Meek and Hildebrand, Field Mus. Publ. Zool., *15*, 1923: 38 (referred to synonymy of *milberti*); not *Carcharias* (*Prionodon*) *milberti* Müller and Henle, 1841.

Carcharhinus obscurus Lesueur, 1818

Dusky Shark, Shovelnose

Figures 71, 72

Study Material. One female, 996 mm., and 5 males, 993 to 1,560 mm., from the vicinity of Woods Hole (Harv. Mus. Comp. Zool.); 8 males, 970 to 1,500 mm. long,

range, but seemingly not on any factual basis; nor would it be expected there even as a stray, for Lake Nicaragua drains into the Atlantic and not into the Pacific.

113. We have received several personal communications on this subject.

114. Smith's (Science, *22*, 1893: 166) statement that unnamed sharks are abundant in Lake Nicaragua no doubt refers to this species.

from the vicinity of Woods Hole, not preserved; also measurements of 5 adult females, 3,115 to 3,465 mm., from Englewood, Florida,[115] one of which was gravid; and a newborn specimen, 848 mm., taken off Bay Chaland, Louisiana, in August 1930 (U.S. Bur. Fish., No. 1360, in U.S. Nat. Mus.).

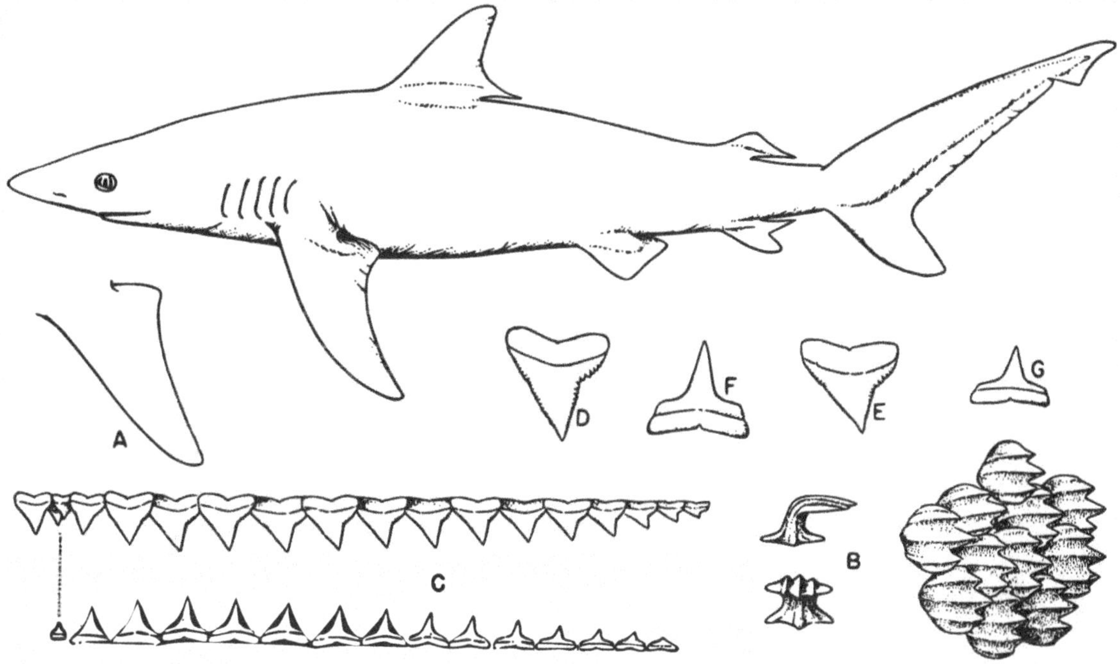

FIGURE 71. *Carcharhinus obscurus*, female, 996 mm. long, from Woods Hole, Massachusetts (Harv. Mus. Comp. Zool., No. 35312). *A* Right-hand pectoral fin. *B* Dermal denticles, about 16 x. *C* Upper and lower teeth, left-hand side, 1.2 x. *D* Third upper tooth. *E* Ninth upper tooth. *F* Fourth lower tooth. *G* Tenth lower tooth. *D–G*, about 2.6 x.

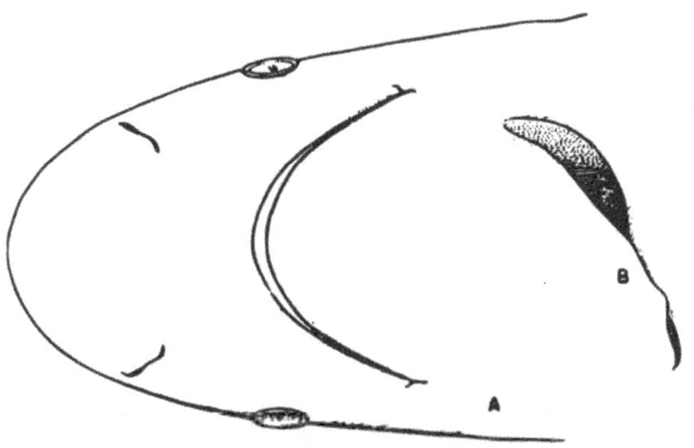

FIGURE 72. *Carcharhinus obscurus*, pictured in Fig. 71. *A* Anterior part of head from below, about 0.4 x natural size. *B* Left-hand nostril, about 2.3 x.

115. Contributed by Stewart Springer.

Distinctive Characters. Among ridge-backed members of the genus, *obscurus* has frequently been confused with *milberti.* But it is easily separable from the latter by the following characters: the first dorsal is somewhat less in vertical height than the distance from the eye to the first gill opening and it originates over or a little posterior to the inner corner of the pectoral (much larger and originating over the axil in *milberti*); the free rear corner of its second dorsal is about 1.5 times as long as the vertical height (only about as long as the vertical height in *milberti*); its dermal denticles are regularly overlapping and with toothed margins. It is easily separable from the other ridge-backed species *falciformis* and *floridanus* by the shape of the second dorsal and anal fins (see Key, p. 323); from *longimanus* by the shape of the first dorsal and by the remoteness of the tip of its anal from the caudal.

It most closely resembles the newly described *springeri*, Bigelow and Schroeder, 1944 (p. 404), but differs from it in a number of respects, most noticeably: (a) in a considerably smaller eye relative to the lengths of the gill openings; (b) that the anterior margin of its nostril is not definitely lobed; (c) the distance from the tips of its pelvics to the origin of its anal is relatively longer (about 1.3 times as long as the base of the anal in *obscurus*, but only about 0.7 that long in *springeri*); (d) the outer corner of its pelvic is considerably more obtuse than a right angle, or about 115° (only about a right angle in *springeri*); and (e) its denticles usually have fewer teeth than those of *springeri* at corresponding stages in growth. The teeth of the two species also differ somewhat in detail (cf. Fig. 71 C–G with 78 E–I).

Description. Proportional dimensions in per cent of total length. Male, 993 mm., from Buzzards Bay, Mass. (Harv. Mus. Comp. Zool., No. 35371). Female, 996 mm., same locality (Harv. Mus. Comp. Zool., No. 35312).

Trunk at origin of pectoral: breadth 11.9, 13.7; height 13.3, 12.9.

Snout length in front of: outer nostrils 3.5, 4.0; mouth 7.9, 7.9.

Eye: horizontal diameter 1.9, 1.7.

Mouth: breadth 7.9, 9.2; height 4.7, 4.9.

Nostrils: distance between inner ends 6.2, 6.3.

Gill opening lengths: 1st 2.6, 2.9; 2nd 2.8, 3.4; 3rd 3.2, 3.5; 4th 3.2, 3.5; 5th 2.7, 2.9.

First dorsal fin: vertical height 7.6, 8.6; length of base 9.8, 9.6.

Second dorsal fin: vertical height 2.3, 2.6; length of base 4.0, 3.5.

Anal fin: vertical height 3.1, 3.4; length of base 3.9, 4.5.

Caudal fin: upper margin 27.4, 28.8; lower anterior margin 11.2, 11.4.

Pectoral fin: outer margin 17.4, 18.5; inner margin 5.0, 5.2; distal margin 13.0, 14.1.

Distance from snout to: 1st dorsal 30.7, 32.0; 2nd dorsal 62.2, 61.2; upper caudal 72.6, 71.2; pectoral 22.6, 23.7; pelvics 50.2, 48.5; anal 62.6, 61.2.

Interspace between: 1st and 2nd dorsal 22.7, 21.2; 2nd dorsal and caudal 6.4, 6.1; anal and caudal 6.1, 5.9.

Distance from origin to origin of: pectoral and pelvics 27.8, 26.1; pelvics and anal 12.7, 12.8.

Proportional dimensions in per cent of total length. Averages of 3 females, 3,195 to 3,465 mm., from Englewood, Florida (calculated from measurements by Stewart Springer).

Snout length in front of: mouth 6.3.
Eye: horizontal diameter 1.0.
Mouth: breadth 10.6; height 4.7.
Nostrils: distance between inner ends 6.2.
Gill opening lengths: 1st 3.4; 5th 2.7.
First dorsal fin: length of base 9.2.
Second dorsal fin: length of base 2.8.
Anal fin: length of base 3.9.
Caudal fin: upper margin 26.5; lower anterior margin 12.0.
Pectoral fin: outer margin 20.6; inner margin 5.4; distal margin 18.6.
Distance from snout to: 1st dorsal 32.9; upper caudal 73.5; pectoral 23.6.
Interspace between: 1st and 2nd dorsal 23 9; 2nd dorsal and caudal 6.1.

Trunk more slender than in *milberti*, its height at 1st dorsal about 1/5 its length to origin of caudal. A low but unmistakable dermal ridge along midline of back between the 2 dorsal fins. Body sector to cloaca about 1.2 times as long as tail sector. Upper precaudal pit deep, subrectangular, the lower less deeply marked than upper. Dermal denticles overlapping so regularly that skin is entirely concealed, their blades but little arched longitudinally, 3 to 5 low ridges in small specimens but perhaps most often 7 in adults, with a corresponding number of strong pointed marginal teeth, the median considerably the largest and the outermost pair very small when there are more than 5.

Head to 5th gill opening about 1/4 of total length, its width nearly or quite 80% as great at eyes as at origin of pectorals. Snout moderately thick anteriorly, broadly rounded, its length in front of a line connecting outer ends of nostrils about 3/5 as great as distance between inner ends of latter, and length in front of mouth about 1/3 as great as that of head. Eye circular, its diameter only a little more than 1/2 (50 to 60%) as long as 3rd gill opening, or a little more than 1/4 (27%) as great as distance between nostrils in young specimens and still smaller relatively in adults; its anterior margin about opposite front of mouth. Gill openings about evenly spaced, slightly oblique and weakly concave in outline, the 2nd to 4th (longest) a little less than 1/2 (about 46%) as long as distance between nostrils or about twice as long as diameter of eye, the 1st and 5th about 0.8 to 0.9 as long as 2nd to 4th, the 4th over margin of pectoral. Nostril strongly oblique, its inner end about 2/3 as far from mouth as from tip of snout, its anterior margin slightly sinuous only without definite expansion. Mouth broadly ovate, about twice as broad as high, occupying a little less than 3/4 (about 72%) of breadth of head. Upper labial furrow only about 0.3 times as long as diameter of eye and almost or entirely concealed when mouth is closed.

Teeth $\frac{14\text{ or }15\text{—}1\text{ to }3\text{—}14\text{ or }15}{14\text{—}1\text{ to }3\text{—}14}$; upper teeth broadly triangular, their edges serrate, most coarsely so toward base, the 1st two erect, symmetrical, with nearly straight or very slightly concave edges, but subsequent teeth weakly oblique, their outer margins considerably more deeply concave in subangular contour, the 1st to 9th or 10th about equal in size, the 10th to 14th or 15th successively smaller and more deeply notched, the outermost 1 or 2 very short; lower teeth erect, symmetrical, with narrow triangular cusps on broadly expanded bases, serrate from tip to base but more finely so than uppers, the 2nd to 6th or 7th a little the largest, the outermost 2 or 3 much the smallest; 1 to 3 small teeth at symphysis in each jaw.

First dorsal much smaller than in *milberti*, *leucas* or *longimanus*, its base appreciably shorter than from eye to 1st gill opening, its origin about over inner corner of pectoral or a little posterior to latter, its anterior margin moderately convex toward tip, its posterior margin much more deeply concave than in *milberti* or in *springeri*, its apex subacute, its free rear corner slender and nearly ½ as long as base, the midpoint of its base about 1.6 times as far from origin of pelvics as from axils of pectorals. Second dorsal only about ⅓ as long at base as 1st and relatively much lower, its free rear tip rather slender and about as long as base or about 1½ times as long as the vertical height, its extreme length from origin to tip about 2.5 its vertical height and therefore considerably greater relatively than in *springeri;* its origin over origin of anal or a little anterior to it. Caudal between ⅓ and ¼ of total length (27 to 29%), its terminal sector a little less than ¼ the length of the fin and noticeably slender with narrowly rounded tip and moderately concave lower posterior outline, the lower lobe (expanded lower anterior corner) about ⅓ to ⅖ (about 40%) as long as upper and thus somewhat shorter relatively than in *springeri*, with moderately and evenly convex anterior margin, nearly straight posterior margin and very narrowly rounded or subacute tip; the re-entrant corner (included by the 2 lobes) moderately rounded. Distance from origin of caudal to tip of anal about ⅔ as long as base of latter, or about as long as its free rear corner. Anal about 1.2 times as long at base as 2nd dorsal and almost 1.5 times as long on anterior margin but only a little higher vertically, its posterior margin much more deeply incised, its free rear corner between ⅔ and ¾ as long as base, its tip a little anterior to tip of 2nd dorsal. Distance from origin of anal to tips of pelvics about 1.3 times as long as base of anal. Pelvics a little longer at base than anal, their outer corners considerably more obtuse than a right angle (about 115°). Pectoral about 9/10 (78 to 91%) as long as head, a little less than ½ as broad as long, with subacute or very narrowly rounded corners, usually with only weakly convex outer margin and weakly concave distal margin, as is the right-hand fin on specimen illustrated, but sometimes with much more strongly convex outer edge and concave distal edge, as is the left-hand fin on this same specimen (Fig. 71).[116]

Color. Back and upper sides, including upper surfaces of pectorals, bluish or leaden

116. A striking example of variability in this genus, illustrating the danger of basing specific distinctions on small differences in the shape of a given fin.

gray in the fresh-caught specimens we have seen, but also described as dirty gray, or very pale, perhaps as a result of living over a white sand bottom; lower parts white; lower surfaces of pectorals grayish and sooty toward tips; pelvics and anal grayish white.

Size. The fact that embryos up to 965 mm. have been reported, as well as a free-living specimen hardly larger (see Study Material, p. 382), suggests that this is about the usual size at birth. Although it is generally recognized that this is a considerably larger shark than *milberti*, the only exact length records of large adult specimens, identified beyond question as this species, are nine females from southern Florida that ranged from 10 feet 4 inches to 11 feet 8 inches[117] and one specimen from Georges Bank 11 feet long.[118] It is reputed to reach 14 feet, although perhaps not from any exact evidence.

Developmental Stages. Embryos have not been described; ten have been recorded in one Florida female and embryos up to 965 mm. long in another.

Habits. Although *obscurus* has been known to science since 1818 and is common enough to be caught occasionally close to Woods Hole and reputedly more often near New York, our only information regarding its diet is that it is a fish-eater; off the east coast of Florida portions of other sharks have been found in its stomach as well as various reef fishes, such as groupers, lizardfish (*Trachinocephalus*), flatfishes (*Citharichthys*), red goatfish (*Mullus*) and cusk eels (*Ophidion*). The wide distribution of the localities where positively identified specimens of *obscurus* have been taken show it to be much more pelagic in habit than are either *milberti* (p. 372) or *leucas* (p. 341). On the other hand, the record of captures proves that it comes closer inshore, even into very shoal water, than does *longimanus* (p. 359). All captures in the northern part of its range have been for the warm months, whereas it is present the year round along eastern Florida but only in the winter off southwestern Florida, which is evidence that some *obscurus* migrate northward along the United States coast in spring or early summer, as do various other warm-water sharks.

All that is definitely known of its breeding habits, beyond what is stated above, is that adult females containing embryos of 575 to 965 mm. are reliably reported off southwestern Florida, and that free-living specimens, so small that they had evidently been born only shortly previous, have been taken off southeastern Florida in late winter,[119] and off Louisiana and near Woods Hole in August (see Study Material, p. 382). It seems, therefore, that *obscurus* may produce young anywhere within its geographic range and over a long season.

Relation to Man. Obscurus is not taken in large enough numbers to be of any commercial importance anywhere, although such as are caught in the shark fishery in southern Florida or in the West Indian region are utilized for leather, etc., as in the case of other large sharks.[120]

117. Springer, Proc. Fla. Acad. Sci., *3*, 1939: 25.
118. Firth, Bull. Boston Soc. nat. Hist., *61*, 1931: 9; Bigelow and Schroeder, Bull. U.S. Bur. Fish., *48*, 1936: 321.
119. Personal communication from Stewart Springer.
120. It seems more likely from the context that a recent account of the actions of a Dusky Shark, when hooked on rod and reel (Wise, Tigers of the Sea, 1937: 262), may have referred to *milberti*.

Range. Tropical and warm-temperate waters on both sides of the Atlantic. In the west *obscurus* is reliably recorded from southern Massachusetts and Georges Bank to southern Florida, Louisiana, and the Bahamas; and as far south as southern Brazil by name. In the eastern Atlantic reports apparently referable to *obscurus* include the Mediterranean coast of Spain,[121] Madeira, Senegal, the Canaries and Cape Verdes, as well as the vicinity of Sierra Leone, Ascension Island, St. Helena, and Table Bay, South Africa.[122] But final decision as to whether or not the "*obscurus*" of the two sides of the Atlantic are identical must await comparison of specimens from the two ocean areas.

Occurrence in the Western Atlantic. Obscurus has been characterized repeatedly in the past as common or even very common along the coast of the United States from New Jersey to Cape Cod. However, it has been proved recently that most of these reports were actually based upon *C. milberti* (p. 374) or on *leucas* in some instances, and that North Carolina records of *obscurus* similarly refer either to *limbatus* or to *Negaprion brevirostris*.[123] In fact, it is only around southern Florida that positively identified specimens of *obscurus* have been taken recently in any numbers. Along southwestern Florida it is present in winter, while off the southeastern coast it is common throughout the year. Also, it has been taken off the coast of Louisiana (one specimen, see Study Material, p. 383), but is not reported otherwise from the western waters of the Gulf of Mexico. To the northward its distribution presents a puzzling picture, for we find no reliable record of it anywhere along the coast between Florida and Delaware Bay. But it has been taken off the mouth of Delaware Bay and repeatedly on the coast of New Jersey, at Long Island, New York and at Woods Hole, where twelve specimens have come into our hands in recent summers, six of them during August 1944 (see Study Material, p. 383) in addition to others reported in earlier years. There is at least one record for Nantucket and another for Georges Bank, which, while by name only, seem referable to this species and not to *milberti* because of the large sizes of the specimens in question (11 to 12 feet). However, these last appear to be the most northerly and easterly of the reliable records of the species on this side of the Atlantic, for while *obscurus* has been reported by name at three localities in the Gulf of Maine,[124] at least one of these records[125] was almost certainly based on *Prionace glauca* (p. 282), and the others probably were the same. Also, reliable reports from New Jersey northward rest on odd specimens only, showing that the numbers of individuals that visit any part of the coast north of Florida are very small as compared with *milberti*, although printed references to "*obscurus*" for Long Island and for southern New England would suggest the reverse. Its recorded appearances in the northern part of its range are limited to the warm months, chiefly August and September.

Information on its occurrence south of Florida is even more scanty, *i.e.*, nominal

121. *Carcharhinus commersonii* Rey (Fauna Iberica Peces, *1*, 1928: 342), identifiable as *obscurus* by the presence of a mid-dorsal ridge, position of first dorsal and teeth.

122. A stuffed specimen in South African Museum (Barnard, Ann. S. Afr. Mus., *21* [1], 1925: 26).

123. Radcliffe, Bull. U.S. Bur. Fish., *34*, 1916: 255, 257.

124. Bigelow and Welsh, Bull. U.S. Bur. Fish., *40* (1), 1925: 30.

125. Storer, Mem. Amer. Acad. Arts Sci., N. S. *9*, 1867: 219, pl. 36, fig. 2.

records for the Bahamas (hence perhaps not actually based on that species at all), Trinidad, British Guiana and Brazil. It is also reported as taken well offshore around Bermuda, as is to be expected from its pelagic nature.

Synonyms and References:[126]

Squalus obscurus Lesueur, J. Acad. nat. Sci. Philad., *1*, 1818: 223, pl. 9 (descr., good ill., N. Amer. but no precise loc.); Gill, Proc. Acad. nat. Sci. Philad., Addend., 1861: 59 (ref. to Lesueur, in list, N. Amer.).

Carcharias falcipinnis Lowe, Proc. zool. Soc. Lond., 7, 1839: 90; Proc. zool. Soc. Lond., *11*, 1848: 93; Trans. zool. Soc. Lond., *3*, 1849: 18 (specs. later ident. by Günther, 1870, as *obscurus*, Madeira).

Carcharias (Prionodon) obscurus Müller and Henle, Plagiost., 1841: 46 (ref. to Lesueur, N. Amer.); Duméril, Hist. Nat. Poiss., *1*, 1865: 371 (descr., from Lesueur, 1818, N. Amer.); Günther, Cat. Fish. Brit. Mus., *8*, 1870: 366 (descr., specs., Madeira, St. Helena); Metzelaar, Trop. Atlant. Visschen, 1919: 187 (refs., both coasts N. Atlant.).

Carcharias obscurus Bory de St. Vincent, Dict. Class. Hist. Nat., *15*, 1829: 597 (ref. to Lesueur, 1818); Storer, Mass. Zool. Bot. Surv., Rep. on Fish., 1839: 184 (ref. to Lesueur only); Boston J. nat. Hist., *2*, 1839: 533 (descr. after Lesueur, but teeth from Mass. Bay more likely *Prionace glauca*); DeKay, Zool. N. Y., 4, 1842: 350, pl. 61, fig. 201 (descr., ill. after Lesueur, no loc.); Rep. St. Cab. nat. Hist. N. Y. (1855), *8*, 1858: 64 (listed for N. York); Mellis, "St. Helena Pisces," 1875 (not seen); Günther, "Challenger" Rep., Zool., *1* (6), 1880: 5 (name only, near St. Helena); Jordan and Gilbert, Bull. U.S. nat. Mus., 16, 1883: 872 (ref. to Lesueur); Proc. U.S. nat. Mus., *5*, 1883: 244 (coast of U.S.); Bean, T. H., Rep. U.S. Comm. Fish. (1882), 1884: 343 (Woods Hole); Günther, "Challenger" Rep., Zool., *31* (2), 1889: 5 (name only, near Sierra Leone and Ascension I.); Nelson, Rep. St. Geol. N. J., *2* (2), 1890: 660 (N. Jersey, not common); Vincinguerra, Atti Soc. ital. Sci. nat., *34*, 1892: 30 (ref. to *obvelatus* Valenciennes, 1844, Canaries); Jordan, Manual Vert. Anim. NE. U.S., 1929: 10 (general); Jordan, Evermann and Clark, Rep. U.S. Comm. Fish. (1928), 2, 1930: 10 (N. and mid. Atlant.); Beebe and Tee-Van, Field Bk. Shore Fish. Bermuda, 1933: 28 (descr., Bermuda); Belloc, Rev. des Trav. Pêches Marit., 7 Fasc. 2, 1934: 132 (ill. after Lesueur; Canaries, Madeira).

Prionodon obvelatus Valenciennes, in Webb and Berthelot, Hist. Nat. Canaries, 1844: 103, pl. 26 (descr. applicable to *obscurus*, but ill. not recognizable; Canaries spec.).

Squalus (Carcharinus) obscurus Gray, List. Fish. Brit. Mus., 1851: 47 (N. Amer.).

Platypodon obscurus Gill, Proc. Acad. nat. Sci. Philad., 1864: 262 (east. Amer.).

Carcharias (Prionodon) obvelatus Duméril, Hist. Nat. Poiss., 1, 1865: 376 (redescr., applicable to *obscurus* except perhaps teeth; Canaries).

Carcharhinus (Eulamia) obscurus Jordan, Rep. U.S. Comm. Fish. (1885), 1887: 795 (name, N. Amer.).

Carcharinus obscurus (at least in part) Jordan and Gilbert, Bull. U.S. nat. Mus., 16, 1883: 22 (descr. after Lesueur, N. Atlant.); Bean, T. H., Rep. For. Comm. N. Y., 1901: 377; Bull. N. Y. St. Mus., *60*, Zool. 9, 1903: 25 (descr., perhaps confused with other species; reported abund., N. Jersey, no doubt *milberti*); Fowler, Rep. N. J. Mus. (1906), 1907: pl. 72 (ill., after Lesueur); Sharp and Fowler, Proc. Acad. nat. Sci. Philad., *56*, 1907: 505 (Nantucket spec., ident. accepted because of large size, 11 ft. 6 in.); Garman, Mem. Harv. Mus. comp. Zool., *36*, 1913: 130 (descr., Buzzards Bay spec.); Sumner, Osburn and Cole, Bull. U.S. Bur. Fish., *31* (2), 1913: 735 (probably partly *obscurus* because of loc., Woods Hole, but refers no doubt to *milberti* chiefly); Nichols, Bull. Amer. Mus. nat. Hist., *37*, 1917: 873 (discuss.); Copeia, 140, 1922: 21 (compares with *milberti*, from photographs, but mid-dorsal ridge said to be absent, Woods Hole); Meek and Hildebrand, Field Mus. Publ. Zool., *15* (1), 1923: 46 (descr., N. and mid. Atlant., no definite loc.); Bigelow and Welsh, Bull. U.S. Bur. Fish., *40* (1), 1925: 29 (descr. and ill., but Gulf of Maine records probably were *Prionace glauca*); Barnard, Ann. S. Afr. Mus., *21* (1), 1925: 25 (spec., Table Bay, S. Afr.); Nichols, Copeia, 140, 1925: 21 (comparison with other species); Thorne, Bull. N. Y. zool.

126. Owing to the fact that other species, especially *C. milberti*, have frequently been reported as *obscurus*, the following list is limited to such references as can be referred to the latter with some confidence from included evidence.

Soc., *31*, 1928: 114 (two reported harpooned near Long Island, N. York, ident. acceptable because distinguished from *milberti* similarly taken); Nichols and Breder, Zoologica, N. Y., *9*, 1929: 15 (states confused with *milberti*, but young near N. York Sept.–Nov.); Breder, Field Bk. Mar. Fish. Atlant. Coast, 1929: 16 (ill., but distrib. evidently confused with that of *milberti*); Bellon and Mateu, Notas Inst. esp. Oceanogr., (2) *53*, 1931: 15 (ref. to *obvelatus* Valenciennes, 1844, Canaries); Firth, Bull. Boston Soc. nat. Hist., 61, 1931: 9 (11-ft. spec. Georges Bank, ident. probable because of large size); Bigelow and Schroeder, Canad. Atlant. Fauna, 12e, 1934: 7 (ill., Georges Bank, but C. Elizabeth record probably for *Prionace glauca*); Norman, "Discovery" Rep., *12*, 1935: 56 (name, listed for St. Helena, W. Indies, Brazil, Canaries, C. Verde I., W. Afr., S. Afr.); Bigelow and Schroeder, Bull. U.S. Bur. Fish., *48*, 1936: 321 (Georges Bank); Cadenat, Rev. des Trav. Pêches Marit., *10* (4), 1937: 430 (C. Verde; Dakar); Norman and Fraser, Giant Fishes, 1937: 36 (name, N. and mid. Atlant.); Springer, Proc. Fla. Acad. Sci., *3*, 1939: 20, 23–24 (descr., ill., size, season, embryos, Florida); Wise, Nat. Hist. N. Y., *28*, 1938: 323 (recognizable photo, Bahamas); Boos, Contr. biol. Lab. Cath. Univ. Amer., 45, 1943: 10 (chemistry of pancreas, Florida); Springer, Copeia, 1946: 174 (food, off Salerno, Florida).

Carcharhinus (Platypodon) obscurus Jordan and Evermann, Rep. U.S. Comm. Fish. (1895), 1896: 215 (name, N. Atlant.); Bull. U.S. nat. Mus., 47 (1), 1896: 35 (descr., refs., N. and Mid. Atlant.).

Eulamia obscura Jordan and Gilbert, Bull. U.S. nat. Mus., 16, 1883: 60 (name only); Fowler, Rep. N. J. Mus. (1905), 1906: 61 (12-ft. spec. off C. Henlopen, Delaware, ident. probably correct because of large size); Bull. Amer. Mus. nat. Hist., *70* (1), 1936: 49 (list of East Atlant. loc., descr. and ill. of N. Jersey specs.);[127] Bigelow and Schroeder, Guide Comm. Shark Fish., Anglo Amer. Caribb. Comm., Wash., 1945: 82, fig. 25 (descr., habits, range, ill.).

References Probably Based on Some Other Shark:

Carcharias obscurus Storer, Boston J. nat. Hist., *2*, 1839: 533 (teeth); Mass. Zool. Bot. Surv., Rep. on Fish., 1839: 184 (same as preceding); Mem. Amer. Acad. Arts Sci., N. S. *9*, 1867: 219, pl. 36, fig. 2; also Fishes Mass., 1867: 243, pl. 36, fig. 2 (this probably was *Prionace glauca*; see p. 292); Bean, Bull. U.S. Fish Comm., 7, 1889: 132, 152 (this probably *milberti* because of reported abund. in bays of N. Jersey).

Carcharhinus obscurus Smith, Bull. U.S. Bur. Fish., *17*, 1898: 88 (Woods Hole, more likely *milberti* because descr. as "very common"); Linton, Bull. U.S. Bur. Fish., *24*, 1905: 339 (N. Carolina, but apparently *Negaprion brevirostris* from his descr.); Tracy, Rep. R. I. Comm. inl. Fish., 1906: 45 (probably *milberti*, from context); Fowler, Rep. N. J. Mus. (1906), 1907: 256 (N. Jersey, "*commersonii*," i.e., *leucas* Müller and Henle, 1841, according to Fowler, Proc. biol. Soc. Wash., *33*, 1920: 144); Smith, Bull. N. C. geol. econ. Surv., *2*, 1907: 33 (N. Carolina, probably *Negaprion brevirostris*, see Linton, 1905); Kendall, Occ. Pap. Boston Soc. nat. Hist., 7 (8), 1908: 4 (reps. for Gulf of Maine probably *Prionace glauca*; those for Connecticut and Rhode Island *Carcharhinus milberti*); Tracy, Rep. R. I. Comm. inl. Fish., 1910: 59 (probably *milberti*, from context); Thorne, Copeia, 35, 1916: 69 (name only, Long Island, N. York; probably *milberti*); Nichols, Copeia, 35, 1916: 72, 73 (Long Island, N. York spec., actually "*commersonii*," i.e., *leucas* Müller and Henle, 1841; see correction by Nichols, Bull. Amer. Mus. nat. Hist., *37*, 1917: 873); Latham, Copeia, 71, 1919: 53 (20 to 22-in. young, Long Island, N. York, probably not *obscurus* because smaller than recorded embryos of latter, see p. 387); Truitt, Bean and Fowler, Bull. Md. Conserv. Dep., *3*, 1929: 30 (descr. is of *Prionace glauca*).

Carcharhinus (Platyodon) obscurus Pietschmann, Ann. naturh. (Mus.) Hofmus. Wien., *21*, 1906: 99 (Morocco, not *obscurus* because origin of 1st dorsal described as over middle of pectoral base; perhaps *plumbeus* Nardo, 1827).

Western Atlantic Records by Name Only, Without Other Indication as to Identity:

Carcharias obscurus Linsley, Amer. J. Sci., *47*, 1844: 76 (Connecticut); Goode, Bull. U.S. nat. Mus., 5, 1876: 76 (Bermuda); Amer. J. Sci., (3) *14*, 1877: 293 (Bermuda); Goode, Fish. Fish. Indust. U.S., *1*, 1884: 672 (Woods Hole); Wilson, Proc. U.S. nat. Mus., *33*, 1908: 624 (N. Carolina); Nichols, Abstr. Linn. Soc. N. Y., 20–23, 1913: 91 (off N. York); Wilson, Proc. U.S. nat. Mus., *64* (17),

127. Spelled "*obscurus*."

1925: 12 (Woods Hole); Bere, Amer. Midl. Nat., *17*, 1936: 593, 604 (Englewood, Florida); Longley and Hildebrand, Pap. Tortugas Lab., *34*, 1941: 2 (Tortugas, Florida).

Eulamia obscurus Verrill and Smith, Rep. U.S. Comm. Fish. (1871–1872), 1873: 520 (Woods Hole); Bean, Proc. U.S. nat. Mus., *3*, 1881: 115 (Woods Hole); Fowler, Copeia, 30, 1916: 36 (in list); Copeia, 31, 1916: 41 (N. Jersey).

Carcharinus obscurus Rathbun, Proc. U.S. nat. Mus., 7, 1885: 488, 489 (off N. York, Lat. 39° 30′ N., Long. 72° W.); Moore, Bull. U.S. Fish Comm., *12*, 1894: 358 (N. Jersey); Linton, Bull. U.S. Bur. Fish., *13*, 1894: 104 (parasites); Proc. U.S. nat. Mus., *20*, 1897: 424, 452 (no loc.); Means, Bull. Amer. Mus. Nat. Hist., *10*, 1898: 311 (Hudson R., N. York); Linton, Bull. U.S. Bur. Fish., *19*, 1901: 272, 426 (food, parasites, Woods Hole); Hargreaves, Fish. Brit. Guiana, 1904: 14, app. 7 (Brit. Guiana, abundance, attacks on man); Bean, B. A., in Shattuck, Bahama Islands, 1905: 296 (Bahamas); Linton, Bull. U.S. Bur. Fish., *26*, 1907: 122, 123 (parasites, N. Carolina); Sullivan, Bull. U.S. Bur. Fish., *27*, 1907: 13 (name only); Wilson, Proc. U.S. nat. Mus., *33*, 1907: 326, 360, 409, 414, 423, 431, 629 (parasites, Woods Hole region); Linton, Bull. U.S. Bur. Fish., *28* (2), 1910: 1200 (parasites); Vincent, Sea Fish. Trinidad, 1910: 53 (Trinidad, W. Indies, abund.); Rosen, Lunds Univ. Årsberätt., N. S. 7 (5), 1911: 47 (Bahamas); Gudger, Yearb. Carneg. Instn., *12*, 1913: 177 (Tortugas, Florida); Nichols, Copeia, 36, 1916: 81; Rockwell, Brooklyn Mus. Quart., *3*, 1916: 162 (Long Island, N. York); Latham, Copeia, 99, 1921: 72 (Long Island, N. York); Breder, Copeia, 127, 1924: 25 (Sandy Hook Bay, N. York); Linton, Proc. U.S. nat. Mus., *64* (21), 1924: 5, 7, 12, 30, 34, 38, 47, 48, 49, 65, 80, 87, 90 (parasites, Woods Hole and N. Carolina); Wilson, Bull. U.S. nat. Mus., 158, 1932: 463, 464, 524 (parasites, Woods Hole region); Nigrelli, Amer. Mus. Novit., 996, 1938: 10 (parasites); Norris, Plagiost. Hypophysis, 1941: 28 (brain); Lunz, Bull. S. Carolina St. Planning Bd., 14, 1944: 27 (Florida).

Carcharias (Prionodon) obscurus Werner, Zool. Jb., Syst. Abt., *21*, 1904: 283 ("Carteret County, New York," N. Carolina perhaps intended).

Eulamia obscura Bean, T. H., Field Mus. Publ. Zool., 7 (2), 1906: 29 (Bermuda).

Carcharinus sp. (probably *C. obscurus*) Norris, Plagiost. Hypophysis, 1941: pl. 1, fig. 2 (brain).

Carcharhinus oxyrhynchus Müller and Henle, 1841

Figure 73

Study Material. Stuffed dried skin of a female, about 443 mm. long, from Surinam (?) (Acad. Nat. Sci. Philad., No. 34635).[128]

Distinctive Characters. This species is sharply marked off from all other Atlantic members of the genus by the large number of teeth and by the great length of the very narrow snout anterior to a line connecting the outer ends of nostrils, this being about 1½ times as long as the distance between the inner ends of nostrils, both in the original illustration of the species and in the specimen listed above. In this respect, and in the narrowness of its teeth, it seems to be most nearly related to *C. velox* Gilbert, 1898, of the west coast of Central America, but it is separable from *velox* by its relatively much shorter nostrils and by the fact that the outer margins of the upper teeth toward the corner of the mouth are not notched.

Description. Proportional dimensions are not available.[129]

Trunk moderately slender, the back without trace of mid-dorsal ridge between

128. Specimen recorded by Fowler (Proc. Acad. nat. Sci. Philad., *60*, 1908: *65*).

129. The specimen listed above was obviously so greatly lengthened in the process of stuffing that measurements taken from it would be only roughly approximate.

1st and 2nd dorsal fins. Upper precaudal pit well marked, subangular, a little less than a right angle, the lower pit only faintly indicated. Dermal denticles overlapping only partly, with the skin visible between them here and there, their blades broader than long, usually with 3 (rarely 5) strong longitudinal ridges and as many moderately strong teeth, the median a little the longest; pedicels very short.

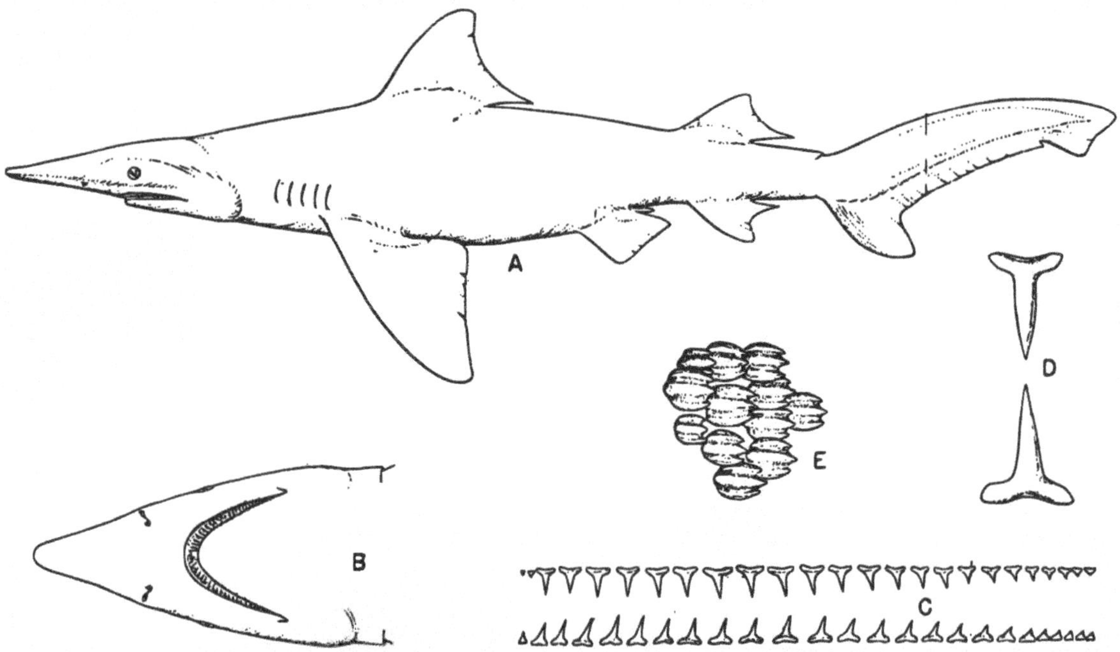

FIGURE 73. *Carcharhinus oxyrhynchus. A, B* Immature male, about 18 inches long, from Surinam, somewhat emended after Müller and Henle. *C* Upper and lower teeth, after Müller and Henle. *D* An upper and a lower tooth, much enlarged, after Müller and Henle. *E* Dermal denticles from a female, about 443 mm. long, from Surinam, much enlarged (see Study Material, p. 391).

Head with snout relatively much longer than in other local species of *Carcharhinus*, its length to origin of pectoral about 40% of length of trunk to origin of caudal in the original illustration. Snout narrow, wedge-shaped, contracted anterior to nostrils, with narrowly rounded tip, its length in front of a line connecting outer ends of nostrils about 1½ times as great as distance between inner ends of latter, its length in front of mouth about twice as great as distance between nostrils or about ⅓ as great as length of head to origin of pectoral.[130] Anterior margin of eye very little posterior to front of mouth. Gill openings about twice as long as diameter of eye, the space between 4th and 5th over origin of pectoral.[131] Nostril only slightly oblique, its inner end only about ½ as far from front of mouth as from tip of snout, its anterior margin only slightly expanded,

130. This is according to the original illustration; in the dried skin the head has obviously been lengthened in the process of stuffing.

131. In the stuffed skin the 5th is a little anterior to the origin of the pectoral, but this is probably due to distortion.

in obtusely subangular outline. Mouth narrow-ovate, about 75% as high as broad. Upper labial furrow about 1/3 as long as diameter of eye.[132]

Teeth in type specimen about $\frac{23-4-22}{24-2-23}$; similar in the 2 jaws, symmetrical, erect, with slender cusps on broadly expanded bases, the edges of cusps coarsely serrate in uppers but very finely so in lowers,[133] the edges of bases smooth; 4 minute teeth at symphysis in upper jaw and 2 in lower, the 3 or 4 outermost teeth in each jaw also very small.

Origin of 1st dorsal a little anterior to axil of pectoral in specimen seen, but over mid-base of pectoral in original illustration, its base a little more than 1/3 as long as head to origin of pectoral by original illustration, its anterior margin moderately convex and increasingly so toward apex, the posterior margin moderately concave, apex subacute or very narrowly rounded, the free rear tip broad, about 1/3 to 1/4 as long as the base, its vertical height a little less than 1/2 as great as length of pectoral, the midpoint of base less than 1/2 as far from axil of pectoral as from origin of pelvics. Second dorsal a little more than 1/2 as long at base as 1st, its origin about over origin of anal,[134] its apex narrowly rounded, rear margin weakly concave, its free rear tip moderately slender and about 1/2 as long as base. Caudal about 1/4 of total length, its upper margin moderately convex, its tip very narrowly rounded or subacute, the terminal sector about 1/4 the length of fin in original illustration (about 1/5 in dried skin), the lower lobe a little more than 1/3 as long as upper (about 36% in original illustration but about 33% in dried skin), its anterior margin strongly convex toward apex, the re-entrant corner (included by the 2 lobes) moderately rounded. Distance from origin of caudal to tip of anal 2/3 (original illustration) to 4/5 (dried skin) as long as base of latter. Anal a little shorter at base than 2nd dorsal but about equally high, its anterior margin only slightly more convex and rear margin more concave than in 2nd dorsal, apex narrowly rounded, free rear corner only about 1/2 as long as base. Distance from origin of anal to tips of pelvics about 1/2 (in original illustration) to 2/3 (dried skin) as long as base of anal. Pelvics with nearly straight margins, about as long at base as anal. Pectoral about 2/3 as long as head in original illustration but only about 1/2 that long in dried skin, about 1/2 as broad as long, with moderately convex outer margin but only very slightly concave distal margin and nearly straight inner margin, rounded inner corner and subacute tip.

Color. Described as yellow-gray above, white below.

Size. The greatest length so far definitely reported for it is about five feet (1,520 mm.), although it has been said to attain six to eight feet. The fact that embryos of 14 inches have been reported, and free-living specimens 17 to 18 inches, suggests that a length of about 15 to 16 inches is usual at birth.

132. Günther (Cat. Fish. Brit. Mus., 1870: 375) credits it with a short labial furrow on the lower jaw as well as on the upper. Actually, however, the dried skin shows that the upper alone is present, as in other members of the genus.

133. Müller and Henle (Plagiost., 1841: 41) state that the upper teeth are serrate only toward their tips, the lowers smooth. But their illustration (pl. 13) shows the uppers as serrate from tip to base, and the lowers as very finely serrate toward their tips.

134. A little anterior to latter in original illustration, but a little posterior to it in dried skin.

Developmental Stages. Development is viviparous, the (four) embryos having been described as attached to the mother by a placenta.[135]

Habits. All that is known of its habits is that it has been described (if identified correctly) as often entering estuaries and river mouths and feeding on small fish, ravaging schools of clupeids and sciaenids in particular.[136] Apparently it is a littoral species.

Relation to Man. While not considered very desirable as food, some are sold in the markets of Trinidad and no doubt in the Guianas as well.

Range. Western tropical Atlantic. The few records of this species are from: Surinam, Dutch Guiana; Demerara, British Guiana; French Guiana, where it has been described as rather common; Trinidad.[137]

Synonyms and References:

Carcharias (Prionodon) oxyrhynchus Müller and Henle, Plagiost., 1841: 41, pl. 15 (descr., ill., Surinam); Müller and Troschel, in Schomburgk, R. H., Reisen Brit. Guiana, 1840–44, *3*, 1848: 642 (Brit. Guiana, said to grow to 6 to 8 ft.); Duméril, Hist. Nat. Poiss., *1*, 1865: 356 (descr., Cayenne, Surinam); Günther, Cat. Fish. Brit. Mus., *8*, 1870: 375 (descr., no loc.).

Isogomphodon oxyrhynchus Gill, Ann. N. Y. Lyc., 7, 1862: 410 (name).

Carcharhinus (Isogomphodon) oxyrhynchus Jordan and Evermann, Rep. U.S. Comm. Fish. (1895), 1896: 216 (name, Surinam); Bull. U.S. nat. Mus., 47 (1), 1896: 40 (descr., Surinam).

Carcharinus oxyrhynchus Hargreaves, Fish. Brit. Guiana, 1904: 14, and append. 8 (name only, Brit. Guiana); Vincent, Sea Fish. Trinidad, 1910: 53 (name only, Trinidad); Garman, Mem. Harv. Mus. comp. Zool., *36*, 1913: 143 (descr., east. S. Amer.); Puyo, Bull. Soc. Hist. nat. Toulouse, *70*, 1936: 85, 87,[138] 250 (ill., but labels transposed, habits, food, vernacular name, French Guiana); White, Bull. Amer. Mus. nat. Hist., *74*, 1937: 128 (in Key).

Eulamia oxyrhynchus Fowler, Proc. Acad. nat. Sci. Philad., *60*, 1906: 65 (meas. of spec. listed above in Study Material, Surinam?); Proc. Acad. nat. Sci. Philad., *67*, 1916: 521[139] (female with embryos, in market, Trinidad); Proc. Acad. nat. Sci. Philad., *71*, 1919: 129 (listed for Surinam).

Carcharhinus porosus Ranzani, 1839

Figures 74, 75

Study Material. Seven small specimens, 330 to 395 mm. long, from Surinam, Pernambuco and Bahia (Harv. Mus. Comp. Zool., No. 307, 526, 721, 1403, 1404); female, 485 mm., and male, 500 mm., from Colón (U.S. Nat. Mus., No. 79317, 79316); male, 831 mm., Pacific Panama (U.S. Nat. Mus., No. 79293); also two females, from Peru (Harv. Mus. Comp. Zool., No. 692).

Distinctive Characters. Porosus differs from all other western Atlantic members of the genus in that the second dorsal originates about over the midpoint of the base of the anal, that the terminal sector of the caudal is relatively smaller and that the outermost four or five lower teeth are strongly asymmetrical with deeply notched outer margins like the uppers.

Description. Proportional dimensions in per cent of total length. Female, 485 mm.,

135. Fowler, Proc. Acad. nat. Sci. Philad., *67*, 1916: 521.
136. Puyo, Bull. Soc. Hist. nat. Toulouse, *70*, 1936: 87.
137. Specimens are in the British Museum and in the museums of Leyden and Paris.
138. Spelled "*Carcharhynus.*"
139. Spelled *oxyrhinchus.*

from Colón, Panama (U.S. Nat. Mus., No. 79317). Male, 500 mm., same locality (U.S. Nat. Mus., No. 79316).

Trunk at origin of pectoral: breadth 12.4, 11.4; height 11.9, 11.8.

Snout length in front of: outer nostrils 4.5, 4.3; mouth 8.9, 8.7.

Eye: horizontal diameter 2.3, 2.1.

Mouth: breadth 8.2, 7.9; height 5.4, 5.3.

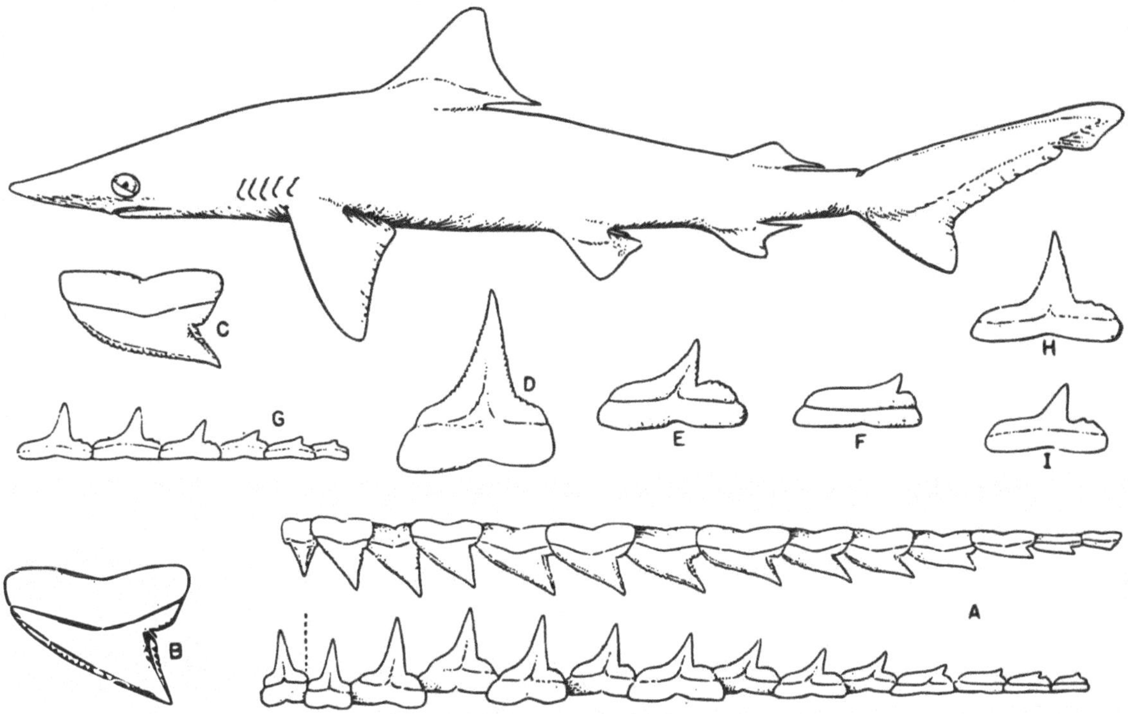

FIGURE 74. *Carcharhinus porosus*, immature male, about 377 mm. long, from Pernambuco, Brazil (Harv. Mus. Comp. Zool., No. 526). *A* Upper and lower teeth, left-hand side, about 4 x. *B* Fifth upper tooth. *C* Ninth upper tooth. *D* Fourth lower tooth. *E* Seventh lower tooth. *F* Tenth lower tooth. *C–F*, about 7.6 x. *G* Eighth to thirteenth lower teeth of a female, 485 mm. long, from Colón, Atlantic Panama (U. S. Nat. Mus., No. 79316), about 3 x. *H* Seventh lower tooth. *I* Tenth lower tooth of same, about 7.6 x.

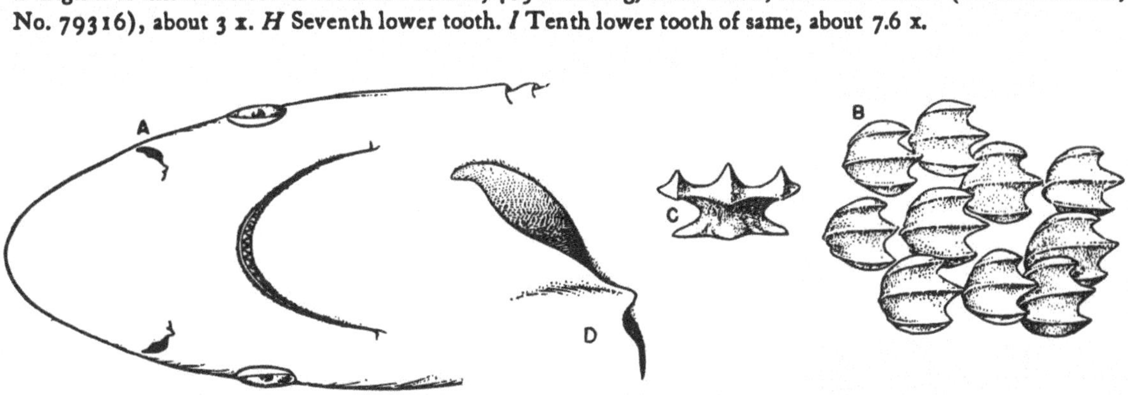

FIGURE 75. *Carcharhinus porosus*, pictured in Fig. 74. *A* Anterior part of head from below, about 6/7 natural size. *B* Dermal denticles, about 50 x. *C* Apical view of dermal denticle, about 100 x. *D* Left nostril, about 6 x.

Nostrils: distance between inner ends 5.8, 5.8.

Gill opening lengths: 1st 2.3, 2.4; 2nd 2.5, 2.6; 3rd 2.7, 2.8; 4th 2.7, 2.8; 5th 2.5, 2.2.

First dorsal fin: vertical height 9.1, 9.4; length of base 11.1, 10.8.

Second dorsal fin: vertical height 2.5, 2.9; length of base 3.4, 3.4.

Anal fin: vertical height 3.4, 3.2; length of base 4.6, 4.6.

Caudal fin: upper margin 26.3, 26.6; lower anterior margin 11.7, 12.4.

Pectoral fin: outer margin 15.5, 15.7; inner margin 5.9, 6.0; distal margin 12.2, 12.4.

Distance from snout to: 1st dorsal 33.0, 32.2; 2nd dorsal 64.3, 62.6; upper caudal 73.7, 73.4; pectoral 23.5, 23.8; pelvics 47.0, 48.6; anal 60.7, 60.4.

Interspace between: 1st and 2nd dorsals 20.7, 20.6; 2nd dorsal and caudal 7.8, 7.0; anal and caudal 7.6, 6.8.

Distance from origin to origin of: pectoral and pelvics 25.2, 26.0; pelvics and anal 13.4, 13.8.

Proportional dimensions in per cent of total length. Male, 831 mm., from Pacific Panama (U.S. Nat. Mus., No. 79293).

Trunk at origin of pectoral: breadth 11.2; height 12.0.

Snout length in front of: outer nostrils 4.1; mouth 7.5.

Eye: horizontal diameter 1.6.

Mouth: breadth 8.2; height 5.1.

Nostrils: distance between inner ends 5.2.

Gill opening lengths: 1st 3.1; 2nd 3.5; 3rd 3.4; 4th 3.4; 5th 3.0.

First dorsal fin: vertical height 9.5; length of base 12.2.

Second dorsal fin: vertical height 2.9; length of base 4.6.

Anal fin: vertical height 3.8; length of base 4.7.

Caudal fin: upper margin 24.7; lower anterior margin 11.9.

Pectoral fin: outer margin 16.0; inner margin 5.8; distal margin 14.2.

Distance from snout to: 1st dorsal 33.3; 2nd dorsal 64.7; upper caudal 75.3; pectoral 23.6; pelvics 48.2; anal 62.5.

Interspace between: 1st and 2nd dorsals 21.9; 2nd dorsal and caudal 7.4; anal and caudal 7.1.

Distance from origin to origin of: pectoral and pelvics 25.3; pelvics and anal 13.5.

Height of trunk at 1st dorsal about 1/6 its length to origin of caudal. Midline of back smooth, with no trace of mid-dorsal ridge. Caudal peduncle about 2/3 as thick as high. Upper precaudal pit strongly marked, semilunar to subangular, the lower similar but less strongly marked. Dermal denticles overlapping so little that the skin is regularly or partially exposed between them, their blades rising rather steeply and varying in size, only a little broader than long, moderately arched longitudinally, with 3 ridges in small specimens but 5 in larger, the median tooth considerably the largest; pedicels very short.

Head about 1/4 of total length, its dorsal profile very weakly convex, about as broad at eyes as at origin of pectorals. Snout ovoid with rather broadly rounded tip, its length in front of a line connecting outer ends of nostrils very nearly equal to distance between inner ends of latter on smallest specimens but only 3/4 that in larger, the length in front of mouth about 1.4 to 1.5 times as great as distance between nostrils and a little less than 2/5 (37 to 38%) as great as length of head to origin of pectorals. Eye approximately circular and relatively large, its diameter a little less than 1/2 as long as distance between inner ends of nostrils in smallest specimens, 1/3 that long in larger, its anterior edge opposite front of mouth or a little anterior to latter. Gill openings very slightly oblique with sinuously concave contours, the 1st about as long as the horizontal diameter of the eye in smallest specimens but increasing relatively to about 1.2 times as long as eye in specimens of 500 mm. and twice the eye at a length of 800 to 900 mm., the 3rd very little longer than 1st, the 5th about as long as 1st, the 4th about over origin of pectoral. Nostrils moderately oblique, about 1/3 as long as the distance between their inner ends, which are nearer to mouth than to tip of snout by a distance about 1/2 as long as between nostrils, the anterior margins somewhat sinuous and expanded near inner end as a short digitate lobe with rounded tip.[140] Mouth broad-ovoid, its height relative to its breadth somewhat greater in smaller specimens (60 to 66%) than in the larger (about 54%), occupying about 2/3 of breadth of head.

Teeth $\frac{13\text{—}1\text{—}13}{12 \text{ or } 13\text{—}0 \text{ or } 1\text{—}12 \text{ or } 13}$; uppers broadly triangular, their edges serrate, most coarsely so basally except for the 2 or 3 outermost which are only slightly irregular, the 1st nearly symmetrical with nearly straight edges, but 2nd and subsequent teeth increasingly oblique, their inner margins slightly sinuous or concave, their outer margins notched more and more deeply toward corners of mouth, the 10th to 13th successively lower, the outermost tooth very low, its cusp hardly discernible, the median upper tooth small and symmetrical; lower teeth with much narrower cusps than uppers, on expanded bases, the 1st to 8th or 9th serrate from tip to base, although somewhat less coarsely so than uppers, but subsequent teeth irregularly wavy at most, the 1st much smaller than 2nd to 6th, those toward center of mouth erect, nearly symmetrical, the outermost 4 or 5 in smallest specimens, but outermost 1–3, only, in larger examples, very oblique, deeply notched outwardly and with very low cusps.

Origin of 1st dorsal about over midpoint of inner margin of pectoral, its anterior margin about 1/2 as long as head, its vertical height about as great as length of snout in front of mouth in small specimens but relatively somewhat greater in larger, its anterior margin only very slightly convex toward apex, its posterior margin moderately and evenly concave, its apex narrowly rounded, its free rear corner about 1/2 as long as base, the midpoint of base about 2/3 as far from axil of pectoral as from origin of pelvics. Second dorsal about 1/3 (30 to 37%) as long at base as 1st, about 1/3 to 1/4 as high vertically and relatively much lower in form, its origin about over midpoint of base of anal, the rear end of its

140. This is clearly shown in Ranzani's illustration (Nov. Comment. Acad. Sci. Inst. Bonon [Bologna], 4, 1839: pl. 2).

base a little posterior to rear end of base of anal, its tip a little posterior to tip of anal, its apex only very narrowly rounded, its free rear corner slender and about as long as base. Caudal about 1/4 of total length, its upper margin slightly to moderately convex, its terminal sector tapering and a little less than 1/3 the length of fin, with narrowly rounded tip and weakly concave lower posterior margin, the lower lobe a little less than 1/2 as long as upper with its tip narrowly rounded, the re-entrant corner (enclosed by the 2 lobes) more obtuse than a right angle and broadly rounded. Distance from caudal to tip of anal about as long as base of anal in smallest specimens but only about 2/3 that long in larger. Anal about 1.0 to 1.3 times as long at base as 2nd dorsal, its anterior margin much more strongly convex and posterior margin much more deeply concave, its apex more rounded, its free corner slender and about as long as base or slightly shorter. Distance from origin of anal to tips of pelvics a little longer than base of anal. Pelvics with nearly straight anterior and posterior margins, about as long at base as anal. Pectoral noticeably small, about 2/3 as long as head or only a little longer than anterior margin of 1st dorsal and a little more than 1/2 as broad as long, the outer margin only very weakly convex, the distal margin weakly and evenly concave, the tip and inner corner very narrowly rounded.

Color. Described as leaden or bluish gray above, the sides sometimes tinged with reddish, the lower surface pale; the pelvics sometimes with reddish tinge toward their bases; edges of lower fins and hind edge of lower caudal lobe white.

Size. The largest specimen so far recorded was 1,235 mm. (49 inches) long.[141] The fact that the claspers in an 831-mm. Pacific specimen are twice as long as the pelvic fins suggests that this shark does not reach a length much greater than perhaps four feet.

Developmental Stages. Embryos have not been described as yet.

Remarks. The specimens listed above can be referred to *porosus* without hesitation, so clearly diagnostic are the original account and illustrations of that species.[142]

Habits. The localities of capture, listed below, show this to be a strictly subtropical-tropical species and probably littoral. But nothing more is known of its habits.

Relation to Man. Saleable for human food in the markets of Colón and Panama.

Range. Western tropical Atlantic; northern Brazil to north shore of Gulf of Mexico; also eastern tropical Pacific, Peru to Panama;[143] represented on the Atlantic coast of North Africa by a form that may finally prove to be identical.[144]

141. Herre, Field Mus. Publ. Zool., *21*, 1936: 22; a Pacific specimen.

142. Ranzani, Nov. Comment. Acad. Sci. Inst. Bonon (Bologna), *4*, 1839: 8, pl. 2.

143. Our comparison of specimens from Payta, Peru and Panama with others from the Atlantic corroborates Meek and Hildebrand's (Field Mus. Publ. Zool., *15* [1], 1923: 49) conclusion that examples from the two sides of the Isthmus of Panama represent only a single species, which they recorded and described as *cerdale* Gilbert, 1898. But we find nothing to separate the latter from the original account and illustrations of *porosus* Ranzani, 1839. And we should perhaps point out that Garman's (Mem. Harv. Mus. comp. Zool., *36*, 1913: 131) account of *porosus* as with broadly rounded snout and with nostril as far from end of snout as from eye, characters used by Meek and Hildebrand in their Key as alternative between *porosus* and *cerdale*, does not fit the West Indian and Brazilian specimens in the Museum of Comparative Zoology, from which Garman's description appears to have been taken.

144. *Carcharias fissidens* Bennett (Proc. zool. Soc. Lond., 1830–1831: 148), in which the origin of the second dorsal is described as over the middle of the anal and the outer edges of the teeth as deeply notched. But the account of it is not detailed enough for decision.

Occurrence in the Western Atlantic. Positive records for *porosus* are from Bahia, Pernambuco, Marajo Island at the mouth of the Amazon; British, Dutch and French Guiana; Trinidad; Colón; and from the north shore of the Gulf of Mexico.[145] Evidently it ranges generally throughout the Gulf of Mexico, Caribbean region, and southward as far as central Brazil.

Synonyms and References:

Carcharias porosus Ranzani, Nov. Comment. Acad. Sci. Inst. Bonon (Bologna), *4*, 1839: 8, pl. 2 (descr., color, excellent ill., Brazil); Boulenger, Ann. Mag. nat. Hist., (6) *20*, 1897: 298 (Marajo I., Brazil); Goeldi, Bol. Mus. Paraense, *2*, 1898: 488 (Marajo I., Brazil); Ribeiro, Ann. Mus. nac. Rio de J., *14*, 1907: 153, 200 (descr., Brazil, refs.); Pellegrin, Poiss. Guyana Franc., Rev. Colon., *67*, 1908: 11 (French Guiana); Ribeiro, Fauna brasil. Peixes, *2* (1) Fasc. 1, 1923: 11 (descr., Brazil).

Carcharias (Prionodon) henlei Müller and Henle, Plagiost., 1841: 46,[146] pl. 19, fig. 6 (descr., ill. of teeth, Cayenne); Müller and Troschel, in Schomburgk, Reisen Brit. Guiana (1840–1844), *3*, 1848: 641 (Brit. Guiana); Duméril, Hist. Nat. Poiss., *1*, 1865: 372 (descr., Cayenne).

Carcharias (Prionodon) porosus Duméril, Hist. Nat. Poiss., *1*, 1865: 373 (Brazil, Cayenne); Günther, Cat. Fish. Brit. Mus., *8*, 1870: 365 (Guiana).

Carcharhinus (Platypodon) henlei Jordan and Evermann, Rep. U.S. Comm. Fish. (1895), 1896: 216 (Brazil to Guiana); Bull. U.S. Nat. Mus., 47 (1), 1896: 37 (descr., Brazil to Guiana).

Carcharhinus cerdale Gilbert, in Jordan and Evermann, Bull. U.S. nat. Mus., 47 (3), 1898: 2746 (descr., Pacif. Panama); Wilson, Ann. Carneg. Mus., *10*, 1916: 58 (Colombia, Ecuador); Meek and Hildebrand, Field Mus. Publ. Zool., *15* (1), 1923: 47 (descr., size, embryos, abund., Atlant. and Pacif. Panama); Herre, Field Mus. Publ. Zool., *21*, 1936: 22 (refs., descr., max. size, Galapagos and at sea).

Carcharias cerdale Gilbert and Starks, Mem. Calif. Acad. Sci., *4*, 1904: 10, pl. 2, fig. 4 (Panama Bay); Starks, Proc. U.S. nat. Mus., *30*, 1906: 762 (Ecuador); Jordan, Evermann and Clark, Rep. U.S. Comm. Fish. (1928), 2, 1930: 16 (Panama).

Carcharinus menisorrah (in part) Garman, Mem. Harv. Mus. comp. Zool., *36*, 1913: 135 (*cerdale* incl. in synonymy); not *Carcharias (Prionodon) menisorrah* Müller and Henle, 1841.

Carcharinus porosus Garman, Mem. Harv. Mus. comp. Zool., *36*, 1913: 131 (descr., W. Indies, Brazil); Meek and Hildebrand, Field Mus. Publ. Zool., *15* (1), 1923: 49 (descr.); White, Bull. Amer. Mus. nat. Hist., *74*, 1937: 126 (in Key).

Eulamia porosus Fowler, Proc. Acad. nat. Sci. Philad., *69*, 1917: 127 (Colón); Proc. Acad. nat. Sci. Philad., *83*, 1931: 391 (Trinidad); Arqu. Zool. Estado São Paulo, *3*, 1942: 128 (listed, Brazil).

Carcharias henlei Jordan, Evermann and Clark, Rep. U.S. Comm. Fish. (1928), 2, 1930: 16.

Carcharhinus lamiella Kumada and Hiyama, Mar. Fish. Pacif. Mexico, 1937: 16, pl. 48 (ident. by ill.); not *Carcharias lamiella* Jordan and Gilbert, 1882.

Eulamia cerdale Fowler, Monog. Acad. nat. Sci. Philad., *2*, 1938: 249 (listed for Galapagos); Beebe and Tee-Van, Zoologica, N. Y., *26*, 1941: 109 (descr., ill., refs., distrib. in Pacif.).

Doubtful References:

Carcharias fissidens Bennett, Proc. zool. Soc. Lond., 1830–31: 148 (Atlant. coast N. Afr., see footnote 144, p. 398).

Carcharhinus henlei Puyo, Bull. Soc. Hist. nat. Toulouse, *70*, 1936: 85, 89, 250 (descr., meas., French Guiana; ident. doubtful since 2nd dorsal is described as "insérée sur le même plan et au même niveau que la nageoire anale.").

145. Stewart Springer, in a personal communication, reports "a nice series of '*cerdale*' " taken off Biloxi, Mississippi, by himself in August 1943.

146. If this page was actually distributed in 1838, as it is stated in Jordan and Evermann (Bull. U.S. nat. Mus., 47 [1], 1896: 37) and in Jordan, Evermann and Clark (Rep. U.S. Comm. Fish. [1928], 2, 1930: 16) the name *henlei* would have priority over *porosus*. Not being in a position to verify this, we credit Müller and Henle's classic work with the date (1841) with which the title page is inscribed.

Carcharhinus remotus (Duméril), 1865

Figures 76, 77

Study Material. Seven specimens from Rio de Janeiro, about 650 to 695 mm. long (Harv. Mus. Comp. Zool., No. 703).[147]

Distinctive Characters. Among the smooth-backed subdivision of the genus, with which it falls, *remotus* is closest to *limbatus* and *maculipinnis* as regards teeth, but it is separated from both of these by its much shorter gill openings (see Key, p. 324), as well as by its lower precaudal pit, which is only faintly indicated, and by its fins which are without conspicuous black markings, at least after preservation.

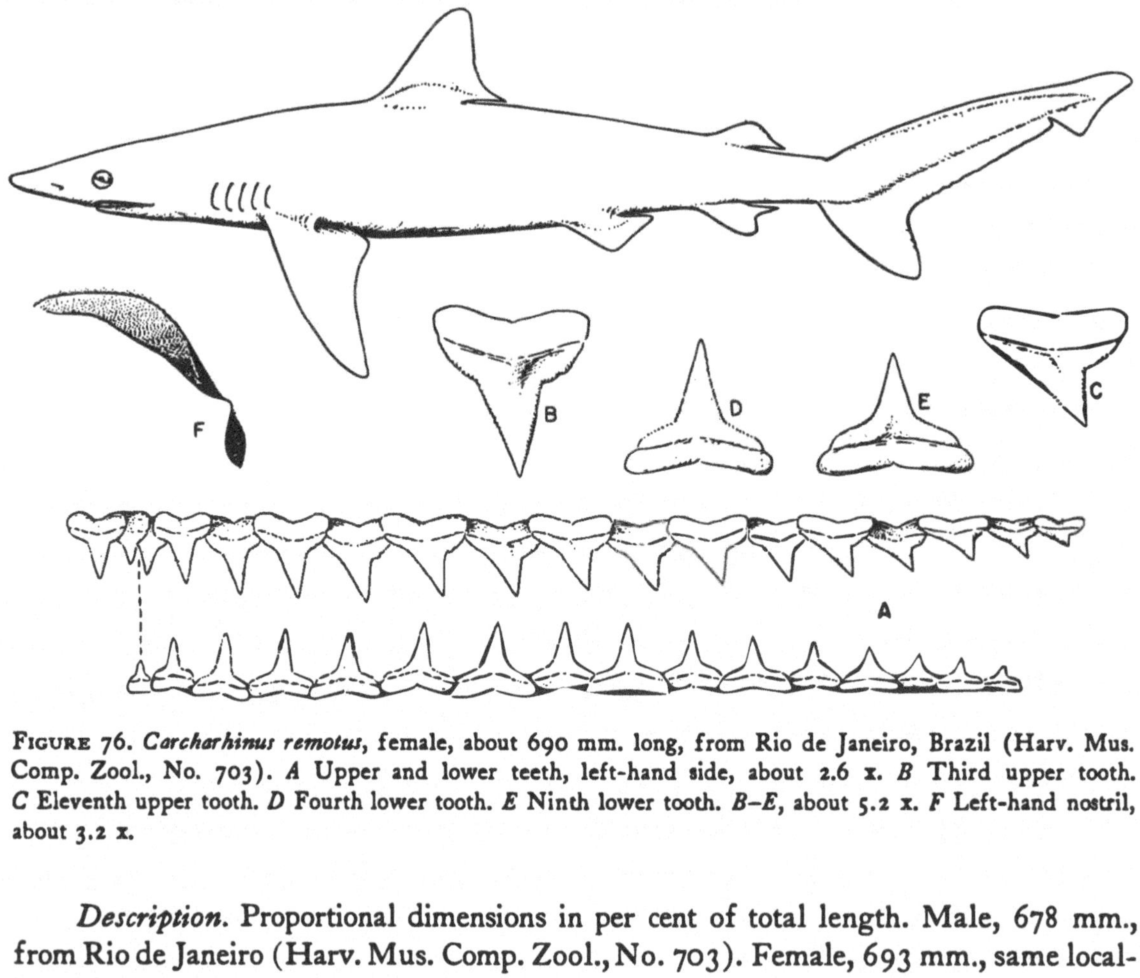

FIGURE 76. *Carcharhinus remotus*, female, about 690 mm. long, from Rio de Janeiro, Brazil (Harv. Mus. Comp. Zool., No. 703). *A* Upper and lower teeth, left-hand side, about 2.6 x. *B* Third upper tooth. *C* Eleventh upper tooth. *D* Fourth lower tooth. *E* Ninth lower tooth. *B–E*, about 5.2 x. *F* Left-hand nostril, about 3.2 x.

Description. Proportional dimensions in per cent of total length. Male, 678 mm., from Rio de Janeiro (Harv. Mus. Comp. Zool., No. 703). Female, 693 mm., same locality and number.

147. While provisional acceptance of Garman's (Mem. Harv. Mus. comp. Zool., *36*, 1913: 138) reference of these specimens to this species seems justified, the original account of it was so brief that the correctness of this identification can be tested only by re-examination of the type specimen, now or formerly in the Paris Museum.

Trunk at origin of pectoral: breadth 9.3, 9.4; height 9.6, 9.5.
Snout length in front of: outer nostrils 3.8, 4.0; mouth 7.2, 7.1.
Eye: horizontal diameter 2.0, 2.0.
Mouth: breadth 7.5, 7.4; height 5.2, 5.1.
Nostrils: distance between inner ends 5.5, 5.5.
Gill opening lengths: 1st 2.8, 2.7; 2nd 3.0, 2.8; 3rd 2.9, 2.7; 4th 2.5, 2.5; 5th 2.4, 2.0.
First dorsal fin: vertical height 8.5, 8.5; length of base 8.9, 9.0.
Second dorsal fin: vertical height 2.5, 2.5; length of base 4.0, 3.9.
Anal fin: vertical height 3.7, 3.6; length of base 4.4, 4.6.
Caudal fin: upper margin 27.0, 27.5; lower anterior margin 11.8, 11.8.
Pectoral fin: outer margin 16.8, 17.0; inner margin 5.3, 5.5; distal margin 13.3, 13.6.
Distance from snout to: 1st dorsal 33.6, 33.4; 2nd dorsal 64.4, 62.7; upper caudal 73.0, 72.5; pectoral 24.2, 23.5; pelvics 50.2, 48.7; anal 62.5, 61.5.
Interspace between: 1st and 2nd dorsals 21.8, 21.1; 2nd dorsal and caudal 6.8, 6.8; anal and caudal 6.2, 6.1.
Distance from origin to origin of: pectoral and pelvics 26.9, 26.4; pelvics and anal 12.5, 12.1.

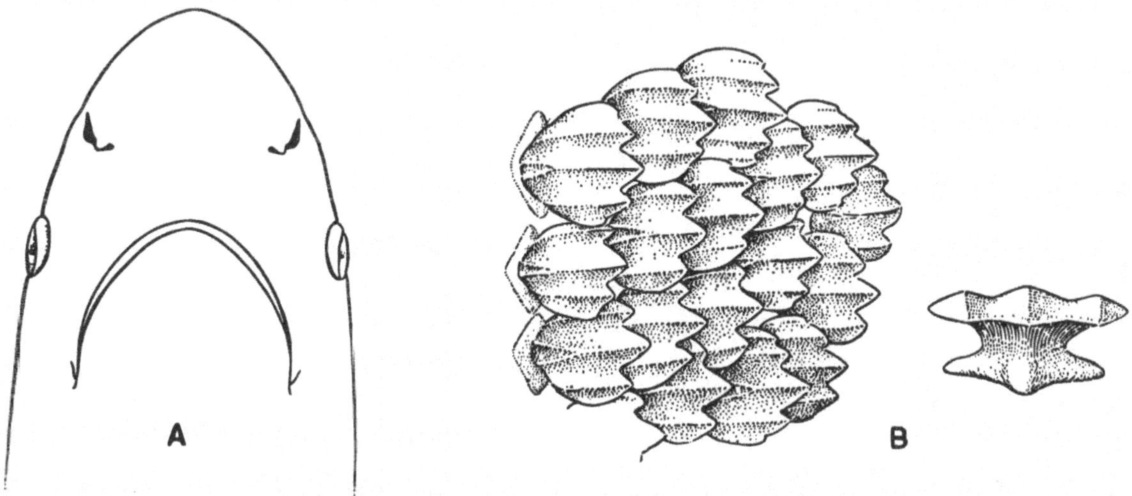

FIGURE 77. *Carcharhinus remotus,* pictured in Fig. 76. *A* Anterior part of head from below. *B* Dermal denticles: general view, about 45 x; apical view, about 90 x.

Trunk slender, its height at 1st dorsal only about ⅙ its length to origin of caudal. Back smooth, without mid-dorsal ridge. Body sector to cloaca about 1.1 times as long as tail sector. Caudal peduncle about ⅔ as thick as deep. Upper precaudal pit strongly developed, subangular and a little less than a right angle in outline, the lower pit only weakly marked. Dermal denticles overlapping so regularly that the skin is exposed only here and there, the blades rising rather steeply, a little broader than high, with 3 to 5 very low ridges and 3 to 5 short, broad teeth, the median considerably the largest and the outermost pair very small when there are 5; pedicels very short.

Head a little less than ⅓ length of trunk to origin of caudal, its dorsal profile sloping evenly forward. Snout moderately thin-tipped, ovate, with broadly rounded tip, its length anterior to a line connecting outer ends of nostrils a little more than ⅔ (about 69 to 73%) of distance between inner ends of latter, its length in front of mouth about 1.3 times distance between nostrils and a little less than ⅓ (about 30%) as great as length of head to origin of pectoral. Eye approximately circular, its anterior margin opposite or very slightly behind front of mouth, its diameter a little more than ⅓ (about 35%) as great as distance between nostrils. Gill openings about evenly spaced, their margins weakly concave, the 1st about ½ (50%) as long as distance between nostrils and a little less than 1.5 times as long as horizontal diameter of eye, the 2nd and 3rd very slightly the longest, the 5th about ¾ (74 to 85%) as long as 1st, the space between 4th and 5th over origin of pectoral. Nostrils strongly oblique, about ¼ as long as distance between their inner ends, which are a little less than twice as far from tip of snout as from mouth, the anterior margin moderately expanded as a low subangular lobe with rounded apex. Mouth ovate, about ⅔ (67%) as high as broad, occupying about ⅔ of breadth of head. Upper labial fold between ⅓ and ¼ as long as diameter of eye.

Teeth $\frac{\text{13 to 15—1 or 2—13 to 15}}{\text{14 or 15—1—14 or 15}}$; uppers with narrow triangular cusps on moderately expanded bases, nearly symmetrical toward center of mouth but slightly oblique toward its corners, both margins concave, the outer margins increasingly so toward corners of mouth in subangular contour, the edges regularly and rather finely serrate from tip to base except for the outermost 1 or 2 teeth, which are only irregular or wavy, the 1st to 10th or 11th teeth of about equal lengths, but subsequent teeth successively shorter, the outermost very short; lower teeth with considerably narrower cusps than uppers, on more broadly expanded bases, erect and symmetrical all along jaw, the edges much more finely serrate than uppers from tip to base, the 2nd to 11th longest and the outermost very short; 1 or 2 small teeth at symphysis in upper jaw and 1 in lower.

Origin of 1st dorsal about over inner corner of pectoral, its vertical height a little more than ⅓ as great as length of head, its anterior margin slightly convex toward apex, its posterior margin moderately concave toward base, its apex moderately rounded, its free rear corner only about ⅓ as long as the base, the midpoint of base only a little nearer to axil of pectoral than to origin of pelvics. Second dorsal slightly less than ½ as long at base as 1st and a little less than ⅓ as high vertically, its apex broadly rounded, its rear margin only very slightly concave, its free rear corner about as long as its base, its origin a little posterior to origin of anal but considerably anterior to midpoint of latter. Caudal a little more than ¼ (27 to 28%) of total length, its upper margin only slightly convex with terminal sector between 1/4 and 1/3 (28 to 29%) the length of fin, slender, with narrowly rounded tip, the lower lobe about 40% as long as upper, its tip narrowly rounded, the re-entrant contour (included by the 2 lobes) well rounded. Distance from origin of caudal to tip of anal about 4/5 as long as base of anal. Anal about 1.1 times as long at base as 2nd dorsal, with slightly more convex anterior and much more deeply concave posterior margins and rounded apex, its free rear corner only about 2/3 as long as base, its tip a little anterior to that of 2nd dorsal. Distance from origin of anal to tips of pelvics a little shorter than base of anal. Pelvics about as long at base as anal, and only about as large as latter in area. Pectoral about 0.7 as long as head, and a little less than 1/2 as broad as long, the outer margin weakly and evenly convex, the distal margin moderately concave, the apex and inner corner both very narrowly rounded.

Color. Preserved specimens are mouse gray or brownish-gray above, paler below, the fins with darker edges, but without conspicuous black markings. The color of fresh-caught specimens has not been recorded.

Size. The fact that two of the present series still show traces of the umbilical scar suggests a length of about 600 to 650 mm. at birth. But any statement as to the size to which *remotus* grows would be pure speculation, the only pertinent information being that the type specimen was 1,200 mm. (about 47 inches), and that an immature male from northern Argentina, probably of this species, was 1,030 mm. (about 41 inches) long.

Developmental Stages. Not known.

Habits. Nothing whatever is known of the habits of this species.

Range. Western tropical and subtropical Atlantic. The few records that can be referred to *remotus* with confidence are for the Antilles (type specimen[148]), Rio de Janeiro (see Study Material, p. 400) and probably northern Argentina.

Synonyms and References:

Carcharias (Prionodon) remotus Duméril, Hist. Nat. Poiss., *1*, 1865: 374 (descr., 1.2-m. spec., Antilles); Günther, Cat. Fish. Brit. Mus., *8*, 1870: 363 (footnote ref. to Duméril, 1865).

Carcharhinus (Platypodon) remotus Jordan and Evermann, Rep. U.S. Comm. Fish. (1895), 1896: 216 (Martinique); Bull. U.S. nat. Mus., 47 (1), 1896: 37 (descr., Martinique).

Carcharinus remotus Garman, Mem. Harv. Mus. comp. Zool., *36*, 1913: 138 (descr. of specs. listed in Study Material, p. 400, Rio de Janeiro); Meek and Hildebrand, Field Mus. Publ. Zool., *15* (1), 1923: 47 (descr., after Garman); White, Bull. Amer. Mus. nat. Hist., *74*, 1937: 127 (in Key).

Carcharias remotus Jordan, Evermann and Clark, Rep. U.S. Comm. Fish. (1928), 2, 1930: 16 (W. Indies and Brazil).

Eulamia remota Fowler, Arqu. Zool. Estado São Paulo, *3*, 1943, 128 (listed, Brazil).

Probable Reference:

Carcharias lamia Lahille, An. Mus. nac. B. Aires, *34*, 1929: 305, pl. 3, lower fig. (north. Argentina, probably referable to *remotus* by proportional dimensions of 1,030-mm. male, by ill. of ventral side of head and absence of black fin markings; but upper teeth, as illustrated, are intermediate between that species and *limbatus*); Pozzi and Bordale, An. Soc. cient. argent., *120*, 1935: 150 (name only, apparently refers to Lahille's reference); not *C. lamia* Risso, 1826, or Müller and Henle, 1841.

148. Martinique, according to Jordan and Evermann (Bull. U.S. nat. Mus., 47 [1], 1896: 37).

Carcharhinus springeri (Bigelow and Schroeder), 1944

Figure 78

Study Material. Female, 805 mm. in total length, from Cozumel, Mexico (U.S. Nat. Mus., type, No. 37141); a somewhat shrivelled skin and head of a female, probably this species, about 1,390 mm. in total length, from Englewood, Florida (Harv. Mus. Comp. Zool., No. 35900). These are the only specimens of the species yet seen.

Distinctive Characters. C. springeri most nearly resembles *C. obscurus,* with which it agrees generally in the relative size and position of fins and the presence of a mid-dorsal ridge. But it differs in a number of features from specimens of *obscurus* of approximately the same size with which we have compared it; (a) its eye is considerably larger relative to

FIGURE 78. *Carcharhinus springeri,* female, 805 mm. long, from Cozumel, Mexico (U. S. Nat. Mus., No. 37141, type). *A* Anterior part of head from below, about 0.3 x. *B* Cross-section of back, midway between the two dorsal fins, to show the mid-dorsal ridge, slightly enlarged. *C* Left-hand nostril, about 1.6 x. *D* Dermal denticles, about 18 x. *E* Upper and lower teeth from right-hand side, about 1.4 x. *F* Fifth upper tooth. *G* Seventh upper tooth. *H* Second lower tooth. *I* Ninth upper tooth. *F–I,* about 2.6 x.

the lengths of the gill openings; (b) the anterior margin of its nostril is expanded as a low, triangular lobe (not lobed in *obscurus*); (c) its first dorsal is relatively larger and more erect, but with the free rear corner relatively shorter and the posterior margin less deeply concave; (d) its second dorsal is larger in area but shorter (from origin to rear tip) relative to its vertical height; (e) the distance from the tips of the pelvics to the origin of the anal is considerably shorter, *i.e.,* about 0.7 of the anal base (1.3 times the anal base in *obscu-*

rus); (f) the outer corners of its pelvics are only about at a right angle (about 115° in *obscurus*); (g) its dermal denticles have a larger number of marginal teeth and ridges in specimens of equal size; and (h) its upper teeth are more strongly oblique, more deeply incised outwardly and more coarsely serrate basally than those of *obscurus*.

Among carcharhinids of the west coast of America, *springeri* resembles most nearly *platyrhynchus* (Gilbert), 1891, in its teeth and fins. But the length of its snout in front of the mouth is considerably less than the breadth of the mouth (a little longer than breadth of mouth in *platyrhynchus*), and its fins show no trace of the white edgings that are so conspicuous in *platyrhynchus*. In combination these differences seem sufficient to demand recognition in nomenclature, especially in view of the geographic discontinuity between the areas of occurrence of the two sharks.

Springeri is similar to *galapagoensis* Snodgrass and Heller, 1905, in the teeth and snout, but separated from it by the shapes of the first dorsal and pectoral fins, much larger eye relative to the lengths of the gill openings, and by the fact that its second dorsal is only very little smaller than its anal in area.

Description. Proportional dimensions in per cent of total length. Female, 805 mm., from Cozumel, Mexico (U.S. Nat. Mus., type, No. 37141).

Trunk at origin of pectoral: breadth 12.0; height 11.4.
Snout length in front of: outer nostrils 3.1; mouth 6.6.
Eye: horizontal diameter 2.1.
Mouth: breadth 8.9; height 5.7.
Nostrils: distance between inner ends 6.5.
Labial furrow length: upper 0.6.
Gill opening lengths: 1st 2.3; 2nd 2.6; 3rd 2.7; 4th 2.6; 5th 2.2.
First dorsal fin: vertical height 10.7; length of base 9.9.
Second dorsal fin: vertical height 3.3; length of base 4.5.
Anal fin: vertical height 3.9; length of base 4.9.
Caudal fin: upper margin 29.8; lower anterior margin 14.2.
Pectoral fin: outer margin 19.9; inner margin 5.2; distal margin 15.6.
Distance from snout to: 1st dorsal 30.7; 2nd dorsal 58.8; upper caudal 70.2; pectoral 20.5; pelvics 47.9; anal 58.8.
Interspace between: 1st and 2nd dorsals 20.7; 2nd dorsal and caudal 7.7; anal and caudal 6.1.
Distance from origin to origin of: pectoral and pelvics 27.2; pelvics and anal 11.0.

General form moderately stout. Trunk height at origin of 1st dorsal (where highest) about ⅕ of length to origin of caudal. Body sector to cloaca somewhat longer than tail sector. Midline of back with a low but unmistakable dermal ridge between dorsals. In the preserved state there is also a similar ridge between 2nd dorsal and caudal, but since this lies along the bottom of a groove of muscular contraction we question whether it is a normal feature. Upper precaudal pit strongly marked, subangular in outline, the lower pit lunate, less distinct than upper. Dermal denticles closely and regularly overlapping, broader than long, with 5 (sometimes 7) short, marginal teeth and an equal number of low ridges: in the larger specimen, probably of this species, the denticles usually have 7 ridges and teeth.

Head to origin of pectoral a little less than 1/3 of length (29%) to origin of caudal, moderately flattened above. Snout broadly rounded, its length anterior to a line connecting outer ends of nostrils a little less than 1/2 as great as length in front of mouth, the length in front of mouth nearly 3/4 (74%) as great as breadth of mouth or about 1/3 of length of head to origin of pectorals. Eye approximately circular, its diameter a little less than 1/3 as long as distance between nostrils and about as long as 1st gill opening. Third gill opening (a little the longest) a little more than 2/5 (42%) as long as distance between nostrils, the 5th about 4/5 as long as 3rd and more oblique than the others, the 3rd gill opening above origin of pectoral. Nostril strongly oblique, its inner corner nearer to mouth than to the tip of snout by a distance a little longer than its own length or diameter of eye, its anterior margin expanded toward inner end as a distinct but low subtriangular lobe. Mouth broad-ovate, a little less than 2/3 (about 62 to 64%) as high as wide.

Teeth $\frac{12-1-13}{12-1-13}$; uppers triangular, 1st and 2nd erect and nearly symmetrical, with concave margins and expanded bases, but subsequent teeth increasingly oblique toward corners of mouth, their inner margins nearly straight, but their outer margins more and more deeply concave in subangular contour, the outermost strongly so, the margins regularly serrate from tip to base, most strongly so outwardly on basal sector; lower teeth nearly erect, except for the outermost 3 or 4, which are moderately oblique, their cusps much more slender than those of uppers, on broadly expanded bases, their margins much more finely serrate than those of uppers; one small symmetrical tooth at symphysis, and outermost 3 to 5 successively smaller in each jaw.

Origin of 1st dorsal a little anterior to inner corner of pectoral, its base nearly as long as from posterior edge of eye to 1st gill opening, its vertical height about equal to its base, its anterior margin only weakly convex toward the apex, the posterior margin only weakly concave (less so than in *obscurus*), apex narrowly rounded, the free rear corner only about 1/3 to 1/4 as long as the base (thus relatively shorter than in *obscurus*), the midpoint of base about 1.7 times as far from origin of pelvics as from axil of pectoral. Second dorsal between 1/2 and 1/3 as long at base as 1st, relatively lower, its vertical height about 2/3 as great as its length at base, its free rear corner a little shorter than its base, its posterior margin weakly concave, its extreme length from origin to rear tip about 2.5 times as great as the vertical height (considerably shorter relative to its height than in *obscurus*), its origin about opposite to that of anal. Caudal between 1/3 and 1/4 (29.8%) of total length, its terminal sector about 1/4 of total length of fin, slender, with narrowly rounded tip and weakly concave lower posterior outline; the lower lobe (expanded lower anterior corner) only a little less than 1/2 (about 47%) as long as upper (somewhat longer, relatively, than in *obscurus*), with weakly convex anterior margin, nearly straight posterior margin, and narrowly rounded tip, the re-entrant corner (included by the 2 lobes) well rounded. Distance from origin of caudal to tip of anal about 2/3 as long as base of latter. Anal a little longer at base than 2nd dorsal and a very little higher vertically, its posterior margin much more deeply concave, its free rear corner about 2/3 as long as base. Distance from origin of anal to tips of pelvics only about 3/4 as long as base of anal. Pelvics about as long at base as base of anal, or a little longer, their outer corners approximately a right angle (less obtuse than in *obscurus*). Pectoral about 4/5 as long as head, a little less than 1/2 as broad as long, with narrowly rounded tip and inner corner, weakly and evenly convex outer margin, and distal margin deeply concave proximally.

Color. After preservation the type specimen is olive gray above, and of a paler shade of yellowish olive below, without any conspicuous fin markings, dark or light.

Size. The fact that the type specimen still shows the umbilical scar, although it is 805 mm. long, suggests that this is one of the larger members of its genus.

Developmental Stages. Embryos have not been seen as yet.

Habits. Nothing is known of the life history of this newly described species.

Range. C. springeri is known only from Cozumel, east coast of Yucatán, and (probably) off the west coast of Florida; see Study Material, p. 404.

Synonyms and References:

Eulamia springeri Bigelow and Schroeder, Proc. New Engl. zool. Cl., *23*, 1944: 30, pl. 9, 10 (descr., ill., Cozumel).

Family *SPHYRNIDAE*

Hammerhead Sharks

Characters. In general the characters are those of the Carcharhinidae (p. 262), except that the anterior portion of the head is much flattened dorso-ventrally and very widely expanded laterally in "hammer" or "bonnet" form, with the eyes at its outer edges; and the skull is modified accordingly, its anterior portion with the olfactory capsules and orbital region being very widely expanded, and the three rostral bars transversely and broadly truncate in front at their union. Development viviparous, with yolk-sac placenta in some species,[1] but perhaps ovoviviparous in others.

Genera. Two, as indicated in the following Key.

Key to Genera

1a. Nostrils closer to midline of snout than to eyes. — *Eusphyra* Gill, 1862. Tropical Indian Ocean, Malaysian region, Indo-China and northern Australia.

1b. Nostrils much closer to eyes than to midline of snout. — *Sphyrna* Rafinesque, 1810, p. 408.

1. For the placenta in *Eusphyra blochii*, see Alcock (J. Asiat. Soc. Beng., *59* [2], 1890: 52).

Genus *Sphyrna* Rafinesque, 1810

Sphyrna Rafinesque, Indice Ittiol. Sicil., 1810: 46, 60; type species, *Squalus zygaena* Linnaeus, 1758, designated by Jordan and Gilbert, Bull. U.S. nat. Mus., 16, 1883: 26.

Generic Synonyms:

Cestracion Klein, Pisc. Natural. Gedoni, *3*, 1742: 12; Neuer Schauplatz Natur., *3*, 1776: 523 (not seen); Walbaum, P. Artedi Genera Pisc. Emend. Ichthyol., 1792: 580; type species, *Squalus zygaena* Linnaeus, 1758, designated by Gill, Ann. N. Y. Lyc., 7, 1862: 403.[2]

Sphyrnias Rafinesque, An. Nature, 1815: 93; substituted for *Sphyrna* Rafinesque, 1810.

Cestrorhinus Blainville, Bull. Soc. philom. Paris, 1816: 121; type species, *Squalus zygaena* Linnaeus, 1758, designated by Fowler, Bull. geol. Surv. N. Jersey, *4*, 1911: 77.

Zygaena Cuvier, Règne Anim., *2*, 1817: 27; type species, *Squalus zygaena* Linnaeus, 1758, preoccupied by *Zygaena* Fabricius, 1775, for Lepidoptera.

Zygoena Risso, Hist. Nat. Europe Merid., *3*, 1826: 125; emended spelling for *Zygaena* Cuvier, 1817.

Sphyrnichthys Thienemann, Lehrb. Zool., 1828: 408; substituted for *Sphyrna* Rafinesque, 1810.

Zygana Swainson, Nat. Hist. Fish. Amphib. Rept., *2*, 1839: 318; emended spelling for *Zygaena* Cuvier, 1817.

Platysqualus Swainson, Nat. Hist. Fish. Amphib. Rept., *2*, 1839: 318; type species, *Squalus tiburo* Linnaeus, 1758.

Sphyra Van der Hoeven, Handb. Dierkunde, *2*, 1858: 68; emended spelling for *Sphyrna* Rafinesque, 1810.

Reniceps Gill, Ann. N. Y. Lyc., 7, 1862: 403, 412; type species, *Squalus tiburo* Linnaeus, 1758.

Generic Characters. Nostrils much closer to eyes than to midline of snout; jaws with or without labial furrows; anterior margin of nostril expanded at inner end as a stiff, triangular flap, hollowed on lower side and overhanging the orifice; 1st dorsal narrow-triangular; 2nd dorsal much smaller than 1st dorsal; caudal with well marked subterminal notch, its lower anterior corner expanded as a definite lobe. Characters otherwise those of the family.

Range. Tropical to warm temperate zones of all oceans, including the Mediterranean.

Fossil Teeth. Upper Cretaceous to Miocene, North America; Miocene, Africa; Miocene to Pliocene, Europe.

Attacks on Bathers. The larger Hammerheads have long borne an unsavory reputation as "man-eaters," partly on the basis of unverifiable rumor, and partly because of the fact that a large specimen taken many years ago off Long Island, New York (hence probably *zygaena*), contained portions of a man in its stomach.[2a] Positive evidence is now at hand that this reputation is deserved, for on Sept. 21, 1913, a Hammerhead about eight feet long (species not determined) attacked and so seriously injured a bather at West Palm Beach, Florida, that the lacerations required some 200 stitches. However, the victim recovered after seven weeks in the hospital.[2b] Attacks on bathers by Hammerheads (nominally, at least, *zygaena*) have also been reported from British Guiana. Hammerheads (probably *S. lewini*) are also considered very dangerous in Australian waters,[3] where shark fatalities are of much more frequent occurrence than anywhere in the western Atlantic.

2. By Ruling 89 of the International Commission on Zoological Nomenclature (Smithson. Misc. Coll., *73*, 1925: 27) Klein's names are not to be taken into account.

2a. Mitchill, Trans. Lit. Phil. Soc. N. Y., *1*, 1815: 482.

2b. Gudger, Bull. Amer. Mus. nat. Hist., *40*, 1937: 417.

3. Coppleson, Med. J. Aust., April 15, 1933: 59.

Species. It was long thought that this was a very monotonous genus including some four or five species at most the world over. However, recent studies have shown that the western Atlantic alone actually supports at least five well marked representatives, separated by well defined and easily detectable characters, but which are so overshadowed by the bizarre appearance of the head that they were largely overlooked in most of the early accounts of the genus. Three new species have been described recently from the eastern tropical Pacific also,[4] while the remaining sphyrnids of the Indo-Pacific region as a whole stand in urgent need of critical revision. Unfortunately many of the older descriptions, other than those of *S. tiburo*, which is the most easily recognizable member of the genus in the Atlantic, omit precisely those characters that have recently been found to be specific. Hence there is no knowing to which species, as now recognized, they actually referred. The case is still further complicated by the fact that opportunity has not yet been offered for a sufficiently extensive comparison of the species now known to exist in the Atlantic with those of the Pacific and Indian Oceans. Consequently, the accompanying Key is restricted to Atlantic species. Fortunately it is clear to which of the Atlantic forms the Linnaean name *zygaena* (type species of the genus) actually referred, because Willughby's[5] illustration of the lower surface of his *zygaena*, to which Linnaeus refers as one of the bases of the species, is an excellent representation of a Hammerhead with head rounded in front, with eyes close to corners of oculo-nasal prominences, and with long caudal peduncle.

Key to Atlantic Species

1a. Anterior contour of midsector of head evenly rounded or nearly straight; not indented or scalloped in median line (Figs. 82 B, 86 A).

- 2a. Contour of head only slightly concave opposite nostrils, if at all (Fig. 82 B); groove from nostril, if any, shorter than horizontal diameter of eye; free tip of 2nd dorsal not longer than its anterior margin; posterior margin of anal only weakly concave; teeth near corners of mouth rounded, without cusps. *tiburo* Linnaeus, 1758, p. 420.
- 2b. Contour of head deeply scalloped opposite nostrils (Fig. 86 A); grooves from nostrils more than twice as long as horizontal diameter of eye; free tip of 2nd dorsal considerably longer than its anterior margin; posterior margin of anal deeply concave; teeth near corners of mouth like those further forward, with cusps. *zygaena* Linnaeus, 1758, p. 436.

1b. Anterior contour of head unmistakably indented or scalloped in midline.

- 3a. Free tip of 2nd dorsal only about as long as its vertical height, and considerably shorter than its anterior margin; teeth serrate on cusps as well as basally. *tudes* Valenciennes, 1822, p. 428.

4. *Vespertina*, *media* and *corona* Springer (Stanford Ichth. Bull., *1* [5], 1940: 161–169).
5. Hist. Pisc., 1686: pl. B, 1.

3b. Free tip of 2nd dorsal considerably longer than its vertical height, and at least as long as its anterior margin; teeth with smooth-edged cusps, serrated only on basal expansions, if at all.

4a. Center of eye opposite or posterior to front of mouth; corner of mouth anterior to outer posterior corner of head (hammer); posterior margin of anal fin deeply concave. *diplana* Springer, 1941, p. 415.

4b. Center of eye considerably anterior to front of mouth; corner of mouth considerably posterior to outer posterior corner of head (hammer); rear margin of anal only weakly concave. *bigelowi* Springer, 1944, p. 410.

Sphyrna bigelowi Springer, 1944

Figures 79, 80

Study Material. Immature male, about 886 mm. long, from Rio de Janeiro, Brazil (Harv. Mus. Comp. Zool., No. 463); young male, about 385 mm., from Uruguay (U.S. Nat. Mus., type, No. 87682).

Distinctive Characters. This newly discovered species falls with *diplana* and *tudes*

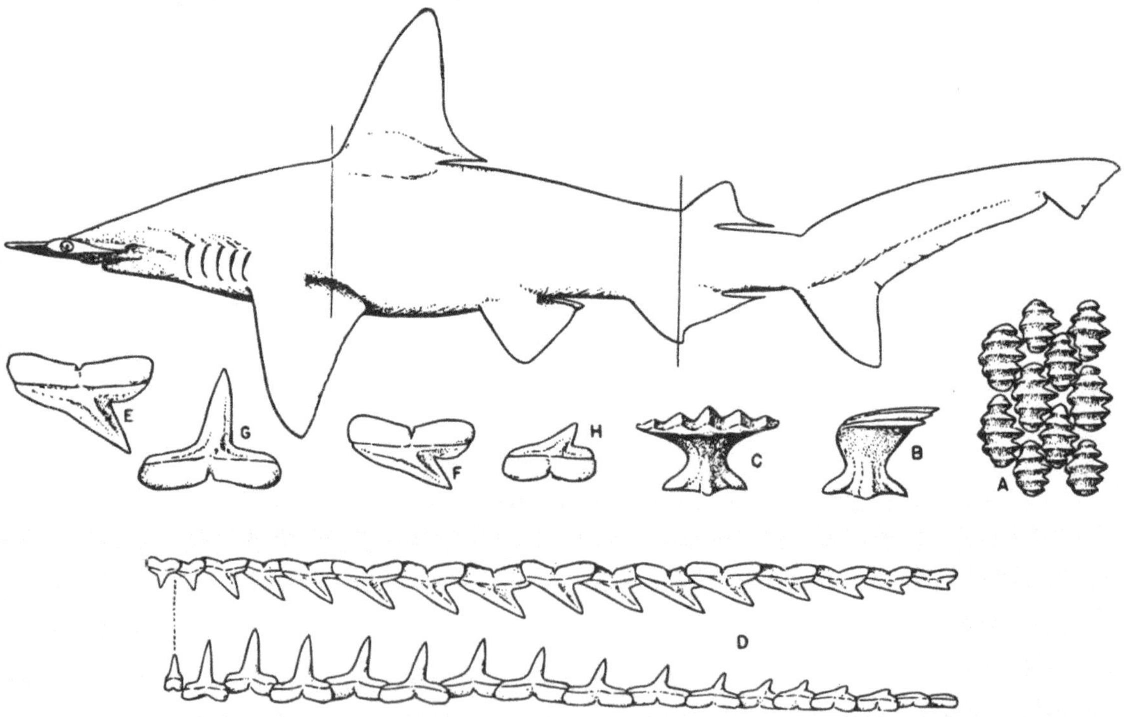

FIGURE 79. *Sphyrna bigelowi*, immature male, about 886 mm. long, from Rio de Janeiro, Brazil (Harv. Mus. Comp. Zool., No. 463). *A* Dermal denticles, about 22 x. *B* Dermal denticle, side view, about 44 x. *C* Dermal denticle, apical view, about 44 x. *D* Upper and lower teeth, left-hand side, about 2 x. *E* Fifth upper tooth. *F* Twelfth upper tooth. *G* Fifth lower tooth. *H* Twelfth lower tooth. *E–H*, about 4 x.

in that the anterior contour of the head is unmistakably indented in the midline. But it differs from *tudes* in the much more definitely marked prenarial groove, more erect first dorsal, and less deeply concave second dorsal, as well as in the fact that the teeth are smooth-edged (serrate in *tudes*) with the lowers noticeably more slender and more erect than the uppers. The most obvious differences from *diplana* are: eyes relatively much smaller, the free rear tip of the second dorsal not longer than the anterior margin of latter (twice that long in *diplana*), origin of pelvics almost under rear tip of first dorsal (considerably behind it in *diplana*) and the anal less deeply concave and considerably longer than the second dorsal. It is separated from *zygaena* by: the relatively long, narrow hammer which is scalloped anteriorly in the midline, the very short free rear corner of the second dorsal, the lower teeth which are much more slender and oblique than the uppers, and the anal fin which is much longer than the second dorsal. The shape of the head separates it from *tiburo*.

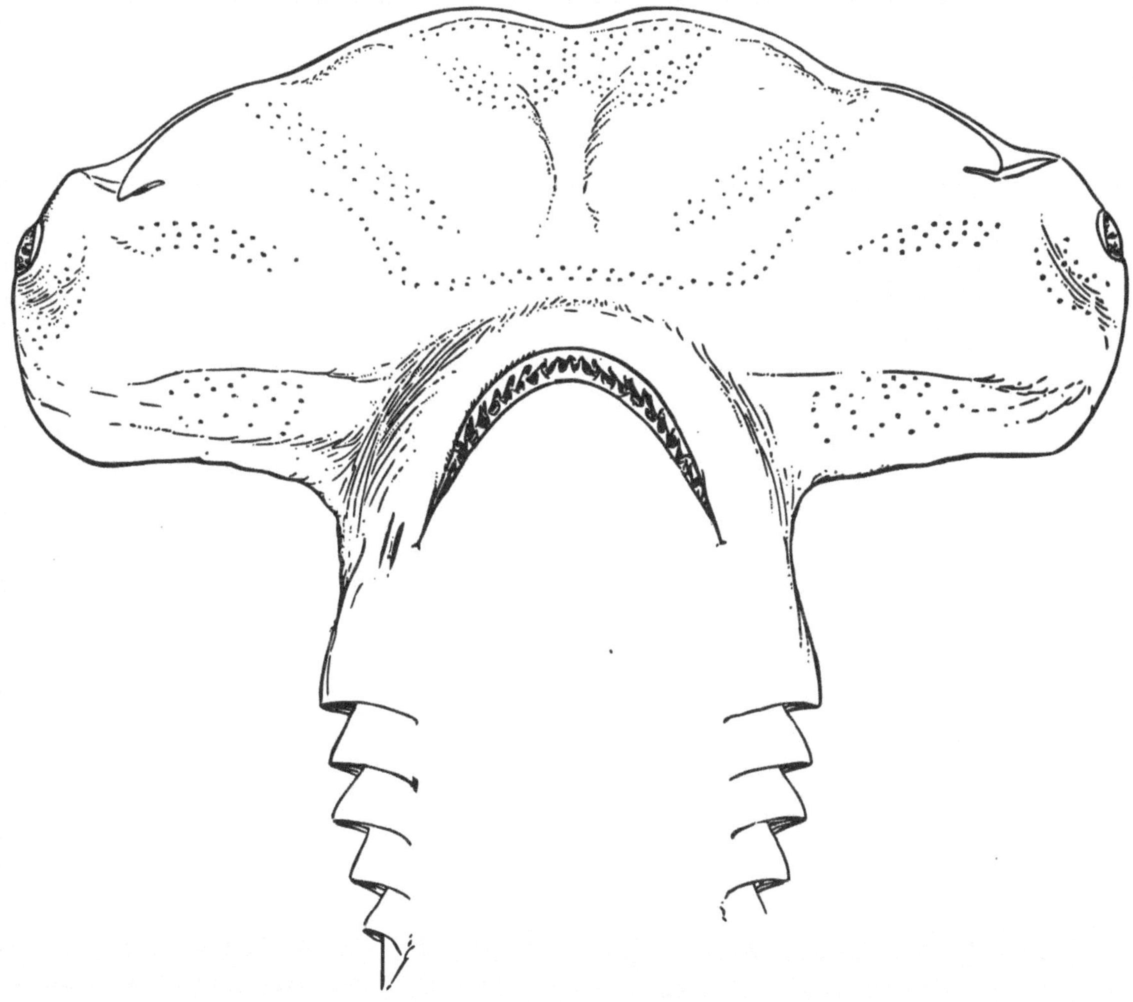

FIGURE 80. *Sphyrna bigelowi*, illustrated in Fig. 79. Head from below, about 0.6 x.

Description. Proportional dimensions in per cent of total length. Male, 886 mm., from Rio de Janeiro (Harv. Mus. Comp. Zool., No. 463).

Trunk at origin of pectoral: breadth 7.9; height 10.7.
Snout length in front of: outer nostrils 3.3; mouth 7.1.
Eye: horizontal diameter 1.0.
Mouth: breadth 6.3; height 3.5.
Nostrils: distance between inner ends 18.1.
Gill opening lengths: 1st 3.2; 2nd 3.4; 3rd 3.7; 4th 3.5; 5th 2.9.
First dorsal fin: vertical height 14.7; length of base 9.8.
Second dorsal fin: vertical height 4.1; length of base 4.1.
Anal fin: vertical height 5.1; length of base 8.5.
Caudal fin: upper margin 29.5; lower anterior margin 11.1.
Pectoral fin: outer margin 13.9; inner margin 5.6; distal margin 14.0.
Distance from snout to: 1st dorsal 27.5; 2nd dorsal 59.5; upper caudal 70.5; pectoral 21.3; pelvics 42.5; anal 55.3.
Interspace between: 1st and 2nd dorsals 21.2; 2nd dorsal and caudal 7.6; anal and caudal 5.8.
Distance from origin to origin of: pectoral and pelvics 20.3; pelvics and anal 14.6.

Trunk about ⅕ as high at origin of 1st dorsal as its length to origin of caudal; more strongly compressed than in other local species. No mid-dorsal ridge. Caudal peduncle about ⅔ as wide as deep, the upper precaudal pit strongly developed as a narrow, transverse, lunate furrow, the lower pit similar but much smaller. Dermal denticles evenly and closely spaced, overlapping but little, the blades thick and rather strongly arched, broader than long, strongly sculptured, usually with 5 (occasionally only 3) high, sharp-topped ridges separated by V-bottomed valleys, the marginal teeth broad and short, the axial only very little the longest; pedicels long, rather slender, the four corners of bases short.

Head in front of pectoral a little less than ⅓ of length of trunk to origin of caudal, its dorsal profile sloping steeply, very thin anterior to eyes, hammer-shaped, the outer posterior margins nearly straight and transverse or even sloping slightly forward (much as in *tudes*); the breadth of head at eyes about 1.2 times its length to origin of pectorals, and about 3.3 times its length at oculo-narial prominence; anterior margin of hammer scalloped with a deep, rounded indentation in midline (as in *diplana*), a shallower indentation opposite each nostril (much shallower than in *diplana* or *zygaena*) and a still more shallow sinuosity between the two others; a well marked groove running inward from corner of nostril along anterior margin of head about ⅓ of distance toward midline (much more strongly marked than in *tudes*). Distance from anterior margin of eye to anterior corner of oculo-narial prominence about equal to diameter of eye; a line connecting the inner ends of nostrils passes about midway between front of mouth and anterior margin of head, one through center of eyes passes anterior to mouth by a distance about twice as long as diameter of eye, and one connecting the corners of the hammer passes a little posterior to the corners of the mouth in very small specimens but a little anterior to them in larger, an alteration that results from an increase in the breadth of the hammer relative to its length with growth, such as takes place in *tudes* also. Mucous pores on lower side of head in midline cover a trapezoidal area (Fig. 80). Eye a little broader than high, much smaller than in the *diplana-zygaena* group, its horizontal diameter only about 1/8 as long as head in front of mouth. Gill openings extending ventrally and spaced nearly equally,[6] the first 1/3 to 1/2 (41%) as long as head in front of mouth and about 3.2 times as long as diameter of eye, the 3rd hardly longer than 1st, and 5th very little shorter and about over origin of pectoral. Nostril transverse. Mouth strongly arched, its breadth about 1.8 times its length, without definite labial furrow on either jaw.

Teeth $\frac{14-2-14}{16-1-15}$ in specimen counted, smooth-edged and narrow-triangular on expanded bases; uppers oblique, deeply notched outwardly and increasingly so toward corners of mouth, the 4th or 5th to 11th largest, the outermost 2 very low but still with definite cusp; lower teeth with much narrower cusps, erect toward center of mouth but somewhat oblique toward corners, the basal expansion somewhat swollen outwardly as a rounded boss or obscure denticle on each side on median tooth and on a few teeth next to the latter;[7] the 2nd to 7th or 8th longest, the outermost 2 very short, rounded, without definite cusp (as in *tiburo*); 2 small symmetrical teeth at symphysis on upper jaw and 1 on lower; 1 or 2 series of teeth functional in alternating rows along sides of upper jaw and 2 to 3 rows along sides of lower.

First dorsal at base a little less than 1/2 as long as head in front of origin of pectorals, its vertical height about 1 1/2 times the base, erect, a perpendicular from its apex falling through rear end of base or a little anterior to latter; its origin about over midpoint of inner margin of pectoral, its anterior margin only slightly convex, rear margin moderately concave toward base, apex subacute, its free rear corner about 1/2 as long as base, the midpoint of base only a little nearer to axil of pectoral than to origin of pelvics with the rear tip about over the latter. Second dorsal at base a little more than 1/3 as long as 1st and a little less than 1/3 as high vertically, its origin about over midpoint of base of anal, its posterior margin moderately concave (much less deeply so than in *tudes*), and apex narrowly rounded, its free rear corner only moderately slender, about 1 1/2 times as long as the base. Caudal a little less than 1/3 (29%) of total length, the terminal sector about 1/4 the length of fin and broadly triangular (considerably larger than in *diplana*, *zygaena* or *tudes*), with nearly straight lower posterior margin and subacute tip; the lower lobe a little more than 1/3 (37%) as long as upper, with weakly convex anterior margin, nearly straight posterior margin and subacute tip; the re-entrant contour (enclosed

6. The 886-mm. specimen is asymmetrical as regards its gill slits, the 4th and 5th coming close together at the lower ends on one side of the head but evenly separated from end to end on the other.

7. On the type specimen the first to fourth lower teeth on each side have these denticles more or less developed; in the 886-mm. specimen the first right-hand and second left-hand lower teeth have one on each side, and the first left tooth has one on one side only.

by the 2 lobes) well rounded and a little more than a right angle. Distance from origin of caudal to tip of anal a little less than ⅓ (30%) as long as base of latter. Anal a little more than twice as long as 2nd dorsal at base but only about as high vertically, its anterior margin weakly convex, its posterior margin deeply concave near apex but nearly straight thence to tip, its apex subacute, its free rear corner about ½ as long as the base. Distance from origin of anal to tips of pelvics a little less than ⅔ (61%) as long as base of anal. Pelvics with nearly straight edges, rather broadly rounded apices and narrowly rounded tips, their origin about under rear tip of 1st dorsal. Pectoral larger than in *diplana*, *tudes* or *zygaena* (*bigelowi* more nearly resembles *tiburo* in this respect), its length about ⅔ (65%) that of head or about equal to vertical height of 1st dorsal, about ⅔ as broad as long, its outer margin nearly straight basally but convex toward apex, the distal margin only very weakly concave or perhaps nearly straight, its apex subacute, the inner corner very narrowly rounded.

Color. The preserved specimens we have studied are grayish brown above and of a paler tint of the same hue below, the fins without conspicuous markings. In the type specimen the anterior margin of the hammer is bordered with yellowish. No information is at hand as to the color of this Hammerhead in life.

Size. The state of sexual development of the 886-mm. specimen, on which the claspers extend a little beyond the tips of the pelvics, suggests that this is a rather small species, perhaps becoming mature when only four to five feet long.

Developmental Stages. Embryos have not been seen.

Habits. Nothing is known of its habits.

Range. So far *S. bigelowi* is known only from Uruguay (type locality) and from Rio de Janeiro, Brazil. Probably it is a tropical species and therefore watch should be kept for it in the West Indian–Caribbean region, in the Gulf of Mexico, and around southern Florida, as well as along the northern and northeastern coasts of South America.

Remarks. This little known species seems closest to *S. corona* Springer, 1940, and *S. media* Springer, 1940, of the Pacific coasts of Central and South America.[7a] It is separated from *corona* by its more broadly rounded mouth, its longer head in front of the mouth relative to the distance between the nostrils (head length in front of mouth about 40 per cent of distance between nostrils in *bigelowi*, but 55 per cent in *corona*), the much more deeply indented anterior outline of its head in the midline, and its much more strongly developed prenarial grooves. It differs from *media*, with which it shares the broadly rounded mouth, in the anterior outline of its head (rounded in *media*), in its more erect lower teeth, and in the fact that the distance between eye and nostril is relatively greater.

Synonyms and References:

Sphyrna tudes (in part) Garman, Mem. Harv. Mus. comp. Zool., *36*, 1913: 159 (Garman's description was based in part on the spec. descr. and pictured here); not *S. tudes* Valenciennes, 1822.

Sphyrna bigelowi Springer, J. Wash. Acad. Sci., *34*, 1944: 274 (descr., ill., Uruguay).

7a. For accounts of these, with illustrations of their heads, see Springer (Stanford Ichthyol. Bull., *1* [5], 1940: 162, 163, fig. 3, 4).

Sphyrna diplana Springer, 1941

Hammerhead

Figures 12 A, 81

Study Material. Eight males and females, 470 to 639 mm. long, and the head of another, about 1,200 mm. (calculated from size of head), from Rio de Janeiro and Rio Grande do Sul, Brazil, and from the vicinity of Galveston, Texas, and male, about 1,340 mm., labelled "Europe" (Harv. Mus. Comp. Zool.). Male, 445 mm., and female, 504 mm., from Galveston, Texas; four specimens, two to three feet, from Colón, and from Charleston, S. Carolina; head of one, about 1,375 mm. (calculated from breadth of head), from Englewood, Florida; jaws of 1,850 mm. and 2,500 mm. specimens from that same locality (U.S. Nat. Mus.). Also, photographs of head, region of caudal peduncle, and second dorsal and anal fins, of a large Florida specimen (from Stewart Springer).

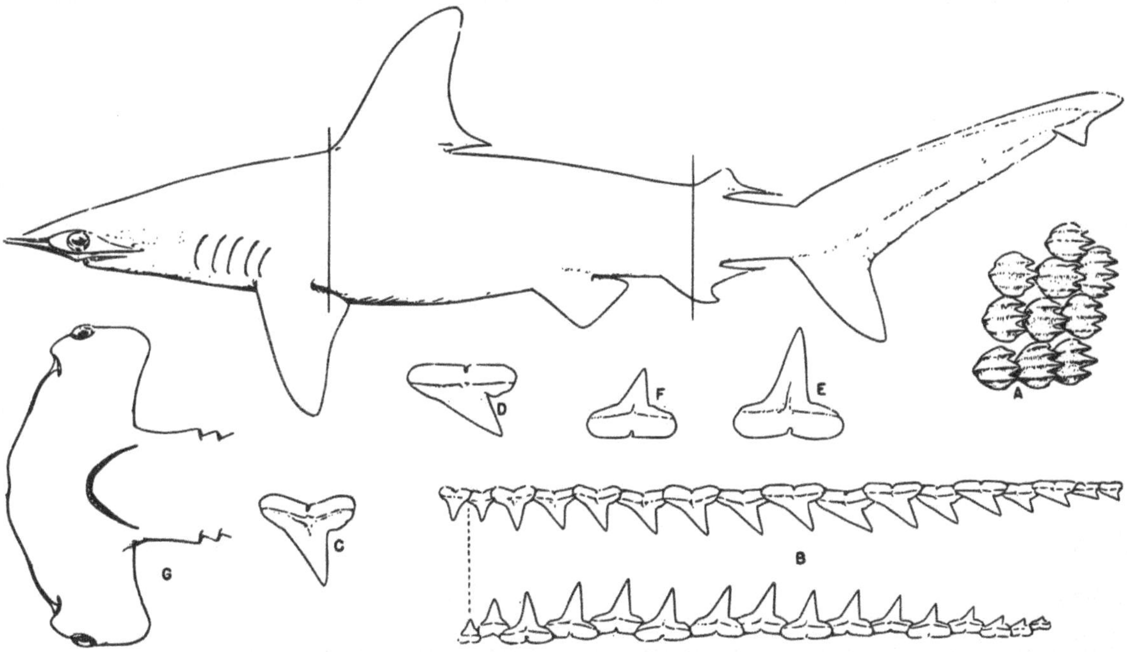

FIGURE 81. *Sphyrna diplana,* female, 639 mm. long, from Rio de Janeiro, Brazil (Harv. Mus. Comp. Zool., No. 462). *A* Dermal denticles, about 55 x. *B* Left-hand upper and lower teeth, about 0.7 x natural size. *C* Fourth upper tooth. *D* Twelfth upper tooth. *E* Third lower tooth. *F* Eleventh lower tooth. *C–F,* about 1.4 x. *G* Head from below.

Distinctive Characters. Diplana has long been confused with *zygaena,* but it is easily distinguished from the latter by the facts that its head is scalloped in front at the midline (rounded in *zygaena*), that its eyes are farther from its nostrils, and by the much shorter interspace between the rear tip of its second dorsal and the origin of its caudal (cf. Fig.

81 with 86 B). It shares a median indentation in the anterior outline of the head with both *tudes* and *bigelowi,* but it is separable from them at a glance by the facts that the corners of its mouth are considerably anterior to the outer rear corners of the head (hammer) and that the free rear tip of its second dorsal is considerably longer than the anterior margin of the fin; it is further separated from *tudes* by its smooth-edged teeth. The hammer shape of its head obviates any danger of confusing it with *tiburo.*

Description. Proportional dimensions in per cent of total length. Female, 554 mm., from Galveston, Texas (Harv. Mus. Comp. Zool., No. 35826). Female, 639 mm., from Rio de Janeiro (Harv. Mus. Comp. Zool., No. 462).

Trunk at origin of pectoral: breadth 9.4, 8.5; height 11.9, 11.6.
Snout length in front of: outer nostrils 4.4, 4.1; mouth 7.4, 7.1.
Eye: horizontal diameter 2.3, 2.4.
Mouth: breadth 6.8, 7.0; height 3.8, 3.8.
Nostrils: distance between inner ends 20.8, 20.2.
Gill opening lengths: 1st 3.6, 3.7; 2nd 3.6, 3.7; 3rd 3.5, 3.7; 4th 3.2, 3.5; 5th 2.5, 2.8.
First dorsal fin: vertical height 13.4, 12.7; length of base 12.4, 11.0.
Second dorsal fin: vertical height 2.9, 3.3; length of base 4.1, 3.1.
Anal fin: vertical height 3.4, 3.3; length of base 6.3, 5.3.
Caudal fin: upper margin 33.4, 31.0; lower anterior margin 11.7, 12.0.
Pectoral fin: outer margin 13.2, 13.0; inner margin 4.9, 5.0; distal margin, 10.0, 11.3.
Distance from snout to: 1st dorsal 26.4, 28.2; 2nd dorsal 56.0, 60.2; upper caudal 66.6, 69.0; pectoral 22.4, 22.9; pelvics 41.3, 46.6; anal 55.0, 57.2.
Interspace between: 1st and 2nd dorsals 20.6, 23.3; 2nd dorsal and caudal 7.4, 6.7; anal and caudal 6.3, 5.6.
Distance from origin to origin of: pectoral and pelvics 20.6, 22.7; pelvics and anal 12.8, 11.7.

Trunk strongly compressed, its height at 1st dorsal about ⅕ its length to origin of caudal, with moderately arched dorsal profile. Back without mid-dorsal ridge. Body sector to cloaca a little longer than tail sector. Caudal peduncle about ⅔ as wide as deep, the upper precaudal pit strongly marked, subrectangular in outline, the lower pit only weakly indicated as a short, semilunar furrow. Dermal denticles much as in *zygaena,* partly overlapping and with skin partly exposed, the blades thin, moderately arched, small specimens usually with 3, but large with 4 or 5, sharp-topped ridges running back about half the length of the blade, the marginal teeth much as in *zygaena,* the axial a little the longest; pedicels very short, moderately slender.

Head about ⅓ of length to origin of caudal, sloping evenly forward to anterior margin, very broadly expanded in hammer form but with posterior edges (outward from the neck) sloping only a little rearward; the breadth at eyes about 1.2 to 1.3 times as great as its length to origin of pectorals, and a little more than 3 times its length at oculo-narial prominence; anterior margin of head scalloped with a deep rounded depression opposite nostril, a somewhat shallower indentation in the median line, and a still shallower sinuosity midway between these two, also with a well marked groove running from the nostril inward along anterior margin of head for about 40% of the distance toward the midline (a little farther than this in *zygaena*). Distance from anterior corner of oculo-narial prominence to anterior margin of eye about as long as diameter of eye (only about ½ to 7⁄10 that long in *zygaena*); a line connecting outer ends of nostrils passes anterior to front of mouth by a distance about as long as horizontal diameter of eye, one connecting centers of eyes passes about through front of mouth, and one through the outer posterior corners of hammer passes a little posterior to corners of mouth. Head (snout) in front of mouth a little less than ⅓ as long as length to origin of pectorals. Rostral cartilage with a median oval hole, the wings of the preorbital processes with an inwardly directed point on anterior margin (this hole and point usually lacking in *zygaena*). Mucous pores in median sector of oral side of head near its anterior margin cover a subrectangular or dumbbell-shaped area (a subtriangular area in *zygaena*). Eye approximately circular, its diameter about ⅓ as long as head in front of mouth. Gill openings noticeably longer than in *zygaena*, the 1st a little more than 1½ (1.6 to 1.7) times as long as diameter of eye, the 5th about 80% as long as 1st, their outlines evenly and moderately concave, the space between 4th and 5th over origin of pectoral. Nostril nearly transverse. Mouth strongly arched, about ½ (49 to 56%) as high as broad. Labial furrows on lower jaw, about ¼ to ⅕ as long as horizontal diameter of eye and concealed when mouth is closed, but none on upper jaw.

Teeth $\frac{15 \text{ or } 16\text{—}0 \text{ to } 2\text{—}15 \text{ or } 16}{15 \text{ or } 16\text{—}1 \text{ or } 2\text{—}15 \text{ or } 16}$, triangular on expanded bases, the cusps smooth-edged, but the bases more or less wavy or fluted on some of the teeth; 1st 3 upper teeth nearly symmetrical and erect, but subsequent teeth increasingly oblique toward corners of mouth, their inner margins more nearly straight, the outer margins more and more deeply notched; the 1st tooth smaller than 2nd, the 15th and 16th very small; lower teeth with somewhat narrower cusps than uppers, similarly oblique and notched outwardly in embryos[8] and in small specimens generally, but with successive series tending to become more erect and their cusps relatively narrower with growth, although there is considerable variation in this respect.[9]

First dorsal erect, a perpendicular from the apex passing close behind rear end of base, its vertical height about ⅔ (57 to 66%) as great as length of head to origin of pectorals with its length at base only a little less, its origin about over midpoint of inner margin of pectoral, its anterior margin rather strongly convex toward apex, posterior mar-

8. Personal communication from Stewart Springer.
9. In the head of a Florida specimen, about 1,375 mm. long (U.S. Nat. Mus., No. 110296, length calculated from the breadth of the head), the first to eighth lower teeth are nearly erect with narrow triangular cusps, and the lower teeth of the type specimen appear to have been similar. In another head of about the same size from Brazil, however, which is otherwise indistinguishable (Harv. Mus. Comp. Zool., No. 845), only the first and second lower teeth are of this type, the third and subsequent teeth being increasingly oblique, and intermediate stages are shown by other specimens. It is not yet known whether the variations in this respect are individual or racial.

gin moderately concave, the free rear tip not very slender, about 1/3 as long as the base; the midpoint of base only about 1/2 as far from axil of pectoral as from origin of pelvics and the rear tip anterior to latter by a distance about 1/3 as long as the base (this distance is considerably greater relatively in *zygaena*). Second dorsal about 1/3 as long as 1st at base but only about 1/5 as high, its origin about over midpoint of base of anal, with narrowly rounded apex and only weakly concave posterior outline, its noticeably slender free rear corner nearly twice as long as base and about twice as long as the vertical height. Distance from tip of 2nd dorsal to precaudal pit a little less than 1/3 as long as base of 2nd dorsal (nearly or quite as long as base of 2nd dorsal in *zygaena*). Caudal a little less than 1/3 (about 31 to 33%) of total length, its upper margin only very slightly convex, the terminal sector about 1/5 the length of fin, narrow-triangular, with subacute tip and rather deeply concave lower posterior margin; the lower lobe a little more than 1/3 as long as upper with weakly convex anterior and posterior margins and subacute tip; the re-entrant contour (included by the two lobes) with rather abruptly rounded corner and a little more than a right angle. Distance from origin of caudal to tip of anal a little less than 1/2 as long as base of anal. Anal about 1.4 to 1.7 times as long at base as 2nd dorsal, about 1.0 to 1.4 times as high vertically, its anterior margin much more convex, posterior margin more deeply concave, apex acute, free rear corner less slender than that of 2nd dorsal, about 2/3 as long as the base, its tip considerably anterior to that of 2nd dorsal. Distance from origin of anal to tips of pelvics only about 1/2 as long as base of anal. Pelvics with nearly straight margins and narrowly rounded corners, and about as long at base as anal. Pectoral a little less than 2/3 (about 57 to 61%) as long as head, a little more than 1/2 (about 56%) as broad as long, the outer margin moderately convex toward tip, distal margin weakly and uniformly concave, the apex and inner corner very narrowly rounded.

Color. Light gray above shading to white below, the pectorals tipped on their ventral surfaces with black in life.

Size. The Study Material (p. 415) suggests that *diplana* may be born at a length no greater than 400 to 450 mm. Males mature at about 1,800 mm. (about 6 feet), and grow to at least 10 feet; it is not known how much larger.

Developmental Stages. Embryos have not been described as yet, except for the teeth (see p. 417), nor have we seen any.

Habits. Recognition of the fact that *diplana* is distinct from *zygaena* is so recent[10] that no attention has yet been devoted to its life history as contrasted with that of *zygaena.* Neither do the few records of Hammerheads that can be positively referred to it contain any pertinent information. It is probable, however, that the account of the habits, food, etc., of *zygaena* (p. 441) applies equally to *diplana,* at least in a general way. Gravid females have been taken in southeastern Florida waters.[11]

Range. Tropical and warm-temperate Atlantic, probably including the Mediter-

10. It was unmistakably pictured by Valenciennes more than a century ago (Mem. Mus. Hist. nat. Paris, 9, 1822: pl. 11, fig. 1) as *Zygaena malleus.*
11. Personal communication from Stewart Springer.

ranean. It is represented in the tropical-subtropical waters of the eastern and western Indo-Pacific by a form (*S. lewini* Griffith, 1834) closely resembling *diplana* in form of head (including arrangement of mucous pores) and in the shape and relative position of the fins. We have not been able to find any significant differences between small specimens from Panama, southern California, Hawaiian Islands or Celebes[12] and the Atlantic series of comparable sizes listed above (p. 415). However, since Whitley[13] describes the teeth of the Australian *lewini* as becoming "finely denticulated" with growth, which is not the case in the Atlantic *diplana*, it seems wise to retain both specific names, awaiting a comparison of adult specimens.

Occurrence in the Atlantic. Diplana was separated from *zygaena* so recently that very few reports of it have yet appeared under its own name. Information as to its occurrence in the eastern Atlantic is confined to the facts that Valenciennes[14] described his *malleus*, which his illustration shows to be the head of *diplana*, as Mediterranean and Atlantic; that Springer[15] reports a head of *diplana* from tropical West Africa (Gold Coast); that the collection of the British Museum contains specimens apparently of this species in addition to *zygaena* from the Mediterranean; and that there is a specimen of it labelled "Europe" in the collection of the Harvard Museum of Comparative Zoology. Locality records that can be referred with certainty to *diplana* in the western Atlantic are Rio Grande do Sul, Rio de Janeiro and probably Pernambuco, Brazil, Colón, both coasts of Florida, South and North Carolina, and a station about 90 miles off Cape May, New Jersey. These are enough to show that its range closely parallels that of *zygaena*. It is so common off south-eastern Florida that we have recently received a report of 19 adult males taken there in a single day. It appears not to range as far north along the United States coast during summer as *zygaena* does. Similarly it is possible that its range may not extend as far to the south in the southern hemisphere, although information of its presence is so scant for the South American coast as to preclude any definite statement in this regard. Neither is any information available as to its abundance relative to that of *zygaena* anywhere off the American Coast.

Synonyms and References:

Shark, no name, Marcgrave, Hist. Nat. Brazil, 1648: frontispiece (this is the earliest illustration that we have found of a Hammerhead with head of the *diplana* shape).

Le Marteau (in part), Duhamel, Traité Gén. Pêches, (2) *3* (9), 1777: 303, pl. 21, fig. 3–7 (ill., apparently this species); Broussonet, Mem. Acad. Roy. (1780), 1784: 661 (by ref. to Duhamel, 1777, as above).

Zygaena malleus (in part) Valenciennes, Mem. Mus. Hist. nat. Paris, *9*, 1822: 223, pl. 11, fig. 1 (descr., ill. of head, Medit., Atlant., but apparently confused with his own *tudes*, because teeth descr. as denticulate); Risso, Hist. Nat. Europ. merid., *3*, 1826: 125 (part, by ref. to Valenciennes, 1822); Storer, Mem. Amer. Acad. Arts Sci., N. S. *2*, 1846: 508 (*malleus* Valenciennes, 1822, incl. in synon.); Günther, Cat. Fish. Brit. Mus., *8*, 1870: 381 (*malleus* Valenciennes, incl. in synon.); Day, Fish. Gt. Brit., *2*, 1880–1884: 294 (*malleus* Valenciennes, incl. in synon.), pl. 154 (ident. by shape of head, no loc.); Doderlein, Man.

12. Specimens in the collection of Harvard Museum of Comparative Zoology, 480 to 740 mm. long.

13. Fish. Aust., *1*, 1941: 121.

14. Mem. Mus. Hist. nat. Paris, *9*, 1822 : 223.

15. Proc. Fla. Acad. Sci., *5*, 1941 : 49.

Ittiol. Medit., *2*, 1881: 46 (*malleus* Valenciennes, incl. in synon.); not *Squalus malleus* Shaw and Nodder, 1796.

Sphyrnias zygaena (in part) Gray, List Fish. Brit. Mus., 1851: 48 (*malleus* Valenciennes, incl. in synon.); White, List Spec. Brit. Mus., Fish., *8*, 1851: 126 (*malleus* Valenciennes, incl. in synon.).

Sphyrna zygaena (in part) Ribeiro, Arch. Mus. nac. Rio de J., *14*, 1907: 202 (*malleus* Valenciennes, incl. in synon.); Nichols and Murphy, Brooklyn Mus. Sci. Bull., *3* (1), 1916: 19 (ill. shows *diplana* shape of head, no loc., but discussion refers to *zygaena*); Springer, Proc. Fla. Acad. Sci., *3*, 1939: 31 (combined with *zygaena* in descr., habits, ident. by shape of head in ill., Florida); Fowler, Bull. U.S. nat. Mus., 100 (*13*), 1941: 217 (*malleus* Valenciennes, incl. in synon.).

Cestracion zygaena (in part) Garman, Mem. Harv. Mus. comp. Zool., *36*, 1913: 157 (*malleus* Valenciennes, incl. in synon.); Radcliffe, Bull. U.S. Bur. Fish., *34*, 1916: 263, pl. 43, fig. 1 (N. Carolina spec., ident. by ill. of head); Meek and Hildebrand, Field Mus. Publ. Zool., *15* (1), 1923: 58 (*malleus* Valenciennes, incl. in synon.; ident. of spec. from Colón, whether *diplana* or *zygaena*, not certain from descr.); not *Squalus zygaena* Linnaeus, 1758.

Hammerhead Shark, Smith, J. Amer. Mus. nat. Hist., *16*, 1916: 348 (photo of 10-foot spec. descr. as *Cestracion zygaena* by Welsh, 1916).

Cestracion zygaena Welsh, Copeia, 38, 1916: 94 (meas. of spec. shown in photo by Smith, 1916, as above; 90 miles off C. May, N. Jersey); not *Squalus zygaena* Linnaeus, 1758.

Sphyrna diplana Springer, Proc. Fla. Acad. Sci., *5*, 1941: 46 (descr., ill., discuss., both coasts of Florida, Mississippi, Louisiana, Texas, trop. W. Afr., and probably Medit.); Bigelow and Schroeder, Guide Comm. Shark Fish., Anglo Amer. Caribb. Comm., Wash., 1945: 136, fig. 50 (descr., range, ill.).

Doubtful References:

Sphyrna tudes Ribeiro, Arch. Mus. nac. Rio de J., *14*, 1907: 157, pl. 5 (ident. probable because of shape of head in photo, Maria Farinha, near Pernambuco, Brazil); Fauna brasil. Peixes, *2* (1) Fasc. 1, 1923: 14 (same as foregoing); not *Zygaena tudes* Valenciennes, 1822.

Sphyrna tiburo (in part) Puyo, Bull. Soc. Hist. nat. Toulouse, *70*, 1936: 82 (head ill. of *diplana* shape, but discuss. apparently of *tiburo*); not *Squalus tiburo* Linnaeus, 1758.

Sphyrna tiburo (Linnaeus), 1758

Bonnet Shark, Shovel Head

Figure 82

Study Material. Seventy-one specimens, embryos to adults, 217 to 1,090 mm. long, including one female containing eight young nearly ready for birth and another with nine, from: Hampton Roads, Virginia; Charleston, South Carolina; Tortugas, Florida; Barrataria Bay and Bayou Fifi, Louisiana; Galveston, Corpus Christi and Harbor Island, Texas; Cuba; Bahia, Pernambuco, Rio Parahyba and Rio de Janeiro, Brazil; also, from Pacific coast of Panama and San Diego, California (Harv. Mus. Comp. Zool. and U.S. Nat. Mus.). Also, paratypes of *S. vespertina* Springer, Pacific Panama, loaned by the Carnegie Museum of Pittsburgh and Stanford University, California.

Distinctive Characters. Tiburo is most obviously marked off from *diplana, bigelowi, tudes* and *zygaena* in having a shovel- (not hammer-) shaped head not indented marginally opposite the nostrils; also, the outermost 4 or 5 teeth in the lower jaw are evenly rounded, without cusps or a definite cutting edge. It is separated from *bigelowi, diplana* and *tudes* by the evenly rounded anterior outline of its head, and further from *tudes* by its smooth-edged teeth.

Description. Proportional dimensions in per cent of total length. Female, 621 mm., from Galveston, Texas (Harv. Mus. Comp. Zool., No. 36157). Male, 812 mm., from Tortugas, Florida (Harv. Mus. Comp. Zool., No. 848).

Trunk at origin of pectoral: breadth 8.7, 8.9; height 10.0, 8.4.

Snout length in front of: outer nostrils 4.0, 4.1; mouth 8.0, 6.6.

Eye: horizontal diameter 1.8, 1.5.

Mouth: breadth 7.2, 6.0; height 3.8, 3.5.

Nostrils: distance between inner ends 12.2, 9.9.

Labial furrow length: lower 0.7, 0.5.

Gill opening lengths: 1st 2.7, 2.2; 2nd 3.2, 2.5; 3rd 3.5, 2.6; 4th 3.2, 2.5; 5th 2.8, 2.1.

First dorsal fin: vertical height 11.6, 11.4; length of base 9.6, 9.2.

Second dorsal fin: vertical height 4.2, 3.9; length of base 5.5, 5.2.

Anal fin: vertical height 3.5, 3.4; length of base 7.6, 6.7.

Caudal fin: upper margin 26.7, 25.7; lower anterior margin 10.1, 10.1.

Pectoral fin: outer margin 13.8, 15.0; inner margin 5.8, 5.8; distal margin 11.6, 12.7.

FIGURE 82. *Sphyrna tiburo.* *A* Female, 367 mm. long, from Rio de Janeiro, Brazil (Harv. Mus. Comp. Zool., No. 90). *B* Head of same, from below. *C* Dermal denticles of same, about 18 x. *D* Lateral and apical views of dermal denticle, about 36 x. *E* Left-hand upper and lower teeth of male, 812 mm. long, from Tortugas, Florida (Harv. Mus. Comp. Zool., No. 848), about 2 x. *F* Fourth upper tooth. *G* Tenth upper tooth. *H* Second lower tooth. *I* Seventh lower tooth. *J* Tenth lower tooth. *F–J,* about 4 x.

Distance from snout to: 1st dorsal 28.9, 29.1; 2nd dorsal 59.4, 60.7; upper caudal 73.3, 74.3; pectoral 21.6, 20.7; pelvics 46.7, 42.7; anal 58.1, 59.2.

Interspace between: 1st and 2nd dorsals 21.6, 22.9; 2nd dorsal and caudal 9.2, 9.0; anal and caudal 6.9, 7.8.

Distance from origin to origin of: pectoral and pelvics 24.0, 23.7; pelvics and anal 12.2, 16.5.

Trunk at origin of 1st dorsal between 1/5 and 1/4 as high as its length to origin of caudal, moderately compressed. No mid-dorsal ridge. Caudal peduncle about 2/3 as wide as deep, the upper precaudal pit strongly developed with its anterior outline oval or subangular, the lower pit only about 1/2 as wide, but of similar form. Dermal denticles relatively somewhat larger than in *zygaena*, varying from closely overlapping to loosely spaced, the blades mostly rather steeply raised, usually with 5 strong ridges and as many sharp-pointed marginal teeth (longer than in other local species), the axial slightly the largest; short, slender pedicels.

Head a little less than 1/3 (average about 28 to 29%) of trunk to origin of caudal, convex dorsally to opposite the angles of mouth but concave thence forward, very thin anterior to the eyes, its anterior part shovel- and not hammer-shaped, the outer posterior margins of the shovel sloping a little rearward with broadly rounded outer corners; the breadth of shovel at eyes about 2.5 times as great as its length in front of mouth in small specimens but nearly 3 times that great in large, 1.3 to 1.6 times as great as its length at outer ends of nostrils[16] and a little shorter than head to origin of pectorals; anterior contour of head an uninterrupted curve from eye to eye, without definite depressions opposite nostrils and merely a slight concavity or sinuosity median to the latter in some cases only, the midsector a little more ovate in adult males where length in front of mouth is almost 2/3 as great as distance between nostrils, while in adult females or in young of both sexes it averages less than 2/3 (almost 57%); a faintly marked groove runs forward from nostril along anterior margin of head in some small specimens for a distance about as great as distance from nostril to eye, but this is hardly visible or wholly obsolete in larger specimens; distance from outer end of nostril to anterior edge of eye a little greater than diameter of latter; a line connecting inner ends of nostrils passes anterior to mouth by a distance about 1 1/2 times the diameter of eye, one connecting the centers of eyes passes about through front of mouth, while one connecting the outer posterior corners of shovel passes posterior to corners of mouth by a distance about equal to diameter of eye. Rostral cartilage without median hole, but preorbital process with a strongly marked inwardly directed spur on anterior margin. Large and small mucous pores on lower surface of anterior part of head distributed in diffuse pattern. Diameter of eye 1/4 to 1/5 as long as length of head in front of mouth. Gill openings extending ventrally and almost evenly spaced, the 1st almost 1.5 times as long as diameter of eye or a little longer, the 5th almost

16. The shovel of *tiburo* may average a little narrower relative to its length in northern specimens than in those of the southern Atlantic. But the probability of local or racial variation is so great that recognition of this difference does not seem essential in nomenclature.

as long as 1st, the 3rd only very little longer, the 5th over origin of pectoral. Nostril approximately transverse, a little shorter than diameter of eye. Mouth strongly arched, about ½ as high as broad. A weakly marked labial furrow around corner of mouth extending a very short distance along lower jaw (concealed when mouth is closed), but none on upper jaw.

Teeth $\frac{12\text{ to }14\text{—}0\text{ or }1\text{—}12\text{ to }14}{12\text{—}1\text{—}12}$, smooth-edged, except somewhat wavy on outer base; 1st upper tooth erect and symmetrical on expanded base, but subsequent upper teeth strongly and increasingly oblique, their inner margins straight or slightly curved, their outer margins deeply notched in angular outline on 2nd to 11th or 12th, the outermost 2 or 3 very low without definite cusps, the 3rd to 10th or 11th considerably the largest; lower teeth shorter than uppers, with narrower cusps and relatively broader bases, the 1st to 3rd erect, symmetrical, the 4th to 7th or 8th slightly oblique with increasingly incised outer margin, relatively broader bases and shorter cusps, the 8th and subsequent teeth without cusps, low, oval, evenly rounded and without definite cutting edge; 3 series functional in front of upper jaw and 1 to 2 along sides; 4 to 6 series functional in front of lower jaw, usually 2 along sides and 3 toward corners of mouth; 1 small symmetrical tooth at symphysis on upper jaw, or none, and 1 on lower.

First dorsal moderately sloping with a perpendicular from its apex falling almost at its rear tip, its origin a little anterior to inner corner of pectoral (or over it), its vertical height a little less than ⅔ (about 60%) and base a little less than ½ (about 44%) of the length of head to origin of pectoral, its apex narrowly rounded, free rear corner about ½ as long as base, the midpoint of base approximately midway between axil of pectoral and origin of pelvics, its rear tip anterior to latter by a distance about as great as diameter of eye. Second dorsal a little less than ½ as long at base as 1st in large specimens (slightly longer in small) and almost ⅓ as high, its anterior margin moderately convex, apex subangular, posterior margin deeply concave, its free rear corner very slender, about 1.2 times as long as the base or about 1.3 as long as the vertical height, its origin a little anterior to midpoint of base of anal. Caudal between ¼ and ⅓ (26 to 29%) of total length, the upper margin nearly straight, its terminal sector about ¼ of the fin, with narrowly rounded apex and concave lower posterior margin; the lower lobe about ⅓ (average 37 to 38%) as long as upper, with weakly convex anterior margin, nearly straight posterior margin, and narrowly rounded or subacute tip; the re-entrant contour (included by the 2 lobes) approximately a right angle with abruptly rounded corner. Distance from origin of caudal to tip of anal about ⅓ as long as base of latter. Anal about 1.2 to 1.5 times as long at base as 2nd dorsal but a little lower vertically, its anterior margin weakly convex, apex subangular, posterior margin weakly concave toward apex, but nearly straight toward rear tip, its free rear corner moderately slender, a little more than ½ as long as base in large specimens or a little less in small. Distance from origin of anal to tips of pelvics about ⅔ as long as base of anal. Pelvics about as long at base as anal, with weakly convex anterior margins, weakly concave posterior margins, and narrowly rounded corners, their origins about as near to inner corners of pectorals as to origin of anal. Pectoral a little more than ⅔ (average about 68%) as long as head, or a little longer than vertical height of 1st dorsal and about ⅔ as broad as long, its outer margin moderately convex, distal margin nearly straight or very slightly concave, apex and inner corner very narrowly rounded.

Color. Gray or grayish-brown above and a paler shade of the same hue below, some specimens with a few small round dark spots on the sides; no conspicuous fin markings.

Size. A female of 1,076 mm. in the Harvard Museum of Comparative Zoology contains eight young of about 300 mm. nearly ready for birth. Sexual maturity is attained at a length of 3½ to 4 feet; it is said to reach 6 feet, but few grow longer than 5 feet.

Developmental Stages. Development is viviparous. The egg, with its developing embryo, is enclosed at first in a tough but elastic shell, iridescent in appearance, the ends of which are curiously plaited and folded, allowing for the growth of the embryo by their expansion. After the embryo is set free from the eggshell and the yolk has been absorbed, the empty yolk sac becomes attached to the uterine walls of the mother, forming a so-called yolk-sac placenta richly supplied with blood vessels. The embryos are described as lying with their heads toward the anterior end of the uterus. The umbilical cord may be about as long as the embryo and is closely set along its whole length with large villi, some of which are simple but others branched. The number of embryos (6 to 9) is much smaller than in *zygaena*, corresponding to their type of development, with males and females in about equal numbers.[17]

Habits. This species occurs chiefly in shallow water, close inshore, often in bays and estuaries, sometimes coming right up to wharves. It is said to be more sluggish than other Hammerheads. It feeds largely on whatever crabs may be available locally and on other crustacea such as mantis shrimps (*Squilla*), shrimps, isopods and even barnacles. But its recorded diet also includes bivalve mollusks, cephalapods (*Octopus*), small fish, and even seaweed, the latter no doubt taken incidentally with crabs, etc. And it has been described as burrowing under coral masses in search of small fish and crustacea in southern Florida waters. It takes a hook readily on almost any kind of bait and is often said to follow fishing boats to pick up any fish or other scraps that may be discarded.

Relation to Man. It is of no commercial value except for a few that may be sold in the fish markets. It is entirely harmless.

Range. Tropical to warm-temperate belt of the Atlantic from southern Brazil northward regularly to the southern part of North Carolina and as a stray to southern New England and Massachusetts Bay in the west; it apparently occurs also in tropical West

17. For an excellent illustration of the embryo with yolk stalk and large yolk sac before the disappearance of the external gills, see Leuckart (Unters. Auss. Kiemen Rochen Hayen, Stuttgart, 1836: 22, pl. 3); for further accounts of the early development of *tiburo*, see Gudger (Science, N. S. *35*, 1912: 466; Proc. biol. Soc. Wash., *25*, 1912: 143), Radcliffe (Bull. U.S. Bur. Fish., *40*, 1916: 266) and Longley and Hildebrand (Pap. Tortugas Lab., *34*, 1941: 3).

Africa in the east (probably not the Mediterranean[19]) as well as on Pacific coasts of America from southern California to Ecuador.[20] *Tiburo* has also been reported by name from China[21] and from the Philippines,[21] but without supporting evidence as to the identity of the particular specimens in question; and since it is not included in any of the descriptive surveys of the sharks of China, Australia, the Philippines or India that have appeared so far, although it is not likely to be overlooked if at all common, we think its presence unlikely in eastern Pacific or Indian waters.

Occurrence in the Western Atlantic. This shark is known from so many localities from southern Brazil to Florida, including the South American coastline, the West Indian–Caribbean region as a whole, and the Gulf of Mexico,[22] and so often reported as common as to prove it not only generally distributed throughout this entire area but in fact one of the more plentiful of littoral sharks there. No doubt it is also a year-round resident throughout this belt, as it certainly is around southern Florida, except perhaps for midsummer. To the north of Florida, however, it occurs on the Atlantic coast only as a summer visitor, common in the warm months between June and October on the coast of South Carolina, where it has been characterized as one of the more abundant of summer sharks, and as far as Beaufort, North Carolina. But few pass Cape Hatteras; it is taken only occasionally in the pound nets in Chesapeake Bay, and it occurs only as a stray farther north, being recorded once or twice in New Jersey waters, once at Newport, Rhode Island, and once in Massachusetts Bay. We have also received a reliable report that about six specimens, about four feet long, were taken in Nantucket Sound among the other species caught in the shark fishery that was carried on there during the summer of 1918.[23]

Synonyms and References:

Squalus tiburo Linnaeus, Syst. Nat., 1, 1758: 234 (descr., refs., Amer.); Syst. Nat., *1*, 1766: 399 (same as foregoing); Bonnaterre, Tabl. Encyc. Meth. Ichthyol., 1788: 9 (descr., S. Amer.); Gmelin, in Linnaeus, Syst. Nat., *1* (3), 1779: 1495 (same as Linnaeus, 1758, Amer. seas); Walbaum, P. Artedi Genera Pisc. Emend. Ichthyol., *3*, 1792: 516 (diagn., refs.); Shaw and Nodder, Naturalist Misc., 7, 1795: pl. 229 (good ill., S. Amer. seas); Bloch and Schneider, Syst. Ichthyol., 1801: 131; Bosc., Nouv. Dict. Hist. Nat., *21*, 1803: 189 (diagn.); Latreille, Nouv. Dict. Hist. Nat., *24*, 1804: 72 (in table of

19. The brief description by Rochebrune (Act. Soc. linn. Bordeaux, [4] *6*, 1882: 44) of the head of the Senegambian shark, reported by him as *Zygaena leeuwenii*, suggests that he was actually dealing with *tiburo*. But we think it improbable that it occurs in the Mediterranean, for while it is credited tentatively to the local fish fauna by Doderlein (Man. Ittiol. Medit., *2*, 1881: 50), the few reports of it there or for southern Spain lack positive evidence as to the specific identity of the specimens concerned.

20. The eastern Pacific form has recently been made a separate species, *vespertina* Springer (Stanford Ichthyol. Bull., *1* [5], 1940: 161) on the basis of a supposedly wider shovel than in *tiburo* relative to its length, longer relative distance between nostrils, higher first dorsal, shorter pectoral and longer caudal. But comparative measurements show that there is no discontinuity in any of these respects between specimens from San Diego, California or Panama and the extensive Atlantic series with which we have compared them.

21. Günther, Cat. Fish. Brit. Mus., *8*, 1870: 382; Elera, Cat. Fauna Filipp., *1*, 1895: 615.

22. Localities of definite record from south to north are: Santos, Rio de Janeiro, Bahia, Pernambuco, Rio Parahyba and Natal in Brazil; French Guiana; Trinidad; Colón; Progreso, Yucatán; Belize, British Honduras; Turks Island; Cuba; the Bahamas; many localities along both coasts of Florida; several along the coasts of Louisiana and Texas.

23. Personal communication by R. H. Bodman, who operated the fishery.

contents); Cuvier, Règne Anim., *2*, 1817: 127, footnote (diagn.); Nardo, Prod. Ittiol. Adriat., 1827: 9 (name only, Medit.); Gray, Cat. Fish. Coll. Descr. by L. T. Gronow in Brit. Mus., 1854: 6 (diagn., Amer. seas).

Le Pantouflier, Broussonet, Mem. Math. Phys. Acad. Sci. Paris, 1780: 662 (not Squale Pantouflier, Lacépède, 1798, in Sonnini, 1802, which is *tudes*; see p. 434).

Round Headed Zygaena, Shaw and Nodder, Naturalist Misc., 7, 1795: pl. 229 (good ill., S. Amer. seas).

Squale carolinien, Bosc, Nouv. Dict. Hist. Nat., *21*, 1803: 190 (diagn., Carolina).

Heart Headed Shark, Shaw, Gen. Zool., *5*, 1804: 355, pl. 154, upper fig. (good ill., S. Amer. seas).

Sphyrna tiburo Rafinesque, Indice Ittiol. Sicil., 1810: 47 (name, for *Squalus tiburo* Linnaeus, 1758);[24] Müller and Henle, Plagiost., 1841: 53 (descr., Brazil); Jordan, Proc. U.S. nat. Mus., *5*, 1882: 345 (Pensacola, Florida, abundant); Jordan and Gilbert, Proc. U.S. nat. Mus., *5*, 1882: 581 (commonest summer shark, Charleston, S. Carolina); Jordan, Rep. U.S. Comm. Fish. (1885), 1887: 796 (name, S. Atlant. and Gulf States of U.S.); Nelson, Rep. St. Geol. N. J., *2* (2), 1890: 661 (N. Jersey, size); Carus, Prod. Fauna Medit., *2*, 1889–1893: 514 (occur. in Medit. questioned); Henshall, Bull. U.S. Fish Comm., *9*, 1891: 384 (Florida); Lönnberg, Overs. Vet. Akad. Forhandl., *51*, 1894: 111 (S. Florida); Henshall, Bull. U.S. Bur. Fish., *14*, 1895: 210 (size, abund., Florida Keys and west coast); Evermann and Bean, Rep. U.S. Comm. Fish. (1896), 1898: 239 (reported Indian R., Florida); Jordan and Evermann, Bull. U.S. nat. Mus., 47 (3), 1898: 2748 (Mazatlan, Mexico); Evermann and Kendall, Rep. U.S. Comm. Fish. (1899), 1900: 44 (W. coast Florida); Jordan and Evermann, Bull. U.S. nat. Mus., 47 (4), 1900: pl. 5, fig. 19 (ill.); Wilson, Amer. Nat., *34*, 1900: 355 (common, N. Carolina); Bean, T. H., Rep. For. Comm. N. Y., 1901: 378 (occasional, N. York); Gregg, Where to Catch Fish, 1902: 16 (ill., descr., size, E. Florida); Bean, T. H., Bull. N. Y. St. Mus., *60*, Zool. 9, 1903: 30 (descr., N. Jersey); Gilbert and Starks, Mem. Calif. Acad. Sci., *4*, 1904: 13 (comp. with other species, Mazatlan, Mexico); Bean, B. A., in Shattuck, Bahama Islands, 1905: 296 (off Abaco, Bahamas); Jordan and Thompson, Bull. U.S. Bur. Fish., *24*, 1905: 232 (habits, food, Tortugas, Florida); Linton, Bull. U.S. Bur. Fish., 24, 1905: 344 (abund., embryos, food, parasites, N. Carolina); Cole and Barbour, Bull. Mus. comp. Zool. Harv., *50*, 1906: 155 (Yucatán); Schreiner and Ribeiro, Arch. Mus. nac. Rio de J., *12*, 1906: 79 (Bahia, Rio de J.); Starks, Proc. U.S. nat. Mus., *30*, 1906: 763 (Guayaquil, Ecuador); Gudger, Science, N. S. *25*, 1907: 1006 (food); Ribeiro, Arch. Mus. nac. Rio de J., *14*, 1907: 156, 201 (descr., Brazil, refs.); Smith, Bull. N. C. geol. econ. Surv., *2*, 1907: 35 (abund., embryos, food, N. Carolina); Wilson, Proc. U.S. nat. Mus., *33*, 1907: 431 (parasites, N. Carolina); Fowler, Proc. Acad. nat. Sci. Philad., *60*, 1908: 66 (Newport, Rhode Island, Florida specs.); Wilson, Proc. U.S. nat. Mus., *33*, 1908: 626 (parasites, N. Carolina); Rosen, Lund Univ. Årsberätt., N. S. (2) 7 (5), 1911: 47; Kongl. Fysiogr. Sällsk. Hand., N. S. *22* (5), 1911: 47 (Bahamas); Wilson, Proc. U.S. nat. Mus., *39*, 1911: 390 (parasites, N. Carolina); Gudger, Science, N. S. *35*, 1912: 466 (young embryo, yolk stalk, N. Carolina); Proc. biol. Soc. Wash., *25*, 1912: 143 (size, embryos, N. Carolina); Starks, Stanford Univ. Publ. Biol., 1913: 5 (Natal, Brazil); Coles, Proc. biol. Soc. Wash., *28*, 1915: 90 (abund., season, N. Carolina); Fowler, Proc. Acad. nat. Sci. Philad., *67*, 1916: 521 (Trinidad); Copeia, 30, 1916: 36 (listed, mid. Atlant. U.S.); Nichols and Murphy, Brooklyn Mus. Sci. Bull., *3* (1), 1916: 18 (diagn., reported near N. York, but no definite Long Island record); Smith, J. Amer. Mus. nat. Hist., 16, 1916: 347 (Atlant. Coast, U.S.); Wilson, Ann. Carneg. Mus., *10*, 1916: 58 (Guayaquil, Ecuador); Fowler, Proc. Boston Soc. nat. Hist., *35*, 1917: 110 (Newport, Rhode Island); Proc. Acad. nat. Sci. Philad., *69*, 1917: 127 (Colón); Ribeiro, Rev. Mus. paul., *10*, 1918: 707 (Santos, Brazil); Fowler, Proc. biol. Soc. Wash., *33*, 1920: 144 (N. Jersey, size); Nichols, Bull. Amer. Mus. nat. Hist., *44*, 1921: 22 (Turks Island); Fowler, Proc. biol. Soc. Wash., *36*, 1923: 28 (Florida S. Coast); Ribeiro, Fauna brasil. Peixes, *2* (1) Fasc. 1, 1923: 13 (same as Ribeiro, 1907); Borodin, Bull. Vanderbilt Oceanogr. (Mar.) Mus., 1 (1), 1928: 5 (no loc.); Hildebrand and Schroeder, Bull. U.S. Bur. Fish., *43*, 1928: 51 (descr., Chesapeake Bay); Rey, Fauna Iberica Peces, *1*, 1928: 370 (descr., occur. in Medit. questioned); Breder, Field Bk. Mar. Fish. Atlant. Coast, 1929: 19 (general); Fowler, Copeia, 1931: 46 (Corpus Christi, Texas); von Bonde, J. Comp. Neurol., *58*, 1933: 381 (ill. head), pl. 2, fig. 4 (ill. brain, no loc.); Borri, Atti Soc. tosc. Sci. nat., *44*, 1934: 91 (Bahia, Brazil); Burton, Sci. Mon. N. Y., 40, 1935: 283 (abund., season, S. Carolina);

24. Spelled "*liburo*," an evident misprint.

Brooks, Parasitology, *26*, 1936: 266 (parasites, Tortugas, Florida); Norman and Fraser, Giant Fishes, 1937: 47 (mentioned); Tortonese, Atti Soc. ital. Sci. nat., 77, 1938: 303 (Rio de J.); Springer, Proc. Fla. Acad. Sci., *3*, 1939: 30 (descr., ill. head, teeth, size, food, abund., Florida W. coast); Stanford Ichthyol. Bull., *1* (5), 1940: 162, 164–166 (comp. with *vespertina*, key, meas., ill. of head); Fowler, Bull. U.S. nat. Mus., 100 (*13*), 1941: 212 (ref. for Atlant., Rhode Island record, but probably not the refs. for China, Philippines); Longley and Hildebrand, Pap. Tortugas Lab., *34*, 1941: 3 (embryos, descr. yolk stalk, Tortugas, Florida); Springer, Proc. Fla. Acad. Sci., *5*, 1941: 5 (comp. with other species, ill. of skull); Norris, Plagiost. Hypophysis, 1941: pl. 11, fig. 40 (brain); Fowler, Arqu. Zool. Estado São Paulo, *3*, 1942: 129 (Brazil); Fish Culturist, *21* (9), 1942: 66 (listed, Cuba); Lunz, Bull. S. C. St. Planning Bd., *14*, 1944: 27 (S. Carolina, Florida); Bigelow and Schroeder, Guide Comm. Shark Fish., Anglo Amer. Caribb. Comm., Wash., 1945: 132, fig. 48 (descr., range, ill.); Fowler, Monogr. Acad. nat. Sci. Philad., 7, 1945: 96 (N. Carolina), 160 (S. Carolina); Gunter, Publ. Inst. Mar. Sci. Univ. Texas, *1*, 1945: 20 (temp., breeding, Texas).

Cestrorhinus tiburo Blainville, Bull. Soc. philom. Paris, 1816: 121 (name only).

Zygaena tiburo Valenciennes, Mem. Mus. Hist. nat. Paris, *9*, 1822: 226, pl. 12, fig. 2^a, 2^b (ill. of head, Brazil); Bory de St. Vincent, Dict. Class. Hist. Nat., *15*, 1829: 597 (general); Cuvier, Règne Anim., *2*, 1829: 394, footnote (diagn., refs.); Cloquet, Dict. Sci. Nat., *60*, 1830: 623 (general); Voigt, in Cuvier, Tierreich, *2*, 1832: 514 (diagn., Brazil); Yarrell, Brit. Fish., Suppl., 1839: 64, 65 (synonyms, ill. of head, Brazil); Hist. Brit. Fish., *2*, 1841: 507, 508 (same as Yarrell, 1839); Cuvier, Règne Anim., ill. ed., 1843: 309, footnote; Baird, Rep. Smithson. Instn. (1854), 1855: 337 (N. Jersey record); Castelnau, Anim. Nouv. Rares Amer. Sud., *2*, 1855: 99 (Bahia, Brazil); Günther, Trans. zool. Soc. Lond., *6*, 1869: 396 (Atlant. coast, Cent. Amer.); Cat. Fish. Brit. Mus., *8*, 1870: 382 (descr., refs., Bahia, Belize, but spec. from China probably not this species); Doderlein, Man. Ittiol. Medit., 2, 1881: 50 (occur. in Medit. questioned); Garman, Bull. Mus. comp. Zool. Harv., *8*, 1881: 232 (Florida Keys); Werner, Zool. Jb., Syst. Abt., *21*, 1904: 284 (Atlant. Oc.); Huber, Z. wiss. Zool., *70*, 1901: 626 (anat. of claspers).

Zygaena (no specific name) Agassiz, L., Poiss. Foss., *3*, 1835: pl. E, fig. 9 (teeth).

Zygaena (*Squalus*) *tiburo* Leuckart, Unters. Auss. Kiemen Rochen Hayen, Stuttgart, 1836: 22 (ill., early embryo, yolk stalk, yolk).

Platysqualus tiburo Swainson, Nat. Hist. Fish. Amphib. Rept., *2*, 1839: 318 (descr.).

Sphyrnias tiburo Gray, List. Fish. Brit. Mus., *1*, 1851: 50.

Cestracion[25] *tiburo* Gill, Proc. Acad. nat. Sci. Philad., Addend., 1861: 59 (in N. Amer. cat.), and later eds.; Duméril, Hist. Nat. Poiss., *1*, 1865: 385 (descr., east coast, N. and S. Amer.); Abbott, Geol. Surv. N. J., Geol. N. J., Append. E, 1868: 829 (N. Jersey); Fowler, Rep. N. J. Mus. (1905), 1906: 65 (N. Jersey); Proc. Acad. nat. Sci. Philad., *58*, 1906: 80 (Florida Keys, abund.); Rep. N. J. Mus. (1906), 1907: 257 (N. Jersey record); Garman, Mem. Harv. Mus. comp. Zool., *36*, 1913: 160, pl. 1, fig. 4, 5 (descr., ill., Massachusetts Bay spec.); Radcliffe, Bull. U.S. Bur. Fish., *34*, 1916: 265, pl. 44, fig. 1, 2, but not fig. 3, which appears to be either *diplana* or *zygaena* (denticles, teeth, food, N. Carolina); Meek and Hildebrand, Field Mus. Publ. Zool., *15* (1), 1923: 60 (descr., Colón, Panama); Breder, Bull. Bingham oceanogr. Coll., *1* (1), 1927: 5 (Florida); Bigelow and Schroeder, Canad. Atlant. Fauna, biol. Bd. Canad., 12^e, 1934: 10 (descr., distrib.); Bull. U.S. Bur. Fish., *48*, 1936: 322 (ref. to Massachusetts Bay spec. descr. by Garman, 1913).

Reniceps tiburo Gill, Ann. N. Y. Lyc., 7, 1862: 412 (name); Rep. U.S. Comm. Fish. (1871–72), 1873: 813 (Cape Cod to Florida); Poey, An. Soc. esp. Hist. Nat., *5*, 1876: 384; Enumerat. Pisc. Cubens., 1876: 188 (listed, Cuba); Jordan and Gilbert, Bull. U.S. nat. Mus., 16, 1883: 25, 874 (descr., size, Atlant. Oc., but rep. for China probably not this species); Bean, T. H., Bull. U.S. Fish Comm., 7, 1888: 152 (old N. Jersey record); Wilson, Proc. U.S. nat. Mus., *33*, 1907: 360 (parasites, St. Mary's R., Florida); Jordan, Manual Vert. Anim. NE. U.S., 1929: 11 (general); Jordan, Evermann and Clark, Rep. U.S. Comm. Fish. (1928), 2, 1930: 17 (Atlant. Oc., Mazatlan, Panama, but rep. for China prob-

25. This name was first used by Klein (Neuer Schauplatz, *3*, 1776: 526), but must date from Gill, 1861, because Klein's names are not applicable by Opinion 89 of the International Commission on Zoological Nomenclature; see footnote 2, p. 408.

ably not this species); Gowanloch, Bull. La. Conserv. Dep., *23*, 1933: 219, 220 (abund. Gulf coast U.S.); Bere, Amer. Midl. Nat., *17* (3), 1936: 583, 605 (parasites, Florida west coast); Whitley, Aust. Zool., *8*, 1936: 190 (*Squalus carolinianus* Latreille, 1804, a synonym of *tiburo*).

Zygaena leeuwenii Rochebrune, Act. Soc. linn. Bordeaux, (4) *6*, 1882: 44; Faune Senegambie, Poiss., *1*, 1883–1885: 21 (ident. probable because of descr. of head; abund. Senegambia; not *Zygaena lewini* Griffith, in Cuvier, Anim. Kingd., *10*, 1834: 640, pl. 50); Metzelaar, Trop. Atlant. Visschen, 1919: 189 (Senegambia by ref. to Rochebrune, 1882).

Reniceps (no spec. name) Adams and Kendall, Bull. U.S. Fish Comm., *9*, 1891: 295 (Florida west coast).

Sphyrna (*Reniceps*) *tiburo* Jordan and Evermann, Rep. U.S. Comm. Fish. (1895), 1896: 217 (Long Island southward and Mazatlan, Mexico, but rep. for China probably not this species); Bull. U.S. nat. Mus., 47 (1), 1896: 44 (descr., Long Island southward, and Mazatlan, Mexico, but rep. for China probably not this species).

Cestracion zygaena Radcliffe, Bull. U.S. Bur. Fish., 34, 1916: pl. 43, fig. 2 (photo. of jaws, evidently mislabelled by accident).

Sphyrna tiburo (perhaps in part) Puyo, Bull. Soc. Hist. nat. Toulouse, *70*, 1930: 83 (ill. of head is *diplana*, but discuss. perhaps covers *tiburo* also).

Sphyrna vespertina Springer, Stanford Ichthyol. Bull., *1* (5), 1940: 161, 164–166 (type descr., comp. with other species, meas., ill. of head, Panama, Guayaquil, Ecuador); Beebe and Tee-Van, Zoologica, N. Y., *26*, 1941: 116 (diagn., Panama, Ecuador).

Not *Zygaena tiburo* Reguis, Ess. Hist. Nat. Provence, Fasc. 1, 1877: 59 (either *diplana* or *tudes* because head descr. as emarginate in midline, Medit.).

Sphyrna tudes (Valenciennes), 1822

Great Hammerhead

Figures 83, 84

Study Material. Male embryo, about 585 mm., a newborn male of 673 mm., another small specimen, and the head of a specimen of about 1,660 mm. as calculated from breadth of head, all from Englewood, Florida (Harv. Mus. Comp. Zool. and U.S. Nat. Mus.); jaws from specimens of 13½ feet and 7 feet 10 inches from Englewood, Florida (U.S. Nat. Mus., No. 110299, 110300); excellent cast of a male, a little more than 10 feet long, from Miami, Florida[26] (Harv. Mus. Comp. Zool.). Measurements of six specimens, male and female, 700 to 3,155 mm., from Englewood, Florida, as well as photographs of a large one on the beach.[27]

Distinctive Characters. The hammer (not shovel) shape of its head places *tudes* with *zygaena*, *diplana* and *bigelowi*; and it falls with the last two because the anterior margin of its head is definitely indented in the midline. But it is easily distinguishable from all three by the edges of both its lower and upper teeth, which are regularly serrate from tip to base (smooth or weakly serrate in *zygaena*), and by its first dorsal fin which is less erect. It is further marked off from *zygaena* and *diplana* by the much shorter free rear corners of its second dorsal and anal fins and by its differently shaped hammer head (cf. Figs. 83, 84 with 81 G, 86 A). Also, its pelvics are much more strongly convex anteriorly and concave posteriorly than those of any other Atlantic member of the genus.

26. Prepared by Al. Pflueger.

27. Contributed by Stewart Springer.

Description. Proportional dimensions in per cent of total length. Male, 673 mm., from Englewood, Florida (U.S. Nat. Mus., No. 106543).

Trunk at origin of pectoral: breadth 10.4; height 10.4.
Snout length in front of: outer nostrils 3.1; mouth 7.0.
Eye: horizontal diameter 2.4.
Mouth: breadth 6.0; height 3.3.
Nostrils: distance between inner ends 16.9.
Gill opening lengths: 1st 2.8; 2nd 2.9; 3rd 3.0; 4th 2.8; 5th 2.4.
First dorsal fin: vertical height 11.9; length of base 10.1.
Second dorsal fin: vertical height 4.4; length of base 5.6.
Anal fin: vertical height 4.0; length of base 6.7.
Caudal fin: upper margin 30.0; lower anterior margin 10.4.
Pectoral fin: outer margin 12.3; inner margin 4.4; distal margin 9.2.
Distance from snout to: 1st dorsal 25.4; 2nd dorsal 57.3; upper caudal 70.0; pectoral 20.2; pelvics 43.7; anal 55.9.

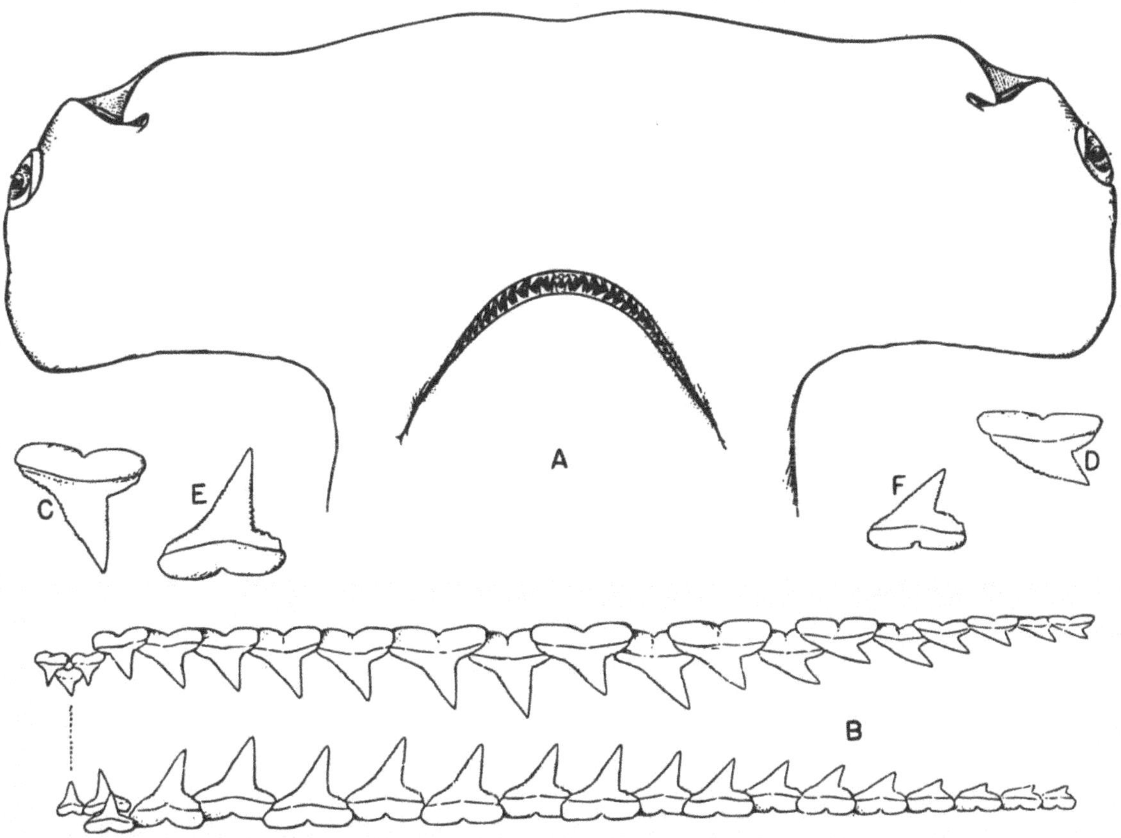

FIGURE 83. *Sphyrna tudes,* about 1,660 mm. long, from Englewood, Florida (U. S. Nat. Mus., No. 108453). *A* Anterior part of head from below. *B* Left-hand upper and lower teeth of same, about 1.3 x. *C* Fifth upper tooth. *D* Twelfth upper tooth. *E* Fifth lower tooth. *F* Eleventh lower tooth. *C-F,* about 2 x.

Interspace between: 1st and 2nd dorsals 20.8; 2nd dorsal and caudal 7.1; anal and caudal 6.8.

Distance from origin to origin of: pectoral and pelvics 21.1; pelvics and anal 12.9.

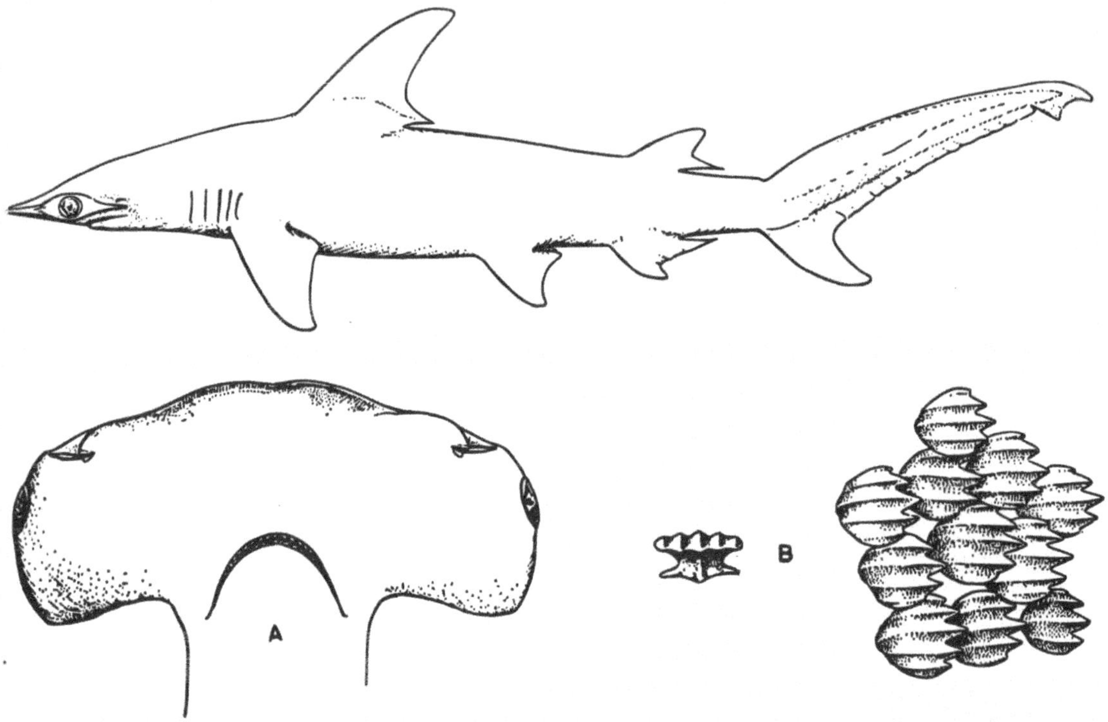

FIGURE 84. *Sphyrna tudes*, new-born male, 673 mm. long, from Englewood, Florida (U. S. Nat. Mus., No. 106543). *A* Anterior part of head from below, about 0.4 x natural size. *B* Dermal denticles, about 65 x.

Two females, 1,745 and 3,155 mm., from Englewood, Florida (calculated from measurements by Stewart Springer).

Snout length in front of: mouth 5.4, 4.6.

Eye: horizontal diameter 1.4, 0.9.

Mouth: breadth 6.6, 7.1; height 2.9, 4.3.

Nostrils: distance between inner ends 16.3, 19.0.

Gill opening lengths: 1st 3.3, 4.0; 5th 2.3, 2.8.

First dorsal fin: length of base 10.0, 10.1.

Second dorsal fin: length of base 4.9, 4.4.

Anal fin: length of base 6.3, 6.0.

Caudal fin: upper margin 31.0, 29.5; lower anterior margin 12.0, 11.7.

Pectoral fin: outer margin 14.1, 15.8; inner margin 3.9, 3.8; distal margin 11.4, 14.4.

Distance from snout to: 1st dorsal 27.8, 28.5; upper caudal 69.0, 70.5; pectoral 20.6, 21.8.

Interspace between: 1st and 2nd dorsals 21.7, 19.8; 2nd dorsal and caudal 8.3, 7.6.

Trunk moderately compressed, as usual in this genus, but somewhat stouter when adult than in the *diplana-zygaena* group, its dorsal profile moderately arched. No trace of mid-dorsal ridge. Caudal peduncle about ⅔ as thick as high, the upper precaudal pit strongly developed as a broadly triangular depression but no definite lower pit. Dermal denticles close-spaced and overlapping regularly (laterally as well as longitudinally) but with skin visible here and there, the blades but little arched longitudinally and not much raised, very thin, about as broad as long, smooth toward base, but with 3 to 5 low rounded ridges toward free margin in small specimen (perhaps 5 to 6 in large specimens), the marginal teeth rather short and broad, the median very little the longest; pedicels short and thick; basal plates with 4 rather long rays.

Head to origin of pectoral about ⅓ of length of trunk to origin of caudal or a little less, its dorsal profile evenly convex posterior to eyes but concave forward to a rather thin tip, hammer-shaped, the outer posterior margins slightly concave and nearly transverse (sloping more strongly rearward in *diplana* and *zygaena*); width of head at eyes about 1.2 times its length to origin of pectoral in small specimens, the width of hammer relative to its length at oculo-narial prominences increasing from about 3:1 in very small specimens to about 4:1 in medium-sized and large; anterior margin of head scalloped with a depression opposite each nostril (not as deep as in the *diplana-zygaena* group), a shallow but unmistakable indentation in the midline, and a shallow sinuosity between these two; the groove running from the nostril inward along anterior margin of head only very faintly marked, but visible upon careful examination nearly halfway to the midline. Distance from anterior margin of eye to anterior corner of oculo-narial prominence about as long as diameter of eye; a line connecting inner ends of nostrils passes anterior to front of mouth by a distance nearly or quite twice as long as diameter of eye (not longer than diameter of eye in the *diplana-zygaena* group), one connecting centers of eyes passes in front of mouth by a distance only about ½ as long as diameter of eye in newborn specimens but about 1½ times that long in larger (it passes close in front of mouth in *diplana*, and a little posterior to it in *zygaena*) and one connecting outer posterior corners of hammer passes a little posterior to corners of mouth in newborn specimens but anterior to it in large by a distance at least ½ as long as diameter of eye (in adults of the *diplana-zygaena* group this line passes posterior to corners of mouth). Head in front of mouth about ⅓ as long as to pectorals in small specimens but decreasing in relative length with growth to only about ⅕ that long in large adults. Rostral cartilage with a median oval or triangular hole, but wings of preorbital processes without inwardly directed points on anterior margin (intermediate in this respect between *diplana* and *zygaena*). Mucous pores in median belt on oral side of front of head covering a subtriangular area, much as in *zygaena*. Eye a little broader than high, its horizontal diameter about ⅓ as long as head in front of mouth in young but increasing so little in size with later growth as to be only about ⅕ as long as head in front of mouth in large specimens (much smaller than in the *diplana-zygaena* group).[28] Gill openings extending ventrally, the 1st to 4th nearly straight

28. Diameter of eye 16 to 18 mm. in specimens of 700 to 725 mm. but only 25 mm. at a length of 1,735 to 1,745 mm., 26 mm. in a specimen of 2,370 mm., and 29 mm. in one of 3,155 mm.

but the 5th slightly concave anteriorly, their relative lengths increasing with growth, the 1st being only a little more than ⅓ as long as head in front of mouth or about 1.3 times as long as diameter of eye in young, nearly ⅔ as long as head in front of mouth or 2.2 to 3.3 times as long as diameter of eye in medium-sized specimens, and nearly 90% as long as head in front of mouth or 4.3 times as long as diameter of eye in largest measured (about 10 ft. 4 in. long); the 3rd gill opening very little longer than 1st, the 5th only a little shorter (about 84%) than 1st, the two spaces between 1st and 3rd about equally broad, but those between 3rd and 5th much narrower, the 4th gill opening about over origin of pectoral. Nostril approximately transverse. Mouth about ½ as long as broad. No trace of lower or upper labial furrows, although there may be wrinkles at corners of mouth when the latter is closed.

Teeth $\frac{17\text{—}2 \text{ or } 3\text{—}17}{16 \text{ or } 17\text{—}1 \text{ to } 3\text{—}16 \text{ or } 17}$, their edges regularly and moderately coarsely serrate from tip to base, except for the outermost 1 or 2, which are nearly or quite smooth; upper teeth triangular, on expanded bases, the 1st tooth erect, symmetrical, but the subsequent upper teeth increasingly oblique toward corners of mouth, their outer margins increasingly notched, inner margins concave toward base except along outer ⅓ of jaw where they are evenly and rather strongly convex; 2nd to 10th or 11th upper teeth considerably the largest and the outermost 2 to 3 very short; lower teeth similar to uppers, except perhaps a little narrower and with inner edges along outer ⅓ of jaw nearly straight instead of convex; 2 or 3 minute teeth at symphysis in upper jaw, and 1 to 3 in lower.

First dorsal narrower toward apex and more sloping than in *diplana* or *zygaena*, a perpendicular from its apex falling posterior to its rear tip in young specimens but a little anterior to its tip in large, its origin about over midpoint of inner margin of pectoral, its vertical height about ½ as great as length of head, the anterior margin moderately convex toward apex but perhaps less so in adult than in young, the posterior margin deeply concave in young but less so in adult, its free rear corner comparatively broad and only about ¼ as long as base both in young and old (more than ⅓ as long as base in *diplana*); the midpoint of base about 1.8 times as far from origin of pelvics as from axil of pectoral in young. Second dorsal at base averaging a little less than ½ as long as 1st and about ⅓ as high, its origin a little anterior to midpoint of base of anal, its anterior margin moderately convex, posterior margin very deeply concave in subangular outline (much more deeply so than in either *diplana* or *zygaena*), apex rounded, its free rear corner about as long as the base (much longer, relatively, in the *diplana-zygaena* group). Caudal a little less than ⅓ (29 to 31%) of total length, its upper margin weakly convex in young and even less so in adult, the terminal sector about ⅙ of the fin, with narrow tip and rather deeply concave lower posterior margin, the lower lobe a little more than ⅓ (35 to 42%) as long as upper, slender, with very narrowly rounded apex, its lower anterior margin rather strongly convex in young but weakly so in adult, its posterior margin nearly straight; the re-entrant contour (included between the 2 lobes) with rounded corner, a little less than a right angle. Distance from origin of caudal to tip of anal a little more than ½ as long as base of latter.

Anal about 1.1 to 1.3 times as long at base as 2nd dorsal, generally similar in shape, deeply incised near the apex much as in *diplana* and in *zygaena*. Distance from origin of anal to tips of pelvics between 2/3 and 3/4 as long as base of anal. Pelvics only about 2/3 as long at base as anal, their anterior margins more strongly convex and the posterior margins more deeply concave than in any other Atlantic member of the genus. Pectoral noticeably small, its length a little more than 1/2 as great as that of head in young but nearly 3/4 (72 to 73%) that length in adults or about as long as vertical height of 1st dorsal, a little more than 1/2 as broad as long, its outer margin strongly convex toward apex in young but less so in adult, its distal margin only moderately concave, apex subacute and inner corner narrowly rounded.

Color. Small specimens are brownish gray above and a paler shade of the same tint below; the dorsals, lower and upper caudal lobes, upper surfaces of the pectorals and lower edge of the caudal are dusky toward the tips. A cast of a large one, colored from the fresh-caught specimen, is dark olive above and pale olive below, without any conspicuous fin markings.

Size. Available information suggests a length slightly less than 700 mm. at birth which corresponds to the large number in a litter. This is the largest of Atlantic Hammerheads, apparently not maturing at a length less than about 10 feet and commonly growing to 13 to 14 feet, with individuals of 15 feet reliably reported.

Developmental Stages. It is not known whether *tudes* is ovoviviparous or viviparous; 30 to 38 embryos have been found in females off southern Florida. Embryos ready for birth differ from adults in the relatively greater length of the hammer relative to its breadth, much larger eyes, longer head in front of mouth, shorter gill openings, more rounded pectoral and caudal, and more oblique and more rounded first dorsal.

Habits. Nothing is known of the habits or diet of *tudes* to set it apart from the *diplana-zygaena* group. Around southern Florida females have been taken in June.

Relation to Man. The Great Hammerheads that are taken in the shark fisheries of Florida and the West Indies are utilized with others for leather, etc. Now that shark liver oil is in demand for its vitamin content, large specimens of *tudes* may prove very valuable, for some (but not all) are extremely rich in Vitamin A. As an example we may cite a 13-foot 10-inch Florida specimen which was recently caught and whose liver oil had a potency of 357,000 units of Vitamin A per gram with a sale value of about $500 at current prices. But this was exceptional. Another of about the same size, caught at the same locality a few days later, had a Vitamin A potency of only about 55,000 units per gram, although it yielded about three times as much oil, its market value thus being only about $150.[29]

Range. Tropical and subtropical Atlantic. Hammerheads have also been reported as *tudes* from the west coast of Central America, the Hawaiian Islands, Australia, Indo-China, the East Indies, the Philippines, India and the Gulf of Arabia.[30] But whether or

29. Personal communication from Stewart Springer.
30. See Fowler (Bull. U.S. nat. Mus., 100 [*13*], 1941: 213) for a list of Australian and East Indian citations.

not the Indo-Pacific form is actually identical with that of the Atlantic remains uncertain, awaiting critical comparison of specimens from the respective seas.[31]

Occurrence in the Atlantic. It is not yet possible to describe the distribution of *tudes* in any detail for either side of the Atlantic, because in very few instances have reports of its presence been accompanied by evidence sufficient to establish the actual identity of the particular specimens on which they were based. On the eastern side of the Atlantic its presence is so far positively established only for the Mediterranean, where it has been taken widely, although never in any numbers. It is also reported off the Atlantic coast of Spain, from the Cape Verde and Canary Islands, and from tropical West Africa (Morocco, Senegambia, Gambia). Positive locality records for the western North Atlantic are: French and Dutch Guiana (Cayenne, Surinam); Cuba; western and southeastern Florida; and North Carolina (Cape Lookout and Beaufort, one specimen each). We have no reason to suppose that it accompanies the other Hammerheads in their yearly migration farther north offshore in the Gulf Stream (p. 442). In the opposite direction there is report of it from Brazil (Santos, Pernambuco), from Uruguay, and from northern Argentina. But the only Brazilian record of *tudes* that is accompanied by an illustration[32] appears actually to have been based on a specimen of *S. diplana* (p. 420), while reports of its presence farther south lack any supporting evidence.

The foregoing, together with the fact that *tudes* is taken in some numbers off both coasts of southern Florida in summer, although large ones at least are unknown there in winter, marks it as characteristically a tropical species.

Synonyms and References:[33]

Le Marteau, Duhamel, Traité Gén. Pêches, (2) *4* (9), 1777: 303 (in part), pl. 21, fig. 8 (ident. by shape of head); Broussonet, Mem. Math. Phys. Acad. Sci. Paris (1780), 1784: 661 (in part, by ref. to Duhamel, 1777).

No species name, Parra, Desc. Piez. Hist. Nat. Havana, 1787: 71, pl. 32 (ident. probable from ill.).

Squale pantouflier, Lacépède, Hist. Nat. Poiss., 4° ed., *1*, 1798: 167, 261, pl. 7, fig. 3 (ident. from ill., Cayenne); in Sonnini, Hist. Nat. Poiss., *4*, 1802: 81 (from Lacépède, 1798).

Squale Marteau, Lacépède, Hist. Nat. Poiss., 4° ed., *1*, 1798: 257 (part, because teeth descr. as serrate, but not pl. 8, fig. 3, which is *zygaena*).

Hammer Headed Shark, Shaw, Gen. Zool., *5* (2), 1804: 354, pl. 154 (ident. from descr. of teeth, and ill., Medit., but report for Indian seas and Tahiti perhaps not this species).

Squalus tiburo Risso, Ichthyol. Nice, 1810: 35 (ident. by descr. of head and ref. to Lacépède, Squale pantouflier, as above, Nice, very rare); not *Squalus tiburo* Linnaeus, 1758.

Zygaena tudes Valenciennes, Mem. Mus. Hist. nat. Paris, *9*, 1822: 225, pl. 12, fig. 1 (type descr., ill. of head, Medit., Cayenne, but Coramandel record perhaps not this species); Risso, Hist. Nat. Europ. Merid., *3*, 1826: 126 (ident. by ref. to Valenciennes, 1822); Bory de St. Vincent, Dict. Class. Hist.

31. The *Zygaena tudes* pictured by Philippi (An. Univ. Chile, *71*, 1887: pl. 2, fig. 4) from Chile was actually of the *diplana* group, to judge from the shape of its head.

32. Ribeiro, Arch. Mus. nac. Rio de J., *14*, 1907: pl. 5.

33. Because of the uncertainty as to whether or not sharks reported from the Pacific and Indian Oceans as *tudes* are actually identical with the Atlantic Hammerhead of that name, the following is limited to citations for the Atlantic. For Indo-Pacific citations of *tudes*, see Fowler (Bull. U.S. nat. Mus., 100 [*13*], 1941: 213) and Beebe and Tee-Van (Zoologica, N. Y., *26* [2], 1941: 115).

Nat., *15*, 1829: 596 (ref. to Valenciennes, 1822); Voigt, in Cuvier, Tierreich, *2*, 1832: 514 (ref. to Valenciennes, 1822); Agassiz, L., Poiss. Foss., *3*, 1835–1843: 91, pl. E, fig. 8 (teeth); Yarrell, Brit. Fish., *2*, 1841: 507, 508, fig. 1 (ill. after Valenciennes, 1822); Günther, Cat. Fish. Brit. Mus., *8*, 1870: 382 (refs. in synon., but ident. of descr. specs. doubtful because of long prenarial groove); Moreau, Hist. Nat. Poiss. France, *1*, 1881: 327 (ident. by descr., Nice, France); Doderlein, Man. Ittiol. Medit., *2*, 1881: 50 (ident. by refs., Medit., occur.);[34] Perugia, Elenc. Pesc. Adriat., 1881: 53 (name only, Adriatic); Rochebrune, Act. Soc. linn. Bordeaux, (4) *6*, 1882: 45; Faune Senegambie, Poiss., *1*, 1883–1885: 22 (refs. in synon., but ident. of W. Afr. specs. doubtful); Belloc, Rev. des Trav. Pêches Marit., 7 (2), 1934: 133 (ill. head), 134 (Morocco to Senegal; Canaries).

Sphyrnichthys zygaena Thienemann, Lehrb. Zool., *3*, 1828: 408 (general, ident. probable because teeth descr. as "gezähnelte").

Sphyrna malleus L. Agassiz, Poiss. Foss., *3*, 1835–1843: 12, 13, pl. E, fig. 7, pl. N, fig. 8 (ident. probable because teeth serrate).

Sphyrna tudes Müller and Henle, Plagiost., 1841: 53 (ident. from ref. to Valenciennes, 1822, Cayenne, but Coramandel record perhaps some other species); Bonaparte, Cat. Pesc. Europ., 1846: 18 (name only, ident. from included refs.); Canestrini, in Cornalia, *et al.*, Fauna d'Ital., *3*, 1871–1872: 47 (ident. by inference, Nice); Doderlein, Atti Accad. Palermo, N. S. *6*, 1878: 30 (name only, Sicily); Jordan, Rep. U.S. Comm. Fish. (1885), 1887: 797 (name only, Europ., W. Indies, but trop. Pacif. ref. perhaps not this species); Carus, Prod. Fauna Medit., *2*, 1889–1893: 544 (name only, Medit.);[35] Ribeiro, Arch. Mus. nac. Rio de J., *14*, 1907: 201 (synon. in part, but not *S. tudes*, p. 157, pl. 5, which appears to be *diplana*, see p. 415); Rey, Fauna Iberica Peces, *1*, 1928: 365, fig. 2 (head, after Valenciennes, 1822, Spain); Borri, Atti Soc. tosc. Sci. nat., *44*, 1934: 91 (name only, Lagos, W. Afr.); Fowler, Bull. Amer. Mus. nat. Hist., *70* (1), 1936: 64 (ident. from descr. of young, Surinam); Puyo, Bull. Soc. Hist. nat. Toulouse, *70*, 1936: 84 (name only, French Guiana); Norman and Fraser, Giant Fishes, 1937: 47 (general); Springer, Proc. Fla. Acad. Sci., *3*, 1939: 32 (first adequate descr. since Valenciennes, 1822, size, no. of embryos, occur. SW. Florida); Stanford Ichthyol. Bull., *1* (5), 1940: 162, 164 (comp. with other species, changes in relative proportions with growth); Proc. Fla. Acad. Sci., *5*, 1941: 49 (discuss., ill. of skull); Fowler, Bull. U.S. nat. Mus., 100 (*13*), 1941: 213 (Atlant. citations in synon., but list of citations of *tudes* for Indo-Pacific and ident. of described Pacific specs. doubtful); Arqu. Zool. Estado São Paulo, *3*, 1942: 129 (listed, Brazil); Lunz, Bull. S. Carolina St. Planning Bd., 14, 1944: 27 (listed, Florida); Bigelow and Schroeder, Guide Comm. Shark Fish., Anglo Amer. Caribb. Comm., Wash., 1945: 138, fig. 51 (descr., range, ill.); Fowler, Monogr. Acad. nat. Sci. Philad., 7, 1945: 97 (Beaufort, N. Carolina).

Sphyrna chiereghini Nardo, Sinon. Modern. Spec. descr. Pesci St. Chiereghin, 1847: 111 (name); Atti Ist. veneto., (3) *4*, 1860: 787 (name only, Medit.); Perugia, Cat. Pesc. Adriat., 1866: 7 (name only, Adriatic).

Sphyrnias tudes Gray, List Fish. Brit. Mus., 1851: 50.

Cestracion (*Zygaena*) *tudes* Duméril, Hist. Nat. Poiss., *1*, 1865: 384 (descr. from Valenciennes, 1822, Nice, Algiers, Cayenne, Demerara?).

Squalus zygaena de la Blanchère, Dict. Pêches, 1868: 499 (at least part, because teeth descr. as "dentelées").

Cestracion tudes Ninni, Revist. Critica Pesc. Adriat. descr. Chiereghini, 1872: 10 (not seen); Garman, Mem. Mus. comp. Zool. Harv., *36*, 1913: 159 (synonymy, but Brazilian spec. descr. is *S. bigelowi*, see p. 414).

Sphyrna (*Platysqualus*) *tudes* Jordan and Evermann, Rep. U.S. Comm. Fish. (1895), 1896: 217 (name, W. Indies, Medit., but ref. to Gulf of Calif. and Indian Ocean perhaps not this species); Bull. U.S. nat. Mus., 47 (1), 1896: 44 (refs., descr., Medit., W. Indies, but ref. to Gulf of Calif. and Indian Ocean perhaps not this species).

Cestracion zygaena Radcliffe, Bull. U.S. Bur. Fish., *34*, 1916: 263 in part, pl. 3, fig. 2 (ident. by descr. of teeth of ill. spec. as serrate, 12-ft. 6-in. spec., Beaufort, N. Carolina).

Sphyrna zygaena Coles, Copeia, 69, 1919: 41 in part, pl. 3, fig. 2 (ident. by shape of head in photo. of 13-ft. 10-inch female, C. Lookout, N. Carolina and meas. of same); not *Squalus zygaena* Linnaeus, 1758.

34. See Doderlein for additional Mediterranean locality records in publications not accessible to us.

35. See Carus for additional Mediterranean records in publications not accessible to us.

Platysqualus tudes Howell-Rivero, Proc. Boston Soc. nat. Hist., *41* (4), 1936: 44 (ident. probable because of large size, 12 to 14 ft., Cuba).

Doubtful References:

Zygaena tudes Cuvier, Règne Anim., ill. ed., 1843: pl. 117, fig. 1; Osorio, J. Sci. math. phys. nat. Lisboa, (2) *3*, 1894: 182 (name only, C. Verde Is.); J. Sci. math. phys. nat. Lisboa, (2) *5*, 1898: 200 (name only, C. Verde Is.); Metzelaar, Trop. Atlant. Visschen, 1919: 189 (by ref. to Rochebrune and Osorio as above, trop. W. Afr.).

Sphyrna tudes Müller and Troschel, in Schomburgk, Reisen Brit. Guiana, 1840–44, *3*, 1848: 642 (one reported as seen, Demerara); Machado, Peces Cadiz, 1857: 9 (Spanish coast, name only); Berg, Ann. Mus. nac. B. Aires, (2) *1* (4), 1895: 8 (Montevideo, Muldonado, no evidence for ident.); Fowler, Proc. Acad. nat. Sci. Philad., *60*, 1908: 66 (meas., 8-in. spec., not diagnostic, and ident. doubtful because of small size, Surinam); Ribeiro, Rev. Mus. paul., 1918: 107 (name only, Santos, Brazil); Devincenzi, An. Mus. Hist. nat. Montevideo, (2) *1* (4), 1920: 119 (name only, Uruguay); An. Mus. Hist. nat. Montevideo, (2) (6), 1925: 323 (name only); Marelli, Elenc. Sist. Fauna B. Aires (1922–23), 1924: 546 (name only, Uruguay, Argentina); Pozzi and Bordale, An. Soc. cient. argent., *120*, 1935: 151 (name only, north. Argentina).

Cestracion tudes Meek and Hildebrand, Field Mus. Publ. Zool., *15* (1), 1923: 59 (descr. appears to be partly *tiburo*).

Not *Sphyrna tudes* Ribeiro, Arch. Mus. nac. Rio de J., *14*, 1907: 157, pl. 5 (photo of head appears to be *diplana*, see p. 420, Maria Farinha, Pernambuco, Brazil); Arch. Mus. nac. Rio de J., Fauna brasil. Peixes, *2* (1) Fasc. 1, 1923: 14 (same record as foregoing).

Sphyrna zygaena (Linnaeus), 1758

Common Hammerhead

Figures 85, 86

Study Material. Eleven specimens of both sexes, from 510 to about 1,780 mm. long, from: Muldonado, Uruguay; Rio de Janeiro, Brazil; New York; off Nahant in Massachusetts Bay; Cape Cod; and off Woods Hole, Massachusetts (Harv. Mus. Comp. Zool.). Others from Virginia Beach, Virginia, and Buzzards Bay, Massachusetts (U.S. Nat. Mus.). Specimens from Woods Hole, Massachusetts (U.S. Bur. Fish.). Several others caught in the offing over the continental shelf by members of the Woods Hole Oceanographic Institution and identified by us, and ten small specimens from Peru, Panama, the Galapagos, Lower California and Japan (Harv. Mus. Comp. Zool., No. 382, 383, 421, 441, 515, 1042, and U.S. Nat. Mus., No. 51289, 51291, 71774, 77711).

Distinctive Characters. The very differently shaped head is enough to separate *zygaena* at a glance from *tiburo* (cf. Fig. 86 A with 82 B). It is easily distinguished from *diplana*, *bigelowi* and *tudes* by the evenly convex anterior outline of the head in the midline (not indented); further from *diplana* by the additional differences summarized under that species (p. 415); further from *bigelowi* and *tudes* by the much more slenderly acuminate tip of its second dorsal fin, and further from *tudes* by its much more erect first dorsal fin.

Description. Proportional dimensions in per cent of total length. Male, 524 mm., from New York (Harv. Mus. Comp. Zool., No. 136). Female, 687 mm., from Nahant, Mass. (Harv. Mus. Comp. Zool., No. 1159).

Trunk at origin of pectoral: breadth 8.2, 8.9; height 9.0, 10.2.
Snout length in front of: outer nostrils 5.9, 5.4; mouth 6.9, 6.1.
Eye: horizontal diameter 2.4, 2.2.
Mouth: breadth 7.6, 7.1; height 4.2, 4.4.
Nostrils: distance between inner ends 19.7, 20.6.
Labial furrow length: lower 0.5, 0.6.
Gill opening lengths: 1st 2.2, 2.5; 2nd 2.4, 2.5; 3rd 2.5, 2.6; 4th 2.5, 2.5; 5th 2.0, 2.1.
First dorsal fin: vertical height 11.1, 12.4; length of base 9.7, 9.5.
Second dorsal fin: vertical height 2.4, 2.3; length of base 3.2, 3.3.
Anal fin: vertical height 3.7, 3.5; length of base 5.3, 4.6.
Caudal fin: upper margin 30.2, 30.6; lower anterior margin 11.6, 13.2.
Pectoral fin: outer margin 12.0, 12.5; inner margin 4.4, 4.1; distal margin 9.2, 9.5.
Distance from snout to: 1st dorsal 27.3, 28.3; 2nd dorsal 60.2, 60.5; upper caudal 69.8, 69.4; pectoral 22.1, 21.3; pelvics 47.5, 46.5; anal 57.8, 58.0.

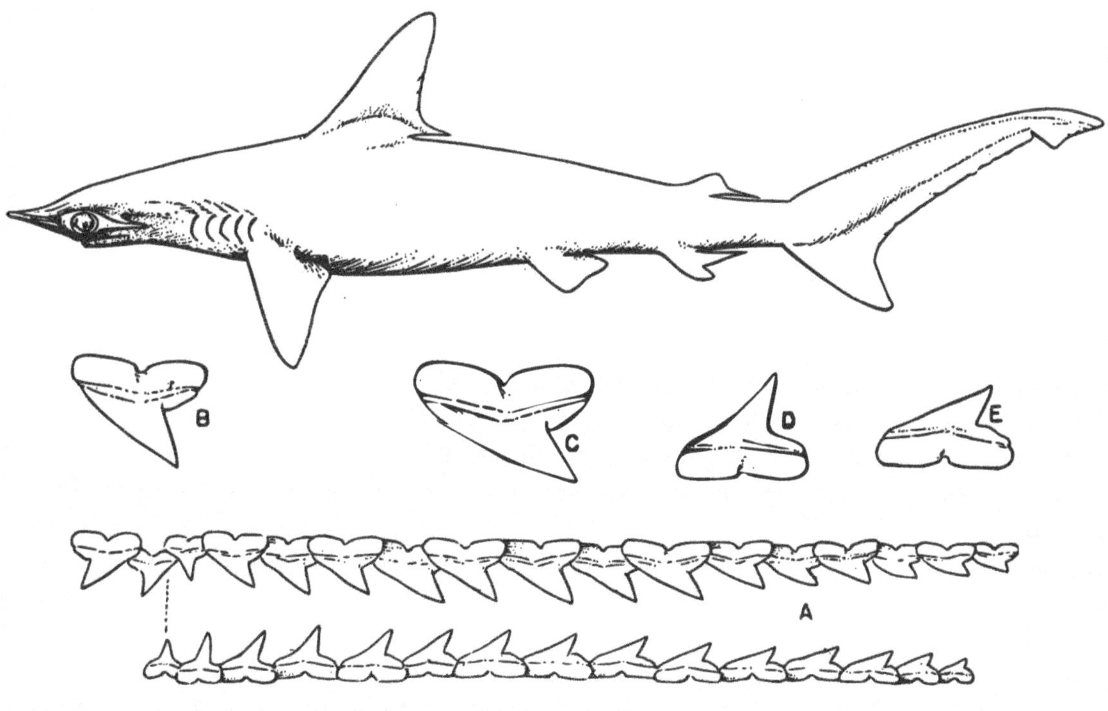

FIGURE 85. *Sphyrna zygaena*, female, 687 mm. long, from Nahant, Massachusetts (Harv. Mus. Comp. Zool., No. 1159). *A* Upper and lower teeth, left-hand side, about 2.5 x. *B* Second upper tooth. *C* Ninth upper tooth. *D* Third lower tooth. *E* Ninth lower tooth. *B–E*, about 5 x.

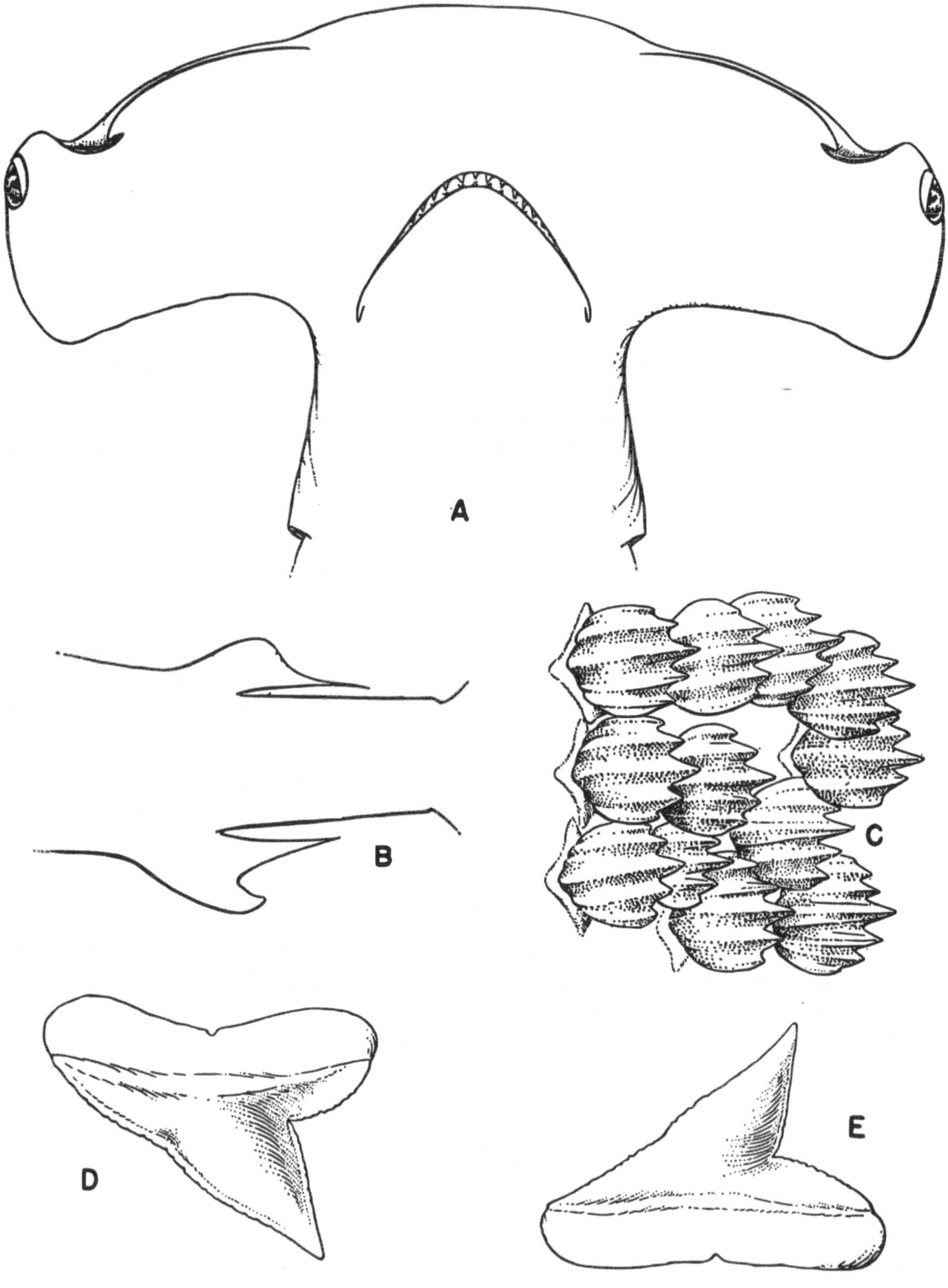

FIGURE 86. *Sphyrna zygaena*, illustrated in Fig. 85. *A* Head from below, about 0.6 x natural size. *B* Second dorsal and anal fins. *C* Dermal denticles, about 60 x. *D* Fourth upper tooth. *E* Fifth lower tooth. *D–E*, about 12 x.

Interspace between: 1st and 2nd dorsals 23.6, 25.3; 2nd dorsal and caudal 7.8, 7.0; anal and caudal 7.8, 7.0.

Distance from origin to origin of: pectoral and pelvics 26.5, 27.0; pelvics and anal 12.0, 12.8.

Trunk strongly compressed, its height at 1st dorsal about 1/5 (18 to 19%) of its length to origin of caudal. Back smooth, without mid-dorsal ridge. Caudal peduncle about 2/3 as wide as high, the upper precaudal pit strongly developed and subtriangular, lower lacking in smallest specimens examined but weakly indicated in larger. Dermal denticles so close-spaced and overlapping that skin is mostly concealed, the blades thin and moderately arched, about as broad as long, small specimens usually with 3 ridges but large ones with 5 and sometimes 7 ridges extending about halfway back from the anterior margin, 3 to 5 marginal teeth, the median considerably the longest; pedicels very short.

Head about 1/3 (37%) of length of trunk to origin of caudal, its dorsal profile slightly and evenly convex from origin of 1st dorsal to eyes, but concave thence forward, its anterior edge very thin, very broadly expanded laterally in hammer form, with outer posterior margins, outward from the neck, directed slightly toward the rear; breadth at eyes about 1.2 to 1.4 times its length to origin of pectoral, and its length at oculo-narial prominence between 1/3 and 1/4 its breadth, much as in *diplana;* anterior margin of head scalloped with a deep depression opposite each nostril, and with a shallow concavity midway between latter and the median line but with the midsector evenly and rather strongly convex (indented in *diplana*, p. 415); a well marked groove from nostril inward along anterior margin of head a little more than halfway toward the midline; distance from anterior corner of oculo-narial prominence to anterior edge of eye only about 1/2 to 7/10 as great as diameter of latter (about as great as diameter of eye in *diplana*); a line connecting outer ends of nostrils passes anterior to mouth by a distance about 1/2 as great as diameter of eye, one connecting centers of eyes passes a little behind front of upper jaw (through front of upper jaw in *diplana*), and one connecting outer posterior corners of hammer passes posterior to corners of mouth by a distance about 1/2–2/3 as great as diameter of eye. Head (snout) in front of mouth a little less than 1/3 (about 26 to 31%) as long as head to origin of pectoral. Rostral cartilage usually without median hole, and wings of preorbital processes without inwardly directed point on anterior margin (there is a median hole and such a point in *diplana*). Mucous pores in median sector on oral side of head cover a subtriangular or irregular area near its anterior margin (a subrectangular or dumbbell-shaped area in *diplana*). Eye a little broader than high and much larger relatively than in *tudes* or *tiburo*, its horizontal diameter about 1/3 as long as head in front of mouth. Gill openings rather strongly concave in outline and about evenly spaced, the 1st about equal to diameter of eye in small specimens (a little more than 1 1/2 times as long as diameter of eye in *diplana* of equal size) but about 1 1/2 times in large, the 3rd only a little longer than 1st, 5th equal to it or a little shorter, the space between 4th and 5th over origin of pectoral. Nostril sloping a little forward from inner end to outer. Mouth strongly arched, a little more than ½ as high as broad, with a very short labial furrow on lower jaw from corner of mouth, but none on upper jaw.

Teeth $\frac{13 \text{ to } 15—0 \text{ or } 1—13 \text{ to } 15}{12 \text{ to } 14—1—12 \text{ to } 14}$ in specimens counted, uppers triangular, the cusps smooth-edged in young, but tending to become weakly serrate with growth,[35a] the bases more or less fluted or wavy on outer side of some teeth but not on others; the 1st upper tooth nearly symmetrical and erect, but subsequent upper teeth strongly oblique, their inner margins straight or slightly convex, their outer margins deeply incised, even the outermost with cusps well developed though very small; lower teeth similar to uppers but a little smaller, smooth or very weakly serrate, the 4 next to the center of mouth with narrower cusps and less strongly oblique; 1 small symmetrical tooth at symphysis of lower jaw, one or none in upper in specimens examined; 1 (in places 2) series functional along sides of mouth and 2 to 3 series toward center in upper jaw; 2 to 3 series all along lower jaw.

First dorsal perhaps averaging a little more sloping than in *diplana*, a perpendicular from its apex falling a little anterior to its tip, its origin a little posterior to axil of pectoral, its vertical height a little more than ½ (50 to 58%) as great as length of head, its anterior margin only slightly convex toward apex, posterior margin moderately concave toward base, apex very narrowly rounded, its free rear corner only moderately slender, a little less to a little more than ⅓ as long as the base, the midpoint of its base about 2.5 times as far from origin of pelvics as from axil of pectoral. Second dorsal about ⅓ as long at base as 1st, its origin about over midpoint of base of anal, its apex abruptly rounded, rear margin moderately concave, its free rear corner very slender, a little longer than the base. Caudal nearly ⅓ of total length, its upper margin varying from very weakly convex to a little more strongly so toward base and tip; the terminal sector a little more than ⅕ the length of fin, slender, with narrowly rounded tip, its lower posterior margin only weakly concave (less deeply so in specimens seen than in *diplana*), the lower lobe a little more than ⅓ as long as upper with weakly convex anterior margin, nearly straight posterior margin, and narrowly rounded tip, the re-entrant contour, enclosed by the 2 lobes, approximately a right angle, with rather broadly rounded corner. Anal about 1⅓ to 1½ times (1.3 to 1.6) as long at base as 2nd dorsal and a little higher, its anterior margin more convex, apex acute, rear margin much more deeply incised in angular contour, its free rear corner about as long as the base or a little longer. Distance from origin of anal to tips of pelvics 1⅕ to 1½ times as long as base of former. Pelvics about as long at base as anal. Pectoral a little less than ⅔ (59 to 64%) as long as head, about as long as vertical height of 1st dorsal or a little longer, a little smaller than 1st dorsal in area, a little more than ½ as broad as long, the outer margin nearly straight, distal margin only slightly concave, the apex and inner corner narrowly rounded.

35a. Stewart Springer informs us that in large *zygaena* the upper teeth may have noticeable serrations while the lowers may have extremely fine serrations or none. In a specimen about six feet long (1,780 mm.), caught in August 1946 off Woods Hole, Massachusetts (Harv. Mus. Comp. Zool., No. 36425), the teeth, both uppers and lowers, vary from smooth to very weakly serrate or slightly irregular.

Color. Deep olive leaden or brownish gray above, paler on sides, shading into pure or grayish white below; fins of same color as back or sides, with tips or margins more or less dusky; pectorals black-tipped in some specimens, but not in others.

Size. Reports of embryos of 450 to 460 mm. and of free-living specimens of only 510 to 590 mm. suggest a length of about 500 mm. at birth. Adults seemingly do not mature at less than 7 to 8 feet, are often taken at 9 to 11 feet in length, and occasionally 12 to 13 feet. It seems likely that the still larger Hammerheads that are reported from time to time are in reality the Great Hammerhead (*tudes*). The following weights of Hammerheads of different lengths probably refer to *zygaena:* 1,651 mm., 57 pounds (26 kilo); 3,610 mm., 836 pounds (380 kilo); and 3,810 mm., 900 pounds (409 kilo).[36]

Developmental Stages. Presumably development is ovoviviparous; at least the presence of a placenta has never been reported. As many as 29 to 37 embryos have been reported repeatedly as being found in a single female.

Habits. This is a strong-swimming shark, often seen at the surface with the tips of its first dorsal and caudal fin exposed. But we have never seen or heard of one jumping clear of the water. They are to be met with indifferently far out at sea, or so close in to the shore that considerable numbers are often taken in beach seines or in pound nets. On the West African coast they (or they and *diplana*) are common in salt-water lagoons; they are also reported in similar situations (*e.g.*, Indian River) in Florida. There is even one report of a Hammerhead of some sort in tidal fresh water in Maryland.[37]

The diet of the *zygaenu-diplana* group consists chiefly of fish. Inshore, in the southern part of their range, they feed largely on sting-rays, which they are often seen chasing and which have been recorded frequently from their stomachs. Sting-ray spines are also found embedded in the jaws of Hammerheads, as are the spines of the gaff-topsail catfish (*Felichthys*). Often the stomach contents of the larger ones contain parts of other sharks, or entire small ones, including their own kind. Where net fisheries for sharks are carried on in warm waters it is their common habit to devour the sharks that are entangled in the nets. In the northern part of their range, skates are a major item in their diet; a Hammerhead of 1,780 mm. caught off Woods Hole in August 1944 had in its stomach one seven-inch scup (*Stenotomus*); they are also known to prey on herring and bass in waters of North Europe, and on Spanish mackerel (*Scomberomorus maculatus*) and menhaden (*Brevoortia*) in North America; no doubt they also feed on any other fishes that may be available locally. Their recorded diet also includes shrimp, crabs, barnacles, and crustacea generally, as well as squid.

Relation to Man. The considerable number taken in the shark fisheries of southern Florida and the West Indies are utilized for leather, fish meal and liver oil. They also afford good sport to any angler who chances to hook one, for they bite freely and have been described (we have never taken one on light tackle) as so lively that one has been known to die of exhaustion when hooked.

36. Schultz, J. Mammal., *19*, 1938: 484.

37. Gunter, Amer. Midl. Nat., *28*, 1942: 316.

Range. Tropical to warm-temperate belts of the Atlantic; north regularly to Portugal and occasionally to the English Channel, Welsh coast and Scotland, in the east; Mediterranean; Azores; Madeira; Teneriffe; Canaries; Cape Verde Islands; Morocco; Dakar; tropical West Africa and South Africa. In the west, north commonly to southern New England and as a stray to Massachusetts Bay and Nova Scotia; south to Uruguay and (nominally) to northern Argentina. It is also widespread in the tropical and warm-temperate Pacific[41] and probably in the Indian Ocean as well. But a more precise statement as to the Indo-Pacific range of *zygaena* must await critical study of the Hammerheads as a whole in that region.

Occurrence in the Western Atlantic. It is not possible to present a satisfactory picture of the distribution of *S. zygaena* in the western Atlantic from existing literature because of the recent discovery that many of the older accounts that ostensibly referred to it may also have covered its companion species, *diplana* (p. 419). In fact, the only western Atlantic localities where the presence of *zygaena* (not including *diplana* or *tudes*) is positively established by pertinent information, either verbal or pictorial, as to shape of head, teeth, relative proximity of the tip of second dorsal fin to caudal, or shape of skull, are: Nahant in Massachusetts Bay; Cape Cod; southern Massachusetts in the general vicinity of Woods Hole[42] and the continental shelf in its offing; vicinity of New York; several localities along the New Jersey coast; near Beaufort and Cape Lookout, North Carolina; southern Florida on both coasts; the Virgin Islands; southern Brazil; Uruguay; and (nominally) northern Argentina. But this is enough to prove it widespread all along the American seaboard in low and midlatitudes.

No doubt it is also responsible, at least in part, for the frequent reports of Hammerheads for the West Indian–Caribbean region, *i.e.*, Porto Rico, Haiti, Jamaica, Cuba, Turks Island, Trinidad, Dutch, French and British Guianas, Venezuela, the Atlantic coast of Panama and likewise for Bermuda. Although definite information is lacking for the southern part of the Gulf of Mexico, it is to be expected there generally, and no doubt it visits the northern coast of the Gulf, at least in small numbers, for there are a few records of Hammerheads for Louisiana.

The most spectacular aspect of the occurrence of Hammerheads is their migration northward (often in schools) in summer along the Atlantic seaboard, both inshore and out along the continental shelf. *Zygaena* is chiefly responsible for this seasonal movement, the most northerly records for *diplana* being from the offing of Delaware Bay (well offshore), and for *tudes* from North Carolina (p. 434), whereas many *zygaena* have been positively identified from New Jersey to southern New England. For example, Hammerheads of one sort or another are moderately common during the summer months near Charles-

41. Specimens that we have examined from Peru, Panama, the Galapagos, Lower California and Japan do not differ in any significant respect from the Atlantic specimens with which we have compared them in regard to teeth, proportionate dimensions, shape of head, or fins. Hussakoff (Copeia, 34, 1916: 63) had already reached the same conclusion for Japanese specimens.

42. A nine-foot specimen, taken in a fish trap in Buzzards Bay on August 6, 1934, and identified by a good photograph published in the *Boston Globe*, is one of several well attested records for the region.

ton, South Carolina; they are among the more plentiful of summer sharks in the vicinity of Cape Lookout, where there is record of as many as 65 taken in a single haul of a seine; and numbers of them enter the shoal North Carolina sounds in some summers but only occasionally in others. Although they appear as strays only in Chesapeake Bay, Hammerheads are common summer visitors to the Atlantic coasts of Maryland, New Jersey and New York, where they are present yearly from July until October in varying numbers, not only offshore but in the shallow coastal bays as well. They even enter New York Harbor occasionally; in fact one of the largest Hammerheads on record north of Cape Hatteras (about 11 feet) was taken many years ago at the head of Peconic Bay, Long Island. Farther east along southern New England they appear less often inshore, although there are a few records for Connecticut and Rhode Island, as well as many reports of them at Woods Hole and Nantucket for summer and early autumn (July to October); we obtained a 6-foot female from a trap near Woods Hole in August 1944. But Hammerheads basking at the surface are familiar objects a few miles offshore along this sector, as we can bear witness, and they are brought in from time to time by tuna and other fishermen. There are also a few records for the tip of Cape Cod, one for the inner part of Massachusetts Bay, and for Casco Bay, Maine, where the capture of two small ones, no doubt *zygaena*, has been reported to us.[43] But the longitude of Cape Cod so sharply bounds their coastwise dispersal in this direction that there are only two records of Hammerheads farther east on the continental shelf, one for Halifax, Nova Scotia, the other (a 12-foot specimen caught in August 1928) between Georges and Browns Banks. Outside the edge of the continental shelf, in the sweep of the Gulf Stream, Hammerheads (probably both *zygaena* and *diplana*) are to be expected much farther to the east and north, perhaps even past the Newfoundland Banks.

The great majority of individuals sharing in the summer movement northward are small (less than 6 feet). In fact, many of them are so small as to suggest that they were born only a short time previously. Dozens of little ones, of about 2½ feet, have been seined on the outer shore of Long Island in August. This has led to the suggestion that they are born in northern coastal waters, but on the other hand large ones are seldom taken near shore along our northern coast,[44] making it more likely that whatever production of young there may be in the northern sector of the range takes place well offshore.

Off New York and to the eastward Hammerheads usually disappear when the temperature of the water falls below 67° F., *i.e.*, by late September or early October. Occasionally, however, one lingers into November, and there is even one record for February in Long Island Sound. It is not known if the Hammerheads that reach the shores of the northeastern United States in the summer migrate south again in the autumn, or if they merely move offshore to escape falling temperatures and are then picked up by the sweep of the Gulf Stream and so lost to the parent population.

Information as to the status of *zygaena* coastwise in South American waters south-

43. W. H. Rich saw these in the fish market at Portland, Maine.

44. Exceptions to this rule are specimens 11 feet 1 inch, from North Carolina, of 11 feet from Long Island, New York, and 9 feet 10 inches from Rhode Island.

ward from the Caribbean is much less extensive than for the United States coast. But since *zygaena* is known positively from as far south as Uruguay, it is undoubtedly responsible at least in part[45] for the nominal reports of the species from Santos, Pernambuco and Maceio; *i.e.*, it occurs generally along the coast of Brazil. But whether or not it expands its range yearly to the southward during the warm season in the southern sector of its range, as it does to the northward in the northern sector, is not yet known.

Synonyms and References:

Atlantic:[46]

Squalus zygaena Linnaeus, Syst. Nat., 1758: 234 (by ref. to ill. by Willoughby, 1686; see above, p. 409; type loc. Europe, Amer.); Bonnaterre, Tabl. Encyc. Meth. Ichthyol., 1788: 9, pl. 6, fig. 15 (ident. by ref. to Linnaeus, 1758; descr. and ill. would apply equally to *diplana*; Atlant. and Medit.); Gmelin, Syst. Nat., *1*, 1789: 1494 (ident. by ref. to Linnaeus, 1758); Bloch, Naturg. Ausländ. Fische, *1*, 1785: 29, pl. 117, upper fig. (ill. apparently this species, but lower fig. is *Eusphyra blochii;* N. Europ., Antilles, Medit.); Bloch and Schneider, Syst. Ichthyol., 1801: 131 (ident. by ref. to Linnaeus, 1758); Bosc, Nouv. Dict. Hist. Nat., 21, 1803: 189 (diagn., ident. from included refs.); Risso, Ichthyol. Nice, 1810: 34 (ident. by ref. to Linnaeus, 1758; general, Medit.); Mitchill, Trans. Lit. Phil. Soc. N. Y., *1*, 1815: 482 (11-ft. spec., its stomach containing parts of a man; ident. probable from loc., Riverhead, N. York); Yarrell, Brit. Fish., Suppl., 2, 1839: 62 (ident. by ill. of head, Norfolk, Eng.); Gray, Cat. Fish. Coll. Descr. by L. T. Gronow in Brit. Mus., 1854: 6 (at least in part, ident. by included refs.).

Balance Fish, Brookes, Nat. Hist., *3*, 1763: 31, pl. 28, fig. 31; also 3rd ed., 1772: 18, pl. 19, fig. 18 (ident. by ill., general).

Le Marteau, Broussonet, Mem. Acad. Math. Phys. Sci. Paris (1780), 1784: 661 (at least in part, by included refs.).

Squalus malleus Shaw and Nodder, Naturalist Misc., *8*, 1796: 375, pl. 267 (ident. by ill., Medit.).

Le squale marteau, Lacépède, Hist. Nat. Poiss., 4° ed., *1*, 1798: 156, 257, pl. 8, fig. 3, in Buffon, Hist. Nat. (in part, ident. by ill. of head, but also includes *tudes*, see p. 434, general); in Sonnini, Hist. Nat. Poiss., *4*, 1802: 74, pl. 8, fig. 1, 2 (after Lacépède, 1798).

Sphyrna zygaena Rafinesque, Indice Ittiol. Sicil., 1810: 46 (ident. by included refs., Sicily); Bonaparte, Icon. Faun. Ital., *3*, 1840: pl. not numbered, fig. 1, and index (descr., ident. by ill., arrangement of pores on head, Italy); Müller and Henle, Plagiost., 1841: 51 (descr., ident. by proportional meas., Medit., France, Brazil, India); Cuvier, Règne Anim., ill. ed., *2*, 1843: 368, pl. 117, fig. 1 (ident. by ill., but ref. in text is to pl. 116, fig. 1, which is a torpedo, *Narcacion*); Bonaparte, Cat. Pesc. Europ., 1846: 18 (ident. by included refs.); Fitzinger, Bild. Atlas Naturg. Fische, 1854: fig. 170 (ill., ident. by shape of head); Nardo, Atti Ist. veneto, *3*, 1860: 787 (at least in part from refs. and loc., near Venice); Brito Capello, J. Sci. math. phys. nat., Lisboa, (1868–1869), *2*, 1869: 141 (at least in part, ident. by inference from incl. refs., Portugal); Canestrini, in Cornalia, *et al.*, Fauna d'Ital., 1870–1874: 47 (at least in part, ident. by inference from included ref.); Gegenbaur, Unters. Vergl. Anat. Wirbelt., *3*, 1872: 23, pl. 9, fig. 1 (ident. by ill. of skull); Perugia, Elenc. Pesc. Adriat., 1881 (at least in part, ident. by included refs.); Jordan and Gilbert, Bull. U.S. nat. Mus., 16, 1883: 26 (at least in part, ident. from range, C. Cod, southward); Kingsley, Stand. Nat. Hist., *3*, 1885: 81 (general, ident. from ill.); Jordan, Rep. U.S. Comm. Fish. (1885), 1887: 797 (at least in part from range, N. Atlant. states to W. Indies, and California southward); Bean, Bull. U.S. Fish Comm., 7, 1888: 151 (at least in part, because of descr. as abund., N. Jersey); Carus, Prod. Fauna Medit., *2*, 1889–1893: 513 (at least in part, because of included refs., Medit.); Nel-

45. Probably *diplana* also.

46. The following list is limited to citations that include information, verbal or pictorial, sufficient to make it reasonably certain that they were based at least in part on *zygaena* and not on *diplana*, or that refer to localities where *zygaena* is known to be common but not *diplana*. For a list of citations, nominally to *zygaena*, but which have referred to *diplana* in reality, see p. 420. For an extensive list of Indo-Pacific citations that certainly or possibly refer to this species, see Fowler (Bull. U.S. nat. Mus., 100 [*13*], 1941: 217).

son, Rep. St. Geol. N. J., *2* (2), 1890: 661 (occasional, N. Jersey, ident. probable because of loc.); Almeida and Roquete, Mammif. Peix. costa e rios do Algarve Inquerito Indust. Lisboa (1889), *2*, 1892: 374 (at least in part because of loc., Portugal, not seen); Jordan and Evermann, Rep. U.S. Comm. Fish. (1895), 1896: 217 (at least in part from range, C. Cod and C. Conception southward); Bull. U.S. nat. Mus., 47 (1), 1896: 45 (descr. not diagnostic, but *zygaena* at least in part because of range, C. Cod and C. Conception southward); Bean, Bull. Amer. Mus. nat. Hist., *9*, 1897: 329 (at least in part from loc. near N. York, Aug. and Sept.); Mearns, Bull. Amer. Mus. nat. Hist., *10*, 1898: 311 (ident. probable because of loc., Hudson River); Smith, Bull. U.S. Bur. Fish., *17*, 1898: 88 (ident. probable because of loc., Woods Hole); Evermann and Kendall, Rep. U.S. Comm. Fish. (1899), 1900: 49 (at least in part from loc., Indian R. to Key West, Florida); Sharp and Fowler, Proc. Acad. nat. Sci. Philad., *56*, 1901: 505 (ident. probable because of loc., Nantucket); Linton, Bull. U.S. Bur. Fish., *19*, 1901: 272 (food, parasites, ident. probable because of loc., Woods Hole); Bean, T. H., Bull. N. Y. St. Mus., *60*, Zool. 9, 1903: 31 (ident. probable because of loc. near N. York); de Braganza, Result. Invest. Sci. "Amelia," *2*, 1904: 46, 47 (name only, Portugal, ident. probable because of loc.); Tracy, Rep. R. I. Comm. inl. Fish., 1906: 45 (ident. probable because of loc., Rhode Island); Pietschmann, Ann. naturh. (Mus.) Hofmus. Wien, *21*, 1906: 99 (descr. not diagnostic, but ident. probable, in part at least, because of loc., abund., Morocco); Linton, Bull. U.S. Bur. Fish., *26*, 1907: 122 (ident. probable because of loc., Woods Hole, August); Smith, Bull. N. C. geol. econ. Surv., *2*, 1907: 36 (ident. by ill., N. Carolina); Ribeiro, Arch. Mus. nac. Rio de J., *14*, 1907: 157, 201 (at least in part, descr. and loc. north. and south. Brazil, equally applicable to *diplana*, synonyms); Kendall, Occ. Pap. Boston Soc. nat. Hist., 7 (8), 1908: 5 (ident. probable from loc., south. New England); Fowler, Rep. N. J. Mus. (1907), 1908: 127 (ident. probable because at many N. Jersey loc., also Delaware Bay); Proc. Acad. nat. Sci. Philad., *60*, 1908: 66 (ident. probable because of loc., N. Jersey, Nantucket by name, also for Italy, Surinam, Padang); Tracy, Rep. R. I. Comm. inl. Fish., 1910: 59 (Rhode Island and Connecticut, ident. because of loc.); Günther, Encyc. Brit., 11th ed., *24*, 1911: 807 (at least in part, general); Nichols, Abstr. Linn. Soc. N. Y., 20–23, 1913: 91 (ident. probable because of loc. near N. Y.); Sumner, Osburn and Cole, Bull. U.S. Bur. Fish., *31* (2), 1913: 736 (ident. because of loc., Woods Hole); Murphy and Harper, Copeia, 23, 1915: 41 (Montauk, Long Island, ident. probable because of loc.); Coles, Proc. biol. Soc. Wash., *28*, 1915: 90 (at least in part, abund. off C. Lookout, N. Carolina, but probably covers *tudes* also); Hussakoff, Copeia, 34, 1916: 63 (Woods Hole spec. compared with Japanese); Nichols and Murphy, Brooklyn Mus. Sci. Bull., *3* (1), 1916: 19 (occur. near N. York, size, food, frequency, season, ident. by loc., but ill. on p. 19 apparently of *diplana*); Brooklyn Mus. Quart., *3* (4), 1916: 151 (fig.), 153 (general, ident. by ill.); Thorne, Copeia, 35, 1916: 69 (ident. probable because of loc., Long Island, N. York, small size); Latham, Copeia, 43, 1917: 37 (Long Island, N. York, ident. probable because of loc.); Fowler, Proc. Acad. nat. Sci. Philad., *69*, 1917: 109 (Atlantic City, N. Jersey, ident. probable because of loc.); Proc. Boston Soc. nat. Hist., *35*, 1917: 110 (Nantucket, ident. probable because of loc.); Nichols and Mowbray, Copeia, 48, 1917: 78 (at least in part because of loc., "tropical seas"); Latham, Copeia, 57, 1918: 53 (Long Island, N. York, ident. probable because of loc.); Coles, Copeia, 69, 1919: 41 (C. Lookout, N. Carolina, food, meas. of adult, ident. by photo of head in fig. 3, but fig. 2 is of *tudes*, see p. 435); Fowler, Proc. biol. Soc. Wash., *33*, 1920: 144 (N. Jersey records, ident. because of loc.); Latham, Copeia, 87, 1920: 91 (Long Island, N. York, November, ident. probable because of loc.); Fowler, Proc. Acad. nat. Sci. Philad., *74*, 1922: 3, 5, 7 (abund., size, season, N. Jersey and Delaware, ident. probable because of loc.); Nichols, Copeia, 116, 1923: 52 (Long Island, N. York, ident. probable because of loc.); Breder, Copeia, 127, 1924: 27 (seasonal occur. near N. York, relation to temperature, ident. probable because of loc.); Wilson, Proc. U.S. nat. Mus., *64* (17), 1924: 8, 12 (parasites, Woods Hole, ident. probable because of loc.); Barnard, Ann. S. Afr. Mus., *21* (1), 1925: 32 (S. Afr., probably in part, but descr. not diagnostic); Nichols and Breder, Zoologica, N. Y., *9*, 1927: 17 (occur., season, size, south. New Engl., ident. probable because of loc.); Chevey, Cons. explor. Mer. Fauna Medit., 1927: plate not numbered (Medit., ident. by ill.); Hildebrand and Schroeder, Bull. U.S. Bur. Fish., *43*, 1928: 50 (ident. probable by ill. from Garman, 1913, Chesapeake Bay); Rey, Fauna Iberica Peces, *1*, 1928: 364 (Spain, Portugal, ident. by ill.); Breder, Field Bk. Mar. Fish. Atlant. Coast, 1929: 19 (general, ident. by ill.); Truitt, Bean and

Fowler, Bull. Md. Conserv. Dep., *3*, 1929: 28 (young abundant, coast of Maryland and Chesapeake Bay, summer, ident. probable because of loc.); Jordan, Manual Vert. Anim. NE. U.S., 1929: 11 (general, at least in part); Jordan, Evermann and Clark, Rep. U.S. Comm. Fish. (1928), 2, 1930: 18 (in check list); Chevey, Faune Ichthyol. Cons. explor. Mer., 1930: plate not numbered (distrib. E. Atlantic, ident. by ill.); Townsend, Bull. N. Y. zool. Soc., *34* (6), 1931: 167 (Virgin Is., ident. by photo); Wilson, Bull. U.S. nat. Mus., 158, 1932: 423, 439, 440, 454, 456 (parasites, Woods Hole, ident. probable because of loc.); Breder, Copeia, 1932: 31 (abundance off Block I., ident. probable because of loc.); Gowanloch, Bull. La. Conserv. Dep., *23*, 1933: 219, lower ill. (ident. probable from ill., listed for Louisiana); Beebe and Tee-Van, Field Bk. Shore Fish. Bermuda, 1933: 29 (ident. probable, at least in part, because of loc., but descr. not diagnostic and ill. not original, Bermuda); Vladykov, Proc. N. S. Inst. Sci., *19*, 1935: 8 (ident. probable because of loc., Halifax, Nova Scotia); Burton, Sci. Mon. N. Y., *40*, 1935: 283 (occur. near Charleston, S. Carolina, but perhaps also incl. other species as well); Vladykov and McKenzie, Proc. N. S. Inst. Sci., *19*, 1935: 45 (Halifax Harbor, Nova Scotia, same specimen as Vladykov, 1935); Marchand, Mar. biol. Rep. Cape Town, *2*, 1935: 42 (S. Afr., ident. by photo); Fowler, Bull. Amer. Mus. nat. Hist., *70* (1), 1936: 64 (W. Afr., but may also cover *diplana*, for ill. after Garman's of Mass. spec. and descr., not diagnostic); Lübbert and Ehrenbaum, Handb. Seefisch. Nordeurop., *2*, 1936: 276 (occur., N. Europ., ident. by ill.); Sigalas and Budker, Bull. Soc. Sci. Arcachon, *34*, 1937: 59 (Arcachon, north. France, newborn, ident. probable because of loc.); Wise, Tigers of the Sea, 1937: 179 (descr., may include other species); Norman and Fraser, Giant Fishes, 1937: 41, fig. 21 (general, may also include other species); Breder, Bull. N. Y. zool. Soc., *41*, 1938: 28 (N. York Harbor, ident. probable because of loc.); Tortonese, Atti Soc. ital. Sci. nat., 77, 1938: 302 (occur. Medit., may cover other species also); Springer, Proc. Fla. Acad. Sci., *3*, 1939: 31 (in part, combined with *diplana*, Florida, see p. 420); Fowler, Bull. U.S. nat. Mus., 100 (*13*), 1941: 217 (extensive synonymy also incl. *diplana*); Springer, Proc. Fla. Acad. Sci., *5*, 1941: 48 (comp. with other species, skull, N. York, New England, N. Jersey, Italian, Azores specs.); Springer, Stanford Ichthyol. Bull., 1 (5), 1941: 168, fig. 5 (ill., comp. with other species); Gunter, Amer. Midl. Nat., *28*, 1942: 316 (tidal fresh water in Maryland, ident. probable because of loc.); Fowler, Arqu. Zool. Estado São Paulo, *3*, 1942: 129 (Brazil); Bigelow and Schroeder, Guide Comm. Shark Fish., Anglo Amer. Caribb. Comm., Wash., 1945: 134, fig. 49 (descr., habits, range, ill.), Boos, Contr. biol. Lab. Cath. Univ. Amer., 45, 1943: 10 (chemistry of pancreas, Florida); Fowler, Monogr. Acad. nat. Sci. Philad., 7, 1945: 97 (Beaufort, N. Carolina), 160 (Charleston, S. Carolina).

Zygaena zygaena Cuvier, Règne Anim., *2*, 1817: 127 (general, ident. by refs.); Seabra, Bull. Soc. portug. Sci. nat., *5*, 1911: 192 (ident. by refs., Portugal); Noronha and Sarmento, Peixes Madeira, 1934: 107 (by ref., not seen).

Zygaena malleus Risso, Hist. nat. Europ. Merid., *3*, 1826: 125 (in part, but also incl. *diplana* by ref. to *malleus* Valenciennes, 1822, general, Medit.); Voigt, in Cuvier, Tierreich, *2*, 1832: 513 (at least in part, ident. by refs.); Bory de St. Vincent, Dict. Class. Hist. Nat., 15, 1829: 597 (at least in part, ident. by refs.); Jenyns, Manual Brit. Vert. Anim., 1835: 507 (Yarmouth, England, ident. probable because of loc.); Lowe, Trans. zool. Soc. Lond., *2* (3), 1837: 195 (Azores, ident. because of subsequent ill., see Lowe, 1843–1860); Yarrell, Brit. Fish., *2*, 1836: 406 (Yarmouth, England, ident. by ill.); also 2nd ed., *2*, 1841: 504 (Brit. records, number of young, ident. as above); Couch, Cornish Fauna, 1838: 51 (Cornwall); DeKay, Zool. N. Y., 4, 1842: 362, pl. 62, fig. 204 (near N. York, ident. by ill.); Storer, Boston J. nat. Hist., *4*, 1842: 185 (near Woods Hole, ident. by ill.); Lowe, Fish. Madeira, *1*, 1843–1860: 83, pl. 12 (Madeira, ident. by descr. and ill.); Hamilton, Naturalist Libr., Ichthyol., *6*; Brit. Fish., *2*, 1843: 306, pl. 24, fig. 1, also subsequent eds. (Eng. loc., ident. by ill.); Linsley, Amer. J Sci., *47*, 1844: 77 (Long Island Sound, N. York, ident. probable because of loc.); Storer, Mem. Amer. Acad. Arts Sci., N. S. *2*, 1846: 508 (Massachusetts, N. York, Caribbean, Brazil; in part, ident. by loc., and refs., *diplana* also incl. by ref. to Valenciennes, 1822); DeKay, Rep. St. Cab. nat. Hist. N. Y., *8*, 1855: 64 (listed for N. York, ident. by a ref. to DeKay, 1842); Storer, Mem. Amer. Acad. Arts Sci., N. S. *9*, 1867: 238, pl. 38, fig. 3; also sep., Fishes Mass., 1867: 262, pl. 38, fig. 3 (occur. south. Massachusetts, ident. by ill.); Günther, Cat. Fish. Brit. Mus., *8*, 1870: 381, 518 (descr., distrib., refs., but also covers *diplana*, see p. 419); Gervais and Boulart, Poiss., *3*, 1877: 174, pl. 66 (Medit. and Atlant., ident. by ill.);

Doderlein, Man. Ittiol. Medit., *2*, 1881: 46 (occur. in Medit., ident. by refs., but probably covers *diplana* also);[47] Moreau, Poiss. France, *1*, 1881: 324 (occur. French coasts; ident. by ill.); Day, Fish. Gt. Brit., *2*, 1880–1884: 294 (occur. Gt. Brit., ident. by local. but ill., pl. 154, is *diplana*); Vieira, Ann. Sci. nat. Porto, *4*, 1897: 66 (Portugal, ident. by inference from included refs.); Lloyd and Sheppard, Proc. zool. Soc. Lond., 1922: 971 (Wales, neuro-anatomy, skull, ident. by ill. of skull); Jenkins, Fish. Brit. Isles, 1925: 308 (ident. probable because of loc.); Joubin and Le Danois, Mem. Pêches Marit. France, *2*, 1925: 27 (coast of France, Tunis, ident. at least in part from refs., but may also cover *diplana*); Ehrenbaum, in Grimpe and Wagler, Tierwelt N.- u. Ostsee, Lief 7 (12°), 1927: 13 (North Sea, ident. by loc.); Lloyd and Sheppard, Zool. Anz., *80*, 1929: 65 (skelet., same spec. as Lloyd and Sheppard, 1922); Belloc. Rev. des Trav. Pêches Marit., 7, Fasc. 2, 1934: 133 (ill. of head; Morocco; trop. W. Afr. in general; C. Verde Is., Canaries); Noronha and Sarmento, Peixes Madeira, 1934: 107 (not seen).

Zygaena vulgaris Cloquet, Dict. Sci. Nat., *60*, 1830: 621 (at least in part, general).

Zygaena (no spec. name) Storer, Rep. Fish. Rept. Birds Mass., 1839: 200 (south. Massachusetts, ident. probable because of loc.).

Zygaena subarcuata Storer, Proc. Boston Soc. nat. Hist., *3*, 1848: 70 (ident. by descr. of 2-ft. spec., Provincetown, C. Cod).

Sphyrnias zygaena Gray, List Fish. Brit. Mus., 1851: 48 (in part; ident. by refs., but also incl. *diplana*, see p. 420); White, List. Spec. Brit. Mus., Fish., *8*, 1851: 126 (part; ident. by refs., but also incl. *diplana*, see p. 420).

Sphyrna malleus Van der Hoeven, Handb. Dierkunde, 2nd ed., *2*, 1855: 262 (general, Medit., ident. by refs. to Linnaeus); Handb. Zool., Eng. transl., *2*, 1858: 68.

Cestracion subarcuata Gill, Proc. Acad. nat. Sci. Philad., Addend., 1861: 59 (name only, evidently based on *Zygaena subarcuata* Storer, 1848); Abbott, Bull. geol. Surv. N. J., Geol. N. J., Append. E, 1868: 829 (occasional N. Jersey, ident. probable because of loc.).

Cestracion zygaena Gill, Ann. N. Y. Lyc., 7, 1862: 403 (name only, ident. by ref. to Linnaeus, 1758); Proc. Acad. nat. Sci. Philad., 1863: 333 (Massachusetts, ident. by refs.); Duméril, Hist. Nat. Poiss., *1*, 1865: 382 (general, ident. by refs., but also incl. *diplana* by ref. to *malleus* Valenciennes, 1822); Fowler, Rep. N. J. Mus. (1905), 1906: 65 (occur. N. Jersey, ident. probable because of loc.); Fowler, Rep. N. J. Mus. (1906), 1907: 257, pl. 76 (N. Jersey record, ident. by ill.); Garman, Mem. Harv. Mus. comp Zool., *36*, 1913: 157, pl. 1, fig. 1–3 (in part, N. York spec. in Harv. Mus. Comp. Zool., ident. from ill., and by personal examination; but refs. also include *diplana*, see p. 420); Radcliffe, Bull. U.S. Bur. Fish., *34*, 1916: 263 (in part), pl. 43, fig. 3 (N. Carolina; ident. by ill., but pl. 43, fig. 1, clearly *diplana* [see p. 420]; fig. 2, *tiburo*; fig. 4 probably *tudes* because of teeth described as serrate [see p. 435]); Linton, Proc. U.S. nat. Mus., *64* (21), 1924: 5, 7, 21, 31, 37, 38, 40, 41, 42, 44, 96 (parasites, Woods Hole, ident. probable because of loc.); Bigelow and Welsh, Bull. U.S. Bur. Fish., *40* (1), 1925: 31 (Gulf of Maine records); Bigelow and Schroeder, Bull. U.S. Bur. Fish., *48*, 1936: 322 (Halifax, Nova Scotia, record, from Vladykov, 1935; also between Browns and Georges Banks, ident. by loc.); Chabanaud and Monod, Bull. Etud. Hist. Sci. Afric. Occid. Franc., 1926: 229 (at least in part, Morocco; not seen); Uriarte and Mateu, Notas Inst. esp. Oceanogr., (2) *53*, 1931: 51 (Canaries, Teneriffe, name only, but ident. probable at least in part); Cadenat, Rev. des Trav. Pêches Marit., *10* (4), 1937: 430 (Dakar, Canakry, W. Africa).

Hammerhead, Couch, Hist. Brit. Fish., *1*, 1867: 70, pl. 16 (Gt. Brit., ident. by ill.).

Sphyrna (*Cestracion*) *zygaena* von Bonde, J. comp. Neurol., *58*, 1933: 377, pl. 1, 3, 4 (S. Afr., skull and nerves, ident. by photo of head).

Zygaena zigaena Nobre, Fauna Marinha Port., Vert., *1*, 1935: 425 (Portugal, ident. by descr. of head).

47. For additional Mediterranean records in publications not accessible to us, see Doderlein, 1881.

Atlantic, by Name Only:[48]

Squalus zygaena Brünnich, Ichthyol. Massil., 1768: 4 (Medit.); Forskål, Descr. Anim., 1775: xviii (no local.); Bowditch, Excurs. Madeira, 1825: 75 (Porto Santo); Nardo, Prod. Ittiol. Adriat., *1*, 1826: 9 (Adriatic).

Zygène marteau, Cloquet, Dict. Sci. Nat., Atlas. Poiss., 1816–1830: pl. 32 (not diagnostic).

Zygaena malleus Busch, Selach. Ganoid. Encephal., 1848: 22, pl. 2, fig. 5 (brain, no loc.); Hasse, Naturl. Syst. Elasm. besond. Theil., 1882: 273, pl. 39, fig. 21–25 (denticles, vertebrae); Rochebrune, Acte. Soc. linn. Bordeaux, (4) *6*, 1882: 43; Faune Senegambie, Poiss., *1*, 1883–1885: 20 (Senegambia); Graeffe, Arb. zool. Inst. Univ. Wien, *28*, 1886: 446 (Medit.); Vinciguerra, Atti Soc. ital. Sci. nat., *34*, 1892: 300 (Teneriffe, name only); Sicher, Atti Accad. gioenia, (4) *11* (5), 1898: 15 (Medit.); Wilson, Proc. U.S. nat. Mus., *33*, 1907: 396 (Cape Verde Is., name only); Brian, Bull. Inst. oceanogr. Monaco, 110, 1908: 3 (C. Verde, Azores, name only); Metzelaar, Trop. Atlant. Visschen, 1919: 189 (W. Afr., name only).

Sphyrna zygaena Nardo, Sinon. Modern. Spec. descr. Pesci St. Chiereghin, 1847: 111 (name only); Machado, Peces Cadiz., 1857: 9 (name only); Canestrini, Arch. Zool. Anat. Fisiol., *1*, 1861: 266 (Genoa, name only); Bocage and Brito Capello, Peixes Plagiost. Portugal, *1*, 1866: 17 (Portugal, name only); Ninni, Ann. Soc. Hist. nat. Modena, *5*, 1870: 66 (Venice, name only); Miklucho-Maclay, Beitr. Vergl. Neurol. Wirbelt., 1870: 25, pl. 5, fig. 2 (brain); Goode, Bull. U.S. nat. Mus., *5*, 1876: 73 (Bermuda, name only); Goode, Amer. J. Sci., *14*, 1877: 293 (Bermuda, name only); Proc. U.S. nat. Mus., *2*, 1879: 121 (Florida, name only); Yarrow, Proc. Acad. nat. Sci. Philad., *29*, 1877: 217 (N. Carolina); Bean, T. H., Proc. U.S. nat. Mus., *3*, 1881: 115 (East Coast, U.S.); Jordan and Gilbert, Proc. U.S. nat. Mus., *5*, 1882: 581 (name only); von Ihering, Rev. Mus. paul., *2*, 1897: 34 (Muldonado, Montevideo, name only); Hilgendorf, Arch. Naturgesch., *54* (1), 1888: 213 (Azores, name only); Carus, Prod. Fauna Medit., *2*, 1889–93: 513 (Medit., refs. probably include both species); Henshall, Bull. U.S. Fish Comm., *9*, 1891: 384 (Florida, name only); Smith, Rep. U.S. Comm. Fish. (1895), 1896: 175 (Florida, name only); Jordan and Rutter, Proc. Acad nat. Sci. Philad., 1897: 91 (Jamaica, name only); Berg, Commun. Ittiol. Mus. nac. B. Aires, *1* (1), 1898: 9 (Montevideo, not diagnostic); Evermann and Bean, Rep. U.S. Comm. Fish. (1896), 1898: 239 (Florida, name only); Gilbert, Proc. Wash. Acad. Sci., *2*, 1900: 161 (Brazil, name only); Gregg, Where to Catch Fish, 1902: 16 (not diagnostic); Evermann and Marsh, Bull. U.S. Fish Comm., *20* (1), 1902: 63 (Porto Rico, name only); Schreiner and Ribeiro, Arch. Mus. nac. Rio de J., *12*, 1903: 79 (Pernambuco, name only); Hargreaves, Fish. Brit. Guiana, 1904: 14, and Append., 8 (Brit. Guiana, attacks on man, name only); Devincenzi, An. Mus. Hist. nat. Montevideo, (2) *1*, 1904: 120 (Rio de La Plata, name only); Jordan and Thompson, Bull. U.S. Bur. Fish., *24*, 1905: 232 (Tortugas, Florida, name only); Bean, Field Mus. Publ. Zool., 7 (2), 1906: 30 (Bermuda, name only); Wilson, Proc. U.S. nat. Mus., *33*, 1907: 416, 431, 626 (C. Verde Is., N. Carolina, parasites, name only); Gudger, Science, N. S. *25*, 1907: 1005 (N. Carolina, 12½-ft. spec., food, name only); Fowler, Proc. Acad. nat. Sci. Philad., *61*, 1909: 407 (C. May, N. Jersey, name only); Rep. N. J. Mus. (1908–1910), 1910: 352 (N. Jersey record, name only); Proc. Acad. nat. Sci. Philad., *63*, 1911: 5 (Delaware, name only); Roule, Bull. Inst. océanogr. Monaco, 243, 1912: 10 (Medit., name only); Fowler, Copeia, 2, 1913: 2 (Maryland, name only); Pietschmann, Jb. nassau. Ver. Naturh., *66*, 1913: 176 (Cameroons, name only); Fowler, Copeia, 13, 1914: 2 (N. Jersey, name only); Proc. Acad. nat. Sci. Philad., *67*, 1915: 245, 521 (Florida, Trinidad, names only); Copeia, 31, 1916: 41 (N. Jersey, name only); Copeia, 27, 1916: 11 (N. Jersey, name only); Copeia, 30, 1916: 36 (name only); Proc. Acad. nat. Sci. Philad., *69*, 1917: 127 (Colón, name only); Copeia, 43, 1917: 39 (Florida, name only); Nichols, Bull. Amer. Mus. nat. Hist., *37*, 1917: 875 (Florida, name only); Ribeiro, Rev. Mus. paul., *10*, 1918: 708 (Santos, Brazil, name only); Fowler, Proc. Acad. nat. Sci. Philad., *71*, 1919: 128, 129, 146, 293 (Panama, Surinam, Jamaica, N. Jersey, name only); Nichols, Bull. Amer. Mus nat. Hist., *44*, 1921: 22 (Turks I., name only); Lahille, Physis, B. Aires, *5*, 1921: 63 (Argentina, name only); Enumer. Peces Cartilag. Argent., 1921: 14 (Argentina, name only); Fowler, Proc. Acad. nat. Sci. Philad., *72*, 1921: 386 (N. Jersey,

48. References ostensibly to *Sphyrna zygaena*, but by name only, either without locality, or for localities where *diplana* may also be expected to occur, hence which are as likely to have been based on the latter as on the former, or on the two combined.

name only); Ribeiro, Arch. Mus. nac. Rio de J., Fauna brasil. Peixes, *2* (1) Fasc. 1, 1923: 14 (Brazil north and south, descr. not diagnostic); Fowler, Copeia, 143, 1925: 41 (Delaware, name only), 43 (N. Jersey, name only); Proc. Acad. nat. Sci. Philad., *78*, 1926: 249 (Florida, food, embryos, name only); Copeia, 156, 1926: 146 (N. Jersey, name only); Copeia, 165, 1927: 90 (Delaware Bay, name only); Beebe and Tee-Van, Zoologica, N. Y., *10*, 1928: 29 (Haiti, name only); Nichols, Sci. Surv. Porto Rico, N. Y. Acad. Sci., *10* (2), 1929: 184 (Porto Rico, not diagnostic); Fowler, Proc. Acad. nat. Sci. Philad., *80*, 1929: 608 (N. Jersey records, names only); MacCallum, Proc. U.S. nat. Mus., *79* (26), 1931: 6 (parasites, name only); Fowler, Proc. Acad. nat. Sci. Philad., *83*, 1931: 391 (Trinidad, name only); Breder, Pap. Tortugas Lab., *28*, 1932: 16 (Tortugas, name only); Pearson, Invest. Rep., U.S. Bur. Fish., (10) *1*, 1932: 18 (listed, N. Carolina); Fowler, Proc. biol. Soc. Wash., *46*, 1933: 57 (Louisiana, name only); Young and Mazet, "Shark, Shark," 1933: 268 (not diagnostic); Brooks, Parasitology, *26*, 1934: 265 (N. Carolina, parasites); Pozzi and Bordale, An. Soc. cient. argent., *120*, 1935: 151 (N. Argentina, name only); Fowler, Bull. Amer. Mus. nat. Hist., *70* (2), 1936: 1152 (Azores, name only); Smith, H. W., Biol. Rev., *11*, 1936: 64 (Louisiana, name only); Fowler, Proc. Acad. nat. Sci. Philad., *89*, 1937: 304 (75 miles off N. Jersey coast, name only); Longley and Hildebrand, Pap. Tortugas Lab., *34*, 1941: 3 (Tortugas, Florida); Lunz, Bull. S. C. St. Planning Bd., 14, 1944: 27 (S. Carolina, Florida).

Sphyrna lewini Duméril, Rep. Poiss. Afr. Occid., 1861: 261 (Gorrée, probably not *Zygaena lewini* Griffith, Cuvier, Règne Anim., *10*, 1834: 640, pl. 50, Aust.).

Cestracion (*Sphyrna*) *zygaena* Steindachner, S. B. Akad. Wiss. Wien, *61* (1), 1870: 576 (Senegal, name only); Denkschr. Akad. Wiss. Wien, *44*, 1882: 51 (Senegambia, name only).

Cestracion zygaena Poey, An. Soc. esp. Hist. Nat., *5*, 1876: 382; Enumerat. Pisc. Cubens., 1876: 187 (Cuba); Fowler, Proc. Acad. nat. Sci. Philad., *58*, 1906: 80 (Florida, name only); Radcliffe, Trans. Amer. Fish. Soc., *44*, 1914: 38 (N. Carolina, name only); Roule, Result. Camp. sci. Monaco, *52*, 1919: 117 (Morocco, Azores, name only); Monod, Faune Colon. Franc., 1927: 647 (Lokundje, not seen); Nigrelli, Amer. Mus. Novit., 996, 1938: 10 (parasites, name only).

Zygaena vulgaris Reguis, Ess. Hist. nat. Provence, *1* (1), 1877: 58 (Medit.).

Cestracyon zygaena Poey, An. Soc. esp. Hist. nat., *10*, 1881: 348 (Porto Rico, Guiana, Medit., E. Indies, name only); Stahl, Fauna Puerto Rico, 1883: 81, 167 (Porto Rico, not seen).

Sphyrna (*Zygaena*) *zygaena* Imms, Proc. zool. Soc. Lond., *1*, 1905: 43 (pharyngeal denticles, name only).

Hammerhead, Gudger, Amer. Mus. nat. Hist., *40*, 1937: 417–418 (Palm Beach, Florida, attack on girl).

Sphyrna sp. (probably *zygaena*) Norris, Plagiost. Hypophysis, 1941: pl. 1, fig. 3 (brain).

Suborder SQUALOIDEA

Characters. No anal fin; 2 dorsal fins, with or without spines; only 5 gill openings, all anterior to pectorals; snout not beak-like, without lateral teeth or cirri; teeth in front of mouth essentially similar to those toward corners; general form subcylindrical (shark-like); eyes lateral; anterior margins of pectorals not expanded forward past 1st gill opening; inner margins of pelvics entirely separate, posterior to cloaca; nostrils entirely separate from mouth; spiracles present; eyes without nictitating fold or membrane; vertebral column completely segmented throughout its length, its axial canal much contracted in the regions of the well differentiated centra, the notochord greatly constricted segmentally, or even obliterated in the centra, but dilated in the spaces between the concave surfaces of adjoining vertebrae; vertebral centra with calcareous lamellae in a ring around central axis; neural spines not attached to dorsals; cranium with antorbital processes more or less developed, but without separate antorbital bar; upper jaw (palatoquadrate cartilage) attached to cranium by a transverse process at one point only, in the ethmoid region, as well as to hyomandibular arch; rostral cartilage single; propterygial cartilage of pectoral with 1 to several radial elements; pelvics transverse; heart valves in 2 to 4 rows. Development usually ovoviviparous, but probably oviparous in some cases.

Key to Families

1a. Each dorsal fin preceded by a long or short spine with tip exposed or concealed.[1]
Squalidae, p. 450.

1b. Second dorsal fin, and usually the 1st, without a spine.[2]
- 2a. Teeth with only 1 cusp; uppers and lowers unlike, the former narrow, raptorial, the latter expanded widely laterally as a cutting edge (sectorial).
Dalatiidae, p. 499.
- 2b. Teeth with several cusps, uppers and lowers similar, sectorial.
Echinorhinidae, p. 526.

Family SQUALIDAE

Characters. Squaloidea with a spine in each dorsal fin, long in some cases but so short in others as to be easily overlooked; teeth with 1 or several cusps, alike or unlike in the 2 jaws. Characters otherwise those of the suborder.

Genera. Generic distribution of the various members of this family is still in some confusion owing to the fact that all the characters that have been regarded as generically diagnostic by one author or another are intergrading, not strictly alternative. Consequently, the accompanying Key is necessarily tentative.

Tentative Key to Genera

1a. Fin spines originating about at midpoint of bases of dorsal fins and running forward to emerge from anterior margins of latter; trunk very stout, subtriangular, with longitudinal dermal ridges anterior to pelvic fins. *Oxynotus*[3] Rafinesque, 1810.
Eastern Atlantic, Mediterranean, Australian region.

1b. Fin spines originating at origins of dorsal fins and lying along anterior margins of latter; trunk slender, subcylindrical, with dermal longitudinal ridges (if any) confined to sector posterior to pelvic fins.
- 2a. Upper teeth with several cusps.
  - 3a. Teeth similar in the 2 jaws.
Centroscyllium Müller and Henle, 1841, p. 480.
  - 3b. Upper and lower teeth noticeably unlike, the lower with only 1 cusp.
Etmopterus Rafinesque, 1810, p. 487.

1. Even in the genera in which the spines are shortest (e.g., *Centroscymnus*, p. 493) they are easily detected by touch.
2. In *Euprotomicrus* the first dorsal may or may not have a spine.
3. Classed as a separate family (Oxynotidae) by some authors.

2b. Upper teeth with only 1 cusp.

4a. Snout in front of mouth considerably longer than from center of mouth to origin of pectorals; dermal denticles pitchfork-shaped, on tall slender pedicels. *Deania*[4] Jordan and Snyder, 1902.[5]
Eastern Atlantic, South Africa, Japan, Philippines, Australia, New Zealand.

4b. Snout in front of mouth considerably shorter than from center of mouth to origin of pectorals; dermal denticles at most only moderately dentate, on short broad pedicels, or sessile.

5a. Teeth similar in the 2 jaws.

6a. Anterior margin of nostril without long barbel.
Squalus Linnaeus, 1758, p. 452.

6b. Anterior margin of nostril with a barbel reaching past corner of mouth. *Cirrhigaleus* Tanaka, 1912.[6]
Japan.

5b. Teeth noticeably dissimilar in the 2 jaws.

7a. Inner corner of pectoral broadly rounded.

8a. Blades of dermal denticles on trunk behind 1st dorsal smooth, with rounded margins; ridged or striate denticles confined to more anterior part of body.
Centroscymnus Bocage and Brito Capello, 1864, p. 493.

8b. Blades of dermal denticles with 3 to several ridges; with marginal teeth on posterior as well as on anterior parts of trunk.
Scymnodon Bocage and Brito Capello, 1864.[7]
Eastern Atlantic, Straits of Magellan, Japan, New Zealand, Philippines, India.

7b. Inner corner of pectoral angular and more or less produced.
Centrophorus Müller and Henle, 1837.[8]
Eastern Atlantic, Mediterranean, South Africa, Japan, New Zealand, Australia.

4. We agree with Fowler (Bull. U.S. nat. Mus., 100 [*13*], 1941: 237) that this name should take precedence over *Acanthidium* as used by Garman (Mem. Harv. Mus. comp. Zool., *36*, 1913: 215), which is properly a synonym of *Etmopterus* (see also footnote 1, p. 487).

5. Including *Deaniops* Whitley, 1932.

6. Including *Phaenopogon* Herre, 1935.

7. Garman (Mem. Harv. Mus. comp. Zool., *36*, 1913: 206) has proposed a separate genus, *Centroselachus*, for species in which the denticles have numerous ridges, as contrasted with those in which they have only three. But this does not seem to us a sufficiently important difference to be regarded as generic.

8. Including *Lepidorhinus* Bonaparte, 1838, and *Entoxychirus* Gill, 1862. Fowler (Bull. U.S. nat. Mus., 100 [*13*], 1941: 223) redivides this group of species between *Centrophorus* and *Entoxychirus*. But the differences on which this division is based, *i.e.*, the relative degrees to which the inner corners of the pectorals are produced and the shapes of the dermal denticles, do not seem to us sufficient for generic separation. We may note also that three of the species included by him in *Centrophorus* (*rossi* Alcock, 1898; *waitei* Thompson, 1930; and *foliaceus* Günther, 1877) appear to us to belong to *Scymnodon*. The genus *Lepidorhinus* of Bonaparte (Nuov. Ann. Sci. nat. Bologna, 2, 1838: 207) has also been revived by Garman (Mem. Harv. Mus. comp. Zool., *36*, 1913: 211). But its type species, the European *Squalus squamosus* of Bonnaterre (Tabl. Encyc. Meth. Ichthyol., 1788: 12), falls in *Centrophorus* as defined here, the inner corners of its pectorals being angular and at least slightly produced.

Genus *Squalus* Linnaeus, 1758[9]

Squalus Linnaeus, Syst. Nat., *1*, 1758: 233; type species, *S. acanthias* Linnaeus, European Ocean; designated by Gill, Ann. N. Y. Lyc., 7, 1862: 405.

Generic Synonyms:

Squallus Scopoli, Intrcd. Nat. Hist., 1777: 464; type species, *Squalus acanthias* Linnaeus, 1758 (emend. spelling).

Acanthorhinus Blainville, Bull. Soc. philom. Paris, 1816: 121; type species, *Squalus acanthias* Linnaeus, 1758, designated by Jordan, Genera Fish., 1, 1917: 95.

Spinax (in part) Cuvier, Règne Anim., *2*, 1817: 129.

Acanthias Risso, Hist. Nat. Europe Merid., *3*, 1826: 131; type species, *A. vulgaris* Risso, equals *Squalus acanthias* Linnaeus, 1758.

Carcharias Gistel, Naturg. Thier., 1848: 8; to replace *Acanthias* Risso, 1826: not *Carcharias* Rafinesque, 1810.

Flakeus Whitley, Aust. Zool., *9*, 1939: 242; type species, *Squalus megalops* Macleay, 1881.[10]

Koinga Whitley, Aust. Zool., *9*, 1939: 242; type species, *Squalus griffini* Phillipps, 1931.[10]

Generic Characters. Squalidae with well developed dorsal spines, without lateral grooves, originating at origins of fins and lying along anterior margins of latter, at least their terminal ⅓ to ¼ free; trunk slender, rounded, with longitudinal ridges confined to caudal peduncle; caudal peduncle with a precaudal pit above but none below; a labial furrow on each jaw and a voluminous pit at corner of mouth; upper and lower teeth alike, with 1 cusp, deeply notched outwardly, and so oblique that their inner margins form a nearly continuous cutting edge; snout in front of mouth considerably shorter than from center of mouth to origin of pectorals; dermal denticles very small, lanceolate, heart-shaped or tridentate, with sharp tip, but varying in shape on different parts of body; eyes and spiracles large; nostrils far from mouth, without barbels; both dorsals with concave posterior margins and elongate free rear corners, the 1st triangular, the 2nd considerably smaller than 1st, the origin of 1st dorsal over inner margin of pectoral or a little posterior to inner corner of latter, the origin of 2nd dorsal behind bases of pelvics; caudal without subterminal notch, its lower anterior corner expanded as a definite lobe, but much shorter than the upper lobe and smaller in area; luminous organs lacking. Characters otherwise those of the family.

Range. Widely distributed in the North Atlantic, west and east, including the Mediterranean; western South Atlantic (Uruguay and Argentina); South Africa; both sides of the Pacific, from southern Alaska, the Aleutians, Kamchatka, Japan, Korea and China in the north to New Zealand, southern Australia, Tasmania and Chile in the south, including the Hawaiian Islands and Philippines; also southern Indian Ocean (Mauritius); in tropical to subarctic and subantarctic latitudes.

Fossil Teeth. Upper Cretaceous, western Asia; Upper Cretaceous to Pliocene, Europe; Eocene, South Africa; Oligocene, South America; Miocene, North America.

Species. The named representatives of *Squalus* fall into three groups, as defined by the relative locations of the fins and by the shape of the anterior margin of the nostril.

9. Fowler (Bull. U.S. nat. Mus., 100 [*13*], 1941: 255) includes in the synonymy of *Squalus* the fossil genus *Centrophorides* Davis (Trans. roy. Soc. Dublin, [2] *3*, 1887: 478).

10. Whitley's proposed subdivision of the genus *Squalus* into two subgenera (*Flakeus* and *Koinga*) according to the position of the first dorsal fin, size of eye and coloration, does not appear to us acceptable.

A. The *acanthias* group: 1st dorsal spine over or behind the inner corner of the pectoral; midpoint of bases of pelvics much closer to 2nd dorsal than to 1st; distal margin of pectoral at least moderately concave; anterior margin of nostril simple (not bilobed); at least most of the members of the *acanthias* group are white-spotted, while the others are not.

B. The *blainville-fernandinus* group: 1st dorsal spine almost over midpoint of inner margin of pectoral; midpoint of bases of pelvics about midway between the two dorsal fins; inner margin of pectoral nearly straight; anterior margin of nostril bilobed (Fig. 87 E, F).

C. The *brevirostris-cubensis* group resembles the *blainville-fernandinus* group in relative position of fins and in bilobed nasal margin, but is set apart by the distal margin of the pectoral, which is deeply concave with its inner angle noticeably acute.

Group A is represented in the North Atlantic by the familiar Spiny or Piked Dogfish (*S. acanthias*) of temperate and boreal latitudes. A close relative in the North Pacific is usually regarded as specifically distinct (*suckleyi* Girard) but recently has been listed as *acanthias*.[11] According to recent Keys[12] the chief alternative character supposedly separating *suckleyi* from *acanthias* is the position of the first dorsal spine, opposite or a little behind the inner corner of the pectoral in the former, behind and remote from it in the latter. Actually, however, our Study Material shows that this criterion is not tenable; not only is the variation considerable in this respect among both Atlantic and Pacific specimens, but in some of the latter the first dorsal spine stands as far behind the pectoral as it does in any of the Atlantic series.[13] It is even doubtful whether there is any average difference between the two populations in this respect. Nor have we been able to find any other difference to separate them, whether in position of fins, in proportionate dimensions, or in teeth. In short, the North Pacific and North Atlantic populations of the *acanthias* group have not differentiated themselves specifically during the period since their ranges became discontinuous.

It is doubtful whether this group occurs in the equatorial Atlantic, unless accidentally. But it is as widespread in the temperate and boreal belts of the southern as of the northern hemisphere. Thus the Spiny Dogs of the Straits of Magellan,[14] Australia and New Zealand,[15] with those reported from Uruguay and northern Argentina as *Squalus acanthias*[16]

11. Soldatov and Lindberg, Bull. Pacif. Fish. Res. Sta., *5*, 1930: 16; eastern Asia.
12. Garman, Mem. Harv. Mus. comp. Zool., *36*, 1913: 192; Fowler, Bull. U.S. nat. Mus., 100 (*13*), 1941: 257.
13. Especially significant is the fact that among three embryos of a single brood from San Francisco the first dorsal spine occupies the same position relative to the pectoral as is commonly true in Atlantic specimens; in another one it is as usually stated for the Pacific form.
14. *A. lebruni* Vaillant, Miss. Sci. Cape Horn (1882–83), 1885: 13, pl. 1.
15. *Fernandinus* Waite (Rec. Canterbury [N. Z.] Mus., *1*, 1901: 142, pl. 16, fig. 1) appears to be identical with *kirki* Phillipps (N. Z. J. Sci. Tech., *12*, 1931: 361) from New Zealand, and with *whitleyi* Phillipps (N. Zealand J. Sci. Tech., *12*, 1931: 361) from South Australia. It is true that the illustration (McCoy, Prod. Zool. Victoria, 1886: pl. 75, fig. 1b) on which *whitleyi* was based fails to show any labial furrows, but this was probably an oversight, for McCoy not only includes labial furrows in the generic diagnosis but states that he was unable to detect any difference between Australian and British specimens.
16. Berg, An. Mus. nac. B. Aires, (2) *1*, 1895: 6; Devincenzi, An. Mus. Hist. nat. Montevideo, (2) *1*, 1920: 123; Lahille, Physis B. Aires, *5*, 1921: 63.

or as *Acanthias vulgaris*,[17] all fall in the *acanthias* group, so far as the fins are concerned, but may perhaps be set apart by the teeth[18] when adult. A form at least very close to *acanthias* is also known under that name from South Africa and from the Island of Reunion in the southern Indian Ocean. The relationship of these southern hemisphere forms to one another and to the northern *acanthias* is uncertain.

In the North Atlantic group B is confined to the Mediterranean (*blainville* Risso, 1826). But it is widely represented in the southern hemisphere (Mauritius, South Africa, Argentina, Tasmania) and in Philippine, Japanese and Korean waters by forms so closely resembling one another, that while they have formed the basis for at least seven supposed species, most of these have recently been united by Fowler[19] under the oldest name, *fernandinus* Molina, 1782.[20]

Our own comparison of specimens of this group from the Mediterranean with others from the west coast of South America, Juan Fernandez and Tasmania[21] confirms this union in so far as concerns the fins, shape of head, and snout and margin of nostril. But the dorsal fin spines (second as well as first) are considerably shorter in the eastern Pacific and Tasmanian specimens, and also (by published accounts) in the South African[21a] and Argentine[21b] representatives of this group than in the Mediterranean representative or in the Japanese as pictured. Hence, since the length of the spines is fairly constant in the only member of the genus (*acanthias*) of which a large series has been examined, it seems wise to retain the name *blainville* for the Mediterranean form, at least for the present, postponing decision as to how many species the *fernandinus* group includes in the southern hemisphere and in the Pacific until adequate series can be compared from representative localities.

Group C includes four named forms: *cubensis* Howell-Rivero, 1936, from Cuba; *brevirostris* Tanaka, 1917, from Japan; *megalops* Macleay, 1881, from southern Australia and Tasmania;[22] and *griffini* Phillipps, 1931, from New Zealand. Here again decision as to whether or not these are all distinct species, and if so by what alternative charac-

17. Perugia, Ann. Mus. Stor. nat. Genova, (2) *10* (30), 1891: 608.
18. See Devincenzi (An. Mus. Hist. nat. Montevideo, [2] *4* [14], 1939: 4) for a recent discussion.
19. Bull. U.S. nat. Mus., 100 (*13*), 1941: 260.
20. Exceptions are *tasmaniensis* Rivero, 1936, which Fowler retains as distinct, and *mitsukurii* Jordan and Fowler, 1903, which he relegates to the synonymy of *suckleyi* Girard, 1854, *i.e.*, the North Pacific representative of the *acanthias* group. Actually, however, *mitsukurii* was a compound species, a fact which has resulted in much confusion in the nomenclature of Japanese sharks of this genus. Thus the form pictured under that name by its authors, Jordan and Fowler (Proc. U.S. nat. Mus., *26*, 1903: 630, fig. 3), and subsequently by Tanaka (Fish. Japan, *26*, 1917: pl. 130, fig. 368–370), was clearly *acanthias*-like, but the specimen described by Jordan and Fowler on the preceding page (which is therefore the type of the species) was of the *fernandinus* group, as pointed out by Jordan and Hubbs (Mem. Carneg. Mus., *10*, 1925: 105, 106). And this is also true of *japonicus* Ishikawa (Proc. Acad. nat. Sci. Philad., 1908: 71); Tanaka (Fish. Japan, *26*, 1917: pl. 130, fig. 365–367). *Japonicus* is therefore a synonym of *mitsukurii*, and the latter in turn probably a synonym of *fernandinus*.
21. The latter is the type specimen of *tasmaniensis* Howell-Rivero, 1936.

21a. *Acutipinnis* Regan, Ann. Natal Mus., *1*, 1908: 248, pl. 37.

21b. *Fernandinus* Lahille, An. Mus. nac. B. Aires, *34*, 1929: 327, fig. 17.

22. For illustration of *megalops* and *griffini*, see Whitley (Fish. Aust., *1*, 1941: 138).

ters they are separated, must await comparison of specimens from the three geographic regions.

Because of these uncertainties the following Key is limited to the western Atlantic representatives of the genus, with the Mediterranean and eastern Atlantic *blainville* included to facilitate comparison.

Key to Western Atlantic Species

1a. First dorsal spine over or posterior to inner corner of pectoral; midpoint of bases of pelvics much nearer to origin of 2nd dorsal than to rear end of base of 1st dorsal; anterior margin of nostril expanded as a simple lobe.
acanthias Linnaeus, 1758, p. 455.

1b. First dorsal spine about over midpoint of inner margin of pectoral; midpoint of bases of pelvics about midway between rear end of base of 1st dorsal and origin of 2nd dorsal; anterior margin of nostril usually with a small secondary lobe. Fig. 87 F, 89 B.

2a. Inner margin of pectoral deeply concave, its inner corner acutely pointed.
cubensis Howell-Rivero, 1936, p. 473.

2b. Inner margin of pectoral only very weakly concave, its inner corner approximately a right angle.

3a. Second dorsal spine reaches to apex of fin (Fig. 87 I); horizontal diameter of eye longer than distance between nostrils, and nearly ⅔ (60%) as long as snout in front of mouth. *blainville* Risso, 1826.
Mediterranean, Canaries, Portugal, Black Sea.

3b. Second dorsal spine reaches only about ⅔ the way to apex of fin (Fig. 87 E); horizontal diameter of eye a little shorter than distance between nostrils, and less than ½ (40%) as long as snout in front of mouth.
fernandinus Molina, 1782, p. 478.

Squalus acanthias Linnaeus, 1758

Spiny Dogfish, Piked Dogfish

Figures 87 A–D, 88

Study Material. About 120 preserved specimens, of all sizes, from embryos and newborn to large adults, from various localities along the New England coast and Newport News, Virginia, in the western Atlantic,[23] from Bohuslän, Sweden, North Sea and Mediterranean in the eastern Atlantic, and from Siberia, Gulf of Georgia, Puget Sound and California in the North Pacific (Harv. Mus. Comp. Zool.); also many specimens, fresh-

23. Including a female, 272 mm. long, from Cuba, with umbilical scar still visible, the type specimen of *S. barbouri* Howell-Rivero, 1936.

caught, from the Gulf of Maine and the vicinity of Woods Hole, Virginia and North Carolina.

Distinctive Characters. S. acanthias is easily separated from other members of its genus in the North Atlantic by the facts that its first dorsal spine is over or posterior to the inner corner of the pectoral (about over the midpoint of the inner margin of the pectoral

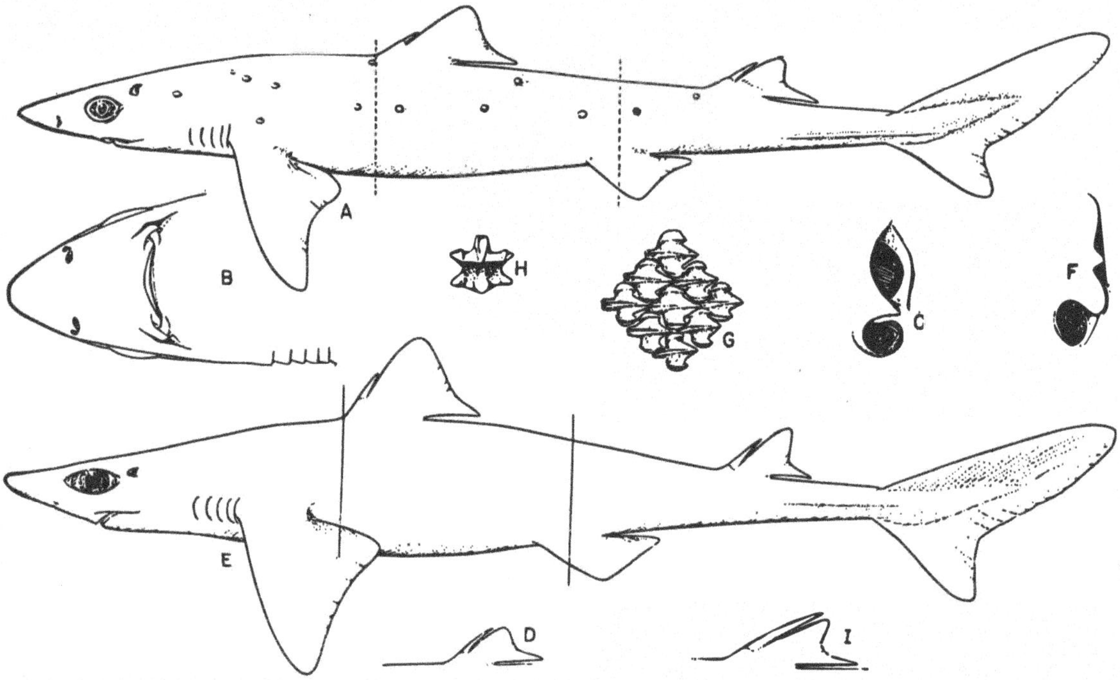

FIGURE 87. *A, Squalus acanthias*, female, about 815 mm. long, from Woods Hole, Massachusetts (Harv. Mus. Comp. Zool., No. 35862). *B* Head of same from below. *C* Right-hand nostril of same, about 2 x. *D* Second dorsal fin of adult male from the same locality to show the length of the spine. *E, Squalus fernandinus*, female, about 914 mm. long, from Island of Juan Fernandez (Harv. Mus. Comp. Zool., No. 841). *F* Right-hand nostril of same, about 2 x. *G* Dermal denticles of same, about 20 x. *H* Apical view of denticle. *I* Second dorsal fin of *Squalus blainville*, female, 570 mm. long, from Italy (?) (U. S. Nat. Mus., No. 28473) to show length of spine.

in *blainville* and *cubensis*), that the midpoint of the bases of its pelvics is much nearer to the second dorsal than to the first (about midway between the two in *blainville* and *cubensis*), by the simple anterior margin of the nostril (bilobed in *blainville* and *cubensis*); by the rounded inner corner of the pectoral (angular in *cubensis*), and by its shorter dorsal spines and white-spotted coloration. It is further separated from *blainville* by the deeply concave distal margin of its pectoral.

Description. Proportional dimensions in per cent of total length. Male, 705 mm., from Buzzards Bay, Mass. (Harv. Mus. Comp. Zool., No. 35864). Female, 814 mm., same locality (Harv. Mus. Comp. Zool., No. 35863).

Trunk at origin of pectoral: breadth 10.9, 11.1; height 9.2, 8.8.
Snout length in front of: outer nostrils 4.1, 4.0; mouth 8.9, 8.6.
Eye: horizontal diameter 3.5, 3.3.
Mouth: breadth 6.9, 6.6; height 1.1, 1.0.
Nostrils: distance between inner ends 3.4, 3.3.
Labial furrow length from angle of mouth: upper 2.4, 2.2; lower 1.1, 1.1.
Gill opening lengths: 1st 1.7, 1.7; 2nd 1.6, 1.5; 3rd 1.6, 1.5; 4th 1.7, 1.7; 5th 2.1, 2.1.
First dorsal fin: vertical height 6.0, 5.7; length of base 7.1, 7.4.
Second dorsal fin: vertical height 3.8, 3.9; length of base 5.4, 6.1.
Caudal fin: upper margin 21.3, 20.3; lower anterior margin 10.2, 11.1.
Pectoral fin: outer margin 15.6, 15.8; inner margin 7.1, 6.8; distal margin 10.1, 11.1.
Distance from snout to: 1st dorsal 33.4, 32.7; 2nd dorsal 63.8, 62.3; upper caudal 78.7, 79.7; pectoral 19.2, 18.2; pelvics 50.7, 51.3.
Interspace between: 1st and 2nd dorsals 23.0, 23.1; 2nd dorsal and caudal 10.7, 11.7.
Distance from origin to origin of: pectoral and pelvics 31.2, 32.7; pelvics and caudal 28.8, 28.3.

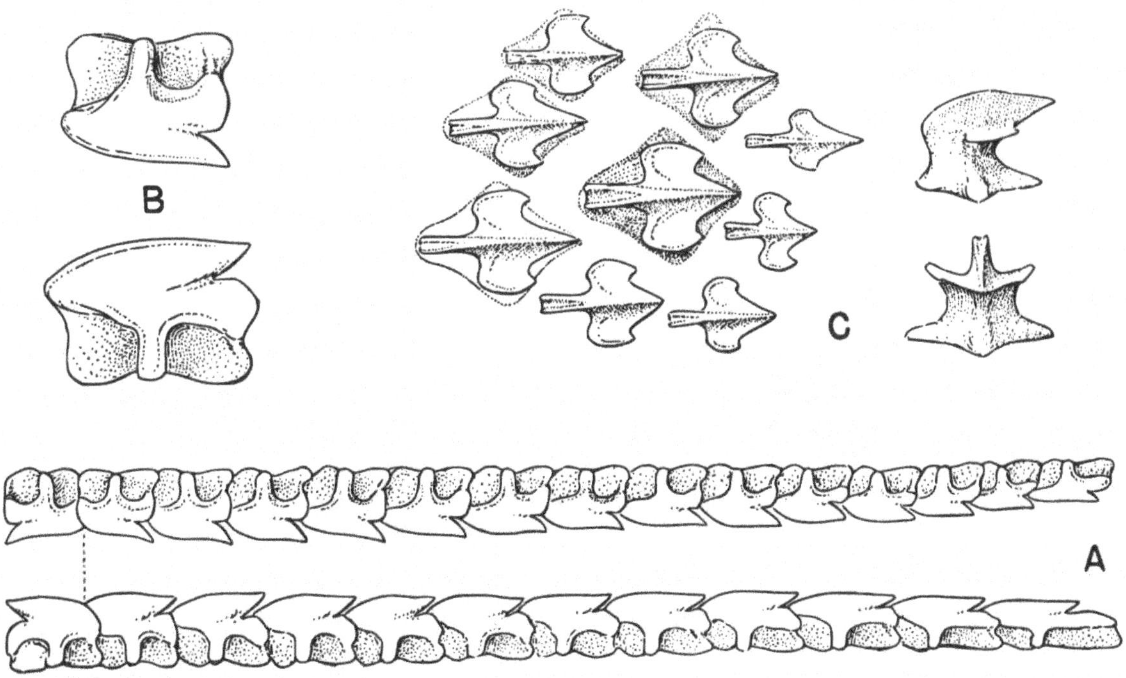

FIGURE 88. *Squalus acanthias*, illustrated in Fig. 87. *A* Upper and lower teeth, left-hand side, about 3.5 x. *B* Third upper and lower teeth, about 4 x. *C* Dermal denticles of another Massachusetts specimen (Harv. Mus. Comp. Zool., No. 842), about 34 x.

Trunk very slender, its height at 1st dorsal only about 1/7 its length to origin of caudal, its dorsal profile sloping forward from 1st dorsal. Body sector to cloaca longer than tail sector by a distance about 2/3 as long as head, and without mid-dorsal ridge. Caudal peduncle flattened below but rounded above, with a low rounded longitudinal dermal ridge along each side a little below the midlevel, which extends from a little behind base of 2nd dorsal to beyond origin of caudal. Upper precaudal pit subrectangular, more or less strongly developed, although some specimens appear to lack it; no lower pit. Dermal denticles loosely spaced and exposing the skin, rising steeply over the trunk as a whole but lying flat and in close contact (overlapping only a little if any) on top of snout and along edges of fins; essentially spine-like but somewhat expanded laterally, the terminal portion mostly a little longer than broad with a strong flat-topped median ridge, but varying considerably in shape on different parts of the body; those on sides and back posterior to 1st dorsal fin as well as on upper surfaces of pectorals more or less definitely tridentate (median tooth much the largest), but interspaced here and there with faintly tridentate forms; those on top of head less strongly tridentate, interspersed with broad-lanceolate; those on lower surface weakly tridentate anteriorly, but mostly very narrow-lanceolate posteriorly, without definite lateral teeth; those on sides of 1st and 2nd dorsals broad-lanceolate and but faintly ridged; those on top of snout and along anterior margins of fins broad-oval and smooth, or very faintly ridged.[24]

Head about 1/4 of trunk to origin of caudal, moderately flattened above. Snout rather thick, ovate, with rounded tip, its length in front of nostrils a little more than 1/3 its length in front of mouth, its length in front of mouth a little less than 1/2 (about 43 to 44%) of length of head. Eye a little longer than high, its upper outline less convex than its lower, its horizontal diameter a little more than 1/3 as long as snout in front of mouth or about the same as distance between nostrils; relatively a little larger in newborn specimens than in adults. Spiracle close behind eye, about 1/3 as long as horizontal diameter of eye, its lower margin about level with upper margin of latter or a little above it. Gill openings low down on sides, the 1st to 4th about evenly spaced, but the 5th closer to 4th; the 1st to 4th about same length, the 5th the longest, the latter about 2/3 as long as horizontal diameter of eye or a little less than 1/4 as long as snout in front of mouth and 1 1/3 times as long as 1st; the 5th close in front of origin of pectoral. Nostril approximately transverse, its inner end a little nearer to tip of snout than to symphysis of upper jaw, its anterior margin expanded as a simple, subtriangular lobe. Mouth only very slightly arched. Upper labial furrow extending inward and forward for a distance about 1/2 as long as diameter of eye, the lower furrow 1/2 to 2/3 as long as upper.

Teeth $\frac{14\text{—}0\text{—}14}{11 \text{ or } 12\text{—}0\text{—}11 \text{ or } 12}$ in specimens counted, essentially similar in the two jaws, smooth-edged, with single sharp-pointed cusp, deeply notched outwardly and so strongly oblique that the inner margins form a nearly continuous cutting edge from one corner of the mouth to the other; the lowers somewhat larger than uppers, and considerably

24. For a more detailed account and discussion of the denticles, see Sayles and Hershlowitz (Biol. Bull. Wood's Hole, 73, 1937: 5).

widest near corners of mouth; either 1 or 2 series in function all along each jaw, or perhaps even 3, depending on their stage in the process of replacement.[25]

Length of 1st dorsal from origin to rear tip nearly 2.3 times its vertical height, which is only a little more than ¼ of length of head, its origin varying in Woods Hole specimens from about over the inner corner of pectoral to posterior to the latter by a distance about as long as horizontal diameter of eye; the spine a little longer in males than in females (cf. Fig. 87 A with 87 D) and reaching at most to the midpoint of anterior margin of fin, the apex rounded, posterior margin moderately concave, free rear corner a little shorter than base, the midpoint of latter nearer to axil of pectoral than to origin of pelvics. Interspace between 1st and 2nd dorsals as long as, or somewhat longer than, head in adult, but only about as long as from tip of snout to 2nd or 3rd gill opening in newborn specimens. Second dorsal nearly as long at base as 1st but only about ⅔ as high vertically, its posterior margin more deeply concave, its free rear corner a little shorter than base, its origin about over tips of pelvics, its spine reaching about ⅔ the way to the apex in females but nearly to the apex in males. Interspace between 2nd dorsal and caudal about twice as long as base of 2nd dorsal. Caudal only about ⅕ the total length, without subterminal notch, its axis only very little raised, its upper margin nearly straight, apex moderately rounded and lower margin somewhat sinuous, the lower lobe about ½ as long as upper, but considerably less than ½ as large in area with narrowly rounded tip; the re-entrant contour between the 2 lobes subrectangular with well rounded corner. Pelvics about as long at base as 2nd dorsal, their anterior margins straight or very slightly convex, posterior margins moderately concave, tips tapering, subangular, midpoint of base nearer to origin of 2nd dorsal than to rear end of base of 1st dorsal by a distance nearly or quite as long as base of latter. Clasper of adult male subdivided at tip into 2 short rounded lobes, the outer soft, the inner cartilaginous, with a sharp recurved hook in its inner edge. Pectoral about ⅘ as long as head and a little more than ½ as broad as long, the outer margin moderately convex toward apex, distal margin moderately and evenly concave, apex and inner corner rather narrowly rounded.

Color. Usually slate-colored above, although sometimes tinged with brown; pale gray, grayish white, or pure white below; a row of small white spots irregularly arranged on each side from above the pectorals to abreast of the pelvics, with a few others in front of and behind the first dorsal, as well as close in front of the second dorsal and scattered on the upper sides of the anterior part of the trunk. These spots are most conspicuous in young specimens up to 12 to 14 inches long, fading with growth, and sometimes entirely lacking in large adults. The upper distal margins of caudal and of first and second dorsals are dusky in at least some newborn specimens, but fade with growth.

Size. Although an occasional Spiny Dog may be born at a length no greater than 165–179 mm. (6½–7 in.), the majority are about 220 to 330 mm. (8⅔–13 in.) at birth. It has also been observed that the smaller parent fishes contained smaller embryos

25. See discussion, p. 65.

than the larger parent fishes.[26] Males mature[27] at about 600 to 800 mm. and females at 700 to 1,000 mm. in different localities. Most of the adult males are from two feet to slightly less than three feet in length, the females from a little less than 2½ feet to almost 3½ feet, averaging about 7–10 pounds in weight; the maximum length attained is about four feet; occasional very large fat females may weigh 15 pounds, and a weight of 20 pounds has been reported.

Developmental Stages. It has been known since Aristotle's day that the Spiny Dog is ovoviviparous. The eggs are large, with much yolk. During the early stages of development those in each oviduct, one to four or more in number, are enclosed in a thin, amber-colored, horny capsule (known as a "candle") which tapers to a fine tip anteriorly but to a blunter end posteriorly; this capsule later breaks down to leave the embryos free in the oviduct. The young have no placental attachment to the uterine wall of the mother, but the latter is complexly folded and has numerous highly vascular papillae. It has been reported that the developing embryo gains about 40 per cent in weight over that of the ripe egg, presumably by absorption of water through the yolk sac, or at least chiefly so.[28] While the embryos are developing, a fresh set of ovarian eggs are growing to take their place. At Woods Hole the number of young in a litter is commonly four to six, sometimes as many as eight to eleven, or as few as two.

According to recent studies the period of gestation is about 18 to 22 months, whether in the Black Sea,[29] the English Channel[30] or the western Atlantic.[31] Accordingly, the adult females taken in summer and autumn in the Gulf of Maine contain either very small embryos that grow to an average length of about 17 mm. by September, or much larger ones ranging from 7 to 11 inches by that month, *i.e.*, nearly full term.

Habits. Spiny Dogs are neither swift swimmers nor very active, putting up little resistance when hooked. They may be either scattered or in schools, and in the latter case it appears that they continue to stay together as they grow. As a rule a given school consists either of small immatures of both sexes in almost equal numbers, of medium-sized mature males and immature females, or of large mature females. It is common knowledge that they are constantly on the move, their appearances and disappearances being so erratic that where there may have been good fishing for cod one day there may be only Dogfish the next, and nothing at all the day after, they having departed in pursuit of the better fish they had driven away. They use their spines for defense, curling around in a bow to strike, and it is probable that the spines are slightly poisonous, the general report to this

26. See Templeman (Res. Bull. Dep. Nat. Resources Newfoundland, 15, 1944: 44) for a detailed account of the life history of the Spiny Dog in Newfoundland waters.
27. English Channel data.
28. See Scammon and Minot (in Keibel, Normaltafeln Entwick. Wirbelt., *12*, 1911) for excellent description and illustrations of embryonic development. For recent accounts of the uterine wall and of the gain in weight of the developing embryo, see Widakowitch (Z. wiss. Zool., *88*, 1907: 499, pl. 30, 31), Ranzi (Pubb. Staz. zool. Napoli, *13* [3], 1934: 372) and Templeman (Res. Bull. Dep. Nat. Resources Newfoundland, 15, 1944: 45).
29. Popovici, in Grig. Antipa. Homm. Oeuvre, Bucharest, 1938: 445.
30. Ford, J. Mar. biol. Ass. U. K., N. S. *12*, 1921: 481.
31. Hisaw and Albert, Biol. Bull. Wood's Hole, *92* (3), 1947: 187.

effect being corroborated by the fact that their concave surfaces are sheathed with a glandular tissue resembling the poison glands of the venomous European weaver (*Trachinus draco*).[32]

They may be anywhere between the surface and bottom to depths as great as 90 to 100 fathoms, and perhaps even deeper. But they have never been found to be pelagic in the ocean basin. On the other hand, while their ability to survive in brackish water has been proved by experiment, and although they have been reported in at least one river[33] in Denmark, they do not normally enter fresh water, and specimens placed in fresh water died within a few hours.[34]

Relationship to Temperature. Spiny Dogfish do not appear on the United States coast in spring until the temperature of the water has risen to about 6° C., and most of them disappear from the inshore belt west and south of Cape Cod by the time the surface has warmed to about 15°, either moving into deeper cooler water nearby or northward to colder seas. Similarly, during their summer stay farther north they are seldom taken in water warmer than about 15° or colder than 6 or 7°, unless temperatures within this range are to be found a few fathoms shallower or deeper. Their northward advance along the coasts of Newfoundland and Labrador clearly appears to follow the vernal warning of the coastal waters, and their autumnal reappearance west and south of Cape Cod coincides roughly with the date when the surface has cooled to 12 to 15°; few if any are seen anywhere along the coast, north or south, after the surface has chilled below about 6° or 7°; and the winter temperature ranges between 6° and 11° on the bottom along the offshore belt, where the majority of the stock is now known to spend the cold months.

The foregoing, added to similar data for European waters, shows that the seasonal migrations north and south, and between shoal water and deep, are chiefly thermal in character, *i.e.*, to avoid extremes of temperature, either lower than about 7° or 8° or higher than about 12° to 15°. The thermal relationship of the Spiny Dog is thus analogous with that of the Mackerel, except that its optimum thermal range is slightly lower.

Breeding Habits. Females, with young nearly ready for birth, are taken in Newfoundland waters in early autumn (October), in the Gulf of Maine in late summer and autumn, near Woods Hole and New York in autumn, off Virginia and North Carolina in January to February. This, with the fact that very young specimens are rarely seen anywhere on the coast, suggests that most of the young are born from late autumn through the winter on the offshore wintering grounds. However, the season of production may extend through the spring in some years, as evidenced by recent captures of young Dogfish with the umbilical scar still discernible, near Woods Hole in June, in the Gulf of Maine in summer, and in schools at the entrance to Long Island Sound in July. It may even extend sporadically into summer, which is proved by a catch of 74 adult females which

32. For details, see Evans (Philos. Trans., [B] *212*, 1923: 27).

33. The fact that they were reported there with cod and *Merluccius* (Feddersen, Naturh. Tidsskr., [3] *12*, 1879: 68, 69, footnote 1; Bean, Amer. Nat., *14*, 1880: 525–526) indicates that the bottom water in the stream in question was salt, or at least brackish and not fresh.

34. For accounts of experiments on the survival of Spiny Dogfish in brackish and fresh water, see Scott (Ann. N. Y. Acad. Sci., *23*, 1913: 30, 60).

gave birth to young in July on capture off Gloucester, Massachusetts.[35] Young are produced throughout the year in the Mediterranean, during autumn in the Black Sea, while in more northern European waters there is wide regional variation, *i.e.*, late summer through autumn into winter in the English Channel, and late April through the summer in the North Sea and in Scandinavian waters.[36] It is probable that pairing takes place shortly after the young are born, but no definite information is available.

Food. The Spiny Dog is as voracious as any fish of its size, and its wanderings on the coast are no doubt chiefly in pursuit of food. Its recorded diet in the western Atlantic includes capelin, herring, menhaden (*Brevoortia*), mackerel, scup (*Stenotomus*), silver hake (*Merluccius*), cod, haddock, pollock, blennies and croakers (*Micropogon*). No doubt it preys on practically any species of fish smaller than itself. Even when newly born they have been seen attacking herring much larger than themselves, as adults do cod and haddock. Fishermen have often described them as harrying schools of mackerel and herring even in the seines, as well as destroying large numbers of cod and haddock in addition to driving them away, and they often bite these and other ground fish from the hooks of long lines. They also prey on squid and to some extent on worms, shrimps, prawns, crabs and amphipods. They occasionally feed on gastropods and jellyfish (*Aurelia*), and even red, brown and green algae have been found in their stomachs.[36a] When they first arrive near Woods Hole in spring they are often full of ctenophores. Probably they feed very little during the winter, for fishermen describe them as thin when they reappear on the coast in spring.

Relation to Man. In northern Europe the Spiny Dogfish is of considerable commercial value as a food fish; in 1931, for example, it fetched the fishermen about 5 cents per pound in the markets of Germany, and the landings for that year came to perhaps 14,000,000 pounds, as estimated from the total catch of sharks of all sorts.[37] Similarly, the British landings for 1923 were 9,597,900 pounds, worth £49,980.[38] However, on the American coast the Spiny Dogfish has never been in demand for the table, although many years ago they were of some value for oil. During more recent periods when they were in great abundance, various efforts were made in America to utilize Dogfish on a large scale both as fertilizer and as a source of oil[39] to combine with cod liver oil (it compares favorably with the cod for Vitamin A, although its Vitamin D content is much lower); it has also been canned for human consumption. And fresh, this is a better food fish than is generally appreciated. On the coasts of the eastern United States and Canada, however, these attempts have been short-lived. Of late years Spiny Dogfish have been so little considered that there is no way of knowing how great a proportion of the total landings of sharks of all kinds

35. McIntire, in Rep. Comm. Fish. Game Mass. (1905), 1906: 108.
36. For details and authorities, see Ford (J. Mar. biol. Ass. U. K., N. S. *12*, 1921: 481, 482).
36a. For a recent list of stomach contents, see Templeman (Res. Bull. Dep. Nat. Resources Newfoundland, 1944: 49).
37. For details, see Lübbert and Ehrenbaum (Handb. Seefisch. Nordeurop., *2*, 1936: 285).
38. Jenkins, Fish. Brit. Isles, 1925: 321.
39. Liver oil in the amount of 176,200 gallons was produced from this species in Canada in 1936 (Hampton, Newfoundl. Fish. Res. Inst. Serv. Bull., 5, 1938: 5).

from the Gulf of Maine to North Carolina may have consisted of them. The United States Bureau of Fisheries reported about 1,250,000 pounds for 1938.

From a practical aspect the Spiny Dog in the western Atlantic is chiefly important because it is undoubtedly more destructive to gear and interferes more with fishing operations than does any other fish—shark or teleost. Its habit of taking the bait is proverbial. In fact, when Dogfish are plentiful, hook and line fishing for cod, haddock and other ground fish is often actually prevented unless cockles (*Lunatia*) are used for bait. Still more serious is the damage they do by tearing and biting nets, biting snoods off long lines, attacking netted or hooked fish and by driving away better fish. It has been estimated that in these ways they do some $400,000 worth of damage annually off the coast of Massachusetts alone, and perhaps much more in their periods of abundance.[40]

Range. Both sides of the North Atlantic, chiefly in temperate to subarctic latitudes; also both sides of the northern Pacific south to California, Japan, northern China and the Hawaiian Islands (as pointed out on page 453, *suckleyi* appears to be indistinguishable from *acanthias*). It is represented in the corresponding thermal belt of the southern hemisphere (South Atlantic, Pacific, Indian Oceans, South Africa) by relatives so close that it is still an open question whether or not any valid specific distinctions can be drawn.[41]

Occurrence in the Eastern Atlantic. The chief center of abundance for the Spiny Dog is from the Atlantic coast of France north to Ireland, Scotland and southern Scandinavia, including the English Channel and the North Sea in general, and as far eastward as the Kattegat. But it rarely enters the Baltic. The Spiny Dogfish is plentiful around the Orkneys, Faroes, and south and east of Iceland in season, but less so to the north and west; it occurs regularly off Norway and as far north and east as the Murman coast. It is also generally distributed in the Mediterranean and in the Black Sea. To the southward it occurs commonly off Morocco and is reported from the Canaries, Madeira and Senegal.

Occurrence in the Western Atlantic. Fishermen are familiar with it in season all along the coast from North Carolina to Nova Scotia and on the southern side of the Gulf of St. Lawrence, as well as offshore on Nantucket Shoals, on Georges and on Browns Banks and on the Nova Scotian and Newfoundland Banks. It is common northward along both coasts of Newfoundland and is known past the Straits of Belle Isle to southeastern Labrador. It is also recorded on the north shore of the Gulf of St. Lawrence from Red Bay. And specimens have been taken on the west coast of Greenland at Sukkertoppen and Holsteinborg, no doubt these being visitors with the summer drift of Atlantic water. But there is no record of it on the east coast of continental North America to the north of Hamilton Inlet.

Southward it is a yearly visitor as far as Cape Lookout, North Carolina. But it is doubtful whether it occurs coastwise any further in that direction in numbers, for while it has been described repeatedly as plentiful in East Florida waters, and around Cuba and Trinidad, the former report seems likely to have referred to some other fish,[42] the latter

40. For detailed records and discussion, see Rep. Comm. Fish. Game Mass. (1905), 1906: 97; (1906), 1907: 20.

41. For discussion of this question, see p. 453.

42. Evermann and Bean (Rep. U.S. Comm. Fish. [1896], 1898: 239) describe it as "probably the most abundant

chiefly to the newly described *S. cubensis* (p. 473). However, at least a few stray as far as southern Florida and Cuba.[43] Offshore its range reaches to the outer edge of the continental shelf; inshore, into the outer reaches of Chesapeake, Delaware, Narragansett and Passamaquoddy Bays and the larger harbors. But it rarely, if ever, enters river mouths, at least on the American coast (but see footnote 33, p. 461).

Seasonal Migrations. The Spiny Dog is a spring and autumn migrant in the southern coastwise section of its range from North Carolina to New York, and mostly so along the southern coast of New England, but it is chiefly a summer visitor to the Gulf of Maine (including Georges Bank) and more northerly waters. South of New York Spiny Dogs are apt to "strike in" nearly simultaneously all along the coast; there are records for New Jersey (March 6, 13, or even earlier) and Chesapeake Bay (March) as early in the season as for North Carolina (April and early May). But the date of their arrival varies considerably from year to year. They depart entirely from Chesapeake Bay and the coast south of it by early May in some years and by late May at the latest; but they do not leave New Jersey waters and the immediate vicinity of New York until early or middle June. In the Long Island and southern New England areas they usually do not appear before late April or early May,[44] and the majority have departed by the end of that month, or by the close of June at the latest. But even in July and August considerable numbers of adults are taken at the mouth of Long Island Sound in deeper water (17 to 24 fathoms) while schools of young are taken inshore; and odd specimens are caught near Woods Hole throughout the summer in some years. On Georges Bank, in the only year of record, a few were taken in late March and April, but not until late June did their numbers sharply increase, the peak of abundance continuing through August. In the western side of the Gulf of Maine they may appear as early as mid-May, as in 1903, or not until well into June, as in 1905 and 1913, when the first big run struck near Cape Ann about the middle of that month. But there may be wide variation in this respect from place to place, as in 1903, when they did not appear at the tip of Cape Cod until early July, although they were numerous a month earlier in Massachusetts Bay, near Cape Ann, and off Penobscot Bay. In most years they have also appeared by June in the eastern part of the Gulf of Maine in general, although not until July in the cold waters of Passamaquoddy Bay, tributary to the Bay of Fundy. Within Massachusetts Bay, where the surface warms to about 18° C. in summer, few are taken between June and September, and the diminution recorded on Georges Bank after July similarly suggests a movement thence into the Gulf of Maine as the water warms.

Shark in the Indian River," Florida, and as a permanent resident there. But it is probable that this record actually referred to some other small shark, or perhaps even to *Amia* (known locally as Dogfish), for the Spiny Dog has never been definitely reported subsequently for the east coast of Florida, either in scientific literature or by fishermen.

43. We have examined the embryos recorded from the Tortugas by Longley and Hildebrand (Pap. Tortugas Lab., 34, 1942: 3) and we have found nothing to separate the type of *S. barbouri* Howell-Rivero, taken off Havana, from young free-swimming *acanthias* of the same size from New England waters.

44. May 17 is the earliest date recorded for the eastern end of Long Island; in 1940 they were first taken at Woods Hole on May 9, or six days after the last were taken south of Delaware Bay (Carolina Biol. Supply Co., Carolina Tips, Elon Coll., N. Carolina, *3* [7], 1940: 25).

Along outer Cape Cod, however, they are present in varying abundance all summer, at least in some years. And this is the case generally thence eastward and northward as far as the species occurs. Ordinarily they appear all along the outer coast of Nova Scotia about as early as in the eastern side of the Gulf of Maine, and a little later (third week in June) along the southeastern coast of Newfoundland, but it is well into July before they are encountered in any numbers in the inner parts of the Gulf of St. Lawrence. In 1942 (the one year of record) they had advanced to the Straits of Belle Isle along both the St. Lawrence and Atlantic coasts of Newfoundland by the second week in July. But they were not reported in southeastern Labrador until the beginning of September.

In general the autumnal withdrawal takes place as early from the western side of the Gulf of Maine as from Labrador, Newfoundland, Nova Scotian waters, Cape Breton or the Gulf of St. Lawrence, the majority ordinarily departing during October, with few caught in November, and as a rule they depart even earlier from the smaller bodies of water (*e.g.*, Passamaquoddy and Massachusetts Bays) than off the outer coast. In some years, however, they may be present in abundance well into November, as in 1903 and again in 1942, or even into December, as in 1913, when large catches were made between Cape Ann and Cape Elizabeth from the fifth to the twelfth,[44b] and again in 1942 when they were reported along the eastern, southeastern and western shores of Newfoundland. Of especial interest is the definite record of a number of Dogfish washed ashore on January 11, 1939, in St. Marys Bay, Newfoundland, after a severe gale;[44a] the implication of this report is discussed on page 466 in relation to their winter home.

Corresponding to this withdrawal from the north they reappear in autumn all along the coast from southern New England to North Carolina, their appearance being successively later from northeast to southwest. And the fact that catches on Georges Bank do not show any marked peak at that time indicates that the autumnal migration route is mostly along shore at first. In the Woods Hole region, near Nantucket and on Nantucket Shoals, they reappear in October, ordinarily disappearing again in November. At the mouth of Long Island Sound the recorded dates of their autumnal arrival in numbers have varied between October 5 and November 7, their subsequent disappearance between November 24 and December 12. Near New York, where they remain into December, they seldom appear in any numbers until November, and similarly along the New Jersey coast, where they may arrive late in October and remain plentiful into the winter.[45] At Cape Charles, at the mouth of Chesapeake Bay, the earliest autumnal record is for November 15. However, they have been reported as early as November 7[46] from North Carolina, and we have ourselves seen them in great numbers, dead on the beach (discarded by seiners) here and there on the coasts of southern Virginia and northern North Carolina late in that month

44a. For this record, for recorded dates of arrival on the Newfoundland coast, and for months when Dogfish were present, see Templeman (Res. Bull. Dep. Nat. Resources Newfoundland, 15, 1944: 56–66, fig. 13–16).

44b. Schools were reported as seen at the surface off Portsmouth, New Hampshire, on February 10, 1882 (Collins, 1883), but there is no proof of identity.

45. Precise information is scanty.

46. Carolina Tips, Elon Coll., N. C., *3* (7), 1940: 26.

and early in December. It is certain that at least some of the Dogfish that summer as far north as Newfoundland journey southward in autumn past the Gulf of Maine, for one tagged near St. John on July 14, 1942, was recaptured on the 23rd of the following November, off Thatcher's Island, Cape Ann, Massachusetts, having travelled a distance of at least 1,000 miles during the interval of 132 days or at an average of about 7.6 miles per day; actually, it probably travelled much farther and faster, for it is not likely to have followed a straight line. An equally interesting case is that of another fish tagged near St. Johns, Newfoundland, in April 1942 which was recaptured in September of the following year at the mouth of the Bay of Chaleur, within the Gulf of St. Lawrence;[47] it is evident that Dogfish that summer off one part of the coast during one year may do so off some other coast many miles distant during another. Additional information in these respects is much to be desired.

Wintering Grounds. It now seems certain that the Spiny Dogfish winter chiefly on bottom in deeper water offshore, from the offing of New York southward, for while none are reported from Georges Bank in February (though a few in January, however), considerable numbers have been trawled on the outer part of the shelf off New York in late November and in January and likewise in depths of 16 to 70 fathoms[48] between the offings of Delaware Bay, of northern Virginia and of Cape Hatteras in February. The fact that Spiny Dogfish have been washed ashore in some numbers on the southwest coast of Newfoundland in mid-January (p. 465) also opens the very interesting possibility that some of those that summer in that general region may survive the winter in the trough of the Gulf of St. Lawrence in temperatures (4°–5° C.; or 39°–41° F.) considerably colder than those that prevail on the southern wintering grounds.

Numerical Abundance. During its periods of abundance this is by far the most numerous local shark; in fact, it is the only one that even remotely rivals the commercially important food fishes in abundance. It has been described repeatedly as "in great abundance," in "schools of thousands," or as being caught as fast as fishermen can haul them in. Unfortunately the statistics of commercial landings do not afford any information in this respect about the Dogfish in American waters. But the foregoing is no overstatement, judging from such records as the following: 690 caught on a 700 hook line at Cape Breton; a Dogfish on nearly every one of 1,500 hooks in the Gulf of Maine; three wagon loads from a single lift of two pound-nets on Long Island; 1,800 pounds in one day in pound nets in North Carolina; more than two tons preserved for use in biological laboratories in a three-weeks' period; or an average trawl catch of 6,000 to 8,000 per trip on Georges Bank during the peak of abundance in 1913. At the time of the 1904–1905 peak of abundance it was estimated, from reports of fishermen's catches, that at least 27,000,000 were taken yearly off the coast of Massachusetts alone.[49] There is, in short, no

47. For these and other tagging records, see Templeman (Res. Bull. Dep. Nat. Resources Newfoundland, 15, 1944: 67, fig. 18).
48. For details, see Bigelow and Schroeder (Bull. U.S. Bur. Fish., *48*, 1936: 323).
49. Rep. Comm. Fish. Game Mass. (1906), 1907: 20.

reason to doubt that the concentrations may be as great in the western side of the Atlantic as in the eastern, where the record catch of 20,000 in a single haul was made many years ago on the Cornwall coast. The foregoing instances also show that Dogfish may be as plentiful, on occasion, off one sector of the coast line as off another from North Carolina to Cape Breton. But in general they are much less plentiful in the bays that they penetrate (*e.g.*, Chesapeake, Delaware, Passamaquoddy) than off the open coast.

The Spiny Dog fluctuates very widely in abundance over periods of years, but there is so much irregularity on different parts of the coast, and the peaks of maximum abundance fall so far apart, that no consistent picture of its ups and downs can yet be offered. Around Newfoundland there seems to have been a period of scarcity for five or six years prior to 1871,[50] but on the other hand a period of abundance in southern New England waters about 1875 to 1880. Available information points next to a pronounced peak about 1904 or 1905. While reports of local fluctuations reflect in part the movements of great schools that may visit one locality in one year and another the next, less complaint has been made of them since about 1913 than previously.

Synonyms and References:[51]

1. North Atlantic:

Squalus acanthias Linnaeus, Syst. Nat., *1*, 1758: 233 (descr., refs., Europ. Oc.); *1*, 1766: 397; Olafsen and Povelsen, Reyse en Island, 1772: 359 (Iceland); Müller, Prod. Fauna Danica, 1776: 3 (Denmark); Olavius, Oecon. Reyse Island, 1780: 80 (Iceland, not seen); Bloch, Fische Deutsch., 3, 1784: 74, pl. 85 (general, early devel., size at birth); Mohr, Forsøg Island Naturh., 1786: 57 (Iceland); Bonnaterre, Tabl. Encyc. Meth. Ichthyol., 1788: 11, pl. 5, fig. 12 (descr., ill.); Blumenbach, Handb. Naturg., 1788: 294 (general) and subsequent eds.; Walbaum, P. Artedi Genera Pisc. Emend. Ichthyol., 1792: 505, pl. 3 (refs.); Forster, in Latham and Davis, Faunula Indica, 1795: 13 (ref. to Bloch, 1784); Vandelli, Mem. R. Acad. Lisboa, *1*, 1797: 70 (not seen); Bloch and Schneider, Syst. Ichthyol., 1801: 135 (descr., refs.); Bosc, Nouv. Dict. Hist. Nat., *21*, 1803: 191 (diagn.); Donovan, Nat. Hist. Brit. Fish., *4*, 1805: 82, pl. 82 (text not seen); Turton, Brit. Fauna, 1807: 114 (descr., Gt. Brit.); de la Roche, Ann. Mus. Hist. nat. Paris, *13*, 1809: 314 (Ivica, Balearic Is. in market); Risso, Ichthyol. Nice, 1810: 40 (general, Medit.); Faber, Fische Islands, 1829: 79 (descr., no. of embryos, Iceland); Nilsson, Prod. Ichthyol. Skand., 1832: 117; Johnston, Hist. Berwicksh. Nat. Cl., *1*, 1834: 17 (Gt. Brit., not seen); Jenyns, Manual Brit. Vert. Anim., 1835: 505 (Brit. coast); Couch, Cornish Fauna, 1838: 51 (Cornwall, abund.); Fries, Ekstrom and Sundevall, Skand. Fisk., 1845: 187, pl. 46 (descr., Scand.); Gaimard, R. Voy. Islande et Groenland, Vol. Zool. Med., 1851: 163 (Iceland); Gray, Cat. Fish. Coll. Descr. by L. T. Gronow in Brit. Mus., 1854: 8 (listed); Knight, Descr. Cat. Fish. N. S., 1866: 8 (Nova Scotia); Jones, List Fish. Nova Scotia, 1879: 9 (common, Nova Scotia); Bean, T. H., Proc. U.S. nat. Mus., *3*, 1881: 116 (New England locs.); Jones, Proc. N. S. Inst. Sci., 1882: 95 (same as Jones, 1879); Jordan and Gilbert, Bull. U.S. nat. Mus., 16, 1883: 16 (descr., value for oil, N. and Mid. Atlantic, U.S.); Bean, T. H., Rep. U.S. Comm. Fish. (1882), 1884: 344 (Woods Hole); Goode, Fish. Fish. Indust. U.S., 1, 1884: 673 (distrib., habits, abund.); Kingsley, Stand. Nat. Hist., *3*, 1885: 76 (general); Rathbun, Proc. U.S. nat. Mus., 7, 1885: 490 (parasites); Whiteaves, Cat. Canad. Pinnep. Cetacea, Fish., 1886: 5 (Gulf of St. Lawrence); Jordan, Rep. U.S. Comm. Fish. (1885), 1887: 793 (distrib.); Nelson, Final Rep. St. Geol. N. J., *2* (2), 1890: 659 (common, N. Jersey);

50. Saxby, Zoologist, (2) *6*, 1871: 2554.

51. Many studies have been made of the anatomy of *Squalus acanthias*, and it has served as the subject of numerous directions for laboratory dissection for the use of students. We have not thought it necessary to burden the present list with citations of these sorts.

Smitt, Hist. Scand. Fish., *2*, 1895: 1159, pl. 52, fig. 1 (descr., habits, fishery); Jordan and Evermann, Rep. U.S. Comm. Fish. (1895), 1896: 219 (Atlant. coast, around Cuba); Bull. U.S. nat. Mus., *47* (1), 1896: 54 (descr., n. and mid. U.S., Cuba); Bean, T. H., Bull. Amer. Mus. nat. Hist., *9*, 1897: 330 (off N. Jersey); Knipowich, Annu. Mus. zool. Acad. St. Pétersb., *2*, 1897: 154 (Murman coast); Bumpus, Science, N. S. *8*, 1898: 850 (breeding season, size of embryos, C. Cod); Smith, Bull. U.S. Fish Comm., *17*, 1898: 89 (abund., season, commercial use, Woods Hole); Jordan and Evermann, Bull. U.S. nat. Mus., 47 (4), 1900: pl. 7, fig. 24, 24a (ill.); Hørring, Rapp. Fisk. Faroerne og Island Fiskberet. Kbh., 1901: 207 (Faroes, Iceland, not seen); Bean, Rep. For. Comm. N. Y., 1901: 381 (season, abund., near N. York); Linton, Bull. U.S. Bur. Fish., *19*, 1901: 274 (parasites, Woods Hole), 430 (stomach contents); Gregg, Where to Catch Fish., 1902: 23 (descr., but no record of occur.); Johansen, Beret. Fisk "Diana" Islands (1898–1900), Fisk Beret. Kbh. (1901–1902), 1903: 250 (Iceland, not seen); Bean, T. H., Bull. N. Y. St. Mus., *60*, Zool. 9, 1903: 43 (descr., season, util., N. York and New England); de Braganza, Result. Invest. Sci. "Amelia," *2*, 1904: 56 (Portugal); Sharp and Fowler, Proc. Acad. nat. Sci. Philad., *56*, 1904: 506 (season, abund., damage to nets, Nantucket); Collet, Norges Fisk., *3*, 1905: 96 (reproduction, Norway); Genzøe, Fiskberet. Kbh. (1903–1904), 1905: 290 (Iceland, not seen); Jordan, Guide to Study Fish., *1*, 1905: 545 (general); Wilson, Proc. U.S. nat. Mus., *28*, 1905: 573 (parasites); Fowler, Rep. N. J. Mus. (1905), 1906: 67 (abund., season, N. Jersey); Pietschmann, Ann. naturh. (Mus.) Hofmus. Wien, *21*, 1906: 74 (abund., Iceland); Tracy, Rep. R. I. Comm. inl. Fish., 1906: 46 (abund., season, Rhode Island); Cornish, Contr. Canad. Biol. (1902–1905), 1907: 81 (abund., north. Nova Scotia); Field, Rep. U.S. Comm. Fish. (1906), Spec. Pap. *6*, 1907: 18, pl. 1, fig. 2 (abund. NW. Atlant., food, possible util.); Jensen, in Schiødte and Hansen, Zool. Danica, *12*, 1907: 308, pl. 28, fig. 3 (descr., depth, abund., food, devel., Denmark); Widakowitch, Z. wiss. Zool., *88*, 1907: 499, pl. 30, 31 (struct. of uterine wall, gain in weight by embryo); Wilson, Proc. U.S. nat. Mus., *31*, 1907: 711 (parasites); Proc. U.S. nat. Mus., *33*, 1907: 452 (parasites, Woods Hole); Fowler, Proc. Acad. nat. Sci. Philad., *60*, 1908: 68 (New England, N. Jersey); Rep. N. J. Mus. (1907), 1908: 128 (season, locs., N. Jersey); Kendall, Occ. Pap. Boston Soc. nat. Hist., 7 (8), 1908: 8 (New England locs.); Regan, Ann. Mag. nat. Hist., (8) 2, 1908: 46 (class., refs., distrib., size); Fowler, Proc. Acad. nat. Sci. Philad., *61*, 1909: 407 (N. Jersey); Saemundsson, Skr. Komm. Havunders. Kbh., *5*, 1909: 115 (Iceland); Gill, Smithson. Misc. Coll., *52*, 1909: 158 (ill.); Tracy, Rep. R. I. Comm. inl. Fish., 1910: 61 (food, size, Rhode Island); Scammon and Minot, in Keibel, Normal Tafeln Entwick. Wirbelt., 1911: 140 pp., 5 pls. (descr., ill., embryonic develop.); Gudger, Proc. biol. Soc. Wash., *25*, 1912: 143 (N. Carolina); Roule, Bull. Inst. océanogr. Monaco, 243, 1912: 11 (Brittany, Morocco, Medit.); Garman, Mem. Harv. Mus. comp. Zool., *36*, 1913: 192, pl. 14, fig. 1–4, pl. 43, fig. 9–10, pl. 59, fig. 1–2 (refs., descr., ill.); Gudger, Proc. biol. Soc. Wash., *26*, 1913: 98 (N. Carolina); Halkett, Check List Fish. Canad., 1913: 41 (Delaware to Davis Strait); Le Danois, Ann. Inst. océanogr. Monaco, *5* (5), 1913: 21 (Engl. Channel); Nichols, Abstr. Linn. Soc. N. Y., 20–23, 1913: 91 (near N. York); Scott, Ann. N. Y. Acad. Sci., *23*, 1913: 30, 60 (physiol., in fresh and brackish water, season near N. York); Sumner, Osburn and Cole, Bull. U.S. Bur. Fish., *31* (2), 1913: 737 (abund., season, embryos, Woods Hole); Jensen, Mindeskr. Jap. Steenstrup, Føds., Kbh., (2) 30, 1914: 7 (W. Greenland, Iceland, Murman Coast, Faroes); Coles, Proc. biol. Soc. Wash., *28*, 1915: 94 (season, N. Carolina); Fowler, Proc. Acad. nat. Sci. Philad., *67*, 1916: 517 (Nova Scotia); Wilson, Proc. U.S. nat. Mus., *47*, 1915: 656 (Maine); Fowler, Copeia, 30, 1916: 36; Copeia, 31, 1916: 41 (in list, N. Jersey); Nichols and Murphy, Brooklyn Mus. Quart., *3* (4), 1916: 150 (abund., season, near N. York); Nichols and Murphy, Brooklyn Mus. Sci. Bull., *3* (1), 1916: 31, pl. 2, fig. 1 (ill., abund., migration, season, devel., util., near N. York); De Nyse, Bull. N. Y. zool. Soc., *19*, 1916: 1424 (as a food fish); Radcliffe, Bull. U.S. Bur. Fish., *34*, 1916: 267 (teeth, denticles, abund., season, N. Carolina); Fowler, Proc. Acad. nat. Sci. Philad., *69*, 1917: 109 (N. Jersey); Proc. Boston Soc. nat. Hist., *35*, 1917: 110 (Gulf of Maine locs.); Latham, Copeia, 41, 1917: 17 (abund., season, Long Island, N. York); Copeia, 57, 1918: 53 (season, young in summer, Long Island, N. York); Huntsman, Trans. roy. Soc. Canada, (3) *12* (4), 1918: 63 (Gulf of St. Lawrence); Fowler, Proc. Acad. nat. Sci. Philad., *71*, 1920: 292 (season); Latham, Copeia, 71, 1919: 54 (abund., season, young, Long Island, N. York); Roule, Result. Camp. sci. Monaco, 1919: 117 (Norway, Gulf of

Gascony, Morocco, food); Fowler, Proc. biol. Soc. Wash., *33*, 1920: 145 (N. Jersey loc.); Latham, Copeia, 87, 1920: 91 (young, abund., Long Island, N. York in summer); Ford, J. Mar. biol. Ass. U. K., N. S. *12*, 1921: 468 (abund., season, commercial util., breeding, habits, Plymouth, Engl.); Latham, Copeia, 99, 1921: 72 (occur., Long Island Sound, N. York); Mavor, Contr. Canad. Biol. (1918–1920), 1921: 129, 130 (occur., Passamaquoddy region, util.); Fowler, Proc. Acad. nat. Sci. Philad., *74*, 1922: 3, 5 (N. Jersey); Huntsman, Contr. Canad. Biol. (1921), 3, 1922: 56 (abund., Bay of Fundy); Drensky, Trav. Soc. bulg. Sci. nat., *19*, 1923: 56, pl. 58, fig. 1 (descr., abund., Black Sea); Hubbs, Copeia, 123, 1923: 101 (Maine); Meek and Hildebrand, Field Mus. Publ. Zool., *15* (1), 1923: 65 (in key); Breder, Copeia, 127, 1924: 26 (season); Linton, Proc. U.S. nat. Mus., *64* (21), 1924: 4, 20, 26, 52, 57, 63, 64, 92 (parasites, Woods Hole); Bigelow and Welsh, Bull. U.S. Bur. Fish., *40* (1), 1925: 44, 551 (habits, abund., Gulf of Maine); Fowler, Copeia, 143, 1925: 43 (N. Jersey); Proc. Acad. nat. Sci. Philad., *78*, 1927: 276 (comp. with Chilean spec.); Manter, Illinois biol. Monogr., *10* (2), 1926: 8 (parasites, Mt. Desert, Maine); Nichols and Breder, Zoologica, N. Y., *9*, 1927: 21 (abund., season, food, also as food fish, N. York to south. New England); Borodin, Bull. Vanderbilt Oceanogr. (Mar.) Mus., *1* (1), 1928: 5 (Newfoundland); Hildebrand and Schroeder, Bull. U.S. Bur. Fish., *43* (1), 1928: 52 (descr., devel., migrations, season in Chesapeake Bay); Rey, Fauna Iberica Peces, *1*, 1928: 424, pl. 4, fig. 1 (descr., ill., Spanish waters); Breder, Field Bk. Mar. Fish. Atlant. Coast, 1929: 24 (general); Jordan, Manual Vert. Anim. NE. U.S., 1929: 14 (general); Truitt, Bean and Fowler, Bull. Md. Conserv. Dep., *3*, 1929: 28 (Maryland); Jordan, Evermann and Clark, Rep. U.S. Comm. Fish. (1928), 2, 1930: 21 (both sides N. and mid. Atlant. U.S., Cuba); Bere, Contr. Canad. Biol., N. S. *5* (13), 1930: 427 (parasites, Passamaquoddy Bay); Schroeder, Bull. U.S. Bur. Fish., *46*, 1930: 11, 72 (enemy of cod, migrations, season, Nantucket Shoals); MacCallum, Proc. U.S. nat. Mus., *79* (26), 1931: 3 (parasites); Schroeder, Copeia, 1931: 41, 42 (winter, depth, off Delaware Bay); Breder, Copeia, 1932: 31 (off Block Island); Anonymous, Newfoundland Fish. Res. Comm. Rep., *1* (4), 1931: 107 (Nova Scotian and Newfoundland Banks, Nfd. coast); Wilson, Bull. U.S. nat. Mus., 158, 1932: 511 (parasites, Woods Hole); Anonymous, Newfoundland Fish. Res. Comm. Rep., *2* (1), 1932: 125 (Nova Scotian Banks, off NE. Newfoundland, Gulf of St. Lawrence); Rep., *2* (2), 1934: 115 (Nova Scotian Banks, Straits of Belle Isle); Belloc, Rev. des Trav. Pêches Marit., 7 (2), 1934: 140 (Morocco to Senegal, Canaries, Madeira); Bigelow and Schroeder, Canad. Atlant. Fauna, biol. Bd. Canad., *12*[e], 1934: 17 (general, east. Canad., W. Greenland); Vladykov and McKenzie, Proc. N. S. Inst. Sci., *19*, 1935: 48 (abund., season, util., Nova Scotia); Bigelow and Schroeder, Bull. U.S. Bur. Fish., *48*, 1936: 322 (breeding, winter habitat); Fowler, Bull. Amer. Mus. nat. Hist., *70* (1), 1936: 69 (refs., descr., ill., embryo, adult, E. Atlant. to Morocco); Lübbert and Ehrenbaum, Handb. Seefisch. Nordeurop., *2*, 1936: 286, pl. 21, fig. 251 (ill., adult, embryos in capsule, general, N. Europe); Howell-Rivero, Proc. Boston Soc. nat. Hist., *41*, 1936: 47, 48 (comp. with *S. cubensis* and *barbouri*); Smith, H. W., Biol. Rev., *11*, 1936: 64 (in fresh water in Scand.); Cadenat, Rev. des Trav. Pêches Marit., *10* (4), 1937: 431 (C. Blanc); Fowler, Proc. Acad. nat. Sci. Philad., *89*, 1937: 297 (N. Jersey); Gudger, Sci. Mon. N. Y., *45*, 1937: 510 (off Chesapeake Bay, Gulf of Maine); Sayles and Hershlowitz, Biol. Bull. Wood's Hole, *73*, 1937: 51, 1 pl. (dermal denticles); Raven and LaMonte, Amer. Mus. Novit., 902, 1937: 1 (eaten by swordfish); Breder, Bull. N. Y. zool. Soc., *41*, 1938: 28 (N. York Harbor); Hampton, Newfoundland Fish. Res. Inst. Serv. Bull., *5*, 1938: 3 (possible util., production of liver oil in Canada); Popovici, in Grig. Antipa, Homm. Oeuvre Bucharest, 1938: 445 (breeding habits, Black Sea); Tortonese, Atti Soc. ital. Sci. nat., 77, 1938: 308 (Medit., Canaries, Massachusetts); Springer, Proc. Fla. Acad. Sci., *3*, 1939: 37 (no definite Florida record given); Popovici, Bull. Acad. Roumaine, *22* (2), 1939: 1 (develop., abund. in Black Sea); Fowler, Bull. U.S. nat. Mus., 100 (*13*), 1941: 257 (N. Atlant. refs.); Hildebrand, Copeia, 1941: 222 (N. Carolina, April, May); Holmgren, Acta zool., Int. Tidskr. zool., 1–3, 1941: 29, 55 (skull); Longley and Hildebrand, Pap. Tortugas Lab., *34*, 1941: 3 (embryos, Tortugas, Florida); Fowler, Fish Culturist, *21* (9), 1942: 66 (listed, Cuba); Lunz, Bull. S. Carolina St. Planning Bd., *14*, 1944: 27 (Florida); Merriman and Warfel, Trans. N. Amer. Wild Life Conf., 1944: 234 (season, east. Long Island Sd.); Templeman, Res. Bull. Dep. Nat. Resources Newfoundland, 15, 1944: 1 (life hist., season, migrations, Newfoundland, Labrador); Bigelow and Schroeder, Guide Comm. Shark Fish., Anglo Amer. Caribb. Comm., Wash., 1945: 140, fig. 52 (descr., ill., habits, range).

Piked Dog, Brookes, Nat. Hist., *3*, 1763: 32, pl., fig. 32 (general); 1772: 18, pl., fig. 18; Pennant, Brit. Zool., *3*, 1769: 77 (descr., weight); Yarrell, Brit. Fish., *2*, 1836: 400 (abund., habits, food, defensive use of spines, Gt. Brit., general); 2nd ed., *2*, 1841: 524; Couch, Hist. Brit. Fish., *1*, 1867: 49, pl. 11 (size, abund., record catch of 20,000, nuisance to fishermen, Gt. Brit.).

Spinax, Duhamel, Traité Gén. Pêches, (2) *4* (9), 1769: 299, pl. 20, fig. 5, 6 (general; ill. as "Aiguillat").

L'Aiguillat, Broussonet, Mém. Acad. Sci. Paris, 1780: 673 (abund., Medit., Atlant. to Scotland).

Squalus spinax Olavius, Oecon. Reyse Island, 1780: 562, pl. 7 (Iceland, not seen).

Squale aiguillat, Lacépède, Hist. Nat. Poiss., 4° ed., *1*, 1798: 270; in Sonnini, Hist. Nat. Poiss., *4*, 1802: 100 (descr.).

Squalus achantias Latreille, Nouv. Dict. Hist. Nat., *24*, 1804: 72 (evident misspelling for "*acanthias*" in table of contents).

Acanthorhinus acanthias Blainville, Bull. Soc. philom. Paris, 1816: 121 (name only).

Spinax acanthias Cloquet, Dict. Sci. Nat., *1*, Suppl., 1816:[53] 91 (abund., defensive spines, fishery, N. Europe); Cuvier, Règne Anim., *2*, 1817: 130 (diagn., abund.); Bory de St. Vincent, Dict. Class. Hist. Nat., *15*, 1829: 598; Bonaparte, Icon. Faun. Ital., *3* (2), 1834: plate not numbered (descr., ill., Italy); Agassiz, L., Poiss. Foss., *3*, 1837: pl. B, fig. 3 (spines);[54] Templeton, Charlesworth Mag. nat. Hist., (2) *1*, 1837: 413 (Ireland); Parnell, Mem. Werner. Soc. Edinb., 7, 1838: 420 (abund., food, habits, Gt. Brit.); Swainson, Nat. Hist. Fish. Amphib. Rept., *2*, 1839: 315 (general); Storer, Rep. Fishes Rept. Birds Mass., 1839: 187 (abund., season, fishery, Massachusetts); Boston J. nat. Hist., *2*, 1839: 538; Nordman, in Demidoff, Voy. Russ. Merid., *3*, 1840: 548 (common, Black Sea); Ayres, Boston J. nat. Hist., *4*, 1842: 288 (near N. York); DeKay, Zool. N. Y., 4, 1842: 359, pl. 64, fig. 210 (descr., ill., abund., season, fishery, C. Cod); Selys-Longchamps, Fauna Belg., 1, 1842: 233 (Belgium); Cuvier, Règne Anim., ill. ed., 1843: 366, pl. 115, fig. 4 (food, ill.); Linsley, Amer. J. Sci., *47*, 1844: 77 (Connecticut); Gosse, Nat. Hist. Fish., 1851: 308 (abund., Brit. waters); Desvaux, Essai Ichthyol. France, 1851: 19 (not seen); Owen, Cat. Osteol. Roy. Coll. Surg., *1*, 1858: 91 (vert., jaws); Van der Hoeven, Handb. Dierkunde, *2*, 1855: 261 (general); Thompson, W., Nat. Hist. Ireland, *4*, 1856: 254 (Ireland); DeKay, Rep. St. Cab. nat. Hist. N. Y. (1855), 8, 1858: 64 (N. York); Doûmet-Adanson, Rev. Zool., (2) *12*, 1860: 446 (Medit.); Schlegel, Naturl. Hist. Nederl. Dieren, 1862: 192, pl. 18, fig. 3 (descr., Netherlands); Gill, Proc. Acad. nat. Sci. Philad., 1863: 333 (Massachusetts); Knight, Descr. Cat. Fish. N. S., 1866: 8 (Nova Scotia); de la Blanchère, Dict. Pêches, 1868: 11 (descr., breeding season); Van Beneden, Poiss. Côtes Belg., 1870: 9; Mem. Acad. R. Belg. (1870), 1871: 9 (food, parasites, Belgium); Saxby, Zoologist, 1871: 2554 (Newfoundland); Reguis, Ess. Hist. Nat. Provence, *1* (1), 1877: 54 (Medit.); Cocco, Nat. Sicil., *3*, 1884: 176 (Messina, not seen); Gröndal, Islenzka Fisk. (Pisces Island) Reykj., 1890–1891 (Iceland, not seen).

Acanthias vulgaris Risso, Hist. Nat. Europe Mérid., *3*, 1826: 131 (Medit.); Müller and Henle, Plagiost., 1841: 83 (descr., Medit., North Sea, Atlant.); Hamilton, Brit. Fish., *2*, 1843: 313 (Gt. Brit.), also later eds.; Bonaparte, Cat. Pesc. Europ., 1846: 15; Nardo, Sinon. Modern. Spec. descr. Pesci St. Chiereghini, 1847: 111 (name only); Gray, List Fish. Brit. Mus., 1851: 70 (in part); White, List Spec. Brit. Mus., Fish., *8*, 1851: 131; Baikie, Zoologist, *11*, 1853: 3846 (Orkneys); Krøyer, Danmarks Fisk., *3*, 1853: 12, 868 (Denmark); Nilsson, Skand. Fauna, *4*, Fisk, 1855: 731 (Scand.); Machado, Peces Cadiz., 1859: 8 (off Cadiz); Jouan, Mém. Soc. Imp. Sci. nat. Cherbourg, 7, 1859 (not seen); Kessler, Bull. Soc. Nat. Moscou, *32* (4), 1859: 473 (abund., Black Sea); Nardo, Atti Ist. veneto, (3) *5*, 1859–1860: 787 (Medit., near Venice); Malmgren, Arch. Naturg., *30* (1), 1864: 346 (N. Norway, west of North Cape); Fitzinger, Bild. Atlas Naturg. Fische, 1864: fig. 177 (ill.); Duméril, Hist. Nat. Poiss., *1*, 1865: 437 (refs., descr., general); Bocage and Brito Capello, Poiss. Plagiost. Portugal, 1866: 21 (Portugal); Günther, Trans. zool. Soc. Lond., *6*, 1869: 396 (mention, but report for Cent. Amer. perhaps not this species); Ninni, Ann. Soc. nat. Modena, *5*, 1870: 66 (near Venice); Brito Capello, J. Sci. math. phys. nat. Lisboa (1868–1869), *2*, 1870: 143 (Portugal); Günther, Cat. Fish. Brit. Mus., *8*, 1870: 418 (refs.,

53. This date has frequently been quoted as 1821, but the copy in the library of the Harvard Museum of Comparative Zoology is dated 1816.

54. This plate, including other species also, is labeled simply *spinax*. But fig. 3 is identified as *S. acanthias* by the Explic. des Planches which appeared in 1843.

descr., includes south. hemisphere records, probably not this species); Miklucho-Maclay, Beitr. Vergl. Neurol. Wirbelt., *1*, 1870: 10, fig. 2, A–C (brain); Saxby, Zoologist, (2) *6*, 1871: 2554 (Newfoundland); Canestrini, in Cornalia, *et al.*, Faun. d'Ital., *3*, Pesc., 1872: 39 (Medit.); Hertwig, Jena Z. Naturw., *8*, 1874: 349, pl. 12, fig. 4, 8, 11, 12, 14, pl. 13, fig. 10, 13, 16 (develop. of dermal denticles, teeth); McIntosh, Ann. Mag. nat. Hist., (4) *14*, 1874: 424 (off W. Scotland); Collett, Norges Fisk., 1875: 211 (Norway); Gervais and Boulart, Poiss., *3*, 1877: 206, 207, pl. 79 (descr., ill., France); Doderlein, Prosp. Metod. Pesci Sicil., 1878: 30 (Medit.); Feddersen, Naturh. Tidsskr., (3) *12*, 1879: 70, footnote (in a Danish stream); Jones, Proc. N. S. Inst. Sci., *5* (1), Append., 1879: 95 (Nova Scotia); Winther, Prod. Ichthyol. Dan. Marin. in Naturh. Tidsskr., 1879–1880: 58 (Denmark); Bean, Amer. Nat., *14*, 1880: 526 (in a stream in Denmark); Günther, Guide to Study Fish., 1880: 331 (general); Stossich, Boll. Soc. adriat. Sci. nat., *5*, 1880: 70 (Adriatic); Day, Fish. Gt. Brit., *2*, 1880–1884: 315, pl. 160, fig. 2 (descr., ill., habits, abund., locs., food, Gt. Brit.); Doderlein, Man. Ittiol. Medit., *2*, 1881: 86 (Medit.);[55] Moreau, Poiss. France, *1*, 1881: 342 (descr., France); Perugia, Elenc. Pesc. Adriat., 1881: 56 (Adriatic); Hasse, Naturl. Syst. Elasm. besond. Theil, 1882: pl. 12, fig. 20–28 (denticles, vertebrae); Marion, Ann. Mus. Hist. nat. Marseille, *1*, 1883: 100 (off Marseilles, not seen); Storm, K. norske Vidensk.-Selsk. Skr. (1883), 1884: 45 (Trondhjem Fjord); Graeffe, Arb. zool. Inst. Univ. Wien, 7, 1886: 3, pl. 1 (29), fig. 1, 2 (develop., descr. of embryos, breeding season, abund., Trieste); Kolombatovic, Cat. Vert. Dalmat., 1888: 28 (abund., Adriatic); Carus, Prod. Fauna Medit., *2*, 1889–1893: 503 (Medit.);[56] Bellotti, Atti Soc. ital. Sci. nat., *33*, 1891: 110 (no. of embryos); Garstang, J. Mar. biol. Ass. U. K., N. S. *3*, 1894: 229 (breeding season, Plymouth, England); Sicher, Atti Accad. gioenia, (4) *11* (5), 1898: 17 (Medit.); Jungerson, Danish "Ingolf" Exped., *2* (1), 1899: 27, pl. 1, fig. 10, 11 (claspers); Lo Bianco, Mitt. zool. Sta. Neapel, *13*, 1899: 538 (breeding season, Naples); Huber, Z. Wiss. Zool., *70*, 1901: pl. 28, fig. 11 (claspers); Steinhardt, Arch. Naturgesch., (69) *1*, 1903: 13, pl. 1, fig. 24–31 (oral and pharyngeal denticles); Schmidt, Fiskeriunders. Island og Faerøerne (1903), 1904: 22 (not seen); Werner, Zool. Jb., Syst. Abt., *21*, 1904: 291 (Medit.); Borcea, Arch. Zool. exp. gén., (4) *4*, 1905: 203, 205 (breeding season, Roscoff); Imms, Proc. zool. Soc. Lond., *1*, 1905: 45 (pharyngeal denticles); Bridge, Camb. nat. Hist., 1910: 455 (general); Günther, Encyc. Brit., 11th ed., *24*, 1911: 804 (general); Murray and Hjört, Depths of Ocean, 1912: 391 (off Morocco, Faroes); Le Danois, Ann. Inst. océanogr. Monaco, *5* (5), 1913: 21, 134 (English Channel); Leigh-Sharpe, J. Morph., *34*, 1920: 259 (claspers); Evans, Philos. Trans., (B) 212, 1923: 27 (poisoning by spines); Jenkins, Fish. Brit. Isles, 1925: 321 (general, commercial catch, Gt. Brit.); Ehrenbaum, in Grimpe and Wagler, Tierwelt N.- u. Ostsee, Lief 7 (12e), 1927: 17 (descr., North Sea, Baltic); Hickling, Rep. Brit. Ass. Adv. Sci., 1927: 331 (bathymetric segregation of sexes, Faroe Shetland Channel); Ehrenbaum, Faune Ichthyol., Cons. explor. Mer., 1929: plate not numbered (general); Borri, Atti Soc. tosc. Sci. nat., *44*, 1934: 99 (Medit.); Ranzi, Pubbl. Staz. zool. Napoli, *13*, 1934: 343, 372 (uterine fluid and wall); Needham, Biochem. Morphogen., 1942: 41 (embryonic weight).

Squalus (*Acantherinus*) *acanthias* Blainville, in Vieillot, Faune Franc., 1825: 57 (descr., France).

Squalus (*Spinax*) *acanthias* Voigt, in Cuvier, Tierreich, *2*, 1832: 510 (descr.); Richardson, Fauna Boreal. Amer., *3*, 1836: 291 (Nova Scotia); van der Hoeven, Handb. Dierkunde, *2*, 1855: 260.

Acanthias acanthias Bonaparte, Mém. Soc. neuchâtel. Sci. nat., *2* (8), 1839: 8 (in synopsis); Poey, An. Soc. esp. Hist. Nat., *5*, 1876: 398; Enumerat. Pisc. Cubens., 1876: 202, in part (discus., refs.); Nobre, Fauna Marinha Port. Vert., *1*, 1935: 445 (descr., Portugal).

Spinax (no spec. name) Owen, Odontogr., 1840–1845: 29, pl. 3, fig. 3 (teeth).

Acanthias americanus Storer, Mem. Amer. Acad. Arts. Sci., N. S. *2*, 1846: 506 (N. York to Labrador and beyond); Boston J. Nat. Hist., *6*, 1857: 270 (Nova Scotia, Labrador); Gill, Proc. Acad. nat. Sci. Philad., Addend., 1861: 60 (name); Leidy, Proc. Acad. nat. Sci. Philad., 1865: 175 (Isles of Shoals, N. Hampshire); Storer, Mem. Amer. Acad. Arts Sci., N. S. *9*, 1867: 232, pl. 38, fig. 1; also as Fishes Mass., 1867: 256, pl. 38, fig. 1 (descr., ill., abund., season, fishery for oil, Massachusetts); Hertwig, Jena. Z. Naturw., *8*, 1874: 349, pl. 12, fig. 5, 9, pl. 13, fig. 15 (devel. denticles, teeth).

55. See Doderlein for additional Mediterranean citations in publications not accessible to us.
56. See Carus for additional Mediterranean citations in publications not accessible to us.

Spinax vulgaris Guichenot, Explor. Alger. Poiss., *5*, 1850: 125 (Algeria).

Squalus americanus Gill, Proc. Acad. nat. Sci. Philad., 1864: 263 (east coast, U.S.); Abbott, Rep. N. J. geol. Surv., App. E, 1868: 824 (common, N. Jersey).

Acanthias linnei Malm, Göteborgs och Bohuslans Fauna, 1877: 624 (commercial catch, Sweden).

Dog Fish, Collins, Bull. U.S. Fish Comm., *2*, 1883: 8 (in winter off N. Hampshire); Gallup, Bull. U.S. Fish Comm., *2*, 1883: 246 (util., Maine); Collins, Bull. U.S. Fish Comm., *4*, 1884: 248 (preying on mackerel); Stevenson, Rep. U.S. Comm. Fish. (1902), 1904: 228 (size, util.); Evermann, Rep. U.S. Comm. Fish. (1903), 1905: 114 (abund., no. of embryos, destroying other fish, Gulf of Maine); Atkins, Bull. U.S. Bur. Fish., *28*, 1910: 253 (possible util.); Cornish, Contr. Canad. Biol. (1906–1910), 1912: 79 (abund., season, Nova Scotia, Prince Edward I.); Beatty, Rep. Fish. Res. Bd. Canad. (1942), 1943: 17 (season, abund., Vitamin A content, Bay of Fundy, Nova Scotia, Gulf of St. Lawrence).

Squalus acanthias or *Acanthias vulgaris* Prince, Rep. Mar. Fish. Canada (1903), 1904: XLV (abund., destructiveness).

Horned Dog Fish, Field, Bull. U.S. Bur. Fish., *28*, 1910: 247 (as a food fish).

Grayfish, Moore, Bull. U.S. Bur. Fish., Econ. Circ., *22*, 1916: (as a food fish, abund.).

Squalus Nichols and LaMonte, Amer. Mus. Novit., 901, 1913: 4 (eaten by swordfish).

Squalus barbouri Howell-Rivero, Proc. Boston Soc. nat. Hist., *41*, 1936: 47, pl. 12 (descr., ill., Cuba, ident. by exam. of type spec.).

Doubtful References:

Spinax acanthias Bowditch, Excurs. Madeira, 1825: 74 (Porto Santo, nominal only, may equally have referred to *S. blainville*).

Acanthias vulgaris Vincent, Sea Fish. Trinidad, 1910: 53 (common, Trinidad, nominal only, more likely to have referred to the newly discovered *S. cubensis*, p. 473).

Probably not *Squalus acanthias* Evermann and Bean, Rep. U.S. Comm. Fish. (1896), 1898: 239 (abund. Indian R., Florida); Evermann and Kendall, Rep. U.S. Comm. Fish. (1899), 1900: 49 (Indian R., Florida).[57]

2. North Pacific:

Acanthias vulgaris Bleeker, Verh. batavia. Genoot., *25* (7), 1853: 21 (name only, Japan); Verh. batavia. Genoot., *26* (4), 1857: 44 (name only, Japan); Günther, Cat. Fish. Brit. Mus., *8*, 1870: 418; Nyström, Bih. Svensk. Vet. Acad. Handl., Stockholm, *13* (4), 1887: 50 (name only, Japan); Ishikawa and Matsuüra, Prel. Cat. Fish. Mus. Tokyo, 1897: 61 (name only, Japan); not *Acanthias vulgaris* Temminck and Schlegel, in Siebold, *et al.*, Fauna Japan Pisces, *4* (15), 1850: 304, pl. 135 (this is probably *fernandinus*; see discussion, p. 454).

Spinax (*Acanthias*) *suckleyi* Girard, Proc. Acad. nat. Sci. Philad., 7, 1854: 196 (descr., Puget Sound).

Acanthias sucklii Girard, Pacif. R. R. Surv. Fish., *10*, 1858: 368 (descr., color, Puget Sound).

Acanthias suckleyi Suckley, Pacif. R. R. Survey, *12* (2), 1860: 367 (descr., size, embryos, abund., util., Puget Sound).

Squalus suckleyi (or *sucklii*)[58] Gill, Proc. Acad. nat. Sci. Philad. (1862), 1863: 499 (refs.); Jordan and Gilbert, Bull. U.S. nat. Mus., 16, 1883: 17 (Pacific coast, U.S.); Jordan and Starks, Proc. Calif. Acad. Sci., (2) 5, 1895: 789 (abund., Puget Sound); Jordan and Evermann, Rep. U.S. Comm. Fish. (1895), 1896: 219 (name only, Aleutian Is. to Santa Barbara, California); Bull. U.S. nat. Mus., 47 (1), 1896: 54 (descr., abund., util., Aleutian Is. to Santa Barbara, California); Jordan and Gilbert, in Jordan, *et al.*, Fur Seals and Fur Seal Industr. N. Pacif. Ocean, 3, 1899: 434, pl. 3 (Bering I.); Evermann and Goldsborough, Bull. U.S. Bur. Fish., *26*, 1907: 228 (SE. Alaska); Starks and Morris, Univ. Calif. Publ. Zool., *3*, 1907: 168 (off San Diego, California, in deep water); Regan, Ann. Mag. nat. Hist., (8) *2*, 1908: 46 (in synopsis); Starks, Ann. Carneg. Mus., 7, 1911: 207 (name only, Puget Sound); Halkett, Check List Fish. Canad., 1913: 41 (Brit. Columbia, Aleutians to California); Thompson, Proc. U.S. nat. Mus., *50*, 1916: 420 (name only, San Diego, California); Hubbs, Copeia, 43, 1917: 37 (stomach contents, California); Starks, Fish Bull. Sacramento, *3*, 1917: 152 (ill., abund. northward, size, depth, California to Alaska); Jordan and Hubbs, Mem. Carneg. Mus., *10*, 1925: 105 (complete list refs.; they point out that ill. by

57. See footnote 42, p. 463.

58. Spelled *sucklii* by some authors, *suckleyi* by others.

Jordan and Fowler, 1903, of their *mitsukurii* was actually *suckleyi*); Hubbs, J. Pan-Pacif. Res. Instn., *3* (3), 1928: 11 (listed for Oregon, Washington); Mori, J. Pan-Pacif. Res. Instn., *3* (3), 1928: 3 (listed for Korea); Quigley, Biol. Bull. Wood's Hole, *50*, 1928: 439 (weight in rel. to length, percentages of sexes; breeding, Straits of Georgia); Fowler, Hongkong Nat., *1*, 1930: 88 (not seen); Jordan, Evermann and Clark, Rep. U.S. Comm. Fish. (1928), 2, 1930: 21 (in check list; Aleutians to Santa Barbara, California; Kamchatka); Fang and Wang, Contr. biol. Lab. Sci. Soc. China, *8*, 1932: 246 (descr., ill., Chefoo, China); Tanaka, Jap. Fish Life Colors, *22*, 1933: (not seen); Daniel, Elasmobranch Fishes, Univ. Calif. Press., 1934: 37, 154, fig. 147 (gill arches and rakers); Walford, Fish Bull. Sacramento, *45*, 1935: 42 (ill., California); Barnhart, Mar. Fish. south. Calif., 1936: 10 (distrib., California).

Squalus acanthias Jordan and Gilbert, Proc. U.S. nat. Mus., *3*, 1881: 458 (listed, Puget Sound to Santa Barbara, California); Bean, Proc. U.S. nat. Mus., *4*, 1882: 261, 269, 272, 274 (records for Brit. Columbia, Alaska); Jordan and Gilbert, Proc. U.S. nat. Mus., *4*, 1882: 33 (habits, food, abund., breeding season); Bean, Proc. U.S. nat. Mus., *6*, 1884: 361 (Brit. Columbia); Eigenmann, Proc. U.S. nat. Mus., *15*, 1893: 129, 132 (near San Diego, California); Bean and Bean, Proc. U.S. nat. Mus., *19*, 1897: 237 (name only, Commander I.); Schmidt, Pisces Mar. Orient. Rossici, 1904: 287 (descr., meas., discuss., E. Siberia, in Russian); Pavlenko, Kazani Turdy Obschestro E. Stest., *42* (2), 1910: 11 (Peter the Great Bay; in Russian); Berg, Faune Russie Poiss., *1*, 1911: 71 (California to Vladivostok; in Russian); Soldatov and Lindberg, Bull. Pacif. Sci. Fish. Inst. Vladivostok, 1930: 16 (Peter the Great and Olga Bays, Japanese Sea, in Russian).

Squalus mitsukurii (in part) Jordan and Fowler, Proc. U.S. nat. Mus., *29*, 1903: 630, fig. 3 (Japan; but as pointed out by Jordan and Hubbs in Mem. Carneg. Mus., *10*, 1925: 103, the descr. was of a member of the *fernandinus* group; see discuss., p. 454); Jordan, Tanaka and Snyder, J. Coll. Sci. Tokyo, *33*, 1913: 18 (ill. and refs. in part).

Squalus mitsukurii Jordan and Evermann, Bull. U.S. Bur. Fish., *23*, 1905: 45, fig. 6 (ill. only; wrongly stated to be "from the type"); Tanaka, Annot. zool. jap., *6*, 1908: 236 (Sakhalin I.); Smith, Proc. U.S. nat. Mus., *41*, 1912: 679 (Japan); Jordan and Metz, Mem. Carneg. Mus., *6*, 1913: 4, fig. 2 (ill., Korea); Tanaka, Fish. Japan, *26*, 1917: 471, pl. 130, fig. 368–370 (descr., ill., Japan); not *Squalus mitsukurii* Jordan and Fowler, Proc. U.S. nat. Mus., *26*, 1903: 629 (descr. of type spec.).

Squalus suckleyi (in part) Garman, Mem. Harv. Mus. comp. Zool., *36*, 1913: 194 (descr., NE. Pacif., but not his var. *mitsukurii*, for *mitsukurii* Jordan and Fowler, 1903: 629, is probably one of the *fernandinus* group, see discussion, p. 454); Fowler, Proc. Pan-Pacif. sci. Congr., *4*, 1930: 495 (California, Brit. Columbia, Japan, Formosa; but refs. for Hawaii probably *fernandinus*); Fowler, Bull. U.S. nat. Mus., 100 (*13*), 1941: 258 (descr., but refs. only in part); not *Squalus suckleyi* Fowler, Mem. Bishop Mus., *10*, 1928: 23 (Hawaiian Is., probably *fernandinus*).

Squalus wakiyae Tanaka, Fish. Japan, *27*, 1918: 475 (to replace *mitsukurii* Tanaka, 1917, because the latter is preoccupied by Jordan and Fowler, 1903, 629).

Squalus sucklii subsp. *mitsukurii* Schmidt, Trans. Pacif. Comm., Leningr., *2*, 1931: 7 (doubts if Japanese and Black Sea specs. are separable).

Squalus cubensis Howell-Rivero, 1936

Cuban Dogfish

Figures 89, 90

Study Material. Male, 524 mm. (Harv. Mus. Comp. Zool., type, No. 1458); female, 672 mm. long (Harv. Mus. Comp. Zool., No. 1461); very young male and female, each about 280 mm., with umbilical scars still evident, and male embryo about 203 mm. long, with yolk sac attached, all from Havana, Cuba (Harv. Mus. Comp. Zool., No. 1459, 1460, 1462).

Distinctive Characters. S. cubensis differs most sharply from *acanthias* and *blainville* in having the inner corner of its pectoral acutely angular (broadly rounded in the other two species); it differs further from *acanthias* in that its first dorsal originates only a little posterior to the axil of its pectoral (over or behind the inner corner of the pectoral in *acanthias*), the midpoint of its pelvics are only very little nearer to its second dorsal than

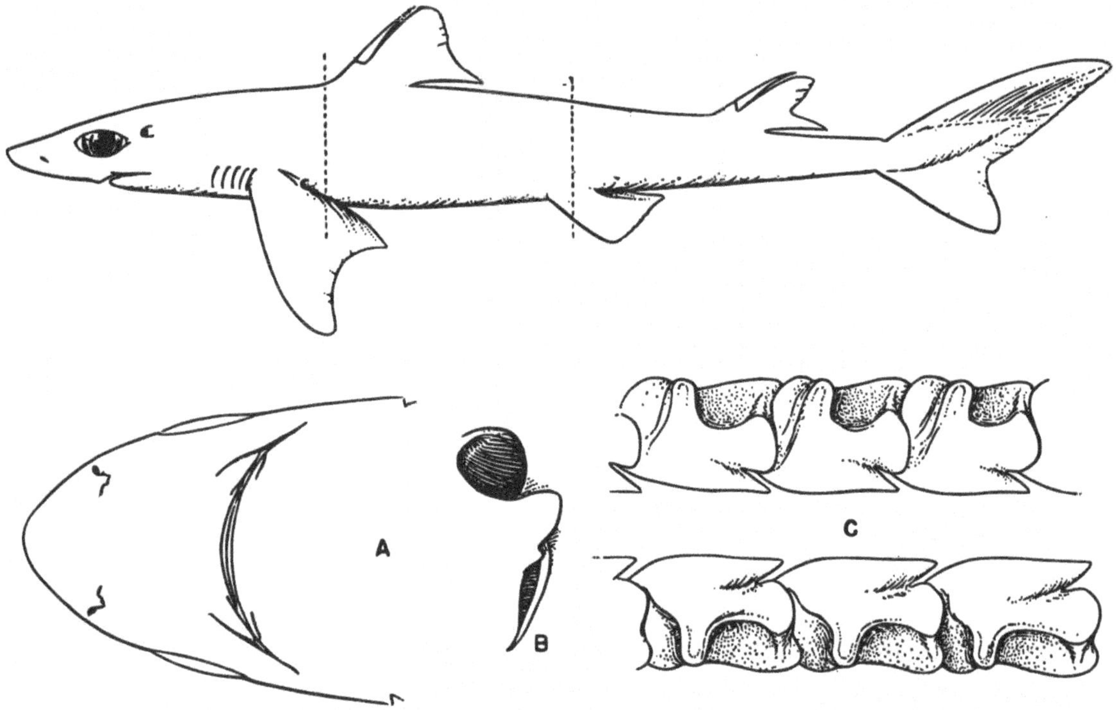

FIGURE 89. *Squalus cubensis*, female, about 672 mm. long, from near Havana, Cuba (Harv. Mus. Comp. Zool., No. 1461). *A* Anterior part of head from below, ½ natural size. *B* Left-hand nostril, about 4 x. *C* Second to fourth upper and lower teeth, about 8.5 x.

to its first (much nearer to the second dorsal than to the first in *acanthias*), it has longer fin spines and its coloration is plain; it differs further from *blainville* in having the distal margin of its pectoral deeply concave.

Description. Proportional dimensions in per cent of total length. Male, 524 mm., from Havana, Cuba (Harv. Mus. Comp. Zool., holotype, No. 1458). Female, 672 mm., same locality (Harv. Mus. Comp. Zool., paratype, No. 1461).

Trunk at origin of pectoral: breadth 12.4, 12.9; height 9.4, 9.5.
Snout length in front of: outer nostrils 3.0, 3.7; mouth 9.5, 9.5.
Eye: horizontal diameter 4.8, 4.8.
Mouth: breadth 6.7, 7.1; height 1.3, 1.2.
Nostrils: distance between inner ends 4.0, 4.6.
Labial furrow length from angle of mouth: upper 2.7, 2.2; lower 1.5, 1.6.

Gill opening lengths: 1st 1.3, 1.9; 2nd 1.1, 1.6; 3rd 1.1, 1.6; 4th 1.2, 2.0; 5th 1.9, 2.4.
First dorsal fin: vertical height 7.8, 8.0; length of base 8.6, 9.2.
Second dorsal fin: vertical height 5.3, 4.9; length of base 6.9, 6.0.
Caudal fin: upper margin 21.8, 21.4; lower anterior margin 12.0, 12.0.
Pectoral fin: outer margin 14.7, 15.9; inner margin 11.2, 11.6; distal margin 12.4, 10.4.
Distance from snout to: 1st dorsal 29.6, 27.2; 2nd dorsal 61.3, 64.1; upper caudal 78.2, 78.6; pectoral 21.7, 21.4; pelvics 46.0, 46.8.
Interspace between: 1st and 2nd dorsals 24.0, 27.5; 2nd dorsal and caudal 10.7, 9.8.
Distance from origin to origin of: pectoral and pelvics 25.3, 27.1; pelvics and caudal 31.3, 30.0.

Trunk slender, as in other members of the genus, its height at 1st dorsal only about 1/7 its length to origin of caudal. Body sector only a little longer than tail sector (much longer than tail sector in *acanthias*), without mid-dorsal ridge. Caudal peduncle strongly flattened below, with an obscure longitudinal ridge low down on each side (much as in *acanthias*). Upper precaudal pit strongly marked, but no lower pit. Dermal denticles close-

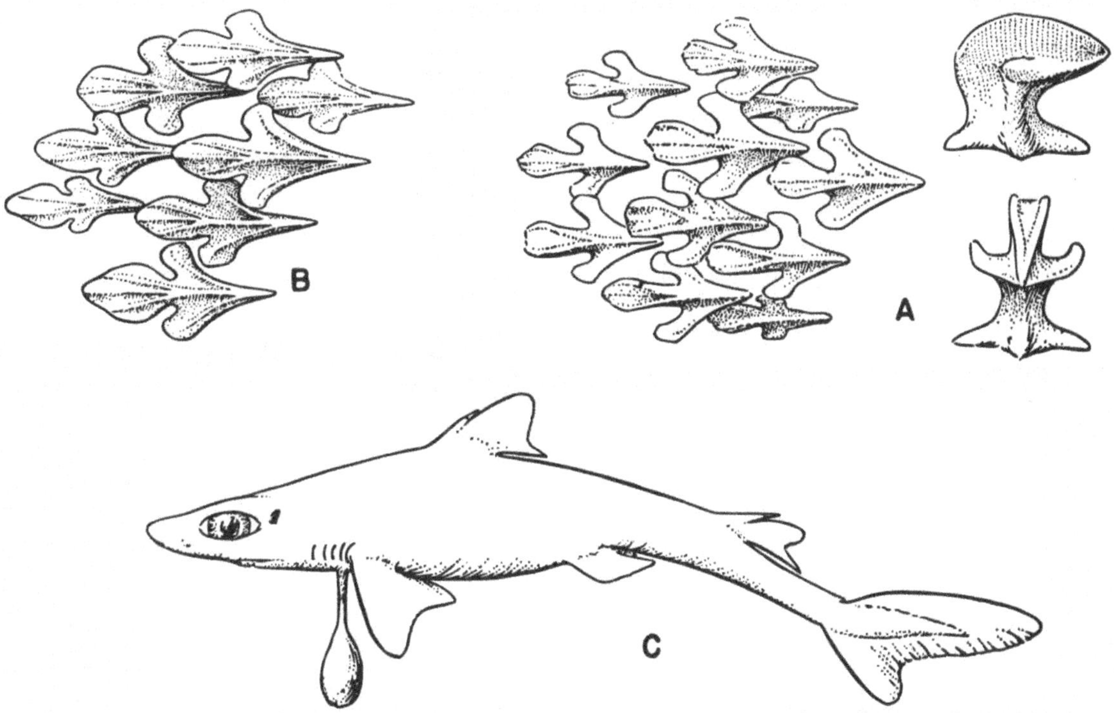

FIGURE 90. *A* Dermal denticles of *Squalus cubensis*, illustrated in Fig. 89, about 80 x. *B* Dermal denticles of adult female, from Cuba (Harv. Mus. Comp. Zool., No. 1458), about 70 x. *C* Embryo, about 203 mm. long, from near Havana, Cuba (Harv. Mus. Comp. Zool., No. 1459), about 0.6 x.

spaced, but not overlapping, varying considerably in shape on different parts of the body and at different ages; in the adult, those on upper sides along midsector of trunk lanceolate, sharp-tipped, with very high median crest dividing anteriorly into 2 or 3 definite ridges, and with a broad wing-like expansion either side opposite the median axis of the pedicel; farther forward, on the trunk, and also rearward and downward, these give place to flatter forms without lateral wings, those on top of snout being ovoid, those on anterior margins of pectoral and dorsal fins either ovoid or lanceolate, with the primary median ridge dividing anteriorly into 2 or 3; those on sides of caudal peduncle either evenly ovoid or weakly tridentate, with 3 separate longitudinal ridges; those on lower surface generally lanceolate, the median ridge either simple or dividing into 2 anteriorly.

Head a little more than ¼ (about 28%) of trunk to origin of caudal, its dorsal outline weakly convex to eye, but flat or slightly concave thence forward. Snout thicker at tip, more broadly ovate and relatively shorter than in *acanthias*, its length in front of nostrils about ⅓ its length in front of mouth, and the length in front of mouth only a little more than ⅓ (about 39 to 40%) of length of head. Eye oval, its outline more convex below than above, its horizontal diameter about ½ as long as snout in front of mouth (only about ⅓ that long in *acanthias*) or about as long as distance between nostrils. Spiracle about ¼ as long as horizontal diameter of eye, about on a level with upper margin of eye and behind latter by a distance about ⅓ as long as horizontal diameter of eye. Gill openings low down on the sides, the 5th slightly the longest, about ½ as long as horizontal diameter of eye and 1.3 times as long as 1st. Nostril very slightly oblique, a little more than ½ as long as horizontal diameter of eye, its inner end nearer to tip of snout than to mouth by a distance ⅔ to ¾ as long as horizontal diameter of eye, its inner margin expanded as a broad, subtriangular lobe, which may or may not have a small subsidiary lobe.[59] Mouth only very slightly arched, occupying about ⅔ of breadth of head. Upper labial furrow about ½ as long as diameter of eye and running obliquely forward, the lower furrow less than ½ as long as upper.

Teeth $\frac{13-0-13}{13-0-13}$, so closely resembling those of *acanthias* that the illustration (Fig. 89 C) is sufficient, 1 or 2 series functional, depending on the stage in the process of replacement.

First dorsal somewhat less sloping than in *acanthias*, its extreme length from origin to rear tip about twice its vertical height (about 2.3 times in *acanthias*), its vertical height about ⅓ as great as length of head, its origin a little anterior to midpoint of inner margins of pectorals, its spine reaching nearly to apex in male and apparently in female also, its anterior margin weakly convex, posterior margin only slightly concave, its apex broadly rounded, the free rear corner about as long as base or a little shorter, the midpoint of base only about ⅓ as far from axil of pectoral as from origin of pelvics. Interspace between 1st and 2nd dorsals about as long as from tip of snout to axil of pectoral. Second dorsal about

59. In the adult male specimen (the type), the left-hand nostril lacks this lobe, but it is present on both nostrils in the female specimen (Fig. 89 B).

4/5 as long at base as 1st and about 2/3 to 4/5 as high vertically, its posterior margin much more deeply concave, free rear corner about as long as base, the spine extending nearly or quite to apex in female as well as in male. Interspace between 2nd dorsal and caudal about twice as long as base of 2nd dorsal. Caudal about 1/5 of total length (as in *acanthias*), its tip perhaps averaging a little narrower than in *acanthias*, its lower lobe a little less than 2/3 (about 60%) as long as upper, the general posterior contour, as included by the 2 lobes, much as in *acanthias* (cf. Fig. 87 with 89). Pelvics about as long at base as 2nd dorsal, the posterior margins more nearly straight and posterior corners perhaps averaging a little less slenderly pointed than in *acanthias*, the midpoint of base in female nearer to origin of 2nd dorsal than to rear end of base of 1st dorsal by a distance only about 1/2 as long as horizontal diameter of eye and about midway in male (much nearer 2nd dorsal than 1st in both sexes of *acanthias*). Claspers of adult male not subdivided at tip (subdivided in *acanthias*, p. 459). Pectoral a little more than 2/3 (69 to 70%) as long as head (about 4/5 in *acanthias* or even a little more), its outer margin rather strongly convex toward apex, distal margin deeply and evenly concave (more deeply so than in *acanthias*), the apex rounded, but inner corner considerably produced and acutely angular, this being the most distinctive feature of the species.

Color. Dark gray above, paler gray below, the young paler than adult; described, when fresh-caught, as having upper lobes of both dorsals black, the caudal, pelvics and pectorals edged with white and the iris green; after preservation the fin markings are much stronger in the young than in the adults.

Size. The only available information is that the male listed above has large claspers, suggesting that maturity is reached at a length not much greater than 500 mm., *i.e.*, somewhat smaller than in the case of *acanthias*, although it may be as large as the latter at birth.

Developmental Stages. Probably the early stages in development (not yet described) correspond to those in *acanthias*. Embryos with a yolk sac already show the bilobed nostril, the shape of the pectoral, the relative positions of dorsal and pelvic fins and the long fin spines characteristic of the adult.

Remarks. This Shark was first described and unmistakably pictured as long ago as 1787 by Parra; although this Dogfish is very different from the northern *S. acanthias*, it was overlooked until 1936, when it was named by Howell-Rivero (see Synonyms, p. 478).

Habits. The fact that all specimens so far taken (about 20 in number) have been from depths greater than 75 fathoms suggests that it is a deep-water species. Nothing further is known of its habits, its breeding season or its diet.

Range. S. cubensis has been reported under this name only from the vicinity of Havana and Matanzas, Cuba. But no doubt most of the Cuban reports of *acanthias* actually refer to it,[60] for specimens of all sizes, including pregnant females, are caught commonly at mid-depths off the north coast of Cuba.[61] While none have yet been reported from the south coast, this is probably due merely to the failure of local fishermen to

60. See p. 463 on this point.

61. Personal communication by Luis Howell-Rivero.

report them, for they appear to be plentiful around Trinidad also.[62] It is to be expected along the northeast coasts of South America generally, for the Spiny Dogfish reported by Ribeiro for Rio de Janeiro appears to have been of this species.[63] No information is at hand regarding its distribution in the Gulf of Mexico, if it occurs there at all.

Synonyms and References:

Galludo, Parra, Desc. Piez. Hist. Nat. Havana, 1787: 84, pl. 34, fig. 1.

Squalus spec. dubia, Poey, Synop. Pisc. Cubens., Havana, 1868: 454 (spec. preserved in Havana and identified as *cubensis* by Howell-Rivero).

Acanthias acanthias Poey, Enumerat. Pisc. Cubens., 1876: 202; An. Soc. esp. Hist. Nat., *5*, 1876: 398, in part (Cuban spec., as above).

Squalus blainvillei Ribeiro, Ann. Mus. nac. Rio de J., *14*, 1907: 168; Fauna brasil. Peixes, *2* (1) Fasc. 1, 1923: 25, pl. 8 (descr. good ill.; ident. as *cubensis* by shape of pectoral, Rio de Janeiro).

Squalus acanthias Sanchez-Roig, Revist. Agric. Peces Cubana *9*, Commerc. Trabaj., 1931: 18, part (Cuba, based on Poey, 1876, not seen).

Squalus cubensis Howell-Rivero, Proc. Boston Soc. nat. Hist., *41* (4), 1936: 45, pl. 10, 11 (type descr., ill., adult and embryo, Cuba); Fowler, Fish Culturist, 21 (9), 1942: 66, fig. 9 (listed, Cuba); Bigelow and Schroeder, Guide Comm. Shark Fish., Anglo Amer. Caribb. Comm., Wash., 1945: 142, fig. 53 (descr., ill., range).

Probable References:

Squalus acanthias Jordan and Evermann, Bull. U.S. nat. Mus., 47 (1), 1896: 54 (probably in part, because reported from Cuba).

Squalus blainvillei Schreiner and Ribeiro, Ann. Mus. nac. Rio de J., *12*, 1903: 79 (name only, Rio de Janeiro, but probably same specs. described under this name by Ribeiro, 1907, 1923); Ribeiro, Pescas "Annie," Bol. Soc. Nat. Agric., Rio de J., 1904: 18 (near Ilha Rassa, Brazil).

Acanthias vulgaris Vincent, Sea Fish. Trinidad, 1910: 53 (common, mud banks, Trinidad).

Genus *Squalus*, Addendum

We include a brief account of *S. fernandinus*, present in southern Argentine waters, but which seems not to range farther north than that in the western South Atlantic.

Squalus fernandinus Molina, 1782

Figure 87 E–H

Study Material. Female, 914 mm. long, from Island of Juan Fernandez (Harv. Mus. Comp. Zool., No. 841).

Distinctive Characters. See Description.

Description. Proportional dimensions in per cent of total length. Female, 914 mm., from Juan Fernandez (Harv. Mus. Comp. Zool., No. 841).

Trunk at origin of pectoral: breadth 12.9; height 10.9.

Snout length in front of: outer nostrils 4.1; mouth 8.9.

62. *Acanthias vulgaris* reported by Vincent (Sea Fish. Trinidad, 1910: 53).

63. Reported by Ribeiro (Ann. Mus. nac. Rio de J., *14*, 1907: 168; Fauna brasil. Peixes, *2* [1] Fasc. 1, 1923: 25, pl. 8) as *S. blainville*, but probably *cubensis*, its pectoral being pictured as with acute inner corner and concave distal margin.

Eye: horizontal diameter 4.3.
Mouth: breadth 6.5; height 0.7.
Nostrils: distance between inner ends 4.6.
Labial furrow length: upper 2.6; lower 1.9.
Gill opening lengths: 1st 1.7; 2nd 2.1; 3rd 2.1; 4th 2.1; 5th 2.4.
First dorsal fin: vertical height 7.7; length of base 8.1.
Second dorsal fin: vertical height 4.6; length of base 6.5.
Caudal fin: upper margin 21.2; lower anterior margin 12.6.
Pectoral fin: outer margin 17.7; inner margin 10.0; distal margin 14.3.
Distance from snout to: 1st dorsal 29.8; 2nd dorsal 62.8; upper caudal 78.7; pectoral 20.2; pelvics 48.2.
Interspace between: 1st and 2nd dorsals 26.0; 2nd dorsal and caudal 10.4.

S. fernandinus is sharply marked off from the *acanthias* group by the positions of the first dorsal spine about over the midpoint of the inner margin of the pectoral (over or posterior to the inner corner of the pectoral in *acanthias*) and of the midpoint of the bases of the pelvics about midway between the two dorsals (much nearer to the origin of the second dorsal than to the rear end of the base of the first dorsal in *acanthias*); and by the more or less noticeably bilobed anterior margin of the nostril (cf. Fig. 87 F with 87 C).

It falls with *cubensis* in the relative positions of the fins and in the contour of the anterior margin of the nostril. But it differs very obviously from *cubensis* in that the distal margin of its pectoral is only very slightly concave (deeply concave in *cubensis*) and its inner corner rounded (acute in *cubensis*). Its teeth are indistinguishable from those of *acanthias*, but its dermal denticles differ noticeably.

S. blainville of the Mediterranean resembles it so closely that it has sometimes been considered a synonym of *fernandinus;* but it appears to be distinct for the reasons given on p. 454.

Color. Described as dark grayish or brown above, paler below, with the tips of the dorsals more or less dusky, the caudal blackish medianly, its lobes pale yellowish gray. *Fernandinus* lacks the white spots so characteristic of the *acanthias* group.

Range. Circumpolar and very widely distributed in boreal and cool temperate latitudes of the southern hemisphere (for localities, see p. 454); it is also represented in the Philippines and Japan by forms so closely allied that it is a question whether they are separable from it (see discussion, p. 454).

Occurrence in the Western Atlantic. S. fernandinus is listed for Argentina, from Lat. 56° S. to Lat. 35° S.,[64] but the only positive record for Atlantic Argentine waters with which we are acquainted is that of a small specimen 370 mm. long taken from the stomach of an albatross at Lat. 34° 44′ S., Long. 53° W.[65]

64. Pozzi and Bordale, Ann. Soc. cient. argent., 120, 1935: 151.
65. Lahille, An. Mus. nac. B. Aires, 34, 1928: 327; identification is made positive by the excellent illustrations that show in particular a very long second dorsal spine.

Synonyms:[66]

Squalus fernandinus Molina, Saggio Stor. Nat. Chili, 1782: 229 (Chile).

Squalus acanthias var. *fernandinus* Walbaum, P. Artedi Genera Pisc. Emend. Ichthyol., 1792: 505 (Chile).

Acanthorhinus fernandinus Blainville, Bull. Soc. philom. Paris, 1816: 121 (name only).

Acanthias blainvillii Gray, List. Fish. Brit. Mus., 1851: 70 (Cape Seas, S. Afr.); not *Acanthias blainville* Risso, 1826 (see discussion, p. 454).

Spinax fernandezianus Guichenot, in Gay, Hist. Chile, Zool., *2*, 1848: 365; Perez Canto, Estud. Escual. Chile, 1886: 10 (descr., Juan Fernandez).

Acanthias blainvillei Bleeker, Natuurk. Tijdschr. Ned. Ind., *21*, 1860: 50, 58, 80 (Cape of Good Hope); not *Acanthias blainville* Risso, 1826 (see discussion, p. 454).

Acanthias fernandezianus Philippi, Tiburones Chile, 1887: 27, pl. 4, fig. 3; An. Univ. Chile, *71*, 1887: 559, pl. 4, fig. 3 (descr., ill., Juan Fernandez).

Squalus blainvillei Delfin, Rev. Chil. Hist. Nat., *4*, 1900: 110 (Chile); not *S. blainville* Ribeiro, Fauna brasil., Peixes, Mus. nac. Rio de J., *2* (1) Fasc. 1, 1923: 25, pl. 8 (Brazil); equals *Squalus cubensis* Howell-Rivero, see p. 478.

Squalus fernandezianus Delfin, Cat. Peces Chile, 1901: 21 (refs., Juan Fernandez).

Squalus acutipinnis Regan, Ann. Natal Mus., *2*, 1908: 248, pl. 37 (descr., ill., Mauritius, S. Afr.).

Squalus tasmaniensis Howell-Rivero, Occ. Pap. Boston Soc. nat. Hist., *8*, 1936: 267, pl. 10, fig. a–e (descr., ill., Tasmania).

Doubtful Synonyms:

Squalus philippinus Smith and Radcliffe, Proc. U.S. nat. Mus., *41*, 1912: 677, pl. 51, fig. 1 (descr., ill., Philippines); not *S. philippinus* Shaw, which is a Heterodont.

Squalus mitsukurii Jordan and Fowler, Proc. U.S. nat. Mus., *26*, 1903: 630 (descr., but not the ill., fig. 3, Japan; see discuss., footnote 20, p. 454).

Squalus japonicus Ishikawa, Proc. Acad. nat. Sci. Philad., *60*, 1908: 71; Tanaka, Fish. Japan, *26*, 1917: pl. 130, fig. 365–367 (Japan).

Squalus montalbani Whitley, Aust. Zool., *6*, 1931: 310 (substitution for *S. philippinus* Smith and Radcliffe, 1912, preoccupied by Shaw, 1804, for a Heterodont).

References for Western South Atlantic:

Squalus fernandinus Lahille, Enum. Peces Cartilag. Argent., 1921: 16 (listed for Argentina); Physis B. Aires, *5*, 1921: 63 (listed for Argentina); Ann. Mus. nac. B. Aires, *34*, 1928: 327 (descr. and good ill. of spec. from stomach of an albatross, Lat. 34° 44′ S., Long. 53° W.); Pozzi and Bordale, Ann. Soc. cient. argent., 120, 1935: 151 (listed for Argent., Lat. 35° S. to 56° S.); not *Squalus fernandinus* Fowler, Arch. Zool. Estado São Paulo, *3*, 1941: 128 (Brazil, by ref. to *S. blainvillei* Schreiner and Ribeiro, 1903, which appears to have been *cubensis*; see discussion, p. 478).

Probable Reference:

Squalus blainvillei Lahille, Enum. Peces Cartilag. Argent., 1921: 16 (listed for Argentina).

Genus *Centroscyllium* Müller and Henle, 1841

Centroscyllium Müller and Henle, Plagiost., 1841: 191: type species, *Spinax fabricii* Reinhardt, 1825, monotypic. West Greenland.

66. *Squalus megalops* Macleay (Proc. Linn. Soc. N. S. W., *6*, 1881: 367 and subsequent authors, Aust.) is referred by Garman (Mem. Harv. Mus. comp. Zool., *36*, 1913: 195) and by Fowler (Bull. U.S. nat. Mus., 100 [*13*], 1941: 261) to the synonymy of *fernandinus*, but it appears to belong to another division of the genus; see discussion, p. 454.

Generic Synonym:

Paracentroscyllium Alcock, Ann. Mag. nat. Hist., (6) *4*, 1889: 379; type species, *P. ornatum* Alcock, monotypic. Bay of Bengal.

Generic Characters. Squalidae with well developed dorsal spines, their anterior edge sharp, posterior flattened, each side with two shallow longitudinal furrows, originating at origins of fins and lying along anterior margins of latter, their tips well exposed; trunk slender, subcylindrical; caudal peduncle with or without faintly marked longitudinal dermal ridges and without precaudal pits; teeth (most distinctive generic character) similar in the 2 jaws, with 3 to 5 cusps, the median much the largest; snout in front of mouth shorter than from center of mouth to origin of pectorals; dermal denticles thorn-like on stellate bases, or granular; eyes and spiracles large; nostrils oblique, their anterior margins expanded as triangular lobes; mouth but little arched, a voluminous triangular pit at corner of mouth extending as a short labial furrow on each jaw; posterior margins of dorsals weakly concave, if at all so, the free rear corners not slenderly elongate; origin of 1st dorsal behind axil of pectoral; 2nd dorsal at least as large as 1st, its origin over or behind bases of pelvics; caudal with only weakly defined lower lobe, its axis raised only a little; pectoral brush-shaped, its inner corner rounded and not at all produced; some species, at least, with dermal thickenings that are presumably luminous.

Range. North Atlantic; Falkland Islands; eastern tropical Pacific; vicinity of Hawaiian Islands; Japan; Indian Ocean, including Bay of Bengal and Arabian Gulf.

Species. The few species so far known are deep-water forms from widely distributed localities which closely resemble one another in general appearance and in their black or very dark brown coloration. While closely allied, they appear sufficiently differentiated from each other to deserve separate names.

Key to Species[1]

1a. First dorsal spine over or in front of tips of pectorals when latter are laid back.

2a. Tip of caudal truncate, with definite subterminal notch. *nigrum* Garman, 1899. Eastern Pacific off Panama; near Galapagos Islands; near Cocos Island and off Hawaiian Islands.[2]

2b. Tip of caudal pointed, without definite subterminal notch. **ornatum* Alcock, 1889. Bay of Bengal and Arabian Gulf.[3]

1. Species marked * have not been seen by us.
2. For list of references and localities of captures, see Fowler (Bull. U.S. nat. Mus., 100 [*13*], 1941: 252) and Beebe and Tee-Van (Zoologica, N. Y., *26*, 1941: 120).
3. It is probable that Burckhardt's (Ann. Mag. nat. Hist., [7] *6*, 1900: 567, fig. 7) figure more nearly represents the normal shape of the caudal than does Alcock's (Ill. Zool. "Investigator," 1894: pl. 8, fig. 3) original illustration of it. For list of references, see Fowler (Bull. U.S. nat. Mus., 100 [*13*], 1941: 254).

1b. First dorsal spine behind tips of pectorals when latter are laid back.

3a. Interspace between 2nd dorsal and caudal as long as from eye to origin of pectoral, or longer.

4a. Second dorsal spine over midbase of pelvics, and extending considerably beyond apex of 2nd dorsal fin; interspace between 2nd dorsal and caudal about as long as from eye to axil of pectoral. **granulosum* Günther, 1880.
Falkland Islands.[4]

4b. Second dorsal spine posterior to rear ends of bases of pelvics, and extending only ½ to ⅔ the distance toward apex of 2nd dorsal; interspace between 2nd dorsal and caudal only about as long as from eye to origin of pectoral.
ritteri Jordan and Fowler, 1903.
Japan.

3b. Interspace between 2nd dorsal and caudal only about as long as from eye to 1st gill opening. *fabricii* Reinhardt, 1825, p. 482.

Centroscyllium fabricii (Reinhardt), 1825

Black Dogfish

Figure 91

Study Material. Two females, 640 and 727 mm. long, one from the continental slope off Browns Bank, the other from off Nova Scotia in 300 fathoms (Harv. Mus. Comp. Zool., No. 35702 and 743).

Distinctive Characters. Among West Atlantic members of its family this species falls with *Squalus acanthias*, *S. cubensis* and *Etmopterus hillianus* in its long and very prominent fin spines. But it is easily separable from the first two by its teeth, which have several cusps, by its dermal denticles, and by its color. It differs most obviously from *Etmopterus hillianus* in that its lower teeth have several cusps like the uppers, and that its second dorsal originates over the bases of the pelvics.

Description. Proportional dimensions in per cent of total length. Female 640 mm., from 240 miles ESE. of Boston Lightship (Harv. Mus. Comp. Zool., No. 35702). Female, 727 mm., from off Nova Scotia (Harv. Mus. Comp. Zool., No. 743).

Trunk at origin of pectoral: breadth 11.8, 11.8; height 7.8, 8.2.
Snout length in front of: outer nostrils 1.4, 1.8; mouth 7.5, 7.4.
Eye: horizontal diameter 4.7, 4.3.
Mouth: breadth 8.1, 8.2; height 3.3, 3.7.
Nostrils: distance between inner ends 3.9, 4.2.
Labial furrow length from angle of mouth: upper 1.9, 2.2; lower 1.9, 1.9.
Gill opening lengths: 1st 3.1, 3.0; 2nd 3.1, 3.0; 3rd 3.1, 3.0; 4th 3.1, 3.0; 5th 2.7, 2.6.

4. Diagnosis based on Burckhardt's (Ann. Mag. nat. Hist., [7] *6*, 1900: 567, fig. 8) illustration of the type specimen.

First dorsal fin: vertical height 3.9, 4.0; length of base 9.8, 12.1.
Second dorsal fin: vertical height 5.3, 5.5; length of base 11.2, 11.0.
Caudal fin: upper margin 24.2, 20.9; lower anterior margin 13.1, 11.8.
Pectoral fin: extreme length 9.7, 10.6; extreme breadth 5.5, 5.8.
Distance from snout to: 1st dorsal 30.5, 27.2; 2nd dorsal 58.6, 58.0; upper caudal 75.8, 79.1; pectoral 18.8, 21.3; pelvics 54.7, 55.3.
Interspace between: 1st and 2nd dorsals 18.0, 19.5; 2nd dorsal and caudal 7.8, 8.8.
Distance from origin to origin of: pectoral and pelvics 35.2, 35.7; pelvics and caudal 18.3, 22.0.

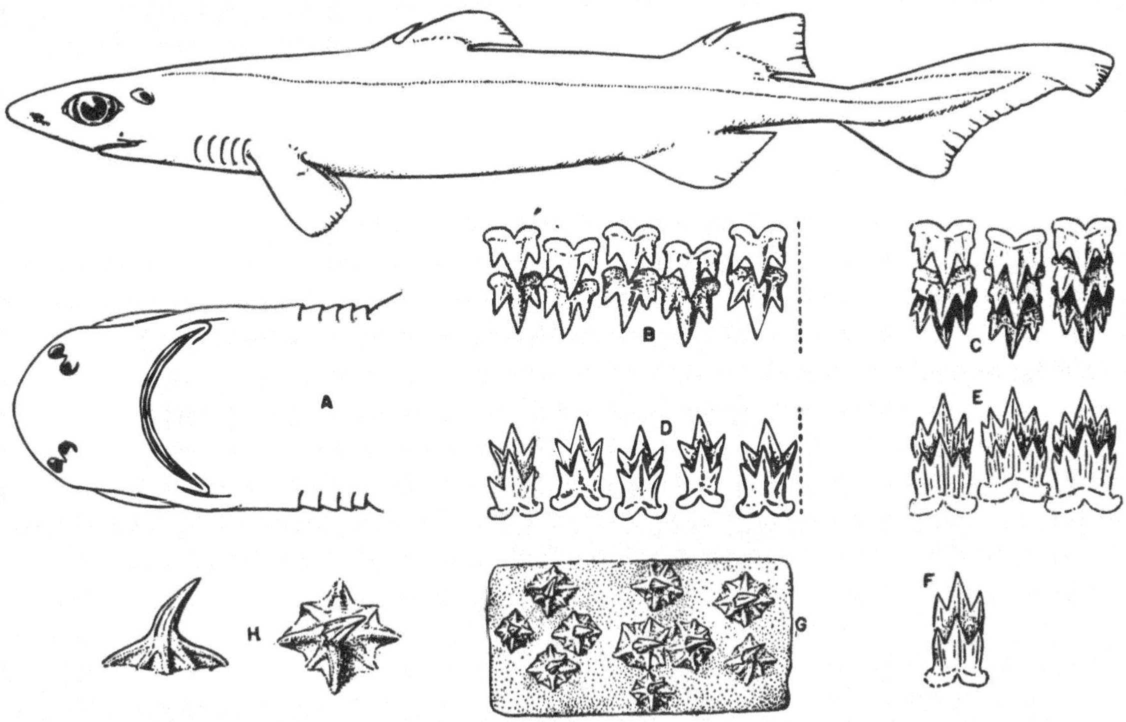

FIGURE 91. *Centroscyllium fabricii*, female, 640 mm. long, from 240 miles ESE. of Boston Lightship (Harv. Mus. Comp. Zool., No. 35702). *A* Head from below, about 0.3 x natural size. *B* First to fifth upper teeth, right-hand side. *C* Twentieth to twenty-second upper teeth. *D* First to fifth lower teeth. *E* Sixteenth to eighteenth lower teeth. *F* Twenty-sixth lower tooth. *B–F*, about 5.4 x. *G* Dermal denticles, about 10 x. *H* Lateral and apical oblique views of denticle, about 20 x.

Trunk slender, its height at 1st dorsal only a little more than 1/6 its length to origin of caudal. Body sector to cloaca about 1.8 times as long as tail sector, without mid-dorsal ridge. Caudal peduncle without longitudinal ridges or precaudal pits either above or below. Dermal denticles minute and very widely spaced, leaving most of the skin bare, on irregularly stellate bases, thorn-like and more or less recurved on the trunk generally, somewhat widest on the tail, those on top of head conical. Skin along upper sides and on top of head with widely scattered and deeply pigmented epidermal thickenings or dots, presumably luminous (p. 485).

Head between ⅓ and ¼ (about 28%) of length to origin of caudal, its dorsal profile weakly and evenly convex. Snout thick, fleshy, broadly-ovate, its length in front of nostrils only about ¼ of length in front of mouth, the distance between nostrils about ½ as long as snout in front of mouth, and length in front of mouth a little more than ⅓ (about 35 to 36%) of length of head. Eye oval, its lower outline the more convex, its horizontal diameter about ⅔ as long as snout in front of mouth or a little longer than distance between nostrils. Spiracle about ⅓ to ¼ as long as horizontal diameter of eye, behind latter by a distance ⅓ to ½ as long as diameter of eye, the lower edge a little above center of eye. Nostril near anterior contour of snout, moderately oblique and about ½ as long as diameter of eye, its anterior margin produced as a well developed, narrow-triangular lobe, its inner corner about midway between mouth and tip of snout. Mouth about ⅖ as high as broad, nearly evenly arcuate. Pit at corner of mouth extending in deep furrows about ¼ of the distance toward symphysis on upper jaw and nearly as far on lower, but only a very short distance rearward.

Teeth about $\frac{34-0-34}{34-0-34}$ in specimen examined (counting difficult), the uppers and lowers similar, with 3 (sometimes 4 or 5) sharp cusps, the median much the longest; 2 to 3 series functional in each jaw.

First dorsal brush-shaped, noticeably small, its vertical height only ⅙ to ⅐ as great as length of head, its anterior margin very sloping, apex very broadly rounded, posterior margin nearly straight, its free rear corner only a little more than ½ as long as base, its origin a little posterior to tips of pectorals (when latter are laid back), the 1st dorsal spine a little more than ⅓ as far from axils of pectorals as from origin of pelvics, with the exposed portion about ½ as long as horizontal diameter of eye and reaching about ⅓ the way along anterior margin of fin. Second dorsal about 1.2 times as long at base as 1st and about twice as large in area, its general contour subrectangular (rounded in 1st dorsal), with nearly straight anterior margin, slightly concave posterior margin and very narrowly rounded apex, its free rear corner a little less than ½ as long as base and only a little more acute than a right angle, its spine a little anterior to midpoint of bases of pelvics, free for about ½ its length, and reaching about midway along anterior margin of fin. Interspace between 2nd dorsal and caudal only about ¾ as long as base of 2nd dorsal. Caudal a little less than ¼ of total length, its upper margin convex, apex broadly truncate, and lower posterior margin with obtuse subterminal notch, thus marking off the terminal sector, its lower anterior corner rounded, only very slightly expanded (not a definite lobe), its anterior margin approximately ½ as long as upper margin of fin. Pelvics about as large as 2nd dorsal, their anterior margins slightly convex, apices broadly rounded, posterior margins nearly straight, tips pointed, their origin a little anterior to origin of 2nd dorsal. Pectoral small, a little less than ½ as long as head, brush-shaped, with transversely truncate tip, rounded corners, the inner not at all produced.

Color. Deep chocolate brown, darkest (almost black) below and on fins generally; inner edge of anterior part of upper eyelid densely pigmented with dark brown or blackish.

Size. Reported by fishermen as growing to about 3½ feet, but adults average only about 2 to 2½ feet, the maximum length of which we find definite record being only about 33 inches (829 mm.). Females average larger than males, at least in Greenland waters.

Developmental Stages. Presumably ovoviviparous like *Squalus*, but the early developmental stages have not been described, although embryos have been reported repeatedly.

Habits. Captures of the Black Dogfish in West Greenland waters range from close to the surface through the ice in winter down to 900 meters at least. Along the Nova Scotian Banks they are most often taken at 200 to 300 fathoms and seldom shoaler than 150 fathoms, often in company with *Centroscymnus coelolepis* (p. 498). Although no regular hook and line fishery is operated there at a greater depth, there is no reason to doubt that they occur as deep in those waters as off Greenland (see above). Records for Icelandic waters (perhaps the chief center of abundance) are mainly from about 500 to about 900 meters; they are recorded down to 1,100 meters off the Faroes, to 1,495 meters off northwest Africa, if the reports of the captures there are well founded (see footnote 5, p. 485).

This, in short, is a deep-water species, normally approaching the surface only in Arctic latitudes and at the coldest, or perhaps darkest, season. The fact that a trawl haul on the south slope of the Davis Strait ridge (in water of 3.12° C.) yielded 42 specimens, while another just north of the ridge took none (from water of 2.47°) is suggestive evidence that it is an inhabitant of Atlantic and not of truly polar waters.

Nothing is known of its life history other than that it bites the hook freely. Cephalopods, pelagic crustaceans, and medusae (*Atolla*) have been found in its stomach, and females containing embryos up to 124 mm. in length have been taken in West Greenland waters in February. The fact that their skins are provided with minute, deeply pigmented papillae, resembling the luminous organs of the brightly luminescent *Isistius brasiliensis* (p. 512), suggests that *C. fabricii* also emits light, although it has not actually been seen to do this so far as we know.

Relation to Man. The Black Dogfish is of no commercial value, hence most of those caught are thrown back.

Range. Both sides of the North Atlantic, chiefly in depths greater than 150 fathoms; Iceland (where most plentiful); Faroe-Shetland Channel and Faroe Bank in the east; also reported from Arguin Bank off Cape Blanco, Northwest Africa;[5] Davis Strait; West Greenland slopes and outer parts of the offshore fishing grounds in the west, from the Grand Banks of Newfoundland to Georges Bank.

Occurrence in the Western Atlantic. To the north the Black Dogfish is at least tolerably common in southwestern Greenland waters, both in the fjords and on the offshore fishing banks; in the southern part of Davis Strait, and northward to the ridge that con-

5. Doubt has been expressed as to the actual identity of the specimen from this locality reported by Vaillant (Exped. Sci. "Travailleur" and "Talisman," Poiss., 1888: 72), since it was not only very small (175 mm.) but in a bad state of preservation.

nects Greenland with Baffin Land. But it has not been found on the northern (polar) slope of the ridge.

It is also taken often by halibut fishermen on long lines all along the edge of the North American continent from the south slope of the Grand Bank southward along the Nova Scotian Banks (Banquereau, Sable Island, Lahave, Browns) to the eastern extremity of Georges Bank. But neither the total number of recorded captures nor the personal reports we have received from fishermen suggest more than a comparatively sparse though widespread population. How far to the westward and southward it may occur regularly is not yet known. Garman[6] credited it with ranging as far as New York, but we find no supporting evidence for this, and while Goode and Bean[7] characterized a young specimen taken in the northern side of the Gulf of Mexico, off the coast of Alabama, in 1885, as "probably this species," the bottom temperature of the water there was so high (67° F., or 19.4° C. at 210 fathoms) as to suggest that some other small shark actually was in hand.

Synonyms and References:

Squalus acanthias Fabricius, Fauna Groenl., 1780: 126 (W. Greenland); not *S. acanthias* Linnaeus, 1758.

Spinax fabricii Reinhardt, Overs. danske Vidensk-Selsk. Forh. (1824–1825), 1825: 3 (W. Greenland, not seen); Overs. danske Vidensk-Selsk. Forh., *3*, 1828: vi (descr., West Greenland, not seen); K. danske Selsk. Naturvid. Math. Afhand., 7, 1838: 116, 132 (W. Greenland).

Spinax acanthias (?) Reinhardt, K. danske Selsk. Naturvid. Math. Afhand., 7, 1838: 116 (W. Greenland, suggests *Squalus acanthias* Fabricius, 1780, equals *Spinax fabricii* Reinhardt, 1828).

Centroscyllium fabricii Müller and Henle, Plagiost., 1841: 191 (descr., Greenland); Krøyer, Danmarks Fisk., *3*, 1852–1853: 888 (note); Gill, Proc. Acad. nat. Sci. Philad., Addend., 1861: 60 and later eds. (name); Duméril, Hist. Nat. Poiss., *1*, 1865: 449, pl. 5, fig. 10 (embryo reported, descr., ill. of scales, Greenland); Günther, Cat. Fish. Brit. Mus., *8*, 1870: 425 (descr., refs., Greenland); Goode and Bean, Bull. Essex Inst. Salem, *11*, 1879: 30 (Banks off Massachusetts); Jones, List Fish. Nova Scotia, 1879: 9 (meas., Nova Scotia Banks); Bean, Proc. U.S. nat. Mus., *3*, 1881: 116 (locs. of capture, abund., Grand Banks and Nova Scotia Banks); Hasse, Naturl. Syst. Elasm. besond. Theil, 1882: pl. 10 (vertebrae); Jones, Proc. N. S. Inst. Sci., *5*, 1882: 95 (same as Jones, 1879); Jordan and Gilbert, Bull. U.S. nat. Mus., 16, 1883: 16 (descr., off Gloucester, Massachusetts); Bean, Rep. U.S. Comm. Fish. (1882), 1884: 344 (off Woods Hole); Günther, "Challenger" Rep., Zool., *22*, 1887: XL, 6 (depths of capture); Jordan, Rep. U.S. Comm. Fish. (1885), 1887: 793 (distrib.); Vaillant, Exped. Sci. "Travailleur" and "Talisman," Poiss., 1888: 72 (off Arguin Bank, Morocco, Lat. about 20° 44′ N.); Goode and Bean, Smithson. Contr. Knowl., *30*, 1895; Mem. Harv. Mus. comp. Zool., *22*, 1896: 11, pl. 2, fig. 7 (descr., ill., offshore banks, and perhaps Gulf of Mexico); Jordan and Evermann, Rep. U.S. Comm. Fish. (1895), 1896: 219 (name, Nova Scotia Banks); Bull. U.S. nat. Mus., 47 (1), 1896: 56 (descr., off Gloucester and Nova Scotia Banks); *ibid.*, 47 (4), 1900: pl. 8, fig. 26 (ill.); Burckhardt, Ann. Mag. nat. Hist., (7) *6*, 1900: 568 (luminescent organs); Werner, Zool. Jb., Syst. Abt., *21*, 1904: 291 (color, embryo, Greenland); Collett, Rep. Norweg. Fish. Invest., *2* (2), 1905: 25 (sizes, embryos, off the Faroes); Fowler, Proc. Acad. nat. Sci. Philad., *60*, 1908: 69 (Georges Bank); Kendall, Occ. Pap. Boston Soc. nat. Hist., 7 (8), 1908: 9 (off Gloucester, Massachusetts); Regan, Ann. Mag. nat. Hist., (8) *2*, 1908: 40 (taxonomic discus., cf. with other species); Murray and Hjort, Depths of Ocean, 1912: 392 (Faroe-Shetland Channel and Faroe Bank); Garman, Mem. Harv. Mus. comp. Zool., *36*, 1913: 231, pl. 10, fig. 5–8 (descr., ill., Greenland to N. York); Halkett, Check List. Fish. Canad., 1913: 41 (off Nova Scotia); Jensen, Mindskr. Jap. Steenstrup, *2* (30), 1914: 4 (discuss., history, size, embryos, food, depth, south. Davis Strait and SW. Greenland); Fowler, Copeia,

6. Mem. Harv. Mus. comp. Zool., *36*, 1913: 232.
7. Smithson. Contr. Knowl., *30*, 1895: 11; Albatross Station 2377.

30, 1916: 36 (in list); Saemundsson, Vidensk. Medd. naturh. Foren. Kbh., *74*, 1922: 167 (depth, occur. Iceland); Bigelow and Welsh, Bull. U.S. Bur. Fish., *40* (1), 1925: 52, 551 (diagn., ill., general, in W. Atlant.); Jensen, Rapp. Cons. explor. Mer., *39*, 1926: 98, 101 (south side of Davis Strait ridge); Breder, Field Bk. Mar. Fish. Atlant. Coast, 1929: 25 (general); Jordan, Evermann and Clark, Rep. U.S. Comm. Fish. (1928), 2, 1930: 21 (listed, Greenland seas, Gloucester, Nova Scotia Banks); MacCoy, Bull. Boston Soc. nat. Hist., 69, 1933: 8 (240 miles ESE. from Boston Lightship); Bigelow and Schroeder, Canad. Atlant. Fauna, biol. Bd. Canad., *12*[e], 1934: 18 (general); Vladykov and McKenzie, Proc. N. S. Inst. Sci., *19*, 1935: 49 (off Nova Scotia); Fowler, Bull. Amer. Mus. nat. Hist., *70* (1), 1936: 83 (descr., ill. of Georges Bank spec., discuss. of ident. of Morocco spec. rep. by Vaillant, 1888); Lübbert and Ehrenbaum, Handb. Seefisch. Nordeurop., 2, 1936: 287 (Iceland).

Not *Centroscyllium fabricii* Lahille, Enum. Peces Cartilag. Argent., 1921: 16 (Falkland Is.); Pozzi and Bordale, An. Soc. cient. argent., 120, 1935: 10 (Argentina, Lat. 52° S.); these doubtless refer to the type specimen of *C. granulosum* Günther, 1880.

Genus *Etmopterus* Rafinesque, 1810

Etmopterus Rafinesque, Carratt. Gen. Nuov. Sicil., 1810: 14; type species, *E. aculeatus* Rafinesque, 1810, equals *Squalus spinax* Linnaeus, 1758.

Generic Synonyms:

Acanthorhinus (in part) Blainville, Bull. Soc. philom. Paris, 1816: 121.

Spinax Cuvier, Règne Anim., *2*, 1817: 129; type species, *Squalus spinax* Linnaeus, 1758.

Acanthias (in part) Risso, Hist. nat. Europ. Merid., *3*, 1826: 132.

Centrina (in part) Lowe, Proc. zool. Soc. Lond., 1833: 144; type species, *C. nigra* Lowe, equals *Squalus spinax* Linnaeus, 1758; not *Centrina* Cuvier, 1817.

Acanthidium Lowe, Proc. zool. Soc. Lond., 1839: 91; type species, *A. pusillum* Lowe, 1839, designated by Goode and Bean (Smithson. Contr. Knowl., *30*, 1895: 10; Mem. Harv. Mus. comp. Zool., *22*, 1896: 10[1]).

Acanthidim Sollas, Zool. Rec., *43*, 1906: 58 (obvious misprint for *Acanthidium*).

Generic Characters. Squalidae with dorsal spines largely exposed, arising at origins of fins and lying along anterior margins of latter; trunk slender, subcylindrical, the peduncle without lateral ridges or precaudal pits; snout in front of mouth is somewhat shorter than from front of mouth to origin of pectorals; upper and lower teeth unlike, the former with several cusps, the latter with only 1 cusp, deeply notched outwardly and so oblique as to form a continuous cutting edge; dermal denticles ranging from bristle-like to scale-like; eyes and spiracles large; anterior margin of nostrils with a long, narrow-triangular lobe; a voluminous triangular pit at corner of mouth and a labial furrow on each jaw; dorsals triangular, their free rear corners elongate but not very slender; origin of 1st dorsal posterior to tips of pectorals; 2nd dorsal larger than 1st, its origin over or posterior to base of pelvics; caudal truncate, with well marked subterminal notch, but without definite lower anterior lobe, its axis only very slightly raised; pectorals brush-shaped, their inner corners rounded and not at all produced; inner edge of anterior part of upper eyelid deeply pigmented; some of the species, perhaps all of them, with luminous organs.[2]

1. We agree with Fowler (Bull. U.S. nat. Mus., 100 [*13*], 1941: 246) that *Acanthidium* Lowe is a synonym of *Etmopterus* Rafinesque, and that the species grouped under *Acanthidium* by Garman (Mem. Harv. Mus. comp. Zool., *36*, 1913: 215) fall properly in *Deania* Jordan and Snyder, 1902. See also footnote 4, p. 451.
2. For a general account of these organs, see Daniel (Elasmobranch Fishes, Univ. Calif. Press, 1934: 29).

Range. Both sides of North Atlantic, Mediterranean, South Africa, Straits of Magellan, southwest coast of South America, East Indies, Philippines, Japan and Hawaiian Islands.

Fossil Teeth. Upper Cretaceous (?) and Miocene, Europe.

Species. The ten or eleven supposed species of these small, deeply pigmented deep water sharks that have been named[3] resemble one another so very closely that a drastic reduction in the number of recognizable species is to be anticipated. But since we lack adequate material from other ocean areas for comparison we limit the accompanying Key to the North Atlantic representatives of the genus.

Key to North Atlantic Species

1a. Interspace between rear end of bases of pelvics and origin of caudal as long as distance from origin of pelvics to tips of pectorals, or longer. *hillianus* Poey, 1861, p. 488.

1b. Interspace between rear end of bases of pelvics and origin of caudal considerably shorter than distance from origin of pelvics to tips of pectorals.

  2a. Dermal denticles bristle- or thorn-like; caudal a little longer than from tip of snout to origin of pectorals. *spinax* Linnaeus, 1758.[4]
  Eastern Atlantic, Mediterranean, South Africa.

  2b. Dermal denticles scale-like; caudal considerably shorter than from tip of snout to origin of pectorals. *pusillus* Lowe, 1839.
  Eastern Atlantic, also Japan, or represented there by a very close ally.[5]

Etmopterus hillianus (Poey), 1861

Figures 92, 93

Study Material. Type specimen, 269 mm. long, from Cuba (Harv. Mus. Comp. Zool., No. 1025); males, 251 and 270 mm. long, from off St. Kitts, West Indies in 208

3. *Brachyurus* Smith and Radcliffe, 1912, Philippines; *frontimaculatus* Pietschmann, 1907, Japan; *granulosus* Günther, 1880, southwest coast of South America; *hillianus* Poey, 1861, western North Atlantic, Florida region; *lucifer* Jordan and Snyder, 1902, Japan, Philippines, East Indies, Natal; *molleri* Whitley, 1939, Australia; *paessleri* Lönnberg, 1907, Straits of Magellan, Argentina; *princeps* Collett, 1904, vicinity of the Faroes (almost certainly a synonym of *spinax*); *pusillus* Lowe, 1839, eastern Atlantic; *villosus* Gilbert, 1905, Hawaiian Islands; and *spinax* Linnaeus, 1758, eastern Atlantic, Mediterranean, South Africa.

4. Including *princeps* Collett, 1904. This was thought by Collett (Forh. Vidensk.-Selsk. Krist., 9, 1904: 3) to be separable from *spinax* because of its somewhat stouter and more thorn-like denticles; but we doubt the validity of this supposed species, based on poorly preserved material.

5. The forms described under this name by Tanaka (Fish. Japan, *5*, 1912: pl. 22; *6*, 1912: 88) and as *E. frontimaculatus* by Pietschmann (Anz. Akad. Wiss. Wien, *44*, 1907: 395; S. B. Akad. Wiss. Wien, 117, 1908: 654, pl. 1, fig. 2, pl. 2, fig. 2), both from Japan, agree with the East Atlantic *pusillus* in the form of the denticles. Whether or not they are actually identical with the latter, as classed by Garman (Mem. Harv. Mus. comp. Zool., *36*, 1913: 228) and Fowler (Bull. U.S. nat. Mus., 100 [*13*], 1941: 249), can be determined only by comparison of specimens from the respective ocean areas.

fathoms, and off northwestern Cuba at Lat. 23° 12′ N., Long. 81° 23′ W. in 375 fathoms (Harv. Mus. Comp. Zool., No. 1024, 1025); also 16 specimens, 225 to 295 mm. long, taken off the north central and northeastern coast of Cuba by the "Atlantis" in March and April 1938, including a female (295 mm.) containing four embryos about 90 mm. long

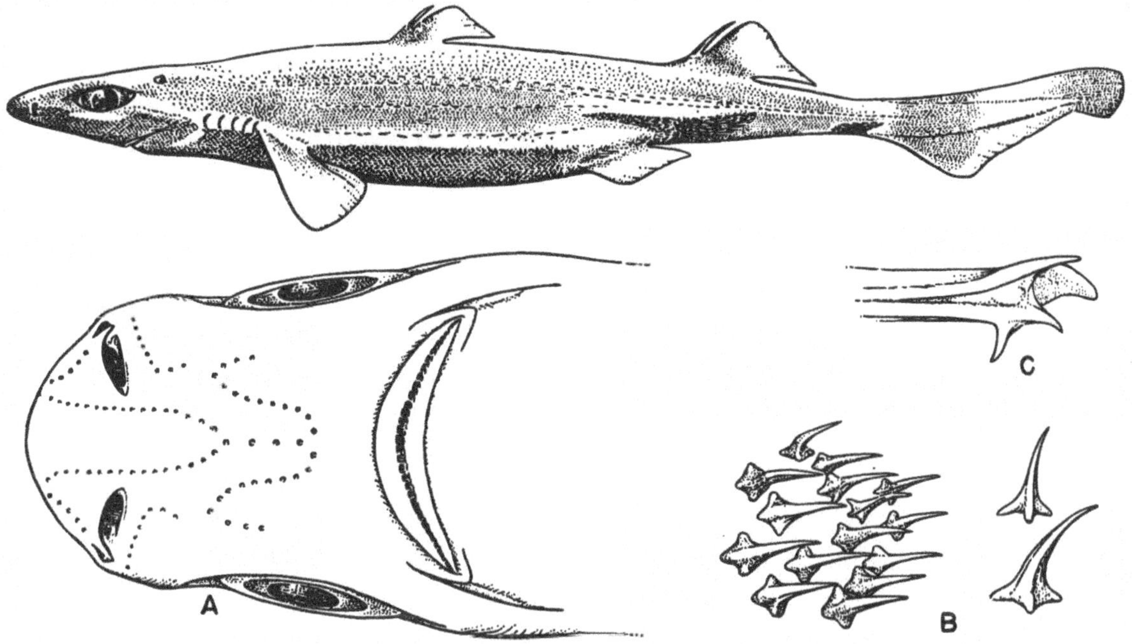

FIGURE 92. *Etmopterus hillianus*, female, 292 mm. long, from off northwestern Cuba (Harv. Mus. Comp. Zool., No. 36112). *A* Anterior part of head from below, about 1.5 x. *B* Dermal denticles, about 30 x. *C* Tip of left-hand clasper of an adult male, 270 mm. long, from dorsal side (Harv. Mus. Comp. Zool., No. 36104).

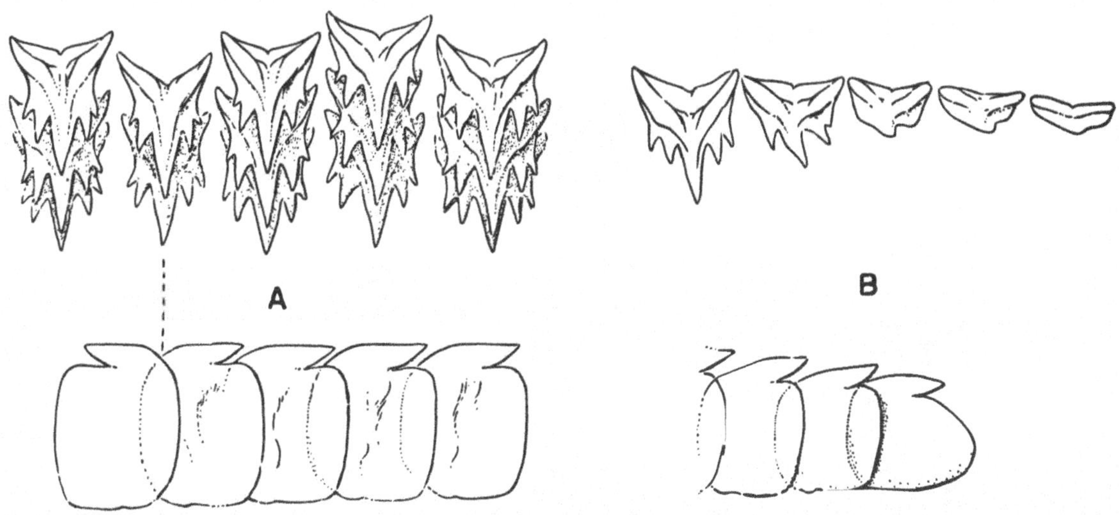

FIGURE 93. *Etmopterus hillianus*, pictured in Fig. 92. *A* Upper and lower teeth at center of mouth. *B* Eighth to twelfth (outermost) upper teeth, and sixteenth to eighteenth (outermost) lower teeth, about 10 x.

apparently ready for birth, and a male, 270 mm. long, with the trifid tips of the claspers fully differentiated (Harv. Mus. Comp. Zool., No. 36104 to 36116); a specimen 214 mm. from the offing of Chesapeake Bay, Lat. 37° 24′ N., Long. 74° 17′ W. in 300 fathoms (U.S. Nat. Mus., No. 26740).

Distinctive Characters. Among northwestern Atlantic members of the family this species falls with *Squalus acanthias*, *S. cubensis* and *Centroscyllium* in its long and conspicuous fin spines. But it is easily distinguishable from all these by the striking dissimilarity of the teeth in its two jaws. It has sometimes been confused with *E. pusillus* of the eastern Atlantic, but it is distinguishable from the latter at a glance by the fact that the interspace between its pelvics and its caudal is at least as long as the distance from the origins of the former to the tips of pectorals (considerably shorter than this in *pusillus* and *spinax*). Its relatively slender caudal further separates it from *spinax*.

Description. Proportional dimensions in per cent of total length. Male, 225 mm., from north coast of Cuba (Harv. Mus. Comp. Zool., No. 36105). Female, 230 mm., same locality (Harv. Mus. Comp. Zool., No. 36104).

Trunk at origin of pectoral: breadth 9.8, 10.4; height 7.1, 7.8.
Snout length in front of: outer nostrils 2.7, 2.2; mouth 10.0, 10.0.
Eye: horizontal diameter 5.6, 6.1.
Mouth: breadth 7.6, 7.6; height 1.3, 1.1.
Nostrils: distance between inner ends 3.5, 3.0.
Labial furrow length from angle of mouth: upper 1.8, 1.4; lower 1.8, 1.2.
Gill opening lengths: 1st 1.3, 1.3; 2nd 1.3, 1.3; 3rd 1.3, 1.3; 4th 1.3, 1.3; 5th 1.3, 1.3.
First dorsal fin: vertical height 3.8, 2.8; length of base 5.8, 5.2.
Second dorsal fin: vertical height 6.0, 5.2; length of base 8.9, 7.4.
Caudal fin: upper margin 21.3, 21.3; lower anterior margin 10.2, 9.6.
Pectoral fin: extreme length 10.2, 9.6; extreme breadth 5.8, 4.8.
Distance from snout to: 1st dorsal 34.7, 34.8; 2nd dorsal 58.0, 60.0; upper caudal 78.7, 78.7; pectoral 24.9, 23.1; pelvics 49.3, 50.0.
Interspace between: 1st and 2nd dorsals 19.1, 20.9; 2nd dorsal and caudal 13.3, 13.9.
Distance from origin to origin of: pectoral and pelvics 24.0, 26.9; pelvics and caudal 28.8, 28.2.

Trunk subcylindrical, moderately slender, its height at 1st dorsal about 1/8 its length to origin of caudal. Body sector to cloaca about 1.3 to 1.4 times as long as tail sector, without mid-dorsal ridge. Dermal denticles minute, close-set, similar over body as a whole, thorn-like, slender, moderately curved, tapering, their bases more or less stellate but concealed in the skin.

Head a little less than 1/3 (28 to 29%) of length to origin of caudal, flattened above. Snout thick and fleshy at tip, its sides slightly concave at eyes, its anterior outline only slightly rounded, its lower surface with large mucous pores arranged in a prominent pattern, its length in front of nostrils only about 1/8 of length in front of mouth, but length in front of mouth only a little less than 1/2 length of head. Distance between nostrils averaging a little less than 1/3 (28 to 35%) of length in front of mouth. Eye oval, its lower outline much more convex than upper, its horizontal diameter about 1/4 as long as head. Spiracle about 1/3 as long as eye, a little above upper margin of latter and behind it by a distance about 1/3 as long as diameter of eye. Gill openings about evenly spaced, all of about equal lengths and very short, about 1/3 to 1/4 as long as diameter of eye, the 5th close in front of the pectoral. Nostril close to anterior margin of snout, about 1/2 as long as horizontal diameter of eye, moderately oblique, its anterior margin with a long narrow lobe near its outer end. Mouth very little arched, in somewhat sinuous contour, about 3/4 as broad as length of snout in front of mouth. Upper and lower labial furrows each a little less than 1/3 as long as to the respective symphysis, the rearward prolongation of pit at corner of mouth a little more than 1/2 as long as horizontal diameter of eye.

Teeth $\frac{12-13}{18-19}$ in specimen illustrated; upper teeth usually with 5 (rarely 3 or 7) cusps, the median cusp longest and the outermost pair very short, except on the outermost 2 teeth, which are much lower and lack definite cusps; lower teeth subquadrate, a little longer than broad and with 1 sharp cusp so oblique that the inner margins are approximately parallel with the jaw, each overlapping the next outermost to form a continuous cutting edge, the outermost tooth of all more broadly expanded than the others basally on outer side; usually 3 series functional all along upper jaw, 1 or 2 on lower, depending on their stage in replacement.

First dorsal with base about 1/4 as long as head, its margins nearly straight, its apex rounded, its free rear corner about as long as base, its origin a little posterior to inner corner of pectoral, its spine exposed for more than 1/2 its length, reaching about 3/4 the way along the fin, the midpoint of its base about 2/3 as far from axil of pectoral as from origin of pelvics. Second dorsal similar to 1st in shape, but nearly twice as high vertically and 1 1/2 times as long at base, its origin a little posterior to rear ends of bases of pelvics, its spine exposed for about 4/5 its length, reaching nearly to apex of fin (thus much longer than 1st dorsal spine, relatively). Interspace between 2nd dorsal and caudal about 1.5 to 1.8 times as long as base of 2nd dorsal. Caudal a little less than 1/4 of total length or about as long as from tip of snout to 3rd gill opening, transversely rounded at tip, with obtuse subterminal notch, the terminal sector about 1/5 of fin, its lower anterior corner rounded, a little more than a right angle, the lower anterior margin between 1/3 and 1/2 as long as upper margin. Interspace between caudal and rear ends of bases of pelvics about as long as from origins of pelvics to tips of pectorals in female and considerably longer in male. Pelvics about as long at base as 2nd dorsal, with nearly straight edges and tapering subacute tips. Pectoral a little less than 1/3 as long as head, about 2/3 as broad as long, brush-shaped with broadly truncate tip and rounded corners, the inner not at all produced.

Color. After preservation the specimens are dark grayish or chocolate brown above, very pale along midzone of back, with a pale spot on top of head and another above posterior part of each eye; lower surface black. In some specimens the gradation from the paler upperparts is gradual, but in others the black of the ventral surface extends in narrow triangular zones forward above the bases of pelvics and backward onto the caudal, but interrupted midway of the peduncle by a pale belt; the posterior portions of the dorsal fins and caudal are pale, the tip of the latter dusky or blackish; the inner surface of the anterior part of the upper eyelid is dark brown and densely pigmented. All the specimens examined also show more or less clearly defined black dots sparsely scattered on the top of head and rearward in a single row along the midline of the back to the origin of the caudal, flanked by others in a scattered belt; also two to four lines of short, very narrow black dashes lower down on each side, one line following the lateral line out onto the caudal. Presumably these are luminous organs, and conditions in the closely allied *E. lucifer* from Japan suggest that in life their centers are of a pearly luster.[6]

Size. Length at birth is a little more than 90 mm.; females mature at a little less than 300 mm., and males by the time they have reached 250 mm. (see Study Material, p. 489). This, with the fact that the maximum length yet reported for it is 315 mm. without caudal fin,[7] shows *E. hillianus* to be one of the smallest of the sharks.

Developmental Stages. Development is ovoviviparous, and females have been taken with as many as five embryos. One in our Study Material contains four young[8] about 80 to 85 mm. long with small yolk sac; these young already show the characters of the adult, including the coloration; two on one side lie with heads forward, the two on other with heads rearward.

Habits. Apparently this is strictly a deep-water species, the recorded depths of capture ranging from 208 fathoms down to 392 fathoms. Nothing else is known of its habits. It is not known positively whether or not it is luminous, as are one of the Japanese representatives of the genus and *E. spinax,* although its coloration suggests that such is the case, for the fine black dots on its back and sides (presumably indicating glandular areas) resemble the luminous spots of other luminescent sharks.[9]

Range. West Indian region and southern Florida to the offing of Chesapeake Bay; probably Bermuda. This little shark is so far known only from Cuban waters, where it is taken quite often on hook and line from deep water[10] (here the "Atlantis" took five specimens on one collecting cruise, Matanzas Bay, vicinity of Havana and off the northwest coast); from near the Island of St. Kitts; from the Tortugas, Florida,[11] and from

6. See Oshima (J. Coll. Sci. Tokyo, *27* [15], 1911: 1–25) for histology of these organs in *E. lucifer.*
7. Longley and Hildebrand, Pap. Tortugas Lab., *34,* 1941: 3.
8. Not in very good condition.
9. For accounts of the luminous organs and phosphorescence of *E. spinax,* see Johann (Z. wiss. Zool., *66,* 1899: 158) and Burckhardt (Ann. Mag. nat. Hist., [7] *6,* 1900: 559); for the Japanese *E. pusillus* and *E. frontimaculatus,* see Oshima (J. Coll. Sci. Tokyo, *27* [15], 1911) and Schmidt (Proc. Pan-Pacif. sci. Congr., [4] *3,* 1929: 461; Trans. Pacif. Comm., Leningr., *2,* 1931: 9).
10. Personal communication from Luis Howell-Rivero.
11. Longley and Hildebrand, Pap. Tortugas Lab., *34,* 1941: 3.

one station in the offing of Chesapeake Bay (Lat. 37° 24′ N., Long. 74° 17′ W., 300 fathoms); probably also from Bermuda.[12]

Synonyms and References:

Spinax hillianus Poey, Memorias, *2*, 1861: 340, pl. 19, fig. 13–14 (descr., teeth, Cuba); Repert. Fisico.-nat. Cuba, *2*, 1868: 454 (Cuba); Regan, Ann. Mag. nat. Hist., (8) *2*, 1908: 44 (St. Kitts, W. Indies, class., depth).

Spinax pusillus (in part) Günther, Cat. Fish. Brit. Mus., *8*, 1870: 425 (*hillianus* incl. in synon.); Jordan and Evermann, Rep. U.S. Comm. Fish. (1895), 1896: 219 (ref. for Cuba); Metzelaar, Trop. Atlant. Visschen, 1919: 190, part (included in synon.); Fowler, Bull. Amer. Mus. nat. Hist., *70* (1), 1936: 81 (ref. for W. Indies, ill. of W. Indian spec.).

Spinax spinax Poey, An. Soc. esp. Hist. nat., *5*, 1876: 399; Enumerat. Pisc. Cubens., 1876: 203 (Cuban spec.).

Etmopterus pusillus Goode and Bean, Smithson. Contr. Knowl., *30*, 1895: 10, pl. 2, fig. 2; Mem. Harv. Mus. comp. Zool., *22*, 1896: 10, pl. 2, fig. 2 (descr., ill., St. Kitts, W. Indies); Jordan and Evermann, Bull. U.S. nat. Mus., 47 (1), 1896: 55 (descr., St. Kitts, W. Indies); Beebe and Tee-Van, Zoologica, N. Y., *13*, 1933: 157 (Bermuda, old record); not *Etmopterus pusillus* Lowe, 1939.

Etmopterus spinax Garman, Mem. Harv. Mus. comp. Zool., *24*, 1899: 27 (thinks type spec. ident. with *spinax*); not *Etmopterus spinax* Linnaeus, 1758.

Etmopterus hillianus Garman, Mem. Harv. Mus. comp. Zool., *36*, 1913: 224, pl. 10, fig. 1–4 (descr., ill., off St. Kitts, W. Indies); Jordan, Evermann and Clark, Rep. U.S. Comm. Fish. (1928), 2, 1930: 21 (Cuba, St. Kitts); Howell-Rivero, Proc. Boston Soc. nat. Hist., *41*, 1936: 49 (recent Cuban specs.); Bull. Mus. comp. Zool. Harv., *82*, 1938: 170 (type spec. in Harv. Mus. Comp. Zool.); Longley and Hildebrand, Pap. Tortugas Lab., *34*, 1941: 3 (Tortugas, Florida, 357 to 392 fathoms, descr.); Fowler, Fish Culturist, *21* (9), 1942: 66 (listed, Cuba); Bigelow and Schroeder, Guide Comm. Shark Fish., Anglo Amer. Caribb. Comm., Wash., 1945: 144, fig. 54 (descr., ill., habits, range).

Genus *Centroscymnus* Bocage and Brito Capello, 1864

Centroscymnus Bocage and Brito Capello, Proc. zool. Soc. Lond., 1864: 263; Mem. R. Acad. Sci. Math. Phys. Lisboa, *3* (2), 1865: 3, extra; type species, *C. coelolepis* Bocage and Brito Capello, 1864, Portugal.

Generic Synonym:

Centrophorus (in part) Günther, Cat. Fish. Brit. Mus., *8*, 1870: 423; Seabra, Bull. Soc. portug. Sci. nat., *5*, 1912: 198; not *Centrophorus* Müller and Henle, 1837.

Generic Characters. Squalidae with dorsal spines arising at origins of fins and lying along anterior margins of latter; their tips either exposed or concealed; trunk subcylindrical, without lateral longitudinal ridges or precaudal pits; snout in front of mouth much shorter than from mouth to origin of pectorals; teeth unlike in the 2 jaws, the uppers with one slender, lanceolate cusp, the lowers approximately quadrate, their outer margins deeply notched and so oblique that the inner margins are nearly parallel to the jaw, forming a continuous cutting edge; dermal denticles scale-like, closely overlapping, with flat or concave blades, smooth or weakly ridged, their margins not toothed, on short, broad pedicels; eyes and spiracles moderate to large; anterior margin of nostrils expanded as a low triangular lobe only; a voluminous triangular pit at corner of mouth; labial furrow on each jaw; origin of 1st dorsal considerably posterior to tips of pectorals; origin of 2nd dorsal over bases of pelvics; caudal with subterminal notch, its lower anterior corner expanded

12. Reported as *E. pusillus* by Beebe and Tee-Van (Zoologica, N. Y., *13*, 1933: 157).

as a weakly defined lobe, its axis raised at an angle of about 30° to 40°; inner corner of pectoral broadly rounded, not at all produced; luminous organs lacking.

Range. Both sides of North Atlantic; South Africa; Japan.

Key to Species[1]

1a. Length of snout in front of mouth considerably less than distance from eye to 1st gill opening. *coelolepis* Bocage and Brito Capello, 1864, p. 494.

1b. Length of snout in front of mouth at least as great as distance from eye to 1st gill opening.

- 2a. Length of snout about as great as distance from eye to 1st gill opening; 1st dorsal about as large as 2nd dorsal. *fuscus* Gilchrist and von Bonde, 1924. South Africa.[2]
- 2b. Length of snout definitely greater than distance from eye to 1st gill opening; 1st dorsal smaller than 2nd dorsal.
  - 3a. Tips of dorsal spines exposed; tip of 2nd dorsal extends back considerably beyond tips of pelvics. *owstoni* Garman, 1906. Japan.
  - 3b. Tips of dorsal spines concealed by skin; tip of 2nd dorsal extends only as far back as tips of pelvics. *cryptacanthus* Regan, 1906. Madeira.

Centroscymnus coelolepis Bocage and Brito Capello, 1864

Portuguese Shark

Figures 94, 95

Study Material. Two adult females, 1,117 and 1,080 mm. long, taken off Banquereau Bank in 200 to 270 fathoms (Harv. Mus. Comp. Zool., No. 35144, 35237); also very young male, about 328 mm. long, from the continental edge south of Nantucket, Lat. 39° 51′ N., Long. 70° 17′ W. (U.S. Nat. Mus., No. 118396).

Distinctive Characters. Among the local members of the suborder, *Centroscymnus* is marked off from *Squalus acanthias, S. cubensis, Centroscyllium* and *Etmopterus* by the following: its fin spines protrude so little from the skin that they are apt to be overlooked (cf. Fig. 94 with 87, 89, 91, 92); while its teeth have only one cusp in each jaw, the lowers and uppers are strikingly unlike. Owing to the inconspicuous nature of its spines it might perhaps be confused with small specimens of *Somniosus,* which it resembles in the general

1. *Centrophorus crepidater* Bocage and Brito Capello, 1864, from Portuguese waters, was referred by Garman (Mem. Harv. Mus. comp. Zool., *36*, 1913: 207) to his new genus *Centroselachus* because of the pluricarinate scales, but by Rey (Fauna Iberica Peces, *1*, 1928: 449) to *Centroscymnus*, but it falls in *Scymnodon* as here defined, likewise *Centroscymnus macracanthus* Regan, 1906, from the Straits of Magellan and Argentina.
2. Known only from the type specimen.

shapes and positions of the fins, as well as in the shape of the teeth, or with *Dalatias*. But even apart from the fin spines, its overlapping, scale-like dermal denticles differ widely in appearance from the thorns with which the skin of *Somniosus* is beset, and its lower teeth differ markedly from those of *Dalatias*.

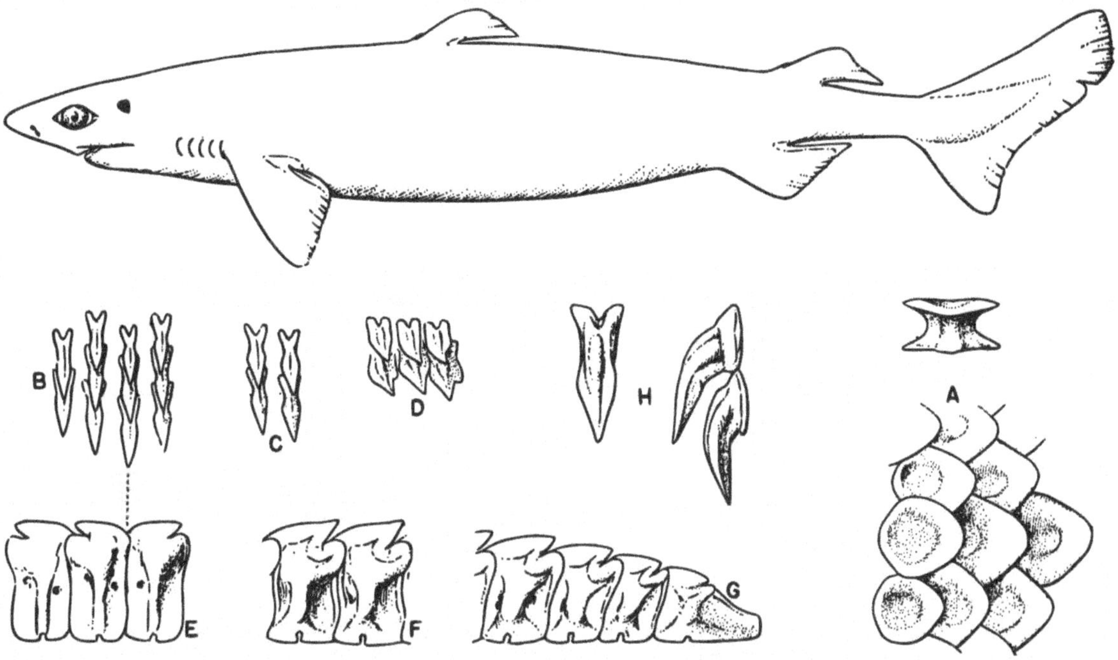

FIGURE 94. *Centroscymnus coelolepis*, female, about 1,080 mm. long, from off Banquereau Bank, Nova Scotia (Harv. Mus. Comp. Zool., No. 35237). *A* Dermal denticles, about 5 x. *B* Upper teeth from center of jaw. *C* Upper teeth from side of jaw. *D* Outermost three rows of upper teeth. *E* Lower teeth from center of jaw. *F* Lower teeth from side of jaw. *G* Outermost rows of lower teeth. *B–G*, about 3 x. *H* Front and lateral views of upper teeth from side of jaw, about 6 x.

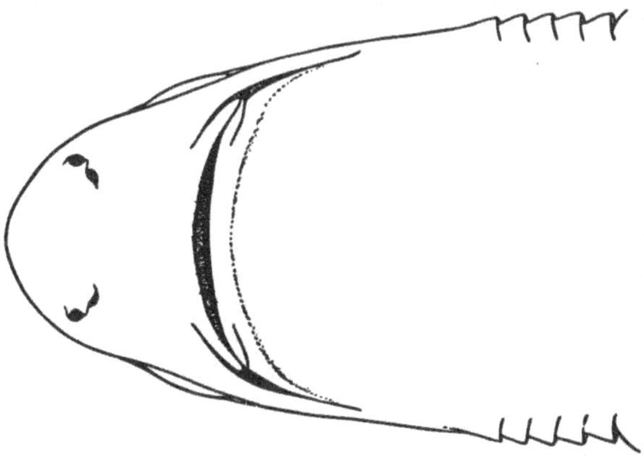

FIGURE 95. Head of *Centroscymnus coelolepis*, pictured in Fig. 94, from below.

Description. Proportional dimensions in per cent of total length. Male, 328 mm., from Lat. 39° 51′ N., Long. 70° 17′ W. (U.S. Nat. Mus., No. 118396). Female, 1,080 mm., from Banquereau Bank (Harv. Mus. Comp. Zool., No. 35237).

Trunk at origin of pectoral: breadth 12.2, 12.6; height 10.7, 11.2.
Snout length in front of: outer nostrils 2.4, 1.9; mouth 8.8, 6.2.
Eye: horizontal diameter 4.9, 3.4.
Mouth: breadth 7.6, 8.1; height 1.1, 1.1.
Nostrils: distance between inner ends 4.3, 3.1.
Labial furrow length from angle of mouth: upper 3.7, 2.4; lower 2.6, 1.4.
Gill opening lengths: 1st 1.2, 1.8; 2nd 1.2, 1.7; 3rd 1.1, 1.5; 4th 1.2, 1.2; 5th 1.7, 1.6.
First dorsal fin: vertical height 3.4, 3.7; length of base 5.2, 4.6.
Second dorsal fin: vertical height 3.8, 4.4; length of base 5.8, 5.7.
Caudal fin: upper margin 25.3, 20.4; lower anterior margin 15.4, 12.8.
Pectoral fin: extreme length 13.4, 13.2; extreme breadth 7.9, 6.3.
Distance from snout to: 1st dorsal 34.7, 36.8; 2nd dorsal 61.3, 67.7; upper caudal 74.7, 79.6; pectoral 23.8, 18.8; pelvics 57.3, 60.7.
Interspace between: 1st and 2nd dorsals 22.0, 24.2; 2nd dorsal and caudal 7.0, 8.0.
Distance from origin to origin of: pectoral and pelvics 35.7, 43.5; pelvics and caudal 15.8, 17.5.

Trunk subcylindrical, moderately stout, its height at 1st dorsal between 1/6 and 1/5 (17 to 18%) its length to origin of caudal, without mid-dorsal ridge. Caudal peduncle without lateral ridges or precaudal pits. Dermal denticles very large, except on fins and lower side of head anterior to gills, and so closely overlapping as to form a continuous armor,[3] the peduncles short and stout, the blades smooth, ovoid, flat or concave, with rounded margins posterior to level of gills, but weakly sculptured with 3 or 5 ridges further forward on head.

Head about 1/4 of trunk to origin of caudal, its dorsal profile weakly and evenly convex, somewhat flattened anteriorly. Snout thick-tipped, very broadly ovate, noticeably short, its length in front of nostrils about 1/4 to 1/3 as great as length in front of mouth, its length in front of mouth a little less than 1/3 of length of head. Distance between nostrils about 1/2 as great as length in front of mouth. Eye oval, its outline about as convex above as below, much smaller than in *Centroscyllium,* its horizontal diameter about 1/2 as long as snout in front of mouth, its center about opposite front of mouth. Spiracle about 1/3 to 1/4 as long as diameter of eye, about level with upper margin of latter. Gill openings much smaller than in *Centroscyllium,* about 1/2 as long as diameter of eye, all of about equal length and evenly spaced, the 5th close in front of pectoral. Nostril

3. The edges of the denticles are so sharp that one must handle *Centroscymnus* carefully, lest one's hands be cut. The closely overlapping denticles are one of the most distinctive features of the genus.

moderately oblique, a little less than ½ as long as horizontal diameter of eye, its inner margin expanded midway of its length as a short, triangular lobe with blunted tip, its inner corner about equidistant from tip of snout and from front of mouth. Mouth only very slightly arched, its breadth about twice as great as distance between nostrils and a little greater than length of snout in front of mouth. Upper labial furrow extending nearly ½ of the way, the lower about ⅓ of the way, toward the respective symphyses. Pit at corner of mouth very voluminous, allowing for considerable expansion when mouth is opened, and prolonged as a narrowing furrow rearward, nearly ½ the way back toward 1st gill opening.

Teeth about $\frac{58}{40}$ on specimen counted ($\frac{70}{42}$ also reported); widely unlike in the two jaws; upper teeth with 1 slender, erect, lanceolate cusp on bifid base,[4] considerably broader toward corner of jaw than toward center, their tips slightly curved outward, the outer margins notched near corner of mouth in some cases, the successive series rather widely spaced; lower teeth quadrate, each overlapping the next on the outer side, their outer margins deeply notched, the 1 broad sharp cusp so strongly oblique that the inner margins form a nearly continuous cutting edge parallel to the jaw; those near center of mouth about twice as high as broad, but the 3 or 4 next to the corner of mouth successively broader, the outermost of all widely expanded basally on outer side, with cusp but weakly outlined; 2 or 3 series regularly functional in upper jaw, 1 or 2 in lower, depending on the stage in replacement.

First dorsal noticeably small, with broadly rounded apex, its length at base only about ¼ to ⅕ as great as length of head, its vertical height a little less than length of base, its rear margin nearly straight, free rear tip a little longer than base, its origin posterior to inner corner of pectoral by a distance about twice as long as horizontal diameter of eye, the midpoint of its base only about ½ as far from axil of pectoral as from origin of pelvics, its spine exposed at tip but so short as to be easily overlooked. Second dorsal similar to 1st in shape, but about 1.2 times as long at base, its origin a little posterior to midpoint of base of pelvics, its rear margin weakly concave, its spine exposed at tip like that of 1st dorsal, but so short as to be apt to escape notice. Interspace between 2nd dorsal and caudal about 1⅓ times as long as base of 2nd dorsal. Caudal ¼ to ⅕ (about 22 to 23%) of total length, noticeably wide, its extreme breadth being about ⅔ its length, truncate posteriorly with deep subterminal notch, its lower posterior contour weakly concave, the lower anterior margin a little more than ½ as long as upper margin. Pelvics a little longer at base than 2nd dorsal, with nearly straight margins, rounded apices and pointed tips. Pectoral a little less than ⅔ as long as head, a little more than ½ as broad as long, with nearly straight outer and distal margins and moderately rounded corners.

Color. Dark chocolate brown below as well as above.

Size. The smallest recorded specimen is about nine inches long (230 mm.); adults

4. The upper teeth resemble those of *Somniosus* in general appearance, but may be recognized by the fact that their cusps do not taper uniformly from base to tip, but are lanceolate.

average 3 to 3½ feet in length, the largest for which actual measurements are available being about 44 inches long (1,117 mm.); see Study Material, p. 494. Twelve kilograms (about 26½ pounds) is the only weight of which we find record.

Developmental Stages. It is no doubt ovoviviparous. All that is known of its early development is that gravid females have been taken with 13 to 16 embryos.

Habits. This is strictly a deep-water shark, as noted below. Apparently it is also a very sluggish one, for those caught in the Portuguese deep-water fishery have been described as falling into the boat entirely inert. But this may be the result of the change in pressure or in temperature to which they are subjected while being hauled in. Off the American coast, this shark occurs mostly at temperatures of 5° to 6° C.; between about 4° and 10° or 11° off Portugal; and at 12° to 13° in the Mediterranean. All that is known of its food is that an argentine (*Argentina silus*) was found in one, suggesting a fish diet. Nothing is known of its breeding habits, except for the number of embryos (see Developmental Stages, p. 498).

Relation to Man. This shark is considered worthless in American waters, but it has been the object of a local deep-water fishery with long lines off Portugal in the past.

Range. Both sides of the North Atlantic, chiefly in depths greater than 200 fathoms, and recorded down to 1,487 fathoms (2,718 meters); taken off Cape Verde, Morocco, Azores, Madeira, Portugal, Faroe Bank, and Iceland in the east, as well as in the western part of the Mediterranean; offing of Nantucket to slopes of the Grand Banks in the west.

Occurrence in the Western Atlantic. Positive records of this deep-water shark in the western Atlantic are from the continental edge off Nantucket, from the deeper slopes of Georges and the Nova Scotian Banks, and from the Grand Banks; a total of perhaps 15 to 20 specimens are recorded at depths ranging from 180 fathoms, which is the shoalest capture of it anywhere, down to 250 fathoms. No doubt, however, an old characterization of it as abundant on the offshore banks at 200 fathoms[5] or deeper presents its status much more correctly than does the meager printed record, for fishermen, long-lining for halibut, take odd specimens all over the halibut grounds in the deep gullies between the offshore banks, usually at least one or two per trip. Since this is the only local type of fishery that is carried on at a depth great enough to take them at all, it would not be astonishing if experimental hook and line fishing on the still deeper slopes, down to 300 or 400 fathoms, were to yield them as plentifully as was the case formerly off Portugal, where there is record of five or six hauled in on a long line with 30 to 40 hooks (baited with fish) after a set of only two hours.[6] As it is of no commercial value only an odd one is brought in as a curiosity.

Synonyms and References:

Centroscymnus coelolepis Bocage and Brito Capello, Proc. zool. Soc. Lond., 1864: 263, fig. 4; Diag. Famil. Squalidae, 1864: 3; also same title in Mem. R. Acad. Lisboa, *3*, 1865: 3 (descr., Portugal); Poiss. Plagiost., 1866: 30, pl. 2, fig. 3 (Portugal, Madeira); Wright, Ann. Mag. nat. Hist., (4) *2*, 1868: 426

5. Goode and Bean, Bull. Essex. Inst. Salem, *11*, 1879: 30.
6. Wright, Ann. Mag. nat. Hist., (4) *2*, 1868: 426.

(Portugal, fishery); Brito Capello, J. Sci. math. phys. nat. Lisboa, *4* (13), 1872: 88 (Madeira); Goode and Bean, Bull. Essex Inst. Salem, *11*, 1879: 30 (abund., Banks off New England); Jones, List Fish. Nova Scotia, 1879: 10 (meas., off Nova Scotia); Bean, Proc. U.S. nat. Mus., *3*, 1881: 116 (list of locs., off New England, Nova Scotia, Grand Banks); Jones, Proc. N. S. Inst. Sci., *5*, 882: 96 (same as Jones, 1879); Jordan and Gilbert, Bull. U.S. nat. Mus., 16, 1883: 17 (descr., Portugal, near Gloucester, Massachusetts); Jordan, Rep. U.S. Comm. Fish. (1885), 1887: 793 (distrib.); Vaillant, Exped. Sci. "Travailleur" and "Talisman," Poiss., 1888: 63, pl. 2, fig. 1 (ill., weight, no. of embryos, temp., off Portugal); Bellotti, Atti Soc. ital. Sci. nat., *33*, 1891: 113 (Medit.); Moreau, Poiss. France, Suppl., 1891: 9 (small no. of embryos, Nice, France); Carus, Prod. Fauna Medit., *2*, 1889–1893: 503 (Medit.); Goode and Bean, Smithson. Contr. Knowl., *30*, 1895: 14, 508, pl. 4, fig. 13; Mem. Harv. Mus. comp. Zool., *22*, 1896: 14, 508, pl. 4, fig. 13 (descr., ill., locs. off New England); Jordan and Evermann, Rep. U.S. Comm. Fish. (1895), 1896: 219 (off Portugal, Massachusetts, Nova Scotia); Bull. U.S. nat. Mus., 47 (1), 1896: 55 (descr., off Portugal, Massachusetts, Nova Scotia); Bull. U.S. nat. Mus., 47 (4), 1900: pl. 8, fig. 25 (ill.); de Braganza, Result. Invest. Sci. "Amelia," *2*, 1904: 86, 106 (abund., depth, off Portugal); Collett, Rep. Norweg. Fish. Invest. *2* (2), 1905: 24 (size, off Faroes); Regan, Ann. Mag. nat. Hist., (7) *18*, 1906: 437 (in Key); Kendall, Occ. Pap. Boston Soc. nat. Hist., 7 (8), 1908: 9 (off Gloucester); Regan, Ann. Mag. nat. Hist., (8) *2*, 1908: 49 (size, distrib.); Murray and Hjort, Depths of Ocean, 1912: 392 (depth, Faroe Bank); Roule, Bull. Inst. océanogr. Monaco, 243, 1912: 13 (Medit. and E. Atlant., list of specs.); Garman, Mem. Harv. Mus. comp. Zool., *36*, 1913: 204, pl. 14, fig. 5–8 (descr., ill., off New England); Halkett, Check List Fish. Canad., 1913: 41 (off Nova Scotia and Massachusetts); Roule, Result. Camp. sci. Monaco, *52*, 1919: 120 (list of E. Atlant. and Medit. specs.); Saemundsson, Vidensk. Medd. naturh. Foren. Kbh., *74*, 1922: 177 (Iceland, Faroe Ridge); Bigelow and Welsh, Bull. U.S. Bur. Fish., *40* (1), 1925: 51 (Gulf of Maine and Nova Scotia, general); Hickling, Ann. Mag. nat. Hist., (10) *2*, 1928: 199 (Atlant. slope off Scotland); Rey, Fauna Iberica, Peces, *1*, 1928: 451 (descr., ill., Portugal); Breder, Field Bk. Mar. Fish. Atlant. Coast, 1929: 25 (off New England); Jordan, Manual Vert. Anim. NE. U.S., 1929: 14 (off C. Ann and fishing banks); Jordan, Evermann and Clark, Rep. U.S. Comm. Fish. (1928), 2, 1930: 21 (Gloucester and Nova Scotia Banks); MacCoy, Bull. Boston Soc. nat. Hist., 69, 1933: 8 (240 miles ESE. of Boston Lightship); Belloc. Rev. des Trav. Pêches Marit., 7 (2), 1934: 144 (ill. after Vaillant; Morocco, Madeira); Bigelow and Schroeder, Canad. Atlant. Fauna, biol. Bd. Canad., *12e*, 1934: 18 (descr., distrib.); Noronha and Sarmento, Peixes Madeira, 1934: 134 (Madeira, not seen); Nobre, Fauna Marinha, Port. Vert., *1*, 1935: 455, pl. 63 (depths, off Portugal); Vladykov and McKenzie, Proc. N. S. Inst. Sci., *19*, 1935: 49 (off Nova Scotia); Fowler, Bull. Amer. Mus. nat. Hist., *70* (1), 1936: 74, 75 (descr., no. of embryos, distrib. in E. Atlant., Madeira specs.).

Centrophorus coelolepis Günther, Cat. Fish. Brit. Mus., *8*, 1870: 423 (at least in part, descr., Portugal, Madeira);[7] "Challenger" Rep., Zool., *22*, 1887: XL, 5 (depth of capture); Seabra, Bull. Soc. portug. Sci. nat., *5*, 1912: 198 (Portugal).

Family DALATIIDAE

Characters. Squaloidea with 2nd dorsal, and in most cases the 1st also, lacking a spine; teeth with 1 cusp only, but widely unlike in the 2 jaws, the uppers slender and conical, the lowers broad and blade-like, each overlapping the next outermost, their edges serrate or smooth.

7. Fowler (Bull. Amer. Mus. nat. Hist., *70* [1], 1936: 75) includes this citation in the synonymy of *Centroscymnus cryptacanthus* Regan, 1906. But Günther's brief account with the cited references and localities no doubt covers both that species and *coelolepis*.

Key to Genera

1a. Rear end of base of 1st dorsal over or posterior to origins of pelvics; interspace between 1st and 2nd dorsals shorter than between 2nd dorsal and caudal.

Isistius Gill, 1864, p. 508.

1b. Rear end of base of 1st dorsal considerably anterior to origins of pelvics; interspace between 1st and 2nd dorsals considerably longer than between 2nd dorsal and caudal.

2a. Lower teeth erect, triangular, nearly symmetrical, serrate.

Dalatias Rafinesque, 1810, p. 500.[1]

2b. Lower teeth strongly asymmetrical with cusps directed outward, not serrate, their outer margins notched.

3a. First dorsal larger than 2nd, its apex angular; tip of pectoral angular.

Heteroscymnus Tanaka, 1912.
Japan.

3b. First dorsal at least no larger than 2nd, its apex rounded; tip of pectoral rounded.

4a. Second dorsal only about as large as 1st; snout in front of mouth about ½ as long as from eye to origin of pectoral.

Somniosus Lesueur, 1818, p. 514.

4b. Second dorsal considerably larger than 1st; snout in front of mouth nearly or quite as long as from eye to origin of pectoral.

5a. Second dorsal twice as long at base as 1st dorsal, or more; 1st dorsal with a spine either partly free or entirely hidden in the skin.

Euprotomicrus Gill, 1864.[2]
Indian Ocean; Philippines; New Zealand; Pacific between Hawaii and California; Madeira.

5b. Second dorsal not more than 1½ times as long at base as 1st; 1st dorsal without spine. *Heteroscymnoides* Fowler, 1934.
Natal, South Africa.

Genus *Dalatias* Rafinesque, 1810

Dalatias Rafinesque, Carratt. Gen. Nuov. Sicil., 1810: 10; type species, *D. sparophagus* Rafinesque, 1810, equals *Squalus licha* Bonnaterre, 1788, designated by Jordan, Tanaka and Snyder, 1913.[3]

1. Including *Pseudoscymnus* Herre, 1935. The differences in the denticles, cited by Herre (Copeia, 1935: 124), do not seem sufficient for generic, and perhaps not even for specific, separation; see p. 501.
2. Including *Squaliolus* Smith and Radcliffe, 1912.
3. Jordan, Tanaka and Snyder's (J. Coll. Sci. Tokyo, *33*, 1913: 22) designation of *sparophagus* Rafinesque, 1810, as the type of *Dalatias* is not invalidated by Swainson's (Nat. Hist. Fish. Amphib. Rept., *1*, 1838: 160) accidental (?) limitation of that genus to *D. nocturnus* alone in one connection, for on an earlier page (129) in the same publication ~~Swainson~~ included *D. sparophagus* Rafinesque in it as well. Therefore, Jordan and Evermann's (Genera Fish., *1*, 1917: 77) subsequent designation of *nocturnus* Rafinesque, 1810, as the type species is not tenable. For the rather confused history of the case, see Gill (Proc. U.S. nat. Mus., *18*, 1896: 191), who arrived at the conclusion that *Dalatias* is a synonym of *Squalus*, hence that the correct generic name for the shark now under consideration is *Scymnorhinus* Bonaparte, 1846.

Generic Synonyms:

Scymnus Cuvier, Règne Anim., 2, 1817: 130; type species, *Squalus americanus* Gmelin, 1789, equals *Squalus licha* Bonnaterre, 1788, but preoccupied for insects by Kugelman, 1794.

Scymnium Cuvier, Règne Anim., ill. ed., Poiss., 1838–1843: pl. 115; type species, *Squalus nicaense* Cuvier, equals *Squalus licha* Bonnaterre, 1788.

Dalatius L. Agassiz, Nomencl. Zool. Index, 1845: 21; emended spelling for *Dalatias.*

Scymnorhinus Bonaparte, Cat. Pesc. Europ., 1846: 16; type species, *Squalus americanus* Gmelin, 1789, equals *Squalus licha* Bonnaterre, 1788.

Barborodes Gistel, Naturg. Tierreich, 1848: X; proposed to replace *Scymnorhinus* Bonaparte, 1846.

Pseudoscymnus Herre, Copeia, 1935: 124; type species *P. boshuensis* Herre.

Generic Characters. Dalatiidae without dorsal spines; snout very short; caudal peduncle without lateral ridges or precaudal pits; a labial fold on each jaw and a voluminous pit at corner of mouth; upper teeth slender, thorn-like, in several functional series; lower teeth broad-triangular with regularly serrate edges; dermal denticles low, ridged, their margins more or less definitely toothed; 2nd dorsal somewhat larger, and pelvics much larger, than 1st dorsal; 1st dorsal far anterior to pelvics; rear end of base of 2nd dorsal considerably posterior to origin of pelvics; interspace between 1st and 2nd dorsals considerably longer than between 2nd dorsal and caudal; caudal noticeably large, with rounded corners, its terminal sector sharply marked off, but lower anterior corner not expanded as a definite lobe; skin without luminous organs. Characters otherwise those of the family.

Range. Both sides of North Atlantic; Mediterranean; South Africa; Japan; Australia; New Zealand.

Fossil Teeth. Upper Cretaceous, western Asia and North America; Eocene, North Africa; Eocene to Pliocene, Europe; and Miocene, North America.

Species. The Australian–New Zealand and South African representatives of the genus have recently been separated[5] from the well known *D. licha* of the North Atlantic and Mediterranean. But we find nothing in the several accounts or illustrations of the Australian form (see Synonyms, p. 508) to separate it from *licha.* It is equally doubtful whether the supposed differences noted by the author of *brevipinnis*, i.e., lower teeth more oblique in adult, smaller fins and less pronounced lower caudal lobe, will prove sufficient for specific separation when critically tested.

New generic and specific names (*Pseudoscymnus boshuensis*) have also been proposed by Herre[6] for a Japanese form, no doubt the same as one earlier reported as *Scymnus lichia*,[7] as *Dalatias americanus*[8] and as *D. licha*,[9] the separation being based on its denticles, which differ in shape between the lower side of the snout and the trunk in general, and its

5. As *phillippsi* Whitley (Aust. Zool., *6*, 1931: 310) and as *brevipinnis* Smith (Trans. roy. Soc. S. Afr., *24*, 1936: 1) respectively.
6. Copeia, 1935: 124.
7. Ishikawa and Matsuura, Prel. Cat. Fish. Mus. Tokyo, 1897: 61.
8. Jordan and Snyder, Annot. zool. jap., *3*, 1901: 129.
9. Jordan and Fowler, Proc. U.S. nat. Mus., *26*, 1903: 637; also Izuka and Matsuura, Cat. Zool. Tokyo Mus. Vert., 1920: 188.

lower teeth, which are serrate. But it has long been known that all this applies to the Atlantic form, and our own comparison of a Japanese specimen with one from the Atlantic coast of the United States shows no significant differences in proportional dimensions, shape or position of fins, teeth, or denticles.

Dalatias licha (Bonnaterre), 1788

Figures 96, 97

Study Material. Female, 1,470 mm. long, from Georges Bank (Amer. Mus. Nat. Hist., No. 14056[10]); 4 specimens of about 367 to 1,080 mm., from Nice, France, and an embryo of 245 mm. from the same locality (Harv. Mus. Comp. Zool.); also immature male, 1,114 mm. long, from Japan (Harv. Mus. Comp. Zool., No. 1116).

Distinctive Characters. The serrate margins and triangular shape of its lower teeth mark *D. licha* off from all other North Atlantic members of its suborder. It is further separated very obviously from the species of *Squalus*, *Centroscyllium* and *Etmopterus* by its lack of fin spines, and from *Isistius* by the position of its first dorsal fin farther forward.

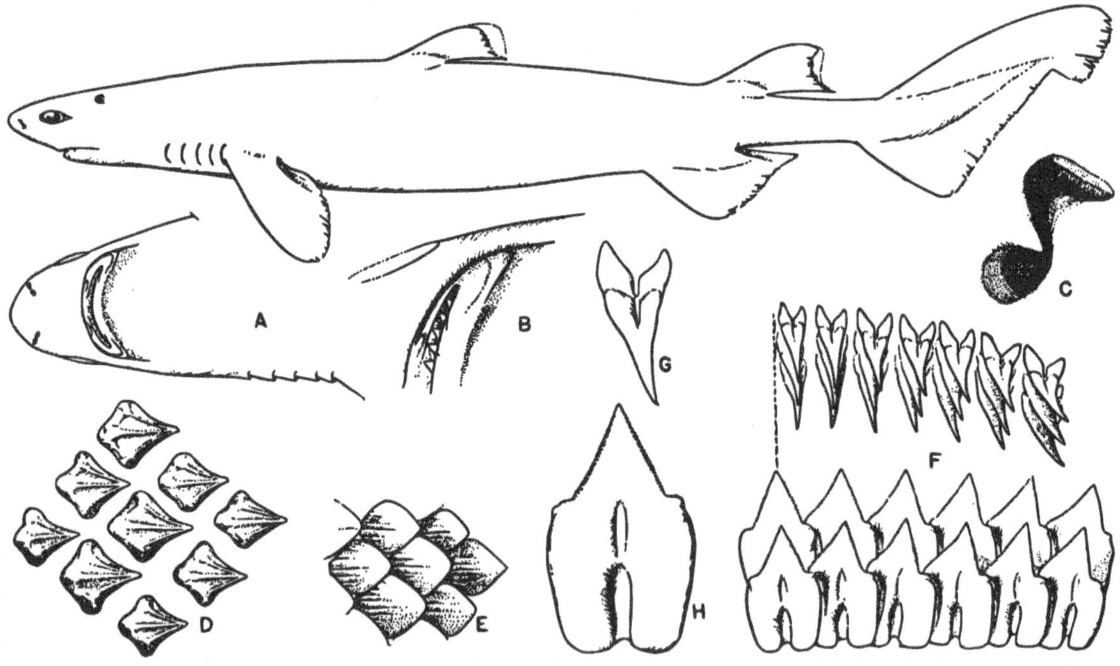

FIGURE 96. *Dalatias licha*, female, 1,470 mm. long, from Georges Bank, Gulf of Maine (Amer. Mus. Nat. Hist., No. 14056). *A* Head from below. *B* Left-hand corner of mouth to show labial furrows, about 0.4 natural size. *C* Right-hand nostril, about 1.2 x. *D* Dermal denticles from side, below first dorsal fin, about 12 x. *E* Dermal denticles from ventral surface of snout, about 9 x. *F* First to seventh upper teeth, and median and first to fifth lower teeth from left-hand side, about 1.2 x. *G* Fourth upper tooth. *H* Median lower tooth. *G–H*, about 2.4 x.

10. Reported five feet one inch (approximately 1,550 mm.) long (Nichols and Firth, Proc. biol. Soc. Wash., 52, 1939: 85); but now only 1,470 mm. by the system of measurement here employed (p. 61).

Description. Proportional dimensions in per cent of total length. Female, 531 mm., from Nice, France (Harv. Mus. Comp. Zool., No. 948). Female, 1,470 mm., from Georges Bank (Amer. Mus. Nat. Hist., No. 14056).

Trunk at origin of pectoral: breadth 11.3, 11.7; height 10.9, 9.0.

Snout length in front of: outer nostrils 1.1, 0.6; mouth 5.4, 4.6.

Eye: horizontal diameter 3.9, 2.1.

Mouth: breadth 4.5, 5.0; height 0.8, 1.3.

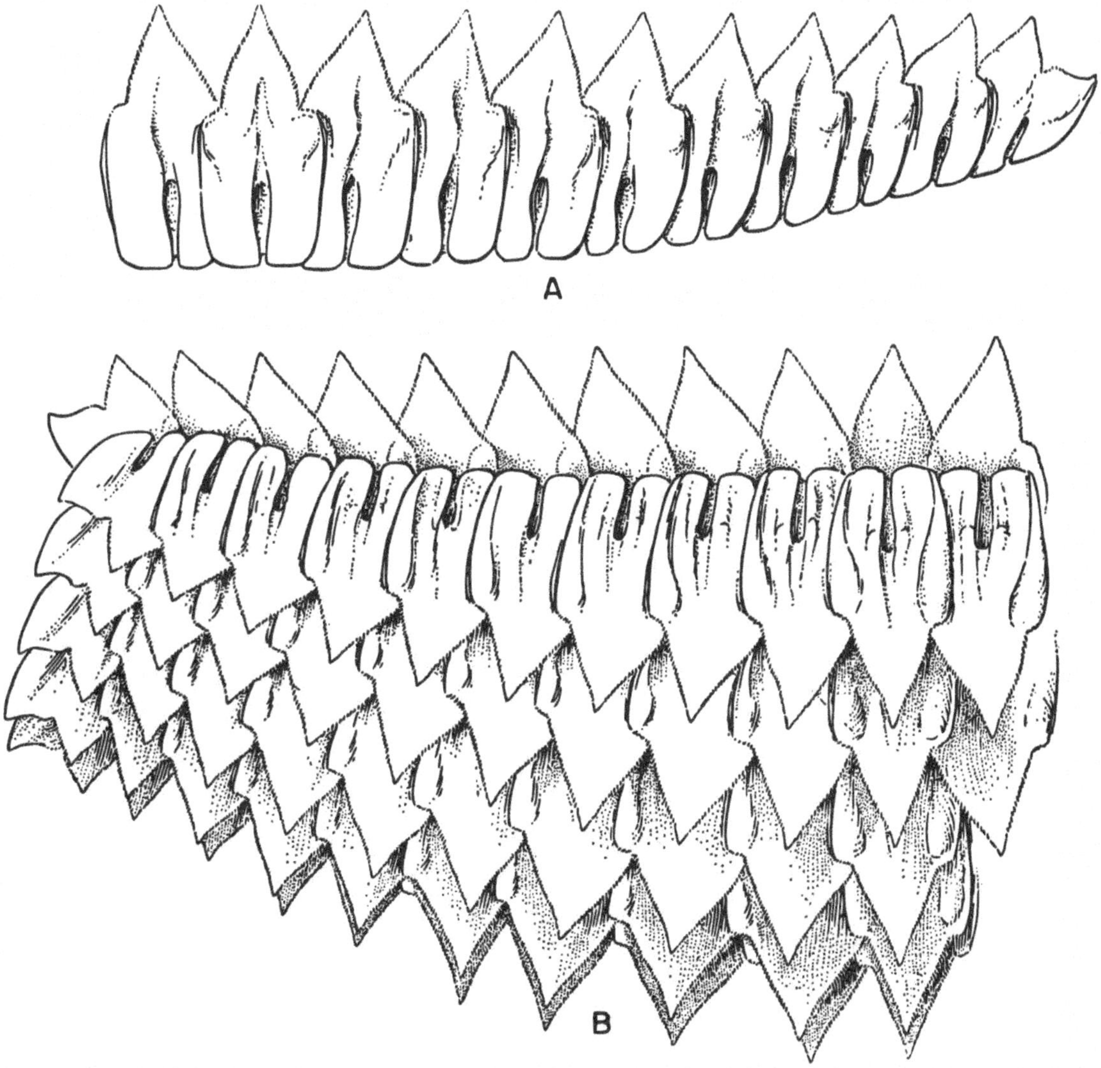

FIGURE 97. *Dalatias licha*, female (dried skin), about 1,080 mm. long, from Europe (Harv. Mus. Comp. Zool., No. 664). *A* Left-hand lower teeth, viewed from without. *B* Dentition of right-hand lower teeth viewed from within the mouth to show the one series of teeth in function, with five replacement series still occupying the reversed position with their points directed downward and inward, about 2.3 x natural size.

Nostrils: distance between inner ends 3.2, 2.6.
Labial furrow length from corner of mouth: upper 1.8, 1.8; lower 2.0, 1.6.
Gill opening lengths: 1st 1.6, 1.8; 2nd 1.6, 1.8; 3rd 1.6, 1.9; 4th 1.9, 1.9; 5th 2.1, 2.1.
First dorsal fin: vertical height 5.3, 4.4; length of base 4.9, 5.2.
Second dorsal fin: vertical height 6.2, 5.9; length of base 6.2, 6.8.
Caudal fin: upper margin 25.2, 21.8; lower anterior margin 12.1, 11.3.
Pectoral fin: extreme length 14.1, 13.0; extreme breadth 6.2, 6.3.
Distance from snout to: 1st dorsal 35.3, 34.2; 2nd dorsal 60.2, 63.0; upper caudal 74.8, 78.2; pectoral 22.1, 19.7; pelvics 55.4, 57.6.
Interspace between: 1st and 2nd dorsals 20.6, 23.5; 2nd dorsal and caudal 10.0, 9.5.
Distance from origin to origin of: pectoral and pelvics 36.8, 39.2; pelvics and caudal 17.9, 18.2.

Trunk slender, subcylindrical, its height at 1st dorsal a little less than 1/6 (15 to 16%) its length to origin of caudal. Body sector to cloaca a little less than twice as long as tail sector, the back without mid-dorsal ridge. Caudal peduncle without lateral ridges or precaudal pits. Dermal denticles, over trunk as a whole, small, loose-spaced, scale-like, their blades close to the skin, thick, quadrate, with 3 weak ridges uniting posteriorly at the margin in a tooth that varies in length and in acuteness from denticle to denticle; pedicels thick and short. Denticles on lower side of snout overlapping, ovate, without marginal teeth, but usually with 3 weakly marked longitudinal ridges.

Head about 1/4 (24 to 25%) of length to origin of caudal, strongly flattened above. Snout thick, fleshy, broadly rounded or slightly ovate anteriorly, very short, its length in front of mouth being only 1/4 to 1/5 (about 22%) as great as that of head. Eye oval, its horizontal diameter 2/3 to nearly 4/5 as long as snout in front of mouth in late embryos and newly born specimens, but decreasing in relative size with growth to only about 1/2 as long as snout in front of mouth and thus only about 1/8 as long as head in adult, its midpoint a little anterior to front of mouth. Spiracle on dorsal side of head a little above level of eyes, transverse, about 1/2 as long as horizontal diameter of eye. Gill openings small, the longest about 1/2 as long as horizontal diameter of eye in newborn, but about as long as eye in adult, low on the sides, the 5th close in front of pectoral. Nostril close to anterior margin of snout, oblique, about 1/2 as long as horizontal diameter of eye, its anterior margin with a low, triangular lobe, rounded at the tip. Mouth only very slightly arched, its breadth about as great as length of snout in front of mouth. Lips noticeably thick and fleshy, but without special cartilaginous supports near corner of mouth, the lower lip free, but the upper joined to gum along central 1/3 of jaw. Upper labial fold extending about 1/3 of distance toward symphysis, but lower less than 1/2 that far. Pit at corner of mouth extremely voluminous, but its rearward prolongation extending only about 1/5 or 1/6 of the distance toward the 1st gill opening.

Teeth $\frac{8 \text{ or } 9—8 \text{ or } 9}{8 \text{ or } 9—1—8 \text{ or } 9}$, widely unlike in the 2 jaws; upper teeth thorn-like, on broad bifid bases, curved rearward, erect toward center of mouth but moderately oblique toward corners, the 1st tooth small; lower teeth blade-like, with quadrate bases and broad-triangular cusps, the latter with regularly serrate edges, erect toward center of mouth, but oblique and decreasing in size toward corners to a degree apparently depending on age and perhaps on individual variation; the median lower tooth as large as others, symmetrical, weakly notched on both edges at junction of cusp with base and overlapping its neighbor basally on either hand, the lateral lower teeth notched only on outer side and each overlapping the next outermost tooth; 3 or 4 series functional in upper jaw, and 1 or 2 series functional in lower jaw, depending on the stage in replacement.

First dorsal only about ¼ as long at base as head, brush-shaped with broadly rounded apex, its posterior margin nearly straight and perpendicular, its free rear corner, or free lower margin, about as long as base, its origin posterior to tips of pectorals by a distance about as long as horizontal diameter of eye when pectorals are laid back. Second dorsal a little larger than 1st, its origin about over middle of bases of pelvics, its distal margin concave and rear corner acute, thus differing from 1st dorsal, its free rear tip about as long as its base. Interspace between 2nd dorsal and caudal a little more than 1½ times as long as base of 2nd dorsal. Caudal a little more than ⅕ of total length, obliquely truncate terminally with broadly rounded apex, its lower margin deeply incised subterminally in rectangular outline and thus sharply marking off the terminal sector, its lower anterior corner rounded, about a right angle, the lower anterior margin a little less than ½ as long as upper margin. Pelvics about 1½ times as long at base as 2nd dorsal, with nearly straight margins, broadly rounded apices and tapering rear corners. Pectoral about ⅔ as long as head, paddle-shaped, with very broadly rounded tip, weakly convex outer margin and more strongly convex distal margin, the transition from distal to inner margin gradual, there being no definite inner corner.

Color. After preservation, uniformly dark chocolate or cinnamon brown below as well as above; also described as sometimes violet brown with poorly defined blackish spots in life, the fins with pale or whitish margins and caudal black-tipped.

Size. The young are born at a length of approximately 300 mm., but most of those caught are between 1,000 and 1,500 mm. (40 to 60 inches) long; the longest of which we have found definite measurement was 1,820 mm. (72 inches).[11] A specimen of about five feet weighed about 23½ pounds gutted.[11a] Females are larger than males, as is commonly the case among sharks.

Developmental Stages. Development is ovoviviparous. Gravid females are reported as containing 10 to 16 young. An embryo of 270 mm., still with the large yolk sac, already shows all the diagnostic characters of the adult except for the teeth, denticles and relatively larger eyes (horizontal diameter about ⅔ as long as snout in front of mouth).[12]

11. Duméril, Hist. Nat. Poiss., *1*, 1865: 452.
11a. Nichols and Firth, Proc. biol. Soc. Wash., *52*, 1939: 85.
12. For account of the uterine wall of a gravid female, see Ranzi (Publ. Staz. zool. Napoli, *13*, 1934: 366).

Habits. In its centers of abundance in the eastern Atlantic this shark is taken most often in at least moderately deep water. Off Nice, on the Mediterranean coast of France, it was long ago described as commonly caught at 1,000 meters depth; other depth records are from 300 to 600 meters and many have been taken on the Irish Atlantic slope between 200 and 350 fathoms (366 to 640 m.).[13] But since the Georges Bank specimen was taken in only 50 fathoms, with report of at least one other on the beach at Madeira,[14] it is apparent that it is not confined exclusively to deep water, and its New Zealand representative is also occasionally washed ashore. Gravid females are taken throughout the year in the Mediterranean. Nothing whatever is known of its feeding habits, although the nature of its teeth, and the fact that Mediterranean ones have been commonly caught on hook and line, suggest a fish diet.

Relation to Man. It is of relatively little commercial importance at present, but in the Azores it is the object of a special fishery for leather; formerly its skin was prized as an abrasive by cabinet makers and jewelers.

Range. In the eastern Atlantic from Rio de Oro, the Canaries, Madeira, Morocco, Azores and western Mediterranean north to the Irish Atlantic slope; plentiful locally (perhaps periodically) off the Mediterranean coasts of France and Portugal, as well as on the fishing grounds west of Ireland. There is but one record for the western Atlantic (see below). It is represented off South Africa, in the New Zealand–Australian region and in Japanese waters by allies so close that they appear to be identical with the Atlantic form (p. 501).

Occurrence in the Western Atlantic. The only record of the capture of this shark in the western Atlantic is the female here pictured (Figs. 96, 97), about five feet long, taken on the northern edge of Georges Bank in 50 fathoms on August 19, 1937.[15]

Synonyms and References:

1. North Atlantic:

La liche, Broussonet, Mem. Math. Phys. Acad. Sci. Paris, 1780: 677 (C. Breton, France); Duhamel, Traité Gén. Pêches, *4*, 1782: 301, 328 (descr.).

Squalus licha Bonnaterre, Tabl. Encyc. Meth. Ichthyol., 1788: 12 (descr., size).

Squalus americanus Gmelin, in Linnaeus, Syst. Nat., *1* (3), 1789: 1503 (descr., named on incorrect assumption that Cap du Breton specimen of Broussonet, 1780, was taken at the Nova Scotian loc. of that name); Walbaum, P. Artedi Genera Pisc. Emend. Ichthyol., *3*, 1792: 521 (descr.); Bloch and Schneider, Syst. Ichthyol., 1801: 136 (descr.); Bosc, Nouv. Dict. Hist. Nat., *21*, 1803: 192 (diagn.); Latreille, Nouv. Dict. Hist. Nat., *24*, 1804: 72 (in contents); Shaw, Gen. Zool., *5*, 1804: 347.

Le squale liche, Lacépède, Hist. Nat. Poiss., 4° ed., *1*, 1798: 279, pl. 10, fig. 3 (descr., ill., C. Breton), also other eds.; in Sonnini, Hist. Nat. Poiss., *4*, 1802: 117 (same as Lacépède, 1798).

Squalus nicaensis Risso, Ichthyol. Nice, 1810: 43, pl. 4, fig. 6 (descr., occur. near Nice, France).

Dalatias sparophagus Rafinesque, Carrat. Gen. Nuov. Sicil., 1810: 10, pl. 13, fig. 2 (descr. and ill. recognizable, though spiracle said to be lacking); Indice Ittiol. Sicil., 1810: 44 (Sicily).

Scymnus americanus Cuvier, Règne Anim., *2*, 1817: 130 (genl., notes earlier error regarding type loc., C. du

13. Fraser-Brunner, Proc. R. Irish Acad., *42*, 1935: 320.
14. Bowditch, Excurs. Madeira, 1825: 74.
15. Nichols and Firth, Proc. biol. Soc. Wash., *52*, 1939: 85.

Breton, not in America), also later eds. and translations; Bory de St. Vincent, Dict. Class. Hist. Nat., *15*, 1829: 598; Bonaparte, Mém. Soc. neuchâtel Sci. nat., *2* (8), 1839: 9 (in synopsis).

Scymnus vulgaris Cloquet, Dict. Sci. Nat., *25*, 1822: 433 (general); Reguis, Ess. Hist. Nat. Provence, *1* (1), 1877: 56 (descr., Medit.).

Scymnus licha Bowditch, T. E., Excurs. Madeira, 1825: 74 (Madeira); Roule, Result. Camp. sci. Monaco, *52*, 1917: 123 (off Lisbon).

Scymnus nicaensis Risso, Hist. Nat. Europe Merid., *3*, 1826: 137, pl. 2, fig. 4 (descr., size, color, Medit.); Lowe, Trans. zool. Soc. Lond., *2* (3), 1837: 194 (Madeira); Cuvier, Règne Anim., ill. ed., 1843: pl. 115, fig. 5 (jaws).

Squalus (Acanthorhinus) americanus Blainville, in Vieillot, Faune Franc. Poiss., 1825: 63, pl. 15, fig. 2 (descr., ill., France).

Squalus scymnus Voigt, in Cuvier, Tierreich, *2*, 1832: 512.

Scymnus lichia Bonaparte, Icon. Faun. Ital., *3*, 1835–1836: plate not numbered (descr., ill., Italy); Agassiz, L., Poiss. Foss., *3*, 1843: 367, pl. 40B, fig. 5 (vertebrae); Busch, W., Selach. Ganoid. Encephal., 1848: 22, pl. 2, fig. 3, 4 (brain); Van der Hoeven, Handb. Dierkunde, (2) *2*, 1855: 260 (general); Canestrini, Arch. Zool. Anat. Fisiol. Bologna, *1*, 1861: 267 (off Genoa); Fitzinger, Bild. Atlas Naturg. Fische, 1864: fig. 179 (good ill.); Bocage and Brito Capello, Poiss. Plagiost., 1866: 34 (Portugal); Brito Capello, J. Sci. math. phys. nat. Lisboa, *2*, 1869: 145 (Portugal); Günther, Cat. Fish. Brit. Mus., *8*, 1870: 425 (descr., refs., Medit., Madeira); Miklucho-Macklay, Beitr. Vergl. Neurol. Wirbelt., *1*, 1870: 11, pl. 1, fig. 3, 4 (brain); Canestrini, in Cornalia, *et al.*, Fauna d'Ital., 1872: 41 (Medit.); Gegenbaur, Unters. Vergl. Anat. Wirbelt., *3*, 1872: 23, 24, pl. 1, fig. 3, pl. 7, fig. 3, pl. 11, fig. 1, pl. 17, fig. 4, pl. 19, fig. 2, pl. 22, fig. 5–7 (anat.); Hertwig, Jena. Z. Naturw., *8*, 1874: 349, pl. 12, fig. 10, 13 (develop. of denticles, teeth); Gervais and Boulart, Poiss., *3*, 1876: 210, pl. 81 (Medit., Portugal); Doderlein, Prosp. Metod. Pesci Sicil., 1878–1879: 30 (near Sicily); Man. Ittiol. Medit., *2*, 1881: 100 (west. Medit.);[16] Moreau, Poiss. France, *1*, 1881: 358 (descr., ill., French coasts); Hasse, Naturl. Syst. Elasm. besond. Theil, 1882: 65, pl. 9 (skelet., dermal denticles); Carus, Prod. Fauna Medit., 1889–1893: 501 (W. Medit.);[16] Collett, Bull. Soc. zool. Fr., *15*, 1890: 219 (Madeira); Almeida and Roquete, Mammif. Peix. Costa e rias do Algarve Inquerit. Indust. Lisboa (1889), 1892: 374 (Portugal, not seen); Vieira, Ann. Sci. nat. Porto, *4*, 1897: 138 (Portugal); Sicher, Atti Accad. gioenia, (4) *11* (5), 1898: 18 (Medit.); Jungerson, Danish "Ingolf" Exped., *2* (1), 1899: 31 (claspers); Huber, Z. wiss. Zool., *70*, 1901: 604, pl. 27, fig. 3, 39 (claspers); Bridge, Camb. nat. Hist., *8*, 1904: 455, and subsequent eds. (general, Medit., neighboring Atlantic); de Braganza, Result. Invest. Sci. "Amelia," *2*, 1904: 90, 106 (Portugal); Helbing, Nova Acta Leop. Carol., *82* (4), 1904: 523 (comp. with *Somniosus*, skelet.); Seabra, Cat. Poiss. Port., 1911: 199 (Portugal); Roule, Bull. Inst. océanogr. Monaco, 243, 1912: 17 (west. Medit., off Morocco, Madeira); Jenkins, Fish. Brit. Isles, 1925: 326 (descr.); Leigh-Sharpe, J. Morph., *42*, 1926: 314 (claspers); Ranzi, Pubbl. Staz. zool. Napoli, *13* (3), 1934: 340, 366 (uterine liquid and wall); Needham, Biochem. Morphogen., 1942: 41 (embryonic weight).

Scymnus (no specific name) Agassiz, L., Poiss. Foss., *3*, 1835–1837: 12, pl. f, fig. 7 (teeth).

Scymnium niciense Cuvier, Règne Anim., ill. ed., 1843: pl. 115, fig. 5 (ill., jaws).

Scymnus (Scymnus) lichia Müller and Henle, Plagiost., 1841: 92 (descr., Medit., French coast); Duméril, Hist. Nat. Poiss., *1*, 1865: 452 (descr., distrib.).

Scymnorhinus lichia Bonaparte, Cat. Pesc. Europ., 1846: 16 (in list); Goode and Bean, Smithson. Contr. Knowl., *30*, 1895: 7, pl. 2, fig. 4; Mem. Harv. Mus. comp. Zool., *22*, 1896: 7, pl. 2, fig. 4 (descr., ill., W. Medit., Madeira); Regan, Ann. Mag. nat. Hist., (8) *2*, 1908: 57 (specs. in Brit. Mus.); Belloc, Rev. des Trav. Pêches Marit., 7 (2), 1934: 148 (ill. after Lozano Rey; Morocco, Rio de Oro, Canaries); Borri, Atti Soc. tosc. Sci. nat., *44*, 1934: 88 (Medit.); Lübbert and Ehrenbaum, Handb. Seefisch. Nordeurop., *2*, 1936: 290 (W. Medit., Portugal); Nichols and Firth, Proc. biol. Soc. Wash., *52*, 1939: 85 (Georges Bank spec., length, weight).

Dalatias licha Gray, List. Fish. Brit. Mus., *1*, 1851: 75 (Madeira); Fowler, Proc. Pan-Pacif. sci. Congr.,

16. See Doderlein, 1881, and Carus, 1889–1893, for additional records for the Mediterranean in publications not accessible to us.

4, 1930: 497 (Atlant.); Bull. Amer. Mus. nat. Hist., *70* (1), 1936: 86 (descr., ill., Medit. spec.); Cadenat, Rev. des Trav. Pêches Marit., *10* (4), 1937: 432 (Azores; special fishery for leather); Fowler, Bull. U.S. nat. Mus., 100 (*13*), 1941: 267 (descr., refs., Atlant. and Pacif.).

Squalus (*Scymnus*) *lichia* Van der Hoeven, Handb. Dierkunde, *2*, 1855: 260.

Scymnorhinus licha Garman, Mem. Harv. Mus. comp. Zool., *36*, 1913: 237 (descr.); Rey, Fauna Iberica Peces, *1*, 1928: 474, pl. 5, fig. 1 (Iberian coasts); Uriarte and Mateu, Notas Inst. esp. Oceanogr., (2) *53*, 1931: 26 (Canaries); Fraser-Brunner, Proc. R. Irish Acad., *42*, 1935: 320 (Irish Atlant. slope, color); Nobre, Fauna Marinha Port. Vert., *1*, 1935: 460, pl. 64, fig. 200 (descr., off Portugal); Tortonese, Atti Soc. ital. Sci. nat., 77, 1938: 311 (Medit.).

Scymnorhinus lichi (*Scymnus lichia*) Holmgren, Acta zool., *22* (1–3), 1941: 24 (skull).

2. South Africa and Pacific; apparently referrable to *D. licha*:[17]

Scymnus lichia Parker, Trans. Proc. N. Z. Inst., *15*, 1883: 223, pl. 31, 32 (descr., ill., N. Zealand); Ishikawa and Matsuura, Prel. Cat. Fish. Mus. Tokyo, 1897: 61 (Japan).

Dalatias americanus Jordan and Snyder, Annot. zool. jap., *3*, 1901: 129 (Japan).

Dalatias licha Jordan and Fowler, Proc. U.S. nat. Mus., *26*, 1903: 637 (Japan); Izuka and Matsuura, Cat. zool. Mus. Tokyo, 1920: 188 (Japan); Fowler, Proc. Pan-Pacif. sci. Congr., *4*, 1930: 497 (Aust., N. Zealand, Japan, Atlant.); Bull. U.S. nat. Mus., 100 (*13*), 1941: 267 (descr., listed Japan, also Atlantic and Medit.).

Scymnorhinus licha McCulloch, Res. "Endeavour," *2* (3), 1914: 81, pl. 14 (ill., Aust.); Waite, Trans. Proc. N. Z. Inst., *46*, 1914: 128, pl. 4, fig. 1 (ill., N. Zealand); Rec. S. Aust. Mus., *2*, 1921: 24, fig. 33; Handb. Brit. Sci. Guild, Fish. S. Aust., 1923: 43; Phillipps, N. Z. J. Sci. Tech., *10*, 1928: 224 (N. Zealand).

Scymnorhinus phillippsi Whitley, Aust. Zool., *6*, 1931: 310; Fish. Aust., *1*, 1940: 150 (ill., Aust., N. Zealand).

Pseudoscymnus boshuensis Herre, Copeia, 1935: 124 (descr., Japan); Fowler, Bull. U.S. nat. Mus., 100 (*13*), 1941: 275 (descr., Japan).

Scymnorhinus brevipinnis Smith, Trans. roy. Soc. S. Afr., *24*, 1936: 1 (descr., S. Afr.).

Dalatias phillippsi Fowler, Bull. U.S. nat. Mus., 100 (*13*), 1941: 268 (Aust., N. Zealand).

Dalatias brevipinnis Fowler, Bull. U.S. nat. Mus., 100 (*13*), 1941: 268 (S. Afr.).

Genus *Isistius* Gill, 1864

Isistius Gill, Proc. Acad. nat. Sci. Philad., 1864: 264; type species, *Scymnus brasiliensis* Quoy and Gaimard, 1824, monotypic.

Generic Synonyms:

Scymnus (in part) Quoy and Gaimard, Voy. "Uranie," Zool., 1824: 197; not *Scymnus* Cuvier, 1817.

Dalatias (in part) Gray, List. Brit. Mus., *1*, 1851: 76; not *Dalatias* Rafinesque, 1810.

Squalus Johann, Z. Wiss. Zool., *66*, 1891: 152; for *S.* (*Scymnus*) *fulgens* Bennett, 1840; not *Squalus* Linnaeus, 1758.

Leius Kner, Denkschr. Akad. Wiss. Wien., *24*, 1865: 9; type species, *L. ferox* Kner, equals *Scymnus brasiliensis* Quoy and Gaimard, 1824.

Generic Characters. Dalatiidae without dorsal fin spines; snout in front of mouth much shorter than from front of mouth to origin of pectorals; caudal peduncle without lateral ridges or precaudal pits; pit at corner of mouth prolonged below upper lip and rearward as a narrow furrow; expanded lips at corners of mouth with special cartilaginous supports; teeth widely unlike in the 2 jaws, the uppers slender, thorn-like, widely spaced, the lowers with triangular smooth-edged or partly serrate cusp and quadrate base, each

17. See page 501.

overlapping the next outermost; dermal denticles low, with depression in the crown; rear end of base of 1st dorsal about over origin of pelvics; 2nd dorsal and pelvics only a little larger than 1st dorsal; interspace between 1st and 2nd dorsal much shorter than between pelvics and caudal; caudal with axis approximately in continuation of main axis of trunk, very broad relative to its length, with shallow subterminal notch, its lower anterior corner expanded as a well defined lobe; pectoral small, paddle-shaped; skin sprinkled with strongly luminescent, glandular points. Characters otherwise those of the family.

Range. Tropical and subtropical belts of Atlantic, Pacific and Indian Oceans.

Fossil Teeth. Upper Cretaceous to Eocene, Africa; Eocene to Miocene, Europe.

Species. Only one species is known.

Isistius brasiliensis (Quoy and Gaimard), 1824

Figures 98, 99

Study Material. Immature male, 383 mm. long, taken by the Research Vessel "Atlantis," Station 2947, north of the Bahamas at Lat. 25° 11′ N., Long. 77° 19′ W., where the depth was 1,000 fathoms[1] (Harv. Mus. Comp. Zool., No. 36039); two fe-

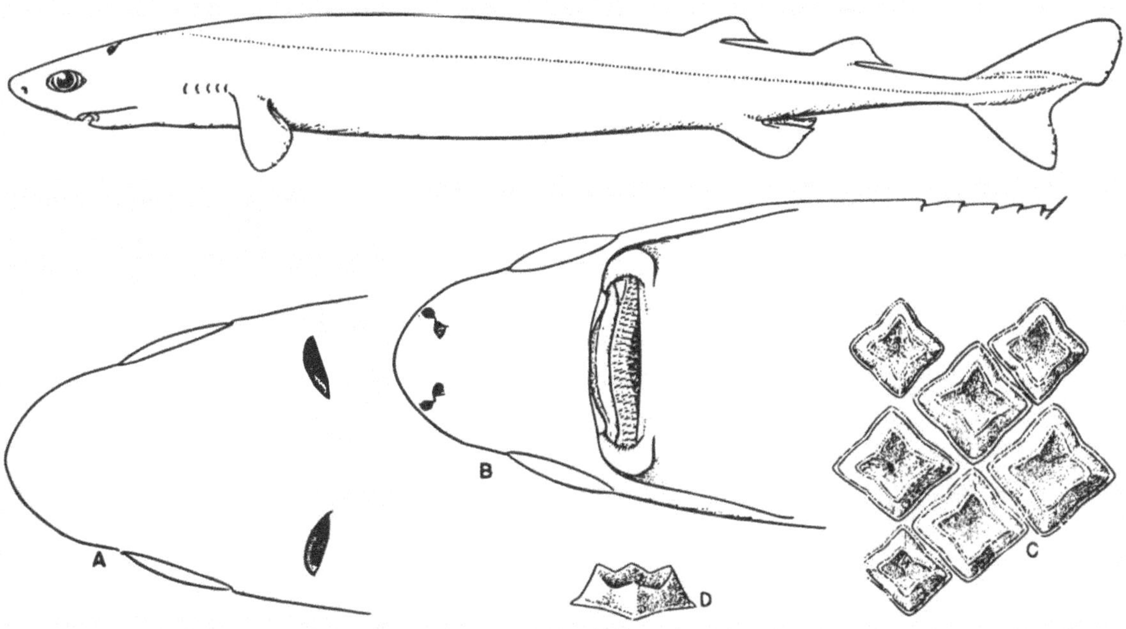

FIGURE 98. *Isistius brasiliensis*, immature male, about 383 mm. long, from north of the Bahamas (Harv. Mus. Comp. Zool., No. 36039). *A* Anterior part of head from above to show spiracles, about 1.1 x. *B* Head from below, about 1.1 x. *C* Dermal denticles, about 45 x. *D* Lateral oblique view of dermal denticle, about 45 x.

1. But not necessarily from so great a depth, as the nets were brought up open to the surface.

males, 465 and 501 mm. long, from Japan (Harv. Mus. Comp. Zool., No. 1368, 1245); female about 485 mm. long, from the vicinity of Albemarle Island, Galapagos, Lat. 2° 34′ N., Long. 92° 06′ W., in a trawl haul from 1,360 fathoms, Albatross Sta. 3413 (Harv. Mus. Comp. Zool., No. 1005).

Distinctive Characters. Isistius is separated from all other northwestern Atlantic sharks of its suborder except *Dalatias* by the triangular shape of its lower teeth; the posi-

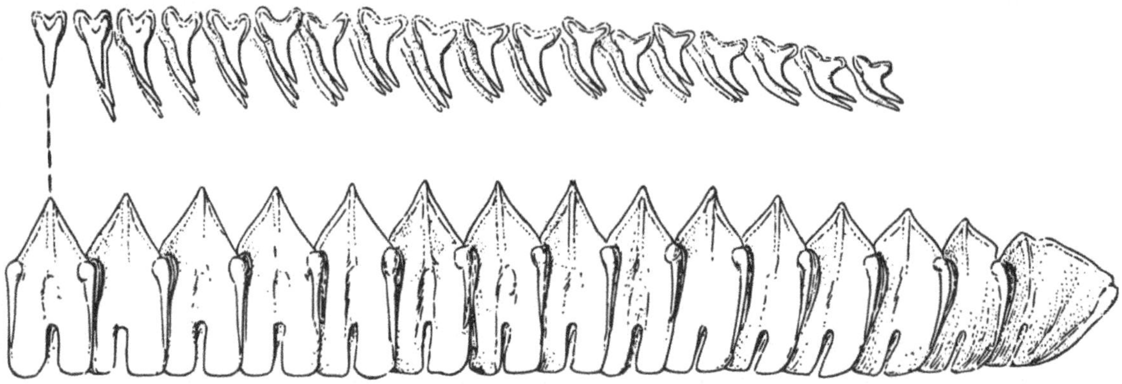

FIGURE 99. *Isistius brasiliensis*, pictured in Fig. 98. Upper and lower teeth, left-hand side, about 4.4 x.

tion of its first dorsal fin far rearward marks it off at a glance from *Dalatias*. It shares this last character with *Echinorhinus*, but there is no danger of confusing it even with new-born specimens of the latter, for the shape of its caudal, its teeth and its dermal denticles are very different and its gill openings much smaller.

Description. Proportional dimensions in per cent of total length. Male, 383 mm., from Lat. 25° 11′ N., Long. 77° 19′ W. (Harv. Mus. Comp. Zool., No. 36039). Female, 501 mm., from Tokyo, Japan (Harv. Mus. Comp. Zool., No. 1245).

Trunk at origin of pectoral: breadth 10.2, 11.0; height 8.4, 11.0.

Snout length in front of: outer nostrils 0.8, 0.8; mouth 7.0, 6.6.

Eye: horizontal diameter 3.4, 3.2.

Mouth: breadth 4.7, 4.2; height 0, 0.

Nostrils: distance between inner ends 1.6, 1.4.

Labial furrow length from angle of jaw: upper 4.2, 5.6.

Gill opening lengths: 1st 0.8, 0.8; 2nd 0.8, 0.8; 3rd 0.8, 0.8; 4th 0.8, 0.8; 5th 0.8, 0.8.

First dorsal fin: vertical height 3.4, 2.6; length of base 3.1, 3.2.

Second dorsal fin: vertical height 2.9, 2.4; length of base 3.9, 3.8.

Caudal fin: upper margin 14.6, 15.3; lower anterior margin 11.2, 10.2.

Pectoral fin: outer margin 7.8, 7.3; inner margin 4.4, 5.6; distal margin 4.2, 4.2.

Distance from snout to: 1st dorsal 59.5, 59.3; 2nd dorsal 70.5, 71.8; upper caudal 85.4, 84.7; pectoral 19.1, 17.9; pelvics 60.1, 62.8.

Interspace between: 1st and 2nd dorsals 8.5, 9.3; 2nd dorsal and caudal 10.7, 9.6. *Distance from origin to origin of:* pectoral and pelvics 43.3, 44.1; pelvics and caudal 22.9, 21.9.

Trunk subcylindrical and very slender, its greatest height only about 1/8 its length to origin of caudal, the dorsal profile only weakly arched and the ventral profile nearly straight. Body sector to cloaca more than twice as long as tail sector. Caudal peduncle without lateral ridges or precaudal pits, and without mid-dorsal ridge. Dermal denticles small, closely spaced, but with skin exposed between them, highly diagnostic in shape, being very low with no distinction between pedicel and blade, transversely truncate apically, quadrilateral in outline, but occasionally polygonal, with concave margins and rounded corners, the crown with a quadrate depression.

Head a little less than 1/4 (22 to 23%) of length to origin of caudal. Snout thick, fleshy, broadly-ovate and very short, its length in front of mouth a little less than 1/3 of length of head. Eye oval, its horizontal diameter almost 1/2 as great as length of snout in front of mouth. Spiracles on top of head, slightly oblique, oval, about 1/2 as long as horizontal diameter of eye and posterior to latter by a distance a little greater than their own length. Gill openings very small, only about 1/3 to 1/4 as long as horizontal diameter of eye, rather high on the sides, the 5th close in front of origin of pectoral. Nostril close to anterior margin of snout, about 1/3 as long as horizontal diameter of eye, oblique, its anterior margin expanded as a short broadly-triangular lobe rounded at the tip. Mouth very little arched. Lips fleshy, adnate to gum along midsector of upper jaw but elsewhere free and widely distensible, with rounded wing-like expansions at corners of mouth, enclosing an extensive funnel-like cavity that extends inward along upper jaw on either side; upper lip overlaps lower as a thin skin fold rearward from corner of mouth for about 1/2 of the distance to 2nd gill opening.

Teeth $\frac{15\text{ or }16—1—15\text{ or }16}{12\text{ to }15—1—12\text{ to }15}$, lowers increasing in number with growth; upper teeth slender, thorn-like, increasingly curved outward toward corners of mouth; lowers much larger, erect, symmetrical, their bases subquadrate with a shoulder on each side at point of transition to the triangular, sharp-pointed cusp, their edges mostly smooth, partly wavy, or even showing faint indication of serrations, the central lower tooth overlapping its neighbor on either side basally, with each subsequent tooth correspondingly overlapping the next outermost; the outermost lower tooth widely expanded basally on the outer side; 3 to 4 series functional in upper jaw, one in lower.

First dorsal very small, its base between 1/5 and 1/6 as long as head, sloping, with broadly rounded apex, its posterior outline very weakly concave, its free rear tip slender and about as long as the base, the rear end of base about over origin of pelvics. Second dorsal similar to 1st, but a little longer at base and larger in area, its origin about over rear tips of pelvics. Interspace between 1st and 2nd dorsals about as long as between 2nd dorsal and caudal. Caudal only about 1/7 the total length, about 2/3 as broad as long, the tip broadly rounded, its posterior outline deeply concave in angular contour, marking off the terminal sector, the lower lobe broadly triangular with narrowly rounded tip, its anterior margin about 2/3 as long as upper margin of fin. Pectoral about 2/5 (40–45%) as long as head to 5th gill opening.

Color. Dark brown above, paler brown or brownish white below, except for a conspicuous dark collar around the neck in the region of the gill openings; fins brown, the pectorals, dorsals and pelvics with pale distal margins, the upper and lower lobes of the caudal with dusky or darker brown tips; the inner side of upper eyelid not pigmented. Except in the region of the dark collar the lower surface is closely, and the sides sparsely, sprinkled with black dots, presumably luminous;[2] these also occur in patches on the sides of the head, on the dorsal and caudal fins, and on the basal parts of the pectorals, with a few along the back.

Size. Recorded specimens have ranged in length from about 140 mm. (5½ inches) to about 495 mm. (19½ inches); females are mature at a length of 18 inches.

Developmental Stages. Presumably development is ovoviviparous, but the only available definite information is that a female has been reported as containing six large eggs.[3]

Habits. This is a pelagic species, the majority of recorded specimens having been taken either from small depths or at the surface at night. And while a few have been recorded from deep hauls,[4] it is likely that they were picked up by the net on its way down or up. Nothing is known of its diet, nor of its breeding habits.

This is the most brilliantly luminescent of sharks. According to an eyewitness account[5] the entire lower surface of its trunk, with the exception of the dark collar, its paired fins and its caudal, emits a vivid greenish light. While the luminescence apparently is not under nervous control, since it is not affected by handling, it is not a constant characteristic of the species, for one specimen taken alive failed to show any trace of it.[6]

Range. The localities of capture include the Gulf of Guinea, the offings of Sierra Leone and Cape Verde in the eastern Atlantic, as well as Brazil, the Bahamas, and north of the Bahamas in the western Atlantic; also the vicinity of the Galapagos, Hawaiian Islands, Japan, Fiji, central equatorial Pacific west of Christmas Island, equatorial belt north of New Guinea, Lord Howe Island off New South Wales, Australia, Mauritius, and between Java and western Australia. These localities are dispersed widely enough to prove this shark cosmopolitan in the tropical and subtropical belts of all three oceans. Records for the western Atlantic are: off Rio de Janeiro (one specimen), among the Bahamas

2. The distribution of these has been described in detail by Burckhardt (Ann. Mag. nat. Hist., [7] *6*, 1900: 565, 566).
3. Bennett, Narr. Whaling Voy., *2*, 1840: 255.
4. See Study Material, p. 509; also Garman (Mem. Harv. Mus. comp. Zool., *24*, 1899: 40) and Parr (Bull. Bingham oceanogr. Coll., *3* [7], 1937: 1).
5. F. D. Bennett's (Narr. Whaling Voy., *2*, 1840: 255) account has been quoted repeatedly.
6. Duncker and Mohr, Mitt. zool. StInst. Hamburg, *44*, 1929: 84.

(one specimen), and about 160 miles north of the Bahamas (one specimen). See Study Material, p. 509.

Synonyms and References:

Scymnus brasiliensis Quoy and Gaimard, Voy. "Uranie," Zool., 1824: 198 (brief descr., Brazil).

Squalus (Scymnus) fulgens Bennett, F. D., Narr. Whaling Voy., *2*, 1840: 255 (descr., size, eggs, luminescence, trop. Pacif. near Christmas I.[7]); Bennett, G., Gatherings Nat. Australia, 1860: 66 (luminescence, a second trop. Pacif. spec.).

Scymnus (Scymnus) brasiliensis (incl. var. *torquatus* and var. *unicolor*) Müller and Henle, Plagiost., 1841: 92 (descr., Mauritius, C. Verde at St. Jago, Rio de Janeiro); Duméril, Hist. Nat. Poiss., *1*, 1865: 453 (refs., descr., Mauritius).

Dalatias brasiliensis Gray, List. Fish. Brit. Mus., *1*, 1851: 76 (Isle of France, St. Jago, Rio de Janeiro).

Scymnus torquatus Duméril, Arch. Mus. Hist. nat. Paris, *10*, 1861: 261 (name only, St. Jago, C. Verde).

Isistius brasiliensis Gill, Proc. Acad. nat. Sci. Philad., 1864: 264, footnote (name); Günther, Cat. Fish. Brit. Mus., *8*, 1870: 429 (refs., descr., S. Pacif. and Gulf of Guinea specs.); Peters, W. C. H., Monatschr. Berlin Akad., 1876: 853 (Indian Oc., Lat. 14° 23′ S., Long. 118° 16′ E.); Rochebrune, Act. Soc. linn. Bordeaux, (4) *6*, 1882: 48; Faune Senegambie, Poiss., *1*, 1883–1885: 26 (C. Verde); Sauvage, in Grandidier, Hist. Phys. Nat. Madagascar, *16*, 1891: 5, 511 (spec. in Paris Mus.); Garman, Mem. Harv. Mus. comp. Zool., *24*, 1899: 34, pl. 1, fig. 1, pl. 2–3, pl. 69, fig. 2 (descr., size, no. of teeth, anat., near Galapagos); Johann, Z. wiss. Zool., *66*, 1899: 152 (luminescence); Burckhardt, Ann. Mag. nat. Hist., (7) *6*, 1900: 565 (luminescence, luminous organs); Waite, Rec. Aust. Mus., *3*, 1900: 195, fig. 1, 2 (no. teeth, largest recorded spec., Lord Howe I.); Mangold, Pflüg. Arch. Ges. Physiol., *119*, 1907: 583 (luminescence); Ribeiro, Arch. Mus. nac. Rio de J., *14*, 1907: 169, 205 (descr., refs., Brazil); Brauer, Wiss. Ergebn. 'Valdivia,' *15*, 1908: 133, pl. 2 (luminescence); Regan, Ann. Mag. nat. Hist., (8) *2*, 1908: 55 (class., size); Garman, Mem. Harv. Mus. comp. Zool., *36*, 1913: 238 (descr.); Jordan, Tanaka and Snyder, J. Coll. Sci. Tokyo, *33*, 1913: 23 (old loc. records, also Fiji, Japan); Lampe, Deutsche sudpolar Exped., *15*, Zool., 7, 1914: 214, pl. 2 (no. of lower teeth, sizes, 3 spec. off Sierra Leone, W. Afr.); Metzelaar, Trop. Atlant. Visschen, 1919: 191 (refs.); Ribeiro, Fauna brasil., Peixes, Mus. nac. Rio de J., *2* (1) Fasc. 1, 1923: 26 (same as Ribeiro, 1907); Fowler and Ball, Bull. Bishop Mus., *26*, 1926: 5 (footnote that *Echinorhinus brucus* Fowler, 1923, from Hawaii, was actually *I. brasiliensis*); Whitley, J. Pan-Pacif. Res. Instn., *2* (1), 1927: 3 (Fiji); Fowler, Mem. Bishop Mus., *10*, 1928: 23 (descr., refs., Honolulu); Duncker and Mohr, Mitt. zool. StInst. Hamburg, *44*, 1929: 84 (size, depth of capture, Equatorial Pacif. near New Guinea, Lat. 00° N., Long. 146° 5′ E.); Fowler, Proc. Pan-Pacif. sci. Congr., (4) *3*, 1930: 497 (distrib.); Mem. Bishop Mus., *11*, 1931: 314 (refs.); Bull. Amer. Mus. nat. Hist., *70* (1), 1936: 87 (refs., descr.); Parr, Bull. Bingham oceanogr. Coll., *3* (7), 1937: 1 (north of Bahamas); Beebe and Tee-Van, Zoologica, N. Y., *26*, 1941: 121 (ref.); Fowler, Bull. U.S. nat. Mus., 100 (*13*), 1941: 270 (refs., descr., distrib., luminescence); Arqu. Zool. Estado São Paulo, *3*, 1942: 129 (Brazil); Bigelow and Schroeder, Guide Comm. Shark Fish., Anglo Amer. Caribb. Comm., Wash., 1945: 146, fig. 55 (descr., ill., habits, range).

Leius ferox Kner, Denkschr. Acad. math. Natur. Wien, *24*, 1865: 10, pl. 4, fig. 2 (descr., ill., Australia); Schmeltz, Cat. Mus. Godeffroy, *3*, 1866: 13 (South Seas, not seen); Johann, Z. wiss. Zool., *66*, 1899: 152 (luminescence); Whitley, Fish. Aust., *1*, 1940: 149 (descr., ill., luminosity, Aust.).

Isistius ferox Schmeltz, Cat. Mus. Godeffroy, *2*, 1865: 10 (South Seas, not seen).

Isistius braziliensis Macleay, Proc. linn. Soc. N. S. W., *6*, 1881: 368 (ref., Aust.).

Squalus fulgens Johann, Z. wiss. Zool., *66*, 1899: 152 (luminescence).

Leius brasiliensis Günther, J. Mus. Godeffroy, *6* (17), 1910: 490 (refs., depth, distrib.).

Apristurus spongiceps Jordan and Jordan, Mem. Carneg. Mus., *10*, 1922: 3 (Hawaii); not *Catulus spongiceps* Gilbert, 1905.[8]

7. A second specimen reported from Lat. 55° N., Long. 110° W.; either the latitude or longitude is evidently an error, for this would place it far inland in Canada.

8. A cast in Bishop Museum, later identified by Fowler (Mem. Bishop Mus., *10*, 1928: 23) as *I. brasiliensis*.

Echinorhinus brucus Fowler, Occ. Pap. Bishop Mus., *8* (7), 1923: 375 (Honolulu); not *Squalus brucus* Bonnaterre, 1788.[9]

Genus *Somniosus* Lesueur, 1818

Somniosus Lesueur, J. Acad. nat. Sci. Philad., *1*, 1818: 222; type species, *S. brevipinna* Lesueur, Massachusetts, monotypic.

Generic Synonyms:

Squalus Gunnerus, Drontheim Gesellsch. Schr. Leipzig, *2*, 1776: 299; for *S. carcharias* Gunnerus; not *Squalus* Linnaeus, 1758.

Acanthorhinus (in part) Blainville, Bull. Soc. philom. Paris, 1816: 121: for *S. norwegianus* Blainville.[1]

Scymnus (in part) Fleming, Hist. Brit. Anim., 1828: 166, for *S. borealis* Fleming, equals *Squalus borealis* Scoresby, 1820; not *Scymnus* Cuvier, 1817.

Laemargus (subgenus) Müller and Henle, Plagiost., 1841: 93; type species, *S. borealis* Fleming, 1828, equals *Squalus borealis* Scoresby, 1820, Spitzbergen.

Leiodon Wood, Proc. Boston Soc. nat. Hist., *2*, 1846: 174; type species, *S. echinatum* Wood, monotypic.

Dalatias (*Somniosus*) Gray, List Fish. Brit. Mus., *1*, 1851: 76, for *Scymnus* (*Laemargus*) *borealis* Müller and Henle, 1841, equals *Squalus borealis* Scoresby, 1820; not *Dalatias* Rafinesque, 1810.

Scimnus Van Beneden, Bull. Acad. Sci. Roy. Bruxelles, *20* (2), 1853: 258; emended spelling for *Scymnus*.

Rhinoscymnus Gill, Proc. Acad. nat. Sci. Philad., 1864: 264, footnote 5; type species, *Scymnus rostratus* Risso, 1826, monotypic.

Generic Characters. Dalatiidae without dorsal fin spines; snout in front of mouth much shorter than from front of mouth to origin of pectorals; midline of back with a faint dermal ridge; caudal peduncle with faint lateral ridges, at least in some cases, but without precaudal pits; labial furrows and a pit prolonged rearward from corner of mouth; teeth widely dissimilar in the two jaws; the uppers slender, conical, widely spaced; the lowers quadrate, each overlapping the next outermost, the cusps smooth edged and so oblique that their inner margins form a continuous cutting edge, much as in *Squalus* and in *Centroscymnus;* dermal denticles conical to thorn-like, curved rearward; rear end of base of 1st dorsal far anterior to origin of pelvics; 2nd dorsal over or a little posterior to rear end of bases of pelvics; 2nd dorsal only about as large as 1st dorsal, but pelvics considerably larger; interspace between 1st and 2nd dorsals longer than between pelvics and caudal; caudal very wide relative to its length, its lower anterior corner forming a more or less definite lobe, its subterminal margin notched; pectorals with broadly rounded corners; with or without functional luminous organs (see p. 516). Development ovoviviparous in one species, perhaps oviparous in another. Characters otherwise those of the family.

Range. Arctic Atlantic (including White Sea) south to North Sea, Portugal, Mediterranean and Cape Cod; Bering Sea, in North Pacific, south to Japan, southeastern Alaska, and occasionally southern California; also Antarctic (Maquarie Island).

9. Fowler and Ball (Bull. Bishop Mus., *26*, 1926: 5, footnote) point out that specimens in the Bishop Museum on which this record was based are actually *I. brasiliensis*.

1. This is a *nomen nudem*, see footnote 37, p. 523.

Species. It seems sufficiently established that the Greenland Sharks of the sub-Arctic on the two sides of the North Atlantic (including the White Sea) and of neighboring parts of the Arctic seas belong to a single species, described first by Gunnerus in 1766 as *Squalus carcharias*, but which, under the rules of nomenclature, must be called *microcephalus*, Bloch and Schneider, 1801, the name *Squalus carcharias* having been used previously by Linnaeus, 1758, for a very different shark (p. 142). It is still an open question what the relationship is between *microcephalus* and the very much smaller form that has been reported repeatedly from the Mediterranean and from the coast of Portugal as *rostratus*. The difficulty, as is so often the case, is that the older portrayals differ widely as regards relative locations of the fins and shape of the head. It is even possible that more than one species may be included among the supposed *rostratus*.[2] But if the more recent illustrations of *rostratus* can be accepted as reliable, it differs from *microcephalus* in a relatively much shorter interspace between the tip of second dorsal and caudal, relatively larger fins, smaller denticles, strongly developed luminous organs, and in various skeletal characters;[3] it also attains maturity when much smaller, and, still more important, it is ovoviviparous.[4]

The North Pacific representative of the genus has usually been considered identical with the North Atlantic form. However, the first dorsal stands considerably farther rearward in a Japanese specimen that we have examined[5] (as shown by Tanaka also[6]) than is ordinarily the case in Atlantic specimens, and its pectoral and caudal fins are larger; the lower anterior and upper posterior margins of its caudal are much more strongly convex, the distance from the tip of its second dorsal to the origin of its caudal shorter relatively, its upper teeth are considerably broader, and the basal outlines of its lower teeth more deeply incised. These differences seem sufficient to mark it off as a separate species. We have therefore proposed the name *pacificus* for it.[7]

In all probability the Greenland Shark of Bering Sea and Alaska belongs to this species. However, no detailed account of it has yet appeared, nor have we adequate material for comparison.

The Antarctic representative of the genus, known from a single specimen only, has also been made the basis of a separate species, *antarcticus* Whitley, 1939. Its first dorsal appears to stand even farther forward than in *microcephalus*, and critical examination may reveal additional differences.[8]

2. Brito Capello's measurements and illustration (J. Sci. math. phys. nat. Lisboa, *2*, 1870: 141, pl. 9, fig. 2) represent both the snout and the caudal peduncle as much longer than in any other *Somniosus*.
3. Burckhardt, Ann. Mag. nat. Hist., (7) *6*, 1900: 559; Helbing, Nova Acta Leop. Carol., *82*, 1904: 347 ff., pl. 8, 9.
4. For a recent report of its embryo, see Borri (Atti Soc. tosc. Sci. nat., *44*, 1934: 101).
5. This is the specimen illustrated by Garman (Mem. Harv. Mus. comp. Zool., *36*, 1913: pl. 15, fig. 1–3) as *brevipinna*.
6. Fish. Japan, *3*, 1911: pl. 13, fig. 32–36.
7. Bigelow and Schroeder, Proc. New Engl. zool. Cl., *23*, 1944: 35.
8. See Whitley (Fish. Aust., *1*, 1940: 152) for illustration of the unique specimen.

Tentative Key to Northern Hemisphere Species

1a. Origin of 1st dorsal much nearer to tip of snout than to tip of caudal; interspace between 1st and 2nd dorsals at least as long as from snout to 1st or 2nd gill opening.

2a. Distance from tip of 2nd dorsal to origin of caudal nearly or quite as long as base of 2nd dorsal; base of 1st dorsal only ¼ to ⅕ as long as distance from tip of snout to origin of pectorals; no evident luminescent organs.
microcephalus Bloch and Schneider, 1801, p. 516.

2b. Distance from tip of 2nd dorsal to origin of caudal only about ½ as long as base of 2nd dorsal; base of 1st dorsal nearly or quite ⅓ as long as from tip of snout to origin of pectoral; adult with luminescent dots. *rostratus* Risso, 1826.
Mediterranean; probably Portugal.

1b. Origin of 1st dorsal almost as near to tip of caudal as to tip of snout; interspace between 1st and 2nd dorsals only about ⅔ as long as from tip of snout to 2nd gill opening.
pacificus Bigelow and Schroeder, 1944.
Japan, probably also Bering Sea and Alaska, south to Puget Sound and occasionally to southern California.

Somniosus microcephalus (Bloch and Schneider), 1801

Greenland Shark, Gurry Shark

Figures 100, 101

Study Material. Excellent mounted specimen, 10 feet long, from Upernavik, West Greenland (Amer. Mus. Nat. Hist.) and old mount of a specimen 1,740 mm. (Harv. Mus. Comp. Zool.); jaws of Gulf of Maine specimen, about 11 feet (Harv. Mus. Comp. Zool.); lower jaw, from a Grand Banks specimen, taken at Lat. 43° 11′ N., Long. 51° 22′ W. (U.S. Nat. Mus., No. 26270); excellent photographs of an 11-foot specimen;[9] also a fresh-caught specimen of 10 to 11 feet from Cape Cod Bay examined by us in the flesh some years ago.[10]

Distinctive Characters. The adult Greenland Shark is so large that it could not be confused with any other Atlantic squalid. Its lack of dorsal fin spines further marks it off from the members of *Squalus, Centroscyllium* and the pigmy genus *Etmopterus* among local genera. Its quadrate lower teeth with strongly oblique cusps separate it sharply from *Dalatias* and *Isistius,* and its dermal armature sets it off from *Centroscymnus* (cf. Fig. 101 B with 94 A). The forward position of its first dorsal, the shape of its teeth in general, its dermal denticles, the shape of its caudal and its much smaller gill openings separate it from *Echinorhinus.*

9. Contributed by F. E. Firth.

10. We have examined the Japanese specimen mentioned above, the lower jaw of one found on the beach near St. Michaels, Alaska, and the head of another from California loaned by the U.S. National Museum.

Description. Proportional dimensions in per cent of total length. Specimen, 2,990 mm. (mounted, Amer. Mus. Nat. Hist., New York).

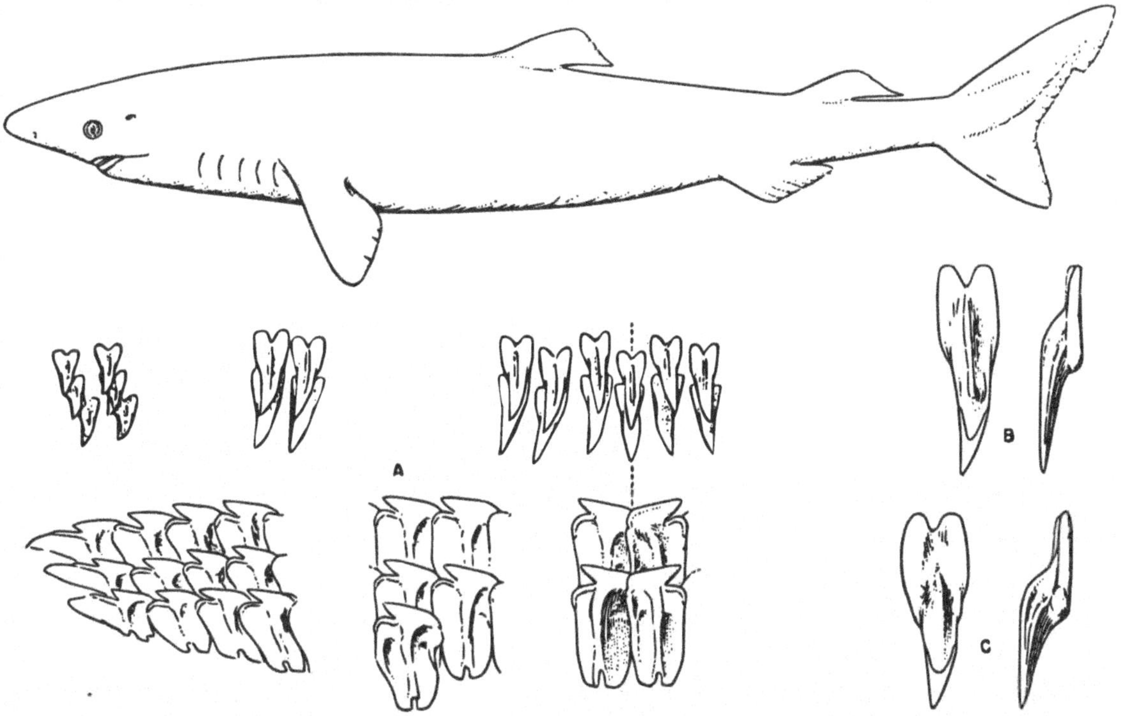

FIGURE 100. *Somniosus microcephalus.* Side view of a female from a mounted specimen 1,740 mm. long (Harv. Mus. Comp. Zool.), from photographs of a fresh 11-foot specimen and from other available information. *A* Upper and lower teeth from right-hand side, at center of mouth, from midway along the jaws and from the corner of the mouth of a specimen about 11 feet long from the Gulf of Maine (Harv. Mus. Comp. Zool., No. 36119)., about 1.8 x. *B* Front and side views of second upper tooth, about 3.5 x. *C* Eleventh upper tooth, about 3.5 x.

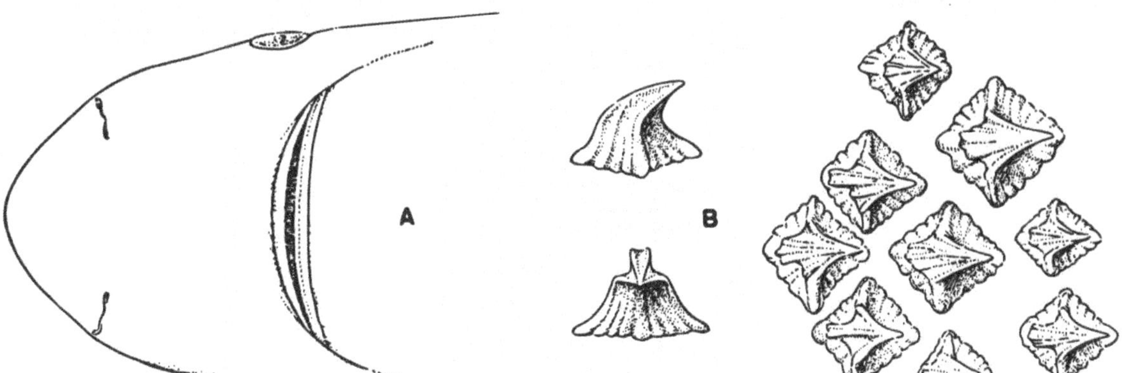

FIGURE 101. *Somniosus microcephalus. A* Anterior part of head from below, obtained from available information. *B* Dermal denticles from a mounted specimen, 1,740 mm. long (Harv. Mus. Comp. Zool.), about 15 x.

Trunk at origin of pectoral: breadth 13.7; height 13.4.
Snout length in front of: outer nostrils 2.1; mouth 8.3.
Eye: horizontal diameter 1.3.
Mouth: breadth 11.7; height 3.5.
Nostrils: distance between inner ends 6.3.
Labial furrow length from corner of mouth: upper 1.3.
Gill opening lengths: 1st 2.2; 2nd 2.8; 3rd 2.8; 4th 2.8; 5th 3.1.
First dorsal fin: vertical height 3.5; length of base 6.6.
Second dorsal fin: vertical height 2.5; length of base 4.9.
Caudal fin: upper margin 19.3; lower anterior margin 13.8.
Pectoral fin: outer margin 12.1; inner margin 6.5; distal margin 6.2.
Distance from snout to: 1st dorsal 40.7; 2nd dorsal 66.6; upper caudal 80.7; pectoral 23.7; pelvics 63.2.
Interspace between: 1st and 2nd dorsals 19.5; 2nd dorsal and caudal 8.5.
Distance from origin to origin of: pectoral and pelvics 39.5; pelvics and caudal 18.9.

Trunk subcylindrical anteriorly, but tapering posteriorly, its greatest height about 1/6 of its length to origin of caudal, somewhat compressed, about 2/3 as thick as high opposite the pectorals and 70 to 80% as thick as high at caudal peduncle, but so soft and flabby that one lying on the dock bears little resemblance to its form when in the water.[11] Back smooth between dorsals, but with a faintly indicated dermal ridge along midline extending rearward from opposite pectorals about to origin of 1st dorsal.[12] Caudal peduncle described as with an indistinct lateral longitudinal ridge, much as in *Squalus,* but this is not visible in the mounted specimens we have examined. Dermal denticles conical, curved rearward, longitudinally fluted, with high axial and lower lateral crests, their bases quadrilateral, moderately closely spaced and of essentially the same form over the trunk as a whole.

Snout thick, fleshy, broadly rounded in front, somewhat contracted between nostrils and eyes, its length in front of mouth about 1/3 of length of head. Eye approximately circular and very small, its diameter only about 1/5 to 1/6 of length of snout, its center about opposite front of mouth. Spiracle a little above level of eye and behind latter by a distance about 1 1/2 times the diameter of eye. Gill openings low on sides of neck, about evenly spaced, very small and all about twice as long as diameter of eye, the 5th, slightly the longest, close in front of origin of pectoral. Nostril nearly transverse and much closer to tip of snout than to mouth, its anterior margin only slightly expanded. Mouth transverse, very little arched, its breadth about equal to length of snout in front of mouth. Labial furrows extending rearward for a considerable distance from corner of mouth and part way along upper and lower jaws:[13]

11. For this reason many of the earlier representations of it are no better than caricatures, for they picture it as enormously stout of body, which is not the case.
12. The exact extent cannot be stated for the Atlantic specimens studied because of their condition.
13. On the specimens examined, the lower labial furrow has been entirely obliterated in the process of mounting, the upper and posterior folds mostly so as well.

Teeth $\frac{49}{50}$ in specimen counted (up to $\frac{52}{52}$ reported), widely unlike in the 2 jaws; upper teeth thorn-like, tapering gradually to tip,[14] moderately stout, very slightly recurved and flattened anteriorly (thus subtriangular in cross-section); lower teeth subquadrate, about ½ as broad as high, except that the outermost 3 or 4 are relatively much broader and lower, each overlapping the next outermost, their cusps smooth-edged, deeply notched outwardly and so strongly oblique that inner margins are nearly parallel with the jaw, forming a continuous saw-like cutting edge (much as in *Centroscymnus*), the basal sectors with a strong longitudinal ridge; about 3 series functional as a rule in upper jaw, 1 or 2 series functional in lower, depending on the stage in their replacement.[15]

Pectorals, dorsals and pelvics all very small. First dorsal brush-shaped, about ¼ as long at base as head, and about 3 to 4 times as long as high, its anterior edge very sloping, apex rounded, posterior margin nearly straight or weakly concave, its free lower posterior corner nearly or quite as long as base, its origin on mounted specimens about midway between tip of snout and origin of caudal[16] or posterior to axil of pectoral by a distance 1 to 2 times as long as its own base. Interspace between 1st and 2nd dorsals about as long as from snout to 1st or 2nd gill opening. Second dorsal nearly as long as 1st at base, but only about ⅔ as high vertically, with more broadly rounded apex, its origin over rear end of bases of pelvics. Distance from tip of 2nd dorsal to origin of caudal about as long as base of 2nd dorsal, or a little longer. Caudal ⅕ to ⅙ of total length, its posterior margin notched subterminally, about opposite end of caudal axis; its lower lobe about ⅔ as long as upper lobe along upper margin of fin, the re-entrant contour included between the two lobes considerably more obtuse than a right angle. Pelvics about as long at base as 2nd dorsal or a little longer and about as high vertically, with broadly rounded apices and tapering tips. Pectoral between ⅓ and ½ as long as head, with nearly straight outer margin, moderately rounded tip, and much more broadly rounded inner corner.

Color. Described as varying in life between coffee brown or black and ashy-gray, purplish-gray, or slaty-gray, below as well as above, changing to bluish-gray if the epidermis is rubbed off, as often happens after capture; the sides are sometimes tinged with violet; the back and sides are crossed by numerous indistinct dark bands, at least in some specimens, or with numerous indistinct whitish spots. Vestigial luminous dots, not functional, are reported[17] as scattered on the top and sides of the head, in a band along the lateral line, and on the caudal.

Size. This is one of the larger sharks, and by far the largest of typically Arctic fishes. It has been said to reach a length of 24 feet, but few, if any, actually grow to so great a size, for the longest of which we find definite record[18] was 21 feet, with specimens of 16 to 18 feet unusual, although occasionally taken. The largest western Atlantic specimens whose

14. They closely resemble those of *Centroscymnus* in general appearance but are recognizable as *Somniosus* by their progressive taper; in *Centroscymnus* they are definitely lanceolate (cf. Fig. 100 A–C with 94 B, C, H).
15. For discussion of this process in squalids with lower teeth of this type, see p. 65.
16. Garman (Mem. Harv. Mus. comp. Zool., *36*, 1913: pl. 15, fig. 4) so pictures it also for a Massachusetts specimen newly caught.
17. Burckhardt, Ann. Mag. nat. Hist., (4) *6*, 1900: 562.
18. Jenkins, Fish. Brit. Isles, 1925: 325.

lengths have been published are one of 16½ feet from the Grand Banks in 1934, one of about 16 feet off Portland, Maine, in 1846, another of about 15 feet off Cape Cod in 1849, a fourth of about that same size caught in an otter trawl north of Cape Ann in February 1931. Perhaps 8 to 14 feet is a fair average for adults, few among the hundreds annually caught around Iceland and Greenland exceeding this size. Females average larger than males, the maximum among 120 specimens brought into Aberdeen, Scotland, being 11 feet 3 inches for the latter, and 15 feet 6 inches for the former. The British specimen of 21 feet, mentioned above, is said to have weighed about 2,250 pounds; an 11-foot specimen from the Gulf of Maine, which we inspected, weighed about 650 pounds; another, of 11 feet 6 inches, taken off Cape Ann in January 1939, weighed about 600 pounds; but one of 13½ feet, found alive on the beach in the estuary of the River Seine many years ago, was only between 300 and 400 pounds. Males of about six feet are still immature; smaller ones of either sex are seldom caught, but there is one record of a free-swimming specimen of only one foot six inches (445 mm.).[19]

Developmental Stages. Adult females have been found repeatedly containing great numbers of soft eggs without horny capsules, with up to as many as 1½ barrels of them in large specimens, these eggs ranging in size up to that of a goose egg. This, combined with the fact that none of the many examined have ever been found with embryos, supports the general belief that this shark, unlike other squalids, is oviparous. If so, it seems likely that the eggs are deposited on the bottom in mud, but eggs naturally laid have not been found as yet.[20] On the contrary, the Mediterranean *Somniosus rostratus* is ovoviviparous, its embryos having been seen by several students.

Habits. Eyewitnesses agree that this is one of the most sluggish of sharks, offering no resistance whatever when hooked or even when drawn up out of the water. An observer of long experience writes that he had driven a boat hook into one larger than himself as it lay basking at the surface and had drawn it easily onto the ice.[21] In view of this passivity it is somewhat astonishing that it is able to capture prey as active as herring, halibut, salmon, and seals which are said to become very scarce when these sharks gather. Experience in the Iceland and Greenland fisheries indicates that they usually lie close to bottom in summer but often swim toward the surface for prey, even in the warm season, and in the winter fishery through the ice of West Greenland they are often lured to the surface by a light.

The diet of the Greenland Shark includes a wide variety of fishes, both large and small, such as skates, herring, salmon, capelin (*Mallotus*), rosefish (*Sebastes*), sculpin (*Myoxocephalus*), lumpfish (*Cyclopterus*), saithe or American pollock (*Pollachius virens*), ling (*Molva*), cod, haddock, wolffish (*Anarrhichas*), and various flatfishes, among them halibut (*Hippoglossus*) and the Greenland halibut (*Reinhardtius*). Seals are a

19. Winther, Prod. Ichthyol. Dan. Mar., 1879: 59.
20. For a summary of evidence on this subject, see Lütken (Vidensk. Medd. naturh. Foren. Kbh. [1879], 1880: 56).
21. Grenfell, Labrador, 1910: 351.

favorite food as well as small cetaceans, the latter perhaps dead when eaten. Sometimes sea birds are captured, and squids, crabs, large snails, and even medusae are devoured. Objects as large as an entire reindeer (without horns), a whole seal, a four-foot ling (*Molva*), a three-foot cod, and a 39-inch salmon have been found in stomachs of the Greenland Shark, which gives some measure of its appetite. They also greedily devour any carrion, such as whale meat, blubber, etc., from whaling operations, and their habit of gathering around whaling stations for this purpose, or when there has been a big killing of narwhals in Greenland waters, is proverbial. Similarly, large numbers are described as haunting the ice fields in spring off the Labrador coast, where sealers have left the carcasses of young seals. But there appears to be no basis for the old story that they attack living whales.

Its depth range is wide. In its centers of abundance it tends to approach the surface in winter, coming right up to the ice off Greenland and along the Labrador coast. In summer, however, it is most often caught at 100 to 300 fathoms, and has been recorded as deep as 660 fathoms. Although it usually lies close to the bottom during the warm season on the Labrador coast, it often becomes entangled in seal nets even then. Its habit of gathering when whales are being cut up was well known during the days of the Arctic Right Whale Fishery. The frequency with which the remains of seals and sea birds are found in its stomach is further evidence of its readiness to swim upward in pursuit of prey. The considerable number that are taken in the North Sea are all caught shoaler than 100 fathoms, irrespective of the season, which applies equally to most of the Gulf of Maine records.[22]

It has been taken in water as cold as minus 0.6° C.,[23] and it is the only shark regularly inhabiting polar temperatures. At the other extreme it is able not only to survive but to feed actively in water at least as warm as 10° to 12° C., as indicated by the repeated capture of specimens in the northern part of the North Sea, their stomachs full of recently eaten fish. In the Gulf of Maine, too, it has been taken in water as warm as about 10° C.[24] But most of the local records have been based on specimens taken when the water temperatures were between 2° and about 7° at the bottom.

Relation to Man. In North American waters the Greenland Shark is of no commercial value. Off northern Norway, however, around Iceland and in West Greenland waters it has long been sought regularly. By the middle eighteen-hundreds the catch off West Greenland was 2,000 to 3,000 sharks yearly, which had risen to 11,000 to 15,000 by the eighteen-nineties, and to upwards of 30,000 by the first decade of the present century. The catch is obtained by hand lines, or on long lines, for the most part in depths of 100 to 200 fathoms, except along the northern part of the West Greenland coast, where

22. Stray specimens have been taken in a few feet of water near the southern boundary of its range, or found stranded on the beach.
23. Murray and Hjort, Depths of Ocean, 1912: 436.
24. Taken in a weir during summer or early autumn in Passamaquoddy Bay.

they are caught chiefly in winter through holes in the ice, close to the surface, either on hook and line, or sometimes even with short-handled gaffs.[25]

They are utilized chiefly for their liver oil, a large specimen sometimes yielding as much as 30 gallons or more (up to about 50% of the volume of the liver). In Greenland the flesh is also dried for dogfood, and in Iceland small amounts have been consumed for human food in the past. But it produces a sort of intoxicant poisoning if eaten fresh, whether by man or by dogs,[26] although it is wholesome when dried. Fishermen regard them as harmless; old tales that they attack Greenlanders in their kayaks appear to be mythical.

Range. White Sea, Spitzbergen, Bear Island, Norwegian Sea, East and West Greenland, south regularly to the North Sea and Kattegat, less commonly to the southern part of the North Sea, accidentally to the mouth of the Seine (one specimen) and perhaps to Portugal, in the eastern Atlantic; regularly to the northern part of the Gulf of St. Lawrence, in the western Atlantic, and less commonly to the Gulf of Maine. Whether its range extends along the Arctic coasts of Eurasia, or to the coasts of Arctic North America westward from Baffins Bay, is not known. It is represented in the North Pacific by a form which, while close, appears to be distinct (for discussion, see p. 515).

Occurrence in the Western Atlantic. This shark is so plentiful along the Greenland side of Davis Strait and in Baffins Bay that in the first decade of the present century the yearly catch there was around 32,000.[27] Similarly, during fishing experiments with long lines off West Greenland by the "Ingolf" Expedition at least 20 per cent of the hooks were bitten off by them. They have been reported from Hudson Strait[28] and are plentiful along the east coast of Labrador, where specimens are often entangled in seal nets and where a recent author also reports catching five through one hole in the ice.[29]

No doubt its normal range includes the outer coast of Newfoundland, although we find no definite statement as to its numbers, there being only one published record for the Grand Banks.[30] It is sufficiently numerous to be a nuisance to fishermen in the Straits of Belle Isle, at least in some years, and inward along the north shore of the Gulf of St. Lawrence;[31] specimens have even been reported from the Saguenay River[32] and from the lower reaches of the St. Lawrence.[33] But while it undoubtedly occurs on the southern side of the Gulf as well, it is at least not numerous enough there for its presence to have occa-

25. For a description of the West Greenland fishery, see especially Jensen (Mindeskr. Jap. Steenstrup. Føds., Kbh., *2* [30], 1914: 15).
26. For accounts, see Jensen (Mindeskr. Jap. Steenstrup. Føds., Kbh., *2* [30], 1914: 12) and Clark (Science, N. S. *41*, 1915: 795).
27. Jensen, Mindeskr. Jap. Steenstrup. Føds., Kbh., *2* (30), 1914: 9.
28. Vladykov, Contr. Canad. Biol., N. S. *8* (2), 1933: 5.
29. Grenfell, Labrador, 1910: 351.
30. Rep. Newfoundland Div. Fisher. Res., Fisher Res. Lab. (1934), 1935: 79.
31. Stearns, Proc. U.S. nat. Mus., *6*, 1883: 123–125.
32. Whiteaves, Cat. Canad. Pinnep. Cetacea, Fish., 1886: 4.
33. At Metis; Dawson, Canad. Rec. Sci., *4*, 1891: 304.

sioned any printed comment; we find but two records from the coast of Nova Scotia, one being from Cape Breton and the other near Halifax.[34]

In view of its apparent scarcity in Nova Scotian waters, it is somewhat astonishing that there is published record of about 27 specimens in the Gulf of Maine up to 1938, with several more reported subsequently by local fishermen. The localities include Passamaquoddy Bay (tributary to the Bay of Fundy), off Eastport, Portland and Cape Elizabeth, Maine; Jeffrey's Ledge,[35] inner part of Massachusetts Bay, Cape Cod Bay, tip of Cape Cod, and the southwestern part of the Gulf of Maine basin. This distribution is wide enough to show that odd specimens are to be expected anywhere in the western side of the Gulf at any time of year. In fact, it is rumored that in early colonial times, when Atlantic Right Whales were still being killed in numbers off the Massachusetts coast, the Greenland Sharks were more abundant there than they have been at any time during the last hundred years. None have been reported either from the Nova Scotian side of the Gulf on the one hand, although this lies in their route from the north, or westward from Cape Cod on the other; but recorded captures in the Gulf of Maine include both small and large specimens (five small ones, from only 39 inches up to four to five feet long, off Portland between 1925 and 1933); furthermore, they have been recorded for November, January, February, March, April, June and August; these facts suggest that the partial enclosure of the comparatively deep western waters of the Gulf by the shoaler banks to the south forms a sort of cul-de-sac for any that may stray that far. Once arrived, they may survive there for years.

Synonyms and References:

Haa-Skierding, Gunnerus, Trondh. Selsk. Skrift., *2*, 1763: 330, pl. 10, 11 (size, food, descr., Norway).

Squalus carcharias Gunnerus, Drontheim Gesellsch. Schr., *2*, 1766: 299, pl. 10, 11 (size, food, descr., Norway); Müller, Prod. Fauna Danica, 1776: 38 (Denmark); Fabricius, Fauna Groenl., 1780: 127 (general account, food, abund. W. Greenland); not *Squalus carcharias* Linnaeus, 1758.

Haa-Kiâerringen, Rosted, K. norske Vidensk.-Selsk. Skr., N. S. *2*, 1788: 203, 1 pl. (fishery, Norway).

Squalus microcephalus Bloch and Schneider, Syst. Ichthyol., 1801: 135 (refs., descr., Arctic Seas); Blainville, in Vieillot, Faune Franc., 1825: 66 (ref. to Bloch and Schneider, 1801).

Squalus squatina Pallas, Zoogr. Rosso Asiat., *3*, 1814:[36] 64 (White and Arctic Seas); not *Squalus squatina* Linnaeus, 1758.

Acanthorhinus norwegianus Blainville, Bull. Soc. philom. Paris, *8*, 1816: 121.[37]

Squalus brevipinna Lesueur, J. Acad. nat. Sci. Philad., *1*, 1818: plate facing p. 222 (ill., spec. from Massachusetts).

Somniosus brevipinna Lesueur, J. Acad. nat. Sci. Philad., *1*, 1818: 222 (descr. of spec. ill. as *Squalus brevipinna*, Marblehead, Massachusetts); Bory de St. Vincent, Dict. Class. Hist. Nat., *15*, 1829: 597 (ref., Massachusetts); Storer, Rep. Fish. Rept. Birds Mass., 1839: 189 (Massachusetts); Boston J. nat. Hist., *2*, 1839: 541 (descr., Massachusetts); Gill, Proc. Acad. nat. Sci. Philad., 1863: 333 (Massachusetts);

34. Storer, Boston J. Nat. Hist., *6*, 1857: 270; Jones, Proc. N. S. Inst. Sci., *5* (1), 1882: 96.

35. Six specimens, which ranged in length from 39 inches upward and were taken in the months of January, February, April, June and August, have been reported to us by W. W. Rich for this general region since 1925.

36. It is generally accepted that 1814 is the date of publication of the part of volume *3* in question; see Cat. Library Brit. Mus., page 1505.

37. Name only, but identification probable by inference.

Garman, Mem. Harv. Mus. comp. Zool., *36*, 1913: 240, in part (Atlant. refs., but descr. and ill., pl. 15, fig. 1–3, are of a Japanese spec.); Rey, Fauna Iberica Peces, *1*, 1928: 479 (in Key); Jordan, Manual Vert. Anim. NE. U.S., 1929: 14 (part); Tortonese, Atti Soc. ital. Sci. nat., 77, 1938: 312 (concludes that *brevipinna* is distinct from *microcephalus*).

Squalus borealis Scoresby, Arctic Regions, *1*, 1820: 538, pl. 15, fig. 3, 4, 5 (descr., habits, size, Spitzbergen Sea); Cordeaux, Zoologist, (2) *1*, 1866: 230 (Dogger Bank, North Sea, stomach contents); Southwell, Zoologist, (2) *10*, 1875: 4424 (England).

Squalus norvegianus Blainville, in Vieillot, Faune Franc., 1825: 61 (refs.).

Scymnus borealis Fleming, Hist. Brit. Anim., 1828: 166 (descr.); Jenyns, Manual Brit. Vert. Anim., 1835: 506 (Scotland); Swainson, Nat. Hist. Fish. Amphib. Rept., *1*, 1839: 145; *2*, 1839: 315 (general); Hamilton, Brit. Fish., *2*, 1843: 315, also subsequent eds. (Gt. Brit.); Busch, Selach. Ganoid. Encephal., 1848: pl. 3, fig. 7 (brain); Smith, Zoologist, *9*, 1851: 3058 (Scotland); Baikie, Zoologist, *11*, 1853: 3846 (Orkneys); Van der Hoeven, Handb. Dierkunde, (2) *2*, 1855: 261 (general); Nilsson, Skand. Fauna, *4*, Fisk., 1855: 724 (Greenland, Iceland, Scand., Spitzbergen); Thompson, Nat. Hist. Ireland, *4*, 1856: 255 (Ireland); Brown, Zoologist, *18*, 1860: 6861 (Scotland); Fiedler, Tidskr. Fiskeri, Jahrg. 1866: 26–33 (Denmark, not seen); Saxby, Zoologist, (2) *6*, 1871: 2553 (Newfoundland); Trail, Scot. Nat., *1*, 1872: 48 (numbers, size, Scotland).

Scymnus gunneri Thienemann, Lehrb. Zool., *3*, 1828: 409.

Scymnus glacialis Faber, Fische Islands, 1829: 23 (habits, eggs, food, fishery, Iceland, Polar Sea); Van Beneden, Sur *Sciaena aquila* et *Scymnus glacialis*, 1852 (not seen).

Scymnus micropterus Valenciennes, Ann. Mus. Hist. nat. Paris, *1*, 1832: 458, pl. 20 (descr., weight, 13-foot spec., estuary of the Seine).

Squalus glacialis Nilsson, Prod. Ichthyol. Skand., 1832: 115 (Norway); Lilljeborg, Vet. Akad. Handl. Stockholm, *2*, 1850: 334 (not seen).

Greenland Shark, Yarrell, Brit. Fish., *2*, 1836: 403 (descr., Scotland); Couch, Fish. Brit. Isles, *1*, 1867: 57, pl. 13 (descr., habits, ill., Gt. Brit.); Buckland, Hist. Brit. Fish., 1881: 213 (Scotland, Iceland, habits, fishery); Norman and Fraser, Giant Fishes, 1937: 38 (general, habits).

Squalus (Scymnus) gunneri Richardson, Fauna Boreal. Amer., *3*, 1836: 313 (Greenland Seas).

Laemargus carcharias Bonaparte, Mém. Soc. neuchâtel. Sci. nat., *2* (8), 1839: 9 (in synopsis).

Scymnus (Laemargus) borealis Müller and Henle, Plagiost., 1841: 93 (descr., refs.); Gaimard, Voy. Islande et Groenland (1835–1836), Atlas *3*, Zool. Med., 1851: pl. 22 (good ill.); Duméril, Hist. Nat. Poiss., *1*, 1865: 456, pl. 5, fig. 3, 4 (refs., descr., comp. with *brevipinna*, ill. of teeth).

Scymnus brevipinna DeKay, Zool. N. Y., 4, 1842: 361, pl. 61, fig. 202 (descr., ill. after Lesueur, 1818, Massachusetts); Storer, Mem. Amer. Acad. Arts Sci., N. S. *2*, 1846: 507 (in synopsis); Kneeland, Boston J. nat. Hist., *5*, 1847: 479 (anat., Massachusetts); Storer, Boston J. nat. Hist., *6*, 1857: 270 (jaw, C. Breton); DeKay, Rep. St. Cab. nat. Hist. N. Y. (1855), *8*, 1858: 64; Storer, Mem. Amer. Acad. Arts Sci., N. S. *9*, 1867: 235, pl. 38, fig. 2; Fishes Mass., 1867: 259, pl. 38, fig. 2 (descr., orig. ill., Massachusetts records).

Laemargus borealis Bonaparte, Cat. Pesc. Europ., 1846: 16 (in Europ. list); Günther, Cat. Fish. Brit. Mus., *8*, 1870: 426 (refs., descr., Brit. spec.); Turner, J. Anat. Lond., 7, 1873: 233 (size, anat., Scotland, North Sea); J. Anat. Lond., *8*, 1874: 285 (anat., stomach contents, meas.); Gervais and Boulart, Poiss., *3*, 1877: 212, pl. 82 (descr., ill.); Turner, J. Anat. Lond., 12, 1878: 604 (corrects earlier statements that oviducts were lacking); Jones, List Fish. Nova Scotia, 1879: 96 (Nova Scotia record, meas.); Hasse, Naturl. Syst. Elasm. besond. Theil, *1*, 1882: 55, pl. 8, fig. 1–6 (vert., dermal denticles); Jones, Proc. N. S. Inst. Sci., *5* (1), 1882: 96 (same as Jones, 1879); Turner, J. Anat. Lond., *19*, 1885: 221 (anat., stomach contents, Scottish spec.); Dawson, Canad. Rec. Sci., *4*, 1891: 304, pl. 4 (ill., meas., size, lower St. Lawrence R.); Dean, Fishes Living and Extinct, 1895: 91 (general); Burckhardt, Ann. Mag. nat. Hist., (7) *6*, 1900: 562 (vestigial luminous organs); Helbing, Nova Acta Leop. Carol., *82*, 1904: 351, 364, 377, 394, 421, 441, 450, 461, 475, 491, 511, pl. 8, fig. 2B, pl. 9, fig. B (skeleton); Günther, Encyc. Brit., 11th ed., *24*, 1911: 808 (general); Ehrenbaum, in Grimpe and Wagler, Tierwelt N.- u. Ostsee, Lief 7 (12[e]), 1927: 19 (general, North Sea).

Leiodon echinatum Wood, Proc. Boston Soc. nat. Hist., *2*, 1846: 174 (descr. of 16-foot spec., off Portland, Maine).

Dalatias (Somniosus) borealis Gray, List. Fish. Brit. Mus., *1*, 1851: 76.

Dalatias microcephalus White, List Spec. Brit. Mus., Fish., *8*, 1851: 132 (Scotland, England).

Scimnus glacialis Van Beneden, Bull. Acad. Sci. R. Bruxelles, *20* (2), 1853: 258 (parasites, Belgium); Bull. Acad. Sci. R. Bruxelles, (2) 1, 1857: 226 (parasites).

Scymnus microcephalus Krøyer, Danmarks Fisk., *3* (2), 1853: 914 (descr., meas., Denmark); Malmgren, Oev. Svensk. Vet. Akad. Forhand. Stockholm (1864), 1865: 536 (N. of Norway); Collett, Forh. Vidensk-Selsk. Krist., 1874: 212 (fishery, depths, Norwegian locs.); Norges Fisk., 1875: 208 (Norway); Malm, Göteborgs och Bohusläns Fauna, 1877: 626 (Kattegat); Lütken, Dyreriget (Laerbog Zool.), *1*, 1881–1882: 379 (Iceland, Greenland, Spitzbergen, fishery); Storm, K. norske Vidensk.-Selsk. Skr., 1884: 44 (Trondhjem Fjord).

Squalus norvegicus Gray, Cat. Fish. Coll. Descr. by L. T. Gronow, in Brit. Mus., 1854: 8.

Laemargus microcephalus Malmgren, Finland Fisk. Fauna, *1*, 1863: 71 (N. Norway, not seen); Arch. Naturgesch., *30* (1), 1864: 345 (transl. of foregoing by Frisch; N. Norway, Greenland, Iceland); Winther, Naturh. Tidskr. Copenhagen, *12*, 1879: 59 (Danish locs., Kattegat, size); Day, Fish. Gt. Brit., *2*, 1880–1884: 320, pl. 162, fig. 1 (genl. account, Gt. Brit.); Calderwood, Rep. Fish. Bd. Scot., App. F, *16*, 1886: 228, pl. 10 (Scotland, size, eggs, stomach contents, numbers); Möbius, S. B. Ges. naturf. Fr. Berlin, 1890: 11 (color, size, Helgoland); White, P. J., Trans. roy. Soc. Edinb., *37* (2), 1893: 287, 2 pls. (skull and visceral skelet.); Jenkins, Fish. Brit. Isles, 1925: 325, pl. 125 (general, size, Gt. Brit.); Beaufort, Tijdschr. ned. dierk. Ver., 1911, (2) *2*, 1911: LXVIII (Holland).

Somniosus microcephalus Gill, Proc. Acad. nat. Sci. Philad., Addend., 1861: 60 (listed); Proc. Acad. nat. Sci. Philad., 1864: 264 (name); Rink, Danish Greenland, 1877: 131 (abund., habits, toxicity of flesh, Greenland); Goode and Bean, Bull. Essex Inst. Salem, *11*, 1879: 31 (off Massachusetts); Lütken, Vidensk. Medd. naturh. Foren. Kbh., 1880: 56 (eggs, reproduction probably oviparous); Jordan and Gilbert, Bull. U.S. nat. Mus., 16, 1883: 15 (descr., C. Cod to Arctic seas); Stearns, Proc. U.S. nat. Mus., *6*, 1883: 125 (Gulf of St. Lawrence, Labrador); Petersen, Vidensk. Medd. naturh. Foren. Kbh. (1884–1886), 1887: 160 (Denmark); Collett, Nyt Mag. Naturv., *29*, 1885: 117 (Banks off Norway); Kingsley, Stand. Nat. Hist., *3*, 1885: 76 (general); Whiteaves, Cat. Canad. Pinnep. Cetacea, Fish., 1886: 4 (weight, Saguenay R. spec.); Jordan, Rep. U.S. Comm. Fish. (1885), 1887: 793 (distrib.); Holm, Medd. Groenland, *2*, 1888: 81 (E. Greenland); Packard, Labrador Coast, 1891: 397 (Gulf of St. Lawrence, Labrador); Goode and Bean, Smithson. Contr. Knowl., *30*, 1895: 7, 507, pl. 3, fig. 8; Mem. Harv. Mus. comp. Zool., *22*, 1896 (descr., ill., distrib.); Jordan and Evermann, Rep. U.S. Comm. Fish. (1895), 1896: 219 (distrib.); Bull. U.S. nat. Mus., 47 (1), 1896: 51 (refs., descr., distrib.); Vanhöffen, in Drygalski, Grönl. Exped. Gesellsch. Erkunde, Berlin, *2* (1), 1897: 124 (N. Greenland); Jungersen, Danish "Ingolf" Exped., *2*, 1899: 1, pl. 1, fig. 1–9 (anat., off W. Greenland); Jensen, Medd. Groenland, *29*, 1904: 276 (abund., E. Greenland); Collett, Vidensk. Selsk. Foren. Christiania, 7, 1905: 100 (Norway); Medd. Norges Fiske, *3*, 1905: 100 (food, size, eggs, Norway); Jordan, Guide to Study Fish., 1905: 547 (in part, but incl. Pacif. form also); Kendall, Occ. Pap. Boston Soc. nat. Hist., 7 (8), 1908: 9 (N. Engl. records); Regan, Ann. Mag. nat. Hist., (8) *2*, 1908: 54 (class., distrib.); Saemundsson, Skr. Komm. Havunders. Kbh., 5, 1909: 119 (eggs, sizes, depths, numbers, Iceland); Grenfell, Labrador, 1910: 350 (abund., captures, yield of oil, Labrador); Berg, Faune Russie (Russian), *1*, 1911: 78 (Murman Coast, White Sea, but Bering I. ref. probably for *pacificus*); Murray and Hjort, Depths of Ocean, 1912: 436 (Norwegian Sea); Halkett, Check List Fish. Canad., 1913: 44 (Labrador); Garman, Mem. Harv. Mus. comp. Zool., *36*, 1913: 241, pl. 15, fig. 4–6 (refs., descr., ill., Massachusetts); Jensen, Mindeskr. Jap. Steenstrup. Føds., Kbh., *2* (30), 1914: 8 (abund., food, devel. fishery, Greenland); Clark, Science, N. S. *41*, 1915: 795 (toxicity of flesh); Smith, J. Amer. Mus. nat. Hist., *16*, 1916: 349 (general); Huntsman, Contr. Canad. Biol. (1921), 1922: 56 (Bay of Fundy, Passamaquoddy Bay); Wollebaek, Norges Fisk., 1924: 24 (Norway, not seen); Bigelow and Welsh, Bull. U.S. Bur. Fish., *40* (1), 1925: 53, 551 (descr., habits, Gulf of Maine); Causey, Parasitology, *18*, 1926: 195 (parasites); Raven, Nat. Hist. N. Y., *26* (6), 1926: 654 (color, food, captures, W. Greenland); Hickling, Ann. Mag. nat. Hist., (10) *2*, 1928: 197, 199 (Faroe–Shetland Channel); Rey, Fauna Iberica Peces, *1*, 1928: 480 (descr. after Garman, 1913, habits, comp. with other species); Breder, Field Bk. Mar. Fish. Atlant. Coast, 1929: 25 (general); Jordan, Manual Vert.

Anim. NE. U.S., 1929: 14 (general); Jordan, Evermann and Clark, Rep. U.S. Comm. Fish. (1928), 2, 1930: 22 (in check list for N. Amer., N. Pacif. form also included); Vladykov, Contr. Canad. Biol., N. S. *8* (2), 1933: 5 (Port Burwell, Hudson Strait); Proc. Pan-Pacif. sci. Congr., *5*, 1934: 3787, 3788 (Hudson Bay region); Bigelow and Schroeder, Canad. Atlant. Fauna, biol. Bd. Canad., 12e, 1934: 19 (descr., range); Brooks, Parasitology, *26*, 1934: 259 (parasites); Piers, Proc. N. S. Inst. Sci., *18*, 1934: 203 (ref. to old Nova Scotia record by Jones, 1882); Vladykov and McKenzie, Proc. N. S. Inst. Sci., *19*, 1935: 49 (old Nova Scotia record); Rep. Newfoundland Fish. Res. Lab. (1934), 1935: 79 (16½-ft. spec., Grand Banks); Bigelow and Schroeder, Bull. U.S. Bur. Fish., *48*, 1936: 323 (addit. Gulf of Maine records); Lübbert and Ehrenbaum, Handb. Seefisch. Nordeurop., *2*, 1936: 288 (size, depths, fishery, N. Europ.); White, Bull. Amer. Mus. nat. Hist., *74*, 1937: 43 (ill. after Garman, 1913); MacCoy, New Eng. Nat., *2*, March 1939: 25 (photos, stomach contents of 11-ft. spec., Gulf of Maine); Fowler, Bull. U.S. nat. Mus., 100 (*13*), 1941: 276 (refs., descr., but N. Pacif. form also included); Holmgren, Acta Zool., Häft. *1–3*, 1941: 28 (skull).

Scymnus (Laemargus) brevipinna Duméril, Hist. Nat. Poiss., *1*, 1865: 456 (descr., ref. to old record for R. Seine estuary).

Scimnus borealis Van Beneden, P. J., Poiss. Côtes Belg. Parasit., 1870: 10 (parasites, Belgium).

Laemargus brevipinna Moreau, Nat. Hist. Poiss. France, *1*, 1881: 361 (descr., ref. to old record for estuary of R. Seine).

Somniosus carcharias Garman, Bull. Mus. comp. Zool. Harv., *17*, 1888: 85, pl. 20 (mucous canals, no loc., but probably Massachusetts).

Laemargus (no specific name) Ewart, Trans. roy. Soc. Edinb., *37* (1), 1893: 59, pl. 1, 2 (sensory canals).

Acanthorhinus carcharias Smitt, Hist. Skand. Fish., *2*, 1895: 1167, pl. 52, fig. 3 (descr., habits, fishery, Skand.); Knipowich, Annu. Mus. zool. Acad. St. Pétersb., *2*, 1897: 154 (Murman Coast, White Sea); Ehrenbaum, Faune Ichthyol., Cons. Explor. Mer., 1930: plate not numbered (descr., ill., distrib. in E. Atlant.).

Family ECHINORHINIDAE

Characters. Squaloidea without dorsal fin spines, the teeth with several cusps and similar in the two jaws.

Genera. One genus, *Echinorhinus.* It is doubtful whether the characters forming the basis for the subgenus *Rubusqualus* Whitley, 1931,[1] are even of specific value.

Genus *Echinorhinus* Blainville, 1816

Echinorhinus Blainville, Bull. Soc. philom. Paris, 1816: 121; type species, *E. spinosus* (no authority quoted, but doubtless Gmelin, 1789), equals *Squalus brucus* Bonnaterre, 1788.

Generic Synonyms:

Scymnus (in part) Cuvier, Règne Anim., *2*, 1817: 131, footnote.

Goniodus L. Agassiz, Poiss. Foss., *3*, 1838: 94; same as *Echinorhinus* Blainville, 1816.

Centrophorus (in part) Swainson, Nat. Hist. Fish. Amphib. Rept., *2*, 1839: 315 (incl. *spinosus*); not *Centrophorus* Müller and Henle, 1837.

Rubusqualus (subgenus) Whitley, Aust. Zool., *6*, 1931: 311; type species, *Echinorhinus spinosus* McCoy, Australia.

Generic Characters. Teeth with 3 to 7 cusps, the median much the largest and so strongly oblique that the inner margins form a continuous cutting edge along the jaw; dermal denticles in the form of tubercles or shields with small central spines, scattered singly

1. Whitley, Aust. Zool., *6*, 1931: 311; eye over anterior mouth, heavier tail, dorsal fins closer together and first dorsal originating over the anterior part of the "anal" (sic), instead of over the middle of that fin.

or in groups and varying greatly in size; a dermal ridge above and one below the lateral line, posterior to gill openings; caudal without precaudal pits; spiracles minute; a labial furrow on each jaw and voluminous pit at each corner of mouth; origin of 1st dorsal over bases of pelvics, and far behind midlength of trunk; caudal without subterminal notch, its lower anterior corner not expanded as a definite lobe; gill openings much larger than is usual in the suborder.

Range. Eastern Atlantic, from Ireland to tropical South Africa, including the Mediterranean; Argentina; California; Japan; Hawaiian Islands; New Zealand; Australia and Tasmania; Arabia; accidental in western North Atlantic.

Fossil. Miocene, North America; Pliocene, Europe.

Species. The South African representatives of the genus, as well as the Australian–New Zealand and Hawaiian representatives, have all been given separate names as supposedly distinct from *E. brucus* of the North Atlantic. By common consent, however, the first of them (*E. obesus* Smith, 1849) has been relegated to the synonymy of *brucus*. Similarly, it has been held recently[2] that the Hawaiian *cookei* Pietschmann, 1928, is merely a variant of *brucus*. A Californian specimen, recently taken, proved to be a typical *brucus*,[2a] and it is at least questionable whether the features that are believed by its author to distinguish the Australian–New Zealand *mccoyi* Whitley, 1931, from *E. brucus*, represent anything more than individual variations.[3] Final conclusions must await critical comparison of adequate series of specimens, however. References for the several geographic regions are therefore segregated in the accompanying synonymy (p. 530).

Echinorhinus brucus (Bonnaterre), 1788

Spiny Shark

Figure 102

Study Material. None.

Distinctive Characters. The following combination makes this Shark easily recognizable among local Squaloidea, should one be taken in the western side of the North Atlantic: dorsal fins without spines, teeth with several cusps in each jaw but so oblique as to form a nearly continuous cutting edge, very large gill openings and peculiar, shield-like dermal denticles.

Description.[4] Trunk subcylindrical, moderately stout. Lateral line lying in a well marked furrow, rearward from opposite 5th gill opening, flanked above and below by a pair of thin, palisade-like dermal ridges, sparsely fringed with small fleshy papillae. Dermal denticles in the form of flat shields, varying greatly in diameter, each with a more or less strongly developed conical, sharp-pointed spine in the center, sometimes 2 spines, from which numerous furrows radiate outward to the margin so that the latter is more or less

2. Fowler, Bull. U.S. nat. Mus., 100 (*13*), 1941: 278; Hubbs and Clark, Calif. Fish Game, *31*, 1945: 65.

2a. Hubbs and Clark, Calif. Fish Game, *31*, 1945: 65.

3. Whitley, Aust. Zool., *6*, 1931: 311.

4. Based on published accounts and illustrations.

denticulate (the larger scales resemble more the bucklers of certain skates than those of other sharks), irregularly distributed, either singly and wide-spaced or in groups of 3 to 5, in which case they may be so closely crowded that their circular outlines are more or less lost, or the adjoining denticles may even be more or less fused.

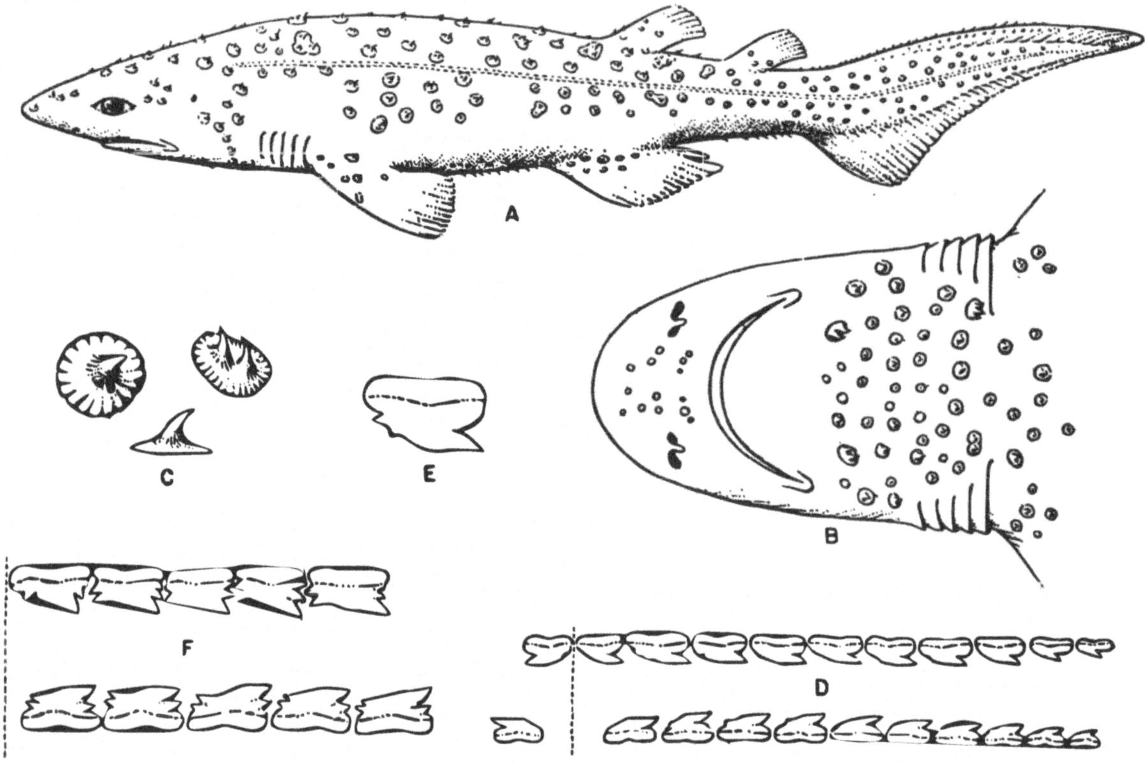

FIGURE 102. *A*, *Echinorhinus brucus*, eastern Atlantic specimen, about three feet long, in British Museum. *B* Head of same from below. *C* Dermal denticles. *D* Upper and lower teeth a little longer than natural size. *E* Third upper tooth, about 2 x. *F* Upper and lower teeth, after *L. Agassiz.*

Head flattened above. Snout ovate, tapering from eyes. Eye opposite front of mouth, approximately circular, its horizontal diameter ⅓ to ½ as long as snout in front of mouth. Spiracle posterior to eye by a distance a little longer than diameter of latter. Gill openings slightly oblique, the 5th about twice as long as 1st or about as long as snout in front of mouth and more than twice as long as diameter of eye (thus much larger than in any other member of the suborder). Nostril about midway between tip of snout and corner of mouth, its anterior margin with a pointed lobe. Mouth crescentic, about ½ as high as broad (thus more strongly arched than in other local Squaloidea). Labial furrows confined to corners of mouth.

Teeth $\frac{20\text{ to }26}{22\text{ to }26}$, alike in the 2 jaws, each with a pointed median cusp usually flanked by 1 small cusp on the inner side and 2 on the outer though described as sometimes[4a] with-

4a. Rey, Fauna Iberica, Peces, *1*, 1928: 485.

out denticles, and so strongly oblique that their inner margins form a cutting edge, but with the teeth separated by distinct interspaces.

Dorsal fins very small, brush-shaped, with broadly rounded corners, the origin of 1st over or slightly anterior to anterior third of bases of pelvics. Second dorsal a little smaller than 1st, its origin pictured as varying from over or a little posterior to rear ends of bases of pelvics to over their rear tips. Interspace between 1st and 2nd dorsals pictured for European specimens as varying from as long as base of 1st dorsal to so short as to be hardly discernible. Caudal extremely characteristic, being broadly scythe-shaped with tapering tip, its posterior contour evenly concave without definite subterminal notch, its lower inferior corner broadly rounded. Pelvics much larger than dorsals, their bases nearly or quite twice as long as bases of latter, with rounded apices and tapering rear corners. Pectoral a little less than ½ as long as head, brush-shaped, with weakly convex outer margin, broadly rounded corners and notably broad base.

Color. This is variously described as dark gray, dull olive or brown above, with reflections of violet, silver, gold or coppery yellow, and with or without obscure darker blackish or reddish blotches; paler brown or gray to white below. The scales have been described as luminescent,[5] but there are no special luminous organs.[6]

Size. The smallest European specimens on record were about three feet in length, the largest about nine feet; the majority of measurements available have ranged between approximately 5 feet and 8½ feet. A specimen of eight feet four inches weighed about 300 pounds, and an eight-foot five-inch example of the New Zealand form about 350 pounds. Females appear to average larger than males.

Developmental Stages. It is not known whether the development is viviparous or ovoviviparous, the latter being much more likely.

Habits. This is described as a ground shark, caught in European waters most often on hook and line. Off the coast of Portugal and in the Gulf of Gascony it is most numerous at depths of about 400 to 900 meters (about 220 to 500 fathoms). But there is also a long list of recorded captures from the shallow waters of the English Channel and North Sea, proving that its choice of depth is not narrow.

Its recorded diet includes smaller sharks (*Squalus acanthias*), other fish, and crabs. Nothing is known of its breeding habits.

Range. Eastern Atlantic, from tropical West Africa to Ireland and the North Sea, including the Mediterranean; apparently it is most numerous in the southern part of the Bay of Biscay and off the coast of Portugal; it is known also from Morocco to Senegal. It is accidental in the western Atlantic. It is also represented off South Africa, Argentina, California,[7] in the Hawaiian, Japanese and Australo–New Zealand regions, and in Arabian

5. Cornish, Zoologist, (2) *10*, 1875: 4501.
6. Burckhardt, Ann. Mag. nat. Hist., (7) *6*, 1900: 568.
7. Hubbs and Clark (Calif. Fish Game, *31* [2], 1945: 64) report the recent capture of a six-foot specimen, off Los Angeles County, California, indistinguishable from the eastern Atlantic *brucus*.

waters by a form (or forms) so close that it probably cannot be distinguished specifically (see discussion, p. 527).

Occurrence in the Western Atlantic. The only reports of it in the western Atlantic are of one that drifted ashore at Provincetown, Massachusetts, at the tip of Cape Cod, in December, 1878,[8] and of a second, 2½ meters long, taken near Buenos Aires more recently.[9]

Synonyms and References:

1. North Atlantic and Mediterranean:

Le bouclé, Broussonet, Mem. Math. Phys. Acad. Sci. Paris, 1780: 672 (descr.).

Squalus brucus Bonnaterre, Tabl. Encyc. Meth. Ichthyol., 1788: 11 (descr., size, N. Atlant.).

Squale bouclé, Lacépède, Hist. Nat. Poiss., 4° ed., *1*, 1798: 167, 283, pl. 3, fig. 2 (descr., ill.); in Sonnini, Hist. Nat. Poiss., *4*, 1802: 123 (from Lacépède, 1798).

Squalus spinosus Gmelin, in Linnaeus, Syst. Nat., *1* (3), 1788: 1500 (by ref. to Broussonet, 1780); Walbaum, P. Artedi Genera Pisc. Emend. Ichthyol., *3*, 1792: 519 (refs.); Bloch and Schneider, Syst. Ichthyol., 1801: 136 (refs.); Bosc, Nouv. Dict. Hist. Nat., *21*, 1803: 192 (diagn.); Latreille, Nouv. Dict. Hist. Nat., *24*, 1804: 72 (in table of contents); Risso, Ichthyol. Nice, 1810: 42 (descr., size, Medit.); Strickland, Ann. Mag. nat. Hist., *4*, 1840: 315 (Yorkshire, England); also, Calcutta J. nat. Hist., *1*, 1840: 285; Gatcombe, Zoologist, (3) *1*, 1877: 108 (size, stomach contents, off Plymouth, England).

Echinorhinus spinosus Blainville, Bull. Soc. philom. Paris, 1816: 121 (name); in Vieillot, Faune Franc., 1825: 66, pl. 16, fig. 1, 2[10] (descr.); Bonaparte, Fauna Ital. Pesc., *3*, 1835: plate not numbered (descr., ill., Medit.); Mém. Soc. neuchâtel. Sci. nat., *2* (8), 1839: 9 (spelled *Echinorrhinus* in synopsis); Yarrell, Brit. Fish., Suppl., 2, 1839: 54 (descr., ill., teeth, denticles, Brit. record); Müller and Henle, Plagiost, 1841: 96 (descr.), pl. 60 (dermal denticles); Yarrell, Brit. Fish., *2*, 1841: 532 (descr., Brit. records); Hamilton, Brit. Fish., *2*, 1842: 317, pl. 28 (descr., Gt. Brit.), also subsequent eds.; Bonaparte, Cat. Pesc. Europ., 1846: 16 (in list); Machado, Peces Cadiz, 1847: 8 (off Cadiz); Cocks, Ann. Mag. nat. Hist., (2) *5*, 1850: 71 (south coast, England); Gray, List Fish. Brit. Mus., *1*, 1851: 78 (refs., Medit., S. Afr.); Smith, Zoologist, *9*, 1851: 3057 (Scotland); White, List. Spec. Brit. Mus., Fish., *8*, 1851: 132; Costa, Fauna Napol. Pesci Cat. Sist. Condrott., in pt. 3, 1854–1857: 6 pp., pl. 16, pl. 17, fig. 1, 2 (descr., best ill., near Naples); Duméril, Hist. Nat. Poiss., *1*, 1865: 459, pl. 12, fig. 16–20 (denticles, descr.; see Pacif. refs., p. 531); Bocage and Brito Capello, Poiss. Plagiost., 1866: 35 (Portugal); Gray, Ann. Mag. nat. Hist., (4) *1*, 1868: 76 (Cornwall); Brito Capello, J. Sci. math. phys. nat. Lisboa, *2*, 1870: 148 (Portugal); Günther, Cat. Fish. Brit. Mus., *8*, 1870: 428 (refs., descr., Cornwall, Medit., see also S. Afr. refs., p. 531); Canestrini, in Cornalia, *et al.*, Fauna d'Ital., *3*, 1871–1872: 42 (Medit.); Jackson and Clarke, J. Anat. Lond., *10*, 1876: 75 (brain and spinal nerves, Cornwall); Trois, Atti Ist. veneto, (5) *3*, 1876–1877: 1179 (Adriatic); Gervais and Boulart, Poiss., *3*, 1877: 214, pl. 83 (descr., ill.); Doderlein, Prosp. Metod. Pesci Sicil., 1878: 30 (Sicily); Goode and Bean, Bull. Essex Inst. Salem, *11*, 1879: 31 (Provincetown, Massachusetts, record); Stossich, Boll. Soc. adriat. Sci. nat., *5*, 1880: 70 (Adriatic); Day, Fish. Gt. Brit., *2*, 1880–1884: 323, pl. 162, fig. 2 (descr., ill., Brit. records); Doderlein, Man. Ittiol. Medit., *2*, 1881: 104 (records from Medit.);[11] Moreau, Hist. Nat. Poiss. France, *1*, 1881: 365 (descr., ill., teeth, denticles, coasts of France); Perugia, Elenc. Pesc. Adriat., 1881: 54 (Adriatic); More, Zoologist, (3) *6*, 1882: 434 (Ireland); Rochebrune, Act. Soc. linn. Bordeaux, (4) *6*, 1882: 48; Faune Senegambie, Poiss., *1*, 1883–1885: 25 (Senegambia); Jordan and Gilbert, Bull. U.S. nat. Mus., 16, 1883: 14 (old C. Cod record); Jordan, Rep. U.S. Comm. Fish. (1885), 1887: 793 (distrib.); Carus, Prod. Fauna Medit., *2*, 1889–1893: 501 (Medit.); Goode and Bean, Smithson. Contr. Knowl., *30*, 1895: 8, pl. 3, fig. 9; also Mem. Harv. Mus. comp. Zool., *22*, 1896 (descr., ill., old C. Cod record); Nobre, Ann. Sci. nat. Porto, *2*, 1895: 225 (Portugal); Jordan and Evermann, Rep. U.S. Comm. Fish. (1895), 1896:

8. Goode and Bean, Bull. Essex Inst. Salem, *11*, 1879: 31.

9. Berg, Com. Ictiol., Comm. Mus. nac. B. Aires, *1* (1), 1898: 10.

10. We have not been able to see this plate.

11. See Doderlein for additional Mediterranean records in publications not accessible to us.

219 (range); Bull. U.S. nat. Mus., 47 (1), 1896: 58 (refs., descr., old C. Cod record); Vieira, Ann. Sci. nat. Porto, *4*, 1898: 138 (Portugal); Ridewood, Anat. Anz., *15*, 1899: 346 (skeleton, Bay of Biscay); Popta, Ann. Sci. nat. (Zool.) Paris, (8) *12*, 1901: 205 (gill arches, France); de Braganza, Result. Invest. Sci. "Amelia," *2*, 1904: 94, 106 (off Portugal); Ninni, Neptunia Venice, *19* (20), 1904: 20 (Adriatic); Borca, Arch. Zool. exp. gén., (4) *4*, 1905: 203, 205 (France); Kendall, Occ. Pap. Boston Soc. nat. Hist., 7 (8), 1908: 10 (old C. Cod record); Regan, Ann. Mag. nat. Hist., (8), *2*, 1908: 42 (class., also S. Afr. and Pacif. refs.); Günther, Encyc. Brit., 11th ed., *24*, 1911: 808 (general); Seabra, Bull. Soc. portug. Sci. nat., *6*, 1911: 200 (not seen); Roule, Bull. Inst. océanogr. Monaco, 243, 1912: 18 (Gulf of Gascony); Le Danois, Ann. Inst. océanogr. Monaco, *5* (5), 1913: 21 (Roscoff, France); Metzelaar, Trop. Atlant. Visschen, 1919: 191 (old records); Jenkins, Fish. Brit. Isles, 1925: 327, pl. 125 (ill., Gt. Brit.); Leigh-Sharpe, J. Morph., *42*, 1926: 313 (claspers); Ehrenbaum, in Grimpe and Wagler, Tierwelt N.- u. Ostsee, Lief 7 (12e), 1927: 20 (general, North Sea); Rey, Fauna Iberica Peces, *1*, 1928: 485 (descr., ill. of teeth, Spanish coasts); Gudger, Sci. Mon. N. Y., *34*, 1932: 417 (food); Belloc, Rev. des Trav. Pêches Marit., 7 (2), 1934: 149 (ill. after Bonaparte, Morocco to Senegal); Borri, Atti Soc. tosc. Sci. nat., *44*, 1934: 100 (Medit.); Nobre, Fauna Marinha Port. Vert., *1*, 1935: 403, pl. 64, fig. 201 (descr., ill., Portugal); Lübbert and Ehrenbaum, Handb. Seefisch. Nordeurop., *2*, 1936: 290 (North Sea); Norman and Fraser, Giant Fishes, 1937: 53 (general).

Scymnus spinosus Cuvier, Règne Anim., *2*, 1817: 131, footnote; Cloquet, Dict. Sci. Nat., *25*, 1822: 434 (general); Risso, Hist. Nat. Europe merid., *3*, 1826: 136 (descr., Medit.); Bory de St. Vincent, Dict. Class. Hist. Nat., *15*, 1829: 598 (general); Cuvier, Règne Anim., *2*, 1829: 368; Reguis, Ess. Hist. Nat. Provence, 1, Fasc. 1, 1877: 57 (Medit.).

Leiche bouclé, Cloquet, Dict. Sci. Nat., Atlas Poiss., 1816–1830: pl. 28, fig. 2.

Squalus (Scymnus) spinosus Voigt, in Cuvier, Tierreich, *2*, 1832: 513 (descr.).

Goniodus spinosus Agassiz., L., Poiss. Foss., *3*, 1835–1838: 94, pl. E, fig. 13 (teeth); Owen, Odontogr., 1840–1845: 31, pl. 4, fig. 4 (teeth).[12]

Spinous Shark, Cornish, Zoologist, *1*, 1866: 102 (Cornwall); Couch, Fish. Brit. Isles, *1*, 1867: 54, pl. 12 (ill., Brit. records); Cornish, Zoologist, (2) *5*, 1870: 2347 (Cornwall); Zoologist, (2) *10*, 1875: 4501 (luminescence, Cornwall); Zoologist, (3) *6*, 1882: 22 (Cornwall).

Echinorhinus (no specific name) Hasse, Naturl. Syst. Elasm. besond. Theil, 1882: 73, pl. 9 (vertebrae, dermal denticles); Burckhardt, Ann. Mag. nat. Hist., (7) *6*, 1900: 568 (luminescent organs lacking).

Echinorhinus brucus Garman, Mem. Harv. Mus. comp. Zool., *36*, 1913: 243 (descr.); Roule, Result. Camp. sci. Monaco, *52*, 1919: 123 (off Brittany); Bigelow and Welsh, Bull. U.S. Bur. Fish., *40* (1), 1925: 55 (ill., old C. Cod record); Jordan, Manual Vert. Anim. NE. U.S., 1929: 14 (general, C. Cod); Breder, Field Bk. Mar. Fish. Atlant. Coast, 1929: 26 (old C. Cod record); Jordan, Evermann and Clark, Rep. U.S. Comm. Fish. (1928), 2, 1930: 22 (ref. to old C. Cod record); de Buen, Faune Ichthyol., Cons. explor. Mer., 1931: plate not numbered; Bigelow and Schroeder, Canad. Atlant. Fauna, biol. Bd. Canad., 12^e^, 1934: 20 (descr., ill., old C. Cod record); Fowler, Bull. Amer. Mus. nat. Hist., *70* (1), 1936: 88 (descr., ill., W. Afr.).

2. South Africa and South Atlantic:

Echinorhinus obesus Smith, A., Ill. Zool. S. Afr., 1849: pl. 1 (descr., ill., S. Afr.).

Echinorhinus spinosus Günther, Cat. Fish. Brit. Mus., *8*, 1870: 428 (Cape of Good Hope); Berg, Com. Ictiol. Comm. Mus. nac. B. Aires, *1* (1), 1898: 10 (near B. Aires); Lahille, Physis B. Aires, 5, 1921: 63; Enum. Peces Cartilag. Argent., 1921: 17 (Argentina); Marelli, Elenc. Sist. Fauna B. Aires (1922–1923), 1924: 546 (Argentina); Barnard, Ann. S. Afr. Mus., *21* (1), 1925: 46; also, *21* (2), 1927: pl. 2, fig. 6 (S. Afr.); Pozzi and Bordale, An. Soc. cient. argent., *120*, 1935: 152 (listed, Argentina).

3. Pacific:

Echinorhinus spinosus Duméril, Hist. Nat. Poiss., *1*, 1865: 459 (refs. for Aust., Tasmania); Parker, Trans. Proc. N. Z. Inst., *16*, 1884: 28 (N. Zealand); McCoy, Prod. Zool. Victoria, Decade *15*, 1887: pl. 144

12. Spelled *Goniadus*.

(descr., meas., good ill., Aust.); Lucas, Proc. roy. Soc. Vict., N. S. *2*, 1890: 44 (Aust.); Jordan, Tanaka and Snyder, J. Coll. Sci. Tokyo, *33*, 1913: 23 (Japan).

Echinorhinus brucus Waite, Rec. Canterbury [N. Z.] Mus., *2* (1), 1913: 17 (size, weight, N. Zealand); Phillipps, N. Z. J. Sci. Tech., *10*, 1928: 221 (N. Zealand); Fowler, Proc. Pan-Pacif. sci. Congr., (4) *3*, 1930: 497; Bull. U.S. nat. Mus., 100 (*13*), 1941: 277 (Pacif. refs.); Hubbs and Clark, Calif. Fish Game, *31* (2), 1945: 64, fig. 16, 17 (descr., photos, California).

Echinorhinus cookei Pietschmann, Anz. Akad. Wiss. Wien, 27, 1928: 297 (descr., Hawaiian Is.); Bull. Bishop Mus., *73*, 1930: 3 (descr., Hawaiian Is.).

Echinorhinus (*Rubusqualus*) *mccoyi* Whitley, Aust. Zool., *6*, 1931: 311 (descr., Victoria); Mem. Qd. Mus., *10* (4), 1934: 200; Fish. Aust., *1*, 1940: 151 (descr., ill., Aust.).

Suborder PRISTIOPHOROIDEA

Characters. No anal fin; 2 dorsal fins without spines; either 5 or 6 gill openings, all of them anterior to origin of pectoral; snout greatly elongate, blade- or beak-like; each edge of snout and of head, anterior to mouth, armed with a row of sharp transverse, tooth-like structures and with a long fleshy barbel; oral teeth small, numerous, with 1 cusp, similar in front and sides of mouth, with several rows functional; trunk subcylindrical, except that head and snout are flattened dorso-ventrally; anterior margins of pectorals not expanded forward past 1st gill opening; nostrils entirely separate from mouth; eye without nictitating fold or membrane; spiracles present; inner margins of pelvics entirely separate posterior to cloaca; vertebral column completely segmented, with well developed centra, the notochord segmentally constricted correspondingly; skull with a separate antorbital bar of spongy cartilage extending rearward from nasal capsule, past orbit, as far as corner of mouth;[1] upper jaw (palatoquadrate cartilage) attached to hyomandibular, and also firmly articulated by a short, narrow transverse process to lower side of cranium in post-orbital region and by a ligament to the antorbital bar;[2] rostral cartilages united as a single elongate, blade-like bar, occupying the entire breadth of the snout to its tip; propterygial cartilage of pectoral bears 1 radial element only; pelvic transverse; heart valves in 3 rows. Development ovoviviparous.

Remarks. The sharks of this group have usually been placed among the Squaloidea, with which they agree in lacking an anal fin. But we believe they should rank as a distinct suborder (see p. 77), for they differ not only as regards their beak-like snout with its lateral teeth, in which they are unique among modern sharks, but in the presence of a separate antorbital bar, from which the upper jaw is suspended by a broad ligament in addition to the articulation to the cranium.

The saw-like beak makes them resemble superficially the sawfishes (Pristidae) among the Batoidei. But they are true sharks because of their free upper eyelids, their pectorals not united with the sides of the head, and their lateral, not ventral, gill openings.

Range. Indo-Pacific; South Africa; Tasmania; Australia; Philippines; Korea; Japan. Fossil pristiophorids are known from the Cretaceous, Miocene, and Tertiary.

Families. One family only, Pristiophoridae, with characters of the suborder.

1. This bar, like the rostrum proper, is armed with lateral teeth.

2. Description based on original dissection.

Key to Genera

1a. Five gill openings. *Pristiophorus* Müller and Henle, 1837.
South Africa, Australia, Tasmania, Philippines, Japan, Korea.

1b. Six gill openings. *Pliotrema* Regan, 1906.
South Africa.

Species. *Pristiophorus* includes three known species, *Pliotrema* only one.[3]

Suborder SQUATINOIDEA

Characters. No anal fin; 2 dorsal fins without spines; only 5 gill openings, all anterior to origin of pectorals; snout not beak-like, without lateral teeth or cirri; teeth in front of mouth essentially similar to those toward its corners; general form skate-like rather than shark-like, with the trunk very much flattened dorso-ventrally and expanded laterally anterior to cloaca, but tapering thence rearward; eyes dorsal; anterior margins of pectorals extending forward past gill openings[4] and partly concealing them; pelvics also very broad, wing-like, their inner margins entirely separate posterior to cloaca; nostrils separate from mouth; spiracles present, dorsal; eyes without nictitating fold or membrane; vertebral column completely segmented throughout its length, its axial canal much contracted in the region of the centra, which are well differentiated, with the notochord greatly constricted segmentally in its passage through them; vertebral centra with 2 or more series of concentric calcified lamellae; neural spines attached to dorsals;[5] rostral cartilage single, very short; skull without antorbital processes or separate antorbital bars; upper jaw (palatoquadrate cartilage) articulated to hyomandibular bar, with a long transverse process that is attached to the ethmoid region of the cranium by a ligament;[6] propterygial cartilage of pectoral elongate, directed anteriorly, corresponding to forward expansion of the fin, but bearing much fewer radials than the metapterygium;[7] pelvis curved rearward, corresponding to the expanded pelvic fins; heart valves in 6 to 7 rows (an exceptionally large number for sharks, but characteristic of skates and rays).[8] Development ovoviviparous.

Remarks. Although very skate-like in appearance, in number of heart valves, and in some skeletal characters, the squatinoids are usually classed with the sharks because they have free eyelids, pectorals with anterior margins not attached to the sides of the head, and gill openings that are not confined to the lower surface but extend up onto the sides of the neck as well.

Their method of swimming, also shark-like, is by a sculling motion of the tail, little

3. For descriptions of the species of the two genera, with lists of references, see Fowler (Bull. U.S. nat. Mus., 100 [*13*], 1941: 280, 283).
4. But not attached to sides of head.
5. Mivart, Trans. zool. Soc. Lond., *10*, 1879: pl. 77, fig. 5.
6. For account and excellent illustration of the skull, see Gegenbaur (Unters Vergl. Anat. Wirbelt., *3*, 1872: 190, pl. 11, fig. 2, pl. 12, fig. 4).
7. Gegenbaur, Unters. Vergl. Anat. Wirbelt., *2*, 1865: pl. 9, fig. 10.
8. For account of the vertebrae and dermal denticles, see Hasse (Naturl. Syst. Elasm. besond. Theil, 1882: 126, pl. 17, fig. 1–10); for heart valves, see Marples (Trans. roy. Soc. Edinb., *58* [3], 1936: 817).

use being made of the wing-like pectorals, which, on the contrary, are the effective swimming organs of skates and rays.[9]

Only one family is known.

Family SQUATINIDAE

Characters. Snout very broad and short; eyes dorsal, without nictitating membranes; spiracles large; nostrils terminal, entirely separate from mouth, their anterior margins with barbels which are variously lobed; gill openings lateral, but extending onto lower surface; mouth protrusible at corners with well developed labial cartilages, nearly terminal but separated from front of snout by a deep transverse furrow which is edged by a thin, variously-lobed fold of skin; lower jaw with deep labial furrows near corners; teeth numerous, similar in the 2 jaws, with single thorn-like cusp on broad base, 3 or 4 series functional; caudal axis not raised at all above main axis of trunk; lower lobe of caudal longer than upper. Development ovoviviparous; embryo with very large yolk sac.[10]

Genera. Only one genus is known.

Genus *Squatina* Risso, 1810

Squatina Risso, Ichthyol. Nice, 1810: 45; type species, *Squatina vulgaris* Risso, equals *Squalus squatina* Linnaeus, 1758.[11]

Generic synonyms:

Rhina Schaeffer, Epist. Stud. Ichthyol., 1760: 20; Klein, Neuer Schauplatz, 2, 1776: 587; Walbaum, P. Artedi Gen. Pisc. Emend. Ichthyol., 1792: 580, not available;[12] not *Rhina* Bloch and Schneider, Syst. Ichthyol., 1801: 352; Cuvier, Règne Anim., 2, 1817: 133; Müller and Henle, Plagiost., 1841: 110; a ray which was subsequently named *Rhamphobatis* by Gill, Ann. N. Y. Lyc., 7, 1862: 408.

Generic Characters. Head broadly rounded, with wing-like lateral expansions; spiracles behind eyes and at the same level as the latter; lips with well developed supporting cartilages, widely protrusible at corners of mouth but not centrally; anterior margins of pectorals expanded anteriorly as narrow shoulder-like extensions, lying below the lateral expansions of the sides of the head, partly concealed by the latter, and more or less overlapping the pelvics rearward; pelvics originating anterior to posterior corners of pectorals; dorsals much smaller than pectorals or pelvics; caudal triangular, its posterior con-

9. For a recent discussion of the affinities of the suborder, based chiefly on the skull, see Holmgren (Acta Zool., *22*, 1941: 79). They are classed (as Angeliformes) among the skates and rays by Le Danois (Rev. des Trav. Pêches Marit., *13*, 1945: 67) because of the nature of their vertebral calcifications.

10. Garman, Mem. Harv. Mus. comp. Zool., *36*, 1913: pl. *61*, fig. 9–11.

11. The name *Squatina* was first proposed by Valmont (Dict. Rais. Univ. Hist. Nat., *1*, 1768: 117). But by ruling of the International Commission on Zoological Nomenclature (Smithson. Misc. Coll., *73* [3], 1925: 27, Opinion 89) his names are not taken into consideration. It was next mentioned by A. M. C. Duméril (Zool. Anal., 1806: 102), but without inclusion of any particular species, so that, as a generic name, it must date from Risso, 1810.

12. Schaeffer did not include any species in his *Rhina*; and the generic names proposed by Klein and republished by Walbaum are not to be taken into account, according to Opinions 21 and 89 of the International Commission on Zoological Nomenclature (Opinions Rendered, Smithson. Publ., No. 1938, 1910: 51; Smithson. Misc. Coll., 7 [3], 1925: 27).

tour truncate or concave, its lower anterior margin longer than the upper anterior margin (a relationship the reverse of that which obtains in all other sharks); caudal peduncle with a faintly defined longitudinal ridge along each side; claspers of males extend only a little beyond pelvics, even in adults, and are attached to inner margins of fins nearly to tips of latter;[13] dermal denticles on dorsal side conical, on broad bases, those on ventral side flat, with scale-like blades. Characters otherwise those of family and suborder.

Range. Continental waters on both sides of the Atlantic, including the Mediterranean, north to the Shetlands and Cape Cod and south to northern Argentina; South Africa (Natal); west coast of America from Chile to southern Alaska; Japan, Korea, Australia and Tasmania.

Fossil. From Upper Jurassic to Pliocene, Europe; Upper Cretaceous, western Asia; Upper Cretaceous to Miocene, North America; Eocene, Africa.

FIGURE 103. Margins of right-hand nostrils of different species of *Squatina: A, australis* from Australia (Harv. Mus. Comp. Zool., No. 659), about 3 x. *B,* Left-hand nostril of same, to show asymmetry. *C, squatina* from the Irish Sea (Harv. Mus. Comp. Zool., No. 846), about natural size. *D, argentina* from Uruguay (U. S. Nat. Mus., No. 87684), about 1.5 x. *E, japonica* from Japan (Harv. Mus. Comp. Zool., No. 1112), about 2 x. *F, californica* from California (Harv. Mus. Comp. Zool., No. 952), about 2.5 x. *G, dumeril* from off New York (U. S. Nat. Mus., No. 118461), about 1.5 x. *H, armata* from Mejillones Island, Peru (Harv. Mus. Comp. Zool., No. 531), about 1.5 x.

13. For account, see Leigh-Sharpe (J. Morph., 35, 1921: 373).

Species. Recent estimates of the number of species actually represented by the named representatives of this curious genus range from only one[14] to eight or nine,[15] a list to which two more have subsequently been added.[16] Our own examination of specimens from widely separated seas[17] has convinced us that the latter opinion is more nearly correct., *i.e.*, that most of the supposed species are separable from one another by characters that seem precise enough to be accepted as specific, although they all resemble one another so closely in general apearance that identification requires close inspection. The conformation of the nasal margins and barbels (Fig. 103) proves a reliable diagnostic character, but the degree to which the mid-dorsal line of denticles is enlarged is far less so, since this may vary widely with age.

Key to Species

1a. Inner nasal barbel strongly ramose at tip; margin of nostril between barbels deeply fringed (Fig. 103 A, B).

- 2a. Outer nasal lobe strongly fringed. — *oculata* Bonaparte, 1840. Eastern Atlantic, Mediterranean.
- 2b. Outer nasal lobe smooth or only weakly fringed (Fig. 103 A, B).
  - 3a. Outer corner of pectoral little if any more obtuse than a right angle, its inner corner subangular; upper surface with small white and gray spots only. — *australis*[18] Regan, 1906. Southern Australia, Tasmania.
  - 3b. Outer corner of pectoral much more obtuse than a right angle, its inner corner broadly rounded; upper surface with conspicuous brown ocelli as well as small spots. — *tergocellata*[18] McCulloch, 1914. Western and southern Australia.

1b. Inner nasal barbel simple, or at least not strongly ramose; margin of nostril between barbels smooth, or at most feebly fringed (Fig. 103 C–H).

- 4a. Fold along front of head with 2 lobes opposite and in front of corner of mouth. — *nebulosa* Regan, 1906. Japan.[19]
- 4b. Fold in front of head with 1 lobe only, or none.
  - 5a. Fold along front of head expanded as a noticeable triangular lobe outside corner of mouth (Fig. 103 C). — *squatina* Linnaeus, 1758. Eastern North Atlantic; Mediterranean.

14. Ribeiro, Fauna brasil. Peixes, 2 (1) Fasc. 1, 1923: 27.
15. Regan, Ann. Natal Mus., *1*, 1908: 248; Garman, Mem. Harv. Mus. comp. Zool., *36*, 1913: 251.
16. *Tergocellata* from Australia (McCulloch, Biol. Res. "Endeavor," 2, 1914: 84) and *argentina* from Argentina (Marini, Physis B. Aires, *10*, 1930: 5).
17. From off the east coast of United States; Mediterranean; coast of northwestern Europe; Uruguay; S. Peru; California; Japan; Australia.
18. We question whether *tergocellata* is actually distinct from *australis*.
19. This form is known only from the original description (without illustration) of the type specimen; it may prove identical with *japonica*.

5b. Fold along front of head either nearly straight or at most obtusely rounded opposite corner of mouth (Fig. 103 D–H).

6a. Inner nasal barbel broadly spatulate (Fig. 103 D–F).

7a. Distance from eye to spiracle nearly or quite twice as long as horizontal diameter of eye. *argentina*[20] Marini, 1930, p. 544.

7b. Distance from eye to spiracle little if any longer than horizontal diameter of eye.

8a. Distal margin of pectoral nearly straight, its inner corner broadly rounded, its inner margin strongly convex.

japonica Bleeker, 1857.
Japan, Korea.

8b. Distal margin of pectoral weakly concave, its inner corner subangular, its inner margin only slightly convex.

californica Ayres, 1859.
Mexico to southern Alaska.

6b. Inner nasal barbel narrow, tapering (Fig. 103 G, H).

9a. Distal margin of pectoral marked off from inner margin by a definite, subangular corner. *dumeril* Lesueur, 1818, p. 538.

9b. Distal margin of pectoral not marked off from inner margin by a subangular corner.

10a. Distance from anterior corner of pectoral to rear end of its base equals 4/5 or more of its length; posterior margin of lower lobe of caudal, as well as upper, vertically truncate.

armata Philippi, 1887.
Chile, Peru.[21]

10b. Distance from anterior corner of pectoral to rear end of its base equals only about 3/5 of its length; posterior margin of lower lobe of caudal oblique. *africana* Regan, 1908.
Natal, South Africa.[22]

20. A specimen from Uruguay which we have examined in the U.S. National Museum is clearly referable to this species, and probably this also applies to the form reported from Rio de Janeiro by Ribeiro (Arch. Mus. nac. Rio de J., *14*, 1907: pl. 10; Fauna brasil. Peixes, *2* [1] Fasc. 1, 1923: pl. 9) as *Squatina squatina*. Although the inner corners of its pectorals are shown as angular or subangular (rounded in *argentina*), the conformation of the margin of the nostrils agrees with that of *argentina*, in addition to the fact that the outer corners of the pectorals are much more obtuse than a right angle, and that the eyes and spiracles are far apart.

21. The original account of this species is not sufficiently detailed to locate it positively in the key. The characters given above are from a specimen from Mejillones Island, Peru (Harv. Mus. Comp. Zool., No. 531), presumably the type of *philippi* Garman, 1913, which we believe to be a synonym of *armata*.

22. Records under this name from Lagos and Gorée in tropical West Africa (Gilchrist and Thompson, Ann. Durban Mus., *1*, 1916: 284; Metzelaar, Trop. Atlant. Visschen, 1919: 191), and from Cape Blanco, Morocco (Chabanaud and Monod, Bull. Étud. Hist. Sci. Afr. Occid. Franc., 1927) do not include evidence sufficient for specific identification.

Squatina dumeril (Lesueur), 1818

Angel Shark

Figures 103 G, 104, 105

Study Material. Female, 382 mm. long, from the continental slope off New York, Lat. 39° 42′ N., Long. 71° 17′ W., taken Sept. 19, 1887 (U.S. Nat. Mus., No. 118461); male, 1,080 mm. (42½ inches) long, from lower Chesapeake Bay at Lynnhaven Roads, Virginia; mounted specimen, 43 inches, from Martha's Vineyard, Massachusetts (New Eng. Mus. Nat. Hist.).

Distinctive Characters. Its skate-like appearance separates the Angel Shark at a glance from any other shark except for some other members of its own genus. For specific characters within the genus, see the preceding Key (p. 536).

Description. Proportional dimensions in per cent of total length. Female, 382 mm., from Lat. 39° 42′ N., Long. 71° 17′ W. (U.S. Nat. Mus., No. 118461).

Extreme breadth at outer extremity of pectorals: 60.8.
Trunk at origin of pectoral: height 9.0.
Snout length in front of: outer nostrils 0; mouth 0.
Eye: horizontal diameter 2.0; distance between eyes 8.4.
Spiracles: distance between 10.0; from spiracle opening to eye 2.2.
Mouth: breadth 13.6; height 2.1.
Nostrils: distance between inner ends 5.8.
Gill opening lengths: 1st 7.0; 2nd 6.3; 3rd 5.8; 4th 5.5; 5th 5.0.
First dorsal fin: vertical height 5.5; length of base 3.3.
Second dorsal fin: vertical height 5.0; length of base 3.1.
Caudal fin: upper margin 13.6; lower margin 16.7.
Pectoral fin: outer margin 28.0; inner margin 15.7; distal margin 14.9.
Distance from snout to: 1st dorsal 65.7; 2nd dorsal 76.5; upper caudal 86.4; pectoral 17.5; pelvics 37.2.
Interspace between: 1st and 2nd dorsals 7.5; 2nd dorsal and caudal 7.3.
Distance from origin to origin of: pectoral and pelvics 22.2; pelvics and caudal 45.6.

General form of trunk as described for family and genus. Caudal peduncle expanded laterally as a low ridge on either side posterior to 2nd dorsal. Dermal denticles on dorsal surface loosely spaced but generally distributed, conical, the spinous portion weakly recurved, usually with 4 or more longitudinal ridges; bases broad, their outlines more or less radiate; larger denticles in clusters of 5 or 6 beside inner anterior and inner posterior edges of eyes, with a row extending from near inner side of nostril toward eye; young specimens also with a single row of denticles, 3 to 4 times as large as the others, along midline of back from opposite anterior ends of bases of pectorals to origin of 1st dorsal, but in large specimens these are little, if any, larger than the denticles that flank them; adult

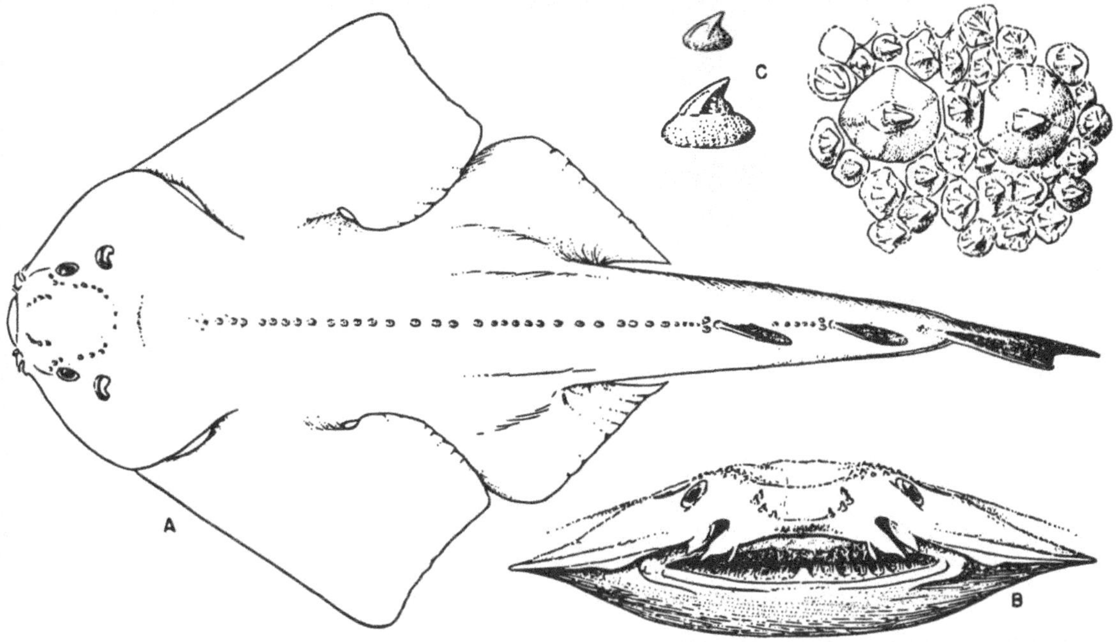

FIGURE 104. *Squatina dumeril*, female, 382 mm. long, from off New York (U. S. Nat. Mus., No. 118461). *A* Dorsal view. *B* Anterior view. *C* Dermal denticles, about 7 x.

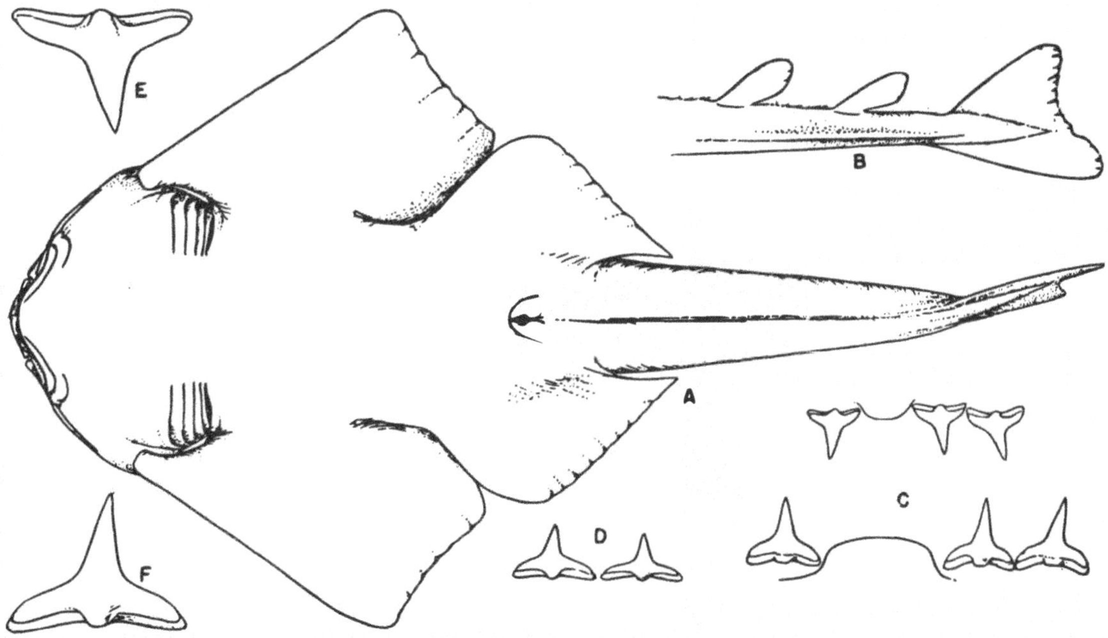

FIGURE 105. *A* Ventral view of *Squatina dumeril* pictured in Fig. 104. *B* Side view of posterior part of trunk with caudal fin. *C* Upper and lower teeth from center of mouth. *D* Eighth and ninth lower teeth, about 3.6 x. *E* Fifth upper tooth, about 7.2 x. *F* Fifth lower tooth, about 7.2 x.

males with larger denticles along anterior margins of pectorals and near their outer corners;[23] denticles on lower surface with flat ovoid blades on very short pedicels, close-spaced on outer parts of paired fins and on lower edge of tail sector of trunk, but abdomen and inner parts of paired fins naked except for patches here and there.

Head broadly rounded anteriorly, but its postero-lateral margin nearly straight, its greatest breadth about 3 times as great as distance between spiracles or ⅓ as great as distance from snout to rear end of base of 2nd dorsal. Horizontal diameter of eye about equal to maximum diameter of spiracle, the distance from eye to spiracle about as long as horizontal diameter of eye. Distance between spiracles longer than between eyes by a distance about equal to vertical diameter of eye. Distance from eye to nostril a little longer than from eye to spiracle. Nostril terminal, its inner anterior margin with 2 barbels, the outer barbel triangular with broad base and slender tip, the inner barbel narrow, widening slightly outward, then tapering to pointed tip and rising from the base of the outer without an intervening lobe. Both barbels smooth-edged or nearly so, the outer anterior margin of nostril expanded as a smooth-edged, subtriangular lobe; the posterior margin of nostril smooth.[24] Fold at front of head only slightly expanded in obtusely rounded contours outside corners of mouth. Mouth terminal, its breadth equal to about ¾ of length of head. Lower labial furrow extending about ⅔ of the distance toward center of mouth, but no upper furrow.

Teeth $\frac{10-10}{9-9}$ in specimen counted, alike in the 2 jaws, with erect, conical cusp on broadly expanded base, the outermost teeth slightly the smallest; 3 rows functional in each jaw; a broad gap at symphysis in each jaw.

Dorsals similar in form, brush-shaped, with broadly rounded apex. First dorsal about ⅕ as long at base as head, its vertical height a little greater than length of base, its origin posterior to tips of pelvics by a distance about ½ as great as distance between eyes. Interspace between 1st and 2nd dorsals about as long as between eyes. Second dorsal a little smaller than 1st. Interspace between 2nd dorsal and caudal about as long as between 1st and 2nd dorsals. Caudal between ⅙ and ⅐ of total length, its upper margin only about ¾ as long as its lower margin, both its corners rounded but the lower more broadly so, the posterior margin moderately concave with a shallow obscure notch opposite termination of caudal axis. Pelvics about ⅔ as long as extreme length of pectorals, the anterior margins weakly convex and distal margins nearly straight, their outer corners broadly rounded, their posterior corners tapering with acute tip (about 40°). Transition from distal margin to inner margin of pectoral marked by a definite subangular corner, its narrowly rounded outer corner approximately a right angle, its outer margin nearly straight, distal margin weakly concave, inner posterior margin moderately rounded and definitely notched at axil; distance from anterior corner of pectoral to rear end of its base about ⅔ as great as length along outer anterior margin, as is also the distance from outer corner to rear end of

23. Clearly shown in a photograph of a 42½-inch male from Chesapeake Bay (Hildebrand and Schroeder, Bull. U.S. Bur. Fish., *43* [1], 1928: fig. 30) and mentioned by Duméril (Hist. Nat. Poiss., *1*, 1865: 467, footnote).

24. More or less fringed in *californica*.

base; its anterior corner posterior to a transverse line at rear edges of eyes by a distance about ½ as great as distance between eyes; posterior parts of pectorals overlap anterior parts of pelvics by a distance a little less than distance between nostrils.

Color. Fresh specimens are described as bluish-gray or ashy-gray above, tinted with red on head and margins of fins; white below, with a reddish spot on the throat, a second on the abdomen and a third extending from behind the cloaca to the caudal; the pelvics are bordered below with irregular reddish bands. After preservation in alcohol the upper surface of specimen described is chocolate brown with pale mottlings, the upper surface of anterior corner of pectorals brownish white, the lower surface of the trunk grayish white anterior to the vent, but pale reddish brown on the tail sector, and the paired fins broadly edged with pale reddish brown.

Size. The fact that the claspers were well developed in a male of 42½ inches suggests that maturity is reached at a length of 3 to 3½ feet; the maximum length so far definitely reported is four to five feet. One of four feet weighed about 60 pounds. It is not known whether the American *S. dumeril* ever grows as large as its European relative, *S. squatina*, which often reaches a length of six feet (maximum reported, eight feet) with a weight of 160 to 170 pounds.

Developmental Stages. Embryos of *S. dumeril* have not been seen, but conditions in allied species make it likely that the yolk sac is very large, and that gravid females may contain as many as 13 to 25 embryos.

Remarks. This species has frequently been considered identical with *S. squatina* of the eastern Atlantic. Comparison of American with European specimens has shown, however, that they are easily distinguishable by the following characters: the shapes of the head folds (cf. Fig. 103 G with 103 C), the conformation of the nasal flaps and barbels, the inner margin of the pectoral notched at the axil in *dumeril* but not in *squatina*, the horizontal diameter of eye almost as great as the maximum diameter of the spiracle in *dumeril* but definitely smaller than the spiracle in *squatina*, the maximum breadth of the head only about ¾ as great as the length of the pectoral in *dumeril* but nearly or quite as great as the length of the pectoral in *squatina*, and (in small specimens) the mid-dorsal denticles larger in *dumeril*.

Habits. Most of the specimens so far reported have been taken in depths of only a few feet. However, one of the specimens listed above (p. 538) was taken in September over the continental edge 80 miles from the coast, where the depth was 705 fathoms,[25] while several others were trawled by the United States Fisheries Steamer "Albatross" about 75 miles off Long Island, New York, in 109 fathoms in February 1920, showing that it may stray far offshore and that it may also occur at considerable depths irrespective of season. Fragments of fish and of bivalve mollusks were found in the stomach of one taken in North Carolina, where it is described as "often troublesome, getting snarled in the nets or eating other fish caught therein; it also bites the fishermen if they are not wary."[25a] No

25. "Albatross" Dredging Station 2749.

25a. Smith, Bull. N. Carolina geol. econ. Surv., 2, 1907: 38.

other firsthand observations appear to have been made on the habits of the North American species. Analogy, however, with its close relative *S. squatina* of the eastern Atlantic suggests that it lives on or close to bottom, often burying itself partially in the sand or mud, as do rays and flatfishes (pleuronectids), and that it feeds on a variety of fish, perhaps chiefly on flounders and skates, as well as on crustaceans and gastropod mollusks.[26] Probably the young are born in summer, when the adults are close inshore.

Relation to Man. Squatina is not plentiful enough in American waters to be of any commercial importance. In Europe a certain number are marketed for fried-fish shops. Formerly its skin was in some demand for polishing wood and ivory, and in earlier days its dried flesh was "prescribed as a sovereign remedy for the itch."[27]

Range. East coast of the United States from southern New England to North Carolina and southern Florida, north coast of the Gulf of Mexico, and reported recently from Jamaica.[28]

This is a summer visitor to the mid-Atlantic coast of the United States. In most years it appears at Cape Lookout, North Carolina, in late March or April to remain until about the first of May. To the northward it has been reported between May and October from Chesapeake and Delaware Bays, and from various localities along the coast to the vicinity of New York, as well as in the bays along the southern shore of Long Island. Most of the records for it have been based on odd specimens only. But it has been reported as sometimes common on the outer coasts of Virginia and Maryland, as well as Delaware, suggesting that this section may be a center of abundance for it. Three specimens have been reported from Rhode Island, two from the vicinity of Woods Hole,[29] this last being the most easterly and northerly record for it. Positive knowledge of it southward from North Carolina is limited to reports that it is occasionally taken in summer among the Florida Keys,[30] and that it is caught occasionally in shrimp trawls off the mouth of the Mississippi.[31] There is one record for Corpus Christi, Texas, and a 30-inch specimen has been taken in the harbor of Port Royal, Jamaica.[32] But it has not been reported at all anywhere between Jamaica and Rio de Janeiro, nor has any other member of its genus for that matter, suggesting that its normal range does not extend to the equatorial belt. Although described as quite abundant for a short period in spring off the North Carolina coast,[33] present indications are that it is far less plentiful than is its European relative *S. squatina*, as many as 26 of which have been reported as lying on the beach on the west coast of England at one time.[34] We have seen only one fresh-caught specimen.[35]

26. We wonder whether an old and oft-quoted account of one seen to come to the surface and to seize a living cormorant may not actually have referred to an angler (*Lophius*) which commonly captures sea fowl in this way.

27. Norman and Fraser, Giant Fishes, 1937: 55.

28. Personal communication from Luis Howell-Rivero.

29. Both of them from Menemsha Bight, Martha's Vineyard Island, the one in 1873, the other in September 1921.

30. Fowler (Proc. Acad. nat. Sci. Philad., *58*, 1906: 80). It has also been conjectured that at least one of the specimens on which the species was founded was from Florida. But the wording of Lesueur's original account (Proc. Acad. nat. Sci. Philad., *1*, 1818: 226) suggests, rather, that both of his specimens were studied by him in a fresh condition, *i.e.*, that they were collected not far from Philadelphia.

31. Personal communication from Stewart Springer.

32. Personal communication from Luis Howell-Rivero.

33. Coles, Proc. biol. Soc. Wash., *28*, 1915: 92.

34. Day, Fish. Gt. Brit., *2*, 1880–1884: 32b.

35. In Chesapeake Bay.

The wintering ground of the sparse population of both young and adults that visit the east coast of the United States in summer is not known. It seems more likely that they move out into deeper water than that they migrate southward along the shore, for Angel Sharks are scarce around Florida, while one specimen was trawled about 75 miles off Long Island, New York, in February 1920, and another about 25 miles off Bodie Island, North Carolina, in about 40 fathoms in February 1931.[36]

Synonyms and References:

Squatina dumeril Lesueur, J. Acad. nat. Sci. Philad., *1*, 1818: 225, pl. 10 (descr., no loc.); Bory de St. Vincent, Dict. Class. Hist. Nat., *15*, 1829: 598; DeKay, Zool. N. Y., 4, 1842: 363, pl. 62, fig. 203 (descr., ill., after Lesueur, 1818, N. York); Rep. St. Cab. nat. Hist. N. Y. (1855), *8*, 1858: 64 (listed for N. York); Fowler, Copeia, 30, 1916: 36 (mid-Atlantic states); Proc. biol. Soc. Wash., *33*, 1920: 145 (N. Jersey); Proc. Acad. nat. Sci. Philad., *74*, 1922: 3, 6 (N. Jersey); Hildebrand and Schroeder, Bull. U.S. Bur. Fish., *43*, 1928: 54 (descr., photo, Chesapeake Bay); Jordan, Manual Vert. Anim. NE. U.S., 1929: 15 (general); Fowler, Proc. Acad. nat. Sci. Philad., *80*, 1929: 608 (N. Jersey); Truitt, Bean and Fowler, Bull. Md. Conserv. Dep., *3*, 1929: 28 (Maryland); Jordan, Evermann and Clark, Rep. U.S. Comm. Fish. (1928), 2, 1930: 22 (Newport, Rhode Island); Fowler, Monogr. Acad. nat. Sci. Philad., 7, 1945: 97 (one 42 inches, Beaufort, N. Carolina).

Squatina dumerilii Wils, Observ. *Squatina laevi*, Inaug. Dissert. Lugduni-Batavorum, 1844: 5 (America).

Squatina dumerili Storer, Mem. Amer. Acad. Arts Sci., N. S. *2*, 1846: 256 (in synopsis); Griffith, Proc. Acad. nat. Sci. Philad., *3*, 1847: 246 (Delaware Bay); Leidy, Proc. Acad. nat. Sci. Philad., *3*, 1847: 247 (no loc.); Baird, Rep. U.S. Comm. Fish. (1871–72), 1873: 827 (Woods Hole, Massachusetts); Gill, Rep. U.S. Comm. Fish. (1871–72), 1873: 813 (listed, C. Cod to Florida); Smith, Copeia, 106, 1922: 33 (Martha's Vineyard).

Squatina angelus (in part) Valenciennes, Hist. Nat. Canaries, *2*, 1837–44: 102 (N. York to Brazil); Jordan and Gilbert, Bull. U.S. nat. Mus., *16*, 1883: 35 (Atlant. coast, N. Amer.); Gaskill, Forest and Stream, *36*, 1891: 68 (C. May, N. Jersey, not seen).

Rhina dumerili Gill, Proc. Acad. nat. Sci. Philad., Addend., 1861: 61 (name); Duméril, Hist. Nat. Poiss., *1*, 1865: 467 (descr., spec. from N. York).[36a]

Rhina dumeril Garman, Mem. Harv. Mus. comp. Zool., *36*, 1913: 252 (descr., Newport, Rhode Island); Radcliffe, Bull. U.S. Bur. Fish., *34*, 1916: 268 (N. Carolina); Bigelow and Schroeder, Guide Comm. Shark Fish., Anglo Amer. Caribb. Comm., Wash., 1945: 148, fig. 56 (descr., ill., discuss.).

Rhina squatina (in part) Günther, Cat. Fish. Brit. Mus., *8*, 1870: 430 (*dumerilii* included in synonymy); Day, Fish. Gt. Brit., *2*, 1880–1884: 326 (*dumerilii* incl. in synonymy).

Squatina squatina (in part) Jordan and Evermann, Rep. U.S. Comm. Fish. (1895), 1896: 220 (Atlant. coast U.S.); Bull. U.S. nat. Mus., 47 (1), 1896: 58 (descr., C. Cod southward; considered ident. with European *squatina*); Bean, T. H., Bull. Amer. Mus. nat. Hist., *9*, 1897: 331 (near N. York); Smith, Bull. U.S. Fish Comm., *17*, 1898: 89 (Martha's Vineyard); Evermann and Kendall, Rep. U.S. Comm. Fish. (1899), 1900: 49 (probably Florida); Bean, Rep. For. Comm. N. Y., 1901: 381 (old Martha's Vineyard record); Bull. N. Y. St. Mus., *60*, Zool. 9, 1903: 45 (descr., near N. York); Smith, Bull. N. C. geol. econ. Surv., *2*, 1907: 38 (diagn., food, N. Carolina); Kendall, Occ. Pap. Boston Soc. nat. Hist., 7 (8), 1908: 10 (Woods Hole region); Tracy, Rep. R. I. Comm. inl. Fish., 1910: 61 (Newport and West Passage, Rhode Island); Fowler, Proc. Acad. nat. Sci. Philad., *62*, 1911: 599 (Delaware Bay and N. Jersey); Proc. Acad. nat. Sci. Philad., *63*, 1911: 5 (Virginia, Maryland, Delaware); Proc. Acad. nat. Sci. Philad., *64*, 1912: 35 (N. Jersey); Gudger, J. Elisha Mitchell sci. Soc., *28*, 1913: 166 (N. Carolina); Nichols, Abstr. Linn. Soc. N. Y., *20–23*, 1913: 91 (near N. York); Sumner, Osburn and Cole, Bull. U.S. Bur. Fish., *31* (2), 1913: 737 (Martha's Vineyard); Coles, Proc. biol. Soc. Wash., *28*, 1915: 92 (season, abund., N. Carolina); Nichols and Murphy, Brooklyn Mus. Sci. Bull., *3* (1), 1916: 33 (near New York); Nichols and Breder, Zoologica, N. Y., *9*, 1927: 22 (vicinity of N. York and Woods Hole); Breder, Field

36. This latter was taken by the United States Fisheries Steamer "Albatross II."

36a. Spelled *dumerilii.*

Bk. Mar. Fish. Atlant. Coast, 1929: 26 (general); Jordan, Evermann and Clark, Rep. U.S. Comm. Fish. (1928), 2, 1930: 22 (N. to Cape Cod, but incl. European *squatina*); Breder, Bull. N. Y. zool. Soc., *41*, 1938: 25 (Sandy Hook Bay, near N. York); Parks, Tech. Bull. Stephen Austin State Teachers' Coll., 1 (4), 1939: 1 (Texas; not seen); Gunter, Amer. Midl. Nat., *26* (1), 1941: 197 (Corpus Christi, Texas); not *Squalus squatina* Linnaeus, 1758.

Rhina squatina Fowler, Proc. Acad. nat. Sci. Philad., *58*, 1906: 80 (reported for Florida Keys).

Not *Squatina squatina* Fowler, Arqu. Zool. Estado São Paulo, 2, 1942: 129 (listed for Rio de Janeiro, Brazil; probablv *argentina* Marini, 1930; see p. 546).

Genus *Squatina*, Addendum

The common Squatina of the temperate coasts of the western South Atlantic seems clearly separable from all other members of the genus. A short notice of it therefore follows.

Squatina argentina Marini, 1930

Figure 106

Study Material. Female, 714 mm. long, from Uruguay (U.S. Nat. Mus., No. 87684).

Distinctive Characters. See Description.

Description. Proportional dimensions in per cent of total length. Female, 714 mm., from Uruguay (U.S. Nat. Mus., No. 87684).

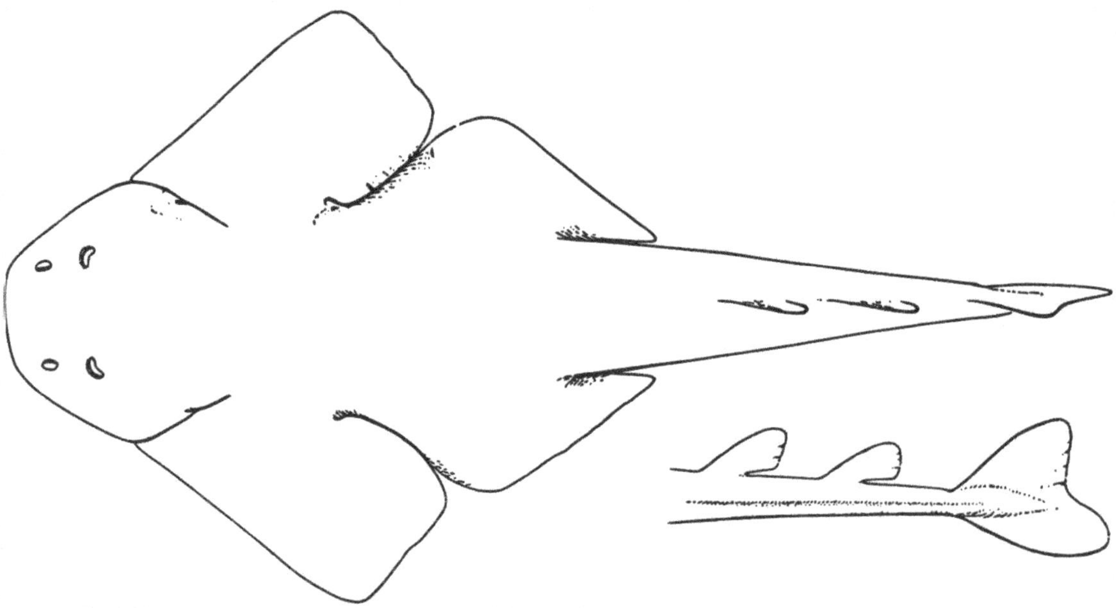

FIGURE 106. *Squatina argentina*, female, 714 mm. long, from Uruguay (U. S. Nat. Mus., No. 87684).

Extreme breadth at outer extremity of pectorals: 54.0.
Trunk at origin of pectoral: height 8.0.
Snout length in front of: outer nostrils 0; mouth 0.
Eye: horizontal diameter 1.5; distance between eyes 8.4.
Spiracles: distance between 8.1; from spiracle opening to eye 2.7.
Mouth: breadth 11.8; height 3.2.
Nostrils: distance between inner ends 4.8.
Gill opening lengths: 1st 7.1; 2nd 6.7; 3rd 6.6; 4th 6.2; 5th 6.0.
First dorsal fin: vertical height 4.9; length of base 4.1.
Second dorsal fin: vertical height 4.5; length of base 4.0.
Caudal fin: upper margin 12.0; lower margin 14.3.
Pectoral fin: outer margin 26.5; inner margin 13.8; distal margin 14.0.
Distance from snout to: 1st dorsal 65.6; 2nd dorsal 76.0; upper caudal 88.0; pectoral 16.8; pelvics 37.8.
Interspace between: 1st and 2nd dorsals 7.0; 2nd dorsal and caudal 8.5.
Distance from origin to origin of: pectoral and pelvics 20.9; pelvics and caudal 49.8.

S. argentina resembles *S. dumeril* so closely in general proportions and in the shape of the pectoral fin[37] that we have found no clear distinction between the two species in most of their proportional dimensions, and the teeth are of the same number ($\frac{10\text{-}10}{10\text{-}10}$) and shape. However, the distance from the spiracle to the eye is only a little more than ½ (55%) as long as the diameter of the eye in *argentina*, but nearly as long as the eye (91%) in *dumeril*. The inner nasal barbel of *argentina* is spatulate (tapering in *dumeril*), while there is a broad quadrate lobe between the two barbels, with a second lobe-like expansion just basal to the outer margin of the outer barbel on the outer side in *argentina*, which is not the case in *dumeril* (cf. Fig. 103 D with 103 G). A further difference if minor is that the dermal fold along the front of the head is not expanded at all opposite the corner of the mouth in *argentina*, but is slightly expanded there in *dumeril* (although much less so than in the European *squatina*; Fig. 103 C).

S. argentina differs from *armata* of the west coast of South America in the contour of the margin of the nostril (cf. Fig. 103 D with 103 H); it differs further from *armata* in that the distal margin of its pectoral is marked off from the inner margin by a definite, narrowly rounded corner, whereas in *armata* the one margin grades insensibly into the other in an even curve.

Color. Described as gray-brown above, marked with many small oval cinnamon-colored spots; white below.[38]

Size. Said to grow to a length of 1.7 meters (5½ feet).

37. In the original illustration of *argentina* (Marini, Physis B. Aires, *10*, 1930: 6) the outer corner of the pectoral is shown as considerably more obtuse than a right angle. But it is only a little more than a right angle in the specimen we have studied, and Ribeiro's (Arch. Mus. nac. Rio de J., *14*, 1907: pl. 10) photograph of a Brazilian specimen shows about a right angle.
38. The only specimen we have seen has not only lost the color pattern, but is now stained red with iron rust.

Range. Temperate latitudes in the western side of the South Atlantic; northern Argentina (the type locality), and apparently common along the coasts of Uruguay and southern Brazil as far north as Rio de Janeiro.[39]

Synonyms and References:

Squatina squatina Schreiner and Ribeiro, Arch. Mus. nac. Rio de J., *12*, 1903: 80 (near Rio de Janeiro); Ribeiro, Peces "Annie," 1904: 18 (off Ilha Rassa, Brazil); Arch. Mus. nac. Rio de J., *14*, 1907: 170, pl. 10 (descr., ill., Rio de Janeiro); Devincenzi, An. Mus. Hist. nat. Montevideo, (2) *1*, 1920: 124 (abund., Uruguay); Ribeiro, Fauna brasil. Peixes, *2* (1) Fasc. 1, 1923: 27, pl. 9 (descr., ill., Rio de Janeiro); Devincenzi and Barattini, An. Mus. Hist. nat. Montevideo, Suppl. Album Ictiol., 1926: pl. 3, fig. 1 (ill., Uruguay); Fowler, Arqu. Zool. Estado São Paulo, *3*, 1942: 129 (listed for Bay of Rio de Janeiro, Brazil); not *Squalus squatina* Linnaeus, 1758.

Squatina angelus Lahille, Physis B. Aires, *5*, 1921: 63 (listed for Argentina); Enum. Peces. Cartilag. Argent., 1921: 17 (listed for Argentina); Pozzi and Bordale, An. Soc. cient. argent., *120*, 1935: 152 (listed for Argentina, Lat. 35°–42° S.).

Rhina angelus Marini, Physis B. Aires, *9* (34), 1929: 422 (listed for north. Argentina); not *Squatina angelus* Blainville, 1816 (sometimes referred, but incorrectly, to Duméril, A. M. C., Zool. Analit, 1806: 102); same as *Squalus squatina* Linnaeus, 1758.

Rhina argentina Marini, Physis B. Aires, *10* (35), 1930: 5, 1 fig. (descr., good ill., meas., Argentina, Lat. 39° S.); Pozzi and Bordale, An. Soc. cient. argent., *120*, 1935: 152 (listed for Argentina, Lat. 39° S.).

Squatina armata Norman, "Discovery" Rep., 16, 1937: 10 (descr., comp. with other sp., Argentina); Hart, "Discovery" Rep., 23, 1946: 260 (Argentina by ref. to Norman, 1937); not *Rhina armata* Philippi, 1887.

39. If our reference to it of the several Uruguayan and Brazilian records for *angelus* and *squatina* be correct.

INDEX OF COMMON NAMES

(Principal references in boldface)

INDEX OF SCIENTIFIC NAMES

(Principal references in boldface)

www.ingramcontent.com/pod-product-compliance
Ingram Content Group UK Ltd.
Pitfield, Milton Keynes, MK11 3LW, UK
UKHW051118290926
15471UKWH00010B/1749

9 781933 789118